LABORATORY PHYSICS

SECOND EDITION

Harry F. Meiners
Professor of Physics
Rensselaer Polytechnic Institute

Walter Eppenstein
Professor of Physics
Rensselaer Polytechnic Institure

Ralph A. Oliva
Manager, World Wide Market Communications
Texas Instruments, Inc.

Thomas Shannon
Manager, Physics Laboratory
Rensselaer Polytechnic Institute

Volker Paedelt
Illustrations
Technical Manager, Composite Laboratories
Rensselaer Polytechnic Institute

With Contributions by:

Kenneth H. Moore
Emeritus
Rensselaer Polytechnic Institute

Gerald Luecke
Manager of Technical Products Development
Texas Instruments Learning Center

JOHN WILEY & SONS
New York Chichester Brisbane Toronto Singapore

*ISBN 0 471 03675 7
Printed in the United States of America*

10 9 8 7 6 5

PREFACE

Laboratory Physics is intended to be a laboratory guide for a calculus based introductory physics course taught primarily to science and engineering students. We believe that the laboratory is an important part of such a course and that very often the experiment bridges the gap between the idealized situations presented in the usual physics text and the real world of the laboratory.

The first six chapters give an introduction to laboratory procedures and instrumentation. How these chapters are used is entirely up to the faculty or staff teaching the course; some of the material may be assigned as reading whereas other topics may be completely left out.

The first chapter presents some of the objectives of the laboratory and makes suggestions for laboratory operations. The second chapter on errors should be studied in detail and at some time during the first physics course the material should be assigned in a formal manner. The ideas of Chapter 3 on graphing will be applied throughout the experiments and should be referred to at the appropriate time. Chapter 4 includes a very brief description of some important pieces of equipment, and students should read these sections before handling the apparatus involved.

Chapters 5 and 6 give a general introduction to the use of calculators and computers in the laboratory as well as some of the principles of digital integrated circuits. This information, usually not found in an introductory laboratory manual, has been included for the benefit of interested instructors and students. A study of these two chapters is not a prerequisite for the experiments in this text.

The rest of the text presents more than 70 experiments, arranged by topic and not by level of difficulty. This gives the instructor a wide choice depending on interest, equipment available, level of difficulty, and the like. Some of the experiments contain detailed instructions while others are - on purpose - rather vague, leaving the details up to the student. The experiments differ greatly in difficulty as well as in length. Some use very simple equipment whereas others use rather sophisticated apparatus. Most of the experiments have several different parts, and the students are not necessarily expected to finish all parts; very often they may have a choice.

This laboratory text should not be regarded as an instruction book that is to be followed word by word; it is meant to be a general guide to various experiments in introductory physics. It is hoped that the instructor and the student will deviate from the directions given and will add their own parts to the experiments. Student creativeness and initiative are stressed by suggesting various projects without going into details. Some experiments suggest further work. In some cases the equipment can be used for "free" experiments not described in this text; examples are the low-friction devices discussed in Chapter 8, the oscilloscope with accessories, the microwave apparatus, and the laser.

Report writing has traditionally been a very time-consuming activity in connection with physics laboratories. We do not want to minimize the importance of well presented experimental results, but this can be accomplished without too much busy work. How the laboratory should be organized and what type of reports, if any, the student should present is left up to the instructor in charge of the course. Generally, students should be encouraged to record all data, including qualitative observations, in a laboratory notebook.

Although this book is based on the text Analytical Laboratory Physics, which was published by Meiners, Eppenstein, Moore, and Nickol in 1956, a great many experiments have been added. Most of the old experiments have been changed so that they are less formal, include more optional parts, and make use of newly available instrumentation. In 1956, for instance, inexpensive air tracks and air tables did not exist, the laser had not yet been invented, and microcomputers were unheard of. All of these relatively recent innovations are now standard equipment in most introductory physics laboratories.

We are grateful to Gerald Luecke of the Texas Instruments Learning Center for writing Chapter 6 on integrated circuits and to M. Dean Lamont of Texas Instruments for his critical review of Chapter 2. We are indebted to Volker Paedelt of Rensselaer Polytechnic Institute for producing most of the art work.

We express our appreciation to the thousands of freshmen and sophomore students at Rensselaer Polytechnic Institute who have tried out all of the experiments and have helped us greatly by their constructive criticism. We acknowledge the many comments we have received from the graduate assistants teaching in our introductory laboratories as well as from faculty members. We express our thanks to the manufacturers of scientific equipment who have supplied us with photographs and literature relating to their products.

June, 1986
Troy, New York

Harry F. Meiners
Walter Eppenstein
Ralph A. Oliva
Thomas Shannon

CONTENTS

Chapter 1

THE LABORATORY

1.1 Introduction

Recent technological advances in all fields of science and engineering have emphasized the microscopic world, the world of the atom and its parts. This emphasis requires every embryo engineer and scientist to develop an individual initiative to see, to question and, if possible, to find out why. This cannot be done quickly but requires a gradual, directed introduction to the fundamental methods of analysis.

The objectives of the physics laboratory are not, as is often imagined, simple or easy to define. They cannot be generalized in a single sentence. The object of the physics laboratory of today should not be the verification of a known law and the blind substitution of data into a formula with the consequent "cranking out" of an answer. The physics laboratory should help to bridge the gap between the idealized laws and postulates discussed in the textbook and the real world. The laboratory is not a separate course in physics. It is an integral part of the study of physics and must be considered as such.

Before the student can see, question and find out why, he or she must master the fundamental tools (or techniques) necessary for the ultimate satisfaction of his or her curiosity. The degree of mastery depends on the attitude of the student toward the laboratory work.

The objectives of this laboratory are:

1. To introduce the student to the significance of the experimental approach through actual experimentation.
2. To apply the theory of the textbook and the recitation class to real-life problems to develop a better understanding of the fundamentals of classical and modern physics.
3. To introduce the student to the methods of data analysis used throughout science and engineering.

4. To develop an "error conscience" so that the engineer and scientist will at least be aware of the relative worth of all measurements, whatever their type. The method of "precision analysis" presented here should not be thought of as the only way of determining a measurement's relative worth. Precision analysis appears in various guises in all branches of business, government and science. The student should become familiar with its many facets and realize that while learning a fundamental technique, he or she is also becoming more proficient in applying calculus.

5. To familiarize the student, by direct contact, with a great many basic instruments and their applications.

6. To make the student realize that such tools as graphing, difference analysis, the use of calculus, etc., are of fundamental importance.

7. To impress on the student that even an experiment which is apparently unimportant to his or her professional future may contribute directly to the student's mental development because of the analytics and mathematics involved.

8. To improve the student's ability of self expression through report presentation.

9. To give the student direct contact with the instructor, and thus the advantages of close direction and personal discussion of ideas and methods.

1.2 Laboratory Operation

1. ASSIGNMENTS

At the first laboratory meeting, working teams of two members will be made up, each team being assigned a number. All assignments will be made in terms of these numbers.

Each laboratory assignment requires the performance of an experiment in the laboratory and the presentation of the data, computations, etc., must be completely worked out in the laboratory notebook. Most of the writeup should be completed in the laboratory on the day the experiment is performed. If not finished, the report in final form must be turned in for inspection and grading AT THE NEXT SCHEDULED LABORATORY MEETING FOLLOWING THE PERFORMANCE OF THE EXPERIMENT.

Although many experiments will be preceded by the necessary theory in the recitation class, this will not always be the case. This should be no disadvantage, however, since laboratory work is usually the forerunner of theory. The student will be expected, in these instances, to study the theory concerned in his theory text, and check thoroughly the content of the experiment.

Each student is expected to prepare in advance for each experiment. He or she should be able, before beginning the experiment, to answer questions based on the general content of the experiment.

2. DATA CHECK

At the completion of the experiment, the laboratory notebooks are to be presented to the instructor to be checked, stamped, and initialed. This permits obvious errors to be found.

3. STUDENT'S RESPONSIBILITIES TOWARD EQUIPMENT

Apparatus is to be treated with respect. Most of the equipment is sensitive and expensive. Damage over and above normal depreciation will be charged to the team involved.

Students must leave their tables and apparatus in good order: i.e., weights put away, instruments returned to cases, water emptied, scrap paper picked up, etc.

1.3 The Report

In most experimental work, records of the work done, data taken, and observations made in the laboratory are kept in a data book. When important research projects are underway, such a book is usually assigned to each team, and the people associated with the project record in it all the data in complete and careful detail. All matters having any conceivable bearing on the problem are carefully recorded and every effort is made to preserve a complete picture of what is going on. All work is carefully dated and signed. Very often, when new effects are noted, the books are signed by witnesses (for possible patent purposes). Progress or final reports are abstracted from such databooks for publication or for presentation to the head of the laboratory or other officials. The completeness or detail of abstracted reports is determined, of course, by the use to which they will be put. In our physics laboratories, all data is to be taken and all work done in laboratory notebooks.

The laboratory report has generally the following eight parts:

1. Purpose of the experiment and preliminary discussion
2. Sketch
3. List of Apparatus
4. Procedure
5. Data
6. Computation Outline
7. Graphs and Results
8. Discussion (or conclusion)

1. PURPOSE

A short statement of the object or objectives should be made. The object may be not only to find an unknown, to establish the linearity of a function, or to verify a known law, but also - more subjectively - to apply a fundamental technique.

The PRELIMINARY DISCUSSION is not a statement of procedure, but an explanation of any special features of the experiment: for example, that it involves the use of precision analysis, or the development of an empirical equation, or the application or fundamental laws to a field problem.

2. SKETCH

A line sketch of the experimental equipment (wiring diagram for electrical experiments) should be made with a straight edge. This need not be an artist's sketch or be drawn to scale. Do not include standard items such as micrometers, meter sticks, stands, timeclocks, clamps, etc. Label with a letter and identify under the list of apparatus.

3. LIST OF APPARATUS

Identify all measuring apparatus. Omit stands, clamps, etc. List:

a) Name of apparatus and manufacturer; Model number and serial number (if available).
b) Range of values covered; Least Count of scale; Instrumental Limit of Error, as guaranteed by the manufacturer or by laboratory tests.

4. PROCEDURE

It is not necessary to copy the detailed instructions of most of the experiments. List the procedure only for those experiments lacking complete instructions.

5. DATA

All measurements must be recorded directly into the laboratory notebook. Data must not be taken on scrap paper and recopied into the report. It must not be erased. If a mistake is made, draw a line through the measurement and place the corrected value above. Data should be presented in a table whenever possible. Place the units of the quantities being measured at the top of the data columns. A typical data table is shown in Fig. 1-1.

6. COMPUTATION OUTLINE

State all formulas. Identify all symbols. Substitute one set of data into each different formula. When stating results, express answers in powers of ten, with proper units and the range of error if the experiment requires precision analysis. Watch your number of significant figures. Number the steps followed so that your approach may be easily understood by the instructor.

7. GRAPHS AND RESULTS

When reporting graphical results, show carefully slope calculations and the values obtained from the axes of the graphs. List the numerical results as found in the computation outline. If the results are qualitative, describe briefly.

Graphical Method:

Mass (gm)	Scale Readings (cm)				mean Scale Readings	Elongation [Y] (cm)
	Run 1	Run 2	Run 3	Run 4		
0	39.28	39.29	39.29	39.28	39.285	
50.0	44.89	44.88	44.88	44.89	44.885	5.600
100.0	50.49	50.48	50.49	50.50	50.490	11.205
150.0	56.10	56.08	56.09	56.10	56.093	16.808
200.0	61.69	61.70	61.68	61.69	61.690	22.405
250.0	62.27	67.26	67.28	67.28	67.273	27.988
300.0	72.85	72.86	72.88	72.89	72.870	33.585
350.0	78.41	78.42	78.43	78.44	·78.425	39.140
400.0	83.98	84.00	84.02	84.01	84.003	44.718

Tabular Difference Method:

Mass (gm) $\Delta\omega = 50$ gms	Elongation Y (cm)	ΔY (cm)	$\Delta^2 Y$ (cm)
0			
50.0	5.600	5.600	
100.0	11.205	5.605	+.005
150.0	16.808	5.603	-.002
200.0	22.405	5.597	-.006
250.0	27.988	5.583	-.014
300.0	33.585	5.597	+.014
350.0	39.140	5.555	-.042
400.0	44.718	5.578	+.023

∴ mean value of ΔY = 5.58(8 2) — Sig. Fig.

Fig. 1-1 Data table from an actual student report.

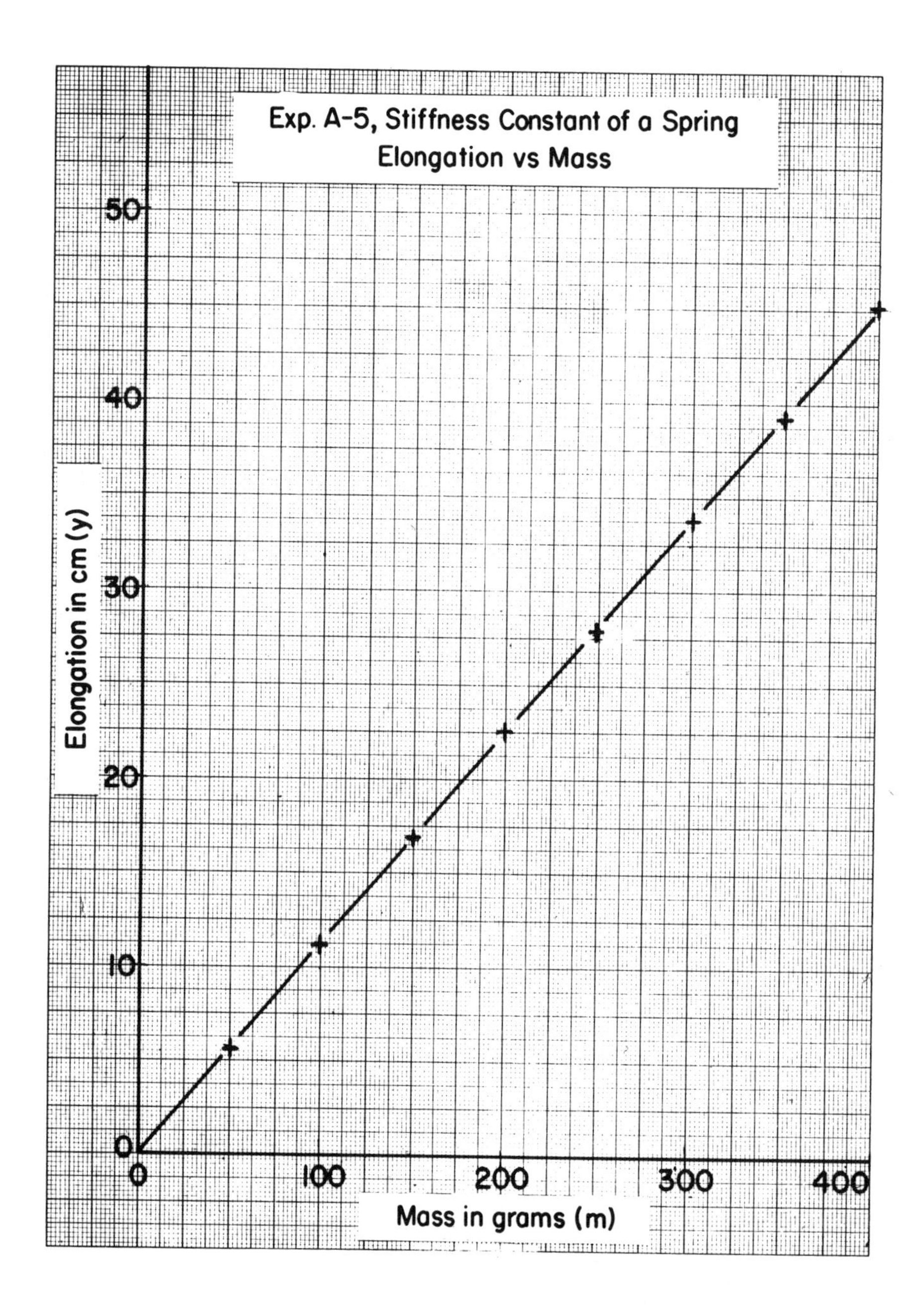

Fig. 1-2 Typical graph from an actual student report.

8. DISCUSSION (or CONCLUSION)

If several methods are used, summarize the worth of the experiment by describing the benefits of a particular analytical approach as compared to another. If only one approach is used, discuss the significance. Discuss the value of a particular result as indicated by the error analysis. Try to determine why the experiment was performed. Discuss the meaning of your graphical results.

The discussion (or conclusion) is the individual's job. No team partner should copy the other member's work. The discussion should be concise and to the point. Do not use over half a page. This, obviously, indicates that the student must think and use judgment before presenting the discussion.

The impossible is not requested of the student in the preparation of a discussion. The suggestions given are designed to help the student to observe, and, as the course progresses, he or she will find this section of the report increasingly easier. The student should, if in trouble, contact the instructor and request help.

1.4 Summary

The report is the individual's responsibility. Data may be secured by a group (or team), but under no circumstances may a student use data which belongs to some different group. Reports which are copied completely or in part from other students' work will receive an "F". An absent student may not copy the partner's data, but must arrange to perform the experiment at the convenience of the instructor.

The average time required to take the data varies greatly, but is usually less than one hour. The balance of the laboratory period must be spent toward completing the report.

A complete report has the seven or eight divisions described above. In special cases, where the taking of data requires an unusually long time, only three divisions are necessary: Data, Computation Outline, and Discussion. Unless the directions for an experiment specify that only three divisions are necessary, a complete report must be submitted.

In requiring adherence to this standard report form, the laboratory is preparing the student for future professional work: almost all industrial and research units have fairly rigid specifications for reports of different types, and engineers and scientists are expected to obey the ground rules.

Chapter 2

MEASUREMENTS AND ERROR

2.1 Introduction

Lord Kelvin once summed up the importance of measurement as part of the very essence of science in a statement which will bear requotation:

> "I often say that when you can measure what you are speaking about and express it in numbers, you know something about it; but when you cannot measure it, when you cannot express it in numbers, your knowledge is of a meager and unsatisfactory kind; it may be the beginning of knowledge, but you have scarcely, in your thoughts, advanced to the stage of science, whatever the matter may be."

Another more recent quote by the popular science fiction writer Stanslow Lem relates to the science of statistics:

> "The mathematical order of the universe is our answer to the pyramids of chaos. On every side of us we see bits of life that are completely beyond our understanding – we label them unusual, but we really don't want to acknowledge them. The only thing that really exists is statistics. The intelligent person is the statistical person."

It is familiar knowledge that the specification of a physically measurable quantity requires at least two items:

1. A number
2. A unit

(Plus, when necessary, a statement of direction for vector quantities and certain tensor quantities.)

3. An indication of the reliability or degree to which we can place confidence in the value stated. This is usually done by specification of a "*precision index.*"

The purpose of this chapter is to develop a technique which may be used to specify the precision index for direct measurements of a single quantity; or indirect measurements resulting from the calculation of a value as a function of one or more direct measurements.

2.2 Types of Measurements

We, therefore, can classify measurements as either (A) direct, or (B) indirect, as follows:

A. **DIRECT MEASUREMENTS** are the result of direct comparison, usually with the aid of instruments, of an unknown amount of a physical entity x with a know or standardized amount of the same entity. Several of the important types of direct comparison are:

(1) Balanced, equality or null measurements

Here, the standard value s is selected or adjusted to be equal to x and the balance value is recorded. Measurements of this sort give extremely precise results.

An example of a null type of measurement is weighing with the old-fashioned, equal-arm balance. An unknown mass (x) is placed on one pan of the balance and standardized masses are added to the other pan until null is again reached. Another is the measurement of electrical resistance with a Wheatstone Bridge.

(2) Ratio measurements

An unknown x is compared with a known s in terms of some fraction or multiple (R).

$$\text{then } x = Rs$$

(R is to be operationally determined)

An excellent example is the linear potentiometer, as used in measuring devices. In this case a standardized electrical potential difference s is applied across a resistance. The latter is usually a long, uniform wire (which may be wrapped up or doubled back) arranged so that a sliding contact can pick off a determinable fraction of its length (R), and therefore the same fraction of its resistance. For uniform wire, this also means that the potential difference picked off also has the ratio (R) to that of s.

The precision with which R is determined (on a fractional or percentage basis) will be a major factor in the reliability of the measured value of x (See Fig. 2-1).

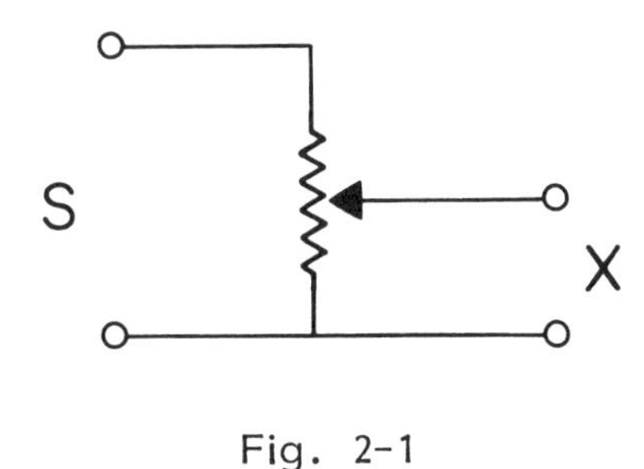

Fig. 2-1

B. A very simple example of an **INDIRECT MEASUREMENT** based on a single direct measurement is the determination of the volume of a sphere from a direct measurement of its radius, or better, its diameter:

$$V = \frac{1}{6}\pi D^3 \quad \text{(see p. 23)}$$

The diameter D, a length, can be measured directly with sufficient accuracy to suit most purposes. The volume V can then be calculated, – any unreliability in D being compounded three times.

Again, the volume of a right circular cylinder of diameter D and length h is

$$V = \frac{\pi}{4} D^2 h \quad \text{(see p. 24)}$$

If D and h are measured by unbiased observers using different instruments, errors in D and h will not necessarily compound. There will, of course, be a compounding effect in D^2.

2.3 Types of Errors

Before determining the precision index for laboratory data, errors which are called systematic errors must be eliminated or at least reduced to a negligible magnitude.

A. SYSTEMATIC ERRORS

Systematic errors are errors which result from consistent observational inaccuracies, such as parallax error. (Parallax error is an optical error that arises when a laboratory meter or similar instrument is consistently read from one side rather than "straight on.") Systematic errors may also be the result of changes in temperature, pressure and humidity. Such changes may adversely affect the accuracy of laboratory equipment. Regardless of the causes, in theory at least, systematic errors can be sufficiently reduced so that they play no part in determining the precision index of an experiment.

The error introduced into a laboratory experiment by the measuring equipment itself is called the "instrumental limit of error" (ILE), and forms one important part of the precision index.

B. INSTRUMENTAL ERRORS

Since the observer and the environment are part of the entire measurement process, it is difficult to separate out purely instrumental errors. It can be done, and instrument manufacturers today stand ready to guarantee that their instruments will have "errors of not over _ _ _ _ _". A statement of this sort gives the residual limit within which the maker guarantees his instrumental errors to fall.

The American Standards Association and the American Society for Testing Materials recommend that the limit of error be made equal to the least count or smallest readable scale division of the instrument. Good examples are the small, panel mounted electric meters which are used for so many purposes.

These meters are usually supplied with fifty divisions (if linear) and are commonly guaranteed to a limit of error of ±2% of full scale value (that is to ± ONE DIVISION). Such instruments are usually repeatable or reproducible in their readings to better than the published limit, but the attainment of anything better than ±2% requires individual and frequent re-calibration and checking.

Enormous improvements in instrumentation have taken place in the past few decades reducing environmental and observation errors. In general, one must rely on the integrity of the instrument manufacturer to maintain his guarantees. Competition is fierce enough and the industry well enough self-policed that serious breaches of error limits are exceedingly rare. Most makers take justifiable pride in the fact that their instruments exceed specifications. There is no longer any secrecy or much vagueness about the limits of error of available instruments.

It is true, of course, that careless or improper use may raise error limits, sometimes drastically, so that check or calibration procedures should be available.

Assuming that you are using the appropriate instrument correctly calibrated, then the resulting error introduced into your experiment due to the instrument will be added to random errors (discussed later) to determine the precision index for the quantity being measured. This is a very conservative procedure since only seldom will random errors and instrumental errors both be maximum in the same direction.

So, a first step in developing a precision index for a given quantity is to study the equipment used in measuring the quantity carefully. Examine it to determine the instrumental limit of error, usually included on the equipment label, in the user's manual, or with accompanying literature. When in doubt, ask your instructor.

C. RANDOM (STATISTICAL) ERRORS

When all systematic disturbances in a measurement process (or in the quantity being measured) have been minimized, balanced out or corrected, there still remains a class of disturbances fully as important as the class of systematic, macroscopic perturbations and often much more so. This is the class of disturbances due to the superposition of the many small, random fluctuations which always abound in any environment, and which are always detectable if the measuring instruments and process are sufficiently sensitive. These random, erratic or accidental errors, while never individually predictable, can be treated by statistical methods applicable to groups of repeated measurements. It should be pointed out at once that such analysis can do NOTHING about unknow, hidden, systematic errors, which can prove and have proved the bane of many otherwise excellent experiments.

We do not live in a stable universe. Erratic variations occur at all levels. In the observable range, measurements of the same quantity made by different individuals will vary. Further, measurements made of any dimension, if done with sensitive instruments, will show variation. Below the observable range, molecules move rapidly in gases, more slowly in liquids, and oscillate in solids. On the nuclear level, radioactive nuclei emit particles and thus change to new elements in an individually unpredictable way. We can analyze these types of behavior by the overall or statistical effect. It may seem hopeless to attempt to set up laws or rules for considering such intangible and individually unmeasurable things as fluctuations. The factor that saves the situation is the fairly definite pattern about the way a series of observations is distributed. The methods of analysis that we use are similar to those used by all scientists and engineers as well as professionals in business.

A nice analogy for systematic and random deviations is provided by the shot patterns obtained when a rifle is fired at a target from a machine rest. The rifle is aimed at the center of the "true" target just as we try to secure experimental data which will provide a "true" value. The patterns shown in Fig. 2-2 represent samples of shots from two different rifles, each set up and aimed at identical targets in the same manner.

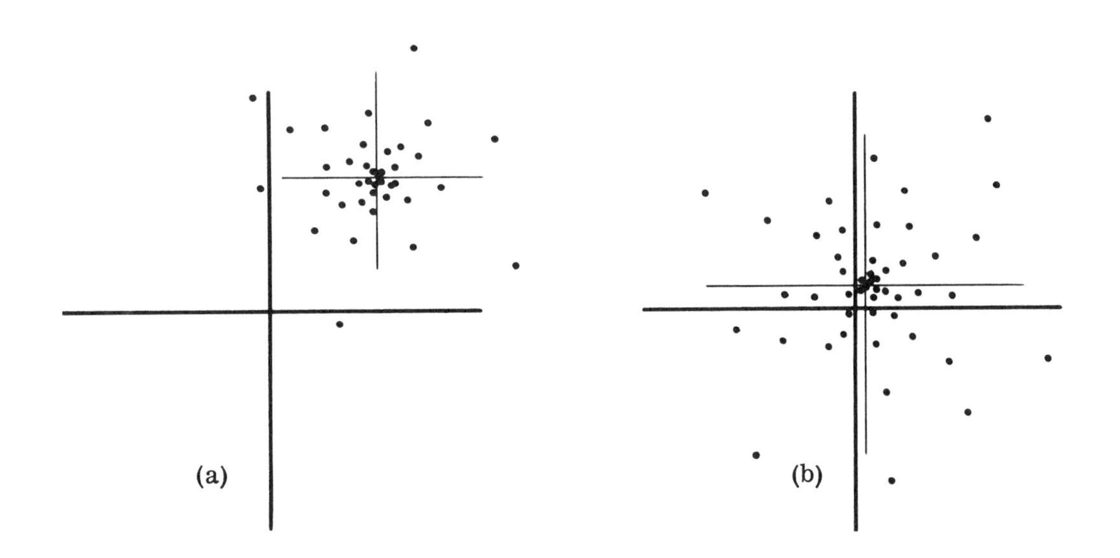

Fig. 2-2

In both cases, the center of fire is systematically displaced from the target center, less in pattern (b) than in (a). The spread or dispersion of the individual "values" (whose individual locations vary randomly and cannot be predicted in advance) is less in (a) than in (b). The word precision has been recommended as an inverse measure of dispersion or scatter. On this basis, the precision of pattern (a) exceeds that of (b).

As an example of fluctuations in the quantity being measured bullets are carefully removed from each of the 50 cartridges in a standard box of 0.22 caliber long-rifle cartridges. Cartridges are removed at random from the box, and the bullets weighed on a balance. With an instrumental limit of error of ±0.00005 g. The balance is of the modern one-pan type, with interval weight addition and digital, optical display of weight values, so that no further errors are introduced. It will be clear from inspection of Table 2-1, which lists the 50 values in the order of their occurrence (not arranged in increasing order), that variations in the quantity being measured are much larger than the negligible fluctuation introduced by the measurement process. While the "mass of a bullet" values that are measured are obviously not those of one particular bullet, the variations will be handled as if they were.

Table 2-1

Mass (grams) of 50 Observations

(1) 2.5696	(18) 2.5625	(35) 2.5586
(2) 2.5725	(19) 2.5776	(36) 2.5745
(3) 2.5693	(20) 2.5819	(37) 2.5700
(4) 2.5780	(21) 2.5666	(38) 2.5678
(5) 2.5735	(22) 2.5595	(39) 2.5865
(6) 2.5816	(23) 2.5608	(40) 2.5730
(7) 2.5658	(24) 2.5637	(41) 2.5712
(8) 2.5788	(25) 2.5768	(42) 2.5587
(9) 2.5613	(26) 2.5713	(43) 2.5693
(10) 2.5658	(27) 2.5769	(44) 2.5778
(11) 2.5713	(28) 2.5669	(45) 2.5578
(12) 2.5715	(29) 2.5747	(46) 2.5632
(13) 2.5612	(30) 2.5687	(47) 2.5746
(14) 2.5694	(31) 2.5746	(48) 2.5690
(15) 2.5681	(32) 2.5643	(49) 2.5745
(16) 2.5660	(33) 2.5685	(50) 2.5742
(17) 2.5523	(34) 2.5513	

When arranged in increasing order, these 50 values extend from 2.5513 gm to 2.5865 gm, with 25 values above and 25 below 2.5693. To examine the pattern of occurrence of various values, we may use a tool called a histogram, which is a plot of the number of times a given value or group of values occurs. To obtain a reasonable histogram, the values may be grouped into 8 classes as shown in Table 2-2, the class interval or difference being 0.0005 gm.

Table 2-2

Measurement in Grams

Central Value	Number of Occurrences	Lying Between
2.550	1	2.5475 - 2.5525
2.555	1	2.5525 - 2.5575
2.560	8	2.5575 - 2.5625
2.565	8	2.5625 - 2.5675
2.570	15	2.5675 - 2.5725
2.575	10	2.5725 - 2.5775
2.580	6	2.5775 - 2.5825
2.585	1	2.5825 - 2.5875

A histogram of the data in the table is shown below:

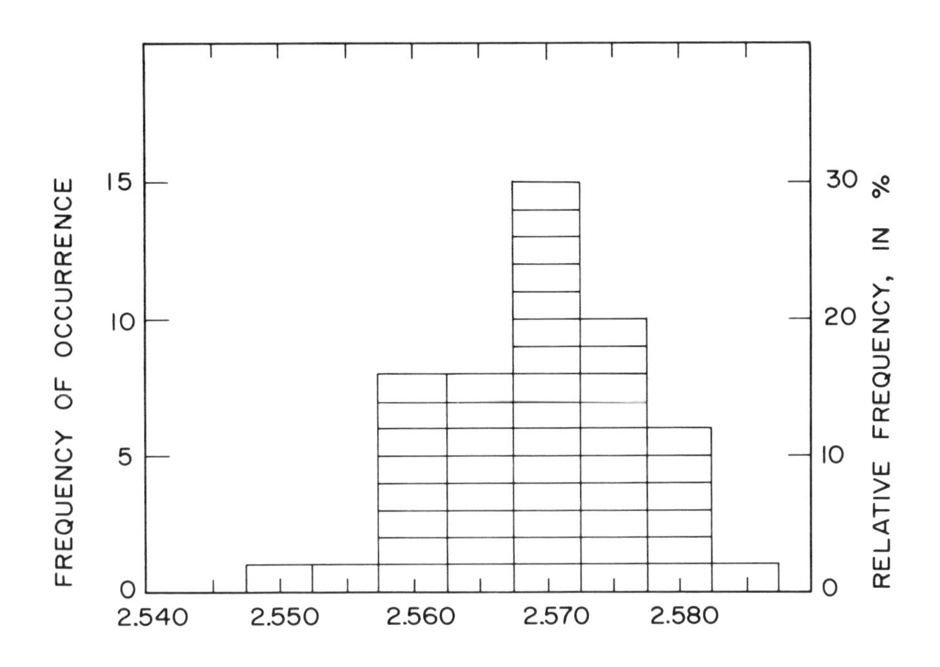

Fig. 2-3

The measurement values arranged in this way constitute a frequency distribution.

While the scope of this chapter is necessarily limited, statistical principles will be applied, with as much emphasis as possible on their physical significance, to develop reliable methods for obtaining a "best" value from a sample (distribution) and an indication of what confidence can be placed on it.

Inspection of the sample distribution indicates a number of simple, common characteristics:

(1) There is definite evidence of clustering about a central value, at or near 2.570 gm (near the mean of the 50 observations).

(2) Values which differ slightly from the central value occur frequently, or have a high probability of occurrence.

(3) Large variations from the central value occur seldom and, therefore, have a low probability of occurrence.

If more and more observations of the mass of the bullet were made and plotted on the histogram, the frequency distribution (due to random variations) would approach the bell-shaped normal curve shown in Fig. 2-4.

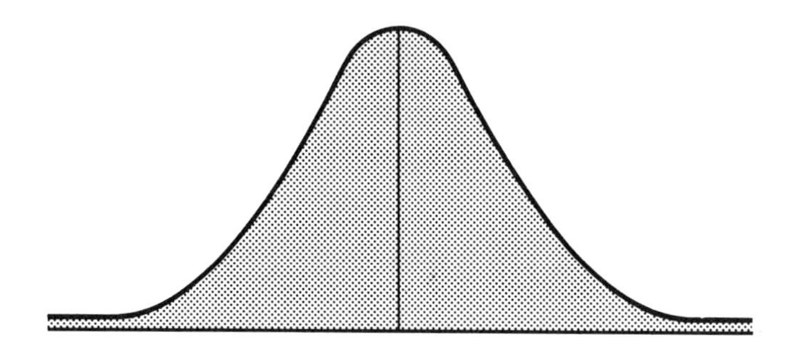

Fig. 2-4

2.4 The Standard Deviation of a Normal Curve

An important quantity called the "standard deviation" plays a key role in determining the portion of the precision index due to random fluctuations.

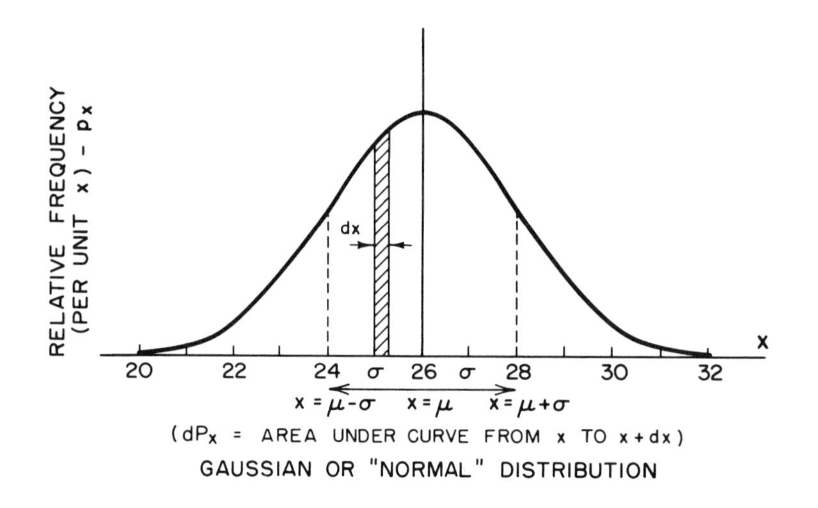

Fig. 2-5

Fig. 2-5 is also a normal curve approached when a very large number of measurements (observations) of a quantity are taken under the condition that the differences between the observations are due to chance or random causes only.

The value of x about which the curve is symmetrical is at $x = y = 26$. The other parameter σ indicates points of inflection at $x = \mu + \sigma$ and $x = \mu - \sigma$. dP_x at $x = 25^+$ is the area beneath the curve that represents the probability of occurrence of a value between $x = 25$ and $x = x + dx$. 68% of the area under the curve lies between $\mu - \sigma$ and $\mu + \sigma$. This means that the probability of a measurement being located within one standard deviation of the mean μ, $\mu \pm \sigma$, is ±68%. Therefore, σ can be used as a measure of the spread or dispersion of the x values included in the distribution.

If some systematic effect (a temperature shift or an instrumental disturbance, for example) occurs, the entire pattern may be shifted to a different central (or μ) value. The magnitude and cause of such systematic shifts can usually be located by a thorough analysis of the physics involved in the measurement process.

If the causes of random fluctuation are reduced by good management or good luck, the spread or scatter of the observations about the central value is reduced. This results in a smaller value of the parameter σ. ($\sigma_2 = 1$ for Curve B in Fig. 2-6)

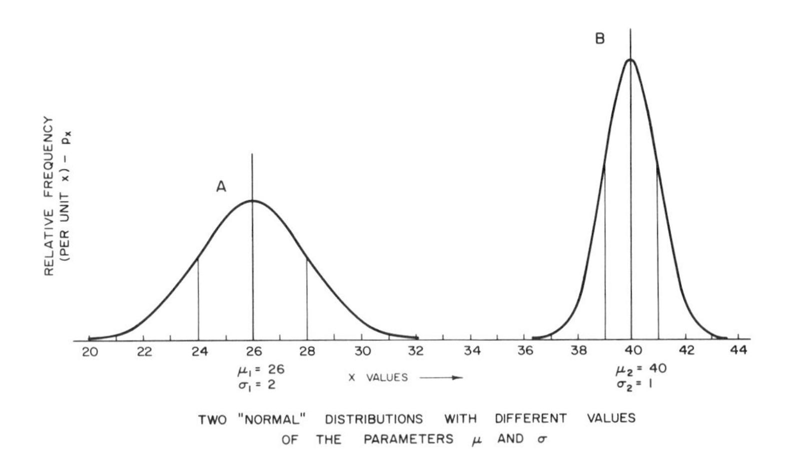

Fig. 2-6

Curve A in Fig. 2-6 is located at $\mu_1 = 26$. Curve B is at $\mu_2 = 40$. Curve A has a wider spread $\sigma_1 = 2$ than Curve B, where $\sigma_2 = 1$. Note that the difference $\mu_2 - \mu_1 = 14$ is nearly five times as large as $\sigma_1 + \sigma_2 = 3$. It is clear that the set of values centered at $\mu_2 = 40$ is statistically more reliable than that at $\mu_1 = 26$, since the spread, or scatter of $\sigma_2 = 1$ is less than $\sigma_1 = 2$.

To determine the standard deviation σ, you can follow the formula shown below, which is explained step by step.

$$\text{Standard Deviation} = \sigma = \text{S.D.} = \sqrt{\frac{\Sigma(x_i - \bar{x})^2}{n - 1}}$$

where $\bar{x}$ is the mean value

x_i are your individual observations

and n is the number of observations made.

The steps to follow are fairly straightforward:

(1) Find the mean of all the sample observations.
(2) Subtract the mean from each observation.
(3) Square each result.
(4) Add all of these squares.
(5) Divide this sum by the number of observations minus 1.
(6) Take the square root.

You'll need to take around 10 to 30 samples (observations) to determine an accurate value for the standard deviation. The greater the deviations from the mean, the greater the number of observations required (up to about 100).

Note that many calculators are either preprogrammed for this computation, or can easily be programmed to perform it. This can save considerable time in performing statistical analysis of your data.

Once the standard deviation is obtained, all normal curves may be divided into the following sections.

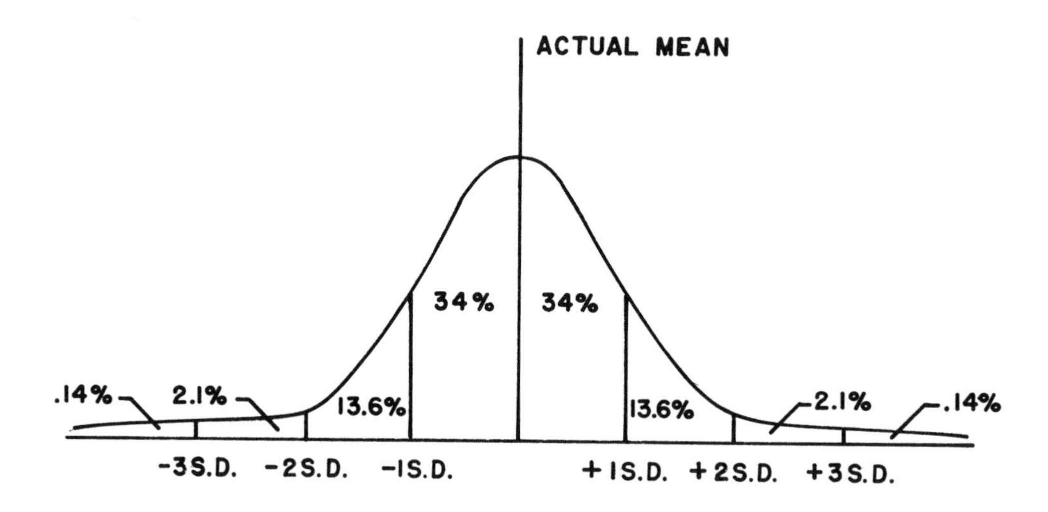

Fig. 2-7

As you can see in Fig. 2.7, very few observations fall outside of 3 S.D. from the mean. (.14% + .14% = .28% or about 3 out of 1000). Stated another way, your chance of taking just one observation and finding it beyond ±3 S.D. of the mean is only 3 out of 1000.

As a rule of thumb, observations which are more than 3 S.D. from the mean should be discarded since they are very likely the result of large sporadic errors.

The more observations you take, the better the average of the observations approximate the correct value for the observed quantity. In other words, the average value of the observations more closely approximates the correct value, as more observations are taken. However, the correct value can never be known (no matter how many observations are taken) more precisely than the instrumental limit of error ILE.

2.5 Precision Index

The formula for the precision index calculated from a certain number of observations (n) is ILE ± (3 S.D./$\sqrt{n}$). We'll refer to 3 S.D./$\sqrt{n}$ as the random error (RE). As you can see the RE can be reduced by taking additional observations, but the overall precision index cannot be reduced below the instrumental limit of error ILE.

Once the precision index is determined it should be expressed to one or at most two significant figures. This precision index should be used as part of your reporting of experimental results. It is one way of expressing the vital third factor you need to completely specify an observed or measured quantity.

A. EXAMPLE: FINDING THE PRECISION INDEX FOR THE 50 BULLETS

Using our previous observations for the masses of the 50 bullets, the sample standard deviation σ or S.D. equals ± 0.00744. We had previously mentioned that the instrumental limit of error (ILE) for the scale is $\pm.00005$ g. Using this data we can compute the precision index, $\pm$(ILE $\pm$ RE)* as: $\pm(.00005 \pm 3(0.00744)/\sqrt{50})$ or ± 0.003206. The precision index should be rounded to one or at most two significant digits, so we'll use 0.0032. As you can see, in this case the RE is by far the most significant portion of the precision index. (This isn't always the case.)

The precision index can also be given as a percent of the mean (2.5693). When the precision index is specified this way, it is referred to as the relative or (%) error. For the 50 bullets the relative error is $.0032/2.5693 = .00124$ or $\pm.12\%$.

Now that you have a feel for the procedure to follow in determining a precision index, here's an outline to follow for reporting your experimental data:

B. OUTLINE OF THE PROCEDURE USED TO DETERMINE THE PRECISION INDEX FOR AN EXPERIMENTAL VALUE

(1) When ready for measurement, take a few observations carefully as a trial or test set (about 4) and observe the deviations. If there are no easily readable deviations or if they are very small compared to the ILE, then a large data set should not be taken because the random error RE is negligible when compared to the ILE. Use the ILE as the precision index.

(2) If the deviations compare with or are larger than the ILE, then take a larger number of observations and determine the sample standard deviation. The larger the deviations the larger the number of observations needed. (Note that 10 to 30 are sufficient in most cases.)

(3) Discard all observations which deviate from the mean value of the data by more than 3 standard deviations (since they are probably the result of drastic errors).

(4) Calculate the precision index using the formula:

*Although we express the ILE as $\pm$ and the RE as $\pm$, we will always calculate the precision index by adding the ILE to the RE. Of course, there is a remote possibility that errors might cancel due to unknown effects. The chances against the total error in the mean $\bar{x}$ being greater than the precision index is thousands to one (i.e., virtually impossible).

$\pm(\text{ILE} + \text{RE})$, where $\text{RE} = \frac{3\ \text{S.D.}}{\sqrt{n}}$ and $\sigma = \text{S.D.} = \sqrt{\frac{(x_i - \bar{x})^2}{n - 1}}$

(Remember $\bar{x}$ is the mean data value
x_i are your individual observations
n is the number of observations)

(5) Your final experimental result should be expressed as follows:

$\bar{x} \pm$ (PRECISION INDEX)* (units)

$\bar{x}$ = mean of data values; PRECISION INDEX = $\pm(\text{ILE} + \text{RE})$

2.6 A Program to Find the Precision Index

The TI-58 or TI-59 calculator is preprogrammed to calculate the mean ([2nd] [X̄]) and the sample standard deviation ([INV] [2nd] [X̄]), so we'll use these keystrokes in a program to find quantities related to the precision index. We'll use label A to enter the observations, label B to calculate the mean, label C to calculate the mean ± 3 S.D., label D to calculate the precision index, and label E to compute the relative error.

A. THE PROGRAM

PURPOSE/FUNCTION	*PROGRAM KEYSTROKES*	*STEP #*
Clear calculator completely, get into "LEARN"	OFF/ON [LRN]	000
Initializes: Clears memories and display	[2nd] [Lb1] [2nd] [A']	002
	[2nd] [CMs] [CLR] [R/S]	005
Enters each observation (and prints it, if printer is attached)	[2nd] [Lb1] [A] [2nd] [Prt]	008
	[2nd] [Σ +] [R/S]	010
Calculates the mean (and prints)	[2nd] [Lb1] [B] [2nd] [X̄]	013
	[STO] 10 [2nd] [ADV]	016
	[2nd] [Prt] [R/S]	018
Calculates the mean plus 3 S.D.	[2nd] [Lb1] [C]	020
	[INV] [2nd] [X̄] [×] 3	024
	[=] [STO] 11 [+]	028
	[RCL] 10 [=]	031
	[2nd] [Prt] [R/S]	033
Calculates the mean minus 3 S.D. (and prints)	[RCL] 10 [-] [RCL] 11	038
	[=] [2nd] [Prt] [R/S]	041

*Note: The term Limit of Error LE is sometimes used in Chapters 7-14 for Precision Index. Therefore, LE = ±(ILE + RE).

To delete an entered value	[2nd] [Lb1] [2nd] [C'] [INV] [2nd] [Σ +] [R/S]	043 046
Enter ILE and calculate precision index	[2nd] [Lb1] [D] [+] [RCL] 11 [÷] [RCL] 03 [√x] [=] [STO] 12 [2nd] [Prt] [R/S]	049 055 058 060
Calculate relative error	[2nd] [Lb1] [E] [RCL] 12 [÷] [RCL] 10 [X] 100 [=] [2nd] [Prt] [R/S]	062 067 072 074
Get out of "Learn", and reset	[LRN] [RST]	0

B. USER INSTRUCTION

OPERATION	*PRESS*	*DISPLAY/COMMENTS*
Turn the calculator OFF, then ON, then enter program keystrokes carefully.		
Clear calculator for a data entry sequence	[2nd] [A']	0
Enter an observation (Repeat for each observation	[A] (NOTE: Be certain to allow enough time between entries for calculation to complete)	Number of observations entered is displayed (observation is printed)
Compute the mean	[B]	Mean displayed (and printed)
Calculate the mean + 3 S.D.	[C]	Mean + 3 S.D. displayed (and printed)
Calculate mean - 3 S.D.	[R/S]	Mean - 3 S.D. displayed (and printed)

NOTE: At this point, each of your individual observations should be examined to see if its within above limits (± 3 S.D.). If not, it should be deleted from your data set.

To delete a value which is outside the ± 3 S.D. limit, enter it and press	[2nd] [C']	Number of observations left is displayed

NOTE: If values are deleted, repeat computation of the mean, etc.

Enter ILE and compute precision index = ILE + $\frac{3\ S.D.}{\sqrt{n}}$	[D]	Precision index displayed (and printed)
Compute relative error (if desired)	[E]	Relative error in % displayed and printed

C. EXAMPLE

A meter stick with an ILE of ±0.02 cm is used to measure the length of a wooden rod with fairly flat ends.

Systematic errors can be minimized by:

(1) Making sure that the meter stick is parallel to the rod
(2) Placing the scale of the meter stick adjacent to the rod
(3) Avoiding the use of the ends of the meter stick

A set of 9 observations is taken with the following results.

96.44
96.12
96.28
96.34
96.45
96.38
96.10
96.38
96.30

What is the precision index for the length of the wooden rod?

Initialize	[2nd] [A']	0
Enter observation	96.44 [A]	1
	96.12 [A]	2
	.	.
	.	.

NOTE: Allow enough time between entries for computation to finish!

	.	.
	.	.
	96.38 [A]	8
	96.30 [A]	9
Find the mean	[B]	96.31
Find the mean ± S.D.	[C]	96.69006578
	[R/S]	95.92993422
(All are within ± RE)		
Enter the ILE	.02 [D]	.1466885946
Relative precision	[E]	.1523087899

The length of the wooden rod measured may be specified as 96.31 ± .15 cm or 96.31 ± .15% cm.

2.7 The Precision Index for Indirect Measurements

A. INDIRECT MEASUREMENTS

Sometimes you will use a measured number in calculating another quantity. Assume that you have found the diameter (D) of a sphere to be 18.13 ± .02 cm. You now wish to compute the volume (V) of the sphere from the measured diameter. Since the volume will be calculated from the diameter, the volume is called an indirect measurement. The question arises, what is the precision index for the volume? The derivative of the volume (V) with respect to the diameter (D) will help in analyzing this situation. For small changes $\frac{dV}{dD} \doteq \frac{\Delta V}{\Delta D}$ so $\Delta V \doteq \frac{dV}{dD}\Delta D$ where ΔD is the precision index for the diameter ($\Delta D = \pm.02$ cm).

To compute the precision index in this case, first calculate dV/dD:

$$V = 4/3\ \pi r^3$$

$$r = D/2, \text{ so } V = 4/3\ \pi D^3/8 \text{ or } (\pi/6)D^3$$

$$\frac{dV}{dD} = \frac{d}{dD}[(\pi/6)D^3] = 3(\pi/6)D^2 = (\pi/2)D^2$$

Now,

$$\Delta V \doteq \frac{dV}{dD}\Delta D \text{ and substituting } \frac{dV}{dD} = (\pi/2)D^2$$

$$\Delta V = (\pi/2)(18.13)^2(\pm.02) = \pm 10.32$$

The volume is $(\pi/6)D^3 = (\pi/6)(18.13)^3 = 3120.268987$, and can be specified as $3120 \pm 10\ cm^3$.

B. RELATIVE INDIRECT CALCULATIONS

Suppose that you are considering a diameter measurement whose precision index is known to be ±.11% (or ± .0011), and you want to know the fractional change in the volume $\frac{\Delta V}{V}$.

We have previously shown that: $\Delta V = (\pi/2)\ D^2\,\Delta D$. Dividing by $V = (\pi/6)\ D^3$ gives

$$\frac{\Delta V}{V} = \frac{(\pi/2)\ D^2\,\Delta D}{(\pi/6)D^3}. \quad \frac{\Delta V}{V} = 3\frac{\Delta D}{D}, \quad \text{so } \frac{\Delta V}{V} = 3(.0011) = .0033 \text{ or } \pm.33\%$$

This result can be generalized to the following: When an indirect measurement involves a direct measurement raised to a power n, the relative error in the indirect measurement is n times as large as the relative error in the direct measurement.

C. INDIRECT (CALCULATED) MEASUREMENT BASED ON TWO OR MORE DIRECT MEASUREMENTS

The quantity Z is computed from the independent measurements a, b, c, ... that have the precision indices Δa, Δb, Δc, ... respectively.

The precision index for Z is found by:

(1) Taking the partial derivative of Z with respect to a, and multiplying by Δa. Note: Treat b, c, etc., as constants when taking the partial derivative with respect to a.

(2) Square the result.
Repeat Step (1) and (2) for each independent variable (b, c, etc.).
(For instance take the partial derivative of Z with respect to b, multiply by Δb and square, etc.)

(3) Sum the squares.

(4) Take the square root.

These steps can be summarized as

$$\pm\ \Delta Z \doteq \sqrt{\left(\frac{\partial Z}{\partial a}\Delta a\right)^2 + \left(\frac{\partial Z}{\partial b}\Delta b\right)^2 + \left(\frac{\partial Z}{\partial c}\Delta c\right)^2 + \dots},$$

where $\frac{\partial Z}{\partial a}$ is the partial derivative of Z with respect to a, etc. This numerical precision index ΔZ is called the probable limit of error and has about the same probability of occurrence as the precision index determined by $\pm$(ILE + RE).

(5) The maximum limit of error for the quantity Z is also computed from the independent measurements a, b, c, that have the precision indices Δa, Δb, Δc. As indicated in Step (1) above, take the partial derivative of Z with respect to a, multiply by Δa; then add the partial derivative of Z with respect to b, multiply by Δb, etc. Therefore, the maximum limit of error for Z becomes:

$$\pm\ \Delta Z \leq \frac{\partial Z}{\partial a}\Delta a + \frac{\partial Z}{\partial b}\Delta b + \frac{\partial Z}{\partial c}\Delta c$$

D. EXAMPLE

What is the precision index for the volume of a cylinder when the diameter (D) is known as 6.32 ± 0.20 cm and the height (h) as 13.46 ± 0.30 cm? The formula for the volume is:

$$V = \pi r^2 h. \quad \text{Since } r = \frac{D}{2},$$

$$V = \frac{\pi}{4} D^2 h.$$

Now the job is to evaluate

$$\Delta V \doteq \sqrt{\left(\frac{\partial V}{\partial D}\Delta D\right)^2 + \left(\frac{\partial V}{\partial h}\Delta h\right)^2}$$

$$\frac{\partial V}{\partial D} = \frac{\pi}{4} \times 2D \times h$$

$$\frac{\partial V}{\partial h} = \frac{\pi}{4} D^2$$

We'll substitute these results into the equation for ΔV and plug in: $D = 6.32$, $\Delta D = \pm 0.2$, $h = 13.46$, and $\Delta h = \pm 0.3$ which gives:

$$\text{(Probable)} \qquad \Delta V \doteq \sqrt{\left(\frac{\pi}{4} \times 2 \times 6.32 \times 13.46 \times 0.2\right)^2 + \left(\frac{\pi}{4} \times 6.32^2 \times 0.3\right)^2}$$

Our result for ΔV is: ± 9.41 or $\pm 9.4\ cm^3$.

The volume, $\frac{\pi}{4} D^2 h = \frac{\pi}{4}(6.32)^2(13.46) = 1136.456322\ cm^3$, and carries the precision index 9.4. The volume may be stated as $1136 \pm 9.4\ cm^3$.

2.8 Significant Figures or Digits

A computed result is no more accurate and no more reliable than the least accurate number entering the computation.

Only those figures or digits of a numerical quantity which are the result of actual measurement, or calculation from an actual measurement, are said to be significant. For example, the quantity (0.0000176/C°) has three significant figures: the 1, the 7 and the 6. The zeros serve only to place the decimal point. Other ways of writing the same quantity would be: 176×10^{-7}, 1.76×10^{-5} (/C°). In every case, there are still 3 significant figures.

As ordinarily used, the zero may or may not be significant, for example:

(1) If it serves only to place the decimal point (as above), or in the value (4,800,000 tons), it is not significant.

(2) If it is the result of actual measurement, it is significant (i.e., 98.0 cm).

(3) The use of the exponential (powers of 10) method of expression will eliminate any question about whether the zero is significant. When a quantity is given as 4,8000,000 tons, it is difficult to determine if any of the zeros have a purpose other than to locate the decimal point. It may indicate 4,800,000 tons as distinct from 4,900,000 tons, 4,800,000 tons as distinct from 4,810,000 tons, or even 4,800,000 tons as distinct from 4,801,000 tons. If we assume that the measurement was made with a final limit of error (LE) of ± 10,000 tons, then the first of the zeros in 4,800,000 ± 10,000 tons is significant while the rest are not. We can avoid the above difficulty by writing the quantity as 4.80×10^6 tons. As a general rule:

All digits placed in front of the power of ten should be significant figures.

In general, the *number* of significant figures gives a rough idea of the reliability of a quantity. Also, one doubtful (but measured, perhaps estimated) figure is always included in the result of any measurement or calculation.

Some useful rules for computations (basis of significant figures) are:

1. Unless some precision index such as the standard deviation (σ) or the precision index is given, the undertainty of the last figure may be interpreted as ± 1. For example, 1.76×10^{-5} without further data, would imply a value lying between 1.75×10^{-5} and 1.77×10^{-5}.

2. In dropping superfluous (non-significant figures) raise the last digit that is retained by 1 if the first of the dropped digits is greater than 5. For example, 32.86 rounded to 3 significant figures would be 32.9, while 32.852 rounded to 3 figures would also be 32.9.

3. When the digit to be dropped is an even 5, the last retained digit is rounded to the nearest *even* value. For example (to 3 significant figures):

   32.65 is rounded to 32.6
   32.75 is rounded to 32.8
   32.85 is rounded to 32.8
   32.95 is rounded to 33.0

4. When *adding*, it is useless to keep any more decimals than are present in the number having fewest decimals. It saves time to drop superfluous figures *before* adding. This is illustrated by the addition (29.32 + 0.01853 + 2.033). In this case there is no need to keep more than 2 decimals in any term.

$$\begin{array}{r} 29.32 \\ 0.02 \\ \underline{2.03} \\ 31.37 \end{array} \quad \text{is the result.}$$

   To show this clearly, the last significant (first doubtful) digit is indicated by a bar above it. Completely unknown digits are represented by x. We now have:

$$\begin{array}{l} 29.3\bar{2}\text{xxx} \\ \;\;0.0185\bar{3} \\ \;\;\underline{2.03\bar{3}\text{xx}} \\ 31.3\bar{7}\text{xxx} = 31.37 \end{array}$$

5. When multiplying or dividing, the result will have essentially the same number of significant figures as occur in the term with the fewest. For example, $(103.\bar{4} \times 0.9\bar{9} = 10\bar{2})$. In this instance, three figures were used, since one part in 99 means an uncertainty of about $\pm 1\%$.

   If we have 1.04π, then $1.0\bar{4} \times 3.14 = 3.27$ and not $1.04 \times 3.1416 = 3.267264$.

2.9 Summary

Real experiments are subject to errors of two general types:

1. Systematic
2. Random or statistical

Systematic errors are due to assignable causes and may, at least in theory, be calculated, compensated or balanced out to an extent which depends on our knowledge of the physics of the experiment. The process may be expensive in time and money.

While reducible, random errors cannot be avoided. They must be treated by statistical methods which enable us to calculate the best representative values (usually means of a set of values) and fairly reliable error limits (the odds are very great that the true value lies within these limits). We have also learned how errors grow or "snowball" in an indirect measurement.

The statistical theory underlying our methods has not been derived. In a few cases, exact theory has been somewhat oversimplified for convenience. Since we have been conservative in all respects, our calculated error limits are wide enough to avoid difficulties due to oversimplification.

QUESTIONS

1. Decide whether each of the following is best classified as a direct or as an indirect measurement and justify your decision. The majority of them are performed in good secondary school physics courses. Completely unfamiliar processes may be omitted; consult appropriate reference to refresh your memory of those about which you may be vague.

   (a) Measurement of force via use of a spring balance.

   (b) Measurement of liquid volume using a pipette.

   (c) Measurement of the specific gravity of a liquid by weighing a metal bob in water and in the liquid.

   (d) Measurement of atmospheric pressure using a:

      (1) A mercury column barometer
      (2) An aneroid (diaphragm type) barometer

   (e) Measurement of relative acidity with litmus paper.

   (f) Measurement of electric current, using an indicating ammeter, with pointer and scale.

   (g) Measurement of the diameter of a dime, using a screw type micrometer caliper.

   (h) Measurement of resistance using a voltmeter and an ammeter.

   (i) Measurement of resistance using a slide-wire bridge.

   (j) Measurement of the luminous intensity of a light source using a Bunsen ("grease-spot") photometer.

(k) Measurement of the focal length of a lens using:

(1) A burning-spot image of the sun.
(2) An optical bench with a linear scale, plus an object and focussed image.

(l) Consider any five measurements, not mentioned above, which you have made at some time or another.

2. For small angles, tan θ and sin θ are often taken to be equal and also replaceable by θ in radians. Find the systematic error introduced by replacing tan θ by θ at $\theta = 2°$ and at $\theta = 10°$. (Find the sign, absolute value and percent.) If tan θ were replaced by sin θ instead of θ, would the error be larger or smaller?

3. At 30°: - $\theta = 0.52360$ radians

$\sin\theta = 0.50000$	$d/d\theta \, \sin\theta = \cos\theta$
$\cos\theta = 0.86603$	$d/d\theta \, \cos\theta = -\sin\theta$
$\tan\theta = 0.57735$	$d/d\theta \, \tan\theta = 1/(\cos\theta)^2$

Suppose that for a measured angle of 30°, there is a systematic error of +1% (+0.3 degrees or +0.005 radians). Find the sign, absolute value and relative (percent) value of the error in sin θ, cos θ and tan θ.

(1) By direct use of trigonometric function tables.
(2) By methods based on calculus.

4. The angular deflection of a galvanometer (and a number of other sensitive instruments) is often determined by reflecting a beam of light from a light mirror attached to the turning system, onto a scale some distance D away. This gives a long, weightless pointer and a free multiplication of 2X, since the light beam turns twice as much as the mirror does. (Why?)

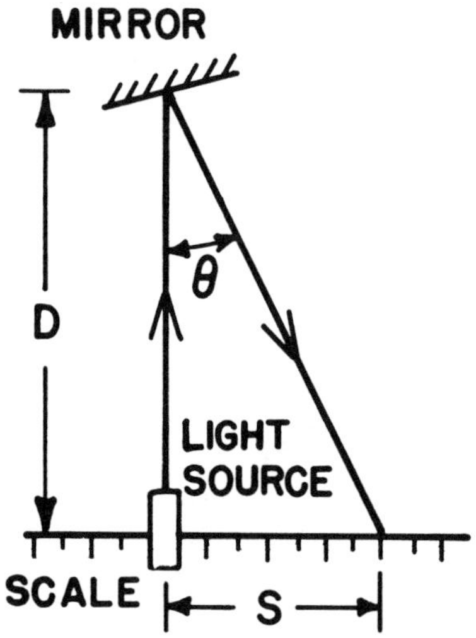

In a well-designed galvanometer, the angle turned is linearly proportional to the current through the instrument.

If a current I_1 produces a deflection $S_1 = 10.0$ cm, and a larger current I_2 produces a deflection $S_2 = 20.0$ cm, find the error made in assuming $I_2 = 2 I_1$. Let D = 100 cm.

5. The theory of special relativity tells us that the mass of a particle is not constant but is a function of the speed of the particle, the experimentally proven relationship being.

$$m = \frac{m_o}{\sqrt{1 - (v/c)^2}}$$

where m_o = the rest mass
v = speed of the particle
c = the velocity of light

In order to measure the ratio of the charge on the electron to its rest mass, this ratio being an important property of charged particles, the electrons must be travelling at an appreciable speed. We actually measure, therefore, the ratio of the charge e to the mass m which the electron has at the experimental speed v. If e/m is the measured ratio, then the ratio of the charge to the rest mass e/m_o is:

$$\frac{e}{m_o} = \left(\frac{1}{\sqrt{1 - (v/c)^2}}\right) \frac{e}{m}$$

PROBLEM: The e/m ratio for electrons is measured with electrons travelling at a velocity of $3(10)^7$ meters/sec (0.1 c). Find the factor by which this value should be multiplied in order to obtain e/m_o. What relative error (%) would be made by omitting this factor?

6. 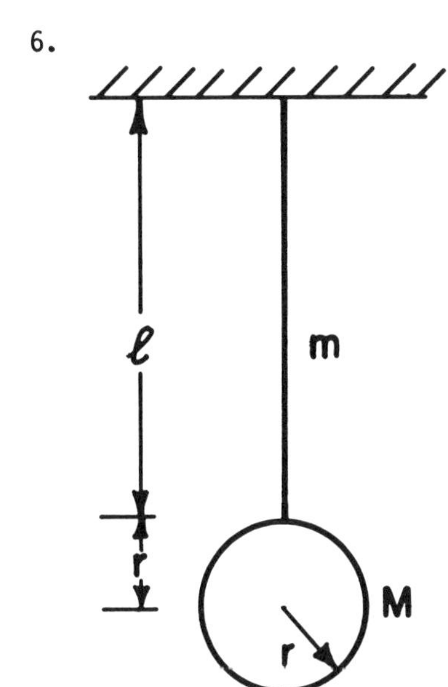

The idealized simple pendulum - a point particle of mass M supported by a weightless, inextensible cord of length L simply does not exist.

A dense metal ball (radius = r), of mass M, supported by a light wire or rod (mass = m) is considered a good approximation.

Suppose that such a pendulum is used to determine "g", the acceleration due to gravity.

Use the familiar equation for the period of a simple pendulum; solved for "g", a first approximation gives:

$$g = 4 \cdot \pi^2 \frac{(\ell + r)}{T^2} \qquad (T = \text{PERIOD})$$

This, of course, neglects the inertia of the supporting wire or rod and the physical size of the ball r.

The accurate equation is:

$$g = \left[4\pi^2 \frac{(\ell + r)}{T^2}\right] \left[\frac{1 + \dfrac{m\ell^2}{3\,M(\ell + r)^2} + \dfrac{2}{5}\dfrac{r^2}{(\ell + r)^2}}{1 + \dfrac{m\ell}{2\,M(\ell + r)}}\right]$$

(as above) [correction factor, > 1]

For the far-out case in which the rod or wire weighs 1/20 as much as the ball ($m/M = 0.05$) and ℓ is only 5 times the radius of the ball ($\ell = 5r$), find the correction factor and the percent systematic error introduced by omitting it.

7. At a particular location, sea level and 45° N. latitude, the value of "g", the acceleration due to gravity is 980.665 cm/sec^2. This value is considered a reference standard, and is often rounded off to 981 or 980 cm/sec^2 (9.81 or 9.8 meters/sec^2). g varies with latitude θ and with altitude above sea level H (as well as with the character of local underground mineral deposits). A semi-empirical equation for "g", involving θ

and H, (Helmert's Formula) is:

$$g = 980.616 - 2.5928 \cos 2\theta + 0.0069 \cos^2 2\theta - 3.086(10)^{-6} H$$

g in cm/sec^2
θ in degrees
H in centimeters

What does this equation predict for "g" at 45° N latitude and sea level? For a point at the equator and sea level? At the North Pole?

8. A steel wire expands in length by just about 10 parts per million (1 in 10^5) for each Celsius (centigrade) degree rise in temperature.

For a simple pendulum: $T = 2\pi\sqrt{\frac{L}{g}}$.

(a) What is the effect % on T (the period) of a rise in temperature of 10 C°?

(b) Is there any way in which a change in "g" could produce the same (or opposite) result? Explain.

9. The resistance of a metal wire can be fairly well stated as:

$$R = \rho \frac{L}{A}$$

R = resistance
L = length
A = cross-sectional area ($\pi/4\ D^2$)
ρ = resistivity, a characteristic of the material, dependent on temperature, among other things.

The resistance (resistivity) of many metals varies by just about 1/273 of the value at 0°C, for each change of 1 Celsius degree.
(Any comments on the significance of this?)

LET the initial temperature of a wire by 0°C, its length L and its diameter D.
NOW: Let the temperature rise by 10 C°.

(a) If the initial resistance were 100 ohms, the change would be?

(b) The change in Part (a) could have been produced, instead, by a change in L of ______? In D of ______? Would the effect of temperature on L and D condition the answer to (b)?

Chapter 3

GRAPHING - TABULAR DIFFERENCES

3.1 Introduction

Graphing is one of the simplest, though to many students the most difficult, of engineering or scientific techniques. The difficulty is caused by the lack of the main requisite for good graphing, JUDGMENT. The use of JUDGMENT and a group of general rules will enable you to construct a good graph.

The engineer or scientist is generally most interested in the quantitative type of graph which shows the relation between two variables in the form of a curve. (The term "curve" is applied to straight lines drawn from point to point or to regular smooth curves.)

Depending on the problem, graphing is done on either millimeter rectangular coordinate paper (Fig. 3-2), paper ruled 20 lines per inch, semi-log (Fig. 3-12), log-log (Fig. 3-14) or other forms, such as polar and triangular, which will not be discussed.

After proper graph paper has been selected, the requirements for good graphing are:

Choice and labeling of the coordinate scales. (Article 3.2)
Plotting of the points representing the data. (Article 3.3)
Fitting a curve to the plotted points. (Article 3.4)
Preparation of the title. (Article 3.5)

3.2 Choice and Labeling of the Coordinate Scales

A. CHOICE OF THE COORDINATE AXES

1. The independent variables are plotted as abscissas parallel to the X-axis and the dependent variables as ordinates parallel to the Y-axis. The independent variable is altered by steps, and the value of the dependent variable is determined for each value of the independent variable.

In the experiment of the VIBRATING SPRING, the independent variable (m = mass of the added weights) is varied and the effect on the dependent variable (T = period) is investigated. An equation relating the two variables (m, T) is expressed as:

$$T = f(m)$$

The choice of the independent variable is determined by the experimental approach or by the character of the data.

An example of the correct and incorrect choice of the independent variable is shown in Fig. 3-1.

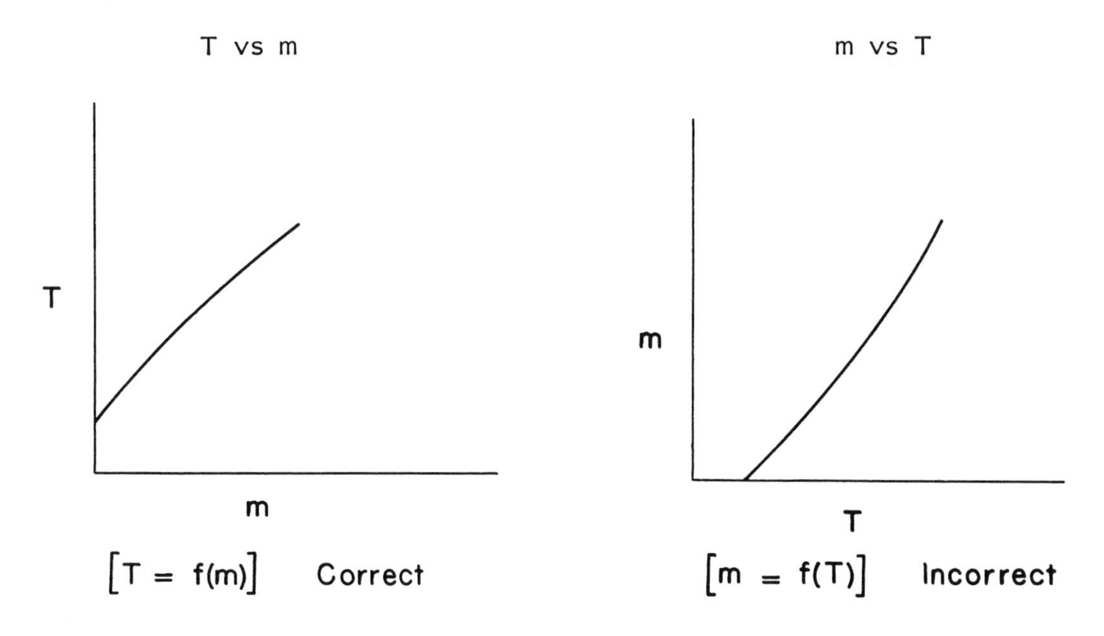

Fig. 3-1. The plot of T = f(m) is correct according to established custom. It is incorrect to plot m = f(T). Note that the independent variable in T = f(m) is plotted along the x-axis and that the period (T) is dependent on (m).

2. The scale is chosen so that one block = 1, 2, 5, 10 (occasionally 4), units. Do not use 3, 7, 9, etc. A graph should be easily readable.

3. Scales usually are not drawn along the boundary between graph and margin. Occasionally, because of the nature of the data and the extent of the graph, this may be necessary. Draw the scales one to two blocks within the graph so that the separation of the coordinate lines may act as guides when labeling (Fig. 3-2). The margin is the blank space around the printed coordinate paper, semi-log, log-log paper, etc.

4. Do not place scales at the top of the graph (Fig. 3-3). See Figs. 3-4 to 3-15 for good placement of scales.

5. Unless impossible, place scales so that if the graph is bound into a report, the scales can be seen.

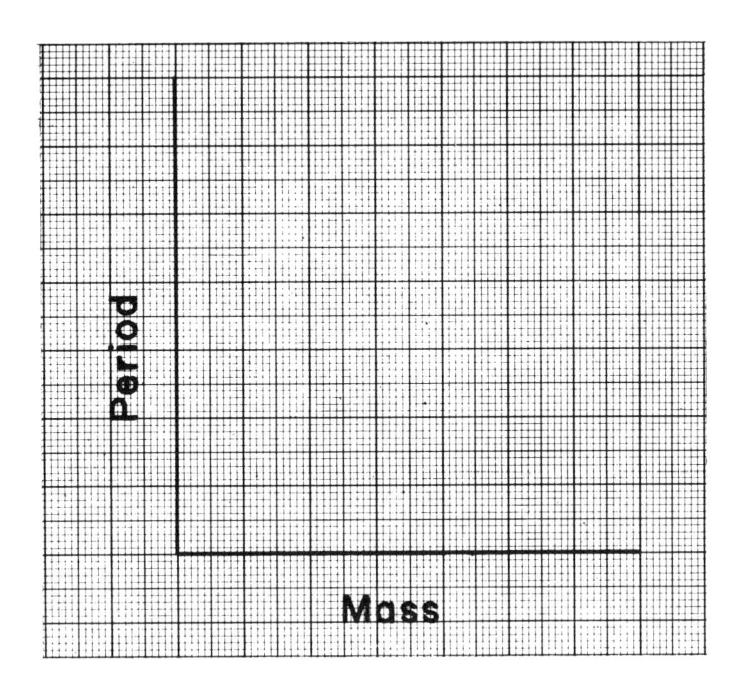

Fig 3-2. Scales drawn within graph margin.

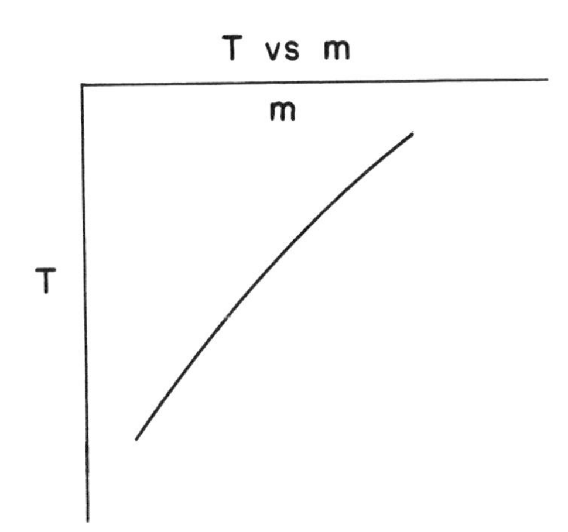

Fig. 3-3.
Poor placement of scales.

6. Scales are usually numbered so that the resultant curve is not confined to a small area of the graph paper (Fig. 3-4). It is not essential that every graph contain the point (0, 0), but if a zero is a significant point, it should appear. Inspect the data carefully and number scales so that each variable begins (when plotting points) near the lowest and highest values in the data. An exception is the plotting of calibration curves for instruments where the point (0, 0) is significant (see Fig. 3-6).

Graph I [Incorrect]

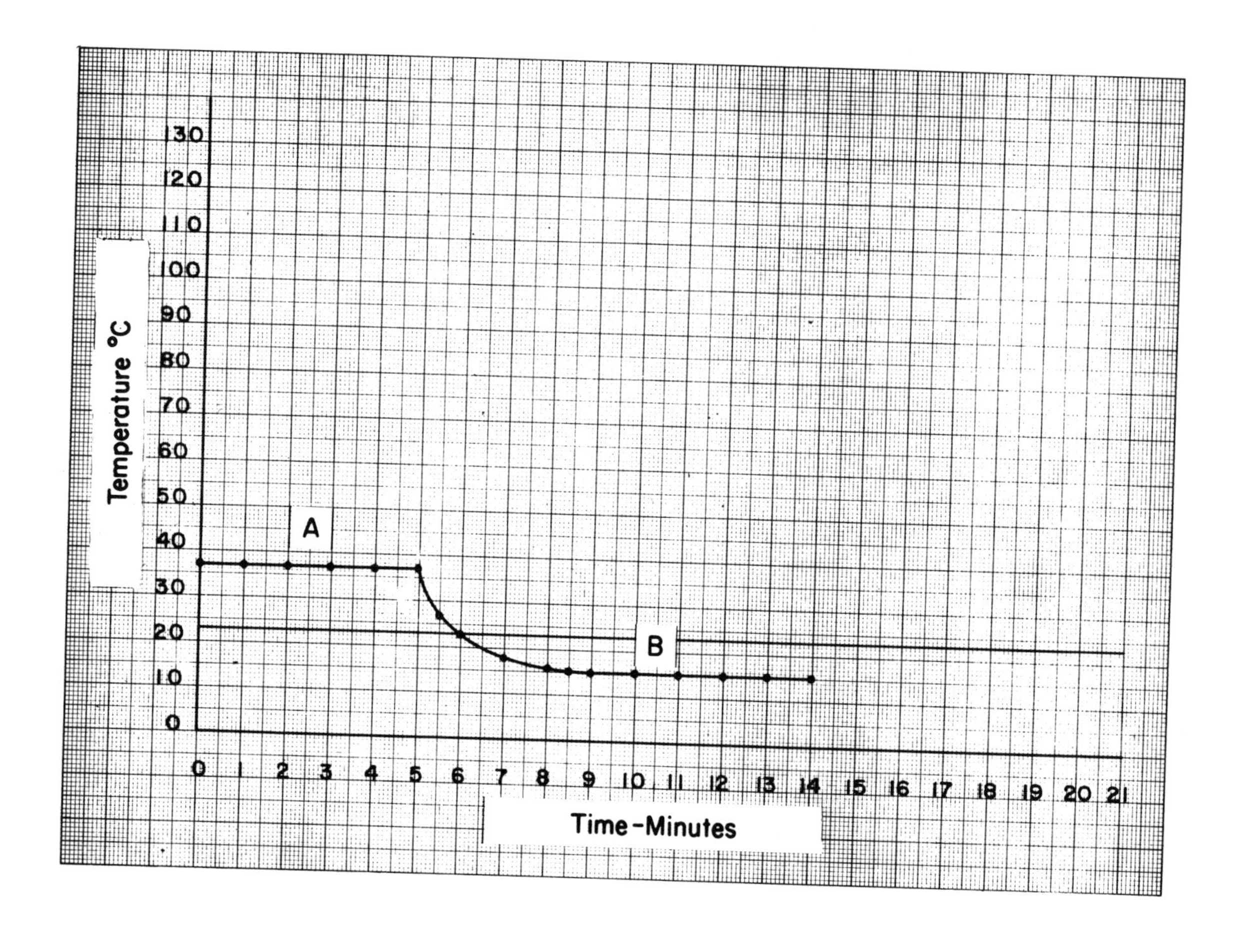

Fig. 3-4. In Graphs I and II, a Temperature vs. Time plot for an experiment on the Latent Heat of Fusion is shown. Graph I is incorrect because the blocks are labeled improperly and the zero for the temperature scale is not significant. Because the graph was confined to a small area, the slopes (A and B) are not sufficient to permit the corrected temperature to be found. (See discussion of temperature corrections in the section on Calorimetry, Chapter IX. In Graph II, the slopes (A' and B') show that the data variations are significant. Temperature corrections can be easily made.

Graph II [Correct]

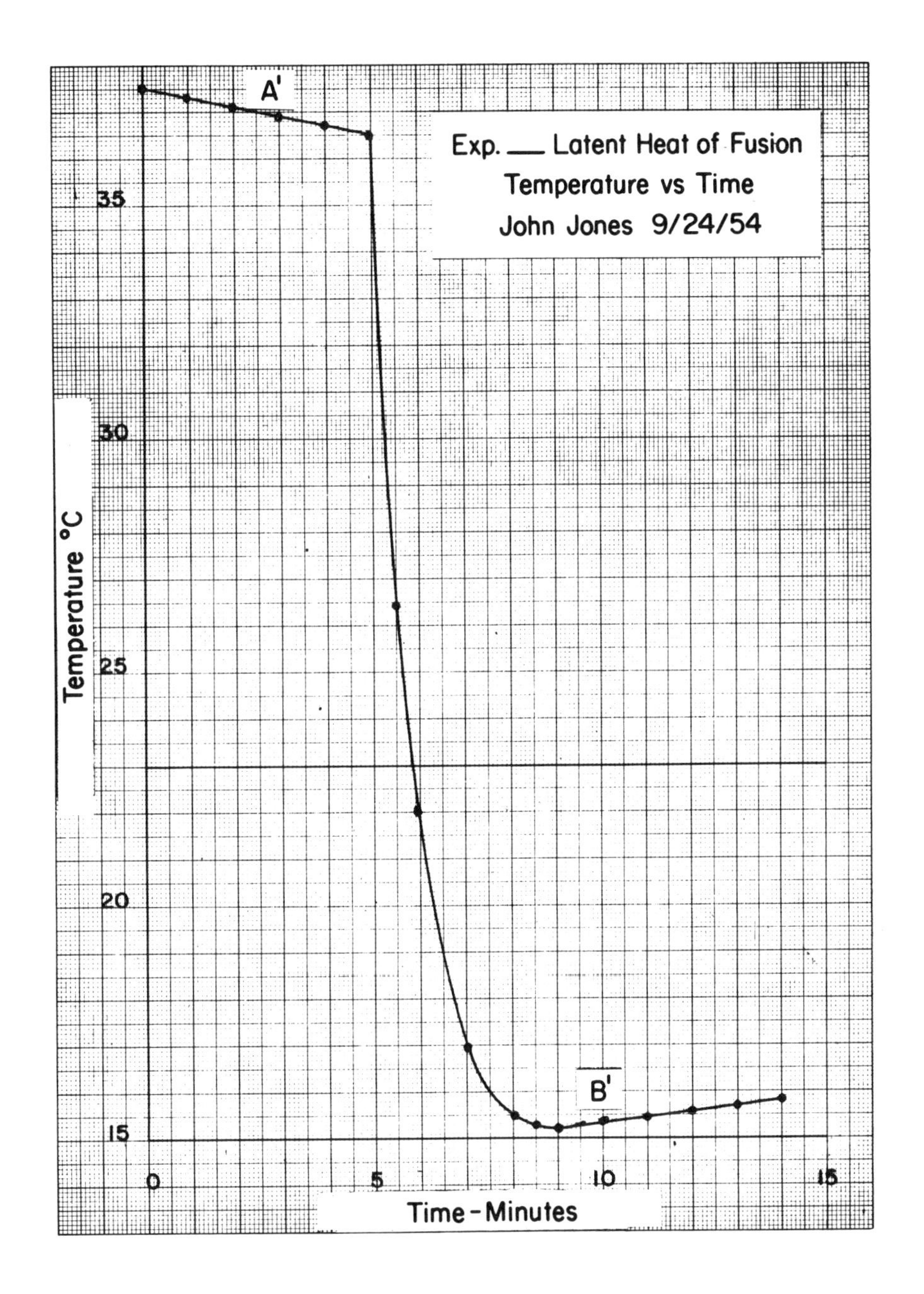

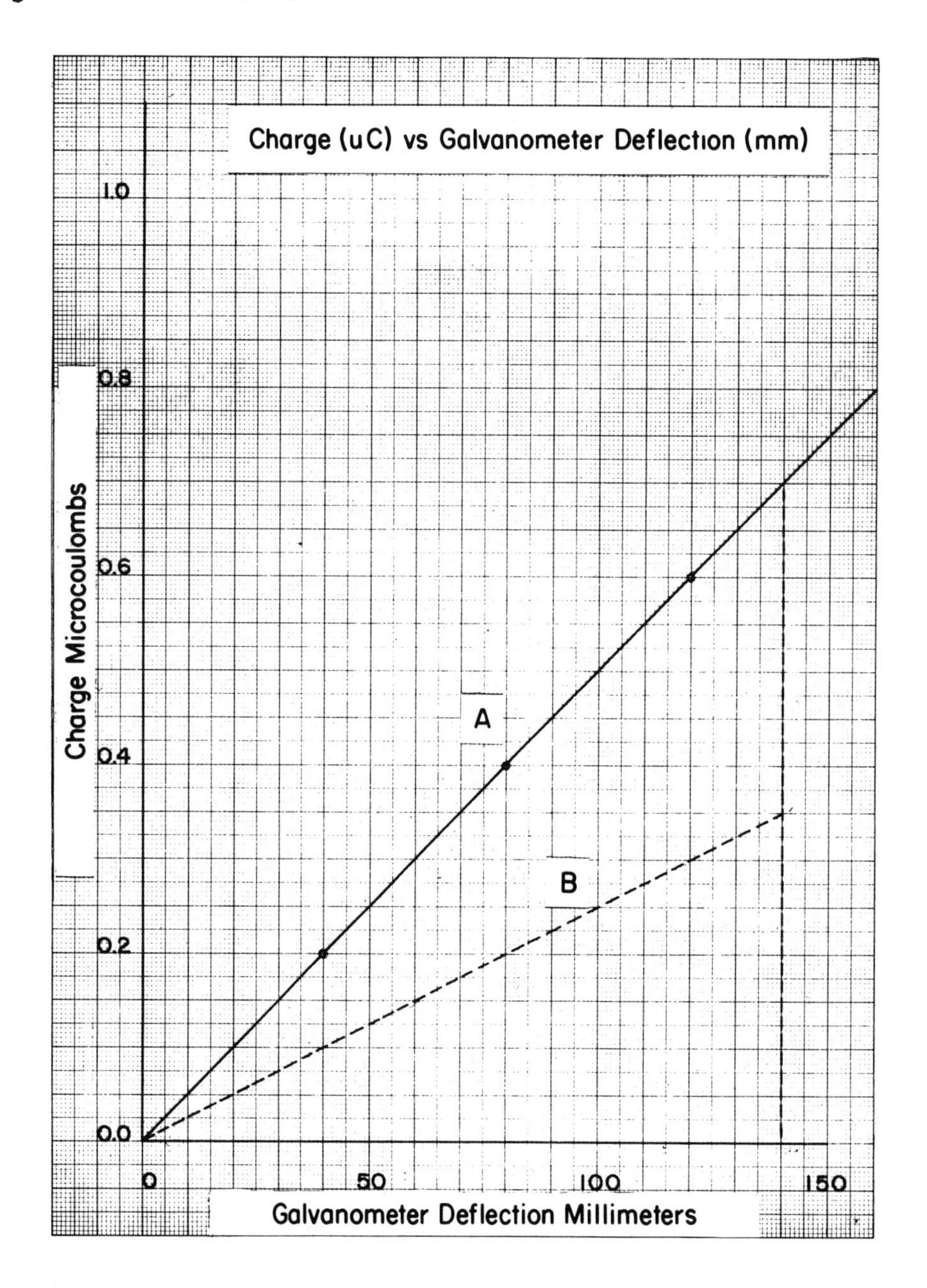

Fig. 3-5. A plot of Charge (microcoulombs) vs Galvanometer Deflection (millimeters). Curve A has a geometrical slope equal to $\frac{14 - 0 \text{ cm}}{14 - 0 \text{ cm}}$ or unity. The actual slope is equal to $\frac{0.7 - 0.0 \text{ microcoulombs}}{140 - 0 \text{ millimeters}} = 5 \times 10^{-3} \frac{\mu C}{mm}$. If the ordinate (Y axis) is changed in value to .1 micrcoulomb per centimeter, then Curve B will have a geometrical slope equal to $\frac{7 - 0 \text{ cm}}{14 - 0 \text{ cm}} = 1/2$, but the actual slope has the same value as Curve A or $5 \times 10^{-3} \frac{\mu C}{mm}$. Therefore, since the same data was plotted in each case, the choice of scale (numbering the blocks) determines the value of the geometrical slope, not that of the actual slope (rate of change of the function). Choice of scale does affect the ease and precision with which the actual slope may be determined. (See, also, Figures 3-4, 3-8 to 3-13 where the slopes are important.)

7. In choosing scales be sure to consider the limit of error (L.E.) of the points being plotted. The smallest graphical division should be less than or at the worst equal to the limit of error of the points being plotted. (See Article 3-4, Paragraph 3, for discussion on "error rectangle.")

8. Whenever possible, curves should have a geometrical slope approaching unity. This implies that the curve make an angle of approximately 45° with the axis of the abscissa. The geometrical slope is that found by drawing a tangent to the curve and then finding the tangent of the included angle. It depends only on the scales chosen and is a pure number.

THE GEOMETRICAL SLOPE IS NOT THE SAME AS THE ACTUAL SLOPE OF THE FUNCTION (dy/dx). The "Actual Slope" requires consideration of the units plotted along each axis (Fig. 3-5). By establishing a geometrical slope of unity, the precision of the plotted points is improved. Any abnormal spread in the data can be easily seen. Conversely, the use of slopes of less than unity conceal variations that might be important.

The use of semi-log and log-log graph paper should be avoided when precision is important.

In many instances, the data will not permit a choice of scale that will lead to a geometrical slope of unity. JUDGMENT is necessary to determine if the selection of a geometrical slope of unity is advisable.

When constructing a slope use as much of the curve as possible (Fig. 3-5). When placing a tangent line to a curve to determine the slope at the point, extend the line as far as possible in order to get a maximum range of values $\frac{y}{x}$. This improves the worth of the calculated value of the slope and also minimizes errors due to plotting technique. See Article 3.2 B, which follows.

B. CONSTRUCTION OF A NORMAL TO A CURVE TO FIND THE ACTUAL SLOPE OF A TANGENT LINE

(1) By use of a small mirror construct a normal (DAF) to the curve (BAC) at a point A. (Place the mirror at right angles to the curve and adjust so that the image of the portion of the curve (BA) when you look into mirror and sight above it is apparently superimposed on (AC). Use the edge of the mirror as a straight edge and draw a normal.

(2) The actual slope (see Fig. 3-5A) which corresponds to the actual slope of a tangent line constructed at A, can be found from:

$$\left(\frac{dy}{dx}\right)_{\text{actual at A.}} = \frac{\left(\frac{\partial y}{\partial x}\right)^2 \rightarrow \text{(Scale Factor)}}{-\left(\frac{dy}{dx}\right) \rightarrow \text{(Actual Slope of Normal)}} \qquad \text{(Eq. ii)}$$

(3) The scale factor for Fig. 3-5A is found by comparing the magnitude of a unit length, say one block along the Y-axis, to a corresponding unit length along the X-axis.

1 block along Y-axis = 100 mm (Hg); 1 block along X-axis = 5°C

$$\left(\frac{\partial y}{\partial x}\right)^2 = \left(\frac{100}{5}\right)^2 = (20)^2 \text{ or } 400 \frac{\text{mm}^2}{\text{deg}^2}$$

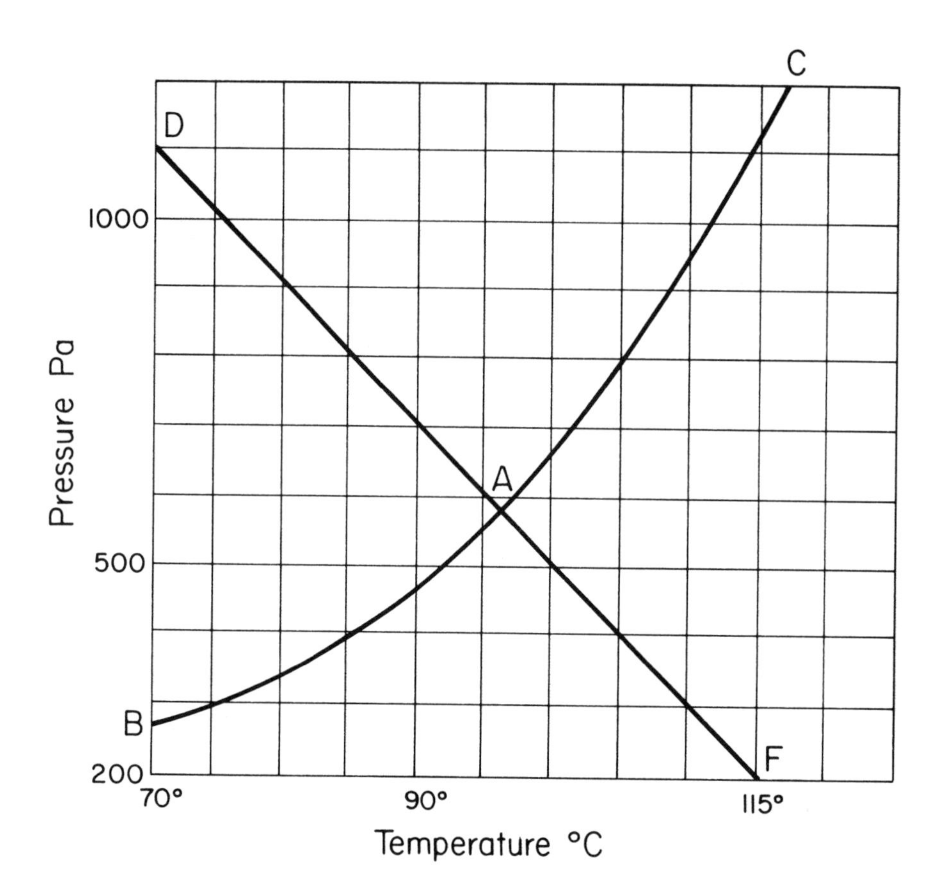

Fig. 3-5A

(4) The actual slope of the normal is:

$$\left(\frac{dy}{dx}\right) = \left(\frac{1100 - 200}{70 - 115}\right) = -\frac{900 \text{ mm}}{45^\circ} = -20 \frac{\text{mm}}{\circ}$$

$\left(\frac{dy}{dx}\right)$ is <u>negative because the normal line has a negative slope</u>.

Substitute the values of the scale factor $\left(\frac{\partial y}{\partial x}\right)^2$ and the actual slope of the normal $\left(\frac{dy}{dx}\right)$ into Eq. ii, as follows,

$$\left(\frac{dy}{dx}\right) = \frac{400 \frac{\text{mm}^2}{\text{deg}^2}}{-(-20) \frac{\text{mm}}{\text{deg}}} = 20 \frac{\text{mm}}{\circ}$$

(5) Please note that once the scale factor is found for a choice of scales, calculation of the actual slope of a tangent line at any point becomes quite simple. This method has been outlined in detail because it permits the actual slope of a tangent line to the curve to be found with good accuracy, if care is taken in construction of the normal.

C. LABELING OF THE COORDINATE SCALES

1. Label coordinates along each axis. Give quantity and units. (Fig. 3-5)

2. Do not label every block. Label every second, fourth, fifth, or tenth block. (Fig. 3-4,I shows incorrect labeling and Fig. 3-4,II correct labeling.)

D. PREPARATION OF THE TITLE

The title should be placed within the margin of the graph paper in a position where it does not interfere with the curve (see Fig. 3-4). There are exceptions to this rule, but in general it is considered poor practice to place captions in the white margin of a graph. The title must include an accurate description of the purpose of the graph. The exact content of the description depends on department or industry policy. The student should use the following form for the title of a graph:

Experiment Number ____________ Name of Experiment ______________________________

Dependent Variable vs. Independent Variable

Name of Student ______________________________ Date ____________________

3.3 Plotting the Points Representing the Data

1. Experimentally determined points can be located by using horizontal and vertical lines in the form of a "+", which permit the consideration of one coordinate at a time. Often the "+" consists of a vertical bar whose length corresponds to the precision index of the measurement in the "y" direction, while the width of the horizontal bar has a similar function in the x direction. See "error rectangle" in Article 3.4.

2. Points can also be located by a small circle ○ (see Figs. 3-12 to 3-15.)

3. When more than one curve is drawn, distinguish between them by using different symbols, dotted or broken lines, or different colored inks.

3.4 Fitting a Curve to the Plotted Points and Empirical Equations

A. CURVE FITTING

1. The principal questions that arise when fitting a curve are:

   (a) Should the drawn curve pass through every point? or
   (b) Should it be drawn smoothly neat, but not necessarily through every point?

The points of a calibration curve (Fig. 3-6) are usually very accurately determined. We know very little of what occurs between points, and therefore draw straight lines between them. When checking a law or other functional relation, there is usually reason to suppose that a uniform curve (or straight line) will result. If one or even two points are quite far from the apparent curve, then check the experimental data to see if a mistake has been made. If none appears, the point may be, in general, disregarded.

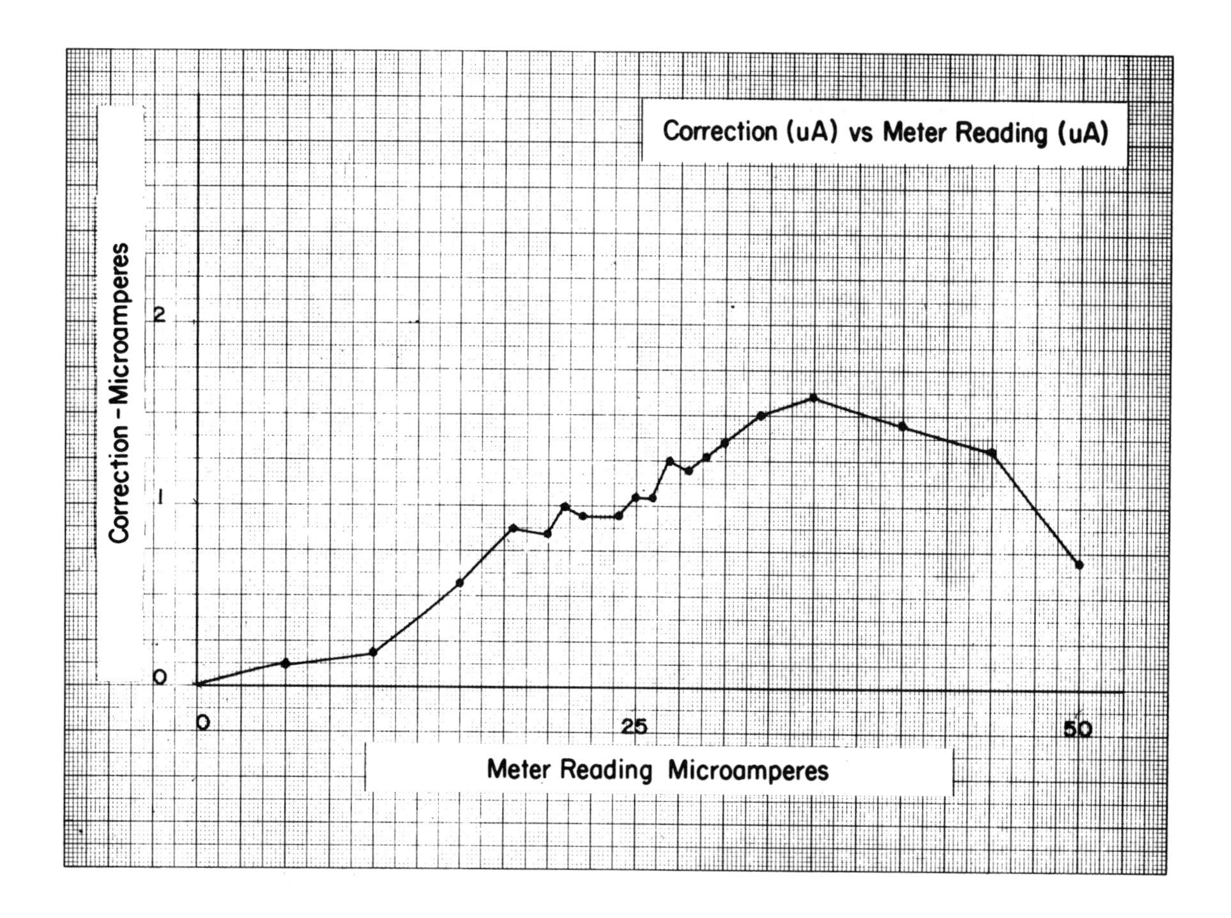

Fig. 3-6. A calibration curve for a 0-50 microampere ammeter. The observed ammeter readings were plotted as abscissas and the corresponding corrections as ordinates. When an ammeter is calibrated, it is joined in series with a standard resistance. The same current must pass through both. This current can be found by connecting a potentiometer across the standard resistance. The current is then compared with the ammeter reading and the difference is the ammeter correction. The corrections for each reading in the general case may be positive or negative or both.

2. Curves should be drawn by use of mechanical aids such as transparent plastic curves, rulers, etc. Abnormal kinks due to poor shifting of the mechanical aid when constructing continuous curves can be easily found by use of the "Sight Test." Hold the graph paper parallel to the eye and sight along the curve. Kinks due to poor construction of the curve should be removed, but use JUDGMENT and make sure that change of slope is not a result of the data.

3. When results are checked and accurate, the curve should pass near enough to the points to fall into the area which would be formed if the error limit values were also plotted around the points (see Fig. 3-7).

A certain amount of JUDGMENT is needed here. If a curve is forced to pass through every point, regardless of the range of error, spurious variations may be interpreted as real. On the other hand, real variations from uniform curves may appear. These variations could be lost if the curve is carelessly plotted as an average.

Although not normally shown on a graph, each point plotted falls within an "error rectangle" (actually an ellipse when statistical techniques are applied). This "error rectangle" is composed of the ± range of the numerical final limit of error (L.E.) for the quantity plotted as an abscissa (independent variable) along the x-axis and for the quantity plotted as an ordinate (dependent variable) along the y-axis (see Fig. 3-7). Each point is the result of data found by measurement techniques. Consequently, each point has an error range extending in both the x and y directions.

B. EMPIRICAL EQUATIONS

1. An empirical equation is found from a set of observed experimental data. The methods used to find an empirical equation (if it exists) can become quite complex. We approach the problem in the simplest manner by plotting the data.

The form of the desired empirical equation can often be found by inspecting the graph. The process of matching an equation to a graph is called "curve fitting." All experimental points may not fall exactly on the curve of the matched "empirical" equation but within the range of their error indices.

2. Curve fitting often requires the assumption of a certain type of equation, such as linear, power law, or exponential. Often, the graph of the data clearly shows the type of equation that should fit.

After graphing a set of data, it is occasionally convenient to graph their logarithms, (log y vs. log x, or y = log x). This sometimes results in a straight line from which the proper equation can easily be found.

3. If data plotted on rectangular graph paper results in a straight line, the variable Y is linear with X. The equation of the graph is a straight line of the form:

$$y = ax + b \qquad 3\text{-}1$$

where a = actual slope b = y-intercept

Example: If the velocity of a body is determined at various times and plotted as a function of time, a straight line indicates a uniform time rate of change in velocity (constant acceleration) and the form is:

$$v = v_o + at \qquad \text{(See Fig. 3-8)}$$

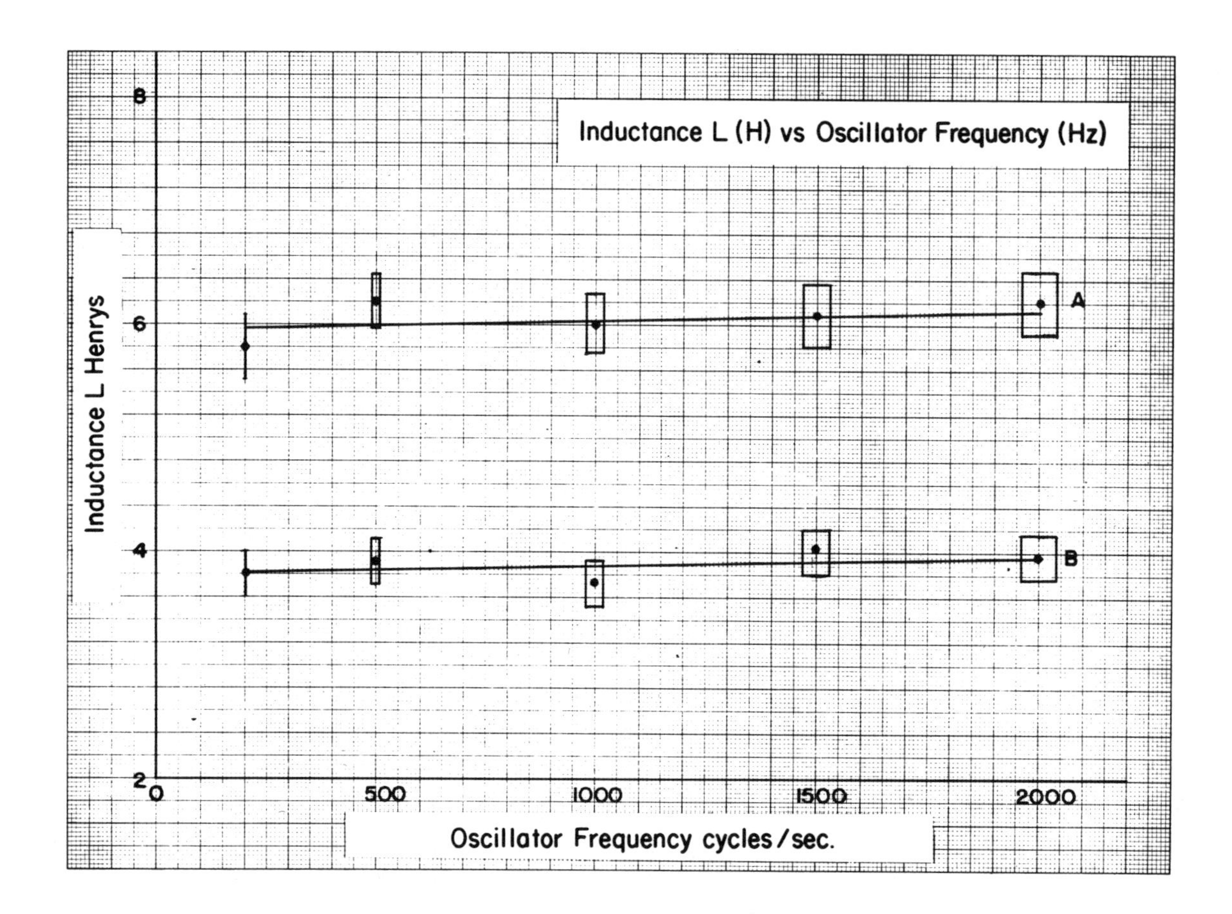

Fig. 3-7. A plot of Inductance vs. Frequency. Curves A, B show values of inductance found by use of a resonance bridge at different oscillator frequency settings. The curves pass through the areas formed by the error limit values which are plotted around each point. The L.E. for any scale setting on the oscillator was ±2%. The L.E. for each value for the inductance was ±5%. The range of error for frequency at each point from 2000 cycles/sec. to 200 cycles/sec. was ±40, 30, 20, 10, 4 cycles/sec. Note that the base of the "error rectangle" varied, but the height (inductance) remained fixed. This was due to the low values of inductance concerned. In practice, plots of range of error often result in a vertical line as seen at 200 cycles/sec. because the error in the dependent Y variable is usually larger than that in the independent X variable.

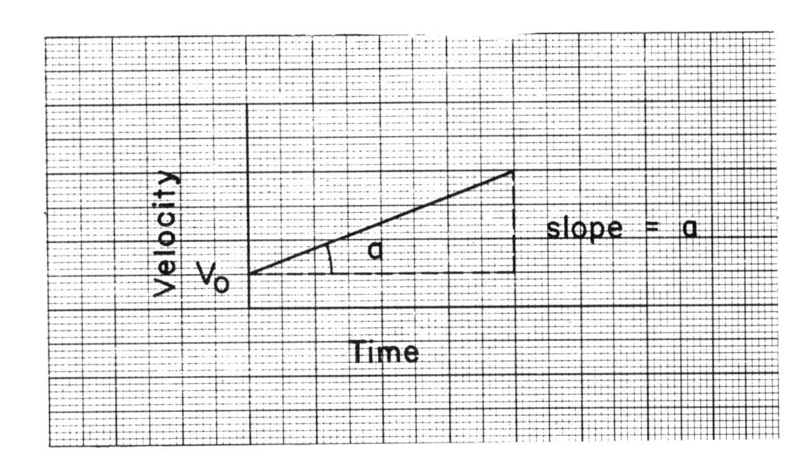

Fig. 3-8. A plot of Velocity vs. Time. Note that a = the actual slope with units. v_o = the y intercept b. In this case, a = constant acceleration.

An equation of the type y = ax + b is a linear function. It is easily recognized and a "best" straight line is readily drawn through or among the experimental points. The two independent constants (a, b) can be determined from the graph.

4. Curves of various degrees of complexity result when y = f(x) is not linear. Some of the curves (parabolas, hyperbolas, simple sine curves, etc.) are easily recognized, but the more complex curves are very hard to analyze.

Many equations which are not linear can be made so by changing the variables which are actually being plotted. This can be shown by a few examples:

(a) If a direct plot y = f(x) of data on a graph looks somewhat like a parabola which has an equation of the form $y = kx^2$ or $y^2 = cx$, we can verify our assumption by replotting the variables.

If we plot y vs x^2, Fig. 3-9, or y^2 vs x, Fig 3-10, and a straight line results, our original assumption is correct. If a straight line does not result, then the problem of analysis becomes more difficult.

The following figures (Fig. 3-9, 3-10) merely indicate a technique. The final empirical equation in each instance would fit or "match" exactly the original curve. The final empirical equation in actual practice only "matches" within the range of the error indices of the experimental points.

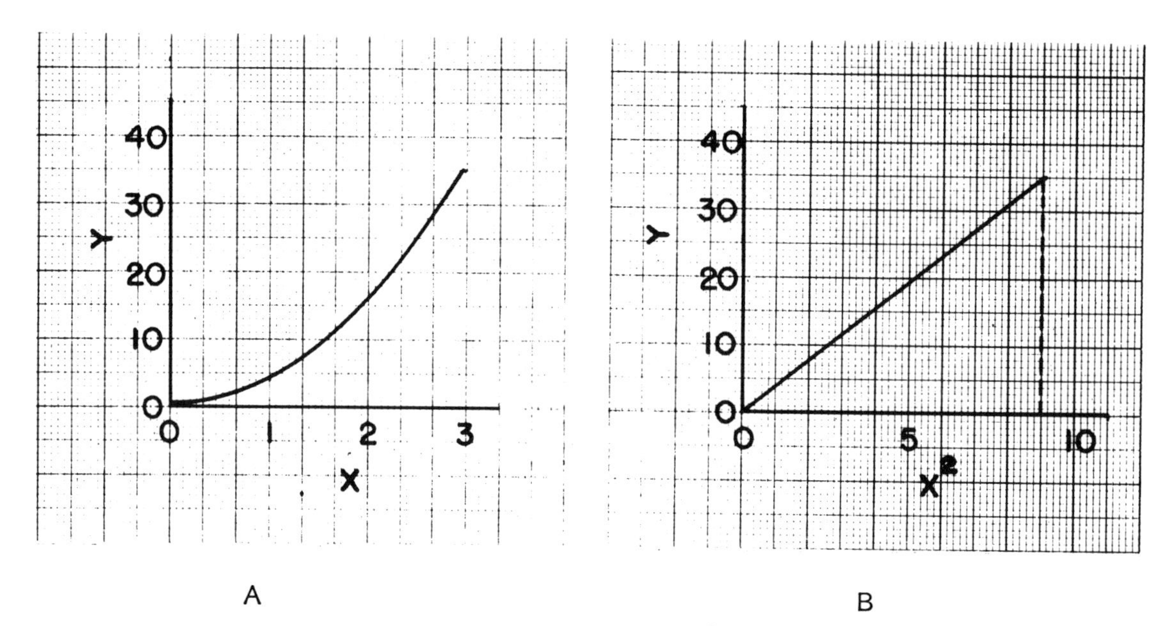

Fig. 3-9. Graph A represents an equation ($y = kx^2$). Graph B is a plot (y vs x^2) from which the constant ($k = 4$) can be found from the actual slope. The final equation (empirical) is $y = 4x^2$.

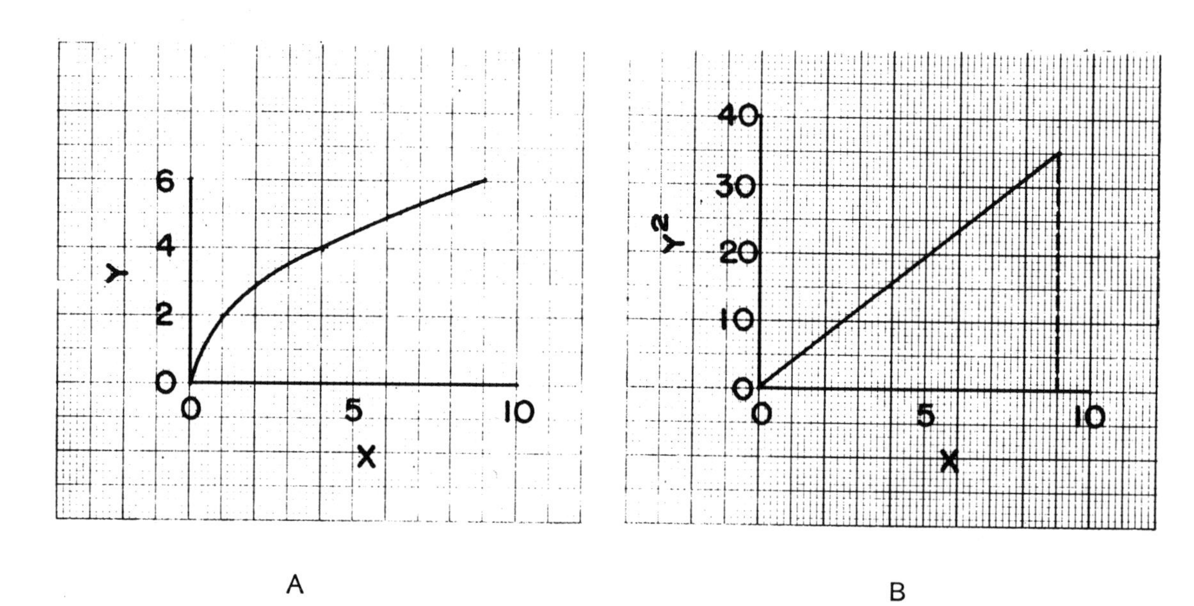

Fig. 3-10. Graph A represents an equation ($y^2 = cx$). Graph B is a plot (y^2 vs x) from which the constant ($c = 4$) can be found from the actual slope. The final equation (empirical) is $y^2 = 4x$.

(b) If a direct plot of data on a graph ($y = f(x)$) looks hyperbolic (see Fig. 3-11A), we can try plotting y vs $1/x$. If the original function is that of a rectangular hyperbola referred to its asymptotes as axes, we will obtain a straight line (see Fig. 3-11B). In general terms, the equation of the rectangular hyperbola is $xy = c$.

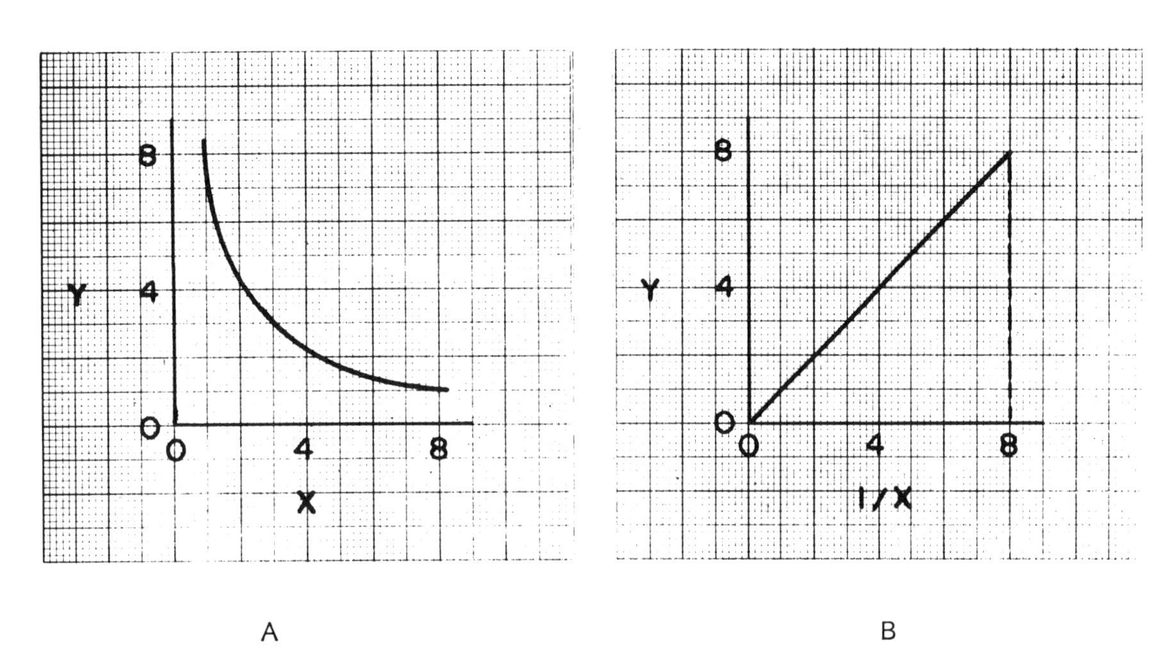

Fig. 3-11. Graph A represents an equation assumed to be of the form $xy = c$. Graph B is a plot (y vs $1/x$). The constant ($c = 8$) can be found from the actual slope. The final equation (empirical) is $xy = 8$.

5. We shall now consider non-linear equations of the form:

$$y = CD^{kx} \qquad \text{(Exponential Law)} \qquad 3\text{-}2$$

$$y = CX^{m} \qquad \text{(Power Law)} \qquad 3\text{-}3$$

An equation of the power law type could also be expressed as:

$$y^{m} = CX \quad \text{or } y = C^{\frac{1}{m}} X^{\frac{1}{m}} \qquad 3\text{-}4$$

(a) Exponential Law

The experiment on Thermal Conductivity is an example of the exponential law form of equation. The equation $t = -\frac{Lmc}{KA}(\ln_\varepsilon i - \ln_\varepsilon i_o)$* can be rewritten in straight line form as:

$$\ln_\varepsilon i = \ln_\varepsilon i_o - \frac{KA}{Lmc} t \qquad 3\text{-}5$$

Since the equation of a straight line is $y = ax + b$, then y corresponds to $\ln_\varepsilon i$, the slope $a = -\frac{KA}{Lmc}$, the independent variable $x = t$ and the intercept on the Y axis $= \ln_\varepsilon i_o$.

In Equation 3-2, D is arbitrary and can be taken as 10. Since $10 = \varepsilon^{2.30+}$, then Eq. 3-2 becomes:

$$y = C\varepsilon^{2.30\,kx} \qquad 3\text{-}6$$

where ε is the base of the natural system of logarithms; ε = an irrational number, 2.71828+.

Equation 3-6 can also be represented as

$$\ln_\varepsilon y = \ln_\varepsilon C + 2.3\,kx \qquad 3\text{-}7$$

Since

$2.3 \log_{10} y$**$= \ln_\varepsilon y$, then Eq. 3-7 is also equal to:

$$2.3 \log_{10} y = 2.3 \log_{10} C + 2.3\,kx$$

or

$$\log_{10} y = \log_{10} C + kx \qquad 3\text{-}8$$

Because the $\log_{10} C$ is still a constant, we can let the $\log_{10} C = c$, and

$$\log_{10} y = c + kx, \qquad 3\text{-}9$$

which is the equation of a straight line.

In summary, we find that the Thermal Conductivity Equation 3-5 has the same form as the semi-log Equation 3-7 expressed in the natural system of logarithms to the

---

* See Chapter 9, Heat, Experiment 6-4, Thermal Conductivity, Eqs. 6-4 to 6-13. In Eq. 3-5, K is the thermal conductivity of a material, m = mass of a cylindrical heat sink, A = cross-sectional area of the cylindrical heat sink, L = thickness of sample slab, c = specific heat of sample slab, i_o = current at time t = 0, is = current at time t.

**Although it is customary to express common logarithms to the base 10 as log., here for added emphasis, the base is given.

base ε. We can transform easily from the log system to the base ε to one of the base 10 (or conversely) by using $2.3 \log_{10} y = \ln_{\varepsilon} y$. Equation 3-5 can be transformed from the base ε to one of the base 10 as follows:

Eq. 3-5
$$\ln_{\varepsilon} i = \ln_{\varepsilon} i_o - \frac{KA}{Lmc} t$$

$$2.3 \log_{10} i = 2.3 \log_{10} i_o - \frac{KA}{Lmc} t$$

$$\log_{10} i = \log_{10} i_o - \frac{KA}{2.3\ Lmc} t \qquad 3\text{-}10$$

An equation of the form 3-9 or 3-10 is plotted upon millimeter rectangular Cartesian coordinate paper or upon semi-logarithmic paper (Fig. 3-12). On semi-log paper the divisions from 1-10, 10-100, 100-1000 along the Y-axis, or in the case of log-log paper along both the X and Y axis, are called cycles.

Because linear distances on log-log paper are proportional to the logarithms of the graduated coordinate scales instead of to the numbers themselves, the actual values of Y and X can be plotted without evaluating the logarithm of each point before plotting. On semi-log paper, the Y axis has a logarithmic scale and the X axis a uniform scale. If data to be plotted lies within one cycle, it can be plotted as shown in Fig. 3-12, otherwise multiple cycle log paper must be used. When plotting logarithms, the origin must be placed at the beginning of a cycle. If it does not appear on the graph due to the nature of the range of values plotted, it can be located by attaching additional sheets.

Figure 3-12 is a graph of the $\log_{10} I$ (galvanometer deflection) plotted against the time t in seconds. The logarithmic scale is numbered in galvanometer deflections and the uniform scale in seconds. The origin is at Y = 10, t = 0. The actual slope of a semi-log plot is found by multiplying the geometrical slope by a ratio determined from the cycle (or fractional cycle) distance on the ordinate (Y axis), compared to the corresponding distance on the abscissa (X axis). The geometrical slope g measured directly by a centimeter rule is found to be:

$$g = \frac{Z(\Delta \log I)}{C(\Delta T)} = -\frac{4.40\ \text{cm}}{15.20\ \text{cm}} = -2.90 \times 10^{-1}$$

The calibrated cycle distance Z, or fractional cycle distance along the ordinate, is expressed in cm/cycle and can be found by measuring the length of the cycle (or fractional cycle) with a centimeter rule. For example:

Length of cycle $\quad 25.4\ \text{cm} = Z \times 1\ \text{(cycle)}$

$$Z = \frac{25.4\ \text{cm}}{\text{cycle}} \quad \text{(or fractional change in cycle)}$$

The calibrated distance C along the abscissa is represented in cm/sec and is also measured with a centimeter rule. The length of this distance is 15.2 cm and C is found to be:

$$15.2\ \text{cm} = C \times 600\ \text{seconds}$$

$$C = \frac{15.2\ \text{cm}}{600\ \text{sec}} = 2.54 \times 10^{-2}\ \frac{\text{cm}}{\text{sec}}$$

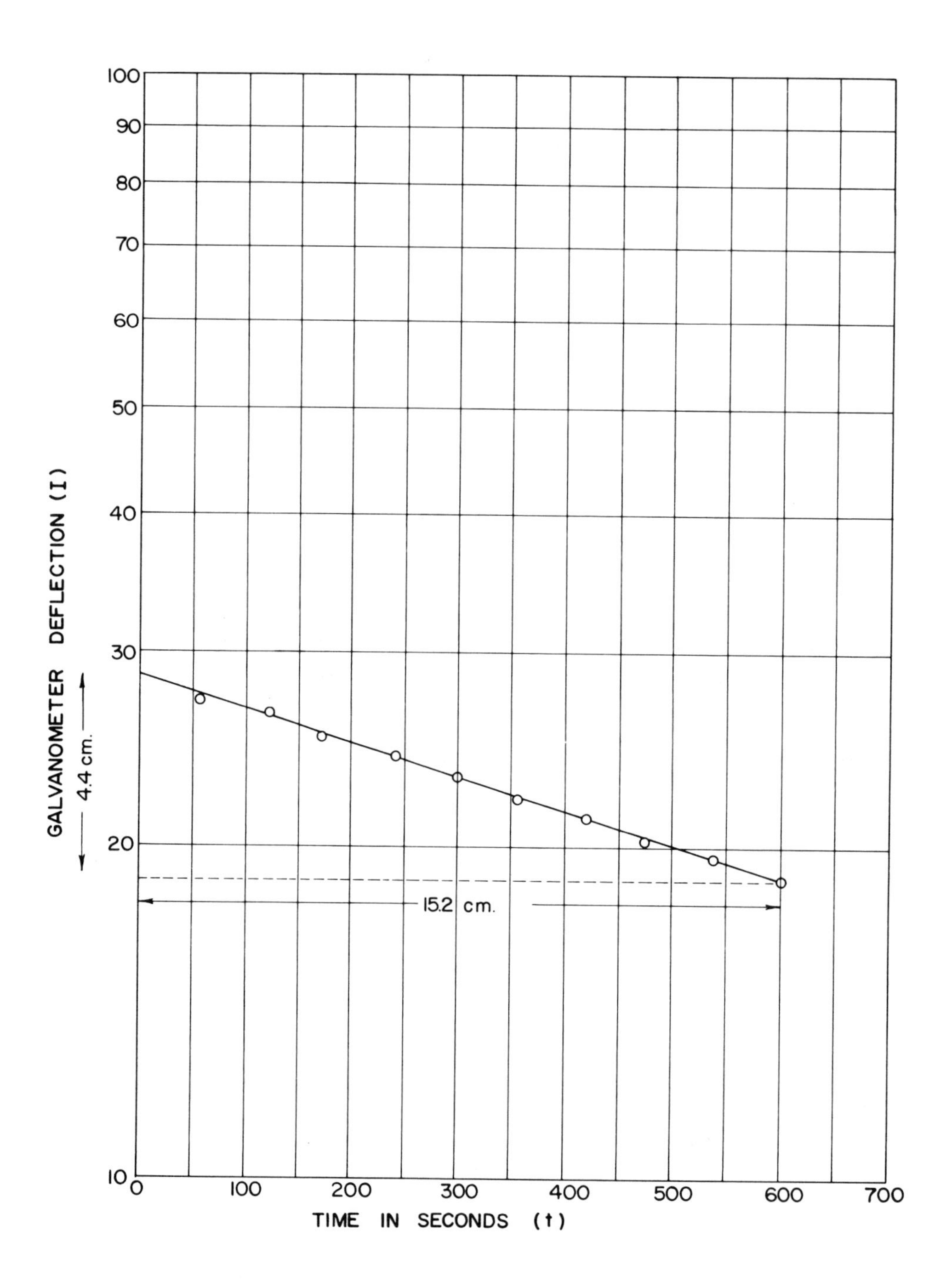

Fig. 3-12. Semi-log graph of Galvanometer Deflection I vs. t. Note that deflections are plotted directly.

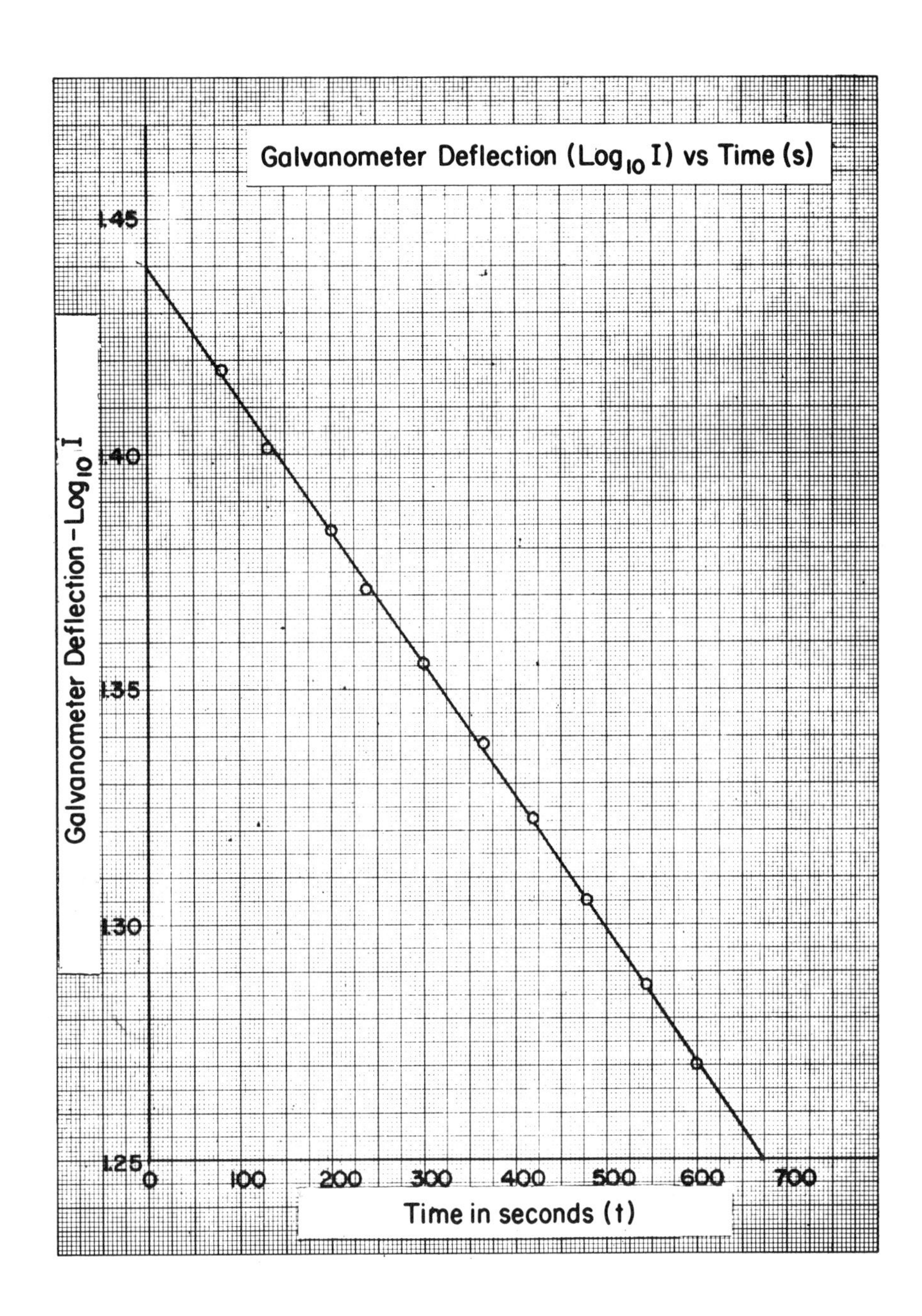

Fig. 3-13. Plot of Log_{10} I Galvanometer Deflection, vs. t. Note that the Log_{10} I must be looked up and plotted on millimeter graph paper.

The actual slope $a = \dfrac{\Delta \log I}{\Delta T}$ can now be calculated:

Since
$$\frac{Z(\Delta \log I)}{C(\Delta T)} = g \qquad 3\text{-}11$$

The actual slope
$$a = \frac{\Delta \log I}{\Delta T} = \frac{C}{Z} g \qquad 3\text{-}12$$

Therefore
$$a = -\frac{2.54 \times 10^{-2} \times 2.90 \times 10^{-1}}{2.54 \times 10^{+1}}$$

$$= -2.90 \times 10^{-4} \frac{1}{\text{sec}} \qquad 3\text{-}13$$

The actual slope in Fig. 3-13 of $\log_{10} I$ vs. t, as read directly from the coordinate scales, provides a simpler and more accurate technique for finding the constants of an exponential equation. Fig. 3-12 and Fig. 3-13 are plots of identical data. The actual slope in Fig. 3-13 is equal to:

$$a = \frac{\Delta \log I}{\Delta T} = \frac{1.44 - 1.25}{670 - 0} \frac{\text{no units}}{\text{seconds}} = 2.83 \times 10^{-4} \frac{1}{\text{sec}} \qquad 3\text{-}14$$

After calculating the actual slope for the semi-log (thermal conductivity) (Eq. 3-10), you obtain the value of the thermal conductivity coefficient K, as follows:

If
$$\log_{10} i = \log_{10} i_o - \frac{KA}{2.3\ Lmc} t, \qquad \text{(Eq. 3-10)}$$

then the slope s is:
$$s = -\frac{KA}{2.3\ Lmc}.$$

Since the actual slope
$$a = \frac{KA}{Lmc},$$

K can immediately be found by multiplying Eq. 3-14 by 2.3 and setting it equal to the actual slope a. Therefore,

$$-\frac{KA}{Lmc} = -2.3 \times \frac{1.44 - 1.25}{670 - 0}, \qquad \text{or}$$

$$\frac{KA}{Lmc} = 6.52 \times 10^{-4}\ 1/\text{sec}, \qquad \text{and}$$

$$K = \frac{6.52 \times Lmc \times 10^{-4}}{A} \frac{\text{cal}}{\text{m}^\circ\text{C sec}}.$$

For the case where $m = 340$ g, $c = 9.23 \times 10^{-2} \dfrac{\text{cal}}{\text{gm }^\circ\text{C}}$, $L = 3.27 \times 10^{-3}$ m and $A = 1.55 \times 10^{-3}\ \text{m}^2$, we find that $K = 43 \times 10^{-3} \dfrac{\text{cal}}{\text{m }^\circ\text{C sec}}$. Can you identify the sample material?

In round numbers, the values of the actual slopes found by two different methods are approximately the same. The difficulties encountered in accurately plotting points on semi-log graph paper and the operations required to compute the actual slope introduce statistical errors which indicate that a semi-log graph should only be used to determine quickly if a proportionality exists in a ($\log_{10}$ y vs x) plot. A replot on millimeter (Cartesian) paper should be made when an accurate slope is desired.

(b) Power Law Equations

Examples of the power law equation ($y = CX^m$) are found in the experiments on the Vibrating Spring ($T^n = km$), and the Period of the Simple Pendulum ($T^n = kL$).

When plotted, data yields a curve that is suspected to be of the power law type, the logarithm of both sides of the equation is taken. If a log-log graph of the data provides a straight line, the slope equals m or 1/m, depending on whether the plotted curve represents Equation 3-3 or 3-4.

When the form of the function is anything like:

$$T^n = kL \qquad 3\text{-}15$$

Then:

$$T = k^{\frac{1}{n}} L^{\frac{1}{n}}$$

which can be expressed logarithmically as:

$$\log_{10} T = \frac{1}{n} \log_{10} k + \frac{1}{n} \log_{10} L$$

or

$$\log_{10} T = C + \frac{1}{n} \log_{10} L \qquad 3\text{-}16$$

Eq. 3-16 is an equation of the form y = ax + b. Therefore, if you plot log T vs. log L, you will obtain a straight line with a slope 1/n.

$$\therefore \frac{1}{n} = \frac{\Delta \log T}{\Delta \log L} \quad \text{(Slope)} \qquad 3\text{-}17$$

When in Eq. 3-13 the $\log_{10} T = 0$ or $T = 1$, then, since $n \neq 0$:

$$\frac{1}{n} \log_{10} k = - \frac{1}{n} \log_{10} L$$

and

$$\log_{10} k = - \log_{10} L$$

$$\therefore |k| = \left|\frac{1}{L}\right| \qquad 3\text{-}18$$

Fig. 3-14 is a graph of the actual values of T (seconds) plotted against L (centimeters) on log-log paper. The ordinate is numbered in seconds and the abscissa in centimeters. If the dependent and independent variables are plotted on the same number of cycles (two in this case), then the slope can be measured directly with a centimeter rule.

If the number of cycles varies, then the procedure for slope determination is the same as that for a semi-log plot. The origin is at Y = .1, X = 1.

Because the number of cycles selected is uniform, the geometrical slope, measured directly, is equal to the actual slope or:

$$g = a = \frac{\Delta \log T}{\Delta \log L} = \frac{2.90 \text{ cm}}{5.70 \text{ cm}} = .508 \quad \text{(no units)} = \frac{1}{n} \qquad \text{3-19}$$

$$\therefore n = 2$$

The constant k is calculated by locating T = 1 second and then taking the reciprocal of the length value at the point of intersection of the straight line with T = 1. Note dotted lines in Fig. 3-14. Therefore:

$$|k| = \left|\frac{1}{25}\right| = .04 \qquad \text{3-20}$$

Since k is a proportionality constant found by log analysis, units for k can be found from $T^n = kL$.

In Fig. 3-14, we chose two-cycle paper to indicate how the size of the slope varies with cycle choice. Usually a log-log plot is planned so that the slope is as large as possible in order to improve the accuracy of the calculated constants. A better choice of cycles would have been 2 for the time and 1 for the length.

From Eq. 3-17 the actual slope a is:

$$a = \frac{1}{n} = \frac{\Delta \log T}{\Delta \log L} \qquad \text{3-21}$$

If the actual (t) time values are substituted, then:

$$a = \frac{1}{n} = \frac{\log 1.8 - \log .90}{\log 80 - \log 20}$$

$$= \frac{0.25 - (-.05)}{1.90 - 1.30} = \frac{0.30}{0.60}$$

$$\therefore a = \frac{1}{n} = 0.50 \text{ and n can be found } (n = 2) \qquad \text{3-22}$$

Fig. 3-15 is a plot of the $\log_{10} T$ vs. $\log_{10} L$ on millimeter (Cartesian) coordinate paper. Note that the $\log_{10} T$ is represented by both positive and negative logarithms. For a value of T = .896 seconds, the negative logarithm of T, which is plotted, is found as follows:

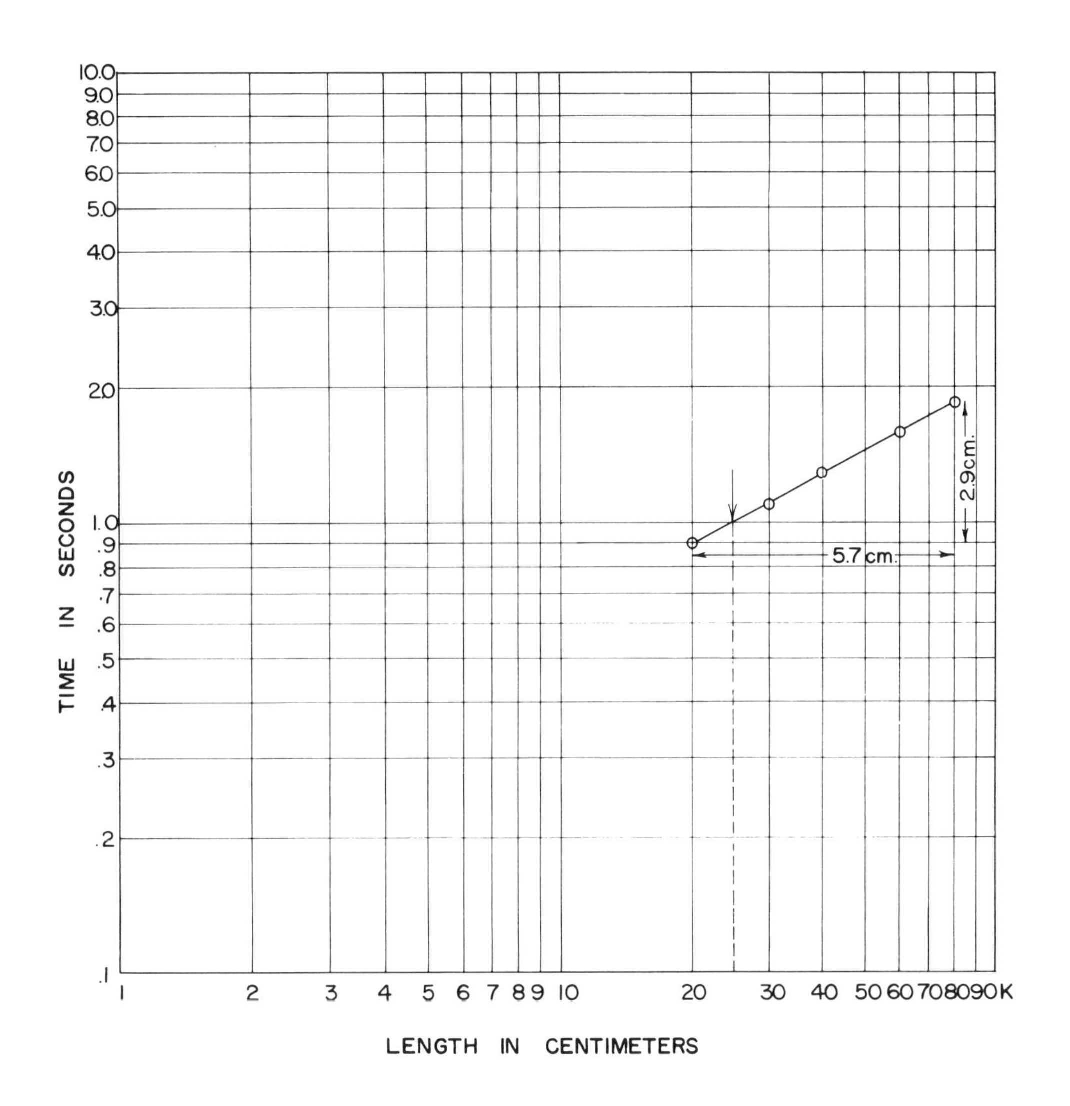

Fig. 3-14. Log-log graph of Time (T) in seconds vs. length (L) in centimeters. Note that T and L are plotted directly.

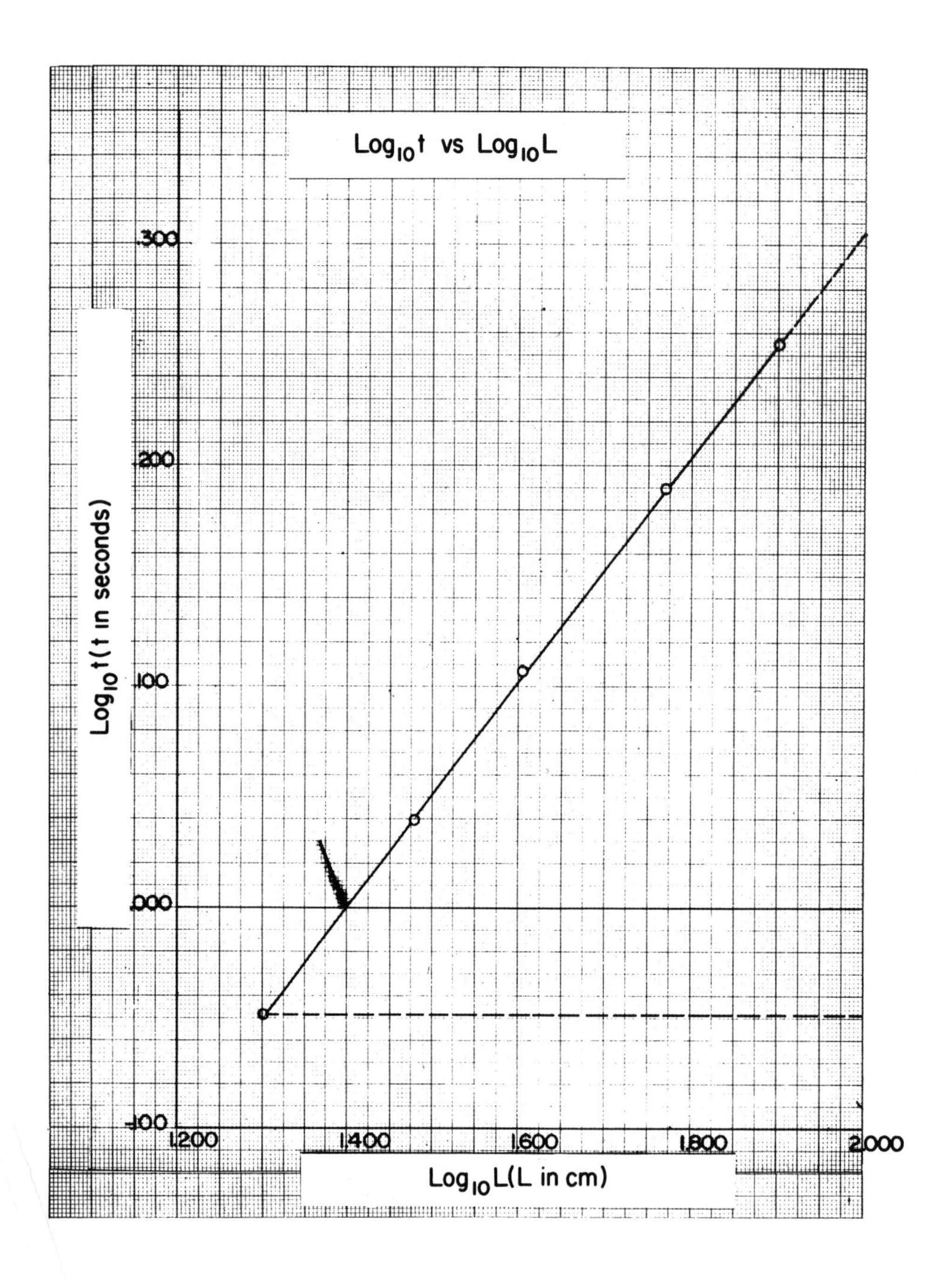

Fig. 3-15. Plot of Log_{10} T vs. Log_{10} L. Note that the logarithms must be plotted.

$$\log_{10} T = \log_{10} .896 = 9.9523 - 10$$

$$\begin{array}{r} 10.0000 - 10 \\ - \underline{9.9523 - 10} \\ - .0477 \end{array} \quad \text{(negative logarithm)} \qquad 3\text{-}23$$

The actual slope in Fig. 3-15 is equal to:

$$a = \frac{.303 - (-.047)}{2.000 - 1.300} = \frac{.350}{.700} = .500 \quad \text{(no units)} = \frac{1}{n}$$

Since the $\log_{10} k = -\log_{10} L$ when the $\log_{10} T = 0$, then k can be found:

$$\log_{10} k = \frac{1}{\log_{10} L} = \frac{1}{\text{anti-log } 1.395} = \left|\frac{1}{24.8}\right| \doteq .040$$

$$\therefore \; k = .040 \qquad 3\text{-}24$$

The actual slopes and the values of k calculated by two different methods are approximately the same. The log-log graph (on log paper) should be used to determine quickly if a proportionality exists in a $\log_{10} y$ vs. $\log_{10} x$ plot.

In summary, if a plot of data on a graph gives a curve that is suspected to be of the exponential form (Eq. 3-2) or the power law type (Eq. 3-3), we can test our assumption by the previously outlined methods. If a straight line results, we can easily determine the unknown constants and find the desired empirical equation. After the empirical equation is found, data can be substituted into it and the resultant curve checked against the original curve to see how well they fit (or match).

The student should exercise care in the interpretation of K. K in the thermal conductivity Eq. 3-10 is the thermal conductivity coefficient. "k" in the simple pendulum Eq. 3-15 is a proportionality constant.

3.5 Tabular Differences

While the graphical methods developed in Articles 3-2 to 3-5 are useful in finding the form and constants of quite a number of functions which describe physical behavior, they are by no means universally applicable. In the case of many complex and even some simple functional relations, the graphical methods so far developed are not very rewarding.

Consider the case of the more general parabolic function:

$$y = A + BX + CX^2 \qquad 3\text{-}25$$

A log-log plot will not give a straight line unless A and B both equal zero (a case already considered in Article 3.4, B-5). This is because logarithms do not simplify sums, but rather products and powers. A plot of y vs. x^2 is also non-linear because of the presence of the term in X^2.

Even the simpler form:

$$y = A + CX^2 \quad (B = 0)$$

is not linear in the log y vs. log x plot unless A is small, although log y vs. log x approaches linearity for large values of X. Why?

For data which fits (or closely approximates) equations of the form:

$$y = A + BX + CX^2 + DX^3 + \ldots\ldots + MX^n \qquad 3\text{-}26$$

a simple TABULAR DIFFERENCE test can be applied. This is convenient because Eq. 3-26 covers a very wide range of functions. Note that the first two terms $(A + BX)$ define a straight line if C, D, etc., are equal to zero. The first three terms $(A + BX + CX^2)$ define a family of parabolas, though not all parabolas. Many complex functions can be closely approximated by POWER SERIES equations like Eq. 3-26 if several terms are used.

Consider the partial (parabolic) form Eq. 3-25. Suppose that X is increased by <u>equal</u> increments ΔX. Then ΔX produces a change ΔY in Y, and the new value of Y, $Y + \Delta Y$, will be:

$$Y + \Delta Y = A + B(X + \Delta X) + C(X + \Delta X)^2$$

$$= A + BX + CX^2 + B\Delta X + 2CX\Delta X + C\Delta X^2$$

$$\therefore \quad Y = (B\Delta X + C\Delta X^2) + 2C\Delta X \cdot X$$

and if ΔX is constant, this amounts to:

$$\Delta Y = a + bx \qquad (\text{or } \Delta Y \text{ varies linearly with } X).$$

If the <u>differences</u> in ΔY, called $\Delta(\Delta Y)$ or $\Delta^2 Y$, are considered for equal values of ΔX, then:

$$\Delta Y + \Delta(\Delta Y) = a + b(X + \Delta X)$$

$$= a + bX + b\Delta X$$

$$\Delta(\Delta Y) = \Delta^2 Y = b\Delta X$$

$$= \text{constant}$$

That is, for equal increments (ΔX) of X, the changes in ΔY or $\Delta^2 Y$ values should be constant if

$$y = A + BX + CX^2$$

fits the data.

If $y = A + BX + CX^2 + DX^3$ fits the data, the changes in $\Delta^2 Y$, i.e., $\Delta(\Delta^2 Y)$ called $\Delta^3 Y$ will be constant, and by extension, if $\Delta^m Y$ = constant for uniform increments of X, the highest power of X needed in Eq. 3-26 is X^m. In many cases, terms beyond X^3 are often not needed for a very decent fit.

Let us consider some actual data. Note that the time t (sec) corresponds to x in Table 3-1.

Table 3-1

Time (t) (sec)	Displacement (y) (cm)	Δy (cm)	$\Delta^2 y$ (cm)	$\Delta^3 y$ (cm)
0.0	0.0			
		6.0		
0.1	6.0		0.6	
		6.6		-0.2
0.2	12.6		0.4	
		7.0		-0.1
0.3	19.6		0.3	
		7.3		0.5
0.4	26.9		0.8	
		8.1		-0.5
0.5	35.0		0.3	
		8.4		-0.3
0.6	43.4		0.0	
		8.4		0.8
0.7	51.8		0.8	
		9.2		-0.8
0.8	61.0		0.0	
		9.2		0.7
0.9	70.2		0.7	
		9.9		-0.7
1.0	80.0		0.0	
		9.9		
1.1	89.9			

Mean Δy = 8.2 cm Mean $\Delta^2 y$ = 0.4 cm

See Fig. 3-16 for a plot of this data.

The data in Table 3-1 was taken on an inclined plane (see Chapter 7, Exp. 7-5) using a hollow cylinder. An empirical equation which fits this data will be derived using the method of tabular differences.

The various differences are computed, remembering that the first value is subtracted from the second: $\Delta y = y_{i+1} - y_i$. The mean of each column is computed (do not neglect the signs of the differences) so that we can determine which difference values, Δy, $\Delta^2 y$, or $\Delta^3 y$, are constant within our limits of error.

If we take the precision index in the y values to be ±0.1 cm, the indices for the difference columns are very conservatively:

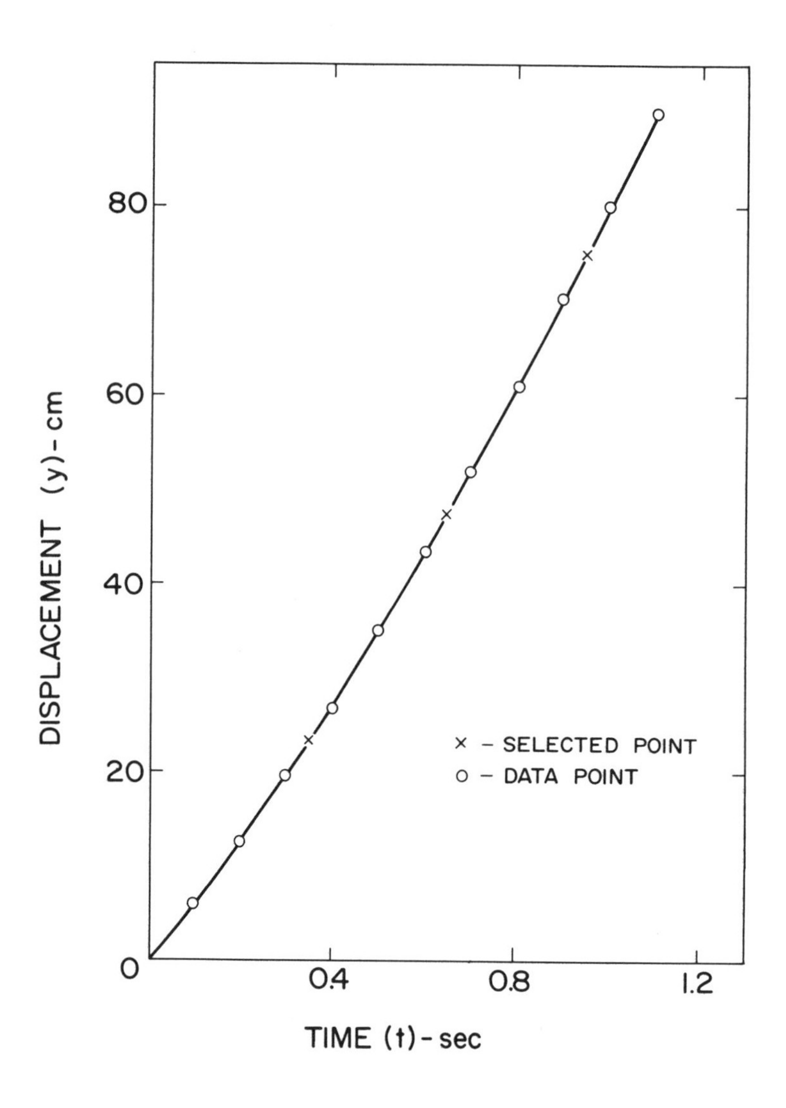

Fig. 3-16

$$PI_{\Delta y} = \pm 0.2 \text{ cm}$$

$$PI_{\Delta^2 y} = \pm 0.4 \text{ cm}$$

$$PI_{\Delta^3 y} = \pm 0.8 \text{ cm}$$

since precision indices add whether we add or subtract two measurements.

The mean of our $\Delta^2 y$ column is +0.4 cm. Since all the $\Delta^2 y$ values equal the mean within the limit of error, we may regard these values as constant; i.e., the $\Delta^2 y$ values are constant because they fall within the interval $0 \leqslant \Delta^2 y \leqslant 0.8$ cm. Note also that the

$\Delta^3 y$ values are zero within the appropriate limit of error, ±0.8 cm. As explained in the text, a constant second-order difference $\Delta^2 y$ implies that our data fits an equation of the form $y = a + bt + ct^2$.

We can immediately simplify this equation by noting that a = 0, since y = 0 when t = 0. We will determine b and c by plugging in pairs of data points and solving the two simultaneous equations for b and c. We will use three pairs of (t, y) data points and average the three values for each constant to obtain a better fit, as follows:

For t = 0.1 sec, y = 6.0 cm and t = 0.7 sec, y = 51.8 cm:

$$6.0 = 0.1b + 0.01c$$
$$51.8 = 0.7b + 0.49c$$
$$b = 57.7 \text{ cm/sec} \qquad c = 23.3 \text{ cm/sec}^2$$

For t = 0.3 sec, y = 19.6 cm and t = 0.9 sec, y = 70.2 cm:

$$19.6 = 0.3b + 0.09c$$
$$70.2 = 0.9b + 0.81c$$
$$b = 59.0 \text{ cm/sec} \qquad c = 21.1 \text{ cm/sec}^2$$

For t = 0.5 sec, y = 35.0 cm and t = 1.1 sec, y = 89.9 cm

$$35.0 = 0.5b + 0.25c$$
$$89.9 = 1.1b + 1.21c$$
$$b = 60.2 \text{ cm/sec} \qquad c = 19.6 \text{ cm/sec}^2$$

Averaging the three values of b and the corresponding values of c, we have

$$b = 59.0 \text{ cm/sec} \quad \text{and} \quad c = 21.3 \text{ cm/sec}^2$$

and

$$y = 59.0t + 21.3t^2 \qquad 3\text{-}27$$

To test our equation, let us determine y at t = 0.6 sec.

$$y = (59.0)(0.6) + (21.1)(0.36) = 42.1 \text{ cm}$$

This deviates from the measured value of 43.4 cm by 3%.

In summary, the equation $f(y) = A + BX + CX^2 + DX^3 + \ldots\ldots$ fits even more cases. f(y) might be Y^2, Y^3, log y, etc. These relations can be tested by making difference tables for f(y) with constant (ΔX). Advanced textbooks develop tests for more complicated relations.

3.6 Graphical (Selected Point) Method

The inherent precision of any graph and therefore any graphical procedure based on it is definitely limited. A limit of error of 0.2 to 0.5 percent has been estimated by a number of careful investigators. Within such limits, however, the graphical method is straightforward, easy to use and relatively rapid. It does not require complicated calculations.

(a) Plot a careful precision graph of the curve y = f(x), balancing out deviations of the measured points (x, y values) as much as possible. This is an averaging process.

(b) Select k pairs of x, y values from the graph (k "points", k = number of constants), choosing points which are well separated, not data points (except by chance), and not extremum values.

(c) Insert the k pairs of values in the equation $y = f(x)$ and solve the k equations for the constants, as shown in the example which follows:

GRAPHICAL (SELECTED POINT) EXAMPLE

We pass a parabola through our data points and read from Fig. 3-16. k = 3 data points for our equation,

$$y = a + bt + ct^2.$$

$$\begin{aligned} y &= 23.5 \text{ cm when } t = 0.35 \text{ sec} \\ y &= 47.4 \text{ cm when } t = 0.65 \text{ sec} \\ y &= 75.0 \text{ cm when } t = 0.95 \text{ sec} \end{aligned}$$

We solve

$$\begin{aligned} 23.5 &= a + 0.35b + 0.12c \\ 47.4 &= a + 0.65b + 0.42c \\ 75.0 &= a + 0.95b + 0.90c \end{aligned}$$

for a, b, and c, obtaining

$$\begin{aligned} a &= 0.3 \text{ cm} \\ b &= 59.1 \text{ cm/sec} \\ c &= 20.6 \text{ cm/sec}^2 \end{aligned}$$

and

$$y = 0.3 + 59.1t + 20.6t^2$$

At $t = 0.7$ cm our equation gives

$$y = 0.3 + (59.1)(0.7) + (20.6)(0.49) = 51.8 \text{ cm},$$

which equals the measured value of 51.8 cm.

We could have simplified our work slightly by noting that $y = 0$ at $t = 0$, implying that $a = 0$. Only two simultaneous equations for b and c would then need to be solved.

The student should realize that the selected points method will give accuracy suitable for nearly all the data he will encounter in introductory laboratories. We cannot manufacture precision beyond that of the basic data; attempts to do so are a waste of time.

The data being tested in these examples is of low precision and requires only graphical treatment. The method of least squares will now be applied, however, for comparison.

3.7 Method of Least Squares

The method of least squares will determine the coefficients of an empirical equation so that the sum of the squares of the vertical (y) distances from each point to the resultant curve is a minimum.

Since we want to minimize the sum of the squares of the distances between our data points (y', x) and our functional curve, $y = f(x)$, we minimize $\sum_{i=1}^{n} (y_i' - y_i)^2$, where the sum is taken over the n measurements we have taken.

For example, let us try to fit the straight line $y = a + mx$ to the set of data points (y', x). We have to determine a and m such that

$$\sum_{i=1}^{n} (y_i' - y_i)^2 = \sum_{i=1}^{n} (y_i' - (a + mx_i))^2$$

is a minimum where a and m are our variables.

From calculus, we therefore must solve

$$\frac{\partial}{\partial a} \left[\sum_{i=1}^{n} (y_i' - y_i)^2 \right] = 0 \quad \text{and} \quad \frac{\partial}{\partial m} \left[\sum_{i=1}^{n} (y_i' - y_i)^2 \right] = 0$$

to obtain the values of a and m that give a minimum for the sum of the deviations. (Note: $\frac{\partial}{\partial a}$ represents the partial derivative with respect to a. Crudely, this means that we take the derivative with respect to a, treating all other variables as if they were constants. Your calculus book will provide examples and more rigorous definitions.)

By forming the derivatives, we obtain:

$$an + m \sum_{i=1}^{n} x_i - \sum_{i=1}^{n} y_i' = 0$$

$$a \sum_{i=1}^{n} x_i + m \sum_{i=1}^{n} x_i^2 - \sum_{i=1}^{n} x_i y_i = 0$$

Solving the two equations simultaneously for a and m will yield the equation of the least squares line for our data points (y_i', x_i). An example follows:

METHOD OF LEAST SQUARES EXAMPLE

While the data previously treated does not warrant application of the elegant and time-consuming method of least squares, this technique will be applied for comparison. We will again use an equation of the form $y_i = a + bt_i + ct_i^2$ and we will designate our data points by (y_i', t_i).

Our problem is then to minimize

$$\sum_{i=1}^{n} (y'_i - y_i)^2 = \sum_{i=1}^{n} (y'_i - a - bt_i - ct_i^2)^2$$

for the variables a, b and c.

' e form the partial derivatives of the expression on the right with respect to a, b and c and equate them to zero.

$$\frac{\partial}{\partial a}\left[\sum_{i=1}^{n} (y'_i - y_i)^2\right] = 2na + 2b\sum_{i=1}^{n} t_i + 2c\sum_{i=1}^{n} t_i^2 - 2\sum_{i=1}^{n} y'_i$$

$$= 2na + bn(n-1)\Delta t + \frac{1}{3}c\,[n(n-1)(2n-1)](\Delta t)^2 - 2\sum_{i=1}^{n} y'_i = 0$$

$$\frac{\partial}{\partial b}\left[\sum_{i=1}^{n} (y'_i - y_i)^2\right] = 2a\sum_{i=1}^{n} t_i + 2b\sum_{i=1}^{n} t_i^2 + 2c\sum_{i=1}^{n} t_i^3 - 2\sum_{i=1}^{n} y'_i t_i$$

$$= an(n-1)\Delta t + \frac{1}{3}b\,[n(n-1)(2n-1)](\Delta t)^2 + \frac{1}{2}c\,[n^2(n-1)^2](\Delta t)^3$$

$$-2\Delta t\sum_{i=1}^{n} (i-1)y'_i = 0$$

$$\frac{\partial}{\partial c}\left[\sum_{i=1}^{n} (y'_i - y_i)^2\right] = 2a\sum_{i=1}^{n} t_i^2 + 2b\sum_{i=1}^{n} t_i^3 + 2c\sum_{i=1}^{n} t_i^4 - 2\sum_{i=1}^{n} y'_i t_i^2$$

$$= \frac{1}{3}a\,[n(n-1)(2n-1)](\Delta t)^2 + \frac{1}{2}b\,[n^2(n-1)^2](\Delta t)^3$$

$$+ \frac{1}{15}c\,[n(n-1)(2n-1)(3n^2-3n-1)](\Delta t)^4 - 2(\Delta t)^2\sum_{i=1}^{n} (i-1)^2 y'_i = 0$$

We have assumed that the t_i are spaced at equal intervals Δt (that is, $t_{i+1} - t_i = \Delta t$) and that our first time reading, t_1, is zero. We have also used the following relations:

$$\sum_{i=1}^{n} t_i = \Delta t\sum_{i=1}^{n} (i-1) = \frac{1}{2}n\,(n-1)\Delta t$$

$$\sum_{i=1}^{n} t_i^2 = (\Delta t)^2\sum_{i=1}^{n} (i-1)^2 = \frac{1}{6}n(n-1)(2n-1)(\Delta t)^2$$

$$\sum_{i=1}^{n} t_i^3 = (\Delta t)^3 \sum_{i=1}^{n} (i-1)^3 = \frac{1}{4}[n(n-1)]^2(\Delta t)^3$$

$$\sum_{i=1}^{n} t_i^4 = (\ t)^4 \sum_{i=1}^{n} (i-1)^4 = \frac{1}{30} n(n-1)(2n-1)(3n^2-3n-1)(\Delta t)^4$$

To analyze the data used in the tabular differences example (a hollow cylinder on an inclined plane), we use:

$\Delta t = 0.1$ sec $\qquad n = 12 \qquad t_1 = 0$

$$\sum_{i=1}^{12} y'_i = 0 + 6.0 + 12.6 + \ldots + 89.9 = 496.4$$

$$\sum_{i=1}^{12} (i-1)y'_i = 0 + (1)(6.0) + (2)(12.6) + \ldots + (11)(89.9) = 3904.3$$

$$\sum_{i=1}^{12} (i-1)^2 y'_i = 0 + (1)(6.0) + (4)(12.6) + \ldots + (121)(89.9) = 34{,}106.9$$

Our equations for a, b, and c become

$$24a + 13.2b + 10.12c - 992.8 = 0$$
$$13.2a + 10.12b + 8.712c - 780.86 = 0$$
$$10.12a + 8.712b + 7.9948c - 682.138 = 0$$

(Four decimal places were kept throughout the solution calculations to prevent round-off error masking the improved fit of the least squares curve. Our final equation will, of course, reflect the appropriate significant digits.)

Solving this system by Cramer's rule, we have:

$$a = -0.2168 \approx -0.2 \text{ cm}$$
$$b = 60.6559 \approx 60.7 \text{ cm/sec}$$
$$c = 19.5009 \approx 19.5 \text{ cm/sec}^2$$

Our final equation is

$$y = -0.2 + 60.7t + 19.5t^2$$

To test the equation roughly, let us calculate y at $t = 0.6$ sec and at $t = 0.1$ sec.

At $t = 0.6$ sec

$$y = -0.2 + (60.7)(0.6) + (19.5)(0.6)^2$$
$$= 43.2 \text{ cm}$$

measured $y = 43.4$ cm: 4.6% deviation

At $t = 1.0$ sec

$$y = -0.2 + (60.7)(1.0) + (19.5)(1.0)^2$$
$$= 80.0 \text{ cm}$$

measured $y = 80.0$ cm: no deviation

3.8 Test for Fit

Although we have gained a rough estimate of the accuracy of our equations by plugging in various time values and comparing the calculated y value with the actual y' value, we can best estimate the closeness of the fit of our equations by computing the sum of the deviations between the observed and calculated y values for each observed t value. The smaller the magnitude of the sum of the deviations, the better the fit we have produced.

The table below summarizes these results for our two examples. In this table the deviation between the observed and the calculated y values is $\delta_y = y_i' - y_i$. The sum of the deviations is $\Sigma\delta_y$, the sum of the squares of the deviations is $\Sigma\delta_y^2$ and $\bar{\delta} = \frac{1}{12}\Sigma|\delta_y|$, the average absolute deviation per point.

Method	Equation	$\Sigma\delta_y$	$\bar{\delta}$	$\Sigma\delta_y^2$
selected points	$y = 0.3 + 59.1t + 20.6t^2$	-1.50	0.21	0.7996
least squares	$y = -0.2 + 60.7t + 19.5t^2$	-0.53	0.13	0.3035

Notice that $\Sigma\delta^2$, the sum of the squares of the deviations, is smaller for the least squares method than for the selected points (graphical)method, as we would expect.

3.9 Summary

The object of the preceding articles were to show how an empirical equation could be obtained from a set of observed data. After a set of data is plotted, we design an equation that will fit (or match) the curve. If we are fortunate, the type of equation can be found from familiarity with the curves of known functions. We can then use the methods suggested to find a straight line equation from which we can find the coefficients of the assumed equations. After the coefficients (constants) of the empirical equation have been found, we check to determine if it will be satisfied by the coordinates of the points plotted from the observed data.

In general, the graph will not pass through the original points, but will pass near (in a region determined by the range of error of the respective coordinates). Curve matching can become quite complex. When a proper equation "form" cannot be readily found, we must use a more exacting analytical approach, such as the tabular difference method, or the method of least squares.

Chapter 4

LABORATORY APPARATUS

4.1 Introduction

In this chapter the characteristics and operation of some typical laboratory apparatus will be discussed. Not all the details of the instruments will be given, nor is all laboratory apparatus included, but only some of the widely applicable measuring devices are discussed. Specialized instruments are described in later chapters along with the experiments requiring their use.

Most apparatus in the introductory physics laboratory is relatively simple. Although the theory on which it operates can often be understood without any advanced work, when in doubt the student should consult both reference books and this chapter.

The descriptions are kept general to apply to most standard models of each device. Particular models described are given as examples only and represent other instruments serving the same purpose.

Some instrumental errors are given in the discussion of the instruments. Manufacturer's specifications should be consulted whenever possible. If no instrumental error is given, the least count of the instrument is used.

4.2 Length and Area

THE METER STICK

The wooden or plastic meter stick is usually used to measure length when the accuracy desired is not very high. Meter sticks are read to one tenth or two tenths of the smallest division - corresponding to 0.1 mm to 0.3 mm in the metric system. Since the instrumental limit of error is between ±0.2 and ±0.4 mm, the meter stick cannot be used in measuring very small lengths or where high accuracy is required. In these cases the micrometer or the vernier calipers are used.

VERNIER PRINCIPLE

A vernier consists of a scale which can move next to a stationary scale. This vernier scale has divisions which are slightly smaller than a division on the stationary main scale along which it slides. If the zero mark of the vernier is on any main scale division, then n divisions of the vernier correspond to n-1 divisions on the main scale. Therefore, each vernier division is $\frac{n-1}{n} = 1 - \frac{1}{n}$ of a main scale division. Consequently, the vernier division is $\frac{1}{n}$ shorter than a main scale division. This quantity $(\frac{1}{n})$ of a main scale division is called the "Least Count" of the vernier. Always determine the "Least Count" of a vernier before attempting a measurement.

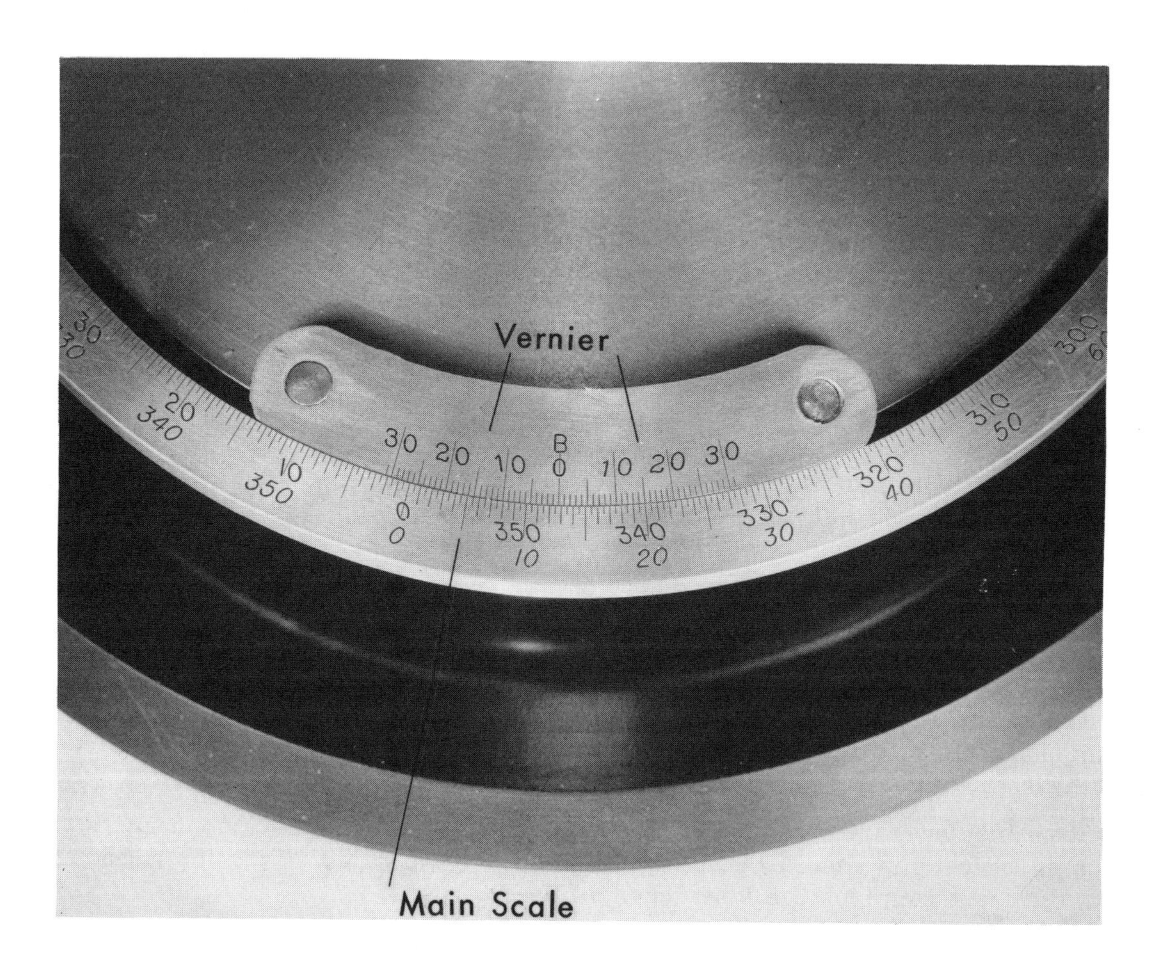

Fig. 4-1. The Angular Vernier. (Courtesy of W. & L. E. Gurley Instrument Co., Troy, N. Y.)

THE ANGULAR VERNIER

In many optical instruments an angular vernier is used. In Fig. 4-1 one type of an angular vernier is shown. Twenty-nine divisions of the main scale are covered by thirty of the vernier. The vernier permits the smallest division (1/2°) on the main scale to be read to 1/30 or 1'. The observer, therefore, can read the instrument to the nearest minute. In using a vernier of this type: (a) find the zero of the vernier and record the main scale reading, (b) find the vernier division coinciding with one of the main scale divisions and add this to the first reading.

The angular vernier shown in Fig. 4-1 can be moved in either a clockwise or counterclockwise (0 - 360°) direction. The vernier contains thirty divisions to the right and to the left of the zero index and since the main scale has two 0 - 360° scales, one clockwise and the other counterclockwise, the vernier can be used in either direction. In Fig. 4-1, using the vernier divisions to the right of the zero index, the reading would be: the main scale reading opposite the zero of vernier = 12° 30' while the vernier division coinciding with the main scale is 15. The final reading is 12° 30' + 0° 15' or 12° 45'. The vernier divisions to the left of the zero index provide as a reading: the main scale reading opposite zero of the vernier = 347° 0' while the vernier division coinciding with the main scale is 15. The final reading is 347° 0' + 0° 15' or 347° 15'. Note that the sum of the two final readings is 360°. The instrumental limit of error (ILE) = ±1'. (Least Count)

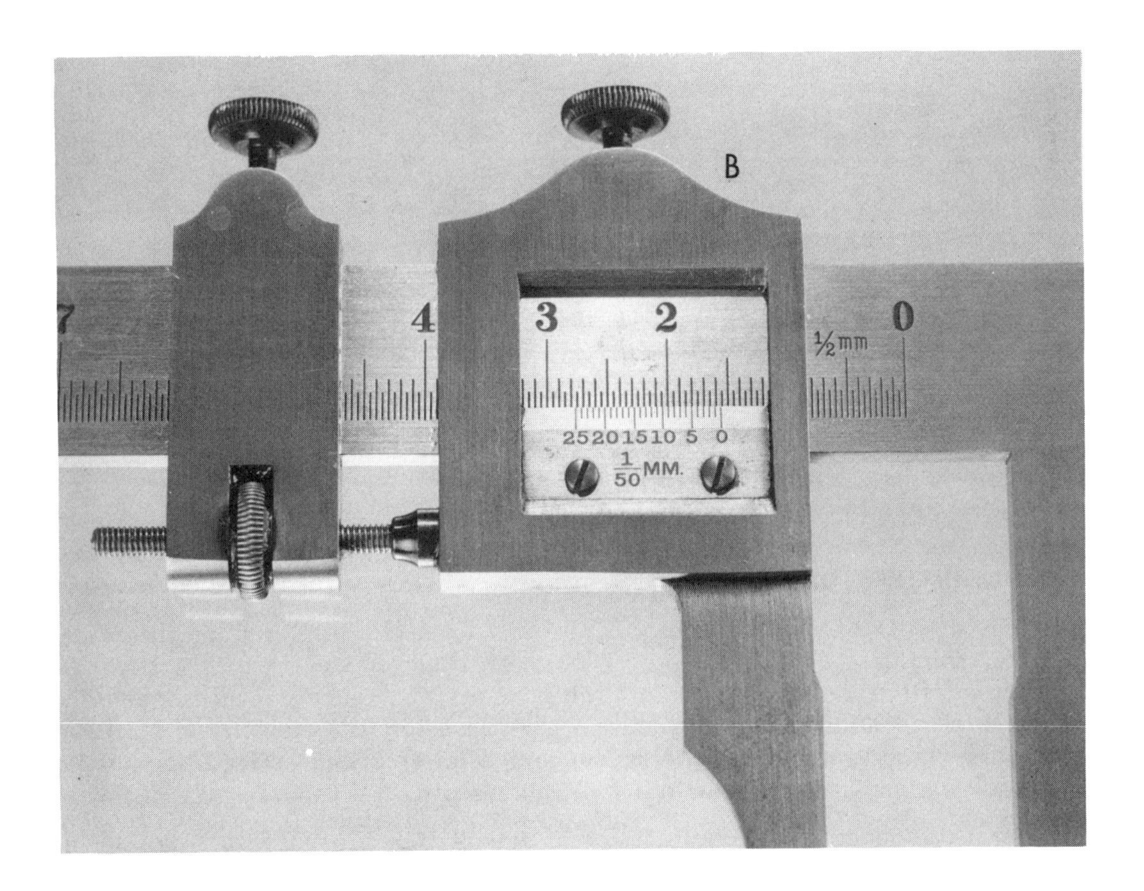

Fig. 4-2. The Metric Vernier Caliper.

VERNIER CALIPERS

In Fig. 4-2, a sample metric vernier is shown. The sliding jaw (B) has a window with the vernier divisions at the bottom. Twenty-five divisions on the vernier correspond in overall length to 24 divisions on the main scale.

The smallest main scale divisions are in half mm (0.5 mm = 0.05 cm). Note in Fig. 4-2 that 25 divisions on the vernier correspond to 24 divisions on the main scale. since 0.5 mm is divided into 25 parts, the smallest vernier division = 0.002 cm. In Fig. 4-2 a measurement of 1.55 + cm is shown on the main scale. The sixth division of the vernier is in alignment with a main scale division. The reading from the vernier, therefore = 0.012 cm. The final reading on the calipers is 1.550 + 0.012 = 1.562 cm. The instrumental limit of error (ILE) = ±0.002 cm. (Least Count)

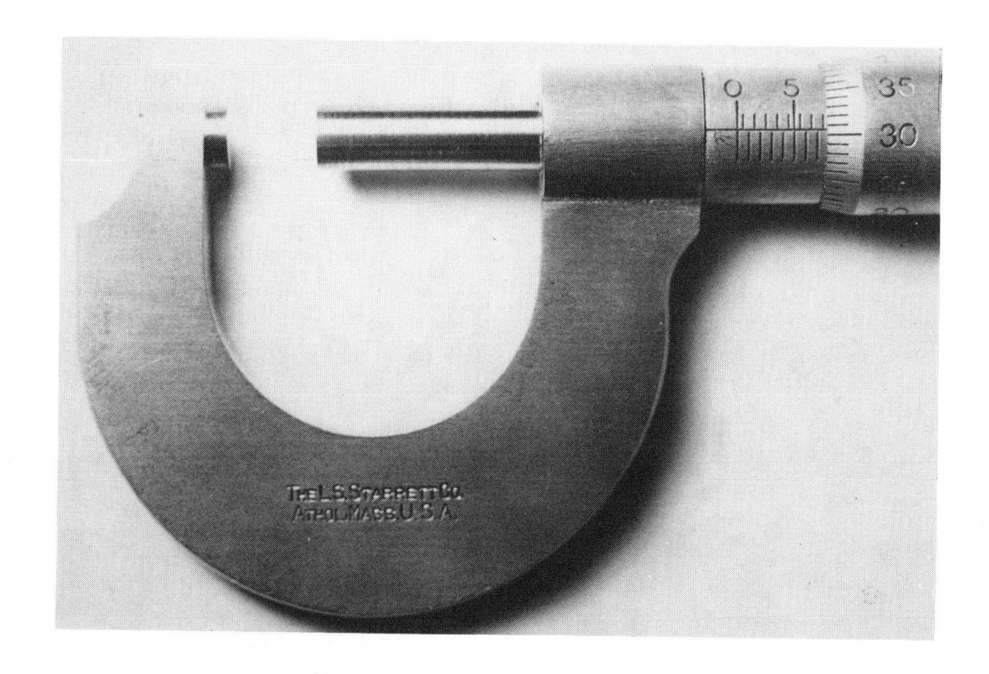

Fig. 4-3. The Micrometer Caliper.

THE MICROMETER CALIPER

This instrument is used for the accurate measurement of short lengths, i.e., the diameter of wire, thickness of sheets, etc. The metric micrometer caliper (shown in Fig. 4-3) has a spindle with a pitch of 0.5 mm. This means that every time the spindle is turned through one revolution, it will advance 0.5 mm along the main scale. The main scale has two sets of divisions, an upper scale and a lower scale. The upper scale division divides the lower millimeter scale into half millimeters.

In order to advance the spindle one mm, it will be necessary to turn the spindle through two revolutions. For example (Fig. 4-3): a measurement of 0.700 cm is taken directly from the main scale on the band of the micrometer. The spindle shows that almost three-fourths of the eighth division of the main scale is exposed. This means that the spindle has already turned over one revolution 0.5 mm along the band and since the spindle has only 50 divisions on it, add 0.05 cm to the main scale reading and the measurement is now 0.750 cm. The horizontal line on the axis of the micrometer falls between the 30th and 31st division on the spindle. The spindle can now be read as 0.030 + cm. By interpolation 0.006 cm are estimated as being shown between the 30th and 31st division on the spindle. The final reading is therefore: 0.750 + 0.0306 cm or 0.7806 cm (ILE = ±0.0002 cm).

Always check the micrometer caliper for <u>zero correction</u>. This must be added or subtracted from any reading.

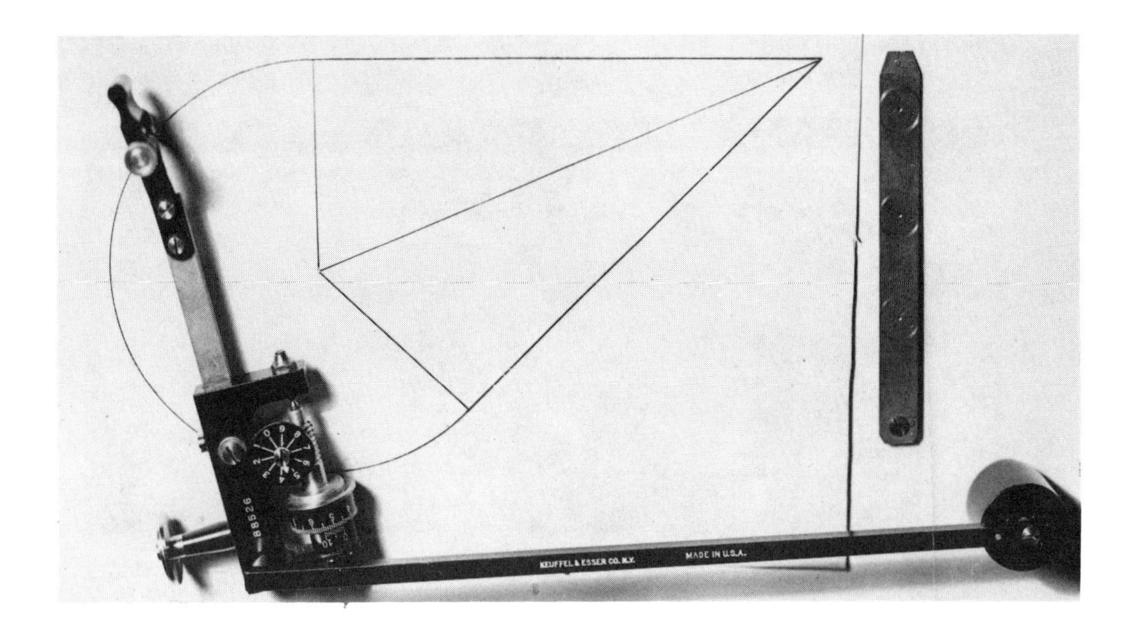

Fig. 4-4. The Polar Planimeter.

POLAR PLANIMETER - INTRODUCTION

The polar planimeter is used to measure irregular areas. It is constructed so that areas can be measured by tracing around the figure in question, reading the drum attached to the instrument and multiplying the turns made by the drum with a constant. The theory of the planimeter involves a knowledge of translational and rotational motion, but its operation and experimental calibration is very simple.

In Fig. 4-4 a planimeter is shown in use. BWA and AC are two rigid arms that are hinged together with a pivot pin. C is fixed to the table with a sharp point. The point B is used to trace around the area. At W is a wheel whose edge rolls on the table top. A revolution counter and divider drum on W turns in such a way that:

$$A \propto N \qquad\qquad A = KN \tag{4-1}$$

(N = Number of revolutions of the wheel around the area)

The pointer B must trace completely around the area in question in a clockwise direction and return to its starting point. The fixed point C must lie outside the area.

Before Eq. 4-1 can be used to determine any unknown area, the constant K must be found. The value of K depends on the length AB, but for a fixed length it remains constant. Therefore, the value of K can be determined by tracing around a known area of a calibration disk, or a calibrated radius arm (Fig. 4-4, D) and measuring N. Hence:

$$K = \frac{A(\text{known})\ \text{cm}^2}{N(\text{measured})\ \text{rev.}} = \frac{A_s}{N_s} \tag{4-2}$$

Any unknown area A_x can then be determined from:

$$A_x = KN_x = (\frac{A_s}{N_s})\, N_x \qquad 4\text{-}3$$

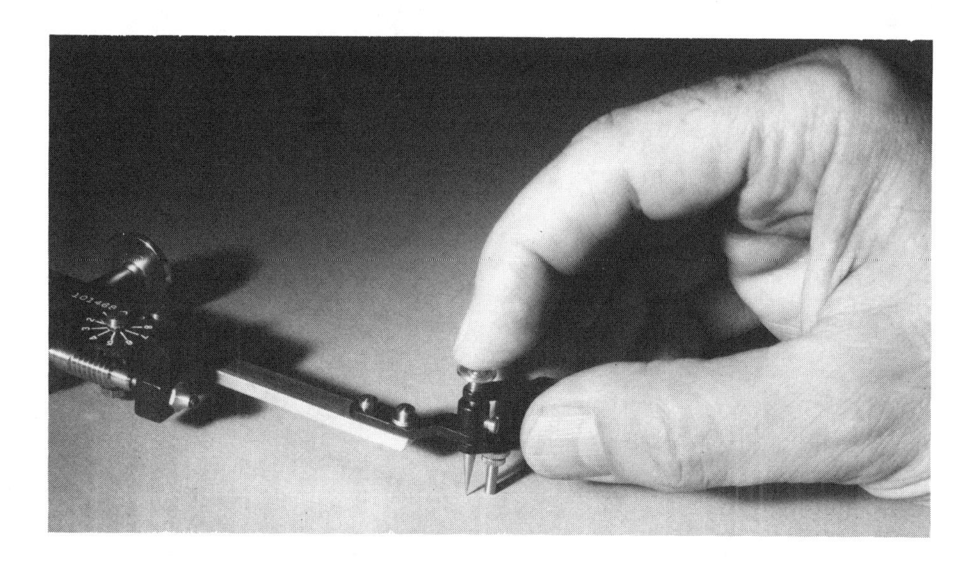

Fig. 4-5. Tracer Point and Stop. (Courtesy of Keuffel and Esser Co.)

OPERATION OF POLAR PLANIMETER (Courtesy of Keuffel and Esser Co.)

The stop (Fig. 4-5) should be adjusted so the tracer point will just clear the paper. Swing the carriage around the tracer point. Note that the measuring wheel turns rapidly, and the dial turns slowly. Now move it parallel to the tracer arm and note that the measuring wheel slides without turning. This characteristic sliding or turning of the measuring wheel is the key to the planimeter performance.

Test the limits of the motion of the tracer from side to side and note that it can be moved freely whenever the tracer arm (Fig. 4-4 AB) and pole arm (AC) make an angle within the wide arc between about 15° and 165° (see Fig. 4-6). The pole in Fig. 4-6 is point C in Fig. 4-4. The shaded area in Fig. 4-6 represents the limit of measurable areas-pole-outside figure.

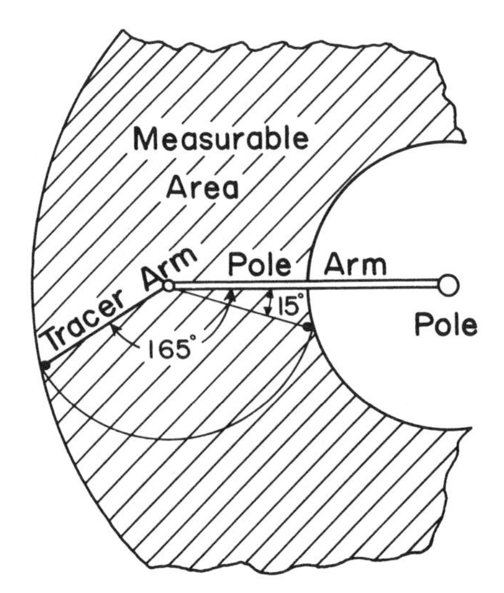

Fig. 4-6. Limit of Measurable Areas. (Courtesy of Keuffel and Esser Co.)

Move the tracer in a series of clockwise circles and, watching the dial and the wheel, stop the motion when the pointer on dial reads 0, and the 0 on the wheel is exactly opposite the 0 on the vernier. This is the zero position (Fig. 4-7). Note that the 10 on the vernier is opposite the 9th graduation on the wheel, so that ten vernier graduations are opposite nine on the wheel. Now move the tracer very slightly to the left, stopping before the vernier 0 has reached the first wheel graduation. You will see that one and only one graduation on the vernier coincides with a wheel graduation. If the 3rd vernier graduation thus coincides, you know that the wheel has turned 3/10 the distance from its 0 to its first graduation. If it is the 4th, it has turned 4/10 the distance, etc. (Fig. 4-8). Thus the vernier is used just to tell you more accurately than you can judge by eye how to divide one division on the wheel into 10 equal parts.

Fig. 4-7. (Courtesy of Keuffel and Esser Co.)

Now move the tracer a little further to the left until the 0 on the vernier is opposite the first graduation on the wheel. The wheel has now moved 10 "vernier units" or 1 "wheel unit" (Fig. 4-9). When the wheel has turned all the way until the graduation numbered 1 on the wheel is opposite the vernier 0, it has moved 10 "wheel units" or 100 "vernier units". When the wheel has made a complete revolution past 2, 3, 4, etc. and on to the point where the two 0's again coincide, it has moved 100 wheel units or 1000 vernier units. Notice that the dial has now moved so that the pointer is at 1 instead of 0 (Fig. 4-10). Each of the 10 numbers on the dial corresponds to one complete revolution of the wheel. When the dial makes one complete revolution, the wheel makes 10 revolutions, or 10,000 vernier units. Thus in terms of the vernier units we read thousands on the dial, hundreds and tens on the wheel, and units on the vernier.

If the reading in wheel units is in the 80s or 90s, the dial marker will be close to the following number and it is important not to read the dial number too high. Thus, if the wheel and vernier read 927 and the dial pointer is near 6, the correct reading is 5927, not 6927.

Fig. 4-8.
(Courtesy of Keuffel and Esser Co.)

Fig. 4-9.
(Courtesy of Keuffel and Esser Co.)

Fig. 4-10.
(Courtesy of Keuffel and Esser Co.)

Fig. 4-11.
(Courtesy of Keuffel and Esser Co.)

As an example, in the illustration (Fig. 4-11), since the pointer on the dial is between 3 and 4, we know the reading in vernier units is 3000+; since the vernier 0 is between 4 and 5 on the wheel we know the reading is 3400+ vernier units; finally since the 4th graduation on the vernier coincides with a wheel graduation, the complete reading is 3474.

In order to secure the number of revolutions around any area, subtract the final drum reading from the initial drum reading and obtain the net number of revolutions around the area. Example:

Reading after - 4612
Reading before - 3474
1138 vernier units

Note: It is not necessary to set the wheel back to zero each time. The instrumental limit of error (ILE) = ±1 vernier unit or ±0.1%, whichever is larger.

4.3 Mass and Weight

Balances are used to measure the mass or the weight of an object. The simplest device used for the measurement of weight (or any other force) is the spring balance. It consists of a coil or wire having very uniform elongation for increasing load, and therefore a definite "scale factor," or elongation-per-unit-force. The sensitivity of the balance is based on the scale factor, and decreases as the range of the balance is increased. When used to measure mass, the spring balance has the disadvantage that it must be calibrated at the place at which it is going to be used; it will give faulty readings when used somewhere else, because of the change in the gravitational attraction from place to place.

The most common balance in the laboratory is the beam balance. In its simplest form it consists of a rigid beam with a pointer pivoted in the middle and with a pan hinged at both sides. The beam balance therefore compares the weight of an object with a known weight. Since the acceleration of gravity is the same on both sides of the balance, the beam balance compares mass rather than weight, although it must be understood that its operation is based on the gravitational force of attraction acting on both the unknown object and the known mass. The measurements of a beam balance are therefore independent of the locality at which the balance is being used.

Beam balances come in different forms and shapes with many different ranges and sensitivities. Trip balances are used very often for the rapid weighing of relatively large masses. The maximum load of the balance shown in Fig. 4-12 is 1110 grams without the attachment weight hanging at the end of the beam, and 2110 with the attachment weight. The sensitivity of the balance depends on the load; it is 0.1 gm, for very light loads and goes up to 0.5 gm for a load of 2000 gm. The main advantage of the trip balance over the ordinary beam balance is that it does not require any loose weight up to 1110 grams. Measurements may be made very rapidly. The instrumental limit of error of trip balance is usually ±0.1 gm.

The analytical balance is another type of beam balance; it has a much smaller range than the trip balance but a far greater sensitivity. Since the analytical balance is a high precision instrument, great care should be taken when using it. Before any measurements can be taken, the balance must be perfectly level. This adjustment is made by

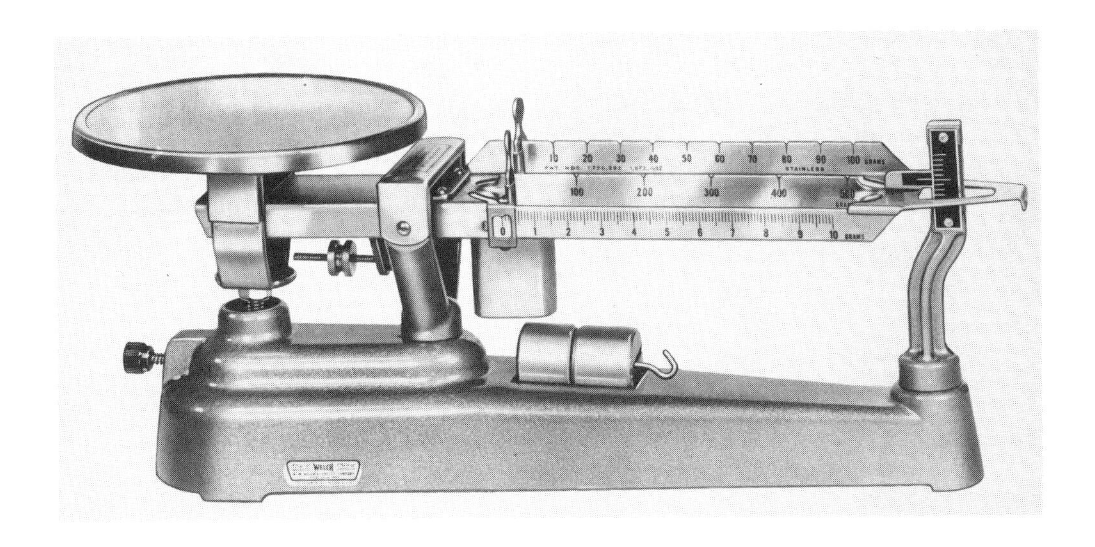

Fig. 4-12. The Trip Balance.
(Courtesy of the Central Scientific Company)

means of the two leveling screws at the front of the balance. A spirit level on the base of the balance aids one in being sure of the adjustment.

The pans should always be locked in position before weights are placed on or taken off. Weights should never be handled with fingers but always with tweezers. The window in front of the balance should be closed while measurements are taken.

There are several methods of weighing; only the Sensitivity Method will be discussed:

The balance is allowed to swing freely without weights. In order to determine the zero point, five successive readings on the scale should be taken - three readings on the left side and two on the right. The average of the readings to the left is then added to the average of the readings to the right and divided by two to obtain the zero point. As an example, let the readings obtained be the following: left -6.5, right +4.5, left -6, right +3.5, left -5.5. The average to the left is -6 and to the right +4. The sum is -2, which divided by 2 gives a zero point of -1, or one division to the left of the center mark.

Now the object to be weighed is placed on the left pan and enough weights are added to the right side of the balance until the pointer vibrates equal distances from the zero point previously determined. Weights of less than 0.01 grams are usually not put on the pan, but a rider is used. The rider weighs 0.01 grams (some balances have 0.012 gm or 0.005 gm riders); they can be put anywhere along the beam of the balance by an attachment operated from the outside. The beam is graduated into 10 large divisions; if the rider is put on the tenth division it acts as though it were placed on the pan; if placed on the fourth division it equals 0.004 grams in the pan.

4.4 Time

MECHANICAL STOPWATCHES. Many watches are equipped with "sweep second" hands, ticking along at 5 beats to the second. These are useful, but are better adapted to measurement if equipped with START, STOP and RESET buttons. A true stopwatch dispenses with hour and minute hands and simply measures time intervals. The least count of a mechanical stopwatch is usually 1/5 sec (0.2 sec.), but 1/10 second movements are available. Beyond 1/10 second, mechanical stopwatches are delicate and hard to keep in adjustment. For accurate work, the movement must be adjusted for proper rate of run, and the start and stop mechanism must be in good order.

DIGITAL WATCHES. Most digital watches have a stopwatch mode and very often a Lap Time function. For details the manufacturer's instructions should be consulted.

ELECTRIC TIMERS are generally driven by synchronous motors. Since the operation of these motors depends on the frequency of the alternating current used, the accuracy of the timers is contingent on the constancy of the 60 cycles/sec line current provided by the power company. A clutch system in the timer consists of an electromechanical unit. This allows the timer hands to move while the clutch is energized and stops the hands when the clutch is de-energized. Most timers are furnished with a manual resetting lever which resets both hands to zero instantaneously. Electric timers are manufactured for many different ranges and accuracies.

SPARK TIMER. The spark timer is used for producing equally timed impulses. It consists of a vibrating bar, the vibration being maintained electrically. The bar has electrical contacts for making and breaking a circuit at equal intervals, the length of one interval being the full period of vibration of the bar. Two sets of contacts are provided. One is used in maintaining the vibration by opening and closing the circuit through an electromagnet; the other set opens and closes a second circuit, such as the primary of a spark coil, for producing the timed sparks. The amplitude of the vibrations can be adjusted by a slow-motion screw on the electromagnet. The frequency of the spark timer shown in Fig. 4-13 is adjustable from 2.5 to 28 vibrations per second.

Fig. 4-13. Spark Timer.
(Courtesy of the Central Scientific Company)

IMPULSE COUNTER. An impulse counter is used to count electrical impulses, mainly in connection with the spark timer discussed above. It consists of an electromagnet, which operates a hand moving over a dial. At each make-and-break of the current in the electromagnet, the hand moves over one division. The impulse counter shown in Fig. 4-14 makes one complete revolution for 100 impulses. Some counters have a small-hand counting the revolutions of the large-hand.

The impulse counter may also be used as a time-measuring device. When connected to a source of current of controlled frequency, the counter becomes a high-speed stopwatch, measuring times to a fraction of a second.

Fig. 4-14. Impulse Counter. (Courtesy of the Central Scientific Company)

4.5 Temperature and Pressure

THERMOMETERS

For measuring non-extreme temperatures and differences in temperature, the mercury thermometer has been the most common type. The sensitivity and accuracy of these thermometers vary according to the requirements of the various experiments.

The <u>sensitivity</u> of a thermometer depends upon: (1) its quickness of response to change in temperature of a body in contact with its bulb; and (2) the amount of apparent linear increase in the mercury column for a given increase in bulb temperature. A thermometer will quickly reach temperature equilibrium with an object it touches - and thus show "quick response" - if it has a small heat capacity and a thin-walled cylindrical bulb. A large rise in the mercury column for a given increase in temperature requires, however, a large original volume of mercury and a small-diameter capillary. Thus a thermometer reading 0°-50°C is usually more sensitive than one whose scale is of equal length but reads 0°-100°C. The first thermometer may not, however, be more accurate.

The <u>accuracy</u> of a thermometer depends upon: (1) the accuracy of its scale calibration; (2) its history since the calibration; and (3) its use in measurements. Aging of the glass causes a thermometer to change its calibration with time. The larger part of this change is a shift of the whole scale. Thus a given thermometer whose calibration has not recently been checked may read differences in temperature more precisely than it reads absolute values of any one temperature.

In using a mercury thermometer the following simple precautions should be observed: (1) The bulb should be in contact with the object whose temperature is to be measured. If the temperature of a liquid is measured, it must be kept well stirred. The thermometer should not be pulled out or pushed in during readings. (2) The necessary level should be observed perpendicular to the length of the thermometer. (3) Great care must be taken that the thermometers are not broken off while in use and that the proper range is used for the temperatures to be measured.

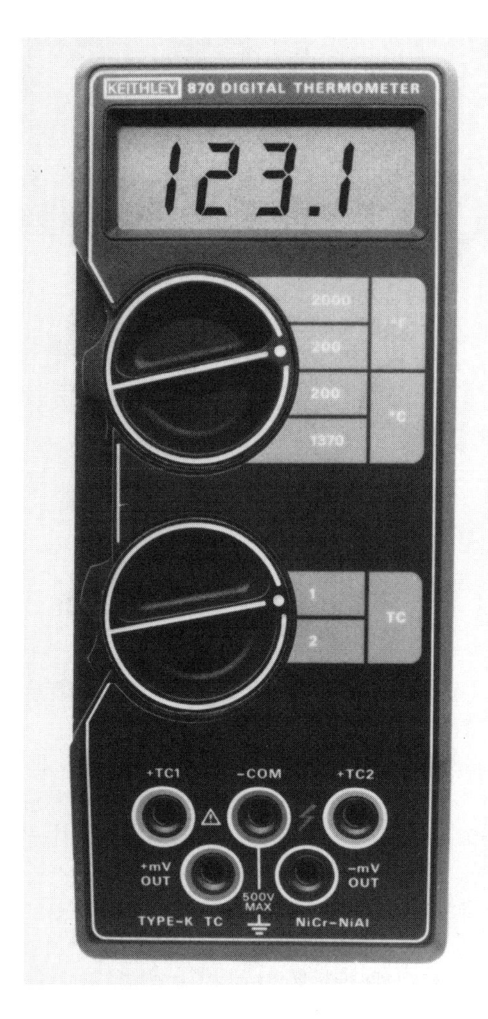

Fig. 4-15. Digital Thermometer.
(Courtesy of Keithley Instruments, Inc.)

Besides the common liquid-in-glass thermometer, many other devices are used to measure temperature or differences in temperatures. Small temperature differences are often measured by means of thermocouples, which are discussed in another section in connection with the experiment on thermal conductivity.

Fig. 4-15 shows a typical thermocouple-based digital thermometer which has a 0.25% accuracy and a 0.1° resolution for temperatures up to 200°. Different kinds of probes are available for use with this digital thermometer, depending on the particular application. Thermistors and various semi-conducting devices are available for the measurement of temperature. The measurement of very high and very low temperatures is an entirely special field.

BAROMETERS

Barometers, instruments for measuring atmospheric pressure, are of two common types: aneroid and mercury.

The aneroid barometer consists of a diaphragm attached to an evacuated chamber. A needle is fastened to the diaphragm by means of a mechanism which will magnify small motions. This instrument is particularly useful for recording barometric pressures continuously. It must be calibrated to the more accurate mercury barometer.

The mercury barometer consists of a glass tube, closed at one end, which has been filled with mercury and inverted in a cup of mercury. The mercury will sink to a level about 76 cm. above the surface of the mercury in the reservoir at normal atmospheric pressure.

A sliding vernier and scale permit accurate readings of the height of the mercury column. The zero point is at the tip of an ivory pointer in the glass reservoir at the bottom of the tube.

To read the barometer the following adjustments must be made:

(a) adjust the instrument until the scale is vertical.

(b) adjust the mercury level in the reservoir until it just touches the ivory pointer.

(c) set the vernier so that its lower edge levels with the top of the meniscus in the tube.

For precise work a number of corrections must be made in order to arrive at an accurate value for barometric pressure. Three of these corrections are listed in order of their importance, the first being the most influential:

Temperature Correction: Since the density of the mercury is a function of temperature and the brass scale will change its dimensions with a change in temperature, all readings are usually reduced to 0°C. Calibration charts very often accompany the barometer. If no such charts are available, the temperature correction to be subtracted from the reading of the barometer may be taken as 0.0123 t cm, where t is the temperature in C°. This temperature correction should be applied whenever a precise reading is desired.

Capillary Correction: Due to the capillary effect the level of the mercury in the tube is slightly lower than it would be for a flat surface. This correction depends on the diameter of the tube, and may be obtained from tables found in handbooks. If the diameter of the barometer tube is more than one centimeter, this correction is so small as to be negligible.

Mercury Vapor Correction: The mercury vapor in the space above the column exerts a pressure which depends upon the temperature. These vapor pressures of mercury may be looked up in any handbook. This correction is very small.

4.6 Electrical Circuits

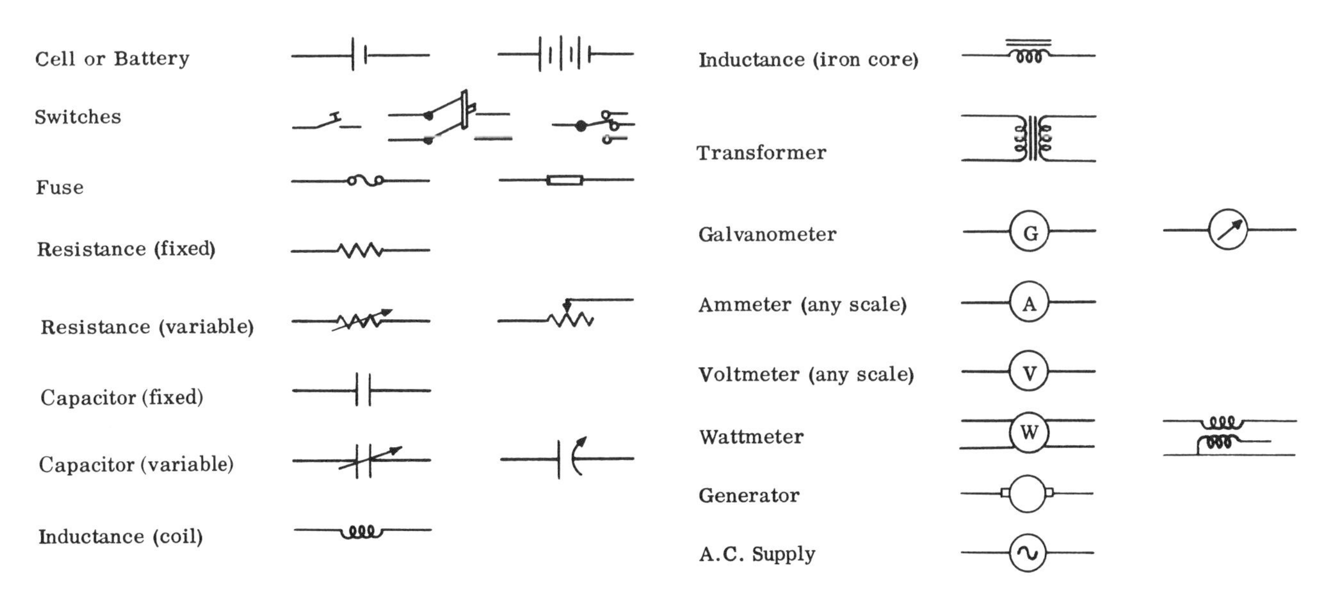

Fig. 4-16. Common Symbols Used in Wiring Diagrams.

Electrical circuits may be very complex, but most are made up of relatively simple components. Fig. 4-16 lists most of the common symbols. Specialized items such as transistors, thermistors, etc., have been deliberately omitted. Energy sources (or power supplies) may take many forms and are represented here only by the symbol for cells or batteries.

POWER SUPPLIES

Power sources are divided into two groups, direct current (d.c.) supplies and alternating current (a.c.) supplies. For small voltages the most common d.c. supply is the 1.5-volt dry cell or the 12-volt storage battery. Many laboratories are equipped with their own d.c. supplies, sometimes varying in voltage from 1.5 to 220 volts. The direct current is obtained from either a motor generator, a bank of batteries or a solid state d.c. power supply. If a very stable potential is required, an electronically controlled d.c. power supply may be used.

Any necessary a.c. supply is usually obtained from the 110-volt wall outlets. If higher or lower voltages are required, step-up or step-down transformers may be used, including transformers of variable ratio, allowing voltage adjustment. Line voltages commonly have a frequency of 60 Hz. For higher frequencies electronically controlled oscillators have to be employed; audio oscillators generate frequencies up to 20,000 Hz, or higher, but have a very low power output.

Any source in a circuit should always be followed by a switch which enables us to turn the current on or off. Knife-blade switches may have one pole or two, they may be single-throw or double-throw switches. If currents are required only very briefly, keys are used instead of switches.

STANDARD CELLS

Standard cells are used to provide a reference E.M.F. for calibration. They usually consist of an H-shaped glass vessel with sealed-in platinum wires at the lower ends of the vertical legs for electrical connections. One leg contains mercury and the other cadmium amalgam. The mercury electrode is covered with a mixture of mercurous sulphate and finely ground cadmium sulphate. A layer of cadmium sulphate crystals is placed upon the surface of both the amalgam and mercurous sulphate. The cell is filled above its crossarm with a saturated solution of cadmium sulphate. After filling the open ends of the H-tube are sealed by fusing in a flame to insure air-tightness. (See Fig. 4-17)

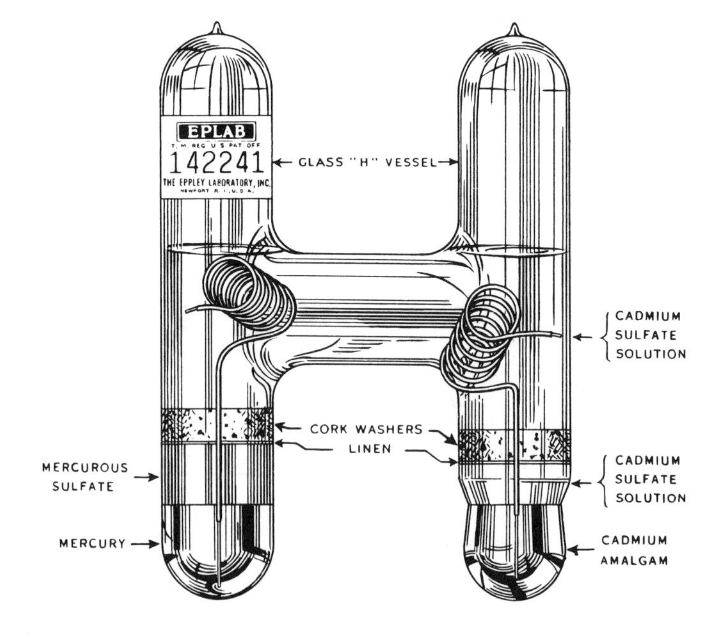

Fig. 4-17. Construction of a Standard Cell. (Courtesy of Eppley Laboratory)

Normal cells accommodate themselves rather slowly to changes of temperature and should be maintained in an oil thermostat, controlled to ±0.02 C° or better, for from 36 hours to one week before measurements are commenced, and also when measurements are being made, if a constancy within ±5 microvolts is to be secured.

The temperature coefficient of a cell is the algebraic sum of the temperature coefficients of the two opposite signs: that of the mercurous sulphate is positive, and that of the cadmium amalgam is negative. Therefore, it is important that both electrodes be held at the same temperature.

Precision, unsaturated, standard cells are tested for temperature coefficient not greater than -0.00001 volt per degree Centigrade.

Standard cells are standards of electromotive force and cannot supply much current without deterioration.

Tests show that current of the order of 100 microamperes may be drawn for six minutes without affecting the cells permanently.

Short circuits for half an hour require eleven hours for recovery to within 0.07% of the original value and 36 days to come back to within 0.007%.

Voltmeters should not be used to determine the voltage of a standard cell. The current-drain is detrimental to the cell. Standard cell determinations should always be made by a null-point method.

RESISTORS

Resistors (or resistances) may have a variety of forms and shapes, some of which will be discussed briefly. A very common type of variable resistance is the <u>rheostat</u>. It may be used for the control of current or voltage. The rheostat consists of resistance wire the effective length of which may be varied, thus changing the resistance. Rheostats are particularly useful for use as voltage dividers. Fig. 4-18a shows a rheostat used as a variable resistance, and Fig. 4-18b shows it connected as a voltage divider. In using rheostats, care must be taken not to exceed the rated voltage. If the rheostat is overloaded it gets too hot and may eventually burn out.

Fig. 4-18. Rheostat used as (a) a variable resistance and (b) as a voltage divider.

Wire-wound rheostats are available in many different ranges: from a few ohms and a current capacity of more than ten amperes to resistances of 5000 ohms or more and current ratings of about 0.25 amperes. Before using a rheostat, the proper range should be determined.

The carbon-pile rheostat is used for controlling large currents. It consists of a number of square carbon blocks mounted between two plates. When the blocks are pressed together, the resistance of the pile decreases. The range through which the resistance can be varied is usually small, 1 to 50 ohms on the average. The current capacity, however, may be as high as 50 to 80 amperes.

Fixed Resistors. The accuracy and corresponding cost of fixed resistors depends upon the use to which they are put. For instance, when a fixed resistor is to be used in a voltmeter of good quality, the value of the resistor must be accurate to within 1/2% or better. Such resistors are specifically designed to withstand adverse atmospheric conditions, such as high humidity, high temperature, and salt air. Since they draw very little current, heat dissipation is not a problem and these resistors may be completely sealed.

Precision resistors are specially designed by the National Bureau of Standards. Carefully heat-treated manganin wire is wound on an insulated brass cylinder and shellacked in position. Separate connections are provided for input and potential leads. After extended baking to insure dryness the assembly is placed in moisture-free oil and sealed in its brass case. Such units, properly handled and used, maintain their accuracy within 10 parts per million for a long time. To minimize errors due to contact resistance, inverted U-shaped arms are placed in mercury cups.

Fixed Radio Type Resistors are available in any range desired, from 1 ohm up to 1 million ohms (1 megohm) or higher. Low resistances (from one to several thousand ohms) are often wire-wound, particularly if they are to carry appreciable current. Higher resistances (200 ohms to 10 or 20 megohms) are usually molded of a composition of clay and carbon, the resistance depending on the percentage of carbon, rather than the dimensions of the device. Very high resistors are made by depositing a thin layer of metal or of carbon on a glass fiber or strip. The tolerance of these resistors is usually 5, 10, or even 20%. A color code is used to identify them and to indicate their resistance and accuracy. Most of the radio resistors have four color bands, each of which has a code-meaning. The first two colors give two significant figures, while the third color gives the decimal multiplier, and the fourth color indicates the tolerance, all according to the following code:

COMPONENTS OF ELECTRICAL CIRCUITS

Color	Significant Figure	Decimal Multiplier	Tolerance
Black	0	1	-
Brown	1	10	-
Red	2	10^2	-
Orange	3	10^3	-
Yellow	4	10^4	-
Green	5	10^5	-
Blue	6	10^6	-
Violet	7	10^7	-
Grey	8	10^8	-
White	9	10^9	-
Gold	-	-	5%
Silver	-	-	10%
No Color	-	-	20%

Plug Box Resistances: Variable plug-type resistance boxes are used in circuits where high precision is desired. To insure good electrical contact the plugs should be screwed clockwise into position without excessive pressure. To remove plugs, unscrew them counterclockwise. The plug boxes contain a series of precision resistors connected between heavy brass lugs or bars. The plugs short-out these resistors when they are screwed in. Hence the total resistance in the circuit equals the sum of the resistances in the positions where the plugs have been removed. Usually these resistances are accurate to at least one-tenth of one percent.

Heavy binding posts and bars are used in these resistance boxes because such posts have very small resistance. The current in these boxes should always be less than the maximum indicated by the manufacturer, sometimes a milliampere or less. Overloading, with consequent overheating, can permanently change the resistance values even though the resistors may not burn out.

Fig. 4-19. Plug Resistance Box. (Courtesy of the Central Scientific Company)

The Ayrton Shunt consists of a series of resistors arranged for shunting a galvanometer (see Fig. 4-20) in such a manner that the sensitivity may be varied in accurately known ratios by manipulation of a multi-contact rotary switch engraved to indicate the several multiplier values provided. The shunt facilitates the use of the galvanometer as a deflection instrument overy very wide current ranges. In null measurements it affords a convenient means of rapidly adjusting the sensitivity of the galvanometer to the requirements of the work at hand.

The resistance of the Ayrton Shunt shown and its multipliers are 1, 10^{-1}, 10^{-2}, 10^{-3}, 10^{-4}, 0 and infinity. Its limit of error is ±0.1%. When contact is at 0, none of the line current passes through the galvanometer; when at 1/1000, then 1/1000 of the total current passes through the galvanometer; etc. When the movable contact is placed on the post marked ∞, then the circuit is open; an infinite resistance is in the line. This terminal is provided so that the galvanometer circuit may be broken suddenly. The 1/1000 position is the least sensitive, and 1 is the most sensitive position on the shunt box.

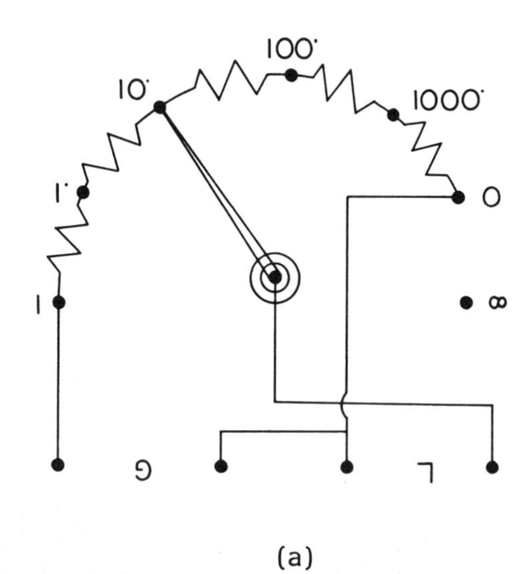

(a)

Fig. 4-20. The Ayrton Shunt. Internal Wiring Diagrams Shown on the Left.

(b)

CAPACITORS AND INDUCTORS

One of the important components of many electrical circuits is the capacitor. Just as we have variable and fixed resistors we also have variable and fixed capacitors. The variable air capacitor is used for tuning in a radio set. It consists of a number of fixed parallel metal plates connected together. By rotating a shaft on which the movable plates are mounted, the second mesh of plates is moved into and out of the first mesh so that the area between the plates, and hence the capacitance, is varied. The value of these capacitors is usually very small - in the picofarad range.

If a high precision is required (about 0.25%), plug box capacitors are used. In the plug box shown in Fig. 4-21 five precision mica capacitors are connected in series. A short heavy metal bar is fastened between each pair of capacitors. These short bars may be connected to two long bars by means of plugs at points 1, 2, 3, etc. These plugs may be arranged to give any series or parallel combination desired. As an example, consider plugs placed at points 1 and 10; we now have two .05 μf capacitors in series. This gives an equivalent capacitance of 0.025 μf. To obtain a capacitance of 0.1 μf, we could place plugs at points 2, 12, and 10 (two .05 μf capacitors in parallel) or at points 3 and 8 (two 0.2 μf capacitors in series). Plugs at 1, 10, and 5 give a total capacitance of 0.125 μf.

Some plug box capacitors work on a somewhat different scheme; always consult the circuit diagram supplied by the manufacturer before using the box.

Make sure never to short-circuit the capacitors by placing two plugs directly across from one another: i.e., on 1 and 12, 4, and 9, etc.

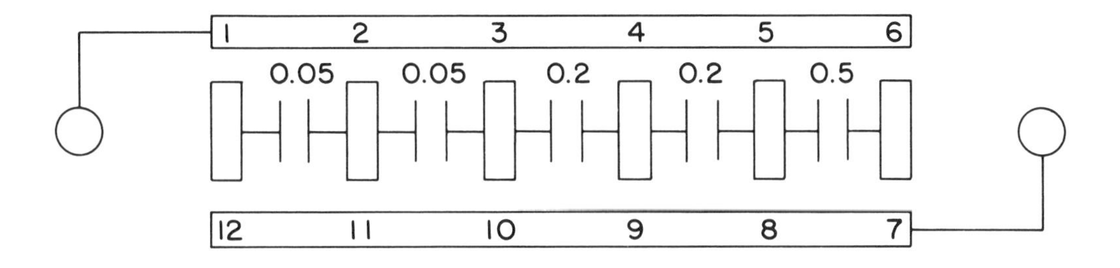

Fig. 4-21. Wiring Diagram for a Plug Box Capacitor.

Fixed Capacitors are made in many different shapes and sizes, depending on the value of capacitance required and on the amount of voltage to be applied across the plates. Most capacitors are made of sheets of tin foil placed as closely as possible with thin layers of mica or paraffine paper in between. Many special types have been developed for very high or very low capacitance values, also for use at high voltages.

Inductors. The inductor is another component of an electric circuit often encountered in practice. An inductor is simply a coil of wire, the center being either air or a magnetic material, such as iron. The most common example of an inductor is an ordinary transformer. Most inductors used in the laboratory have a fixed value of inductance depending on the geometery of the coil, the number of turns, and the material inside the core. Most variable inductors are made up of two coils, the mutual inductance being varied by turning one of the coils.

4.7 Electrical Measuring Instruments

This section describes briefly some common types of instruments used in the introductory physics laboratory to measure current, voltage, resistance, and power. Only general features will be discussed. For constructional and other details the student is referred to the literature published by the manufacturer.

DIRECT CURRENT METERS

With the availability of Digital Meters, galvanometers are not being used very often for general laboratory work. There are applications, however, where galvanometers can be very useful and for this reason some of their properties are described below.

The D'Arsonval Galvanometer. A galvanometer is used to detect very small currents. Practically all galvanometers used today are of the D'Arsonval moving coil or pivoted coil type. By means of suitable external resistors any D'Arsonval galvanometer may be converted into a voltmeter or an ammeter.

The essential parts of a D'Arsonval galvanometer are a permanent magnet, a rectangular coil suspended between the poles of the magnet, and an indicator of the angular position of the coil. The magnet is constructed so that the magnetic field is always perpendicular to the sides of the rectangular coil and is constant in magnitude for the range of the angular deflections used. The wires forming the other two sides of the coil are usually not in the field. The coil itself consists of from 10 to 20 turns of insulated copper wire wound on a rectangular frame. If a current is sent through the wire, a force will act on each wire; this force is proportional to the current. A total torque will be produced which is also proportional to the current sent through the coil.

The suspension of the coil must be such as to provide a restoring torque proportional to the angular displacement of the coil. In the more sensitive instruments this is accomplished by a very fine gold suspension strip, which also serves as one connection to the coil. The other connection is a flexible one to the lower end of the coil. A small mirror is mounted on the coil - as shown in Fig. 3-22 and the angle of deflection is observed with the aid of a beam of light reflected from the mirror. Readings are usually taken by means of a telescope, on a scale mounted 50 cm or one meter in front of the mirror.

The current-sensitivity of this type of meter is of the order of 10^{-8} amperes per thousandth of a radian deflection. This corresponds to a deflection of one millimeter on a scale mounted one meter from the mirror. Current sensitivities are generally expressed in microamperes per millimeter at scales 1 meter from the mirror.

The megohm sensitivity of a galvanometer is the resistance in megohms (10^6 ohms) that must be placed in series with the galvanometer to give a deflection of one millimeter per volt of the applied E.M.F. The megohm sensitivity can be observed directly, while the current sensitivity has to be computed.

For some purposes it is useful to know the sensitivity of a galvanometer to differences of potential applied to its terminals; this is referred to as the voltage sensitivity. It depends mainly on the resistance of the galvanometer, and is expressed in volts per millimeter. It is equal to the produce of the current sensitivity and the resistance of the galvanometer.

Fig. 4-22. (a) The D'Arsonval Current Galvanometer and (b) The D'Arsonval Ballistic Galvanometer. (Courtesy of the Leeds and Northrup Company)

Another constant of the galvanometer is the external critical damping resistance (C D R X). This resistance must be in parallel with the galvanometer to produce the critically damped condition. The motion is said to be critically damped when the coil will move promptly to its new equilibrium position without swinging back and forth.

The Spotlight Galvanometer consists of a compact taut-suspension moving-coil reflecting galvanometer, with built-in lamps and scale. It is highly stable, sturdy, and sensitive.

The moving system of the galvanometer is supported between upper and lower gold suspensions held taut by phosphor bronze springs. Each system is carefully balanced so that the center of gravity is closely in line with the axis of the suspensions thus eliminating the need for careful leveling and giving greater stability to the instrument.

The index is a sharply defined hair line in a circular spot of light completely free of parallax and easily readable to within 0.2 millimeter. The instrument shown in Fig. 4-23 may be used for deflection measurements as well as null measurements. It is provided with a 100-millimeter scale subdivided into 0-100 and 50-0-50 divisions.

The sensitivities of the instruments range from 25×10^{-4} μA to 2×10^{-1} μA.

Fig. 4-23. Spotlight Galvanometer. (Courtesy of Rubicon Company)

A D'Arsonval ballistic galvanometer (see Fig. 4-24) is basically the same as the current galvanometer just discussed; the shape of its coil is so constructed as to give it a higher rotational inertia and therefore a much longer period of swing. The ballistic galvanometer does not require a current to flow through the coil continuously, but will measure a current impulse of very short duration, or a charge, ($q = \int idt$).

Less sensitive D'Arsonval galvanometers have coils mounted on pivots, the torque being provided by a spiral spring similar to the hair-spring of a watch. A mechanical pointer and a circular scale are used for indicating measurements (see Fig. 4-24). If a shunt is put across an instrument of this sort, it then becomes an ammeter. The range of the ammeter can be varied by putting in appropriate shunts and calibrating the scale in amperes directly.

A voltmeter can be made out of the D'Arsonval galvanometer by connecting a high resistance in series with the meter. Again, the range of the voltmeter can be varied to any value by adding the proper series resistance and calibrating the scale in volts.

Fig. 4-24. Permanent Magnet Galvanometer.
(Courtesy of Weston Company)

ALTERNATING CURRENT METERS

When an alternating current is applied to a D'Arsonval galvanometer, the coil tends to reverse its direction of rotation each time the current reverses; because of the moment of inertia of the coil, however, there would be no deflection if a frequency of 60 cycles per second were used. There are several types of alternating current meters, some of which will be described briefly.

The Dynamometer does not contain a permanent magnet. Instead, the magnetic field is provided by a stationary coil. The moving coil of the instrument is similar to that used in d.c. meters. The current in both the stationary and the moving coils alternates at the same frequency (60 cycles per second for ordinary line voltage). Since this causes the magnetic field to reverse whenever the current in the moving coil reverses, the result is a steady deflection.

The dynamometer mechanism is used in a.c. ammeters and voltmeters by using appropriate shunts or series resistance together with the meter itself. The same type of meter is particularly suitable for power measurements. Variations of the dynamometer are used for the measurement of power factors, phase angles or frequency.

While the torque produced in a D'Arsonval type of meter is directly proportional to the current and hence the scale is uniform, this is not the case in a dynamometer. Here the torque is proportional to the current squared, which results in a non-uniform scale, somewhat compressed towards the zero end.

Another type of a.c. meter depends on the Iron Vane Mechanism. The repulsion iron vane mechanism contains two parallel iron vanes. These are located within a solenoid; the outer cylindrical vane is fixed, while the inner one is pivoted and can rotate freely, subject to the restoring force of the spring.

A current passed through the solenoid causes a magnetic field to be set up. Magnetic poles of like polarity are induced in the ends of both vanes. Since like poles repel each other, the moving vane adjusts itself to a position corresponding to the strength of the current flowing through the solenoid, and a pointer indicates readings in amperes or volts.

The torque produced by the magnetic repulsion depends upon the spacing and shape of the vanes, as well as on the magnitude of the current. In general, this torque will not be directly proportional to the current, which results again in a non-uniform scale requiring proper calibration. By using appropriate shunts and series resistances, the iron vane mechanism can also be used for a.c. ammeters as well as voltmeters.

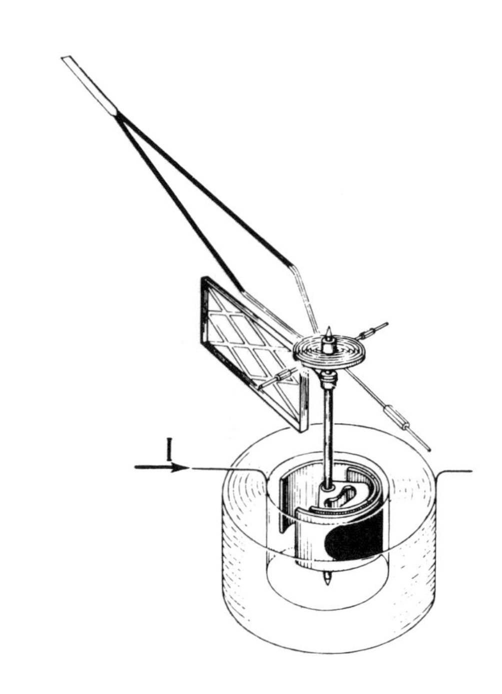

Fig. 4-25. The Iron Vane Mechanism. (Courtesy of Weston)

The dynamometer and iron vane repulsion instruments are satisfactory for the measurement of alternating current when the frequency is relatively low; i.e., not more than a few hundred cycles per second. When the frequency is of the order of hundreds or thousands of cycles per second, as in radio and other communication work, these meters are not satisfactory. In order to measure alternating currents and voltages at high frequencies Thermocouple Meters are used.

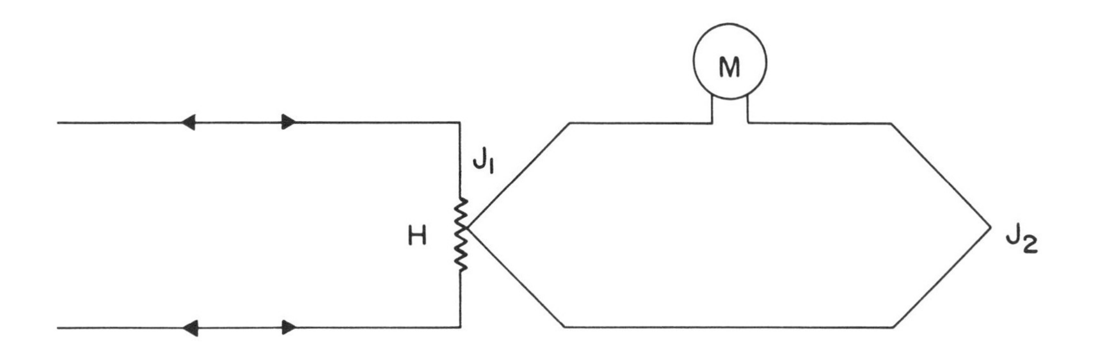

Fig. 4-26. Diagram for Thermocouple Meter.

In a thermocouple meter the permanent magnet-moving coil mechanism is adapted for a wide range of frequencies. A heater element H (see Fig. 4-26) is connected into the line to be measured, and the instrument coil is connected to a thermocouple J. The heat applied to the junction of the two dissimilar metals of the thermocouple causes a difference in potential in the instrument circuit. The instrument is calibrated to measure the current flowing through the heater. Since the reading is proportional to the heating effect of the current, the scale is non-uniform and the readings are independent of the frequency or wave shape. The primary application of this type of a.c. meter is in measuring radio frequency currents.

THE WATTMETER

The principle involved in the dynamometer type of voltmeter and ammeter also applies to the wattmeter.

A magnetic field (F) shown in Fig. 4-27 is produced by a current flowing in the field coil and through the circuit. This field is dependent on the current in the circuit. The instrument's moving coil (C) is composed of many turns of fine wire and is connected in parallel with the load circuit. The current in the moving coil is dependent on the voltage across the circuit.

The torque produced on the moving coil is proportional to the magnetic field, also to the current in the moving coil, and hence is proportional to their product. This means, then, that the torque is proportional to the product of the current and voltage at any instant. Moreover, when the current and voltage are in the same direction, the torque will tend to rotate the coil one way, and when the current and voltage are in opposite directions, the coil will try to rotate the opposite way.

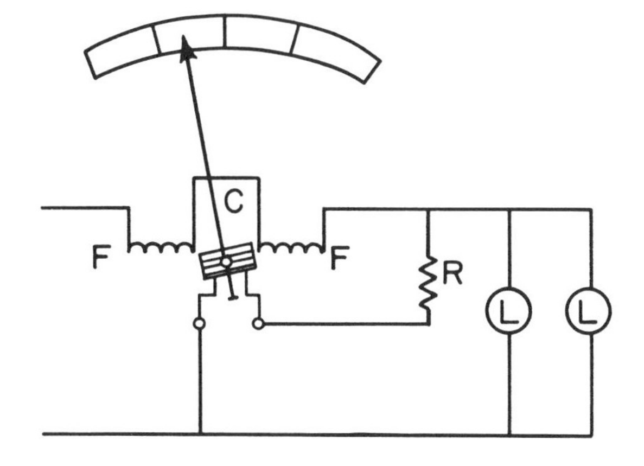

Fig. 4-27. Wiring Diagram for the Wattmeter.

If now, we attach a spring to the coil which provides a restoring torque proportional to the deflection, the rotation of the coil will be proportional to the torque produced by the interaction of the moving coil current with the magnetic field; i.e. to the instantaneous power. The power must be averaged over a complete cycle; i.e., the coil should not fluctuate back and forth as the power goes up and down, but should remain at an average position dependent on the average torque. This is accomplished quite easily by making the coil and pointer system sufficiently massive, so that its rotational inertia prevents it from following the rapid fluctuations within one cycle. Thus the deflection of the coil is directly proportional to the average power in the circuit, and all that remains for completion of the instrument is the calibration of the scale in the appropriate units.

OHMMETERS AND MULTIMETERS

As long as ammeters and voltmeters are available, no separate instrument is required for the measurement of resistance; it can always be calculated from Ohms Law if the currents and voltages are known. Very often, however, it is practical to measure the resistance directly, which can be done with an Ohmmeter. An ohmmeter sets up its own circuit of which the resistance to be measured is a part. The voltage is usually supplied by a number of small dry cells inside the meter housing. The scale is calibrated on ohms directly, so that no computations are required. Readings taken with an ohmmeter are not very accurate, however. If precise values of resistances are required, other methods of measurement have to be used, such as the ammeter-voltmeter method or the Wheatstone Bridge.

Many different types of Multimeters are on the market. By a multimeter (also called analyzer) we mean a combination voltmeter, ammeter, and ohmmeter, all in one housing with one dial. A typical multimeter may be used for the measurement of d.c. and a.c. voltages with scales ranging from 2.5 volts to 1000 volts. Direct currents may also be measured with scales from 100 μa (10^{-4} amperes) to 10 amperes. Ohmmeter scales range from 3000 ohms to 30 megohms. Besides the basic measurements of voltage, current, and resistance, a meter of this kind has many other applications in circuit-testing.

All voltmeters mentioned have a definite resistance (usually from 100 to 20,000 ohms per volt), and therefore draw a current, which must be considered in precise measurements. A Vacuum Tube Voltmeter draws practically no current. The vacuum tube voltmeter (VTVM) shown in Fig. 4-28 is an a.c. meter with 10 voltage ranges from 0.01 to 300 volts RMS full scale and can be used for a wide range of frequencies. It is especially useful in the measurement of microwave intensities when a 1000 cps modulation is used. (See Section 4.10 Microwaves.)

Digital Multimeters, using a liquid crystal display (LCD) have become very popular for general laboratory work. A typical example, shown in Fig. 4-29 has 5 different functions and many ranges for each. D.C. and A.C. (rms) voltages can be read from 100 μV (10^{-4} V) per digit to 1000 V. D.C. and A.C. (rms) currents can be read from 100 nA (10^{-7} A) per digit to 2000 mA (2A). Resistance can be read from 100 mΩ (0.1 Ω) per digit to 20 MΩ ($2 \cdot 10^7$ Ω).

The accuracy of the digital multimeter shown in Fig. 4-29 is as follows:

D.C. voltages - ±0.25 of reading plus 1 digit
A.C. voltages - ±0.75 of reading plus 5 digits
D.C. currents - +0.75 of reading plus 1 digit
A.C. currents - 1.5 of reading plus 5 digits
Resistance - .2% of reading plus 1 digit for 2 KΩ to 2000 KΩ

Fig. 4-28. A.C. VTVM (Courtesy of Heath Company)

Fig. 4-29. Digital Multimeter. (Courtesy of Keitley Instruments, Inc.)

Note that the accuracy is given as a percent plus a digit; sometimes the percentage error is more important while for other readings the uncertainty at the last digit is much larger. This can be illustrated by the following example:

meter reading	0.25%	digit uncertainty	total uncertainty
199.9 mV	0.5 mV	0.1 mV	0.6 mV or 0.3%
1.1 mV	0.003 mV	0.1 mV	0.1 mV or 9%

RECORDERS

In many situations it is desirable to have a continuous and permanent record of quantities and their variation with time. By the use of suitable transducers and appropriate circuitry a chart recorder may be used to record not only voltages and currents, but also such quantities as pressure, speed, temperature, light intensity and many other phenomena.

An example of such a recorder is shown in Fig. 4-30. The Heath/Malmstedt Enke Servo Chart Recorder shown is designed to provide direct readout of signals in the range 10 to 250 millivolts d.c. This particular model is equipped with a multi-speed drive for the chart paper, ranging from 12 inches per minute to 1/2 inch per hour. The chart motor may also be programmed by external signals.

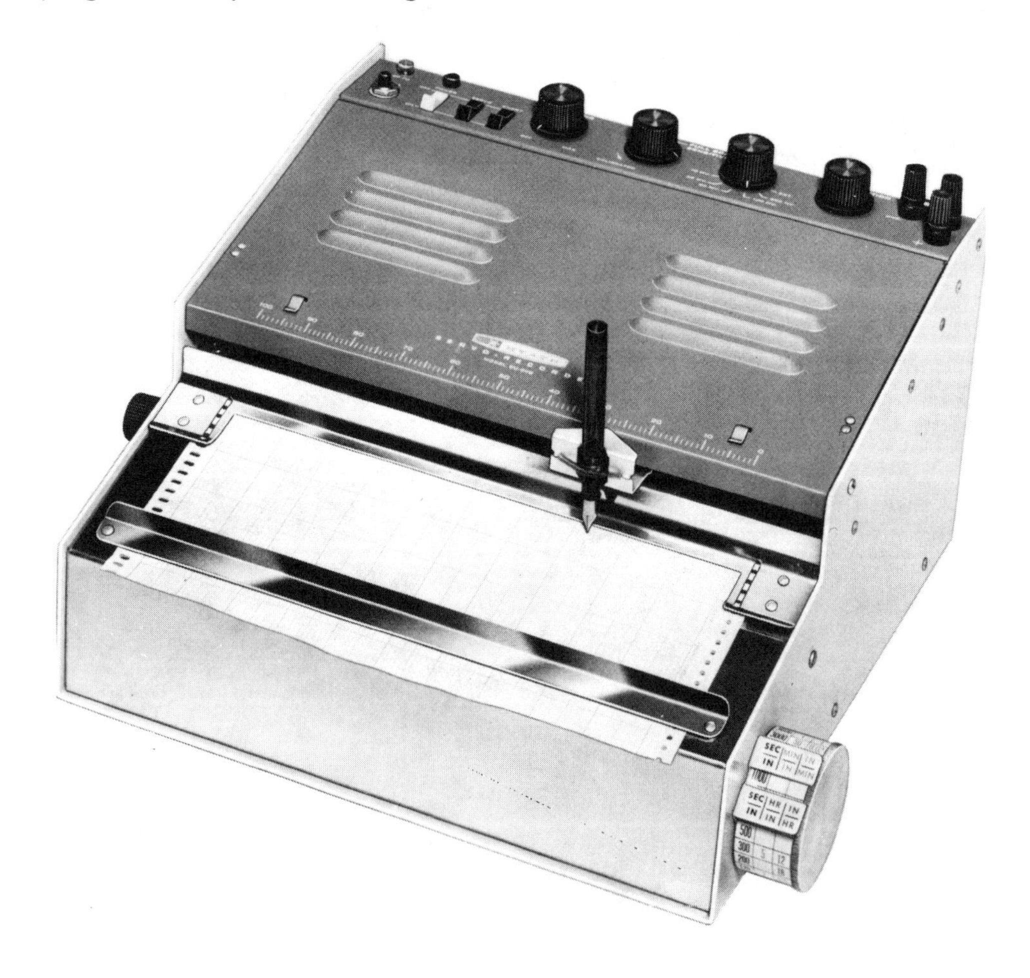

Fig. 4-30. Heath/Malmstedt-Enke Servo Chart. (Courtesy of Heath Company)

CATHODE RAY OSCILLOSCOPE

Most conventional meters (other than digital meters) used for measuring voltages and currents involve mechanical parts, such as coils with pointers, mirrors, etc. The inertia of these moving parts is much too great to permit them to follow a very rapid variation in the applied torque. These instruments, therefore, do not measure instantaneous voltages and currents, but rather average or effective values.

The cathode ray oscilloscope is an instrument that does record instantaneous values of rapidly varying voltages. It can therefore be used to actually observe various wave forms. There is no mechanical moving part in the oscilloscope, but the "moving" is done by a beam of electrons. The inertia is thus negligible, and the pencil of electrons becomes an ideal indicator for any rapidly changing voltage.

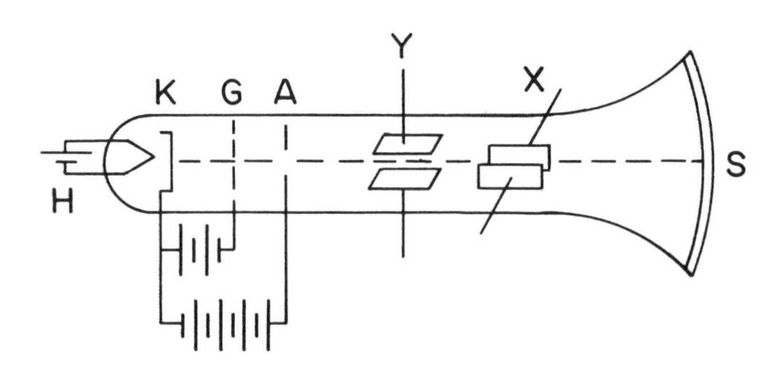

Fig. 4-31. Essential Features of the Cathode Ray Oscilloscope.

Fig. 4-31 shows some of the essential but simplified features of a cathode ray oscilloscope. The heater H heats the cathode K which emits thermionic electrons. These electrons are emitted towards the anode A, which is actually cylindrical, and is maintained at a positive potential of several thousand volts with respect to the cathode. Many of the electrons strike the anode, but a narrow beam is guided through it and eventually strikes the fluorescent screen S. This screen consists of a thin layer of phosphor coated on the evacuated glass envelope of the tube. When the screen is struck by the electrons, it emits light, it fluoresces. The intensity of the light spot on the screen - i.e., the intensity of the beam of electrons is controlled by means of a grid G placed between the cathode and the anode.

Between the anode and the screen the electrons pass between two sets of plates, X and Y. Electric fields across these plates control the horizontal and vertical deflections of the beam. A field across the plates X produce a horizontal deflection, while a field across the vertical plates Y produces a vertical deflection. Because of the very low inertia of the electrons, the electron beam is able to follow nearly instantaneously any variation in the electric fields produced by voltage applied across the plates.

If no voltages are applied to the deflecting plates, a bright spot will be observed at the center of the screen. In using the scope NEVER LET AN EXCESSIVELY BRIGHT SPOT APPEAR ON THE SCREEN BECAUSE IT MAY BURN THE SCREEN AND ALSO DECREASE THE LIFE OF THE CATHODE RAY TUBE. If we now apply a voltage across the vertical plates that makes the upper plate positive, the spot on the screen will move up; if we make the upper plate negative, the spot will move down. If an alternating voltage is applied, the spot will move up and down; the location of the spot at any instant therefore depends on the voltage at the moment. If the voltage applied alternates with a frequency of more than about ten cycles per second, the retentiveness of the screen and of the observer's eye cause the moving spot to appear as a continuous line.

If the wave shape of an applied voltage is to be studied, the spot must be moved horizontally too, so that the pattern may spread out on the screen. This is done by connecting the horizontal deflecting plates to a source of voltage that rises gradually at a constant rate to a maximum value and then suddenly drops back to zero. Such a voltage is said to have a saw-tooth shape. It causes the beam to move horizontally across the screen at a uniform speed and when the voltage suddenly drops to zero, the beam moves back to its original position and begins another horizontal sweep with the "triggering" of the next saw-tooth. The electronic circuit producing this saw-tooth voltage in the oscilloscope is called the sweep circuit. It enables us to obtain a pattern on the screen which is exactly the same as a curve of the varying voltage applied across the vertical plates as a function of time. The only condition that has to be fulfilled is that the saw-tooth voltage must be equal to a multiple of the period of the applied voltage to be studied. The observed curve is really a snapshot of one or more cycles of the voltage being studied. Knowing the sweep frequency, the frequency of the vertical input can be determined.

The oscilloscope may be used for a variety of purposes. In order to allow the study of wave forms of different frequencies, the sweep frequency can usually be changed over a wide range, going up to 60 MHz ($60 \cdot 10^6$ Hz or higher for some of the better oscilloscopes.

To make certain that the sweep frequency is of exactly the same frequency as the voltage being studied, the "trigger" of the oscilloscope is used.

Besides the controls already mentioned, there are some other knobs whose function should be understood. The intensity of the electron beam is controlled as well as the focus on the screen. The pattern may be centered both vertically and horizontally. The gain, or amplification, of the signal may be varied for the horizontal and the vertical plates. ALWAYS MAKE SURE YOU UNDERSTAND WHAT A CONTROL DOES BEFORE YOU USE IT. Different oscilloscopes have somewhat different controls and the manufacturer's instruction manual should be consulted if it is available.

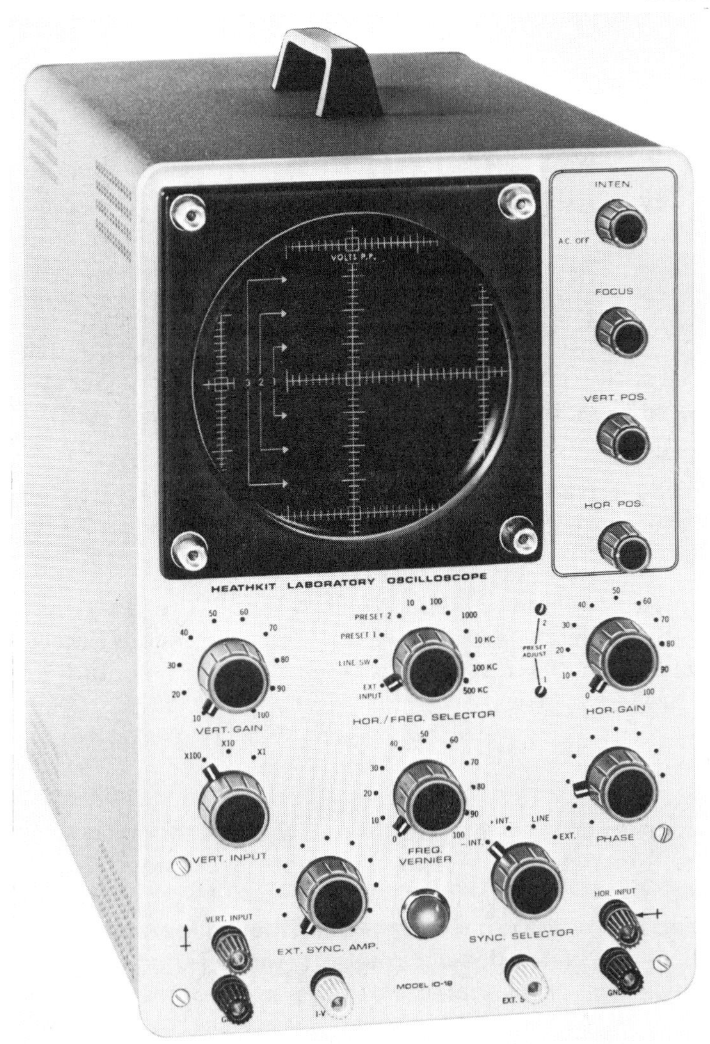

Fig. 4-32. A Simple Laboratory Oscilloscope. (Courtesy of Heath Company)

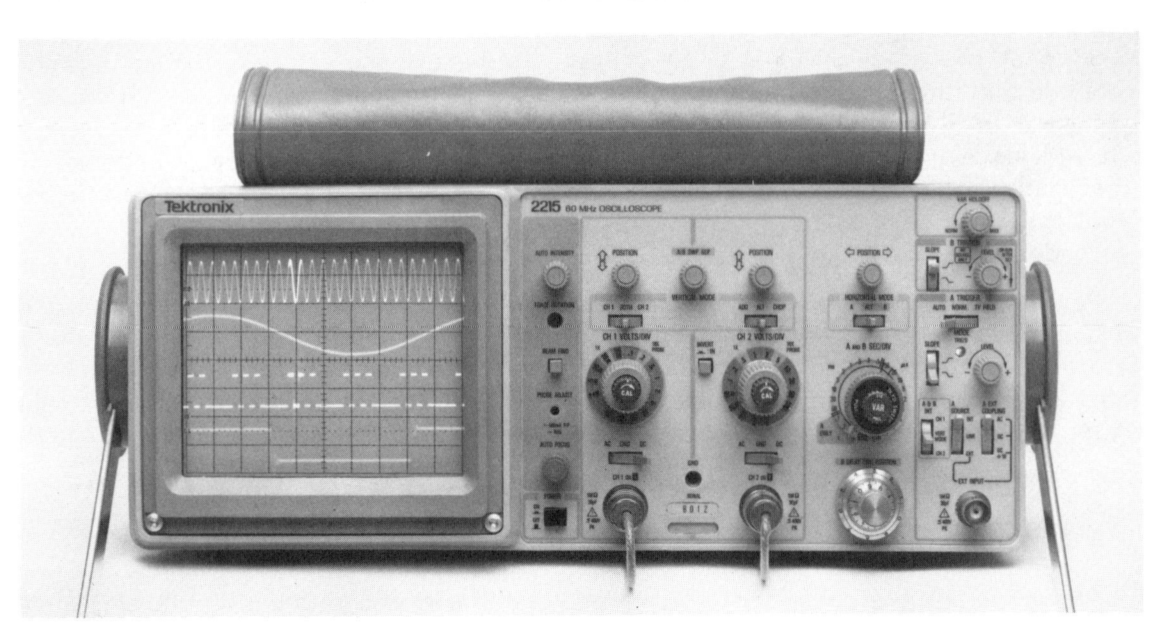

Fig. 4-33. A General Purpose Portable Dual Trace Oscilloscope. (Courtesy of Tektronics, Inc.)

4.8 Optical Instruments

CARE OF LENSES, PRISMS, AND MIRRORS

The accuracy of all optical devices depends upon the condition of the optical surfaces. Dust, grease, and scratches are the common cause of unsatisfactory performance of lenses, prisms, and mirrors.

Dust results from extended exposure to air. It may be largely avoided by covering up the instrument when it is not in use. Dust should be removed only by using a soft cloth, a camel's hair brush, or special lens paper. Other methods may scratch the surface. Wipe with a lifting motion, always bringing fresh tissue in contact with the surface. The abrasiveness in dust must not be scribbed into the glass.

Grease should never be allowed on any optical surface. It is generally found when the surface has been touched with the fingers, which should never happen. To remove grease, rub gently with lens paper.

Scratches on any optical surface are inexcusable. They spoil the sharpness of the image, and sometimes appear in it. Since many metals are harder than glass, keep all objects, particularly metal objects, away from the optical surfaces. Lenses should be handled carefully at the rim and prisms at the top and base.

PARALLAX

By parallax is meant the apparent motion of one object with respect to another when both objects are stationary and the position of the observer is changed. In optical instruments it refers to the apparent motion of the image with respect to the cross-hairs when the eye is moved from side to side. It occurs only when the image and the cross-hairs are not in the same plane. Parallax is used as a test for focusing instruments. When the image is in the same plane as the cross-hairs we have a good focus and no parallax.

To use the parallax test for focusing, simply move your eye from side to side while looking at the image and the cross-hairs. If there is no relative motion between them, they are in the same plane and are in focus. If there is relative motion, adjust the instrument until it is eliminated. When the image is between your eye and the cross-hairs, it will move to the left of the cross-hairs when you move your eye to the right. When the image is beyond the cross-hairs, it will move to the right of the cross-hairs when you move your eye to the right. This enables you to know which way to move the image in order to focus it.

Parallax is important also when measuring the distance between two points with a rule. Have the points and the rule in the same plane (perpendicular to your line of sight) in order to eliminate parallax and the error it involves. The parallax problem also occurs in reading meters with moving pointers. It can be minimized by having a mirror below the pointer, so that the eye will always be perpendicular above the scale and also over the pointer.

THE TELESCOPE

The simplest form of laboratory telescope consists of two lenses, the objectives and the eyepiece or ocular. Both lenses are mounted in a tube with cross-hairs between them. The objective lens forms an image of the object in the plane of the cross-hairs; this image is then viewed through the eyepiece.

To properly adjust the telescope, the eyepiece should be focused on the cross-hairs until they are seen clearly. The correct distance between the eyepiece and the cross-hairs will be different for different observers. When the cross-hairs are properly focused, the image will also be in focus if it coincides with the cross-hairs. This is accomplished by adjusting the distance from the cross-hairs to the objective lens until the image is seen clearly. Some telescopes have a focusing screw for making this last adjustment; in other telescopes the objective lens is mounted in a tube that can be pulled out or pushed in.

As mentioned before, the correct adjustment of the eyepiece will be different for different observers, but the proper distance from the cross-hairs to the objective lens does not depend on the observer and therefore stays the same as long as the same object is being viewed.

The telescope is used in conjunction with a great many scientific instruments, such as the cathetometer, the D'Arsonval galvanometer, the spectrometer, etc. The adjusting procedure, however, is always the same.

THE MICROSCOPE

Fig. 4-34 is a ray diagram of a compound microscope. It consists of two lens groups, the objective and the eyepiece or ocular. The objective lens forms a real, inverted, enlarged image of the object, and the eyepiece forms a virtual, erect, and enlarged image of the object for the eyepiece, which is the image formed by the objective. While both the objective and the eyepiece of an actual microscope are highly corrected compound lenses, they are here shown as simple thin lenses.

The magnification of a microscope is the product of the lateral magnification of the objective and the angular magnification (sometimes called "magnifying power") of the eyepiece.

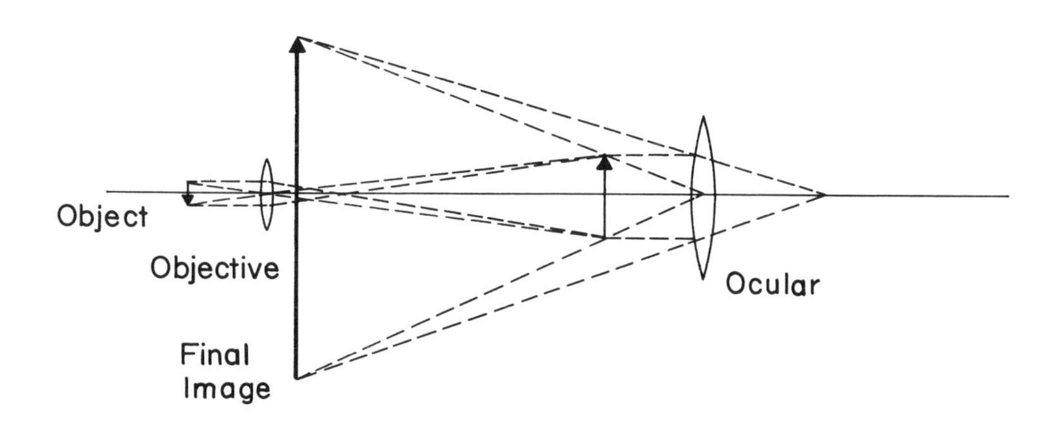

Fig. 4-34. Ray Diagram for a Compound Microscope.

Fig. 4-35 shows the optical and mechanical features of a microscope. Before using the instrument, the purpose of all parts should be understood, particularly the following:

(1) The quick and slow-motion focusing screws. In some microscopes the fine adjustment may also be used for depth measurements.

(2) The illuminating mirror with its plane and curved surfaces, which can be set at any angle.

(3) The condensing lens under the stage, which is used to illuminate the object.

(4) The iris diaphragm lever which can be used to control the amount of light illuminating the object.

(5) The draw tube, which, in some microscopes, can be raised or lowered by pushing it into the lens tube. In many instruments the draw tube is fixed, usually at 160 mm.

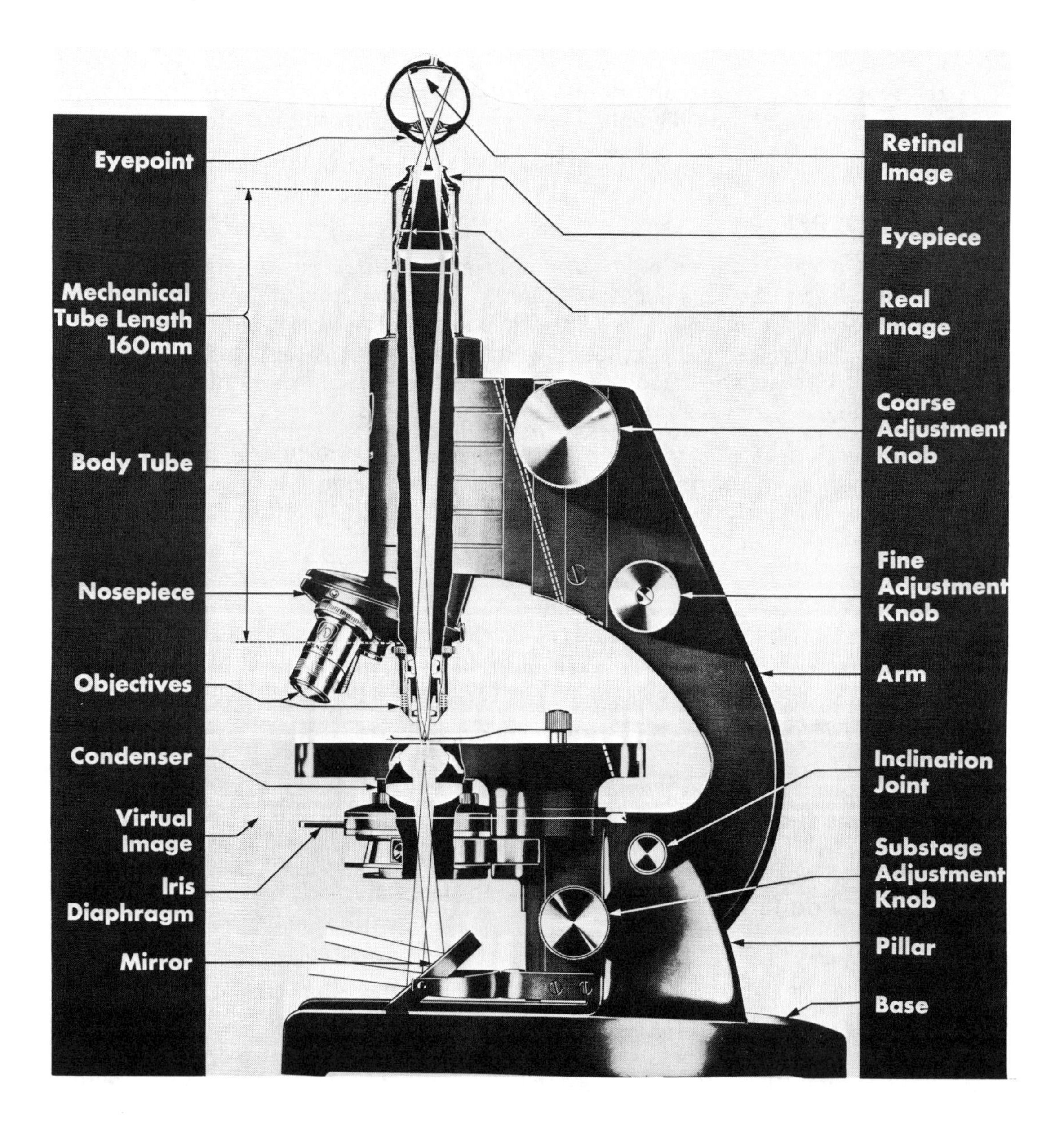

Fig. 4-35. Optical and Mechanical Features of the Microscope. (Courtesy of the American Optical Company)

In use the microscope must be illuminated; this can be accomplished in several ways. Daylight, but not direct sunlight, may be used by turning the mirror so that light is reflected into the microscope. Since daylight is not always available or is of variable quality, artificial light is generally used. Simple lamps may be placed on the table or directly below the microscope stage. In some microscopes, illuminators are built in. Critical microscopic work requires a lamp with a condensing lens and a diaphragm.

Besides magnification and illumination, several other features of the microscope should be mentioned. The Numerical Aperture (N.A.) is the measurement of the angle of the maximum cone of light that may enter the object. It is common to express the numerical aperture by the equation,

$$N.A. = n \sin a$$

where (n) is the lowest index of refraction between the object and the objective, and a is the angle between the extreme ray of the axial bundle and the axis. The greater the N.A., the greater the resolving power of the instrument.

Resolution (or resolving power) is the ability of the microscope objective to separate small details. The primary function of a microscope is not simply to magnify, but to reveal detail not observable with the unaided eye. Even an objective of the highest spherical and chromatic corrections does not yield a "perfect" image because of diffraction. An objective lens should therefore be chosen which has enough Numerical Aperture to insure the resolving power needed.

Another important characteristic of the microscope is its depth of field, sometimes erroneously called depth of focus. This refers to the thickness of the specimen which may be seen in focus. The greater the magnification and the N.A., the thinner is the layer in focus. A lens of longer focal length and less magnification will be more satisfactory for the study of the general arrangement of the specimen. Despite the lower resolving power, a greater depth will be obtained, the field will be larger and the image will be brighter.

Using the Microscope: For proper illumination remove the eyepiece, look into the tube, and adjust the mirror until the beam light coming from the lower lens is clear and unobstructed by any object. Either a plane or concave mirror may be used.

Now mount the object on the stage, secure it with the clips, and then lower the objective lens until it almost touches the object. Center the object carefully under the lens. Look through the eyepiece and focus up slowly with the coarse adjustment knob until the image comes into focus. The fine adjustment knob can now be used for a final setting.

When observing, look into the instrument easily without straining the eyes. Working with both eyes open prevents nervous strain. Do not place the eye too close to the eyepiece because this will restrict the field of view. By raising the eye a certain point will be found from which a maximum amount of the object is visible. This is called the eye point (see Fig. 4-35).

THE SPECTROMETER

Optical instruments fall into two general groups: image-forming instruments, such as the microscope, telescope, etc.; and analyzing instruments, such as the spectrometer. Instruments of the first class serve to form an image of some given object; instruments of the second class are used to analyze light as to its composition, intensity, state of polarization, and other characteristics.

One of the most important of the group of analyzing instruments used in measuring the wavelength of light is the spectrometer. Because of its simplicity and accuracy the spectrometer has been used extensively in the collection of experimental data that is the basis for the theory of the structure of the atom. There are different types of spectrometers in use, which range in accuracy from a limit of error of ±1 minute on a circular vernier to ±1 second or less. The instrument shown in Fig. 4-36 can be read to 1 minute.

Fig. 4-36. The Spectrometer. (Courtesy of the Gaertner Scientific Corporation)

The essential parts of a spectrometer are, first, a narrow slit illuminated by the light to be analyzed. It is located at the first focal point of an achromatic lens, called the collimator. The beam of light leaving the collimator is parallel and falls on the prism, in the case of a prism spectrometer. The light is deviated by the prism and the emerging beam is examined by means of a telescope. In the spectrometer shown, the central shaft and the collimator-support tube are fixed in position on the base plate with the divided circle inscribed on an embossed ring. A vernier is fixed to the telescope support tube, which rotates about the central shaft. A clamp and a tangent screw are provided for delicate settings of the telescope.

The prism table rests on three leveling screws acting against three springs. The prism table rotates freely and may be clamped firmly in any position. The telescope and collimator are adjustable in height to accommodate any prism, grating, etc.

The collimator is focused by hand and may be clamped in any position. The telescope focusing is by rack and pinion. The slit jaws are of nickel silver with accurately ground edges which remain parallel in all positions. The movable jaw closes by spring pressure.

A spectroscope is an instrument used for the visual observation of a spectrum only. If the telescope of the spectrometer is replaced by a camera, the instrument becomes a spectrograph. This device is used for practically all measurements of unknown wavelength today. It permanently records the spectrum for later detailed analysis.

Adjustment of Spectrometer: The slit is adjustable in width. A wide slit admits more light but reduces the precision of the readings. A narrow slit is necessary for exact work, but it may give too faint an image. It is therefore advisable to start observations with a fairly wide slit and to narrow the slit before taking final readings. The collimator will usually be found adjusted. If not, the draw tube should be adjusted until the image of the slit is formed.

Focus the eyepiece of the telescope on the cross-hairs until they are clearly seen. Then adjust the telescope (alone) until an image of a distant object falls in the plane of the cross-hairs without parallax. The telescope is now adjusted to accept parallel rays (effectively from infinity). Now bring the telescope into line with the collimator, and adjust the collimator until an image of the slip appears in the plane of the telescope cross-hairs. The light rays between collimator and telescope are now parallel - an important condition for proper use of the instrument. Focus the telescope on the slit by means of the focusing screw. Follow the directions given for the focusing of telescopes on Page 99. After the collimator and the telescope have been properly focused, make sure that they are in line and that both tubes are horizontal. See that the prism table is at the proper height, but do not readjust the leveling screw.

4.9 Light Sources

POLYCHROMATIC SOURCES

Most light beams are polychromatic, which means that they are a mixture of different wavelengths or colors. These wavelengths may extend through the visible spectrum and beyond. Incandescent light bulbs are the most commonly available source; here a heated filament produces a continuous spectrum. At the temperature usually used in lamps, the light looks white. Its spectrum can easily be investigated with a prism on a diffraction grating.

A line spectrum is produced in a gas discharge tube. The characteristic of the spectrum depends on the gas in the tube. In general a high voltage (about 5000 volts) must be applied across the tube. The voltage is obtained by the use of a transformer. A typical laboratory spectrum tube power supply with a tube is shown in Fig. 4-37.

Fig. 4-37. Spectrum Tube Power Supply. (Courtesy of the Central Scientific Company)

A fluorescent light bulb produces both a continuous spectrum as well as a line spectrum.

MONOCHROMATIC SOURCES

A monochromatic light source emits a beam of one particular wavelength or color. A sodium vapor lamp gives off yellow light that is almost monochromatic within the visible range - the two sodium lines have wavelengths of 588.995 and 589.592 nm respectively. Sodium light can be produced easily by soaking a piece of asbestos in a saturated solution of common salt and wrapping it around a bunsen burner.

Monochromatic light of a number of wavelengths can be produced by placing appropriate filters in front of a mercury light source. This may consist of a very small mercury discharge bulb mounted in a metal shield which is used on a 110-volt a.c. line with a proper ballast. A mercury spectrum tube can be used in the special power supply shown in Fig. 4-37. For higher intensities special mercury vapor lamps are available. A set of filters is needed corresponding to the lines in the mercury spectrum. With appropriate filters the following wavelengths may be obtained: 690.7 nm (red), 578.0 nm (yellow), 546.1 nm (green), 435.8 nm (blue) and 404.6 nm (violet). In the ultraviolet (not visible) the 336.0 nm line may be used.

THE LASER

The continuous visible gas laser[1] has become available at a reasonable cost and it can be used in a great number of experiments as the light source. It has many advantages over the more traditional sources. First of all, a laser emits a high intensity monochromatic well collimated (parallel) beam of light which may be treated as an ideal point source. Since the beam divergence is negligibly small, the intensity of the beam is very high even at large distances. In the second place, the laser light is coherent; this very unusual property may be used to demonstrate phenomena that cannot be seen with ordinary light.

Although different types of lasers exist, such as the optically pumped (ruby) laser and the semiconductor laser, only the helium-neon gas discharge laser will be discussed briefly[2]. This laser emits a continuous beam in the visible range while some other types are pulsed or emit beams in the infrared.

The first helium-neon laser was built by Javan, Bennett and Harriott[3] at the Bell Telephone Laboratories in 1960 with all lines in the infrared. The visible He-Ne laser was discovered by White and Rigden[4] in 1962.

---

[1]The word laser is an acronym from the initial letters of Light Amplification by Stimulated Emission of Radiation.

[2]This brief discussion is very much simplified; for a more detailed description of the gas laser one of the many books recently published should be consulted.

[3]A. Javan, W. R. Bennett, Jr., and D. R. Herriott, Population inversion and continuous optical maser oscillation in a gas discharge containing a He-Ne mixture, Phys. Rev. Lett. 6, 106-110, 1961.

[4]A. D. White and J. D. Rigden, Continuous gas maser operation in the visible, Proc. IRE 50, 1697, 1962.

The principal part of the gas laser is a glass tube filled with helium and neon at a reduced pressure. A high voltage applied across the tube gives rise to a gas discharge of a red-orange color. This color is characteristic of neon spectrum - the helium spectrum by itself appears white.

Fig. 4-38. Helium-Neon Gas Laser. (Courtesy of Spectra-Physics)

The emission of the laser light may be understood by means of the simplified energy level diagram for helium and neon shown in Fig. 4-39. The neon atom in the ground state has 10 electrons and is represented by a 1 s^2 2 s^2 2 p^6 configuration. Excited states are formed when one of the 2 p electrons goes into one of the high energy levels, such as the level labeled 3 s on the energy level diagram. The excited electron may then go to a lower level such as the 2 p one, emitting the excess energy as light. Actually there are many other substates in both, the 3 s and the 2 p states; the only levels shown on the diagram are those giving rise to visible light in the helium-neon laser. Several other transitions result in wavelengths in the infrared

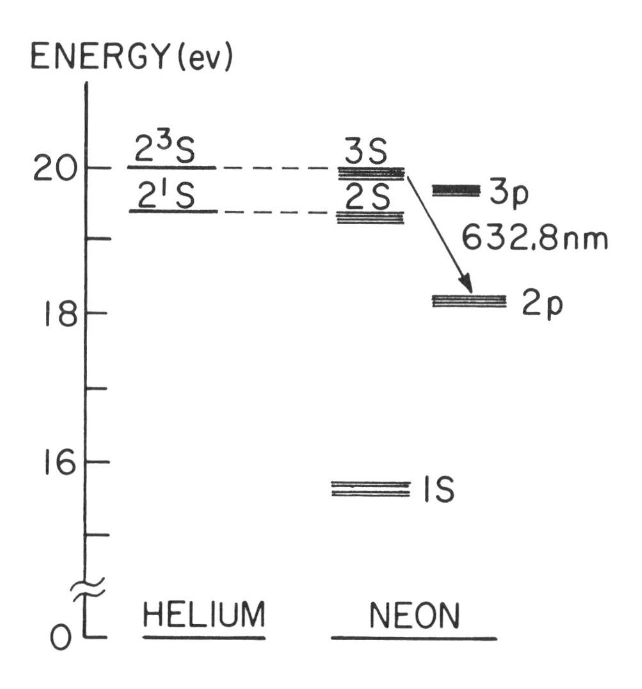

Fig. 4-39. Simplified Energy Level Diagram for the 3 s_2 - 2 p_4 Transition in the Helium-Neon Laser.

region. The wavelength due to the 3 s_2 - 2 p_4 transition is 632.8 nm (red). The neon atom returns from the 2 p_4 level to the ground state.

At thermal equilibrium the population of the 3 s_2 state is somewhat smaller than that of the 2 p_4 state, thus preventing the emission of radiation. It is therefore necessary to over-populate the upper state which is acheived by a population inversion. About 20 ev are required to excite a neon atom to the 3 s_2 level. This may be accomplished by a collision with an excited helium atom since the 2 s metastable state of helium has an energy of about 20 ev. In a collision of the excited helium atom with a neon atom in the ground state, the helium atom transfers its energy to the neon atom and excites it to the 3 s_2 level. The excitation of the helium atom into the 2 s level is accomplished by the potential difference across the tube.

As a result of the population inversion in neon, radiation of wavelength 632.8 nm continues to be emitted without depleting the upper state. The photon emitted spontaneously by the 3 s_2 - 2 p_4 transition can stimulate other neon atoms to emit like photons. These in turn increase the probability of stimulated emission by other atoms resulting in a large buildup of photons all having exactly the same wavelength and phase producing a coherent beam of light. For stimulated emission to occur, the emitted photons must be kept in the vicinity of the radiating atom and should not be lost to the system. This is accomplished by placing mirrors on both sides of the tube. About 99% of the light is reflected back and forth between the mirrors and the rest is emitted as the laser beam. In many lasers great stability is produced by using a plane mirror on one end and a hemispherical one at the other end of the tube.

The laser beam produced is not polarized. When external mirrors are used the end windows are very often oriented parallel to each other and at Brewster's angle to the beam in order to bring the beam out of the tube with negligible loss. In this case, the beam is polarized because of the use of the Brewster window.

The power of gas lasers used as light sources in the laboratory is of the order of a milliwatt. Although this may seem small, it must be remembered that this power is emitted continuously and that it is concentrated in a very small beam.

WARNING: DO NOT LOOK DIRECTLY INTO A LASER BEAM!!

4.10 Microwaves

Microwaves of 3 cm wavelength are particularly well suited for the quantitative investigation of the optical properties of electromagnetic radiation because the radiation is monochromatic, plane polarized and coherent. Since the wavelength is much longer than that of visible light, laboratory experiments in geometrical and physical optics can be performed easily on a convenient observable scale with simple accessories. Although microwave radiation is not identical to that of light in all respects, the use of microwaves will enable the student to understand wave phenomena which are fundamentally the same for each form of radiation.

Microwaves can be produced in a number of different ways. As a low-power source in the frequency range of 3000 to 30 000 Mc/sec the reflex klystron has been found most useful. The tube, usually a 2K25 is connected to the transmitter horn. The cavity of the klystron can be tuned slightly by an adjustment screw which changes the cavity dimensions and the spacing between the cavity grids. A wavelength of approximately 3 cm is normally used for microwave experiments although other wavelengths are also available. Different makes of microwave transmitters are commercially available, consisting of the reflex klystron, the horn and the necessary power supplies for the resonator, the reflector and the filament of the klystron. In some cases the reflector is maintained at a constant d.c. level or it may be modulated at some frequency such as 1000 cps or 60 cps line frequency.

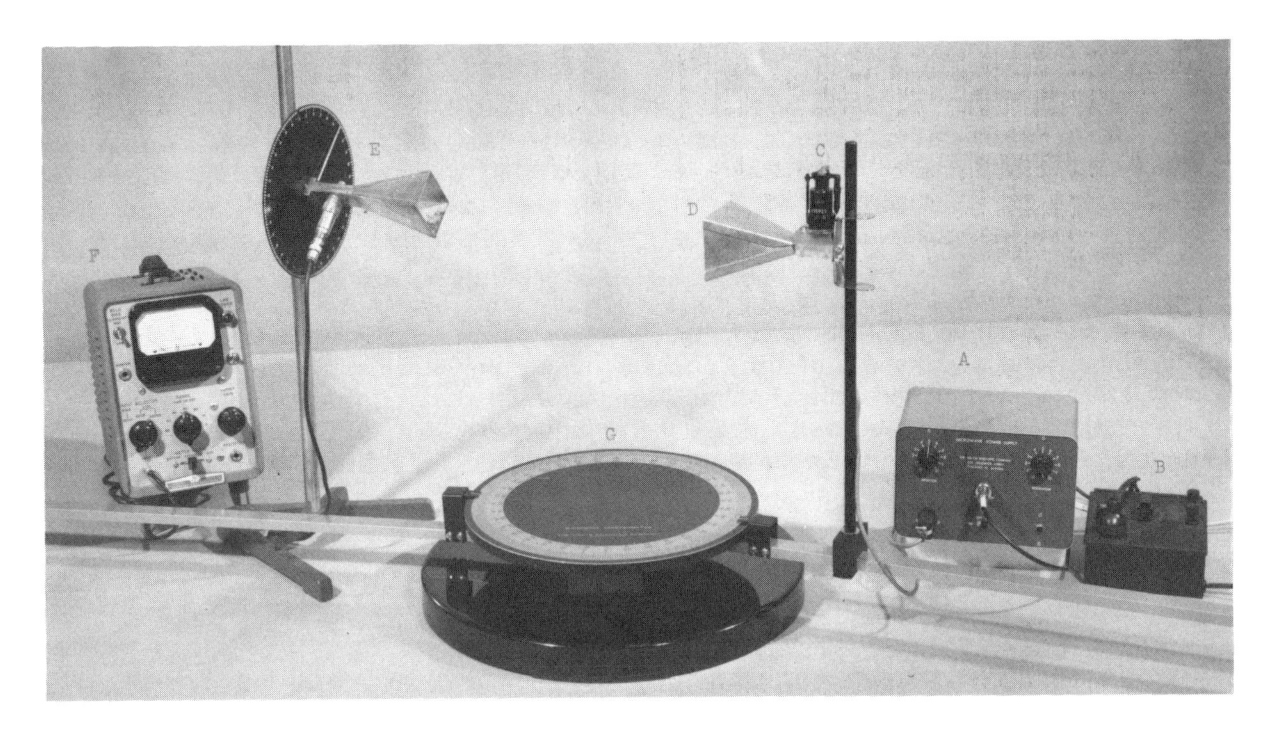

Fig. 4-40. Typical Microwave Setup for Studying Polarization.

A - Power Supply for Klystron
B - External Modulation
C - Reflex Klystron
D - Transmitting Horn
E - Receiver
F - Detector (Standing Wave Indicator)
G - Microwave Spectrometer Table

Microwaves are detected by a receiving horn connected to 1N23B diode. This is an ultra high frequency silicon rectifier mounted in a small cartridge. The output is read on a sensitive meter. If no external modulation is employed, a galvanometer may be used as a detector. The range depends, of course, on the power of the microwaves; a range between 1 and 100 microampere is not uncommon. When modulation is used, an a.c. detector is required, such as a standing wave indicator or a sensitive vacuum tube voltmeter.

4.11 Nuclear Detector

The sealed Geiger-Muller tube (G-M tube) is the most reliable method for detecting and measuring radioactivity and it is the most common detector for beginning studies in nuclear science. Thin-walled side window G-M tubes have typical wall thicknesses of the order of 30 mg/cm^2 while end-window G-M tubes have a window thickness of the order of a few mg/cm^2.*

The basic construction of a typical side window G-M tube is shown in Fig. 4-41. It consists of a glass cylinder with a metal coating on the inside, and a wire - usually tungsten - along the axis. The cylinder contains a gas such as argon at a reduced pressure of about 10 cm of mercury. In a "self-quenching" counter another gas, usually some alcohol, is added. A difference in potential, slightly less than that necessary to produce a discharge, is maintained between the wire and the metal coating on the cylinder. For most commercial counters this potential difference is about 500 or 1000 volts. Too low a voltage will not produce any counts and too high a voltage will cause a continuous discharge, which shortens the life of the counter considerably. Most counters have a "plateau" of several hundred volts - a range over which the counting rate does not change with a variation in the applied voltage. Usually the operating point is selected to be about one-third of the way into the plateau region. Before using a counter the manufacturer's specifications should be consulted.

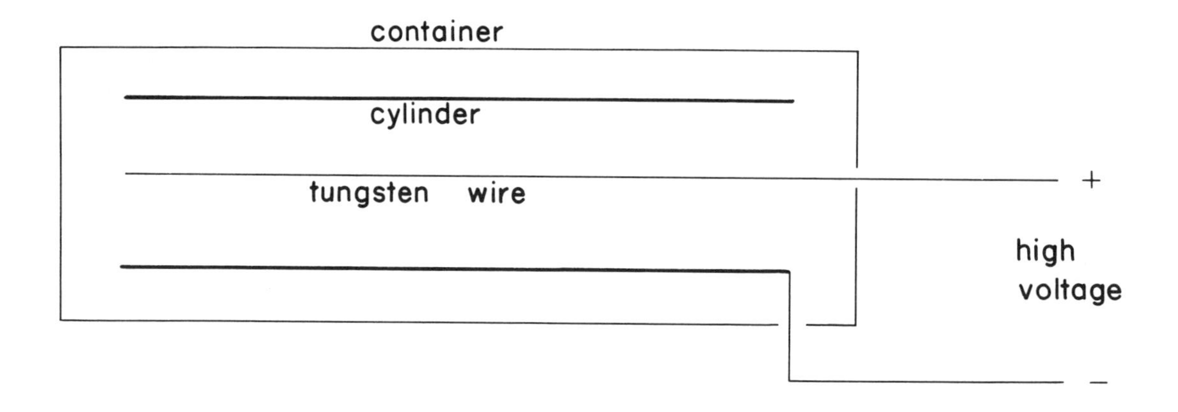

Fig. 4-41. The Construction of a G-M Tube.

Beta particles, or gamma radiation enter the cylinder of the counter and produce ionization of the gas molecules. These ions are accelerated by the electric field and produce more ions by collisions, causing the ionization current to build up rapidly. The current, however, decays rapidly since the circuit has a small time constant. There is, therefore, a momentary surge of current, which is referred to as an electric pulse. Each pulse corresponds to one particle entering the counter.

---

*When dealing with the absorption of radioactivity, the mass per unit area (g/cm^2) of aluminum is generally used instead of the actual thickness of the absorber. The thickness can be found by dividing the mass per unit area by the density of aluminum (2.7 gm/cm^3).

It takes the counter a certain time to recover and be ready for the next particle; this period is called the recovery time and is several hundred microseconds for most counters. If a second particle enters the counter before it has recovered, this particle will not be counted. A "dead-time correction" has to be made if the counting rates are very high.

The electrical pulses from the Geiger counter, after being amplified, must be counted in some way so that their rate can be measured. For very slow counting rates (less than ten counts per second) the pulses could be made to operate an electromechanical recorder directly. For fast rates, however, mechanical devices either jam completely or miss many counts. We therefore use a scaling unit, which is an electronic instrument that may select only certain pulses and pass them on to a mechanical counter or an LED readout. A typical scaler is shown in Fig. 4-42. This particular model features 5 different preset timing intervals (from 0.5 to 10 minutes) and a 5-decade LED readout (99999 counts), all electronic with no mechanical registrar. Also included is a power supply with a range from zero to 2000 volts for various types of detection.

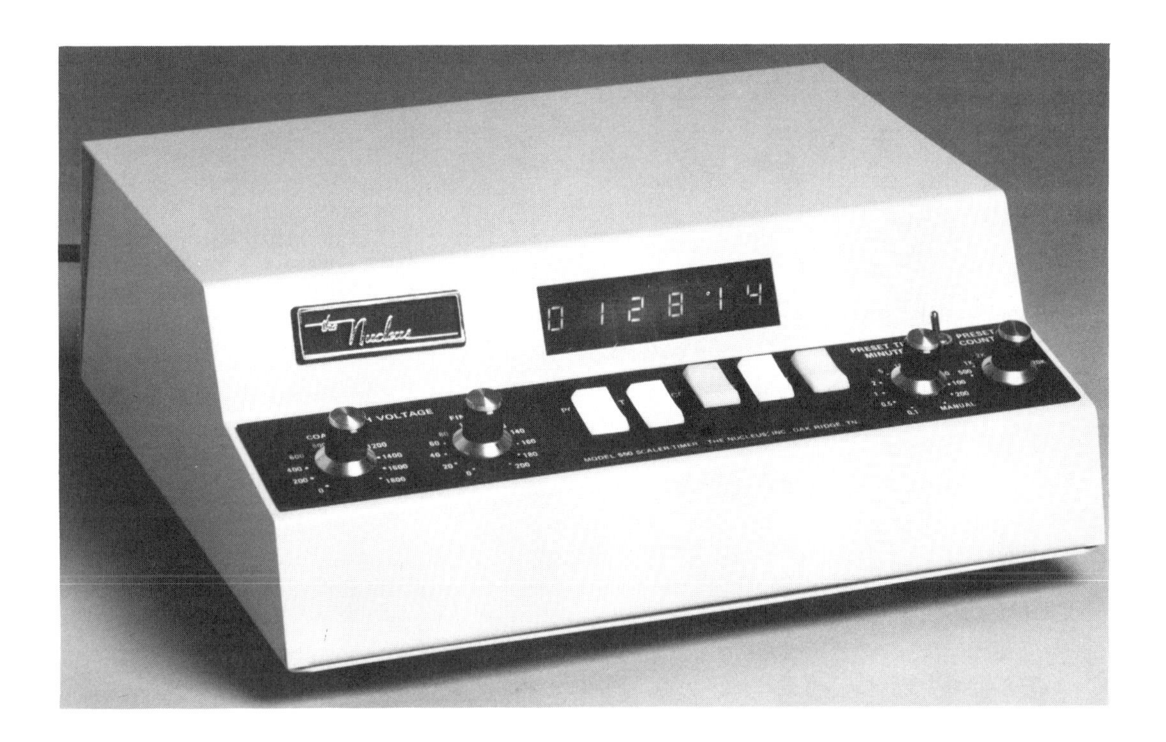

Fig. 4-42. Model 500 Scalar/Timer. (Courtesy of The Nucleus, Inc.)

For the actual operation of the nuclear detector (G-M tube and scaler), the manufacturer's instructions should be consulted because some of the operating procedures differ for different equipment.

Chapter 5

CALCULATOR AND COMPUTER EXPLORATIONS IN BASIC EXPERIMENTAL PHYSICS

5.1 Introduction

The advent of today's affordable personal computers and programmable calculators has brought powerful and easy-to-use new tools to the basic physics laboratory. They provide easily accessible alternatives to larger computers for many applications. Far beyond that, however, they make new sorts of exploration possible that were previously quite difficult to implement in basic courses. In this chapter, we will survey some uses of calculators and personal computers as laboratory tools, and present a series of experiments where these tools themselves are used as the medium of exploration.

For our purposes, we will assume that you have access to a programmable calculator such as the Texas Instruments TI programmable 59, or to a computer capable of running the BASIC programming language. For TI-59 users, we've made provision for those who have access to a printer that works with the calculator. For computer users, we have tried to structure the programs to run on as many BASIC machines as possible. The programs in this chapter were originally developed using Microsoft BASIC on the Texas Instruments Professional Computer, but should run with a minimum of modification on any machine using a standard version of BASIC.

5.2 Calculators and Computers in the Laboratory

The rapid and ongoing decline in cost of microprocessors and memory circuits have made possible whole new families of products, which are having a significant impact on the way scientists and engineers deal with numbers, analyses, and computation. Much greater speed and accuracy – and whole new approaches to problem solving – are possible, particularly in laboratory situations such as:

- Data Gathering
- Statistical Analysis
- Curve Fitting
- Numerical Methods for Differentiation and Integration
- Graphing and Graphic Analysis
- Experiments with Mathematical Models of Systems

... and much more.

In addition, students will find that these tools provide excellent first experiences in programming. Students can gain first-hand experience with the essential features of larger computers, without having to gain access to larger facilities. In addition, building programs around real situations encountered in the laboratory is a stimulating way to gain deeper insight into the function of a theoretical model or discovered laboratory result.

In this chapter, we will concentrate on examples that follow the freshman physics laboratory curriculum, and that illustrate how the calculator or computer can allow study in areas where the sheer weight of mathematics involved previously created barriers. We will be exploring two general areas:

A. GRAPHIC EXPLORATIONS OF A THEORETICAL MODEL

In many areas of study in basic physics, a detailed derivation results in an equation or mathematical model of a system or phenomenon. The student may then ask the questions such as: Now what? How does it work? Why did we do that anyway? Here, the calculator or computer can be used to store the model, and then to "exercise it," and graph it for the student. With a small investment in time, we can examine how the model reacts as various parameters and constants are altered, and plot quick graphs that actually show the model in action.

B. EXPLORATIONS WITH NUMERICAL METHODS

Many times – particularly in early study – the complete exploration of a topic in physics is halted when the theory leads up to a differential equation. Here, unless the equation is particularly simple, the "music usually stops." The detailed solution of the equation would yield valuable results, and provide new insights into the concept being covered; but are usually "beyond the scope of this course." Often a limiting "simplifying assumption" is introduced to allow the solution to proceed in a form that's easier to handle.

A variety of easy-to-use methods exist to handle the solution of differential equations called <u>numerical</u> methods. These methods involve "chopping up" a problem into many "bite-size" parts that are easy to analyze. Simple computational techniques are then applied to each of these parts. The result is that a very nearly exact solution to the equation can be built out of many simpler calculations.

These methods have been around for years, but only rarely used in basic lab. Why? Sheer boredom – that's why! Although each single computation used in a numerical method is by itself quite simple, lots of them are required to see interesting results. Hence, most students would lose track of what they were trying to learn while "grinding out the numbers." Here is where access to a programmable calculator or personal computer comes in. They are ideal for handling the repetitive computations normally involved in implementing numerical methods.

We'll present some general techniques for handling numerical integration and differentiation that have a wide range of applicability, and explore several specific examples.

5.3 Important Notes on this Chapter

As we proceed through this chapter, it will be assumed that the student has access to a programmable calculator such as the Texas Instruments TI-58 or 59; or a personal computer which will run a standard version of the BASIC programming language. As mentioned previously, we have tried to develop the programs to run with minimum modification on any computer running a standard BASIC. The specific language used is Microsoft BASIC running on the TIPC.

Experiments in this chapter are divided into four major sections:

I. An **INTRODUCTION**, briefly describing the area we will be exploring, along with reference materials and suggested areas for further study.

II. A **THEORY REVIEW**, where essential equations will be developed.

III. **THE PROGRAM**, presented in two sections:

IIIa. **THE PROGRAM FOR THE TI-58 or TI-59**: presented in a format which allows for key-by-key entry of the program, or...

IIIb. **THE PROGRAM IN BASIC**: which follows a similar format, allowing the student to carefully enter program keystrokes to begin the exploration.

IV. **EXPLORATIONS/NOTES**, which includes exercises and suggestions on the use of the program in exploring the particular area of physics under study. These explorations are suitable for a typical laboratory period, and will hopefully serve to stimulate further applications of your calculator or computer as a laboratory tool. (Your instructor may assign additional exercises for you to complete during your lab session.)

IF YOU USE A TI-58 OR TI-59:

It is important that you are generally familiar with the calculator, the basics of its operation and the keys involved in programming it. (It is not necessary to be an accomplished calculator "whiz-kid," but important that you know your way around.) The first laboratory session involving the calculator should include at least one hour to let you review the owner's manual, and get some general hands-on experience.

Program information for use with the TI-58 or TI-59 is provided in a special 3-column format that breaks the program into blocks:

Function: Exactly what this group of keystrokes is accomplishing.

Program Keystrokes: To be keyed in exactly into the TI Programmable 58 or 59 while in the learn mode to accomplish the function.

Step Number: Whenever the calculator is in the learn mode, the display takes on a unique 5-digit format:

000	00
The first three digits are the step number.	The last two digit spaces are for key codes. (These will remain zeros as you key in the program.)

The step number code shown in the left column will be the left-most 3 digits appearing in the display as you complete entry of a block of keystrokes. As you enter keystrokes into the calculator, you can use these step numbers as a check to be certain that all keystrokes in a given block have been entered (or that no extra keystrokes have been entered by accident).

Note that when using the calculator to list the program, or when stepping through the program with the [SST] or [BST] keys, the step number shown for any step will be one less than the one shown as it was originally keyed in. This is because as steps are initially entered in learn mode, the calculator "jumps ahead" showing the next available step number in the program memory. (See the Texas Instruments Owner's Manual, "Personal Programming," Pages V-41ff for details.)

In each of the program keystroke sections, asterisked instructions are included to be used in conjunction with the PC-100A printer. Instructions are provided which allow you to observe the results in the calculator display with repeated use of the [R/S] key, or have the results automatically printed on the PC-100A. Also included for each experiment are the following sections:

A. CALCULATOR MEMORIES USED:

A list of calculator memories used in the program, and what's being stored there.

B. USER INSTRUCTIONS:

A three-column form which details how to use the calculator program in the exploration. Operations involved, keys to press, and displayed results are described. (These instructions are used after the program has been completely and accurately entered, and the calculator has been taken out of learn mode.)

The "user instructions" section also includes a set of "Program Check Data," to enable you to be certain that the program is correctly entered and working in your calculator before proceeding. Simply follow the user instructions to enter the test values given, and check to see that you get the expected result, before proceeding to the "Explorations/ Notes" section of the experiment.

One additional note may be helpful concerning the "Fix" feature of the TI Programmable 58 and 59. The programs as originally presented in this chapter are designed to display results to the full limits of calculator accuracy. For use in graphing or laboratory work, two or three significant digits is usually quite satisfactory. You can arrange for data to be displayed to three significant digits by pressing:

[2ND] [FIX] 3

before running any program. (Note that calculations will still be carried out with the full internal accuracy of the machine. Results correctly rounded to three digits will appear in the display.)

IF YOU USE A COMPUTER RUNNING BASIC:

It is important that you have time to become familiar with the particular computer that you use, although extensive programming knowledge is not required for the purposes of this chapter. Your instructor may need to work with you if any minor modifications are required to allow the programs listed here to run on the machine you are using in your lab.

For each experiment, a section is included entitled "The Program in BASIC." The section presents exactly what is needed to key in the program and start viewing results. Note carefully the following components of each "Program in BASIC" section:

A. PROGRAM STATEMENTS

These statements, along with each line number shown, should be carefully and exactly keyed into the computer, once the BASIC programming language is properly up and running on it. (Your instructor should work with you on the appropriate steps involved in starting up and loading BASIC on the machine you are using. As mentioned, some changes to the Program Statements listed here may be required - if you have problems, check with your instructor.)

B. EXPLANATION

For each BASIC program statement you enter, a brief explanation of what function it serves in the program is provided.

C. VARIABLES USED

The names of variables used in the program, each with an explanation of their value (initial value, constant value, or value which will change as the program runs) are provided.

(Note that in some examples arrays of examples are used, and in these cases the names of the arrays and what they represent are also included.)

D. USER INSTRUCTIONS

Here, exact instructions for you to follow after entering and running the program are provided. Each program will typically provide you with messages or "prompts" on the screen which request that a specific piece of information be entered, or that you take some other sort of action. Follow the instructions carefully, and the results described in this section will begin to appear on the computer screen, for your use in the laboratory exercise.

F. PROGRAM CHECK DATA

In this section, data is provided to allow you to check to be sure that you've entered the program correctly, and that all is well before you begin the rest of the laboratory exercise. User prompts you can expect to see on the screen are shown, along with sample values or special instructions for you to use in checking out the program. Note: If the results from your computer do not check with those shown in the book, check with your instructor before proceeding further.

Some additional notes when using BASIC programs:

When handling very large or very small numbers, a computer using Microsoft BASIC will often represent the numbers using scientific notation. Very large or small numbers may be entered into a program using this notation as well. In scientific notation a number is represented by a base number (or "mantissa") followed by an exponent or power of 10 to which the mantissa is raised, represented by an "E," followed by the exponent. Numbers are represented in scientific notation as shown in the examples below:

160,000,000	is represented by	1.6E8
0.08063179	is represented by	8.061379E-02
123.98×10^{10}	is represented by	1.2398E+12

Note also that in microsoft BASIC, occasionally an entered number or calculated result will appear represented with a string of trailing zeros and a one as the least significant digit. (e.g., 9.81 represented as 9.8100001). Alternatively, on occasion a number may be represented with a string of "trailing 9's." (e.g., 4.3 represented as 4.299999). This has to do with the way numbers are represented internally in the computer, does not significantly affect the computations we are doing here, and should be ignored by the student.

SECTION A - GRAPHIC EXPLORATIONS OF A THEORETICAL MODEL

As mentioned, in this section we will be using the programmable calculator or computer to explore some of the equations that result from classic derivations in physics. Some of the results can appear quite static "ends in themselves" when first encountered. In this section we'll begin with several simple examples and examine how they are programmed. We have two ends in mind: First, for those of you unfamiliar with programming, these straightforward examples should serve as a good place to get started. Secondly, once the programs are working, we hope that you will discover some new and dynamic aspects of the basic equations and results derived for you in lecture or in your text.

Experiment 5-1 PROJECTILE MOTION

I. INTRODUCTION

We'll begin with a simple example – the case of an object tossed obliquely into the air subject only to the acceleration of gravity, g. The resulting motion can be easily programmed, which allows us to "feel" the kinematics of ballistic motion at work. A review of the material in Chapter 4 of Resnick & Halliday's physics texts is recommended before you begin.

II. THEORY REVIEW

We'll choose the point at which the projectile is released as the origin of our coordinate system, as shown in Fig. 5-1.

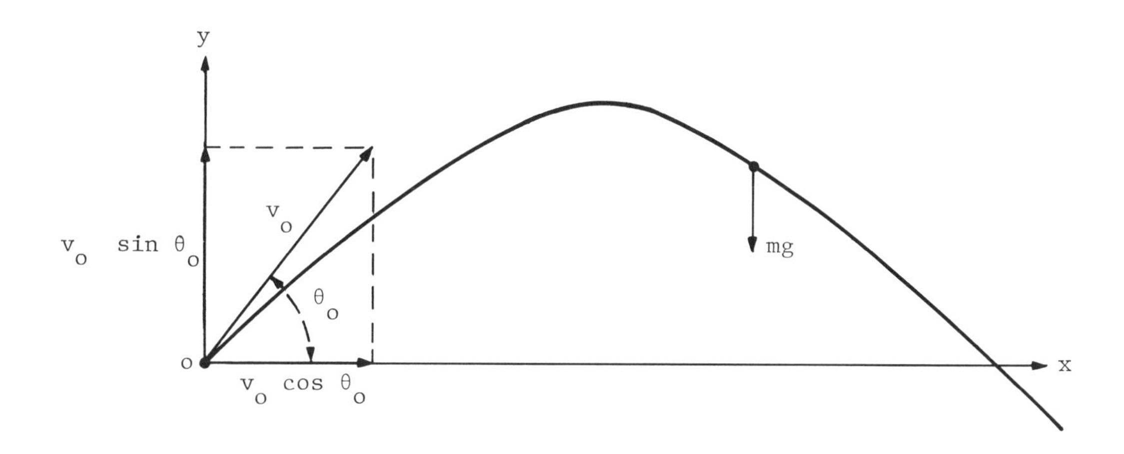

Fig. 5-1. Projectile Motion.

We'll assume that gravity is the only force acting on the particle, as shown. The particle is released at time t = 0, with an initial speed v_o at an angle θ_o with respect to the horizontal. The x component of velocity is $v_o \cos \theta_o$, and since we've assumed that no forces act on the particle in the x direction, this remains constant:

$$v_x = v_o \cos \theta_o$$

In the y direction, gravity acts to accelerate the particle downward, so that:

$$v_y = v_o \sin \theta_o - gt$$

The equations above give rise to the following equations for the position of the particle at any given time. We'll simply be programming these equations to allow us to see how they behave with time.

$$x = (v_o \cos \theta_o)\ t$$

$$y = (v_o \sin \theta_o)\ t - 1/2\ gt^2$$

III. THE PROGRAM

We'll construct our program in several parts. The first part will allow us to store initial values of v_0 and θ_0 for use in our program explorations, as well as a value for Δt, the time increments at which we'll examine the motion. The value of g will be stored separately to allow us to change units if desired. The situation is illustrated in Fig. 5-2.

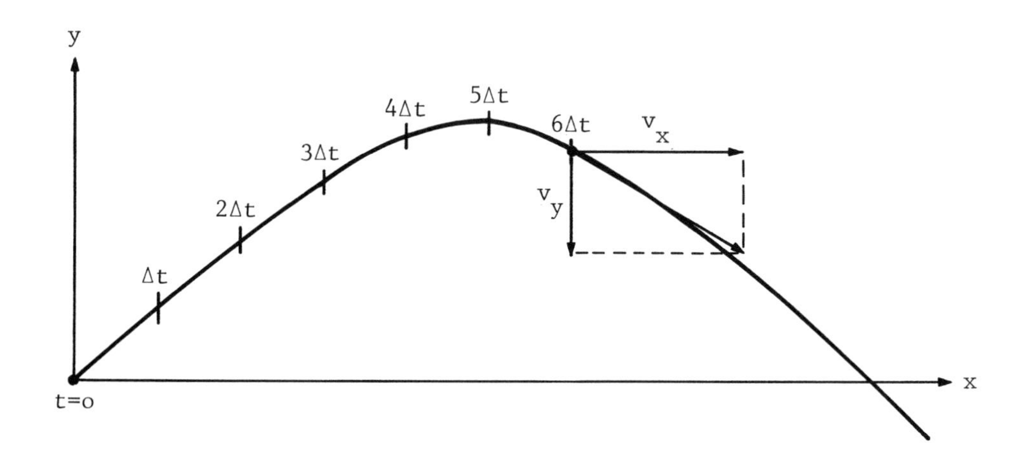

Fig. 5-2. Projectile Motion Divided into Equal Time Segments, Δt.

IIIa. THE PROGRAM FOR THE TI-58 or TI-59

PROJECTILE MOTION

FUNCTION	*PROGRAM KEYSTROKES*	*STEP*
Clear calculator completely, get into "learn mode"	OFF/ON [LRN]	000
Store v_o in memory 10	[2nd] [Lb1] [A] [STO] 10 [R/S]	005
Store θ_o in memory 11	[2nd] [Lb1] [B] [STO] 11 [R/S]	010
Store g in memory 12	[2nd] [Lb1] [C] [STO] 12 [R/S]	015
Store Δt in memory 13	[2nd] [Lb1] [D] [STO] 13 [R/S]	020
(Current time value will be in memory 14)		
Label beginning of calculation	[2nd] [Lb1] [2nd] [A']	022
Display current time value	[RCL] 14 [R/S]*	025

FUNCTION	*PROGRAM KEYSTROKES*	*STEP*
Compute x value = $(t)(\cos\ \theta_o)(v_o)$	[RCL] 14 [X] [RCL] 11 [2nd] [cos]	031
	[X] [RCL] 10 [=] [R/S]*	036
Compute y value = $[-g(t^2)/2]$ $+(v_o)(\sin\ \theta_o)(t)$	[RCL] 12 [+/-] [X] [RCL] 14	042
	[X^2] [÷] 2 [+] [RCL] 10 [X]	049
	[RCL] 11 [2nd] [sin] [X]	053
	[RCL] 14 [=] [R/S]*	057
Increment Δt & resume computation	[RCL] 13 [SUM] 14 [2nd] [A']	062
Reinitialize & start next computation	[2nd] [Lb1] [E] 0 [STO] 14	067
	[2nd] [A']	068
Get out of "learn mode" and reset	[LRN] [RST]	

A. CALCULATOR MEMORIES USED:

Memory	Value
10	v_o
11	θ_o
12	g
13	Δt
14	$t_i = (i \times \Delta t)$

B. USER INSTRUCTIONS:

To use the program, follow these instructions:

OPERATION	*PRESS*	*DISPLAY/COMMENTS*
First, enter program keystrokes carefully, following instructions in "the program" section.		
Enter v_o value	[A]	v_o
θ_o value	[B]	θ_o
g value (as a positive number)	[C]	g
Δt value	[D]	Δt
To start program run - reinitialize	[E]	0 t = 0
Press [R/S] repeatedly to see	[R/S]	0
t_i	[R/S]	0

*Note: For those with access to a printer, [2nd] [Prt] instructions may be substituted for these R/S instructions.

OPERATION	*PRESS*	*DISPLAY/COMMENTS*
x_i		
y_i	[R/S]	$t + \Delta t = t_1$
in succession.	[R/S]	x_1
	[R/S]	y_1
<u>Program Check Data:</u>		
v_o = 180 m/s	[A]	180
θ_o = 40°	[B]	40
g = 9.81 m/sec^2	[C]	9.81
Δt = 0.5 sec	[D]	0.5
	[R/S]	0.
	[R/S]	0.
	[R/S]	0.
	[R/S]	0.5
	[R/S]	68.94399988
	[R/S]	56.62463487
	[R/S]	1.
	.	.
	.	.

IIIb. THE PROGRAM IN BASIC

A. <u>PROGRAM STATEMENTS</u>	B. <u>EXPLANATION</u>
`100 T=0`	Set time to 0
`110 INPUT "Acceleration of gravity";G`	Enter acceleration of gravity
`120 PI=3.141593`	Set PI constant
`130 INPUT "Initial speed";V`	Enter initial speed
`140 INPUT "Initial angle";A`	Enter initial angle
`150 A=A*(PI/180)`	Convert angle to radians
`160 INPUT "Time interval";DT`	Enter time interval
`1000 PRINT`	Display blank line
`1010 PRINT "Time =";T`	Display time
`1020 X=T*COS(A)*(V)`	Compute X
`1030 PRINT"X =";X`	Display X
`1040 Y=((-G*(T^2))/2)+(V*SIN(A)*T)`	Compute Y
`1050 PRINT "Y =";Y`	Display Y
`1060 T=T+DT`	Increment time
`1070 IF INKEY$="" GOTO 1070`	Wait for keypress
`1080 GOTO 1000`	Repeat computation

Note: If you have access to a printer, you can substitute LPRINT instructions for PRINT instructions.

C. VARIABLES USED

Variable	Value
A	Initial angle, with respect to the horizontal (NOTE: You enter the angle in degrees. The program converts the angle to radians because the trigonometric functions in BASIC operate on angles measured in radians.)
DT	Time interval: "delta-T" (seconds)
G	Gravity (9.81 meters per second squared)
PI	Pi (3.141593)
T	Time (seconds)
V	Initial speed (meters per second)
X	X
Y	Y

D. USER INSTRUCTIONS

Enter the program carefully. Enter the command RUN to start the program; then follow the user instructions or "Prompts" for the program.

Prompt	Action
Acceleration of gravity?	Enter the value of g (in appropriate acceleration units).
Initial speed?	Enter the initial speed (in appropriate velocity units).
Initial angle?	Enter the initial angle in degrees.
Time interval?	Enter the time interval in seconds.

The program displays the time (in seconds) and the X and Y values.

Press any key to display the next set of values.

E. PROGRAM CHECK DATA

```
Acceleration of gravity? 9.81
Initial speed? 180
Initial angle? 40
Time interval? .5

Time = 0
X = 0
Y = 0

Time = .5
X = 68.94399
Y = 56.62464

Time = 1
X = 137.888
Y = 110.7968
```

When finished running the program for one set of values, enter a BREAK command. (On most computers this is accomplished by holding down a "shift" or "control" key, and pressing the "Break" or "Pause" key on the keyboard.) To run the program for a new set of values, type in the command RUN, and then press the "ENTER" or "RETURN" key.

IV. EXPLORATIONS/NOTES

Note: When displayed values for y become negative, the projectile has fallen below its initial level of launch.

1. a. Plot the motion of a projectile with the following initial launch conditions, for Δt = 0.5 seconds:
   v_o = 130 m/s
   θ_o = 25° Assume launch occurs from the ground.

   b. When will this projectile hit the ground?

   c. What is its greatest height from the ground?

   d. If θ_o is increased to 30°, when will the projectile hit the ground?

   e. What will be its greatest height from the ground?

2. a. Plot the motion of a projectile with initial velocity v_o = 200 m/s for Δt = 1.0 sec, and for θ_o values from 35° to 55° in 5° increments – on the same set of axes. Which projectile is in the air for the longest time?

   b. Which launch achieved the greatest horizontal range? What is that range?

   c. Which launch achieved the greatest height?

3. a. Plot the path of a projectile launched from 5000 feet in the air at an initial velocity of 350 feet per second in the horizontal direction, as shown in Fig. 5-3. (Use Δt = 1 second.)

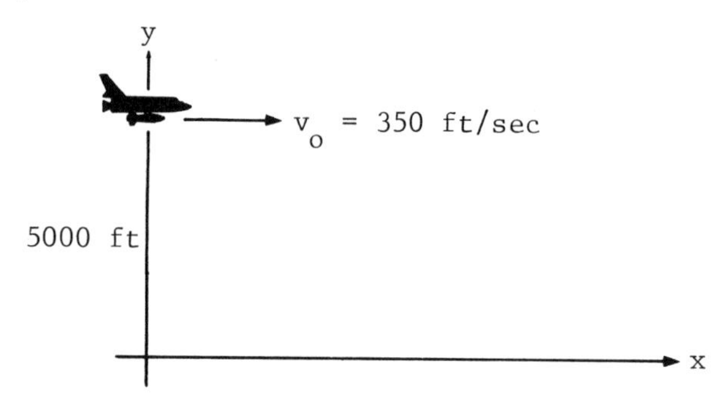

Fig. 5-3. Problem 3.

   b. When will this projectile hit the ground?

   c. What horizontal distance will it have traveled from launch point?

   d. A similar projectile is launched from the same point with 700 ft/sec horizontal velocity. Plot its path.

   e. When will this projectile hit the ground? What horizontal distance will it have travelled?

Experiment 5-2 SIMPLE HARMONIC OSCILLATIONS

I. INTRODUCTION

In this case we will be building a simple "model" of a spring-mass oscillating system, such as that shown in Fig. 5-4. The program will simulate the action of the oscillator, and allow you to study several interesting aspects of this motion simultaneously. A review of the theory of simple harmonic motion, as covered in Resnick and Halliday's physics texts, prior to beginning would be helpful.

II. THEORY REVIEW

We will assume that the oscillating mass under consideration is resting on a horizontal frictionless surface, and attached to a spring of constant k. The mass is pulled from its rest position to the right (positive x direction) a distance A, then released (without any initial velocity). The situation is depicted in Fig. 5-4.

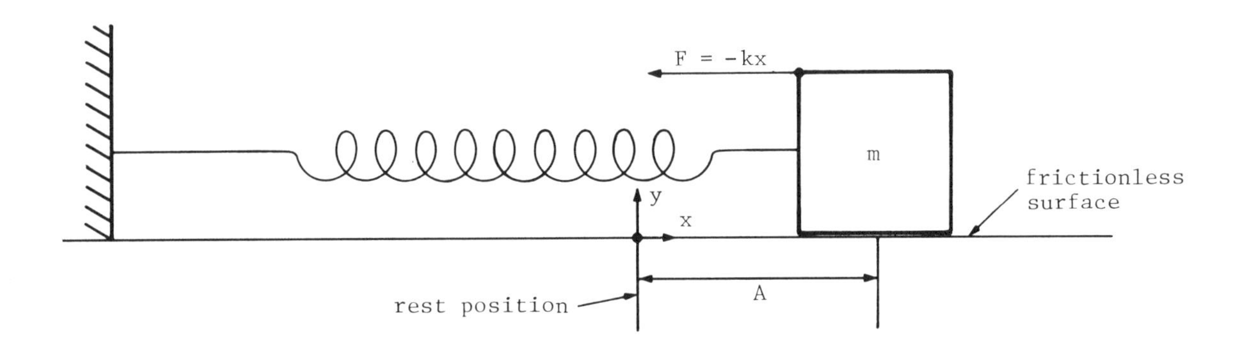

Fig. 5-4. Simple Harmonic Oscillator.

The force on the block at any distance x from the rest position is given by:

$$F = -kx. \qquad 5\text{-}1$$

Applying Newton's law we see:

$$m\frac{d^2x}{dt^2} = -kx; \text{ or}$$

$$\frac{d^2x}{dt^2} + \frac{kx}{m} = 0 \qquad 5\text{-}2$$

The solution to this equation is of the form:

$$x = A \cos(\omega t + \delta), \qquad 5\text{-}3$$

where $\omega = \sqrt{k/m}$ is the "natural frequency" of the system, and δ is a "phase constant" for the system, which describes the system's initial condition. (ω is measured in radians/second.)

The velocity of the mass m at any point in time is given by:

$$v = dx/dt = -A\omega \sin(\omega t + \delta) \qquad 5\text{-}4$$

(For this case we will assume the value of δ to be zero.)

At any point in the motion, the potential energy of the system is given by

$$U = \frac{1}{2} kx^2 , \qquad 5\text{-}5$$

and the kinetic energy by:

$$K = \frac{1}{2} mv^2 \qquad 5\text{-}6$$

The total energy of the system will remain constant during the motion, and is given by:

$$E = K + U \qquad 5\text{-}7$$

III. THE PROGRAM

We will be developing a program that lets us store the critical values for a simple harmonic oscillator system:

A = the initial maximum displacement of the object (at zero velocity)
m = the mass of the object
k = the spring constant

Our program will then simulate the operation of this sytem and give us a look at how the motion unfolds as time goes on. You'll be able to enter an interval of time - Δt - at which you'd like to periodically observe the motion. For each observation point, the program will provide values of:

t = the time value of the observation
x = the position of the mass m
v = the velocity of the mass m
K = the kinetic energy of the mass m
U = the potential energy of the system
E = the total energy of the system

IIIa. THE PROGRAM FOR THE TI-58 or TI-59

SIMPLE HARMONIC OSCILLATOR

FUNCTION	*PROGRAM KEYSTROKES*	*STEP*
Clear calculator completely, get into "learn" mode	OFF/ON [LRN]	000
Store A in memory 10, put calculator in radians mode	[2nd] [Lb1] [A] [STO] 10 [2nd] [Rad] [R/S]	006
Store m in memory 11	[2nd] [Lb1] [B] [STO] 11 [R/S]	011
Store k in memory 12	[2nd] [Lb1] [C] [STO] 12 [R/S]	016
Store Δt in memory 13	[2nd] [Lb1] [D] [STO] 13 [R/S]	021
(Current time value will be in memory 14.)		
Label beginning of calculation	[2nd] [Lb1] [2nd] [A']	023
Display current time value	[RCL] 14 [R/S]*	026
Compute x value, store in memory 16 (& display):	[RCL] 10 [X] [( ] [( ] [Rcl] 15 [X]	034
$x = A(\cos(\omega t))$	[RCL] 14 [ )] [2nd] [cos] [ )] [=] [STO] 16 [R/S]*	043
Compute v value (& display):	[RCL] 10 [+/-] [X] [RCL] 15 [X]	050
$v = -A\omega(\sin(\omega t))$	[( ] [( ] [RCL] 15 [X] [RCL] 14	057
	[ )] [2nd] [sin] [ )] [=] [R/S]*	062
Compute K value, store in memory 17 (& display): (value of v is in display) $K = v^2 (m \div 2)$	[X^2] [X] [( ] [RCL] 11 [÷] 2 [ )] [=] [STO] 17 [R/S]*	074
Compute U value (& display): $U = x^2 (k \div 2)$	[RCL] 16 [X^2] [X] [( ] [RCL] 12 [÷] 2 [ )] [=] [R/S]*	086
Compute E value (& display): $E = U + K$	[+] [RCL] 17 [=] [R/S]*	091
Add Δt to t & resume calculation	[RCL] 13 [SUM] 14 [2nd] [A']	096
Reinitialize and start computation:	[2nd] [Lb1] [E] 0 [STO] 14	101
Store zero in memory 14, compute	[RCL] 12 [÷] [RCL] 11 [=] [$\sqrt{x}$]	108
$\omega = \sqrt{k/m}$ and store in memory 15, Go to A'	[STO] 15 [2nd] [A']	111
Get out of "learn" mode and reset	[LRN] [RST]	0

*Note: For those with access to a printer, substitute [2nd] [Prt] instructions for these [R/S] instructions.

A. CALCULATOR MEMORIES USED:

Memory	Value
10	A
11	m
12	k
13	Δt
14	t
15	ω
16	x
17	K

B. USER INSTRUCTIONS:

OPERATION	*PRESS*	*DISPLAY/COMMENTS*
First, enter program keystrokes carefully, following instructions in "The Program" section.		
Enter initial (zero velocity) displacement A	[A]	A
Enter mass m	[B]	m
Enter spring const k	[C]	k
Enter time interval for the observation, Δt	[D]	Δt
Begin computation	[E]	0 first time value t=0
	[R/S]*	$x_o = A$
	[R/S]*	$v_o = 0$
	[R/S]*	K = initial kinetic energy = 0
	[R/S]*	U = initial potential energy = $1/2 k x_o^2$
	[R/S]*	E = total energy
	[R/S]*	Δt, next time value $t_1 = \Delta t$
	[R/S]*	x_1 = x value at t_1
	[R/S]*	v_1 = v value at t_1
	etc.	etc.
NOTE, TO RE-RUN PROGRAM	Press [CLR] [RST]	0
	Enter new values for m, k, A or Δt; and press [E]	

For those with access to PC100A printer: If [R/S] instructions in program labelled with asterisk are replaced by [2nd] [Prt] instructions, these values will be printed automatically.

Program Check Data

For original displacement,

A = 0.333 ft	[A]	0.333
M = 0.0468 slug	[B]	0.0468
k = 3 lb/ft	[C]	3.
Δt = 0.1 sec	[D]	0.1
	[E]	0.
	[R/S]	0.333
	[R/S]	0.
	[R/S]	0.
	[R/S]	0.1663335
	[ ]	0.1663355
	["]	0.1
	["]	0.23185022
	["]	-1.913757137
	["]	.0857017133
	["]	.0806317867
	["]	0.1663335
	["]	0.2
	[.]	.
	[.]	.

IIIb. THE PROGRAM IN BASIC

A. PROGRAM STATEMENTS / B. EXPLANATION

A. PROGRAM STATEMENTS	B. EXPLANATION
100 T=0	Set time to 0
110 INPUT "Initial displacement";A	Enter initial displacement
120 INPUT "Mass";M	Enter mass
130 INPUT "Spring constant";SC	Enter spring constant
140 INPUT "Time interval";DT	Enter time interval
150 W=SQR(SC/M)	Compute "natural frequency"
1000 PRINT	Display blank line
1010 PRINT "Time =";T	Display time
1020 X=A*COS(W*T)	Compute position
1030 PRINT "Position =";X	Display position
1040 V=-A*W*SIN(W*T)	Compute velocity
1050 PRINT "Velocity =";V	Display velocity
1060 K=(V^2)*(M/2)	Compute kinetic energy
1070 PRINT "Kinetic energy =";K	Display kinetic energy
1080 U=(X^2)*(SC/2)	Compute potential energy

1090 PRINT "Potential energy =";U	Display potential energy
1100 E=U+K	Compute total energy
1110 PRINT "Total energy =";E	Display total energy
1120 T=T+DT	Increment time
1130 IF INKEY$="" GOTO 1130	Wait for keypress
1140 GOTO 1000	Repeat computation

Note: If you have access to a printer, you can substitute LPRINT instructions for PRINT instructions.

C. VARIABLES USED

Variable	Value
A	Initial displacement
DT	Time interval: "delta-T"
E	Total energy
K	Kinetic energy
M	Mass
SC	Spring constant (K)
T	Time
U	Potential energy
V	Velocity
W	"Natural frequency"
X	Position

D. USER INSTRUCTIONS

Enter the program carefully. Run the program and respond to these prompts:

Prompt	Action
Initial displacement?	Enter the initial displacement.
Mass?	Enter the mass.
Spring constant?	Enter the spring constant.
Time interval?	Enter the time interval.

The program displays the time, the position, the velocity, the kinetic energy, the potential energy, and the total energy.

Press any key to display the next set of values.

E. PROGRAM CHECK DATA

```
Initial displacement? .333
Mass? .0468
Spring constant? 3
Time interval? .1

Time = 0
Position = .333
Velocity = 0
Kinetic energy = 0
Potential energy = .1663335
Total energy = .1663335
```

Time = .1
Position = .2318502
Velocity = -1.913757
Kinetic energy = .08570172
Potential energy = .08063179
Total energy = .1663335

Time = .2
Position = -.01015
Velocity = -2.664895
Kinetic energy = .166179
Potential energy = .0001545337
Total energy = .1663335

When finished running the program for one set of values, enter a BREAK command. (On most computers this is accomplished by holding down a "shift" or "control" key, and pressing the "Break" or "Pause" key on the keyboard.) To run the program for a new set of values, type in the command RUN, and then press the "ENTER" or "RETURN" key.

IV. EXPLORATIONS/NOTES

1. Using the program, analyze the behavior of a simple harmonic oscillator, as shown in Fig. 5-4 with the following values:

$$A = 0.16 \text{ m}$$
$$m = 0.75 \text{ kg}$$
$$k = 46 \text{ n/m}$$

   a. Plot the value of x vs t, for $\Delta t = 0.05$ seconds for 1 full period, after release from $x = A$ (0.16 m) at zero velocity.
   b. Plot v vs t for 1 full period, on a separate graph.
   c. Plot K, the kinetic energy for 1 full period, on a separate graph, and also the total energy value.
   d. Plot U, the potential energy, for 1 full period, on a separate graph. Plot the total energy value also.

2. Repeat steps a through d for double the mass value, then 1/2 the mass value, holding all other oscillator values constant.

3. Which of the oscillators above achieves the greatest:

   a. Period?
   b. Velocity?
   c. Kinetic energy?
   d. Potential energy?
   e. Total energy?
   f. Which oscillator achieves its maximum kinetic energy first?
   g. Which oscillator achieves its minimum potential energy first?

4. Now go back and run the Harmonic Oscillator Program again, this time with oscillator values as follows:

   A = 0.16 m (Note: Select a Δt value which allows you to observe
   m = 0.75 kg the motion conveniently.)

   For these three cases:

   (i) k = 46 n/m (ii) k = 23 n/m (iii) k = 92 n/m

   In each case observe one full period of the motion.

   In this situation, we're examining the same oscillator and observing its behavior as the spring constant k is varied: first through half its value; then through double its value. In this situation answer these questions from the data the program provides:

   a. Which oscillator achieves the greatest displacement?
   b. Which has the greatest period?
   c. What is the total energy value in each case?
   d. Which oscillator achieves the greatest velocity?
   e. Which achieves the greatest kinetic energy?
   f. Which achieves the greatest potential energy?
   g. Which achieves its maximum velocity first?
   h. Which achieves its minimum potential energy first?
   i. Which achieves its maximum kinetic energy first?

5. Finally, use the program to examine the behavior of oscillators with the following values:

   m = 0.75 kg
   k = 46 n/m

   For these three cases:

   (i) A = 0.16 m (ii) A = 0.32 m (iii) A = 0.08 m
   (Examine at least one full period of the oscillator in each case.)

   Answer Questions a through h of Question 4 for the three oscillators that are examined.

Experiment 5-3 SUPERPOSITION THEOREM: ELECTROSTATIC OR GRAVITATIONAL FIELD STRENGTH COMPUTATION

I. INTRODUCTION

The superposition theorem is a simple and useful tool for study in physics that relies on large amounts of relatively straightforward computation. These situations are ideal for handling on the programmable calculator or personal computer. The program handles the tedium; you get to examine and explore the results. In this case we'll be providing a program which allows you to compute the field at any point in a plane, created by up to 10 point charges (or masses) in the plane. With the program you will first enter the charge (or mass) magnitude Q_i and the coordinates x_i and y_i for up to ten charges (or point masses) (i = 1 to 10). You'll then enter the coordinates of the point at which you want to compute the field strength. The program will provide the field strength at the point in polar form. With the printer option included, you will be able to go beyond that and build a table which allows you to "map" the field strength throughout the plane.

Through the medium of the program, you are actually using the superposition theorem to explore the nature of fields. You can explore the behavior of fields in the space surrounding various charge arrangements, see where magnitude and direction changes occur, and explore various symmetries in field patterns. Your calculator or computer is actually a probe here – giving you field strength and direction information at any point you wish. The computations performed in this example are actually quite simple, but as you'll see they're often quite lengthy and tedious. With the calculator or computer handling the computation, you are free to see some facets of field behavior you may have never had a chance to examine before. A review of electric fields will be helpful before beginning, as covered in Halliday and Resnick's physics texts.

II. THEORY REVIEW

The program we'll be developing will work for both gravitational and electric field computations, but for now let's focus on the electric field.

Consider computation of the total field at the point P in Fig. 5-5, resulting from Q_1, Q_2, and Q_3. The superposition theorem states that to find E at point P you simply follow these two steps:

a. First, calculate E_i, the field at P due to each charge Q_i, as if it were the only charge present. Then,

b. Add these separately calculated fields vectorially to find the resultant field E at P.

For each computation, Coulomb's Law is used for the charge Q_i and the point P to calculate the magnitude of E:

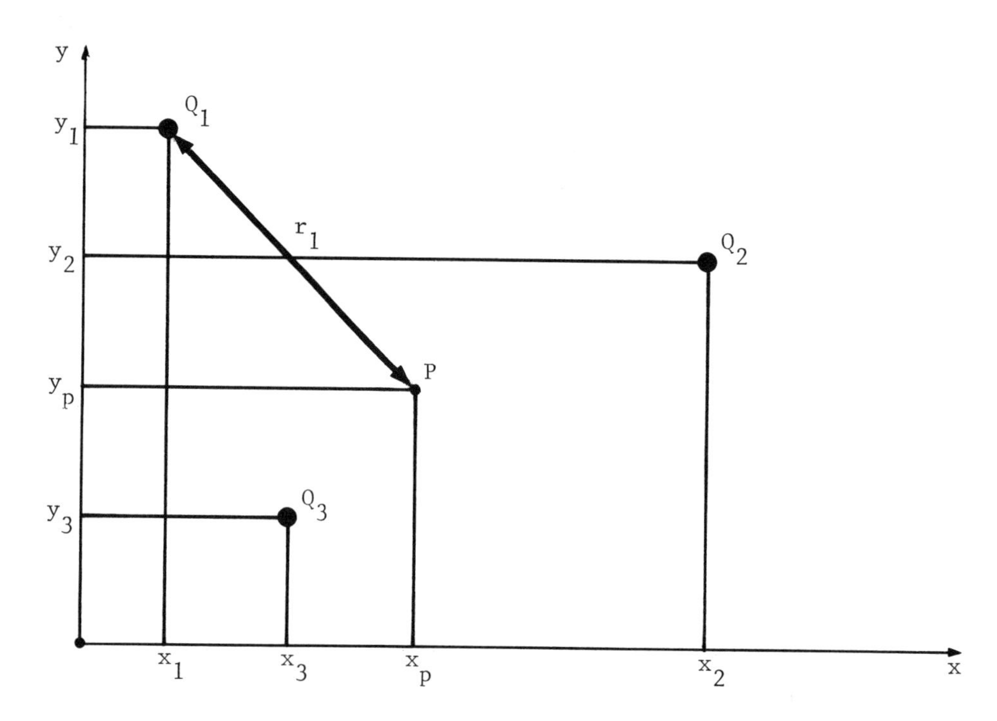

Fig. 5-5. Computation of E at Point P in a Plane with Three Charges Q_1, Q_2, Q_3.

$$E = Q_i/4\pi\varepsilon_o r_i^2 = kQ_i/r_i^2 \qquad 5\text{-}8$$

(Note: Here $k = 1/4\pi\varepsilon_o$ or 9.0×10^9 nt-m^2/coul^2 in mks units. For gravitational fields another constant can be substituted such as $-G = -6.673 \times 10^{-11}$ nt-m^2/kg^2.)

The direction of the field in each case is along the direction of r, either toward Q_i if it's a negative charge, or away from Q_i if it's positive. To complete a superposition computation: you select one charge, compute E from that charge, and reduce it into its components E_x and E_y. You then repeat the procedure for the next charge, and algebraically sum the new field components with those computed for the first charge. You repeat the procedure for all the charges in the system. The final result gives you the components of the net vector field at point P.

III. THE PROGRAM

The program we will develop will allow you to handle up to 10 charges in the x, y plane – and compute the resulting field at any point P. We will arrange it so that you can:

- Enter the magnitude of each charge (positive or negative) Q_i
- Enter the x coordinate of each charge, x_i
- Enter the y coordinate of each charge, y_i.

The calculator or computer will remember where each charge is. You will then enter the coordinates of the point at which you want the field computed, x_p and y_p and the program will handle the computations for you.

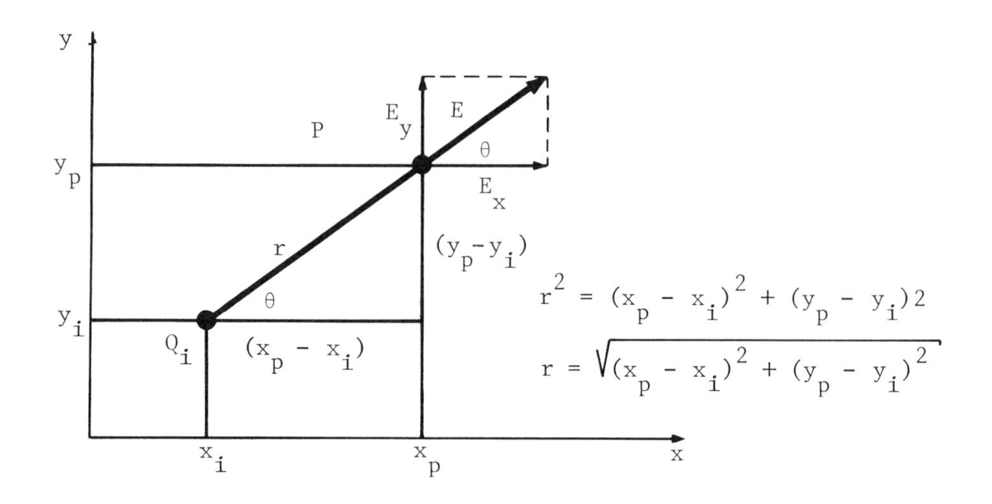

Fig. 5-6. Superposition Theorem Calculation.

The program will first compute $(x_p - x_i)$ and $(y_p - y_i)$ and their squares, and then r^2 and r (as represented in Fig. 5-6). It will then proceed to compute and store the value of $E = Q_i k/r^2$, and then the x and y components of E:

$$E_x = E\,(x_p - x_i)/r \quad (= E \cos\theta) \tag{5-9}$$

$$E_y = E\,(y_p - y_i)/r \quad (= E \sin\theta) \tag{5-10}$$

This method is repeated for each charge in the system, and the final E_x and E_y vector resultants are obtained. To allow for easier graphing and analysis, the program then converts the final result to polar coordinates, giving the field direction in degrees, and magnitude in field units – as shown in Fig. 5-7.

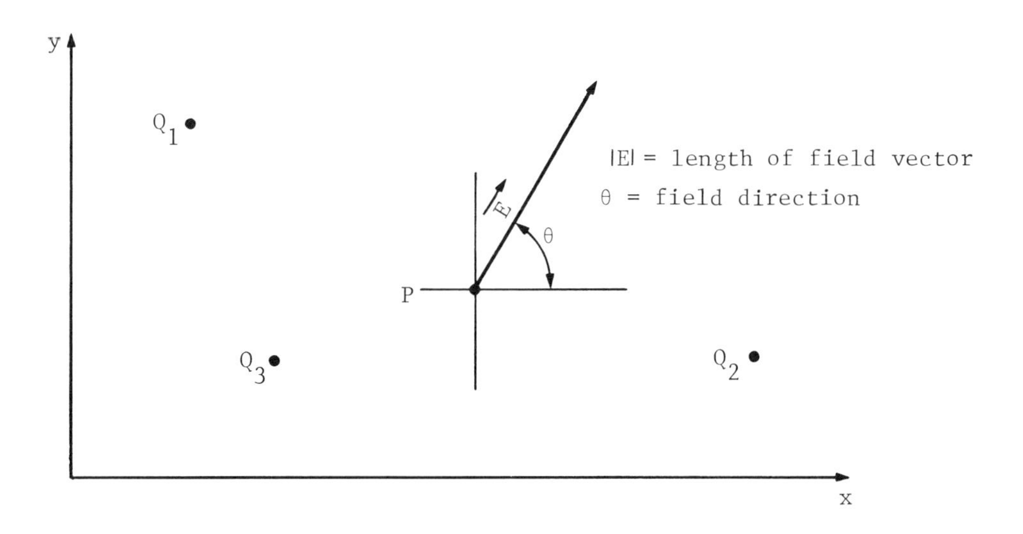

Fig. 5-7. Program Result: Field Direction in Degrees and Magnitude in Field Units

IIIa. THE PROGRAM FOR THE TI-58 or TI-59

Some specific notes about this program when using the TI-58 or TI-59:

1. The angle θ will be displayed as a number between $-90°$ and $+270°$. Angles in the fourth quadrant will be displayed as a negative number. (See "Personal Programming," Pages V-30ff.)
2. If you ask the calculator to compute the field at the location of one of the charges Q_i, the effects of Q_i are ignored in the computation. (The calculation will proceed as if the charge at point P is not present.) In general, these results will not be useful in your work.
3. If E equals zero at the point P, a θ value of $45°$ will be given.

In the program an indirect addressing method is used to move from charge to charge as E is computed. Indirect addressing is a programming technique where a memory is used to store an address or location where data is stored. In this case we will store the charge magnitudes Q_i in memory locations which will be stored in memory 01. Coordinates x_i will be stored in memory locations whose addresses are stored in memory 02, y_i will be stored in memories whose addresses are stored in memory 03. A review of indirect addressing is covered in the T.I. owner's manual "Personal Programming" on Pages IV-84 ff and V-68ff.

SUPER POSITION THEOREM

FUNCTION	*PROGRAM KEYSTROKES*	*STEP*
Clear calculator completely, get into "learn" mode.	OFF/ON [LRN]	000
Initialize - Clear all memories.	[2nd] [Lb1] [2nd] [A'] [2nd] [CMs]	003
Set up memories 01, 02, 03 for use in indirect storage of Q, x and y.	10 [STO] 01 20 [STO] 02	011
	30 [STO] 03 0 [R/S]	017
k will be entered here. Store in memory 07	[2nd] [Lb1] [2nd] [B'] [STO] 07 [R/S]	022
Q_i will be entered here and stored in memory address contained in memory 01. Add 1 to memory 01. Display value of i (1.00 to 10.00).	[2nd] [Lb1] [A] [STO] [2nd] [Ind] 01	026
	1 [SUM] 01 [RCL] 01 [-] 10 [=] [R/S]	036
x_i will be entered here, stored in memory address contained in memory 02. Add 1 to memory 02, display i.	[2nd] [Lb1] [B] [STO] [2nd] [Ind] 02	040
	1 [SUM] 02 [RCL] 02 [-] 20 [=] [R/S]	050
y_i will be entered here, stored in memory address contained in memory 03. Add 1 to memory 03, display i.	[2nd] [Lb1] [C] [STO] [2nd] [Ind] 03	054
	1 [SUM] 03 [RCL] 03 [-] 30 [=] [R/S]	064

FUNCTION	*PROGRAM KEYSTROKES*	*STEP*
Enter coordinates at which field is to be determined: x_p	[2nd] [Lb1] [D] [STO] 05 [R/S]	069
y_p	[2nd] [Lb1] [E] [STO] 06	073
Label to allow for print option.	[2nd] [Lb1] [X]	075
Zero memories which will contain E_x, E_y.	0 [STO] 47 [STO] 48	080
Reset memory addresses for Q_i, x_i, y_i.	10 [STO] 01 20 [STO] 02 30 [STO] 03	092
Recall y_p value just entered and halt (or return).	[RCL] 06 [INV] [SBR]	095
Start field calculations at Label C', zero t memory.	[2nd] [Lb1] [2nd] [C']	097
Compute $(x_p - x_i)$, store in memory 40. Square it and get ready to add.	0 [x:t]	099
	[( ] [RCL] 05 [-] [RCL] [2nd] [Ind] 02 [ )]	106
	[STO] 40 [X^2] [+]	110
Compute $(y_p - y_i)$, store in memory 42, square it and complete addition. r^2 now in memory 44.	[( ] [RCL] 06 [-] [RCL] [2nd]	
	[Ind] 03 [ )]	117
	[STO] 42 [X^2] [=] [STO] 44	123
Compute r, store in memory 45. If r = 0 (we are at a charge location), skip field computations.	[$\sqrt{x}$] [STO] 45	126
	[2nd] [x=t] [X^2]	128
Compute $Q_i\,k/r^2$, store in 46 (Value of E now in 46.)	[( ] [RCL] [2nd] [Ind] 01 [X] [RCL] 07 [ )]	135
	[÷] [RCL] 44 [=] [STO] 46	141
Multiply E by $(x_p - x_i)/r$ to compute E_x, sum to memory 47. Multiply E by $(y_p - y_i)/r$ to compute E_y, sum to memory 48.	[X] [RCL] 40 [÷] [RCL] 45 [=]	148
	[SUM] 47	150
	[RCL] 46	152
	[X] [RCL] 42 [÷] [RCL] 45 [=]	159
	[SUM] 48	161
Bypass label: if r = 0 program skips to here.	[2nd] [Lb1] [X^2]	163
Increment memory addresses for recall of next Q and its coordinates.	1 [SUM] 01 [SUM] 02 [SUM] 03	170
Check to see of all Q's have been used in superposition computation, if not repeat computation.	[RCL] [2nd] [Ind] 01	172
	[2nd] [X=t] [$\sqrt{x}$]	174
	[2nd] [C']	175

FUNCTION	*PROGRAM KEYSTROKES*	*STEP*
Superposition computation	[2nd] [Lb1] [√x]	177
complete label change final field to polar coordinates	[RCL] 47 [X:t] [RCL] 48	182
$\theta(E_\theta)$ displayed (degrees)	[INV] [2nd] [P→R]	184
$\|E\|$ (E_r) displayed (field units)	[INV] [SBR]	185
	[2nd] [X:t] [INV] [SBR]	187
(Note: If printer not used, program complete to this point. Get out of learn and reset.)		

Printer Option

FUNCTION	*PROGRAM KEYSTROKES*	*STEP*
Δx entered here - stored in memory 08	[2nd] [Lb1] [2nd] [D'] [STO] 08 [R/S]	192
Δy entered here - stored in memory 09	[2nd] [Lb1] [2nd] [E'] [STO] 09 [R/S]	197
"Begin print" label	[2nd] [Lb1] [2nd] [Prt]	199
(When running program number of Δx's and Δy's desired must be stored in memories 49 and 50.) Zero memories for x_p, y_p, 50 and 51.	0 [STO] 05 [STO] 06 [STO] 51 [STO] 52	208
Label this point for loop return.	[2nd] [Lb1] [1/x]	210
Execute "reset" subroutine.	[SBR] [X]	212
Execute field computation for current x_p, y_p. Store E_θ	[2nd] [C'] [STO] 55 [X:t] [STO] 56	218
and E_r, advance printer.	[2nd] [Adv]	219
Print x_p	[RCL] 05 [2nd] [Prt]	222
Print y_p	[RCL] 06 [2nd] [Prt]	225
Print θ (E_θ), degrees	[RCL] 55 [2nd] [Prt]	228
Print E (E_r), field units	[RCL] 56 [2nd] [Prt]	231
Test to see if desired number of Δy's have been computed. If so, go to [Y^x]. If not, increment y_p by Δy, increment Δy counting memory and repeat computation. If all Δy's computed, advance printer and test to see if all Δx's have been computed.	[RCL] 50 [X:t] [RCL] 52 [2nd] [X=t]	237
	[Y^x]	238
	[RCL] 09 [SUM] 06 1 [SUM] 52	245
	[GTO] [1/x]	247
	[2nd] [Lb1] [Y^x]	249
	[2nd] [Adv] [2nd] [Adv]	251
	[RCL] 49 [X:t] [RCL] 51	256
	[2nd] [X=t] [+]	258

FUNCTION	*PROGRAM KEYSTROKES*	*STEP*
If not, zero y_p register and y_p counting memory.	0 [STO] 06 [STO] 52	263
Increment x_p by Δx, increment Δx counting memory. Repeat computation.	[RCL] 08 [SUM] 05 1 [SUM] 51 [GTO] [1/x]	270 272
When computation is complete for N Δx's and N Δy's, halt.	[2nd] [Lb1] [+] [R/S]	275
Get out of "learn" mode and reset.	[LRN] [RST]	0

A. CALCULATOR MEMORIES USED

Memory	Value
01	Address of Q_i (for i up to 10) values 10-19
02	Address of x_i (for i up to 10) values 20-29
03	Address of y_i (for i up to 10) values 30-39
05	x_p coordinates of point at which field is to be determined
06	y_p
07	Field constant, K
08	Δx for printer option
09	Δy
10-19	Values of Q_i (for i up to 10)
20-29	Values of x_i (for i up to 10)
30-39	Values of y_i (for i up to 10)
40	$(x_p - x_i)$
42	$(y_p - y_i)$
44	$r^2 = (x_p - x_i)^2 + (y_p - y_i)^2$
45	$r = \sqrt{(x_p - x_i)^2 + (y_p - y_i)^2}$
46	E value
47	E_x components collected
48	E_y components collected
49	Number of desired Δx values for computation, N_x
50	Number of desired Δy values for computation, N_y
51	Δx counting memories for printer option
52	Δy counting memories for printer option

B. USER INSTRUCTIONS

To enter and use the program, follow these instructions:

OPERATION	*PRESS*	*DISPLAY/COMMENTS*
First, enter program keystrokes carefully, following instructions in "the program" section.		
Initialize - set up program for use.	[2nd] [A']	0.
Enter K value	[2nd] [B']	K value displayed
Next, enter values for point charges on the sheet and their coordinates.		
Q_1	[A]	1.
x_1	[B]	1.
y_1	[C]	1.
Q_2	[A]	2.
x_2	[B]	2.
y_2	[C]	2.
Repeat for up to 10 charges in the sheet.		
To compute field at any point, enter coordinates.		
x_p	[D]	x_p
y_p	[E]	y_p
To start calculation θ value	[2nd] [C']	θ - Field direction
to compute $\|E\|$ value	[R/S]	$\|E\|$ - Field magnitude

NOTE (1)

To rerun for new point, enter coordinates and repeat procedure.

NOTE (2)

To enter new charge distribution, press [2nd] [CMs] [RST] [2nd] [A']. Re-enter K value, enter Q_i, x_i, y_i and proceed.

Printer Option: To have the calculator print out a grid of points automatically after all charges and their coordinates (Q's, x's and y's) have been entered, as shown in Fig. 5-8.

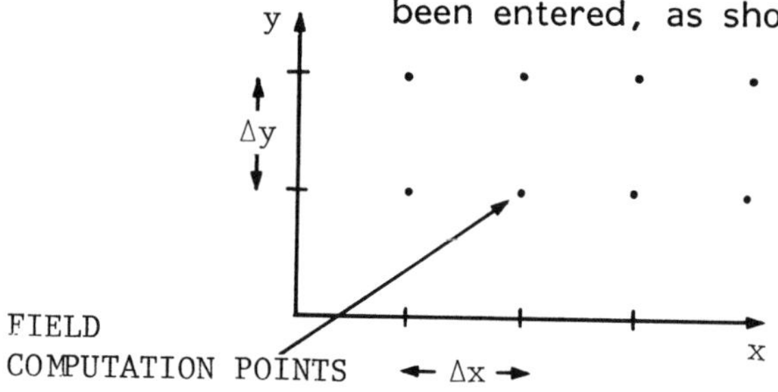

Fig. 5-8.
Grid of Points

OPERATION	*PRESS*	*DISPLAY/COMMENTS*
Enter Δx	[2nd] [D']	x
Enter Δy	[2nd] [E']	y
Enter number of Δx values desired, N_x	[STO] 49	N_x
Enter number of Δy values desired, N_y	[STO] 50	N_y
Start computations and printout of results.	[SBR] [2nd] [Prt]	x_i
		y_i
		E_θ (degrees)
		$\|E\|$ (field units)

Note: Ignore field results computed at charge locations.

Program Check Data

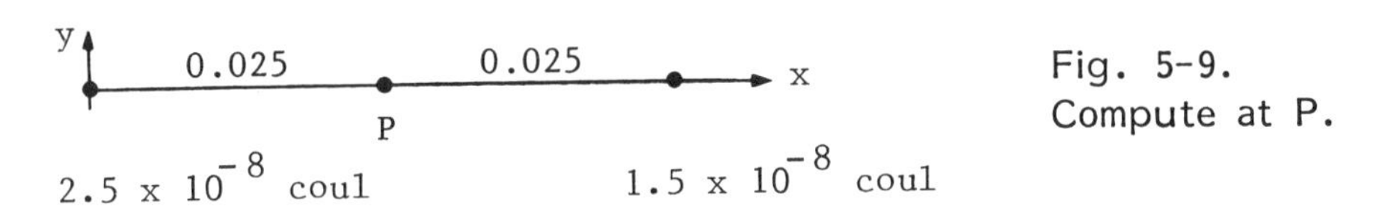

Fig. 5-9.
Compute at P.

To compute the field at the midpoint of the charge distribution in Fig. 5-9,

Initialize	[2nd] [A']	0
Enter $K = 9 \times 10^9$	9 [EE] 9 [2nd] [B']	9. 09
Enter $Q_1 = 2.5 \times 10^{-8}$	2.5 [EE] 8 [+/-] [A]	1.00
$x_1 = 0$	0 [B]	1.00
$y_1 = 0$	0 [C]	1.00
Enter $Q_2 = 1.5 \times 10^{-8}$	1.5 [EE] 8 [+/-] [A]	2.00
$x_2 = 0.05$	0.05 [B]	2.00
$y_2 = 0$	0 [C]	2.00
Enter $x_p = 0.025$	0.025 [D]	2.5 -02
$y_p = 0$	0 [E]	0.00
Compute θ value	[2nd] [C']	0.00
Compute $\|E\|$ value		1.44 05

(Field is to the right, 1.44×10^5 nt/coul.)

IIIb. THE PROGRAM IN BASIC

A. PROGRAM STATEMENTS	B. EXPLANATION
100 PI=3.141593	Set PI constant
110 OPTION BASE 1	Set array element base to 1
120 DIM Q(10),X(10),Y(10)	Dimension charge arrays
200 FOR I=1 TO 10	Set charge arrays to 0
210 Q(I)=0	Set magnitude to 0
220 X(I)=0	Set X coordinate to 0
230 Y(I)=0	Set Y coordinate to 0
240 NEXT I	Repeat loop
500 INPUT "Field constant";K	Enter field constant
1000 FOR I=1 TO 10	Enter up to 10 charges
1010 PRINT	Display blank line
1020 PRINT "Charge";I	Display charge number
1030 INPUT "Charge magnitude";Q(I)	Enter charge magnitude
1040 IF Q(I)=0 GOTO 1100	Exit loop if no magnitude
1050 INPUT "Charge X";X(I)	Enter charge X coordinate
1060 INPUT "Charge Y";Y(I)	Enter charge Y coordinate
1070 NEXT I	Repeat loop
1100 PRINT	Display blank line
2000 INPUT "Desired field X";FX	Enter desired field X coordinate
2010 INPUT "Desired field Y";FY	Enter desired field Y coordinate
2030 GOSUB 9000	Perform field computations
3000 PRINT	Display blank line
3010 PRINT "Angle =";A	Display angle
3020 PRINT "Field strength =";F	Display field strength
3100 PRINT	Display blank line
3110 PRINT "Press 1 for new point"	Menu: 1 to enter new point
3120 PRINT "Press 2 for new charges"	2 to enter new charges
3130 PRINT "Press 3 to print map"	3 to print field "map"
3140 PRINT "Press 4 to exit"	4 to exit program
3150 PRINT	Display blank line
3160 K$=INKEY$	Get keypress
3170 IF K$="" GOTO 3160	Wait for keypress
3180 IF K$="1" GOTO 2000	1 -- Enter new point
3190 IF K$="2" GOTO 200	2 -- Enter new charges
3200 IF K$="3" GOTO 4000	3 -- Print field "map"
3210 IF K$<>"4" GOTO 3160	Not 4 -- Get another keypress
3220 END	4 -- Exit program

```
9000 EX=0                              Zero field X component
9010 EY=0                              Zero field Y component
9020 FOR I=1 TO 10                     Compute for all 10 charges
9030 IF Q(I)=0 GOTO 9500                 Exit loop if no magnitude
9040 DX=FX-X(I)                          Compute X difference
9050 DY=FY-Y(I)                          Compute Y difference
9060 R=SQR((DX^2)+(DY^2))                Compute radius vector
9070 IF R=0 GOTO 9110                    Skip computation if radius = 0
9080 E-Q(I)*(K/(R^2))                    Compute field
9090 EX=EX+(E*(DX/R))                    Increment field X component
9100 EY=EY+(E*(DY/R))                    Increment field Y component
9110 NEXT I                              Repeat loop
9500 F=SQR((EX^2)+(EY^2))              Compute field strength
9510 IF F<>0 GOTO 9540                 Field strength = 0?
9520 A=45                                Yes -- Set angle to 45
9530 GOTO 9900                           Subroutine complete
9540 S=ABS(EY/F)                       Compute sine of angle
9550 SX=SQR(-S*S+1)                    Compute temporary variable
9560 IF SX <>0 GOTO 9700               Temporary variable = 0?
9570 IF EY<0 GOTO 9600                   Yes -- Field Y negative?
9580 A=90                                  No -- Set angle to 90
9590 GOTO 9900                             Subroutine complete
9600 A=270                                 Yes -- Set angle to 270
9610 GOTO 9900                             Subroutine complete
9700 A=(ATN(S/SX))/(PI/180)            Compute angle, in degrees
9710 IF EX<0 GOTO 9750                 Field X negative?
9720 IF EY>=0 GOTO 9900                  No -- Field Y positive?
9730 A=-A                                  No -- Quadrant 4
9740 GOTO 9900                             Subroutine complete
9750 IF EY<0 GOTO 9780                 Field Y negative?
9760 A=180-A                             No -- Quadrant 2
9770 GOTO 9900                           Subroutine complete
9780 A=180+A                             Yes -- Quadrant 3
9900 RETURN                            Return to main program
```

Note: If you have access to a printer, you can substitute LPRINT instructions for PRINT instructions.

If you have access to a printer, you can add the following program statements to print a field "map" (refer to Fig. 5-8). You enter values for ΔX (the X increment) and ΔY (the Y increment), as well as the number of steps along the x-axis and y-axis you want to see. The computer displays a table of field strength and angle of direction for each point on the grid.

```
4000 INPUT "X increment";DXP           Enter X increment
4010 INPUT "Y increment";DYP           Enter Y increment
4020 INPUT "Number of X's";NX          Enter number of X's to compute
4030 INPUT "Number of Y's";NY          Enter number of Y's to compute
4040 FOR FX=0 TO DXP*NX STEP DXP       Compute for X coordinate range
4050 FOR FY=0 TO DYP*NY STEP DYP       Compute for Y coordinate range
4060 GOSUB 5000                        Compute and print results
4070 NEXT FY                           Increment Y coordinate
```

4080 NEXT FX	Increment X coordinate
4090 GOTO 3100	Return to menu
5000 GOSUB 9000	Perform field computations
5010 LPRINT	Print blank line
5020 LPRINT "X =";FX	Print desired field X coordinate
5030 LPRINT "Y =";FY	Print desired field Y coordinate
5040 LPRINT "Angle =";A	Print angle
5050 LPRINT "Field strength =";F	Print field strength
5060 RETURN	Return to main program

ARRAYS USED

Array	Value
Q	Charge (or mass) magnitude
X	Charge (or point mass) X coordinate
Y	Charge (or point mass) Y coordinate

C. VARIABLES USED

Variable	Value
A	Angle of field direction (degrees)
DX	X difference between coordinates
DXP	X increment ("delta-X") for printer routine
DY	Y difference between coordinates
DYP	Y increment ("delta-Y") for printer routine
E	Magnitude of field strength for individual charge
EX	Accumulator for field X components
EY	Accumulator for field Y components
F	Field strength at desired point
FX	Desired field X coordinate
FY	Desired field Y coordinate
I	FOR-NEXT loop counter
K	Field constant
K$	INKEY$ key code
NX	Number of X coordinates for printer routine
NY	Number of Y coordinates for printer routine
PI	Pi (3.141593)
R	Radius vector
S	Sine of field direction angle

D. USER INSTRUCTIONS

Enter the program carefully and run the program. Respond to this prompt:

Prompt	Action
Field constant?	Enter the appropriate field constant.

The program displays the charge number for each of up to ten charges (or point masses). Respond to these prompts for each charge:

Prompt	Action
Charge magnitude?	Enter the magnitude of the charge (or mass). (Enter a zero to stop entering charges.)
Charge X?	Enter the X coordinate of the charge.
Charge Y?	Enter the Y coordinate of the charge.

Respond to these prompts to enter the coordinates at which you want to determine the field:

Prompt	Action
Field X?	Enter the X coordinate of the desired field.
Field Y?	Enter the Y coordinate of the desired field.

The program displays the field direction angle (in degrees) and the field strength (in appropriate units). (Note that the angle θ will be displayed as a number between $-90°$ and $+270°$. Angles in the fourth quadrant will be displayed as a negative number. If $E = 0$, a θ value of $45°$ will be given.)

Next, the program displays a menu (option list). Press a key to select one of the following options:

1 Enter a new point.
2 Enter new charges.
3 Print a field "map."
4 Exit the program.

If you select option 3 (print field map), respond to these prompts:

Prompt	Action
X increment?	Enter the X increment.
Y increment?	Enter the Y increment.
Number of X's?	Enter the number of X's.
Number of Y's?	Enter the number of Y's.

For the origin (0,0) and for each grid point generated, the program displays the X and Y coordinates for points along the grid you specified with ΔX and ΔY; the angle of field direction (in degrees); and the field strength (in appropriate units) at each point.

E. PROGRAM CHECK DATA (Refer to Fig. 5-9.)

Field constant? 9E9

Charge 1
Charge magnitude? 2.5E-8
Charge X? 0
Charge Y? 0

Charge 2
Charge magnitude? 1.5E-8
Charge X? .05
Charge Y? 0

Charge 3
Charge magnitude? 0

Field X? .025
Field Y? 0

Angle = 0
Field = 144000

IV. EXPLORATIONS/NOTES

Carefully enter the program into the calculator or computer. It can be used in two modes, as mentioned.

i. To probe the field of a charge configuration at any single point you enter, x_p, y_p; or

ii. To prepare a table of data which will allow you to "map" a field: you enter Δx, Δy, N_x, N_y and data is provided on your printer which allows you to prepare a table like Table 5-1.

y \ x	0	Δx	$2\Delta x$	$3\Delta x$.....	$N\Delta x$
0	$\theta_{o,o}$ $\lvert E\rvert_{o,o}$	$\theta_{\Delta x,o}$ $\lvert E\rvert_{\Delta x,o}$			
Δy	$\theta_{o,\Delta y}$ $\lvert E\rvert_{o,\Delta y}$				
$2\Delta y$					
⋮					
$N\Delta y$					$\theta_{N_{\Delta x},N_{\Delta y}}$ $\lvert E\rvert_{N_{\Delta x},N_{\Delta y}}$

Table 5-1

You're encouraged to use the program to explore field configurations of your own design or try the specific exercises below.

Note: For distributions with 4 charges, data point computation takes a little over 30 seconds using the TI-59; for those of you using computers computations will be much faster. The more charges in the distribution, the longer the computation time will be.

1. a. Calculate the electric field E at the center of the charge distribution shown in Fig. 5-10.

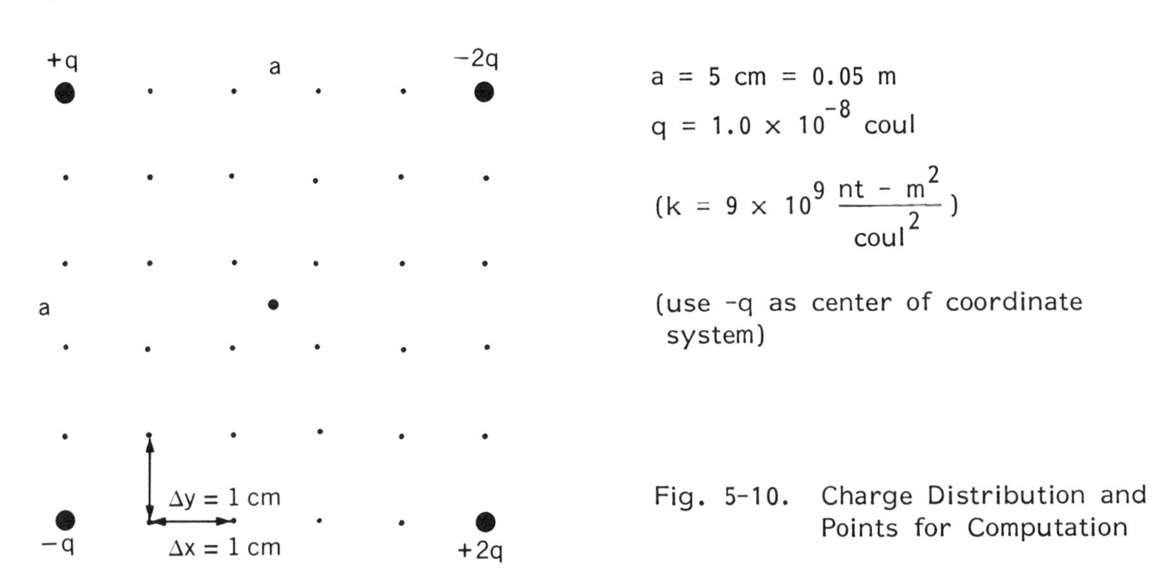

Fig. 5-10. Charge Distribution and Points for Computation

b. Compute the magnitude and direction of the field at all the points indicated in Fig. 5-10 ($\Delta x = \Delta y = 1$ cm = 0.01 m).

c. Draw a "field map" for the points computed in Part b above on a large sheet of graph paper. (Note: The program will compute spurious field values at the change locations themselves. Ignore these values in your work.) On the graph paper either draw in field vectors to scale, or use short lines to indicate field direction and label the field magnitude near the line.

d. Identify and draw any lines of symmetry you find in the field map above.

e. What would you expect the field to look like at large distances from this configuration? Compute the field at 5 meters away in the +x, +y, -x, -y directions:

x_p = 5.0 m y_p = 0.0 m $|E|$ = ________ θ = ________

x_p = 0.0 m y_p = 5.0 m $|E|$ = ________ θ = ________

x_p = -5.0 m y_p = 0.0 m $|E|$ = ________ θ = ________

x_p = 0.0 m y_p = -5.0 m $|E|$ = ________ θ = ________

2. a.

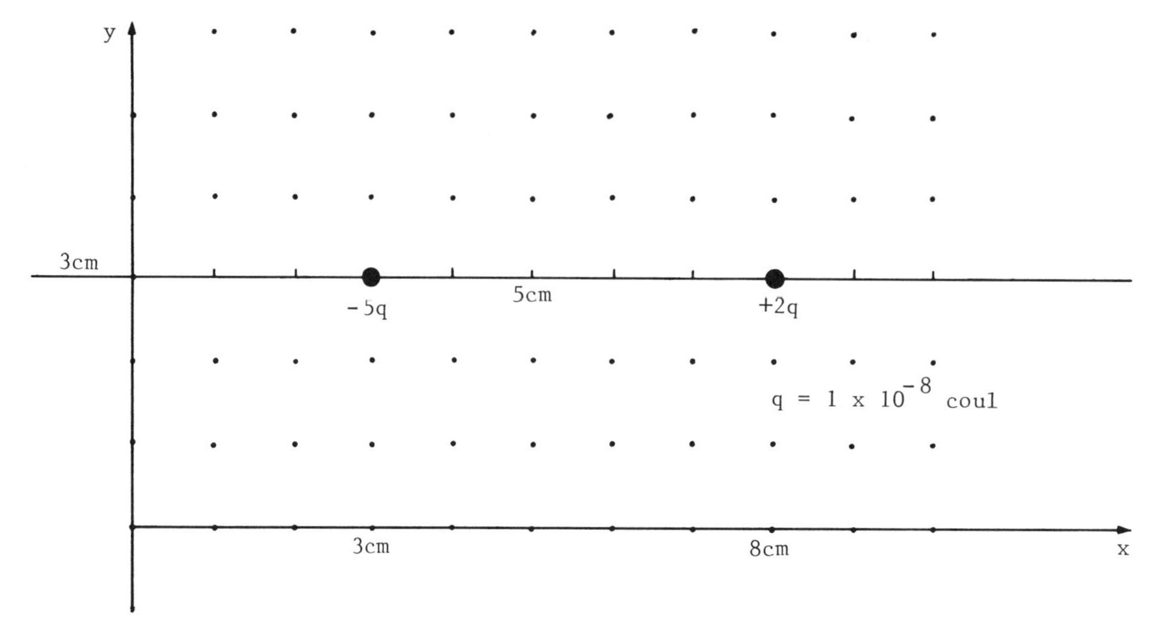

Fig. 5-11

Plot a field map for the charge configuration in Fig. 5-11.

b. Use the program to predict where the field strength equals zero.

Experiment 5-4 ONE-DIMENSIONAL ELASTIC COLLISIONS

I. INTRODUCTION

In this exploration, we will be using the calculator or computer to study the one-dimensional elastic collision. The program will let us focus our attention on the effect of the relative masses of the colliding bodies on the situation. As it turns out, the relative masses of these bodies impose some interesting restrictions on the amount of momentum and kinetic energy that can be transferred during the collision. The program will solve for all important variables before and after each collision. With a calculator or computer, we can quickly observe many "collisions" and examine patterns in momentum and kinetic energy as the relative masses of the colliding bodies are varied.

Here once again, we are actually putting equations into a program which "models" a phenomenon we wish to study. With this model we can vary key parameters and probe for relationships and patterns that result without going through the tedium of repeated computation.

Before beginning this study, a review of the theory of one-dimensional collisions will be helpful, as covered in Resnick & Halliday's physics texts.

II. THEORY REVIEW

We'll be studying the case where 2 point masses undergo a fully elastic ollision while moving along the same line, as shown in Fig. 5-12 below.

Fig. 5-12. One-Dimensional Elastic Collision.

We assume that both momentum and kinetic energy are conserved, and use the relationships that result to solve for the unknown final velocities after the collision. (m_1, m_2, v_{1i} and v_{2i} are assumed known.)

In this case it can be shown that:

$$v_{1f} = \left(\frac{m_1 - m_2}{m_1 + m_2}\right) v_{1i} + \left(\frac{2m_2}{m_1 + m_2}\right) v_{2i} \qquad 5\text{-}11$$

and ...

$$v_{2f} = \left(\frac{2m_1}{m_1 + m_2}\right) v_{1i} + \left(\frac{m_2 - m_1}{m_1 + m_2}\right) v_{2i} \qquad 5\text{-}12$$

III. THE PROGRAM

We will develop a program which will work with or without a printer. At the beginning, the program will allow us to enter and store m_1, m_2, v_{1i}, v_{2i}. The magnitudes of the initial momenta (p_{1i}, p_{2i}) and initial kinetic energies (K_{1i}, K_{2i}) will then be computed and displayed (or printed).

The program will then proceed to compute and display the post collision velocities, v_{1f} and v_{2f}, using the formulae above. With these results, the final moments (p_{1f}, p_{2f}) and kinetic energies (K_{1f}, K_{2f}) will be computed and displayed.

The program includes an option which allows all of these variables to be printed on the printer, and to automatically vary the relative masses of the colliding bodies by an amount Δm – repeating all computations.

IIIa. THE PROGRAM FOR THE TI-58 or TI-59

ONE-DIMENSIONAL ELASTIC COLLISIONS

FUNCTION	*PROGRAM KEYSTROKES*	*STEP*
Clear calculator completely, get into "learn" mode.	OFF/ON [LRN]	000
m_1 will be entered first, store m_1 and halt.	[2nd] [Lb1] [A] [STO] 00 [R/S]	005
m_2 will be entered next, store m_2 and halt.	[2nd] [Lb1] [B] [STO] 01 [R/S]	010
v_{1i} will be entered at this point. Store v_{1i} and halt.	[2nd] [Lb1] [C] [STO] 02 [R/S]	015
v_{2i} will be entered at this point. Store v_{2i} and halt.	[2nd] [Lb1] [D] [STO] 03 [R/S]	020
Computation Label	[2nd] [Lb1] [2nd] [A']	022
Compute p_{1i} (= $m_{1i} \times v_{1i}$) and display it.	[( ] [RCL] 00 [X] [RCL] 02 [ )] [R/S]*	030
Compute p_{2i} (= $m_{2i} \times v_{2i}$) and display it.	[( ] [RCL] 01 [X] [RCL] 03 [ )] [R/S]*	038
Compute K_{1i} (= $1/2\ m_{1i} \times v_{1i}^2$) and display it.	[( ] [RCL] 00 [X] [RCL] 02 [X^2] [÷] [2] [ )] [R/S]*	049
Compute K_{2i} (= $1/2\ m_{2i} \times v_{2i}^2$) and display it.	[( ] [RCL] 01 [X] [RCL]	056
	03 [X^2] [÷] [2] [ )] [R/S]	060
	[2nd] [Adv] [2nd] [Adv]	062

FUNCTION	*PROGRAM KEYSTROKES*	*STEP*
Computation of v_{1f} begins here.	[2nd] [Lb1] [2nd] [B']	064
First compute $(m_1 + m_2)$. Store in memory 04	[( ] [RCL] 00 [+] [RCL] 01 [ )] [STO] 04	073
$(((m_1 - m_2) \div (m_1 + m_2)) \times v_{1i}+$	[( ] [( ] [( ] [( ] [RCL]	
	00 [-] [RCL] 01 [ )] [÷]	084
	[RCL] 04 [ )] [X] [RCL]	
	02 [ )] [+]	092
$(((2 \times m_2) \div (m_1 + m_2)) \times v_{2i}))$	[( ] [( ] [( ] [2] [X]	
	[RCL] 01 [ )] [÷]	101
	[RCL] 04 [ )] [X] [RCL] 03	
	[ )] [ )]	109
Store v_{1f} in memory 05 and display it.	[STO] [05] [R/S]*	112
Computation of v_{2f} begins here.	[2nd] [Lb1] [2nd] [C']	114
$((((2 \times m_1) \div (m_1 + m_2)) \times v_{1i}) +$	[( ] [( ] [( ] [( ] [RCL] 00	
	[X] 2 [ )] [÷]	124
	[RCL] 04 [ )] [X] [RCL] 02 [ )] [+]	132
$(((m_2 - m_1) \div (m_1 + m_2)) \times v_{2i}))$	[( ] [( ] [( ] [RCL] 01 [-]	
	[RCL] 00 [ )] [÷] [RCL] [04]	144
	[ )] [X] [RCL] 03 [ )] [ )]	150
Store v_{2f} in memory 06 and display it.	[STO] 06 [R/S]*	153
Compute p_{1f} (= $m_1 v_{1f}$), and display it.	[( ] [RCL] 00 [X] [RCL] 05	
	[ )] [R/S]*	161
Compute p_{2f} (= $m_2 v_{2f}$), and display it.	[( ] [RCL] 01 [X] [RCL] 06	
	[ )] [R/S]*	169
Compute K_{1f} (= $1/2\ m_1 v_{1f}^2$), and display it.	[( ] [RCL] 00 [X] [RCL] 05	
	[X^2] [÷] [2] [ )]	179
	[R/S]*	180
Compute K_{2f} (= $1/2\ m_2 v_{2f}$), and display it.	[( ] [RCL] 01 [X] [RCL] 06	
	[X^2] [÷] [2] [ )]	190
	[2nd] [Prt] [INV] [SBR]	192

FUNCTION	*PROGRAM KEYSTROKES*	*STEP*
Printer option: need not be added if PC-100A printer is not used.		
Δm_1 entered	[2nd] [Lb1] [E] [STO] 07	196
Print option loop label	[2nd] [Lb1] [2nd] [E']	198
Δm_1 added to m_1	[RCL] 07 [SUM] 00	202
Recall m_1, m_2, v_{1i}, v_{2i} and print	[RCL] 00 [2nd] [Prt]	205
	[RCL] 01 [2nd] [Prt]	208
	[RCL] 02 [2nd] [Prt]	211
	[RCL] 03 [2nd] [Prt]	213
	[2nd] [Adv]	
Execute computation	[SBR] [2nd] [A']	217
Advance printer	[2nd] [Adv] [2nd] [Adv] [2nd] [Adv]	220
Repeat loop	[GTO] [2nd] [E']	222
Get out of "learn" mode and reset	[LRN] [RST]	0

*Note: When using PC-100A printer replace asterisked instructions by [2nd] [Prt] instructions.

A. CALCULATOR MEMORIES USED

Memory	Value
00	m_1
01	m_2
02	v_{1i}
03	v_{2i}
04	$m_1 + m_2$
05	v_{1f}
06	v_{2f}
07	Δm

B. USER INSTRUCTIONS

PURPOSE/FUNCTION	*PROGRAM KEYSTROKES*	*STEP*
First, enter program keystrokes carefully, following instructions in "The Program" section.		
Enter m_1	[A]	m_1
Enter m_2	[B]	m_2
Enter v_{1i}	[C]	v_1
Enter v_{1f}	[D]	v_{1f}

PURPOSE/FUNCTION	*PROGRAM KEYSTROKES*	*STEP*
Begin computation	[2nd] [A']	p_{1i}
	[R/S] **	p_{2i}
**Note: If printer option is used, these results will be printed automatically.	[R/S] **	K_{1i}
	[R/S] **	K_{2i}
	[R/S] **	v_{1f}
	[R/S] **	v_{2f}
	[R/S] **	p_{1f}
	[R/S] **	p_{2f}
	[R/S] **	K_{1f}
	[R/S] **	K_{2f}

At this point, new values can be loaded for m_1, m_2, v_{1i} and/or v_{2i} and the computation repeated. (Press [2nd] [Cms] when loading complete new computation.)

OTHER OPTIONS IN THIS PROGRAM

To compute v_{1f} directly	[2nd] [B']	v_{1f} displayed
To compute v_{2f} directly	[2nd] [C']	v_{2f} displayed

TO USE PRINTER OPTION

Automatically computes all parameters for collision, adds m to m_1 and repeats until manually halted with [R/S] key:

First, enter m_1, m_2, v_{1i}, v_{2i} as above.

Next, enter Δm, and press	[E]	m_1 + m
		m_2
	Computation	v_{1i}
	repeats	v_{2i}
	automatically	p_{1i}
	until	p_{2i}
	halted	K_{1i}
	with	K_{2i}
	[R/S]	v_{1f}
	key	v_{2f}
		p_{1f}
		p_{2f}
		K_{1f}
		K_{2f}
		m_1 + 2 m etc.

Program Check Data

To check program, enter values for the collision shown in Fig. 5-13.

v_{1i} = 5m/sec v_{2i} = 1m/sec

M_1 = 15kg M_2 = 10kg

Fig. 5-13. Collision between Mass M_1 and Mass M_2.

PURPOSE/FUNCTION	*PROGRAM KEYSTROKES*	*STEP*
Enter $m_1 = 15$	[A]	15.
$m_2 = 10$	[B]	10.
$v_{1i} = 5$	[C]	5.
$v_{2i} = 7$	[D]	1.
Begin computation for p_{1i}	[2nd] [A']	75.
p_{2i}	[R/S]	10.
K_{1i}	[R/S]	187.5
K_{2i}	[R/S]	5.0
v_{1f}	[R/S]	1.8
v_{2f}	[R/S]	5.8
p_{1f}	[R/S]	27.
p_{2f}	[R/S]	58.
K_{1f}	[R/S]	24.3
K_{2f}	[R/S]	168.2

IIIb. THE PROGRAM IN BASIC

A. PROGRAM STATEMENTS	B. EXPLANATION
`100 INPUT "Mass #1";M1`	Enter 1st mass
`110 INPUT "Mass #2";M2`	Enter 2nd mass
`200 INPUT "Initial velocity #1";V1`	Enter 1st initial velocity
`210 INPUT "Initial velocity #2";V2`	Enter 2nd initial velocity
`300 GOSUB 9000`	Compute results
`500 PRINT`	Display blank line
`510 PRINT "Initial momentum #1 =";P1`	Display 1st initial momentum
`520 PRINT "Initial momentum #2 =";P2`	Display 2nd initial momentum
`530 PRINT "Init kinetic energy #1 =";K1`	Display 1st init kinetic energy
`540 PRINT "Init kinetic energy #2 =";K2`	Display 2nd init kinetic energy
`550 PRINT`	Display blank line
`560 PRINT "Final velocity #1 =";VF1`	Display 1st final velocity
`570 PRINT "Final velocity #2 =";VF2`	Display 2nd final velocity
`580 PRINT "Final momentum #1 =";PF1`	Display 1st final momentum

```
590 PRINT "Final momentum #2 =";PF2          Display 2nd final momentum
600 PRINT "Final kinetic engy #1 =";KF1      Display 1st final kinetic energy
610 PRINT "Final kinetic engy #2 =";KF2      Display 2nd final kinetic energy
1000 PRINT                                   Display blank line
1010 PRINT "Press 1 for new masses"          Menu: 1 to enter new masses
1020 PRINT "Press 2 for new velocities"            2 to enter new velocities
1030 PRINT "Press 3 to print mass range"           3 to print range of masses
1040 PRINT "Press 4 to exit"                       4 to exit program
1050 PRINT                                   Display blank line
1060 K$=INKEY$                               Get keypress
1070 IF K$="" GOTO 1060                      Wait for keypress
1080 IF K$="1" GOTO 100                        1 -- Enter new masses
1090 IF K$="2" GOTO 200                        2 -- Enter new velocities
1100 IF K$="3" GOTO 2000                       3 -- Print range of masses
1110 IF K$<>"4" GOTO 1060                      Not 4 -- Get another keypress
1120 END                                       4 -- Exit program

9000 P1=M1*V1                                Compute 1st initial momentum
9010 P2=M2*V2                                Compute 2nd initial momentum
9020 K1=(M1/2)*(V1^2)                        Compute 1st init kinetic energy
9030 K2=(M2/2)*(V2^2)                        Compute 2nd init kinetic energy
9040 M=M1+M2                                 Compute total mass
9050 VF1=(((M1-M2)/M)*V1)+(((2*M2)/M)*V2) Compute 1st final velocity
9060 VF2=(((2*M1)/M)*V1)+(((M2-M1)/M)*V2) Compute 2nd final velocity
9070 PF1=M1*VF1                              Compute 1st final momentum
9080 PF2=M2*VF2                              Compute 2nd final momentum
9090 KF1=(M1*(VF1^2))/2                      Compute 1st final kinetic energy
9100 KF2=(M2*(VF2^2))/2                      Compute 2nd final kinetic energy
9110 RETURN                                  Return to main program
```

Note: If you have access to a printer, you can substitute LPRINT instructions for PRINT instructions.

If you have access to a printer, you can add the following program statements to print a range of masses. You enter a value for ΔM (the mass increment); the computer displays the new masses and initial and final values of the velocities, moments, and kinetic energies for each new mass.

```
2000 INPUT "Mass #1 increment";DM1           Enter 1st mass increment
2100 M1=M1+DM1                               Increment 1st mass
2110 LPRINT "Mass #1 =";M1                   Display 1st mass
2120 LPRINT "Mass #2 =";M2                   Display 2nd mass
2130 LPRINT "Initial velocity #1 =";V1       Display 1st initial velocity
2140 LPRINT "Initial velocity #2 =";V2       Display 2nd initial velocity
2150 GOSUB 9000                              Compute results
3000 LPRINT                                  Print blank line
3010 LPRINT "Initial momentum #1 =";P1       Print 1st initial momentum
3020 LPRINT "Initial momentum #2 =";P2       Print 2nd initial momentum
3030 LPRINT "Init kinetic energy #1 =";K1    Print 1st init kinetic energy
3040 LPRINT "Init kinetic energy #2 =";K2    Print 2nd init kinetic energy
3050 LPRINT                                  Print blank line
3060 LPRINT "Final velocity #1 =";VF1        Print 1st final velocity
3070 LPRINT "Final velocity #2 =";VF2        Print 2nd final velocity
```

```
3080 LPRINT "Final momentum #1 =";PF1     Print 1st final momentum
3090 LPRINT "Final momentum #2 =";PF2     Print 2nd final momentum
3100 LPRINT "Final kinetic engy #1 =;KF1  Print 1st final kinetic energy
3110 LPRINT "Final kinetic engy #2 =;KF2  Print 2nd final kinetic energy
3120 LPRINT                               Print blank line
4000 PRINT "Repeat (Y/N)?'                Display "repeat" prompt
4010 K$=INKEY$                            Get keypress
4020 IF K$="" GOTO 4010                   Wait for keypress
4030 IF K$="Y" OR K$="y" GOTO 2100          Y -- Repeat
4040 IF K$="N" OR K$="n" GOTO 1000          N -- Return to menu
4050 GOTO 4010                              Not Y/N -- Get another keypress
```

C. VARIABLES USED

Variable	Value
DM1	Mass #1 increment ("ΔM") for printer routine
K$	INKEY$ key code
K1	Initial kinetic energy #1
K2	Initial kinetic energy #2
KF1	Final kinetic energy #1
KF2	Final kinetic energy #2
M	Total mass
M1	Mass #1
M2	Mass #2
P1	Initial momentum #1
P2	Initial momentum #2
PF1	Final momentum #1
PF2	Final momentum #2
V1	Initial velocity #1
V2	Initial velocity #2
VF1	Final velocity #1
VF2	Final velocity #2

D. USER INSTRUCTIONS

Enter the program carefully and run the program. Respond to these prompts:

Prompt	Action
Mass #1?	Enter the 1st mass.
Mass #2?	Enter the 2nd mass.
Initial velocity #1?	Enter the 1st initial velocity.
Initial velocity #2?	Enter the 2nd initial velocity.

The program then displays the initial momenta and kinetic energies, and the final velocities, momenta, and kinetic energies.

Next, the program displays a menu (option list). Press a key to select one of the following options:

1 Enter new masses (and velocities).
2 Enter new velocities.
3 Print a range of masses.
4 Exit the program.

If you select option 3 (print range of masses), respond to this prompt:

Prompt	Action
Mass #1 increment?	Enter the increment for the 1st mass.

The program displays the masses; the initial velocities, momenta, and kinetic energies; and the final velocities, momenta, and kinetic energies.

Next, respond to this prompt:

Prompt	Action
Repeat (Y/N)?	Press Y to increment the mass again and display the new results, or press N to return to the menu.

E. PROGRAM CHECK DATA (Refer to Fig. 5-13.)

```
Mass #1? 15
Mass #2? 10
Initial velocity #1? 5
Initial velocity #2? 1

Initial momentum #1 = 75
Initial momentum #2 = 10
Init kinetic energy #1 = 187.5
Init kinetic energy #2 = 5

Final velocity #1 = 1.8
Final velocity #2 = 5.8
Final momentum #1 = 27
Final momentum #2 = 58
Final kinetic engy #1 = 24.3
Final kinetic engy #2 = 168.2
```

To stop running the program at any time, enter a BREAK command. (On most computers this is accomplished by holding down a "shift" or "control" key, and pressing the "Break" or "Pause" key on the keyboard.) To run the program for a new set of values, type in the command RUN, and then press the "ENTER" or "RETURN" key.

IV. EXPLORATIONS/NOTES

1. a. Begin exploring the one-dimensional collision, using the program to compute the unknown values listed for the situation shown in Fig. 5-14.

$v_{1i} = 2.5$m/sec, $m_1 = 10$Kg $\qquad$ $v_{2i} = 0.5$m/sec, $m_2 = 25$Kg

Fig. 5-14. One-Dimensional Collision Example.

p_{1i} = __________ $\qquad$ v_{2f} = __________

p_{2i} = __________ $\qquad$ p_{1f} = __________

K_{1i} = __________ $\qquad$ p_{2f} = __________

K_{2i} = __________ $\qquad$ K_{1f} = __________

v_{1f} = __________ $\qquad$ K_{2f} = __________

b. Now reverse the relative masses (i.e., m_1 = 25 Kg, m_2 = 10 Kg) with all else as before, and repeat the experiment.

2. Use the calculator program to analyze the collision situation shown in Fig. 5-15.

v_{1i} = 5m/sec $\qquad$ v_{2i} = 0m/sec

m_1 = 1kg $\qquad$ m_2 = 8kg

Fig. 5-15. One-Dimensional Collision, m_2 Initially at Rest.

In this case a 1 Kg mass moving at 5 m/sec collides with an 8 Kg mass initially at rest.

a. Compute all relevant parameters in the collision: P_{1i}, P_{2i}, K_{1i}, K_{2i} and v_{1f}, v_{2f}, P_{1f}, P_{2f}, K_{1f}, K_{2f}.

b. Using the program, rerun all computations for this collision as m_1 is increased in mass from 1 Kg to 35 Kg; in increments (Δm) of 1 Kg. Arrange the data in the form shown in Table 5-2.

Table 5-2

m_1	m_2	v_{1i}	v_{2i}	P_{1i}	P_{2i}	K_{1i}	K_{2i}	v_{1f}	v_{2f}	P_{1f}	P_{2f}	K_{1f}	K_{2f}
Kg	Kg	M/se	M/sec	Kg M/sec	Kg M/sec	Kg M/sec	Kg M/sec	M/sec	M/sec	Kg M/sec	Kg M/sec	Kg M/sec	Kg M/sec
1	8	5	0	5	0	12.5	0	-3.89	1.11	-3.89	8.89	7.56	4.94
2	8	5	0	..									
3	8	5	0	..									
4	8	5	0	..									
5	8	5	0	..									
.													
.													
.													
.													
35	8	5	0	..									

c. Using the data in Table 5-2 compute the following ratios for each value of m_1 and arrange them as shown in Table 5-3

Table 5-3

m_1	v_{2f}/v_{1i}	K_{2f}/K_{1i}	K_{1f}/K_{1i}
1 Kg	0.22	0.39	0.60
.			
.			
.			
.			
.			
35 Kg			

d. Plot the data obtained in Part 2(c) on the same set of axes (use values of m_1 for the X axis.)

e. Answer the following questions based on observations of the graph obtained in Part 2(d).

(i) What conclusions can be drawn about the final velocity of m_2 (initially at rest), as m_1 gets very large?

(ii) At what point does the maximum kinetic energy transfer take place between the two bodies?

(iii) Describe the behavior of the kinetic energy transfer between the bodies as m_1 gets very large.

(iv) At what value of m_1 does $K_{2f} = K_{1f}$?

SECTION B - EXPLORATIONS WITH NUMERICAL METHODS

5.4 Calculator and Computer Experiments in Motion

In this next series of experiments, we will be using a simple but powerful numerical method to explore physical systems whose detailed solutions are in practice seldom explored by the student. The reason is that solutions of equations of motion where the force on an object varies with position and/or velocity and/or time are extremely difficult (and in some cases impossible). The technique we will describe will enable you to find the velocity $\frac{dx}{dt}$, and the position $x(t)$, in cases where the force operating on the system is varying. A simple technique will let you put the force function into a program and watch it operate the system. With this technique, explorations are possible that are not normally (or easily) treated in undergraduate texts.

5.5 Numerical Methods

To handle the solution of equations of motion of this sort, we will use what are termed "numerical methods." There are a variety of such methods available, and they are similar in that they allow us to arrive at an approximate solution to a complex motion, by proceeding through a series of simple steps. The method we'll be exploring is the "Modified Euler" method. As described by Eisberg[1] in his excellent book, it is a simple method, but it yields results accurate enough in practice to allow for interesting explorations without the need for long complex procedures. Other more accurate methods, such as the Runge-Kutta methods, are also implementable on a programmable calculator or personal computer and the student is encouraged to experiment with these as they are encountered in other texts[2,3,4] or in analysis courses.

5.6 Euler's Method

Euler's method can be used to analyze an equation of motion by chopping time into "bite-size" intervals Δt and treating the motion as if the velocity was constant over each

---

[1] Robert M. Eisberg, Applied Mathematical Physics with Programmable Pocket Calculators (McGraw-Hill, Inc., New York, 1976).

[2] Sourcebook for Programmable Calculators (McGraw-Hill, Inc., New York, 1979).

[3] Jon M. Smith, Scientific Analysis on the Pocket Calculator (John Wiley and Sons, New York, 1977).

[4] H. R. Meck, Scientific Analysis for Programmable Calculators (Prentice-Hall, New Jersey, 1981).

interval. The actual motion is approximated by a series of small constant velocity segments. In the "original" Euler's method, the velocity value at the beginning of each small interval Δt is used to approximate the velocity value throughout the interval. The force, acceleration, velocity, and position is computed from interval to interval using an approximate procedure such as that described in Fig. 5-16.

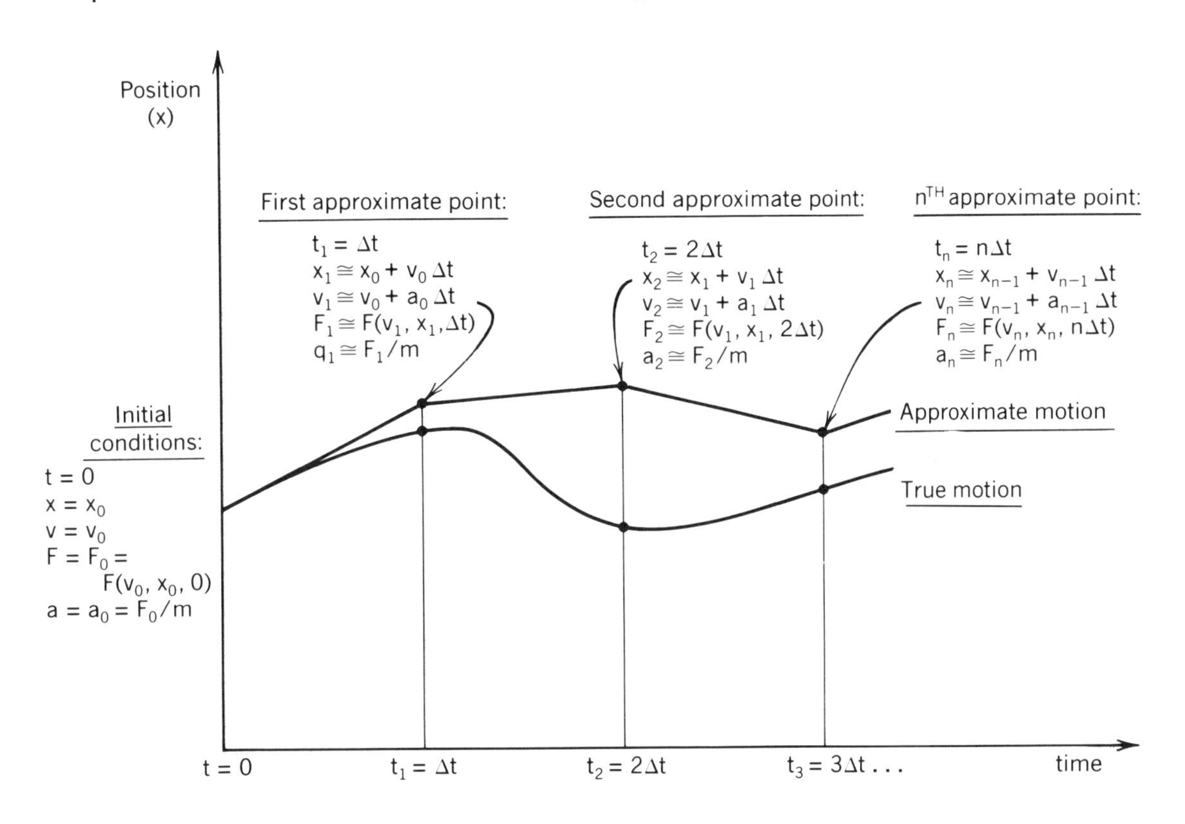

Fig. 5-16. "Original" Euler Method Solution.

Our initial values are given at time $t = 0$, and we'll consider time chopped up into bite-size pieces each Δt in duration. At $t = 0$

$$
\begin{aligned}
x &= x_o \\
v &= v_o \\
F(v_o, x_o t) &= F(v_o, x_o, 0) = F_o \\
a_o &= F_o/m
\end{aligned}
\qquad 5\text{-}13
$$

Now at the end of our first interval, Δt

$$
\begin{aligned}
t &= t_1 = \Delta t \\
x_1 &\cong x_o + v_o \Delta t, \text{ where } v = \frac{\Delta x}{\Delta t} \\
v_1 &\cong v_o + a_o \Delta t, \text{ where } a = \frac{\Delta v}{\Delta t} \\
F_1 &\cong F(v_1, x_1, \ t) \\
a &\cong F_1/m
\end{aligned}
\qquad 5\text{-}14
$$

This procedure continues for the next interval,

$$\begin{aligned} t_2 &= 2\Delta t \\ x_2 &\cong x_1 + v_1\Delta t \\ v_2 &\cong v_1 + a_1\Delta t \\ F_2 &\cong F(v_2, x_2, 2\Delta t) \\ a_2 &\cong F_2/m \end{aligned} \qquad 5\text{-}15$$

and each subsequent interval. In general for the nth interval:

$$\begin{aligned} t_n &= n\Delta t \\ x_n &\cong x_{n-1} + v_{n-1}\Delta t \\ v_n &\cong v_{n-1} + a_{n-1}\Delta t \\ F_n &\cong F(v_n, x_n, n\Delta t) \\ a_n &\cong F_n/m \end{aligned} \qquad 5\text{-}16$$

Using this repeating procedure it is possible to get an approximate picture of the motion, without actually solving the differential equation. The accuracy of this method depends on the size of Δt, better accuracy in general being achieved as Δt is made smaller.

Note that such repetitive computations, although each simple in and of itself, represent quite a tedious arithmetic barrage when seen together. Here's where a programmable calculator or personal computer is ideal -- it can be programmed <u>both</u> with the numerical method <u>and</u> the Force law -- $F(v,x,t)$ -- and carry out the evaluations necessary at each value of Δt. You can then rapidly plot x vs. t or v vs. t quite easily and get a quick picture of the motion.

The method we've outlined above is called an "Euler Method" solution to the motion, also called a "full increment method" or "tangent line" method of solution. In general this method produces only approximate results, and very small Δt values are required -- and hence many iterations -- to achieve reasonable accuracy. Details on how to estimate the errors encountered in a Euler's method solution are covered in many texts on differential equations or numerical methods. It is possible with a simple modification to <u>improve</u> the method significantly.

5.7 Improved or Modified Euler Method

We get a better approximation of the actual motion by following a slightly modified procedure as outlined in Fig. 5-17 -- and documented in many texts.

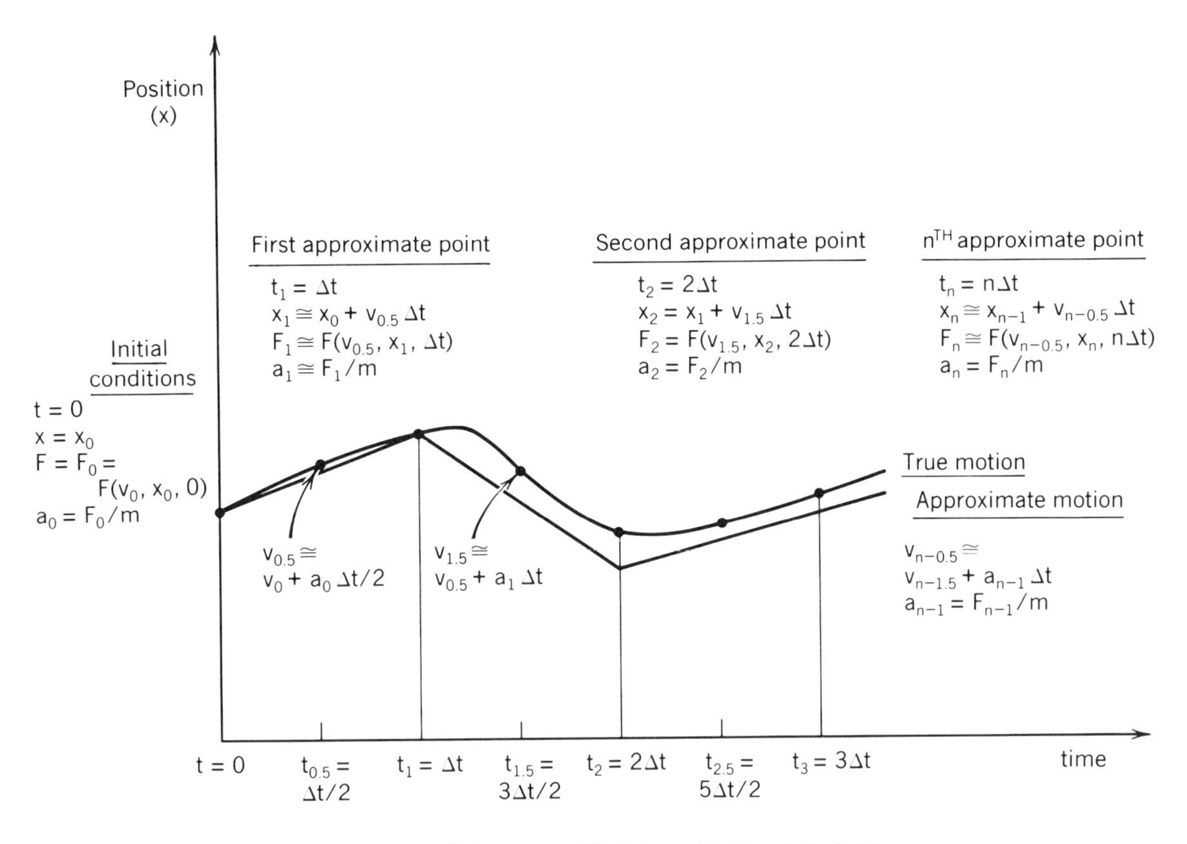

Fig. 5-17. "Improved" Euler Method Solution.

In using the modified Euler method we proceed as before by dividing the problem up into "bite-size" chunks, each of time Δt. We again assume that the velocity is constant over the entire time interval Δt, but we do not use the velocity at the beginning of the interval as that constant. Instead, we compute the velocity value at the midpoint of each interval -- $\Delta t/2$ into the interval -- and use that value to approximate the velocity value throughout the interval. The procedure we will follow is illustrated in Fig. 5-17 and it works like this. We begin with our initial conditions:

$$t = 0$$
$$x = x_o$$
$$F_o = F(v_o, x_o, 0)$$
$$a_o = F_o/m \qquad \text{5-17}$$

Then, we compute the value of the velocity one half way into the first interval, at time $\Delta t/2$:

$$v_{0.5} = v \text{ (at } \Delta t/2) \cong v_o + a_o\,(\Delta t/2)$$

Then using that result, we compute:

$$x_1 \cong x_o + v_{0.5}\Delta t$$

Next, we move on to calculate

$$F_1 \cong F(v_{0.5}, x_1, \Delta t)$$
$$a_1 \cong F_1/m \qquad \text{5-18}$$

Using this result we compute the velocity at the midpoint of the <u>next</u> interval:

$$v_{1.5} \cong v_{0.5} + a_1 \Delta t$$

Using that result we then go on to compute:

$$x_2 \cong x_1 + v_{1.5}\Delta t$$
$$F_2 \cong F(v_{1.5}, x_2, 2\Delta t)$$
$$a_2 \cong F_2/m \qquad \text{5-19}$$

In general, for any interval we compute:

$$F_n \cong F(v_{n-0.5}, x_n, n\Delta t)$$
$$a_n \cong F_n/m$$
$$v_{n+0.5} \cong v_{n-0.5} + a_n \Delta t$$
$$x_{n+1} \cong x_n + v_{n+0.5}\Delta t \qquad \text{5-20}$$

The following programs will allow you to use the modified Euler's method for exploring a wide variety of physical situations. Basically, all you need to do is insert the appropriate force law as a subroutine, select and enter an appropriate value for Δt and other initial conditions, and the program will "take off" and do the rest. (Some minor modifications will be required in certain situations – as will be explained later.)

Note that you'll be able to use this program in explorations that follow even if you do not understand each step completely. The program follows the logic of the modified Euler method described previously, and enables you to apply the method to various force law situations with minimum modifications. In the rest of this chapter we will be examining a selection of interesting applications for this method. It is hoped that after exploring these applications you will find a few of your own.

5.7a THE PROGRAM FOR THE IMPROVED EULER METHOD USING THE TI-58 or TI-59

FUNCTION	*PROGRAM KEYSTROKES*	*STEP*
Clear calculator completely, get into "learn" mode	[OFF/ON] [LRN]	000
Store Δt and Δt/2	[2nd] [Lb1] [A] [STO] 10	004
	[÷] 2 [=] [STO] 15	009
Display Δt	[RCL] 10 [R/S]	012

FUNCTION	*PROGRAM KEYSTROKES*	*STEP*
Store initial position, x_o	[2nd] [Lb1] [B]	014
	[STO] 11 [R/S]	017
Store initial velocity, v_o	[2nd] [Lb1] [C]	019
	[STO] 12 [R/S]	022
Store initial force value, F_o	[2nd] [Lb1] [D]	024
	[STO] 13 [R/S]	027
Store mass, m	[2nd] [Lb1] [E]	029
	[STO] 14 [R/S]	032
Start calculations	[2nd] [Lb1] [2nd] [A']	034
Clear memory 20 - used to store the time	0 [STO] 20	037
Calculate $a_0 = F_o/m$	[RCL] 13 [÷] [RCL] 14	042
Multiply by $\Delta t/2$	[X] [RCL] 15	045
Add v_o	[+] [RCL] 12 [=]	049
Result is $v_{0.5}$, store in memory 22	[STO] 22	051
Compute $x_1 = v_{0.5}\Delta t + x_o$	[X] [RCL] 10 [+]	055
and store in memory 21	[RCL] 11 [=] [STO] 21	060
Repetitive calculation loop label	[2nd] [Lb1] [2nd] [D']	062
Increment t by Δt	[RCL] 10 [SUM] 20	066
Display $t_{n-0.5}$	[RCL] 20 [-] [RCL] 15	071
	[=] [R/S]*	073
Display v at $t_{n-0.5}$	[RCL] 22 [R/S]* [2nd] [ADV]	077
Display t_n	[RCL] 20 [R/S]*	080
Display x at t_n	[RCL] 21 [R/S]* [2nd] [ADV]	084
Direct program to subroutine [E'] where force function is computed	[SBR] [2nd] [E']	086
Compute $F_n/m = a_n$	[÷] [RCL] 14	089
Compute $a_n\Delta t + v_{n-0.5}$	[X] [RCL] 10 [+] [RCL] 22	095
$= v_{n+0.5}$, store in memory 22	[=] [STO] 22	098
Compute $v_{n+0.5}\Delta t + x_n =$	[RCL] 22 [X] [RCL] 10	103
x_{n+1}, store in memory 21	[+] [RCL] 21 [=] [STO] 21	109

FUNCTION	*PROGRAM KEYSTROKES*	*STEP*
Repeat computation	[GTO] [2nd] [D']	111
Label for force subroutine	[2nd] [Lb1] [2nd] [E']	113

–Force subroutine is entered next–

*Notes: If you're using a PC-100A printer, [R/S] instructions marked with an asterisk should be replaced with [2nd] [Prt] instructions.

If only velocity or position data is required, [2nd] [Prt] or [R/S] instructions may be replaced by [2nd] [NOP] instructions. This will cause the calculator to skip these steps and proceed.

To rerun program for new values, press [2nd] [CMs], reload all initial values and press [2nd] [A'].

A. CALCULATOR MEMORIES USED

Label	Memory	Variable
A	10	Δt
	15	$\Delta t/2$
B	11	x_o
C	12	v_o
D	13	F_o
E	14	m
	20	t_n
	21	x_n
	22	$v_{n-0.5}$

B. USER INSTRUCTIONS

Turn the calculator OFF, then ON, then enter program keystrokes carefully.

Enter the subroutine that computes value of force.

In constructing subroutine use:

[RCL] 20 for t value
[RCL] 21 for x value
[RCL] 22 for v value

Value of F should be in display after computation is complete.

Last keystrokes of subroutine should be: [INV] [SBR] [LRN]

Enter initial conditions:		*PRESS*	*DISPLAY/COMMENTS*
Enter	Δt	[A]	Δt
	x_o	[B]	x_o
	v_o	[C]	v_o
	F_o	[D]	F_o
	m	[E]	m
To begin computation		[2nd] [A']	$t_{n-0.5}$
		[R/S]	$v_{n-0.5}$
		[R/S]	t_n
		[R/S]	x_n
		[R/S]	results displayed sequentially

Additional Program Note

It is difficult to predict the accuracy of the improved Euler Method, but it is almost always true that as Δt is made smaller the accuracy improves. In cases where you are unsure of the accuracy of your results, try decreasing Δt and compare results. If the program seems to be taking too long to run, try doubling Δt and compare results. Rapidly varying results with great curvature will require a smaller Δt for good results.

5.7b THE PROGRAM IN BASIC

A. PROGRAM STATEMENTS	B. EXPLANATION
`100 T=0`	Set time to 0
`110 INPUT "Time interval";DT`	Enter time interval
`120 DT.5–DT/2`	Compute half time interval
`130 INPUT "Initial position";X`	Enter initial position
`140 INPUT "Initial velocity";V`	Enter initial velocity
`150 INPUT "Initial force value";F`	Enter initial force value
`160 INPUT "Mass";M`	Enter mass
`200 V.5=((F/M)*DT.5)+V`	Compute mid-interval velocity
`210 X1=(V.5*DT)+X`	Compute new position
`300 T=T+DT`	Increment time
`310 PRINT`	Display blank line
`320 PRINT "Mid-interval time =";T-DT.5`	Display time at mid-interval
`330 PRINT "Mid-interval velocity =";V.5`	Display velocity at mid-interval
`340 PRINT "Time =";T`	Display time
`350 PRINT "Position =";X1`	Display new position
`360 IF INKEY$="" GOTO 360`	Wait for keypress
`400 GOSUB 1000`	Execute subroutine
`500 V.5=((F1/M)*DT)+V.5`	Compute mid-interval velocity
`510 X1=(V.5*DT)+X1`	Compute new position
`520 GOTO 300`	Repeat computation

Notes: If you have access to a printer, you can substitute LPRINT instructions for PRINT instructions.

If only velocity or position data is required, unwanted PRINT or LPRINT statements can be deleted; or, they can be changed into REM (remark) statements so that they will not be executed.

C. VARIABLES USED

Variable	Value
DT	Time interval: "delta-T"
DT.5	Half of time interval
F	Initial force
F1	Force computed in subroutine
M	Mass
T	Time
V	Initial velocity
V.5	Velocity at mid-interval
X	Initial position
X1	New position

D. USER INSTRUCTIONS

Enter the program carefully. Note that the program is designed to run only when you enter a subroutine (beginning with line 1000) that computes the value of the force (F1).

Enter the subroutine that computes the value of the force. The first line of the subroutine must be line 1000; the subroutine should end with a RETURN statement in line 1999. Before entering the subroutine, you can erase a previously entered subroutine with the following command:

```
DELETE 1000-1999
```

Use the following variables in the subroutine:

Variable	Value
F1	Force computed in subroutine
T	Time
V.5	Velocity at mid-interval
X1	New position

Run the program and respond to these prompts:

Prompt	Action
Time interval?	Enter the time interval.
Initial position?	Enter the initial position.
Initial velocity?	Enter the initial velocity.
Initial force value?	Enter the initial force value.
Mass?	Enter the mass.

For each time increment, the program displays the mid-interval time and velocity and the new time and position.

Press any key to display the next set of values.

When finished running the program for one set of values, enter a BREAK command. (On most computers this is accomplished by holding down a "shift" or "control" key, and pressing the "Break" or "Pause" key on the keyboard.) To run the program for a new set of values, type in the command RUN, and then press the "ENTER" or "RETURN" key.

Experiment 5-5 THE "SKYDIVER" PROBLEM - FREE FALL WITH VISCOUS DAMPING

I. INTRODUCTION

In this experiment we will consider the case of a body falling freely through a viscous medium – much as is the case when a skydiver takes to the air. Two forces work on such an object – gravity pulling straight down, and viscous damping retarding the fall upwards, as shown in Fig. 5-18.

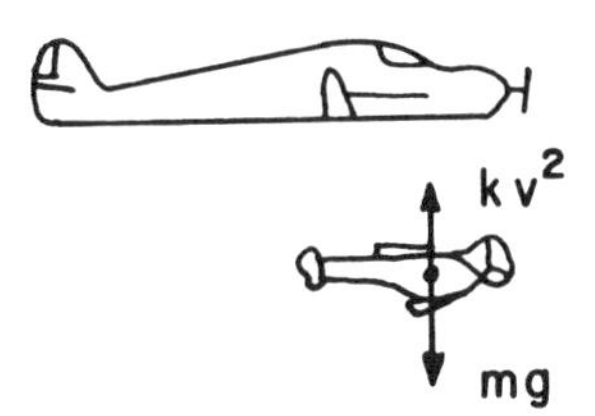

Fig. 5-18. Skydiver Problem

II. THEORY REVIEW

Most typically this damping force can be approximated as being proportional to the skydiver's speed squared. The total force acting on the object can then be written:

$$F = mg - kv^2 \qquad 5\text{-}21$$

In this case k is a constant that describes the damping force. As the object begins falling, assuming zero initial velocity, it begins accelerating rapidly downward. The upward viscous damping force then begins acting to slow the downward acceleration, until finally the body reaches a constant "terminal velocity," at which point the net force acting on it is zero. At that point:

$$mg = kv^2_{term}, \text{ so we can compute}$$

$$v_{term} = \sqrt{\frac{mg}{k}} \qquad 5\text{-}22$$

The exact solution of this motion in practice is quite difficult – usually a general quantitative explanation is all we get to see. The application of Euler's method to this motion, however, lets us study the behavior of this system in detail – and quite easily. We can watch the object approach equilibrium, see the effects of varying the value of k, or see the effect of assuming different "models" for the damping force.

IIIa. THE PROGRAM FOR THE TI-58 or TI-59

We can use the improved Euler method to study this motion by simply constructing a subroutine for the program previously discussed in Section 5-7, which solves for the total force on the skydiver, and inserting it as subroutine [E'] in the program (as specified in the user instructions). The complete process is outlined here:

> Turn calculator OFF, then ON, then press [LRN].
> Enter the improved Euler program described in Section 5-7.
> Next, enter the following subroutine for $F = mg - kv^2$.

PURPOSE/FUNCTION	*SUBROUTINE KEYSTROKES*	*STEP*
Recall $v_{n-0.5}$, square it	[RCL] 22 [X^2]	116
Multiply by k (to be stored in memory 30)	[X] [RCL] 30 [=]	120
Change sign and add to mg	[+/-] [+] [RCL] 14	124
(g to be stored in 31)	[X] [RCL] 31 [=]	129
$F = mg - kv^2$ now in display. End subroutine.	[INV] [SBR]	130

Finally, press [LRN] and [RST]

To run the program and actually "watch" the progress of the skydiver as time goes on – you simply enter all of the appropriate initial conditions and press [2nd] [A'], as described earlier.

Program Test Data

Once your program and subroutine is entered as shown above, try the following example to be certain it's working correctly:

> A skydiver leaves a plane in vertical freefall from v = 0, with damping constant k = 0.005. Use Δt = 0.5 seconds, m = 80 kg, $F_o = mg$ = 80 x 9.8 = 784 N.

Enter Δt	0.5	[A]	0.5
x_o	0.	[B]	0.
v_o	0.	[C]	0.
F_o	784.	[D]	784.
m	80.	[E]	80.
k	0.005	[STO] 30	0.005
g	9.8	[STO] 31	9.8

Start computation	[CLR] [2nd] [A']	
	t/2	0.25
	v at t/2	2.45
	t	0.5
	x at Δt	1.225
	3Δt/2	0.75
	v at 3Δt/2	7.349812422
	2Δt	1.0
	x at 2Δt	4.899906211
	$t_{n-0.5}$	1.25
	v at $t_{n-0.5}$	12.2481243
	t_n	1.5
	v at t_n	11.02396836
	.	.
	.	.
	.	.
	.	.

IIIb. THE PROGRAM IN BASIC

A. PROGRAM STATEMENTS / B. EXPLANATION

A. PROGRAM STATEMENTS	B. EXPLANATION
1000 G=9.8 (Note: or other value – e.g. 32.2 ft/sec)	Set gravity constant
1010 IF T<>DT GOTO 1030	First time? (Time - Interval)
1020 INPUT "Damping constant";K	Yes -- Enter damping constant
1030 F1=(M*G)-(K*(V.5^2)	Compute force
1999 RETURN	Return to main program

C. VARIABLES USED

Variable	Value
DT	Time interval: "delta-T"
F1	Force computed in subroutine
G	Gravity (9.8 meters per second squared; or 32.2 ft/sec^2)
K	Damping constant
M	Mass
T	Time
V.5	Velocity at mid-interval

D. USER INSTRUCTIONS

Enter the Improved Euler Method program and this subroutine carefully, and run the program. Respond to the prompts generated by the program, as explained previously; the program displays the mid-interval time and velocity and the new time and position for the first interval. Press any key to continue.

Respond to the prompt generated by the subroutine:

Prompt	Action
Damping constant?	Enter the damping constant.

For each time increment, the program displays the mid-interval time and velocity and the new time and position.

Press any key to display the next set of values.

E. PROGRAM CHECK DATA

A skydiver leaves a plane in vertical freefall from v = 0, with damping constant k = 0.005. Use Δt = 0.5 seconds, m = 80 Kg, F_o = mg = 80 x 9.8 = 784 N.

```
Time interval? .5
Initial position? 0
Initial velocity? 0
Initial force value? 784
Mass? 80

Mid-interval time = .25
Mid-interval velocity = 2.45
Time = .5
Position = 1.225
Damping constant? .005

Mid-interval time = .75
Mid-interval velocity = 7.349813
Time = 1
Position = 4.899906

Mid-interval time = 1.25
Mid-interval velocity = 12.24812
Time = 1.5
Position = 11.02397
```

When finished running the program for one set of values, enter a BREAK command. (On most computers this is accomplished by holding down a "shift" or "control" key, and pressing the "Break" or "Pause" key on the keyboard.) To run the program for a new set of values, type in the command RUN, and then press the "ENTER" or "RETURN" key.

IV. EXPLORATIONS/NOTES

1. a. Using either a calculator or BASIC program as described in the preceding pages, plot the distance fallen versus time, and velocity versus time for a 1-Kg body dropped from rest into free fall in a vacuum; i.e. $k = 0$. Use $\Delta t = 0.5$ seconds. (Plot values for $t = 0$ through 15 seconds.)

   b. On the same axes, plot the distance fallen vs. time and velocity vs. time of the same object dropped under identical conditions into a medium with a damping constant $k = 0.005$.

   c. Compute the terminal velocity of the object considered in Part (b) above. Is this consistent with your graphic results? Plot this value on the graph for Part (b) above.

2. a. The terminal velocity of an 170-lb. skydiver is clocked at 210 mph (on a certain day, for a certain body attitude). Compute k for this skydiver and situation.

   b. Plot distance and velocity for this skydiver for $t = 0$ to $t = 30$ seconds. Use $\Delta t = 2$ sec. Plot the terminal velocity on an appropriate graph. (Note: In your program be sure the correct value of the acceleration due to gravity is used.)

   c. On the same axes plot the distance and velocity behavior for skydivers weighing 100 and 200 lbs. Compute and plot terminal velocities for each skydiver on an appropriate graph.

3. a. Plot the distance versus time and velocity versus time behavior as above for an object in freefall with the following conditions:

   $m = 10$ Kg
   $k = 0.005$
   $x_0 = v_0 = 0$
   (Use $\Delta t = 2$ sec, plot from $t = 0$ to $t = 30$ seconds)

   b. Modify the subroutine in the improved Euler program used above to change the damping force from:

   $F \text{ damping} = -kv^2$
   to
   $F \text{ damping} = -kv$ (i.e., a linear damping force)

   Plot the distance versus time and velocity versus time behavior on the same axes as above.

   c. Modify the subroutine once again – this time to a cubic damping force:

   $F \text{ damping} = kv^3$

   Once again plot the distance versus time and velocity versus time behavior on the same axes. (Note: In this case use $\Delta t = 0.5$ seconds.)

   d. Which object reaches terminal velocity first? Which attains the highest velocity?

Experiment 5-6 SIMPLE HARMONIC OSCILLATIONS EULER METHOD

In this experiment (and the next) we will be using the Euler Method to investigate systems that oscillate. We have already examined the behavior of a simple harmonic oscillator where we assumed the solution to the motion. In the four cases that follow, the Euler Method will be used to approximate the motion directly – with no "assumptions." This allows the calculator or computer to actually model or simulate the behavior of a system whose actual motion may be quite complex. The cases we will consider below are among the most important encountered in physics, namely:

a. Undamped, Undriven Oscillation
b. Damped, Undriven Oscillation
c. Damped, Driven Oscillation
d. Undamped, Driven Oscillation

Case a - Undamped, Undriven Oscillation

I. INTRODUCTION

In the first case, we will consider the situation as diagrammed in Fig. 5-19.

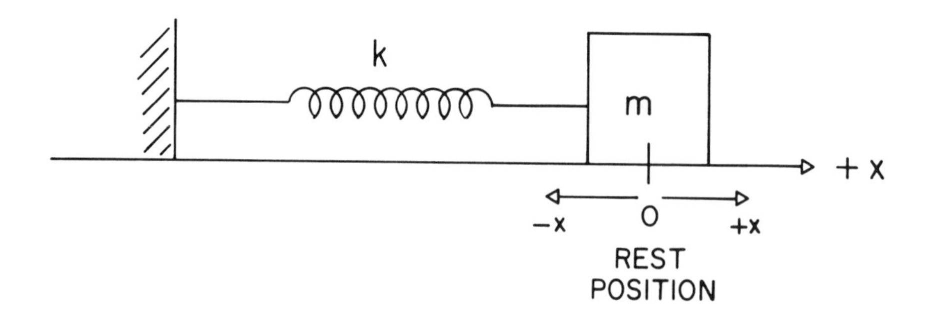

Fig. 5-19. Undamped, Undriven Oscillation.

II. THEORY REVIEW

This is a system we considered using a different approach earlier in the chapter. The mass m is resting on a frictionless surface, subject only to forces exerted by the spring. If the mass is moved to the right a certain distance and released, it will undergo periodic oscillations about its rest position which we call Harmonic Oscillations. When this motion is usually studied in basic physics, the solution is "assumed," and has the form:

$$X = A \cos (\omega t + \delta)$$

where $\omega = \sqrt{k/m}$ is the "natural frequency" of the system, A is the amplitude (maximum displacement from rest position) and δ is a phase constant (which describes where the motion started).

In this experiment, we will use our program for the improved Euler method to compute the motion behavior of this system directly – actually solving the equations of Newton's second law step by step as the motion unfolds. In this case the only force acting on the system is the spring force itself:

$$F = -kx$$

IIIa. THE PROGRAM FOR THE TI-58 or TI-59

This experiment can be easily run using the Improved Euler Method Program described in Section 5-2. Simply follow these steps:

i. Enter the program for the Improved Euler Method (you can either key it in directly, or enter it from a magnetic card if it was stored previously.) Note that for the experiments that follow, only position vs. time data will be needed, so follow the instructions for modifying the program to display these results only. (This will speed things considerably.) Simply insert [2nd] [NOP] instructions for asterisked [R/S]* instructions that are needed.

ii. Next, enter Subroutine 1 for the force (F = -kx) as shown below. When constructing the program, use [RCL] 21 for the x value, and store k in memory 30. To begin, first be sure your out of learn mode, then enter the subroutine as follows:

Subroutine 1: Undamped, Undriven Harmonic Oscillation

Press	Display
[GTO] [2nd] [E'] [LRN]	113
[RCL] 30 [+/-] [X]	117
[RCL] 21 [=]	120
[INV] [SBR]	121
[LRN]	0

USER INSTRUCTIONS/PROGRAM TEST DATA

To try the program, plot the oscillations for a simple (undamped) spring-mass system such as the one described in Fig. 5-20. Follow the steps shown.

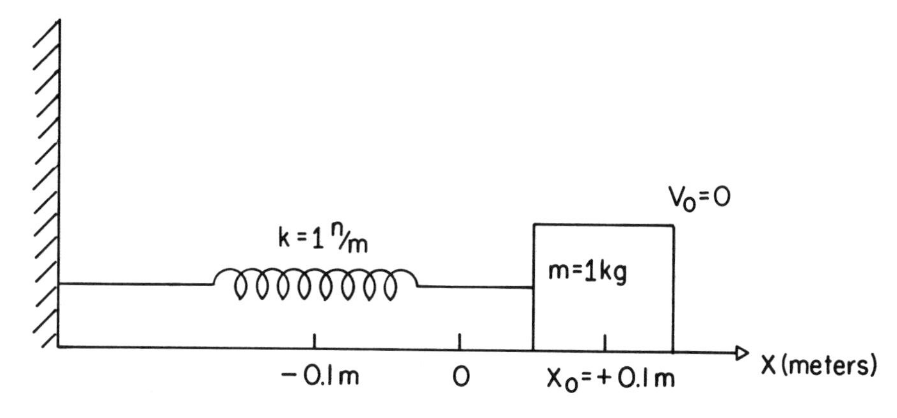

Fig. 5-20

OPERATION		*PRESS*	*DISPLAY/COMMENTS*
Enter:	k = 1 nt/m	[STO] 30	1.
	Δt = 0.1 sec	[A]	0.1
	x_o = 0.1 m	[B]	0.1
	v_o = 0	[C]	0.
	$F_o = -kx_o$ = -0.1 nt	[D]	-0.1
	m = 1 Kg	[E]	1.

To begin running the program:

Press: [2nd] [FIX] 4
[2nd] [A']

The calculator should produce the following motion:

t (sec)	x (m)
0.1000	0.0995
0.2000	0.0980
0.3000	0.0955
0.4000	0.0921
0.5000	0.0877
.	.
.	.
.	.

Press [R/S] to stop the calculation when desired.

Note: To use the program to plot a new motion, first press: [2nd] [CMs] [RST] and [CLR]; then enter all of the parameters required for the new run.

IIIb. THE SUBROUTINE IN BASIC

A. PROGRAM STATEMENTS

```
1000 IF T<>DT GOTO 1020
1010 INPUT "Force Constant";K
1020 F1=-K*X1
1999 RETURN
```

B. EXPLANATION

First time? (Time = Interval)
  Yes -- Enter force constant, k
Compute force
Return to main program

C. VARIABLES USED

Variable	Value
DT	Time interval: "delta-T"
F1	Force computed in subroutine
K	Force constant
T	Time
X1	New position

D. USER INSTRUCTIONS

Enter the Improved Euler Method Program and this subroutine carefully. As this experiment considers only position and time, you can change lines 320 and 330 to REM statements, as follows:

```
320 REM PRINT "Mid-interval time =";T-DT.5
330 REM PRINT "Mid-interval velocity =";V.5
```

(Simply enter these lines exactly as above, and the program will be changed.)

Run the program. Respond to the prompts generated by the program, as explained previously; the program displays the new time and position for the first interval. Press any key to continue.

Respond to the prompt generated by the subroutine:

Prompt	Action
Constant?	Enter the force constant (k).

For each time increment, the program displays the new time and position.

Press any key to display the next set of values.

E. PROGRAM CHECK DATA

Plot the oscillations for the simple, undamped spring-mass system described in Fig. 5-20.

```
Time interval? .1
Initial position? .1
Initial velocity? 0
Initial force value? -.1
Mass? 1

Time = .1
Position = .0995
Constant? 1

Time = .2
Position = 9.800499E-02

Time = .3
Position = 9.552994E-02

Time = .4
Position = 9.209959E-02

Time = .5
Position = 8.774824E-02
```

When finished running the program for one set of values, enter a BREAK command. (On most computers this is accomplished by holding down a "shift" or "control" key, and pressing the "Break" or "Pause" key on the keyboard.) To run the program for a new set of values, type in the command RUN, and then press the "ENTER" or "RETURN" key.

IV. EXPLORATIONS/NOTES (Case a - Undamped, Undriven Oscillation)

1. For the simple harmonic oscillator described above, compute the period of the motion. Plot the motion of this oscillator for 1 full period of the motion – assume the mass m is initially moved to the position $x_0 = 0.1$ m, then released.

2. Consider now the situation where the same oscillator is started from the equilibrium position ($x_0 = 0$), and thrust to the right (+x direction) with an initial velocity v_0. For this oscillator to achieve the same amplitude as in Part 1, what would v_0 have to be?

3. Plot the motion described in Part 2 on the same set of axes used in Part 1. Are the periods of both motions the same? Why?

4. Plot the motion for the same oscillator, released at rest from $x_0 = 0.1$ m, except with a "weaker" spring: $k = 0.5$ nt/m. What is the period for this motion? Is the total energy of this system the same as in the oscillator considered in Exercises 2 and 3? Why?

Case b - Damped, Undriven Oscillation

I. INTRODUCTION

In this case, we will add an additional element to our spring-mass system, as shown in Fig. 5-21.

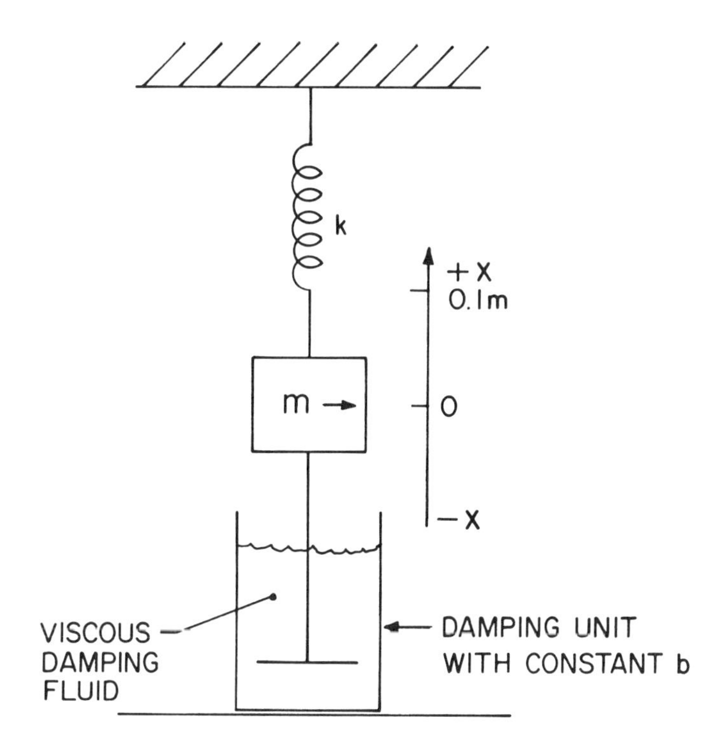

Fig. 5-21. Spring-Mass System with Damping.

The new element we've added is a damping unit, which operates much like the shock absorber found in auto suspensions. This type of device acts to impede the motion of the system, and dissipates the system energy as heat in the damping fluid.

II. THEORY REVIEW

The damping force created by a system such as the damper shown on Page 177 is opposite to the direction of the system's velocity, and to a pretty good approximation is directly proportional to it. We can describe the strength of the damping force with a constant b, and express the damping force as follows:

$$F_d = -bv$$

Note that if the damping unit contains a relatively light fluid like water, b will be lower. If the unit is filled with a heavy, viscous (oily) fluid like gear lubricant, b will be larger.

In the case we are considering, we will assume that the damped spring-mass system is initially hanging at rest, with the pointer on the mass block pointing to 0 on the scale. (In this position, the spring has already stretched just far enough to counter the force of gravity downward. For the rest of the experiment, we will neglect gravitational effects.)

If we take the system mass and raise it above the equilibrium position, then release it with zero initial velocity, the following equation describes the total force on the system:

$$F_t = -kx-bv$$

The development of an actual solution which describes this motion is in practice very difficult. Once again with our Euler method approach, getting an approximate picture which comes very close to the exact motion is "push-button easy."

IIIa. THE PROGRAM FOR THE TI-58 or TI-59

To perform this experiment, our earlier Euler Method Program is modified to include the damping force. The steps to follow are these:

i. First, enter the Euler Method Program discussed in Section 5-7 (either key it in directly, read it in from a magnetic card, or it may already be in place from the previous part of this experiment.) Be sure you're out of learn mode after the program is entered.

ii. Next enter the force subroutine below. We'll store the value of k in memory 30, and damping constant b in memory 31. In this case [RCL] 21 is used for the x value, [RCL] 22 is used for the value of v.

Subroutine 2: Damped, Undriven Harmonic Oscillation

Press	Display
[GTO] [2nd] [E'] [LRN]	113
[RCL] 30 [+/-] [X]	117
[RCL] 21	119
[-] [RCL] 31 [X]	123
[RCL] 22 [=]	126

[INV] [SBR]	127
[LRN]	0.

USER INSTRUCTIONS/PROGRAM TEST DATA

First, consider a system with the following parameters:

$$m = 1 \text{ Kg}$$
$$k = 1 \text{ nt/m}$$
$$b = 0.5 \text{ nt/m/sec}$$

Let's assume the mass block for this system is lifted to a distance $x_o = +0.20$ m from equilibrium, and released with zero initial velocity. To use the program to plot the motion, follow the user instructions below:

OPERATION	*PRESS*	*DISPLAY/COMMENTS*
Enter: $k = 1$ nt/m	[STO] 30	1.
$b = 0.5$ nt/m/sec	[STO] 31	0.5
$\Delta t = 0.2$ sec	[A]	0.2
$x_o = 0.20$ m	[B]	0.2
$v_o = 0$	[C]	0.
$F_o = -kx_o = -0.20$ nt	[D]	-0.2
$m = 1$ Kg	[E]	1.

To begin running the program:

Press; [2nd] [FIX] 4
[2nd] [A']

The calculator should then produce the following motion:

t (sec)	x (m)
0.2000	0.1960
0.4000	0.1846
0.6000	0.1669
0.8000	0.1443
1.0000	0.1182
.	.
.	.
.	.

Press [R/S] to stop the computation when desired.

Note: To run this program for a new set of values, press [2nd] [CMs], [RST], and [CLR]; enter the new set of values for the new run, then proceed.

IIIb. THE SUBROUTINE IN BASIC

A. PROGRAM STATEMENTS

```
1000 IF T<>DT GOTO 1030
1010 INPUT "Force Constant";K
1020 INPUT "Damping constant";B
1030 F1=(-K*X1)-(B*V.5)
1999 RETURN
```

B. EXPLANATION

First time? (Time = Interval)
  Yes -- Enter force constant
         Enter damping constant
Compute force
Return to main program

C. VARIABLES USED

Variable	Value
B	Damping constant
DT	Time interval: "delta-T"
F1	Force computed in subroutine
K	Constant
T	Time
V.5	Velocity at mid-interval
X1	New position

D. USER INSTRUCTIONS

Enter the Improved Euler Method Program and this subroutine carefully. As this experiment considers only position and time, you can change lines 320 and 330 to REM statements, as follows:

```
320 REM PRINT "Mid-interval time =";T-DT.5
330 REM PRINT "Mid-interval velocity =";V.5
```

Run the program. Respond to the prompts generated by the program, as explained previously; the program displays the new time and position for the first interval. Press any key to continue.

Respond to the prompts generated by the subroutine:

Prompt	Action
Constant?	Enter the force constant (k).
Damping constant?	Enter the damping constant (b).

For each time increment, the program displays the new time and position.

Press any key to display the next set of values.

E. PROGRAM CHECK DATA

Use the program to compute the motion for a system such as that shown in Fig. 5-21, with the following characteristics and initial values:

$m = 1$ Kg
$k = 1$ N/m
$b = 0.5$ N/m/sec
$x_o = +0.20$ m
$v_o = 0$
$F_o = -kx_o = -0.2$ N

```
Time interval? .2
Initial position? .2
Initial velocity? 0
Initial force value? -.2
Mass? 1

Time = .2
Position = .196
Force constant? 1
Damping constant? .5

Time = .4
Position = .18456

Time = .6
Position = .1668816

Time = .8
Position = .1442958

Time = 1
Position = .1181967
```

When finished running the program for one set of values, enter a BREAK command. (On most computers this is accomplished by holding down a "shift" or "control" key, and pressing the "Break" or "Pause" key on the keyboard.) To run the program for a new set of values, type in the command RUN, and then press the "ENTER" or "RETURN" key.

IV. EXPLORATIONS/NOTES (Case b - Damped, Undriven Oscillation)

1. Plot the motion of the oscillator above from t = 0 to t = 12 seconds. How does the period of this motion compare to that of an undamped oscillator with the same values of k and m?

2. Now, let's assume we drain the damping unit and fill it with a more viscous fluid, such that b = 3.0. Plot this motion on the same set of axes. Describe how this motion is different from the motion plotted in Exercise 1 above.

3. Run the program with increasing values of b, starting with b = 1. Don't plot these motions, but watch the motion in each case. Look for the case where the motion just barely makes a small single swing below the x axis. This is called the critically damped case. Plot this motion on the same set of axes as above. What value of b would you estimate to yield critical damping?

Case c - Damped, Driven Oscillation

I. INTRODUCTION

Next, we consider the motion of a damped oscillating system being driven with a periodic disturbance, such as that in Fig. 5-22.

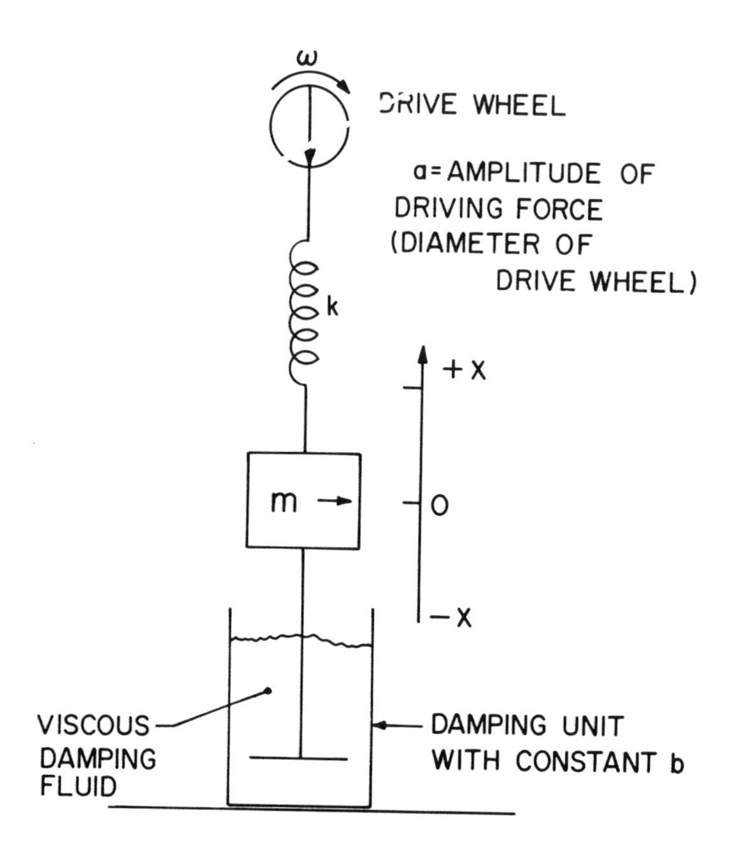

Fig. 5-22. Driven Spring-Mass System with Damping.

II. THEORY REVIEW

Here, a drive wheel attached to the spring-mass system imparts a periodic, sinusoidal disturbance to the system of amplitude a. The rotational frequency of the wheel in radians per second is ω'. This new driving force, the spring force, and the damping force will now all act on the system simultaneously (we'll once again neglect the effects of gravity on this system). The total force on the system is given by:

$$F = -kx - bv + a \sin \omega' t \qquad 5\text{-}23$$

IIIa. THE PROGRAM FOR THE TI-58 or TI-59

Once again we use the Improved Euler Program described in Section 5-7 to allow us to get a look at what the motion of this system will be like, without actually attempting to solve the formidable differential equations that describe it. The steps to follow:

i. Enter the Euler Method Program from Section 5-7 (by keying it in, from a magnetic card, or by previous entry from an earlier experiment.) Be certain you're out of learn mode after the program is entered.

ii. Next, enter the force subroutine as described below. The value of k will be stored in memory 30, the damping constant b in memory 31, the amplitude of the driving force a in memory 32, and the frequency of the driving force ω' in memory 33.

In this case [RCL] 21 is used for the x value, [RCL] 22 for the v value, and [RCL] 20 for t.

Subroutine 3 : Damped, Driven Harmonic Oscillator

Press	Display
[GTO] [2nd] [E'] [LRN]	113
[RCL] 30 [+/-] [X]	117
[RCL] 21	119
[-] [RCL] 31 [X]	123
[RCL] 22 [=] [+]	127
[RCL] 32 [X] [( ]	131
[( ] [RCL] 33 [X]	135
[RCL] 20 [ )]	138
[2nd] [SIN] [ )] [=]	141
[INV] [SBR]	142
[LRN]	

USER INSTRUCTIONS/PROGRAM TEST DATA

For our basic system, in this case we will use the parameters:

m = 2 grams
k = 5 dyne/cm
b = 0.5 dyne/cm/sec

We will assume the system is driven at several frequencies as this experiment proceeds. We will assume that the driving force has an amplitude of 1 cm; and that the system is started at its equilibrium position ($x_0 = 0$), at rest ($v_0 = 0$), such that the initial force acting on the system is zero ($F_0 = 0$).

The resonant frequency for a damped system such as this can be computed, as follows: (Resnick and Halliday, Physics, Part I, Chapter 15)

$$\omega_{res} = \sqrt{\frac{k}{m} - \left(\frac{b}{2m}\right)^2} \quad ; \tag{5-24}$$

$$\omega_{res} = \sqrt{\frac{5}{2} - \left(\frac{0.5}{4}\right)^2} = \sqrt{2.5 - 0.015625} \text{ and}$$

$$\omega_{res} = 1.58 \text{ rad/sec}$$

We will first consider the case when the system is driven at its resonant frequency. To use the program to plot this motion, follow these steps:

OPERATION	*PRESS*	*DISPLAY*
Enter k = 5 dyne/cm	[STO] 30	5.
b = 0.5 cm/m/sec	[STO] 31	0.5
amplitude a = 1 cm	[STO] 32	1.
frequency ω = 1.58 rad/sec	[STO] 33	1.58
Δt = 0.2 sec	[A]	0.2
x_o = 0	[B]	0.
v_o = 0	[C]	0.
F_o = 0	[D]	0.
m = 2g		0.

To begin running the program:

First, press [2nd] [FIX] 4, and then
[**Important: Put the calculator in radians mode by pressing [2nd] [Rad].]

Press [2nd] [A'] to start the program.

In this case, the program will produce the following motion:

t (sec)	x (cm)
0.2000	0.0000
0.4000	0.0062
0.6000	0.0233
0.8000	0.0535
1.0000	0.0958
.	.
.	.
.	.
.	.
10.000	0.8894
.	.

IV. EXPLORATIONS/NOTES (Case c - Damped, Driven Oscillation)

1. a. Plot the motion of the system described above from 0 to 15 seconds.

   b. Describe the behavior of the amplitude of this motion.

   c. Examine also the period of the motion - when does the period of the motion become equal to that of the driving force?

2. Next, run the program for the oscillator when being driven at a frequency below its resonant frequency - $\omega' = 0.50$ radians/sec. For this run, use $\Delta t = 0.5$ sec.

   a. Plot the motion from $t = 0$ to $t = 30$ seconds.

   b. Describe the amplitude and period behavior of this oscillator.

   c. Theoretical calculations (see Resnick and Halliday, Chapter 15) indicate that the low frequency limit of the oscillator's amplitude is F_{max}/k. Does your result indicate this?

3. Next, run the program for the same oscillator when being driven at a frequency above its resonant frequency - $\omega' = 2.5$ radians per second. Use $\Delta t = 0.2$ sec.

   a. Plot the motion from $t = 0$ to $t = 14$ seconds.

   b. Describe the amplitude behavior of this motion, and

   c. the frequency behavior.

   d. Does the motion ever reach a "steady" state during the 14 seconds?

Case d - Undamped, Driven Oscillation

I. INTRODUCTION

In this final case we will consider a driven oscillator with no damping. The physical situation is the same as that illustrated in Fig. 5-22, with the damping unit removed (or damping constant "b" reduced to zero.) The case of undamped driven oscillation can give rist to "extreme" oscillatory phenomena – which can be of critical importance in the study of mechanical and electrical systems.

In this case when the oscillator is driven above its resonant frequency, and when driven below its resonant frequency, its motion will be similar to the case considered in Section c of this experiment. The final equilibrium amplitude achieved in these two cases will be substantially larger than in the damped case, but after several oscillations a steady state amplitude is approached.

II. THEORY REVIEW

The case of interest is where an undamped oscillator is driven at its resonant frequency. To see this behavior in action, we'll use the previous program to simulate the motion of the system being driven at its resonant frequency (ω' = 1.58 radians/sec). In this case, however, we'll set the damping constant b equal to zero.

IIIa/b. THE PROGRAM

The programs in this case (both for the TI-58/TI-59 or in BASIC) are the same as the programs used in Case c; this time run with the following values:

k = 5 dyne/cm
b = 0 (no damping)
a = 1 cm
ω' = 1.58 rad/sec

$x_o = v_o = F_o = 0$
m = 2 g
Δt = 0.2 sec

IV. EXPLORATIONS/NOTES (Case d - Undamped, Driven Oscillation)

1. a. Plot this motion from t = 0 to t = 15 seconds, examine (but do not plot) its motion after that to t = 60 seconds.

   b. Discuss the frequency behavior of this motion.

   c. Examine the amplitude behavior of this system. Does this motion appear to be reaching a steady state? What would you expect to happen to this oscillator for very large values of t?

Experiment 5-7 ORBIT AND CENTRAL FORCE MOTION

I. INTRODUCTION

In this situation we will use a modification of our Euler Method Program to study an important class of phenomena: motion under the influence of a central force. The physical situations we will be examining are basic to the study of astronomy and celestial mechanics, as well as to the study of atomic and nuclear phenomena. In astronomy and celestial mechanics, gravity is the dominant force; and as we know it gives rise to orbital systems and complex motional effects throughout the universe. (What we do not know much about is the nature of gravity itself, which remains a well-guarded secret of nature.) In atomic and nuclear systems, other sorts of force come into play – the electric field force and others which are also quite mysterious.

The specific case we will focus on here is the inverse square law force, which will allow us to generalize our result to both gravitational and electric effects. Our program will be constructed to allow us to examine a wide variety of phenomena with a minimum of modification.

II. THEORY REVIEW

To begin examining the theory involved in this experiment, let's focus for a moment on a gravitational system such as the one shown in Fig. 5-23.

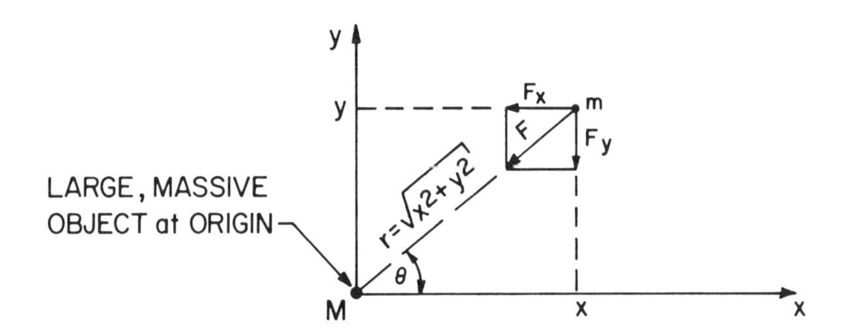

Fig. 5-23. Central Force Motion Situation.

In this case, a small object m is considered to be moving in the field of a much more massive object M, which is at the origin of our coordinate system. M is assumed to remain at rest as the motion proceeds – which is a valid assumption if $M >> m$.

F is the total force exerted by M on m. It acts along a line joining the centers of the two masses, and is attractive – constantly attempting to move the two bodies together. To begin analyzing the situation, we will resolve the total force F into its two components along the x and y axes:

$$F_x = -F \cos \theta$$
$$F_y = -F \sin \theta, \text{ where}$$

$$\cos \theta = \frac{x}{\sqrt{x^2 + y^2}} \quad \text{and} \quad \sin \theta = \frac{y}{\sqrt{x^2 + y^2}} \qquad 5\text{-}27$$

The gravitational force between two masses is in general given by:

$$F = \frac{GMm}{r^2} = \frac{GMm}{(x^2 + y^2)}, \quad \text{where} \qquad 5\text{-}28$$

G is the universal gravitation constant, and r is the distance between M and m.

Our two force components can then be given by:

$$F_x = -F \cos \theta = - \frac{GMm}{(x^2 + y^2)} \frac{x}{\sqrt{x^2 + y^2}} \qquad 5\text{-}29$$

$$F_x = -GMm\, x(x^2 + y^2)^{-3/2}, \quad \text{and} \qquad 5\text{-}30$$

$$F_y = -F \sin \theta = - \frac{GMm}{(x^2 + y^2)} \frac{y}{\sqrt{x^2 + y^2}} \qquad 5\text{-}31$$

$$F_y = -GMm\, y(x^2 + y^2)^{-3/2} \qquad 5\text{-}32$$

Notice carefully that in this case both the x and y components of force acting on mass m are each functions of the coordinates x and y. The acceleration of the mass m is, therefore, also a function of x and y, and its components can be found by dividing the force components by m:

$$a_x(x,y) = \frac{F_x}{m} = -GMx(x^2 + y^2)^{-3/2} \quad \text{and,} \qquad 5\text{-}33$$

$$a_y(x,y) = \frac{F_y}{m} = -GMy(x^2 + y^2)^{-3/2} \qquad 5\text{-}34$$

To keep things general as we develop our program, we introduce the following constants:

$$C = -GM$$

$$P = -3/2 = -1.5$$

$$a_x(x,y) = Cx(x^2 + y^2)^P \qquad 5\text{-}35$$

$$a_y(x,y) = Cy(x^2 + y^2)^P \qquad 5\text{-}36$$

III. THE PROGRAM

We will use these equations in a program which uses the Improved Euler Method to compute the motion in "bite-size" pieces as we have done before. We will build the program step-by-step, following our analysis of a small portion of the motion shown in steps "a" through "e" in Fig. 5-24.

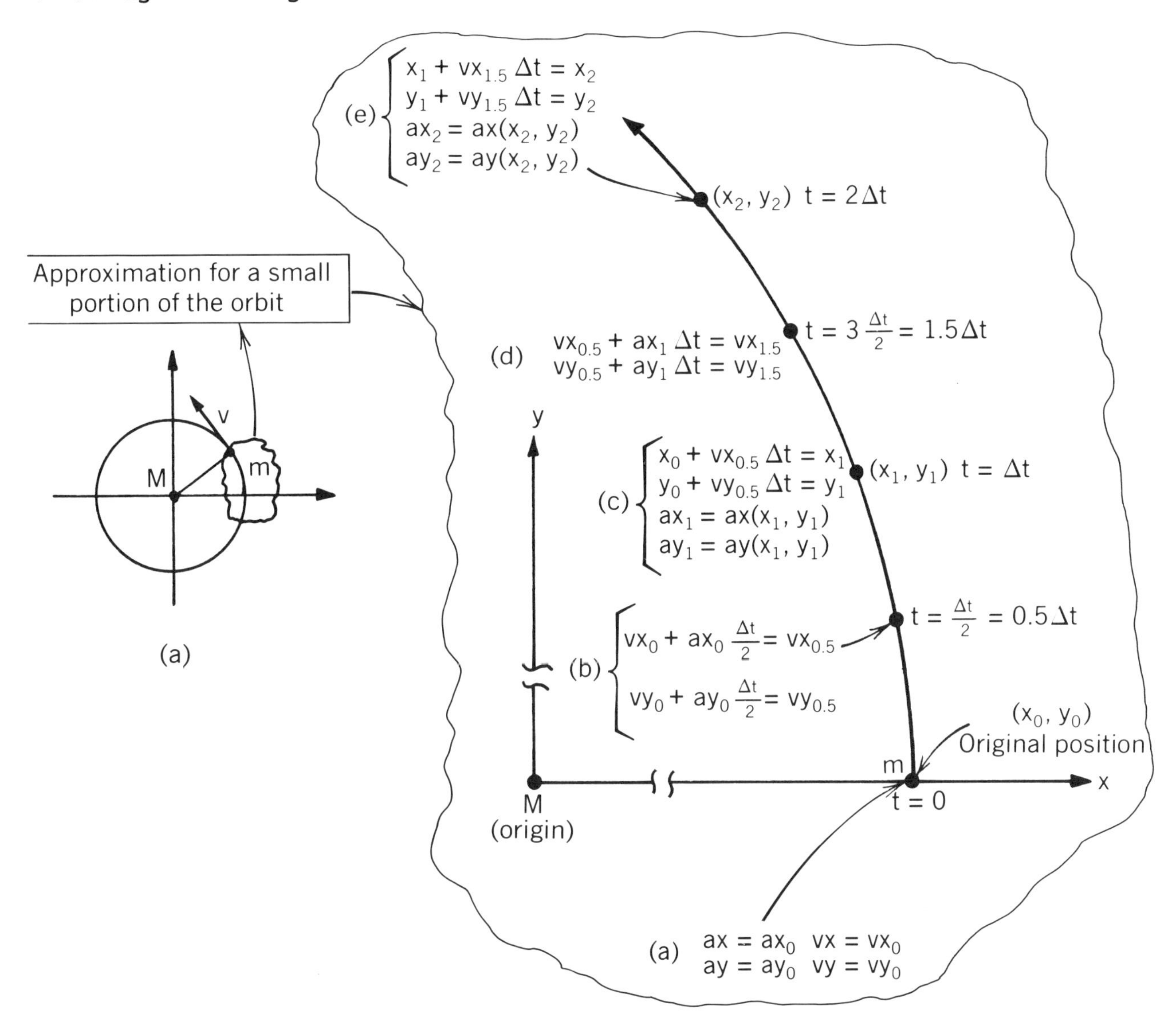

Fig. 5-24. Approximation for a Small Portion of the Orbit.

The motion begins at the original position shown x_0, y_0 at time $t = 0$. (In this case, we've chosen the original position to be on the x axis for convenience. In practice, it could be anywhere.

a) In the original position x_o, y_o at time t_o, the initial velocity components v_{xo} and v_{yo} will be known initial conditions, and the acceleration components a_x and a_y will be computed using our previous formulas:

$$a_{xo} = Cx_o(x_o + y_o)^P \qquad 5\text{-}37$$

$$a_{yo} = Cy_o(x_o + y_o)^P, \quad \text{(where } C = -GM,\ P = -1.5\text{)} \qquad 5\text{-}38$$

b) Next the value of t will be increased to $\Delta t/2$, and new velocity values will be computed from the initial velocity values and the acceleration components just computed:

$$v_{xo} + a_{xo}\,\Delta t/2 = v_{x0.5} \qquad 5\text{-}39$$

$$v_{yo} + a_{yo}\,\Delta t/2 = v_{y0.5} \qquad 5\text{-}40$$

c) Next t is increased to Δt, and the new position coordinates for our mass m are computed from the initial coordinates and the velocity values just completed:

$$x_o + v_{x0.5}\,\Delta t = x_1 \qquad 5\text{-}41$$

$$y_o + v_{y0.5}\,\Delta t = y_1 \qquad 5\text{-}42$$

Using these new coordinates, new acceleration values are computed:

$$a_{x1} = a_x(x_1,y_1) = Cx_1(x_1 + y_1)^P \qquad 5\text{-}43$$

$$a_{y1} = a_y(x_1,y_1) = Cy_1(x_1 + y_1)^P \qquad 5\text{-}44$$

d) Time is advanced to $1.5\Delta t$, and new velocity components are computed from the previous values and the acceleration components just computed:

$$v_{x0.5} = a_{x1}\,\Delta t = v_{x1.5} \qquad 5\text{-}45$$

$$v_{y0.5} + a_{y1}\,\Delta t = v_{y1.5} \qquad 5\text{-}46$$

e) Time advances to $2\Delta t$, and the velocity components just computed are now used (along with the previous position coordinates) to compute the new position coordinates from m (x_2,y_2):

$$x_1 + v_{x1.5}\,\Delta t = x_2 \qquad 5\text{-}47$$

$$y_1 + v_{y1.5}\,\Delta t = y_2 \qquad 5\text{-}48$$

These coordinates are then used to compute the new values for the acceleration components:

$$a_{x2} = a_x(x_2,y_2) = Cx_2(x_2 + y_2)^P \quad 5\text{-}49$$

$$a_{y2} = a_y(x_2,y_2) = Cy_2(x_2 + y_2)^P \quad 5\text{-}50$$

The process then repeats itself as the motion unfolds...

Some Notes on this Approach:

The program we develop will be quite powerful and general, and allow us to simulate central force motion of many sorts. By substituting various values for C and P, different sorts of force laws and phenomena can be studied. We will focus on several cases of interest including:

- bound orbits
- escape trajectories, and
- scattering (by an inverse-square repulsive force)

IIIa. THE PROGRAM FOR THE TI-58 or TI-59

PURPOSE/FUNCTION	*PROGRAM KEYSTROKES*	*STEP #*
Clear calculator completely, get into "LEARN" mode	OFF/ON [LRN]	000
Store Δt in memory 10 and $\Delta t/2$ in memories 11, 31 and 32	[2nd] [Lb1] [A]	002
	[STO] 10 [÷] 2 [=]	007
	[STO] 11 [STO] 31 [STO] 32 [R/S]	014
Store x_o in memory 12, 24	[2nd] [Lb1] [B]	016
	[STO] 12 [STO] 24 [R/S]	021
Store y_o in memory 13, 25	[2nd] [Lb1] [C]	023
	[STO] 13 [STO] 25 [R/S]	028
Store v_{xo} in memory 14, 22	[2nd] [Lb1] [D]	030
	[STO] 14 [STO] 22 [R/S]	035
Store v_{yo} in memory 15, 23	[2nd] [Lb1] [E]	037
	[STO] 15 [STO] 23 [R/S]	042
Store C in memory 16	[2nd] [Lb1] [2nd] [A']	044
	[STO] 16 [R/S]	047

PURPOSE/FUNCTION	*PROGRAM KEYSTROKES*	*STEP #*
Store P in memory 17	[2nd] [Lb1] [2nd] [B']	049
	[STO] 17 [R/S]	052
Start Calculations	[2nd] [Lb1] [2nd] [E']	054
Compute $C(x^2 + y^2)^P$, store in memory 26	[( ] [RCL] 24 [X^2] [+]	059
	[RCL] 25 [X^2] [ )] [Y^x]	064
	[RCL] 17 [X] [RCL] 16 [=]	070
	[STO] 26	072
Compute $a_x = xC(x^2 + y^2)^P$ and store in memory 20	[RCL] 24 [X] [RCL] 26 [=]	078
	[STO] 20	080
Compute $a_y = yC(x^2 + y^2)^P$ and store in memory 21	[RCL] 25 [X] [RCL] 26 [=]	086
	[STO] 21	088
Update t = n-1/2 Δt	[RCL] 11 [SUM] 30 [RCL] 30	094
Compute v_x at (n-1/2)Δt = $v_x + a_x\Delta t/2$, store in memory 22	[RCL] 20 [X] [RCL] 31 [+]	100
	[RCL] 22 [=] [STO] 22	105
Compute v_y at (n-1/2)Δt = $v_y + a_y\Delta t/2$, store in memory 23	[RCL] 21 [X] [RCL] 31 [+]	111
	[RCL] 23 [=] [STO] 23	116
Memory 31 contains Δt/2 for first loop only, after that it contains Δt as required.	[RCL] 32 [SUM] 31 0 [STO] 32	123
Update, then display t = Δt	[RCL] 11 [SUM] 30	127
	[RCL] 30 [R/S]**	130
Compute x at n Δt, store in memory 24, then display it.	[RCL] 24 [+] [RCL] 22 [X]	136
	[RCL] 10 [=] [STO] 24 [R/S]*	142
x (at n Δt) = x (at n-1 Δt) + v_x (at n-1/2 Δt) Δt		
Compute y at n Δt, store in memory 25, then display it.	[RCL] 25 [+] [RCL] 23 [X]	148
	[RCL] 10 [=] [STO] 25 [R/S]*	154
y (at n Δt) = y (at n-1 Δt) + v_y (at n-1/2 Δt) Δt		
Repeat for next Δ t/2 + Δt values	[GTO] [2nd] [E']	156

PURPOSE/FUNCTION	*PROGRAM KEYSTROKES*	*STEP #*
Label "Reinitialize Subroutine"	[2nd] [Lb1] [2nd] [C']	158
Zero all critical memories for new run, reinsert all initial conditions in appropriate locations.	[RCL] 12 [STO] 24	162
	[RCL] 13 [STO] 25	166
	[RCL] 14 [STO] 22	170
	[RCL] 15 [STO] 23	174
	[RCL] 11 [STO] 31 [STO] 32	180
	0 [STO] 30 [STO] 20	185
	[STO] 21 [STO] 26 [R/S]	190
Get out of Learn, and reset fix at 2 places	[LRN] RST] [2nd] [FIX] 3	

*, ** For those with access to a PC-100A printer:
[R/S]* instructions should be replaced with a [2nd] [Prt] sequence.
[R/S]** instructions should be replaced with a [2nd] [Adv] [2nd] [Adv] [2nd] [Prt] sequence.

A. CALCULATOR MEMORIES USED

10	Δt	A
11	$\Delta t/2$	A
12	x_o	B
13	y_o	C
14	v_{xo}	D
15	v_{yo}	E
16	C	A'
17	P	B'
20	a_x	
21	a_y	
22	v_x	
23	v_y	
24	x	
25	y	
26	$C(x^2 + y^2)^P$	
30	t current	
31	$\Delta t/2$ (first loop only, then Δt)	

32	$\Delta t/2$ (first loop only, then zero)

B. USER INSTRUCTIONS

OPERATION	*PRESS*	*DISPLAY/COMMENTS*
Turn the calculator OFF, then ON, then enter program keystrokes carefully.		
Enter Initial Conditions:		
Δt	[A]	$\Delta t/2$
x_o	[B]	x_o
y_o	[C]	y_o
v_{xo}	[D]	v_{xo}
v_{yo}	[E]	v_{yo}
C	[2nd] [A']	C
P	[2nd] [B']	P
Begin Computation	[2nd] [E']	$t = \Delta t$
		x at Δt
		y at Δt
		$t = 2\Delta t$
		x at $2\Delta t$
		y at $2\Delta t$
		.
		.
		.
		$t = n\Delta t$
		x at $n\Delta t$
		y at $n\Delta t$
		.
		.
		.
To Run Program for New Initial Conditions:		
Reloads all original initial conditions, zeros all critical memories	[2nd] [C']	0.

Enter any new initial conditions desired, and press [CLR], and [2nd] [E'] to begin new "RUN".

Program Test Data:

Once the program is entered as shown on the preceding pages, try the following example to be certain it's working correctly. For this example, the initial conditions are set to describe a closed "bound" orbit.

OPERATION	*PRESS*	*DISPLAY/COMMENTS*
Enter $\Delta t = 0.1$	[A]	0.050
$x_o = 1.0$	[B]	1.000
$y_o = 0$	[C]	0.000
$v_{xo} = 0$	[D]	0.000
$v_{yo} = 1.0$	[E]	1.000
C = -1.0	[2nd] [A']	-1.000
P = -1.5	[2nd] [B']	-1.500

To begin the computation: Press [2nd] [E']. If you're using the PC100A printer, the results for t, x and y are printed out consecutively. If not, repeatedly press [R/S] to plot the motion.

The calculator should produce the following motion:

t	x_t	y_t
0.100	0.995	0.100
0.200	0.980	0.199
0.300	0.955	0.296
0.400	0.921	0.390
.	.	.
.	.	.
3.200	-1.004	-0.048
.	.	.
6.300	1.000	-0.004
6.400	0.995	0.096

IIIb. THE PROGRAM IN BASIC

A. PROGRAM STATEMENTS	B. EXPLANATION
100 T=0	Set time to 0
110 INPUT "Time interval";DT	Enter time interval
120 INPUT "Initial X position";X	Enter initial X position
130 INPUT "Initial Y position";Y	Enter initial Y position
140 INPUT "Initial X velocity";VX	Enter initial X velocity
150 INPUT "Initial Y velocity";VY	Enter initial Y velocity
160 INPUT "Force constant";C	Enter force constant
170 INPUT "Power constnat";P	Enter "power" constant
180 DT.5-DT/2	Compute half time interval
190 DTV=DT.5	Set variable time interval
200 X1=X	Set current X position

```
210 Y1 = Y                          Set current Y position
220 VX1=VX                          Set current X velocity
230 VY1=VY                          Set current Y velocity
1000 CXY=C*(((X1^2)+(Y1^2))^P)      Set variable C * (X^2 + Y^2)^P
1010 AX=X1*CXY                      Compute X acceleration
1020 AY-Y1*CXY                      Compute Y acceleration
1030 T=T+DT.5                       Increment time
1040 VX1=(AX*DTV)+VX1               Compute current X velocity
1050 VY1=(AY*DTV)+VY1               Compute current Y velocity
1060 DTV=DT                         Set variable time interval
1100 PRINT                          Display blank line
1110 T=T+DT.5                       Increment time
1120 PRINT "Time =";T               Display time
1130 X1=X1+(VX1*DT)                 Compute current X position
1140 PRINT "X position =";X1        Display current X position
1150 Y1=Y1+(VY1*DT)                 Compute current Y position
1160 PRINT "Y position =";Y1        Display current Y position
1170 IF INKEY$="" GOTO 1170         Wait for keypress
1180 GOTO 1000                      Repeat computation
```

Note: If you have access to a printer, you can substitute LPRINT instructions for PRINT instructions.

C. VARIABLES USED

Variable	Value
AX	X acceleration
AY	Y acceleration
C	Force constant
CXY	Intermediate variable: C * (X^2 + Y^2)^P
DT	Time interval: "delta-T"
DT.5	Half of time interval
DTV	Variable time interval
P	"Power" (exponent) constant
T	Time
VX	Initial X velocity
VX1	Current X velocity
VY	Initial Y velocity
VY1	Current Y velocity
X	Initial X position
X1	Current X position
Y	Initial Y position
Y1	Current Y position

D. USER INSTRUCTIONS

Enter the program carefully and run the program. Respond to these prompts:

Prompt	Action
Time interval?	Enter the time interval.
Initial X position?	Enter the initial X position.
Initial Y position?	Enter the initial Y position.

Prompt	Action
Initial X velocity?	Enter the initial X velocity.
Initial Y velocity?	Enter the initial Y velocity.
Force constant?	Enter the force constant.
Power constant?	Enter the "power" constant.

The program displays the time and the current X and Y positions.

Press any key to display the next set of values.

E. PROGRAM CHECK DATA

Run the program, and respond to the prompts as follows to check it out.

```
Time interval? .1
Initial X position? 1
Initial Y position? 0
Initial X velocity? 0
Initial Y velocity? 1
Force constant? -1
Power constant? -1.5

Time = .1
X position = .995
Y position = .1

Time = .2
X position = .9800504
Y position = .199

Time = .3
X position = .9553017
Y position = .2960104

Time = .4
X position = .9210032
Y position = .3900616
         .
         .
         .
Time = 3.199998
X position = -1.003849
Y position = -4.773731E-02
         .
         .
         .
Time = 6.300006
X position = .9999918
Y position = -4.020751E-03

Time = 6.400007
X position = .9953969
Y position = 9.599852E-02
```

When finished running the program for one set of values, enter a BREAK command. (On most computers this is accomplished by holding down a "shift" or "control" key, and pressing the "Break" or "Pause" key on the keyboard.) To run the program for a new set of values, type in the command RUN, and then press the "ENTER" or "RETURN" key.

IV. EXPLORATIONS/NOTES

1. a. Plot the motion for one complete orbit using the "Program Test Data" initial conditions on a set of axes, as shown in Fig. 5-25.

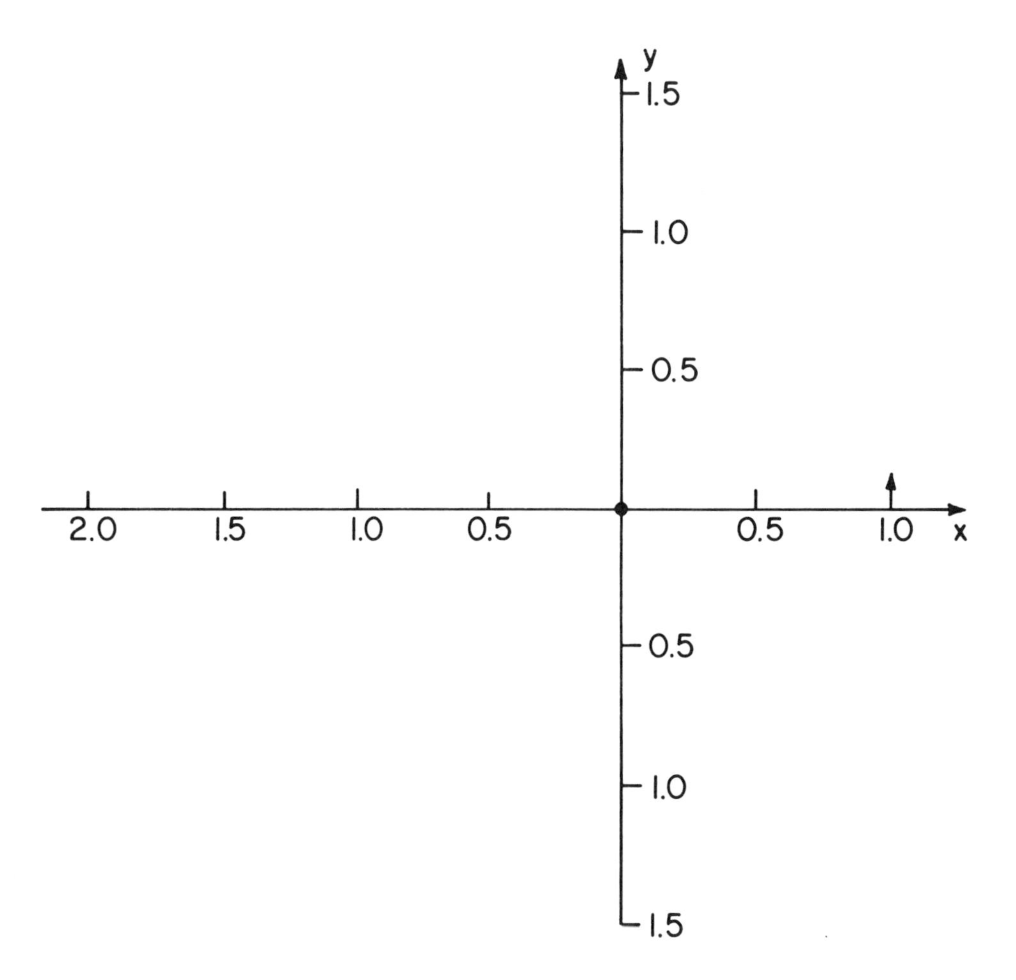

Fig. 5-25. Experiment 7. Orbit and Central Force Motion. Format for Graphing Results for Explorations 1, 2, 3.

b. What is the period of this motion (time for 1 orbit)?

c. What is the shape of this orbit? Use a ruler to measure the dimensions of the orbit along different axes through the center. What do you observe?
Note that a plot such as this can give you an idea of the speed of the orbiting body at different parts of its orbit. Since each point plotted is equally spaced in time from the next, the relative distance between the points gives you an

indication of the speed of the body. The farther apart the points are in a given region, the faster the orbiting body is moving. The closer together the points are, the slower it is moving.

d. Describe the speed behavior for the orbiting body for this set of initial conditions.

2. a. Use the Euler program to plot the motion of one orbit for a system with all of the same initial conditions as case 1, except for a higher initial velocity in the y direction: $v_{yo} = 1.15$. For this run, use $\Delta t = 0.2$ seconds. (Plot the motion on the same axes used in case 1).

   (To use the TI-58 or TI-59 program to observe this motion, first press [2nd] [C']; enter the new Δt value and press [A], enter the new v_{yo} value and press [E], then press [CLR] and [2nd] [E'] to begin the run.)

   b. What is the period of this motion?

   c. Describe the shape of this orbit, using a ruler as described in Exercise 1.

   d. Describe the speed behavior of this motion. Where is the orbiting body moving fastest, where slowest?

3. a. Plot the motion of the same system discussed in Explorations 1 and 2, but this time with a reduced initial velocity in the y direction: $v_{yo} = 0.85$. (Put your plot on the same set of axes used previously, and for this run use $\Delta t = 0.1$ sec).

   b. What is the period of this motion?

   c. Describe the shape of this orbit, using a ruler as you did earlier.

   d. Describe the speed behavior for the orbiting body. Where is this body moving fastest? Where is it slowest?

4. a. Now plot the motion of this system for a much higher initial velocity, say $v_{yo} = 2.0$. Use $\Delta t = 0.1$, but plot the motion every 1/2 second only. Use a new set of axes for this plot, and plot the motion from $t = 0$ to $t = 5$ sec.

   b. Describe the motion of this system. (This is called an escape trajectory, where the orbiting body actually leaves the region of the central force and moves out into space.)

5. Next let's consider the case of an entirely different system. In this situation, we'll assume a central repulsive inverse square law force. This force constantly acts to repel the two bodies, so a bound or orbital solution to this sort of motion does not exist. This motion is classically the sort considered in scattering situations, where light particles (such as alpha particles or other light projectiles) are scattered off a nucleus. Nuclei are many times more massive than alphas, and possess an intense positive charge. (Examinations of this phenomenon provided the first evidence that nuclei existed.)

   To examine scattering behavior using our Euler program, we can use the same inverse square law analysis but reverse the sign of the force constant. If the sign of C is positive, but we change nothing else, our program will simulate a situation where the force on the two bodies:

- acts to <u>repel</u> the two bodies
- acts along a line joining their centers
- follows an inverse square law behavior (P = -1.5).

Let's use our program to simulate the motion of this system in the four situations shown in Fig. 5-26.

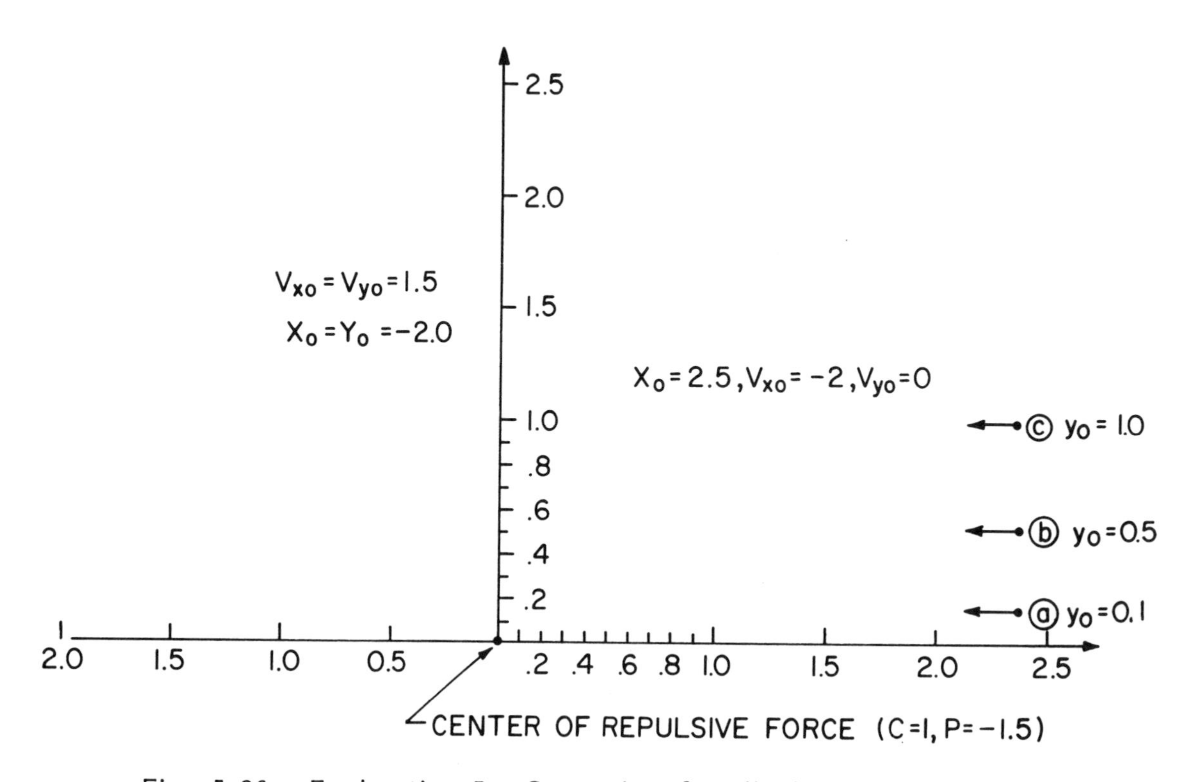

Fig. 5-26. Exploration 5. Scattering for all Plots, $\Delta t = 0.1$.

a. Plot the motion of a body released at $x_o = 2.5$, $y_o = 0.1$

   with the initial velocity: $v_{xo} = -2.0$, $v_{yo} = 0$

   For t = 0 to t = 2.5 seconds.

   For this run (and those that follow) use: C = +1
   P = -1.5, and
   $\Delta t = 0.1$

   Examine the motion carefully, and describe its velocity behavior.

b. Next plot the motion for the exact same situation, except that the object released toward the center of force is moved farther out in the y direction to $y_o = 0.5$. Again examine the motion and describe its velocity behavior.

c. Repeat the exploration for the same initial conditions, except this time use $y_o = 1.0$.

d. For the final case, examine the behavior of the motion for a "head-on" collision. Start with the initial conditions: $x_o = y_o = -2.0$; $v_{xo} = v_{yo} = 1.5$

   Examine (but do not plot) the motion from t = 0 to t = 2.5 seconds. Describe the velocity behavior and note the "distance of closest approach" to the repulsive force center.

Chapter 6

BASIC EXPLORATIONS OF LINEAR AND DIGITAL INTEGRATED CIRCUITS[1]

Introduction

Most all of the electronics equipment we see in our homes and laboratories incorporates one or more integrated circuits. These devices combine active circuit elements, resistors, capacitors and interconnections all on the surface of a "chip" of silicon. By enabling the execution of complex electronic functions in a small, low cost, low power consumption package, integrated circuits – or I.C.'s – are forming the backbone of today's revolutions in consumer electronics. This series of experiments is designed to introduce several common integrated circuit families and allows you to gain familiarity with how they are used in a variety of typical applications.

The first set of four experiments deals with Linear Integrated Circuits – which amplify or control electronic signals that vary in a smooth and continuous fashion. The most common of these devices is called the operational amplifier, which provides an extremely useful tool in the laboratory that is easy to use – and quite affordable.

The second set of experiments deals with digital circuits – which handle electronic signals that assume discrete levels only. These circuits are commonly used to handle computation, or in circuits that "make decisions." Logic circuits or "Gates" form a common family of I.C.'s and many useful configurations are explained and explored in the last four experiments of this chapter.

---

[1]This chapter was written by Gerald Luecke, Manager of Technical Products Development, Texas Instruments Learning Center.

Experiment 6-1 LINEAR CIRCUITS, SMALL SIGNAL (DC OR AC OPERATION)

PURPOSE

This experiment introduces small signal linear amplifiers. It discusses the symbology, some definitions, some techniques for interconnecting amplifiers to obtain dc or ac operation.

DISCUSSION

Fig. 6-1 is a symbolic representation of an amplifier. In our case it is a circuit consisting of resistors, capacitors, transistors, diodes and interconnections that make up an electronic amplifier. This is just like a record or tape amplifier, or the amplifier in a television set that produces the audio output.

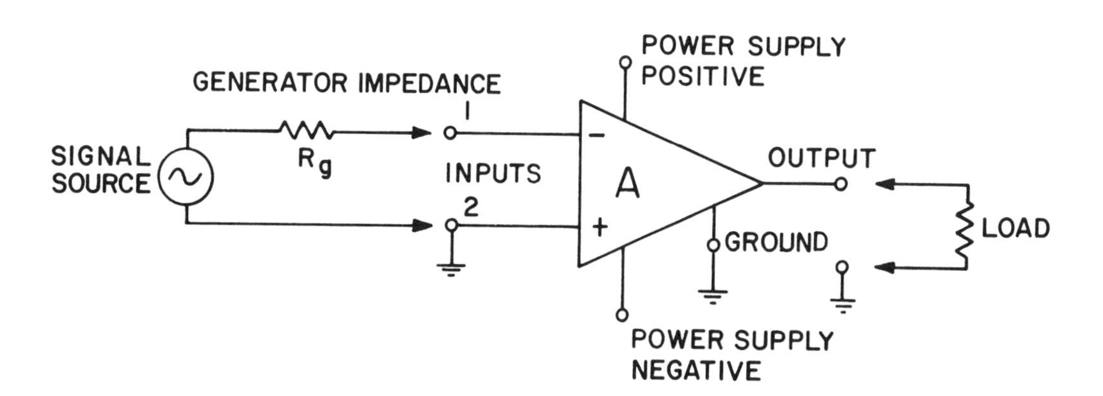

Output is detected across a load connected between the output terminal and ground.

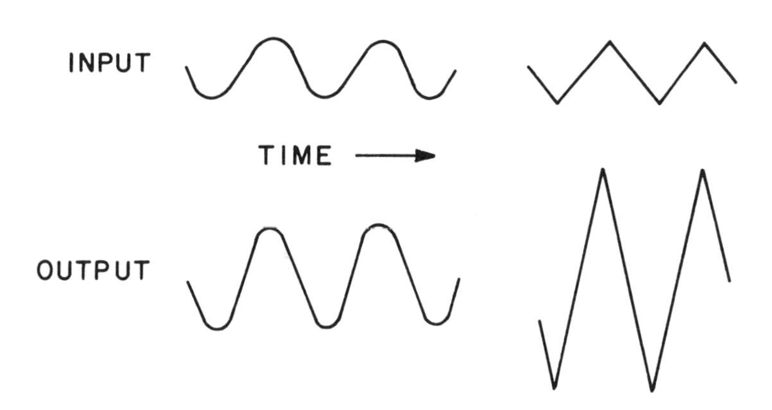

Fig. 6-1. Electronic Amplifier.

Such a circuit is called an amplifier because, as the name implies, it provides a "gain" in the sense that something is being made larger. As Fig. 6-1 implies, the input signals are being amplified and made larger at the output.

In Fig. 6-1, there are two input terminals and two output terminals, connections to a power supply or supplies, and a common connection to a "ground" terminal indicated by the ⏚ symbol. As shown in the figure, the input signal is applied between terminals #1 and #2 with a signal generator. Terminal #2 is connected to the common terminal (ground). The amplified signal is detected, measured and used by connecting a "load" across the output – between the output terminal and the common terminal (ground).

Note that in Fig. 6-2, the output signal is an exact reproduction of the input signal with no distortion, except it is larger in amplitude. An amplifier which performs this way is called a linear amplifier. It is designed to handle signals that are continuously time varying, both in amplitude and frequency and reproduce these signals as accurately as is practical, amplified in amplitude or in power level depending on the design of the amplifier. The gain is normally greater than 1, but does not necessarily have to be. It can be less than 1.

Next examine Fig. 6-2. This shows an amplifier which has a gain control (like a volume control on your radio). If we have a constant input signal and increase the gain, the output signal gets larger and larger in amplitude. However, there are limits. Increasing the gain too far causes the output signal not to follow on certain portions of the signal. It hits boundaries determined by the particular amplifier design. The output waveform becomes distorted and is no longer a reproduction of the input voltage. When this condition exists, we say that the output signal is exceeding the dynamic range of the amplifier.

With linear amplifiers, the operating mode is always within the dynamic range; otherwise, the reproductions are not pure. Amplifiers, where the amplified signal levels are many times smaller than the dynamic range, are called "small-signal" amplifiers. Amplifiers that use a large portion of the full dynamic range are called "large-signal" or, more commonly, "power amplifiers."

Electronic amplifiers can be constructed by using vacuum tubes or solid-state devices (transistors and diodes) in combination with discrete components such as resistors, capacitors, and inductors to produce the "gain." However, solid-state technology has advanced to the state where such amplifiers are constructed all at the same time in integrated circuit form. This experiment uses such integrated circuit amplifiers.

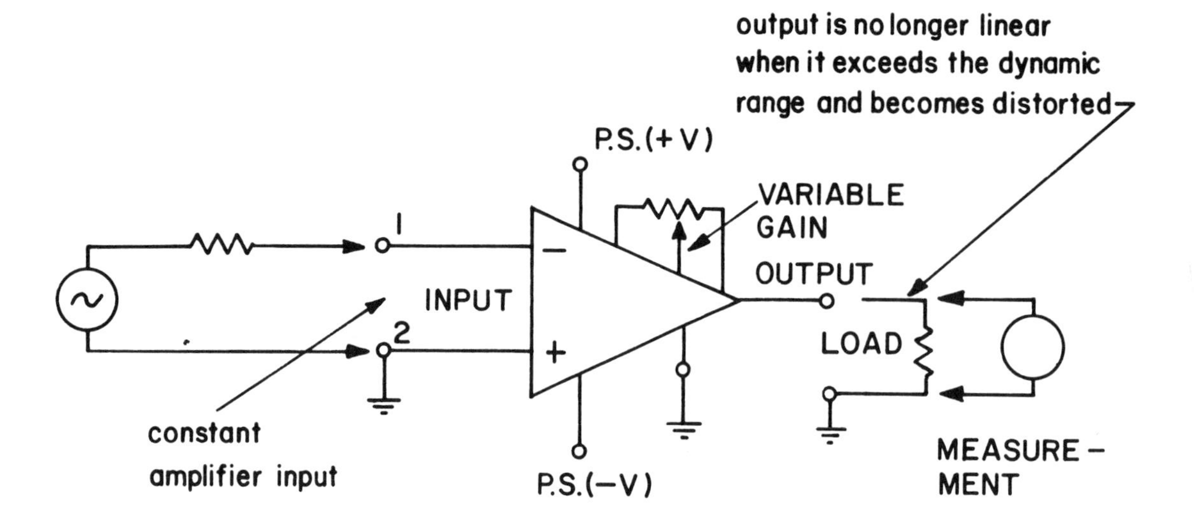

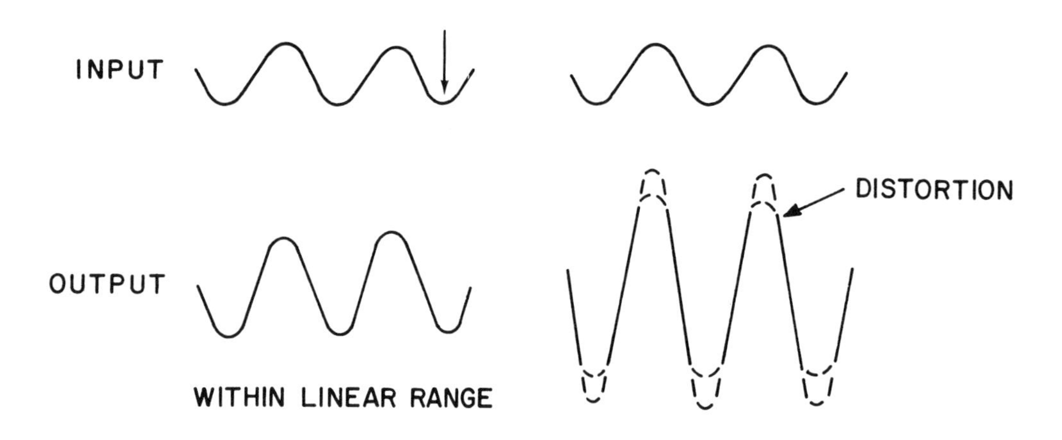

Constant amplitude input

Fig. 6-2. Dynamic Range.

A. EQUIPMENT REQUIRED

Amplifier Card with Amplifier Circuit
Oscillator
Oscilloscope - dc to 10 MHz Bandwidth Dual Channel
S1, S3 Switching Network and Resistors
-15V Power Supply - 100 ma
+15V Power Supply - 100 ma

B. OPERATING CONDITIONS

The amplifier to be used in this experiment will be represented by the symbol used in Fig. 6-1. However, it actually is an amplifier within an amplifier. The internal amplifier has additional components, some of which you will use to learn about the characteristics of linear amplifiers.

Interconnect the amplifier, as shown in Fig. 6-3.

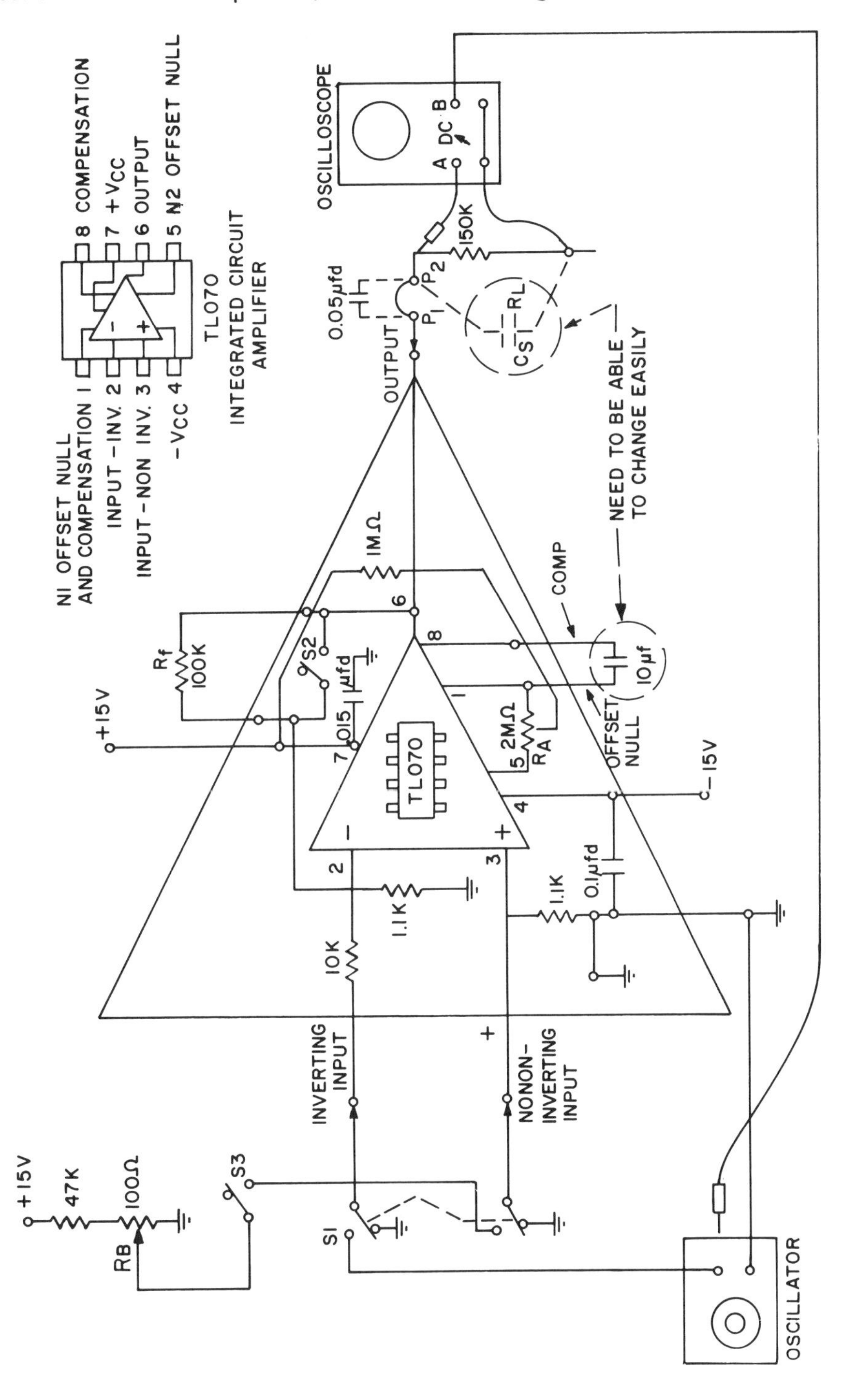

Fig. 6-3. Amplifier Circuit.

Power supplies of +15V and -15V are used so that the output voltage is balanced around ground (0 volts). Such integrated circuit amplifiers normally are dc amplifiers. Without voltage on the inputs, the output voltage should be zero.

Connect the equipment, as shown; the oscillator at the inputs through the network with S1 and S3 and the oscilloscope (Channel A) at the output across R_L. Make sure P1 is shorted to P2.

I. DC QUIESCENT OUTPUT OFFSET VOLTAGE

1. Turn off the oscillator and place S1 in the position where both inputs are connected to ground. Keep S2 open. Set the oscilloscope on dc amplification and place the amplification of Channel A on a sensitive range. Short out the inputs to the oscilloscope and determine the position of the beam scan line representing zero volts. Adjust the vertical position to place the zero volt reference scan line in the center of the screen. The oscilloscope time sweep should be at least 1 millisec/cm.

2. Measure and record the voltage across R_L at the output of the amplifier.

   DC Quiescent Output Offset Voltage ____________ volts.

   This is called offset voltage because it is a voltage displacement caused by a differential input voltage resulting from a mismatch of the amplifier's input stages. The input voltage mismatch is multiplied by the amplifiers gain to cause the output offset.

3. Adjust RA until the DC Quiescent Output Offset Voltage is zero. A small current is injected internally in the amplifier by the network of R_A to compensate the input voltage mismatch.

II. DC OPERATION - TYPICAL

1. Make sure S2 and S3 are open. Connect the amplifier inputs to the oscillator with S1. Set the oscillator frequency at 1,000 Hertz. Turn down the voltage output control of the oscillator to zero.

2. Set the oscilloscope vertical amplifier sensitivity of Channel A to 0.5 volts/cm. Set the time sweep at 0.5 millisec/cm. Synchronize the sweep trigger on the Channel A signal.

3. Increase the oscillator output voltage until the oscilloscope voltage waveform is 2 volts peak to peak. Plot 2 cycles of the waveform on Fig. 6-4 indicating the voltage measured at the zero amplitude points of the sine wave and the peak amplitude points.

   When the amplifier is correctly biased, its normal dc operation is with the output voltage centered around zero volts.

4. With the oscilloscope synchronized as before, measure the input voltage from the oscillator with Channel B and plot 2 cycles of the waveform with dotted lines on Fig. 6-4. Increase the vertical sensitivity of Channel B so that the waveform amplitude appears as one-half the amplitude of the output voltage measured with Channel A. Keep the time sweep synchronized with the Channel A signal.

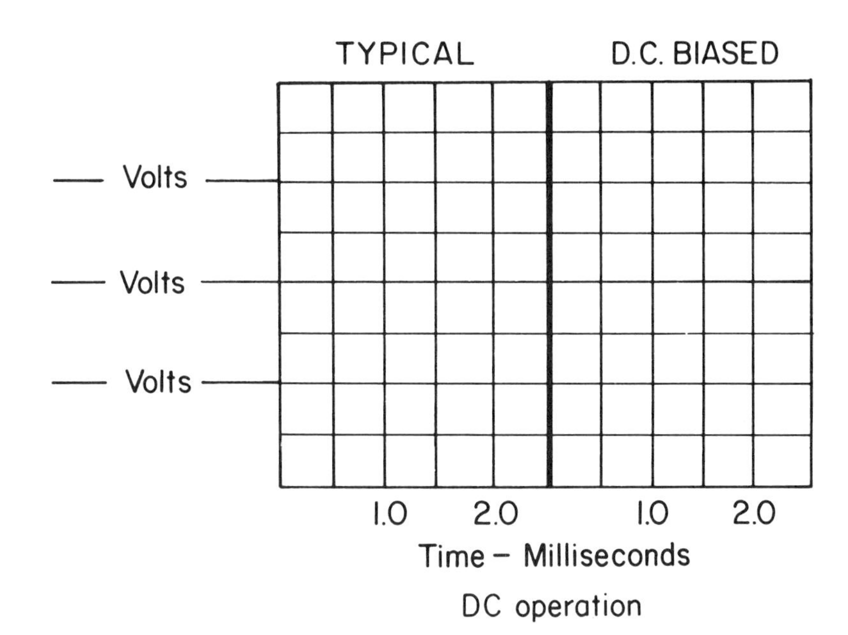

Fig. 6-4.
DC Operation.

What is the phase relationship of the output to the input?

(a) In phase ____________?

(b) 180° out of phase ____________?

Signals applied at the inverting input of the amplifier (-) produce an output that is inverted from the input (180° out of phase).

III. DC OPERATION - BIASED

1. Turn down the output of the oscillator. Close S3. Measure the output voltage with Channel A as before. Adjust R_B so the output voltage is displaced by 1 volt.
2. Increase the oscillator output voltage until the oscilloscope voltage waveform again is 2 volts peak to peak. Plot 2 cycles of the waveform as before.

   What is the phase relationship of the output voltage displacement in relationship to the input bias applied at the (+) input?

   (a) In phase ____________?

   (b) 180° out of phse ____________?

   Signals at the output will be displaced by dc voltages applied at the inputs to amplifiers operating as dc amplifiers. Signals applied at the non-inverting input of the amplifier (+) produce a non-inverted (in phase) output.

IV. AC OPERATION

AC operation of amplifiers means that signals are coupled from one amplifier to another without including the dc component of the signal. Only time varying signals are transmitted. Steady-state quiescent dc levels are not coupled from stage to stage. This is accomplished by coupling amplifier stages together with capacitors.

1. To demonstrate this on the amplifier circuit of Fig. 6-3, replace the shorting strip between P_1 and P_2 with a 0.05 μF capacitor.
2. Repeat Steps 1, 2, 3 and 4 of DC Operation - Typical.

   Plot the waveforms on Fig. 6-5.

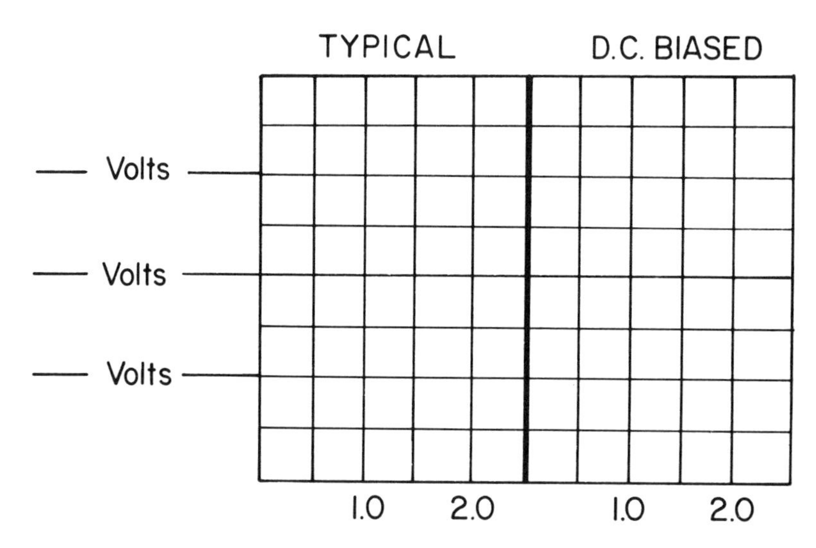

Fig. 6-5.
AC Operation.

3. What is the phase relationship of the output to the input?

   (a) In phase ______________?

   (b) 180° out of phase ______________?

4. Repeat Steps 1 and 2 of DC Operation - Biased.

   In Step 1, when adjusting R_B for the 1-volt displacement, measure the output voltage at terminal P (the same place as in Step 1 for DC Operation - Biased). However, for Step 2 measure the output voltage at terminal P2.

   Plot the waveform on Fig. 6-5.

   (a) In phase ______________

   (b) 180° out of phase ______________

   Other ____________________

V. GAIN

1. As mentioned previously, linear amplifiers are designed to handle signals that are continuously time-varying both in amplitude and frequency, and reproduce these signals as accurately as is practical, amplified in amplitude. The amplification of the amplitude is called the "gain" of the amplifier. As implied, the voltage measured at the output is larger in amplitude than that applied at the input.

   With S2 and S3 open on the amplifier circuit of Fig. 6-3 and the oscillator applied to the inputs of the amplifier through S1, measure the output voltage across R_L with Channel A of the oscilloscope and measure the input voltage with Channel B. You may leave the 0.05 μF capacitor connected between P1 and P2 for ac operation or you may short P1 and P2 for dc operation, whichever you prefer.

   The "gain" of the amplifier is the ratio of the output voltage over the input voltage, $\frac{V_o}{V_i}$. Adjust the output voltage across R_L to 2 volts peak to peak, as before. Measure the input voltage by increasing the sensitivity of Channel B. Record the values.

   (a) V_o = ________________ peak to peak volts

   (b) V_i = ________________ peak to peak volts

   Gain = $\frac{V_o}{V_i}$ = ____________________

2. Gain normally is measured in db, which means "decibels." Voltage gain is expressed in decibels as:

$$\text{Gain db} = 20 \log_{10} \frac{V_o}{V_i}$$

   Remember, the logarithm of a number is the exponent to which the base must be raised in order to get the number.

   If $\frac{V_o}{V_i}$ = 1,000 , you would have to raise the base 10 to 10^3 in order to get 1,000. Therefore, the Gain in db = 20 x 3 = 60 db. Record the gain in db.

   The gain of the amplifier is ________________ db.

3. Amplifier gain might typically be between 20 db and 100 db. What magnitude of gain does an amplifier have for 100 db of gain?

   $\frac{V_o}{V_i}$ = ____________________

VI. DYNAMIC RANGE - DISTORTION

1. Increase the output voltage V_o of the oscillator at the input of the amplifier, and observe the output on Channel A of the oscilloscope which is measured across R_L. Reduce the sensitivity of Channel A so that the output waveform amplitude can always be observed on the oscilloscope face. As the input voltage V_i is increased, a point will be reached where the output waveform will become distorted. Measure the peak to peak amplitude of the output waveform V_o at the point where distortion just begins.

   V_o = ____________________ peak to peak volts

2. Compute the gain in db at this point. (Measure the input voltage.) This is a measure of the dynamic range of the amplifier expressed in db.

   $$\text{Gain db} = 20 \log_{10} \frac{V_o}{V_i} = \text{\_\_\_\_\_\_\_\_\_\_\_\_\_\_\_\_\_\_\_\_ db}$$

   V_i = ____________________ volts

Experiment 6-2 LINEAR CIRCUITS, SMALL SIGNAL (FREQUENCY RESPONSE)

PURPOSE

In this experiment we will examine a linear amplifier's frequency response, showing the variations that occur and the causes of these variations. Terms such as bandwidth, and mid-band frequency response, and low-frequency and high-frequency corner frequencies are defined.

DISCUSSION

The gain of an amplifer varies with the frequency of the signal it is amplifying. There is a range of signal frequencies where the gain is relatively constant. This is called the mid-band frequency range. The constant gain is called the mid-band frequency gain. It is identified as A in Fig. 6-6. The frequency above the mid-band frequencies at which the gain is 0.707 of the mid-band gain is called the high-frequency corner frequency f_2 as shown in Fig. 6-6.

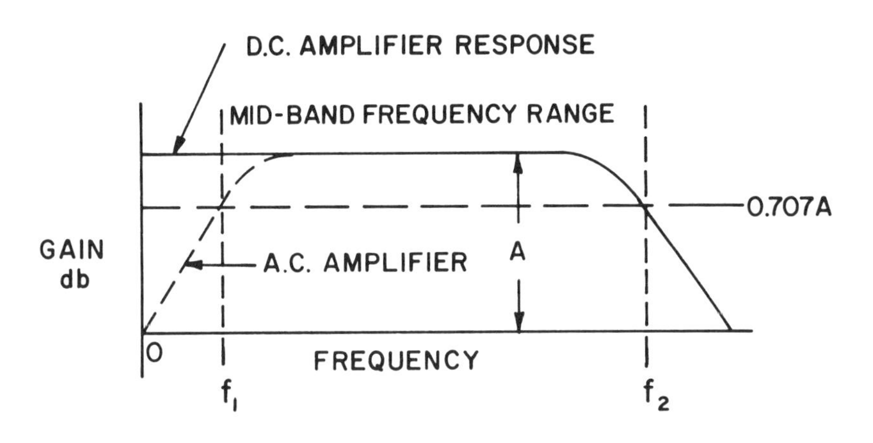

Fig. 6-6. Frequency Response of an Amplifier.

For dc amplifiers, the mid-band frequency gain usually is constant down to dc ($f=0$). However, for ac amplifiers there is a low-frequency corner frequency (f_1 of Fig. 6-6) at which the gain is reduced to 0.707 of its mid-band frequency value (as with the high-frequency corner). Such a low-frequency corner occurs because of the coupling capacitors between stages.

I. LOW FREQUENCY CORNER, f_1

1. To demonstrate the low-frequency characteristics of a typical ac amplifier, as shown in Fig. 6-6, and to see how f_1 is determined, we will use the amplifier shown in Fig. 6-3. (The ac mode will be used with the 0.05 μF capacitor inserted between P_1 and P_2.) With the oscillator at 1,000 Hertz, adjust the input voltage until the output voltage measured by Channel A across R_L is 2.84 volts peak to peak. (In this step of the experiment, it is important that the input voltage amplitude remain constant as the input signal frequency is varied. Measure the input voltage amplitude with Channel B. If the input voltage should vary as the frequency is varied, keep it the same by adjusting the oscillator output control.)

   Now reduce the frequency of the oscillator and note the output voltage measured by Channel A. When the output voltage is 2 volts ($2.84 \times 0.707 = 2$ volts), record the frequency of the oscillator. This is the low-frequency corner frequency f_1 of the amplifier. Adjust the output to 2.84 volts at 1,000 Hz and set the mid-band frequency amplitude at a convenient value for this measurement. The mid-band frequency gain is the same as measured in Experiment 1.

   Low-frequency corner frequency f_1 = ______________________ Hz.

2. The low-frequency corner frequency occurs because the capacitive reactance of the coupling capacitor of Fig. 6-7 equals the load resistor R_L. The combination series circuit acts just like a voltage divider. Prove for yourself that when $jX_C = R_L$, THE OUTPUT VOLTAGE IS 0.707 OF THE VALUE when $jX_C << R_L$ (the mid-band frequency case).

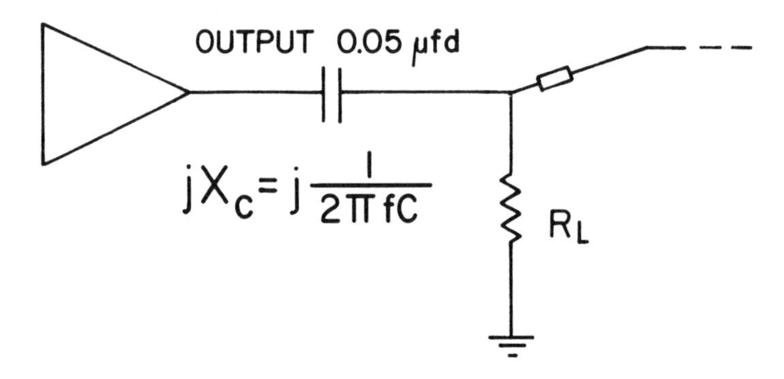

Fig. 6-7. Low Frequency Corner.

II. HIGH FREQUENCY CORNER, f_2

As with the low-frequency corner frequency, the high-frequency reduction of amplifier gain is also due to capacitance. However, in this case it is the shunt (or parallel) capacitance that causes the problem rather than series capacitance, as with the low-frequency corner.

For example, an amplifier stage output circuit can be represented, as shown in Fig. 6-8. R_o is the output impedance of the amplifier, which is in series with the voltage generator E_o.

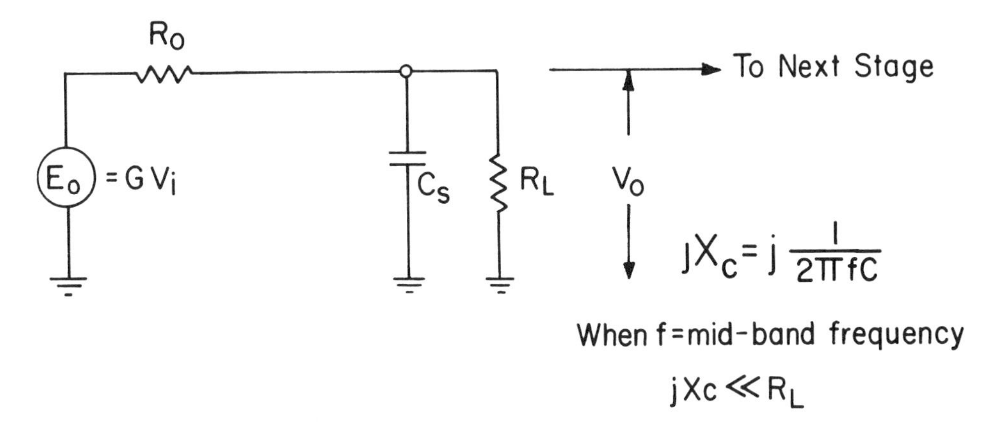

Fig. 6-8. Amplifier Output.

E_o is equal to the mid-band frequency gain of the amplifier times the input voltage. R_L is the load on the amplifier. C_s is a shunt capacitance across the load which is caused by the next stage that is coupled to the output. Suppose that in the mid-band frequency region of Fig. 6-6, $jX_C >> R_L$ and $R_L = R_o$. Then V_o would be equal to 1/2 GV_i because

$$V_o = \frac{R_L}{R_o + R_L} GV_i$$

and with $R_L = R_o$

$$V_o = \frac{GV_iR_L}{2R_L} = \frac{GV_i}{2}$$

R_L is normally much, much greater than R_o when amplifier stages feed each other, and hence the reduction in output voltage is due primarily to the reduction in the capacitive reactance of C_s as frequency increases. As a result, the circuit of Fig. 6-8 basically behaves as shown in Fig. 6-9 at higher frequencies.

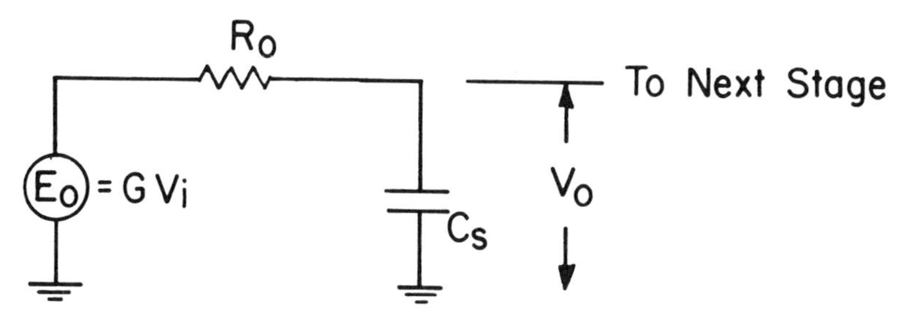

Fig. 6-9. High Frequency Corner.

When the capacitive reactance $-jX_C$ equals R_o, the output impedance of the amplifier stage V_o will be 0.707 of the mid-band frequency value E_o.

1. The integrated circuit amplifier used in these experiments is composed of several amplifier stages, and has a very low output impedance. To simulate an amplifier stage which has a much higher R_o, a 150K resistor is connected between P1 and P2 of the amplifier shown in Fig. 6-3. As shown in Fig. 6-10, a 20 pF capacitor is also connected from P2 to ground. This simulates the shunt capacitance of the next stage. With the oscillator set at 1,000 Hz, adjust the input voltage until the output voltage measured by Channel A across C_S is 2.84 volts peak to peak. Again, it is important that the input voltage amplitude be kept constant as the input frequency is varied. Measure the input voltage with Channel B and keep it constant as you did in the low-frequency corner experiment.

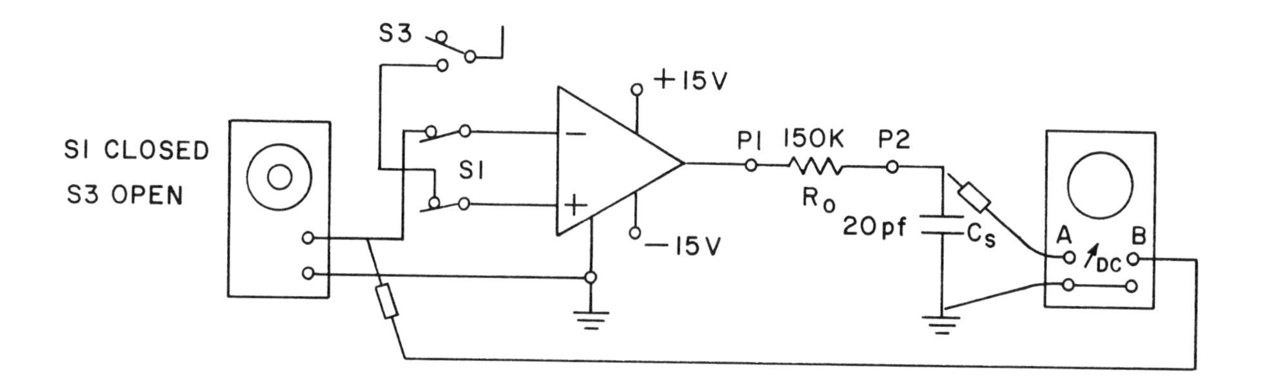

Fig. 6-10. Integrated Circuit Amplifier.

Increase the frequency of the oscillator and note the output voltage measured by Channel A. When the output voltage has fallen to 2 volts peak to peak, record the frequency of the oscillator. High-frequency corner f_2 = ________________

2. Prove for yourself that when $jX_c = R_o$, the output voltage V_o is 0.707 of the value when $jX_c >> R_o$ (the mid-band frequency case).

   (a) At what frequency does $jX_c = R_o$? ____________ Hz

   (b) Is this the same as the high-frequency corner frequency f_2? ____________

3. Next, insert a 0.05 μF coupling capacitor in the circuit as shown in Fig. 6-11. Leave R_o (150K) in place, and add a load resistor R_L of 1.5 M connected as shown. Measure V_i and V_o at selected frequencies from f_1 to f_2, convert to db and construct a plot that looks like Fig. 6-6. The gain should be relatively "flat" over this frequency range. The range of frequencies from f_1 to f_2 is called the "bandwidth" of the amplifier.

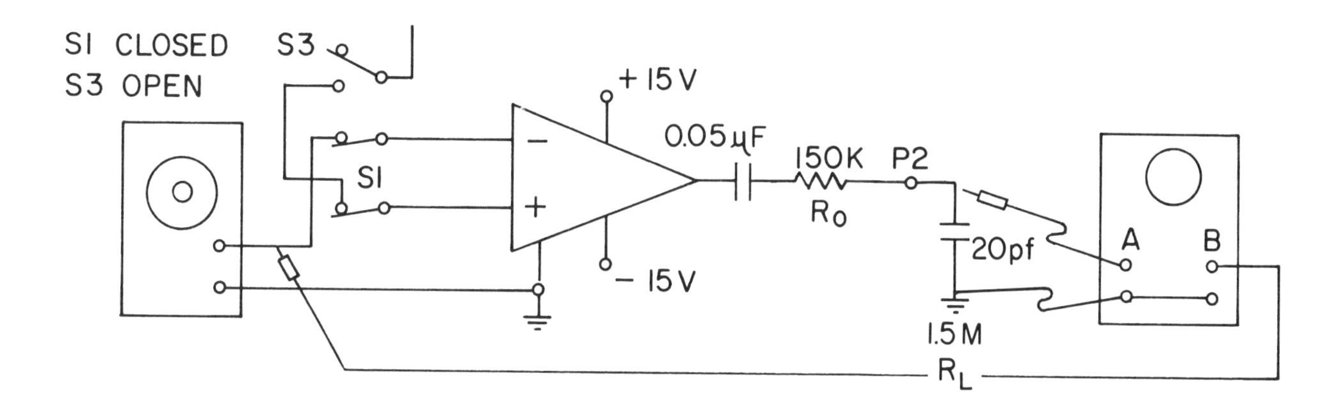

Fig. 6-11. Mid-Band Frequency.

Experiment 6-3 LINEAR CIRCUITS, SMALL SIGNAL (OPERATIONAL AMPLIFIERS)

PURPOSE

This experiment introduces operational amplifiers. It discusses their characteristics, terminology, definitions, and shows how they can be used to provide amplifiers with differing gain and frequency behaviors.

DISCUSSION

Operational amplifiers receive their name from amplifiers used in analog computers to perform mathematical operations. Today the term "Op Amp" means an amplifier that is direct-coupled, has high-gain, high input impedance, low output impedance, wide bandwidth, and whose characteristics can be varied by using external components. In this experiment, we'll focus in particular on the effects of feedback in the circuit – i.e., where a portion of the output signal is fed back to the input through a feedback resistor which we'll label Rf.

An operational amplifier can be represented by the network shown in Fig. 6-12. The amplifier normally provides a linear output voltage V_O that is proportional to the difference voltage between the two input terminals. The output voltage having the same polarity as the voltage at the non-inverting (+) input with respect to the voltage at the inverting (-) input. If the non-inverting input is more positive than the inverting input, then V_O is positive with respect to ground. V_O will swing negative when the non-inverting input is negative with respect to the inverting input. The definitions of the various elements shown in Fig. 6-12 are as follows:

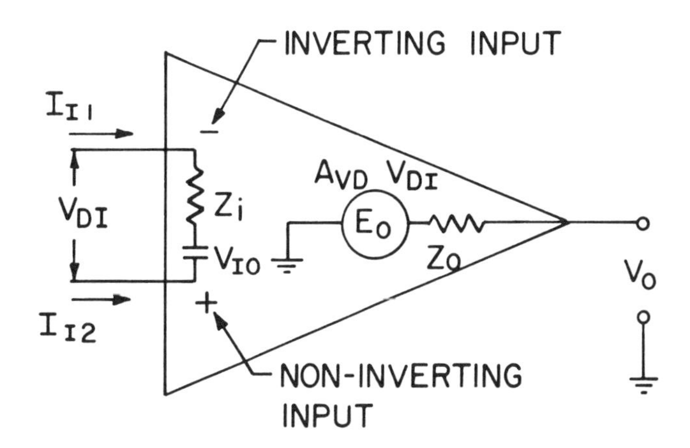

Fig. 6-12. Operational Amplifier.

a. I_{I1} and I_{I2} are input currents.
b. V_{DI} is the differential input voltage.
c. Z_i is the input impedance.
d. V_{IO} is the input offset voltage.
e. A_{VD} is the "open loop differential voltage gain."
f. Z_o is the output impedance.
g. V_o is the output voltage.

An ideal operational amplifier has the following characteristics:

Z_i is very high, approaching infinity	$Z_i = \infty$
Z_{IO} is zero	$V_{IO} = 0$
A_{VD} is very high, approaching infinity	$A_{VD} = \infty$
Z_o is zero	$Z_o = 0$
In addition, the bandwidth approaches infinity	$f_2 - f_1 = \infty$

If we assume these ideal characteristics exist for a moment, we can drive several of the fundamental relationships that describe operational amplifier behavior. If we assume that in Fig. 6-13 below, the gain A_{VD} is infinity, then we may assume that the input differential voltage V_{DI} is very close to zero. With $V_{DI} = 0$, the current into the inverting input of the amplifier – labeled I_{IN} – will also approach zero. Applying Kirchoff's current law at the inverting input then gives:

$$I_1 + I_2 = 0$$

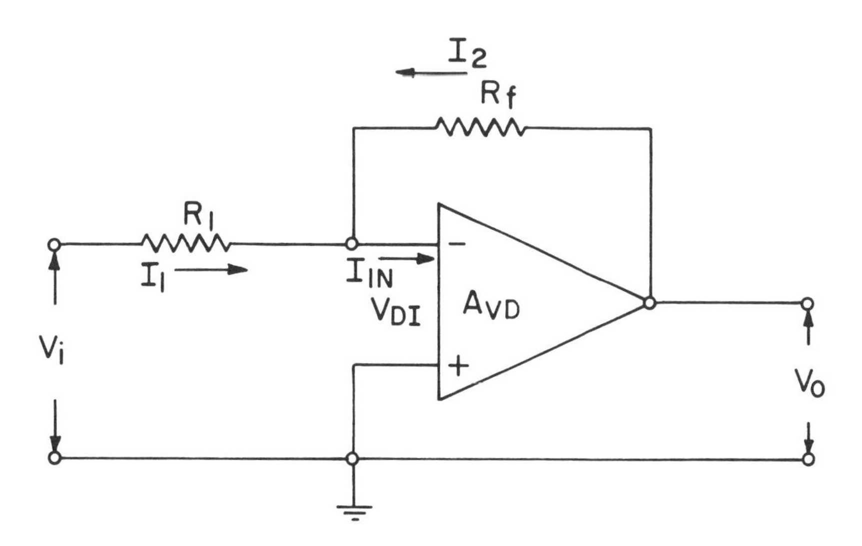

Fig. 6-13. Ideal Op Amplifier.

Now,

$$I_1 = \frac{V_i - V_{DI}}{R_1} = \frac{V_i}{R_1}$$

and

$$I_2 = \frac{V_o - V_{DI}}{R_f} = \frac{V_o}{R_f}$$

since

$$V_{DI} = 0$$

Therefore,

$$\frac{V_i}{R_1} + \frac{V_o}{R_f} = 0$$

$$\frac{V_i}{R_1} = -\frac{V_o}{R_f}$$

and

$$\frac{V_o}{V_i} = -\frac{R_f}{R_1}$$

Now, the ratio $\frac{V_o}{V_i}$ for an amplifier with feedback is called the feedback voltage gain A_{vf}, and represents the amplification provided to the input signal when feedback is present in the circuit. A_{vf}, as shown, is equal to the ratio of the feedback resistor R_f to the input resistor R_1. The minus sign indicates that the output voltage is 180° out of phase from the input voltage, and consequently the circuit of Fig. 6-13 is called an inverting amplifier.

Of course, real operational amplifiers actually have finite open loop gain, finite input impedance and finite output impedance, but their real characteristics are such that the approximations we have used result in an error of much less than 0.5% when the amplifier is operated up to mid-band frequencies. In Experiment 6-1, the effect of offset voltage and its adjustment to zero were discussed. Occasionally, compensation must be made for the input bias currents I_{I1} and I_{I2} shown in Fig. 6-12; otherwise, besides gain variations with frequency, other parameters of operational amplifiers are quite stable and contribute little error in behavior from that predicted using ideal characteristics.

I. VARIATION OF GAIN

1. Use the same circuit setup as for Fig. 6-3. Repeat the steps you followed in Experiment 6-1 to adjust the offset voltage on the output to zero. After adjustment make sure S2 is open. Continue to use the dc gain arrangement with P1 shorted to P2, R_L = 150K and C_s = 0.

   With R_f = 100K, the dc gain of the amplifier measured in Experiment 6-1 was 100.

2. Now change R_f to 10K and measure the gain with a 1,000 Hz oscillator signal. Increase the input signal until the output is 2 volts peak to peak.

   Gain with R_f = 10K ____________________

3. Now change R_f to 4.7K and measure the gain again. The input voltage level of the oscillator will have to be increased again to make the output 2 volts peak to peak.

   Gain with R_f = 4.7K ________________

4. Now change R_f to 1K and measure the gain again. Again, the input level of the oscillator will have to be increased to make the output 2 volts peak to peak. If the oscillator output is limited, set the output at 1 volt peak to peak and measure the gain.

   Gain with R_f = 1K ________________

   Operational amplifier characteristics are such that the gain can be adjusted very easily by choosing the ratio of two resistors R_f and R_1. R_1 in Fig. 6-3 is determined by Thevinen's theorem, as shown in Fig. 6-14.

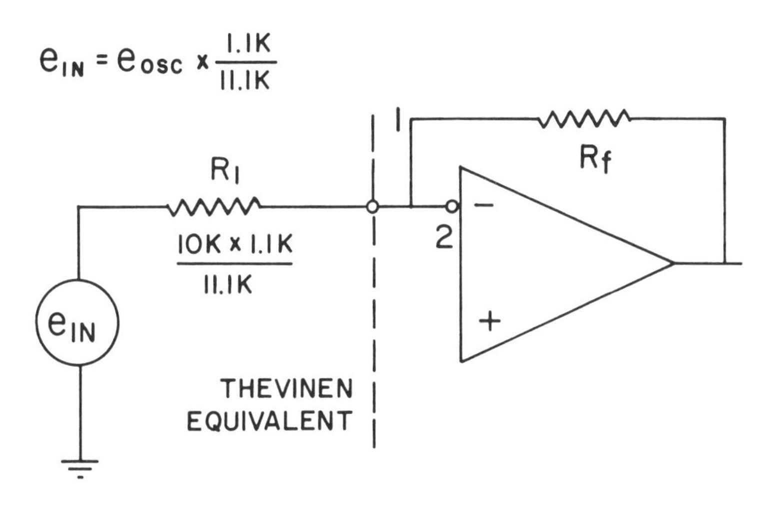

Fig. 6-14. Thevinen Equivalent.

II. VARIATION OF GAIN WITH FREQUENCY

Operational amplifiers also provide a way to easily vary the frequency response of the overall amplifier circuit. Fig. 6-15 shows the frequency response of the differential voltage amplification A_{VD} of the circuits shown in Figures 6-12 and 6-13. This is the gain when R_f is equal to infinity or the "open-loop" gain. There is no feedback. With feedback the amplifier gain vs. frequency will fit under the response curve envelope of Fig. 6-15. For example, if the overall gain of the amplifier is equal to 1, then the frequency response of the amplifier will have an f_2 = 3MHZ. If, however, the overall gain with feedback is 40, then the frequency at which the gain starts falling off with frequency from the mid-band gain is approximately 100KHZ). This gain vs. frequency response curve is shown as a dotted line in Fig. 6-15.

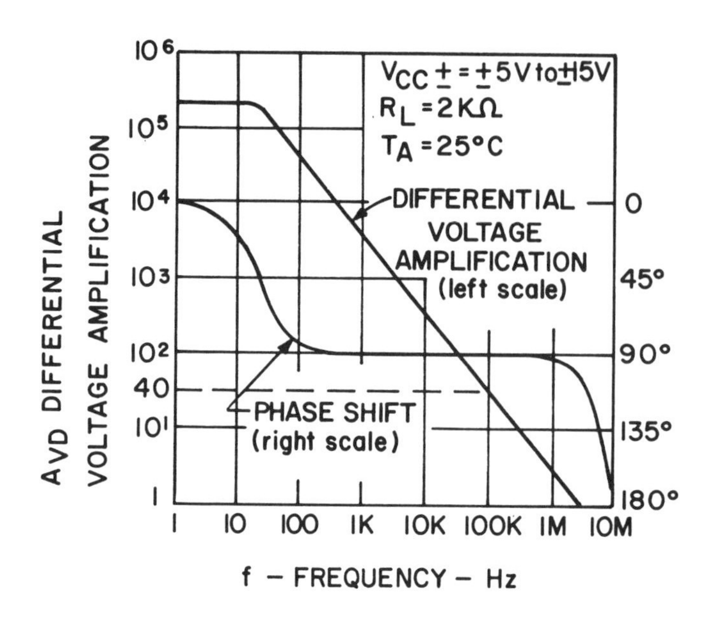

Fig. 6-15. Amplifier Frequency and Phase Shift Response.

When adjusting the gain with feedback, caution must be exercised to make sure that the signal phase shift caused by the load and feedback network does not exceed 180° while the gain is still greater than 1. If it does, the amplifier circuit will oscillate. This is shown in Fig. 6-16. If oscillations occur, it means that the output signal is being fed back to the input to reinforce the input. The frequency of oscillation will be at the frequency point in the response curve where the phase shift is at 180° and the gain is greater than 1. Therefore, networks that are used in the load and in the feedback path to compensate for frequency response must be carefully selected so that the phase shift and gain criterion is not violated. Otherwise, instability occurs and the circuit oscillates instead of amplifying.

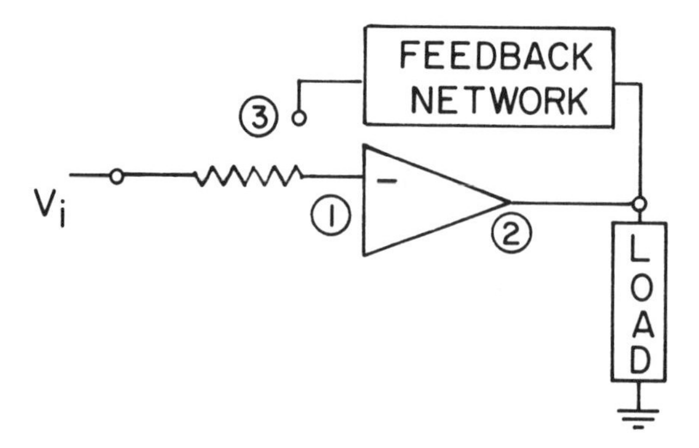

Fig. 6-16. Phase Shift of Feedback Signal.

Many operational amplifiers have external connections called "compensation" which are to be used to add external capacitance to make sure the amplifier is stable. These two terminals are shown in the amplifier circuit of Fig. 6-3 with C_c = 10 pF connected to them. The amplifier manufacturer usually specifies what this value should be in order to provide a stable amplifier over all gain variations from open loop to unity gain. Many amplifiers have internal compensation, and these connections are not available.

The external terminals (COMP and OFFSET NULL/COMP in this case) can be used to adjust the frequency response of the amplifier and make the gain fall off much faster with frequency than it normally would with 10 pF compensation inserted.

1. With the same circuit setup shown in Fig. 6-3 (used in Part I) make R_f = 39K. Adjust the input frequency to 1kHz, 50kHz, 100kHz and 500kHz. Measure the gain at each frequency.

   (a) Gain at 1kHz ________ (c) 100kHz ________

   (b) 50kHz ________ (d) 500kHz ________

   Plot these on a frequency response curve similar to Fig. 6-15.

2. Change the 10 pF compensation capacitance of the circuit of Fig. 6-3 to <u>20 pF</u> and repeat Step 1.

   (a) Gain at 1kHz ________ (c) 100kHz ________

   (b) 50kHz ________ (d) 500kHz ________

   Plot this response on the same plot as for Step 1 to compare the change in frequency response of the amplifier. If the application of a linear amplifier in a certain system requires specific frequency response characteristics, operational amplifiers can be used rather easily to obtain these characteristics.

III. 1. Integrated circuit amplifiers of the type we are using can detect and amplify signals that are applied as "differential" signals. This means that the <u>difference</u> of the signal applied at the inverting input and the non-inverting input is what is amplified by the amplifier and observed at the output. If the non-inverting input is more positive than the inverting input, the output signal will be more positive, and vice-versa.

Change the input network to the amplifier of Fig. 6-3, as shown in Fig. 6-17. Check to make sure that R_f = 100K and C_c = 10 pF. If they are different values, change them.

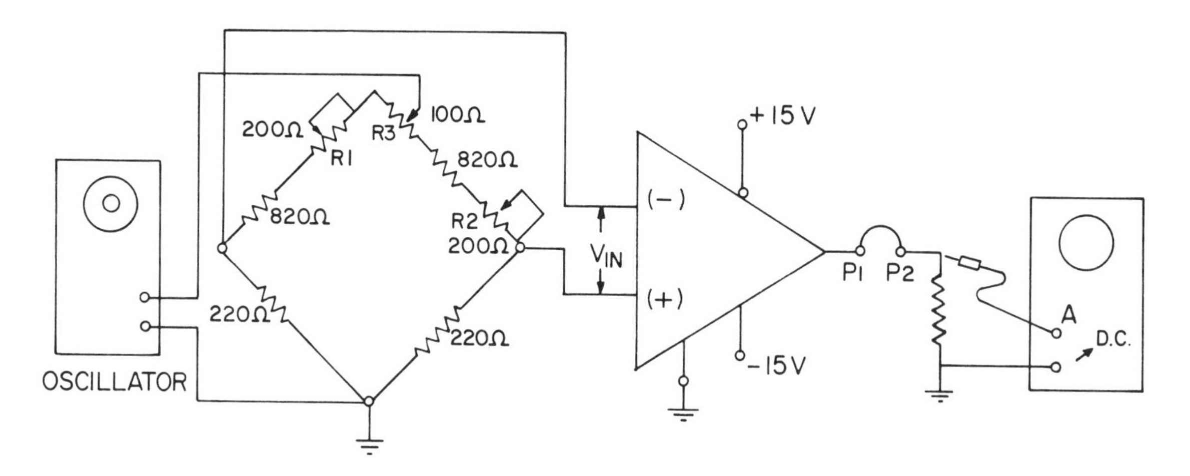

Fig. 6-17. Differential Gain.

2. Place R1 at its minimum value and R2 at its maximum value. Adjust the output of the 1,000-Hz signal from the oscillator until the output voltage is 2 volts peak to peak on the oscilloscope. Is the output voltage in-phase or out-of-phase with the oscillator voltage?

   (a) In-phase ____________

   (b) Out-of-phase ____________

   Place R1 and R2 at their minimum value and adjust R3 until the output voltage is as close to zero as possible.

3. Place R1 at its maximum value and R2 at its minimum value. Is the output voltage in-phase or out-of-phase with the oscillator voltage?

   (a) In-phase ____________

   (b) Out-of-phase ____________

   Is it opposite to what it was in Step 1?

   Place both R1 and R2 at their minimum value. What is the output signal amplitude?

   ____

   Probably it is apparent to you that if the exact same signal is placed on the inputs at the same time, there will be no differential signal. Amplifiers of this type are said to have "common-mode rejection." They reject the same signal applied to both inputs and amplify only the differential signal.

Experiment 6-4 LINEAR CIRCUITS, LARGE SIGNAL AMPLIFIERS

PURPOSE

This experiment points out the characteristics of large signal amplifiers and how operational amplifiers can be extended to obtain "power" amplifiers. Control to prevent thermal runaway is also discussed.

DISCUSSION

In Experiment 6-1 the dynamic range of an amplifier was discussed and demonstrated. Recall that when the input voltage was not reproduced exactly because the output waveform became distorted, the dynamic range of the amplifier was being exceeded. Small signal amplifier operation is always well within the dynamic range. Large signal amplifiers are using the full dynamic range.

Even though the full dynamic range is used for an operational amplifier such as the ones used in Experiment 6-3, the output current is limited because of the small-signal amplifier design. Since amplifiers of this type normally draw only 1 to 2 ma, it is difficult for them to supply a great deal of current to the load. Therefore, in order to supply watts of power to a load, additional "power" amplifier stages are used.

When amplifier stages are designed as power amplifiers, they must have these characteristics:

1. Good dynamic range - This usually means relatively high breakdown voltage for the output transistors so that sufficient voltage swing can be produced to develop the power.
2. Good current gain - Since power equals V x I, and voltage swing may be restricted because of the low load impedance, power amplifiers are required to supply large currents. They must have good current gain at these large currents.
3. Good frequency response - Even though they provide large currents to low impedances, they must maintain good frequency response.
4. Low distortion - Signals amplified in power level must still be maintained with good fidelity. The power amplifier is to boost the power level without adding distortion.
5. Low thermal resistance - Since a large amount of power is being delivered to a load, significant power is dissipated in the amplifier stage itself. Therefore, the heat generated by the power dissipated in the amplifier must be adequately "heat sinked." Otherwise, a safe semiconductor junction operating temperature will be exceeded and the amplifier will be damaged.

I. THERMAL OPERATING CONDITIONS

In Fig. 6-18 is shown a thermal model of a semiconductor chip mounted in a package and the package has some form of additional heat sink. As power is dissipated in the semiconductor chip, the junction temperature T_J of its semiconductor devices rises above the temperature of the exterior of the package which is at ambient temperature T_A. The amount of temperature rise above ambient for a given power dissipation can be expressed as a thermal resistance $R_{\theta JA}$. For example, $R_{\theta JA}$ = 100°C/W means that the junction temperature will rise 100°C for every watt that is dissipaed in the chip.

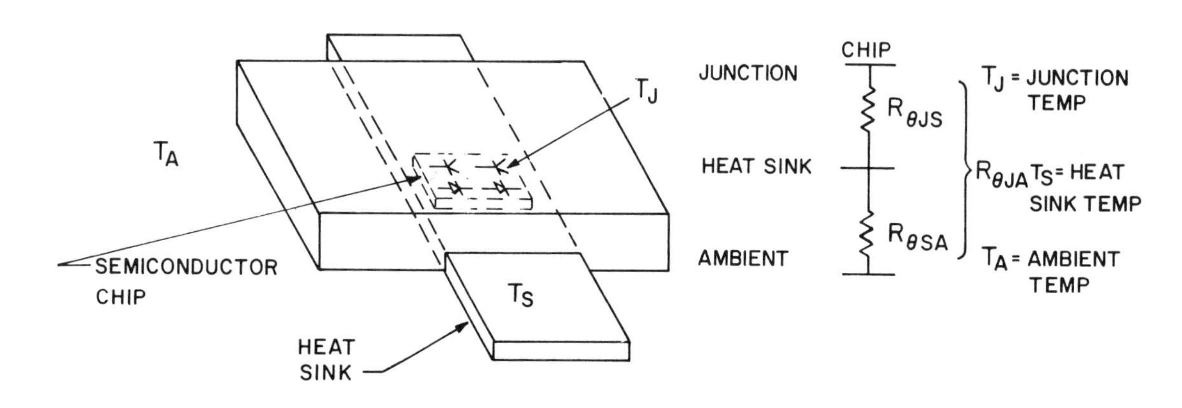

Fig. 6-18. Thermal Resistance.

The manufacturer specifies the maximum power that a given power amplifier will dissipate and the operating free-air temperature range. T_J = 150°C is usually considered as the maximum junction temperature for a silicon integrated circuit. If this T_J is exceeded, the integrated circuit devices will be damaged.

A. MAXIMUM SAFE AMBIENT TEMPERATURE WITH A GIVEN POWER DISSIPATION (WITHOUT HEAT SINK)

A power amplifier has the following manufacturer's specifications:

1. Free Air Operating Temperature: -25°C to +85°C
2. Maximum Power Dissipation: 1 Watt @ 25°C
3. Thermal Resistance $R_{\theta JA}$ = 75°C/Watt

Determine the maximum junction temperature T_J when the amplifier is operating at maximum rated power dissipation if the surrounding ambient is:

(a) T_A = 45°C T_J Max ______________

(b) T_A = -25°C T_J Max ______________

(c) T_A = 85°C T_J Max ______________

(d) Are any of these operating ambient temperatures going to damage the device?

Yes ____________

No ____________

(e) If so, which one?

a. ________

b. ________

c. ________

(f) If the answer to d above is Yes, what maximum ambient temperature would be acceptable?

T_A Max = ______________

If the maximum free-air operating temperature range ambient T_A plus the temperature rise due to the thermal resistance $R_{\theta JA}$ causes the T_J Max = 150°C to be exceeded, then the $R_{\theta JA}$ must be reduced. One way of accomplishing this is with a heat sink. Recall that $T_A + P_D \times R_{\theta JA} = T_J$.

B. MAXIMUM SAFE AMBIENT TEMPERATURE WITH A GIVEN POWER DISSIPATION (WITH HEAT SINK)

As shown in Fig. 6-19, attaching the heat sink of the package to a printed circuit board with 1 square inch of copper laminate has reduced the $R_{\theta JA}$ to 38°C per watt.

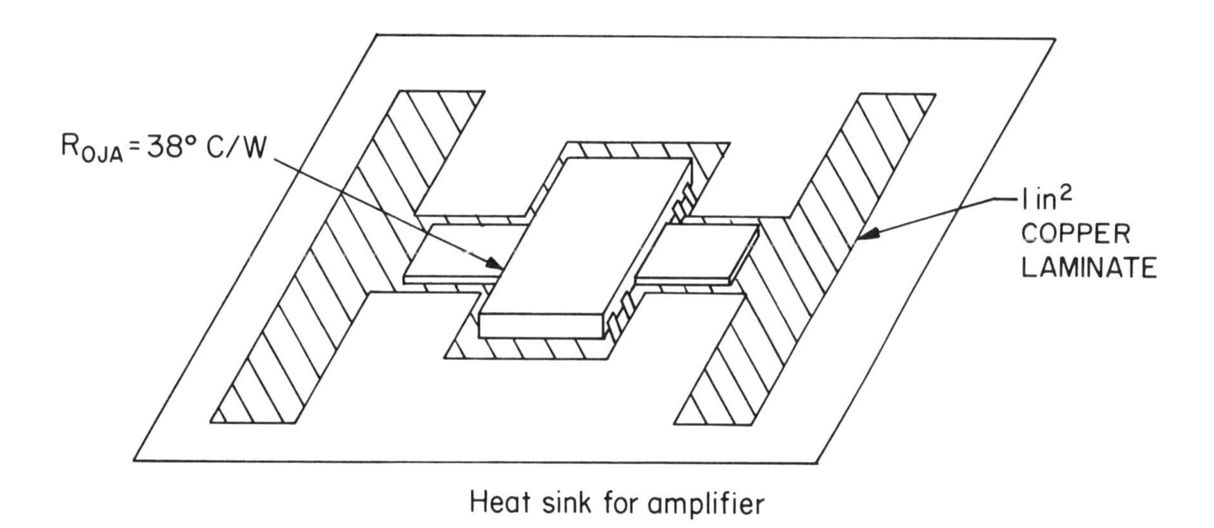

Fig. 6-19. Heat Sink for Amplifier.

Repeat the calculations of Part A.

(a) T_J Max = ______________

(b) T_J Max = ______________

(c) T_J Max = ______________

(d) No ________ Yes ________

(e) a. ________ b. ________

c. ________ None ________

(f) What maximum ambient temperature is acceptable with the heat sink?

T_A Max = ________________

C. POWER DERATING

For some power amplifiers with high thermal resistance, in order to operate over the full free-air operating temperature range, the amount of power that can be dissipated in the amplifier must be reduced from the rated maximum power. This is called "derating."

A power amplifier has the following specifications:

1. Free Air Operating Temperature: -55°C to +125°C
2. Maximum Rated Power Dissipation: 1 Watt at 25°C Rated $P_D = P_{DR}$
3. Thermal Resistance $R_{\theta JA}$ = 125°C/Watt

What is the maximum T_A at which the amplifier can be operated at rated power without a heat sink and without exceeding T_J = 150°C?

(a) T_A Max ________________

To determine safe operating conditions when an amplifier must be derated, the maximum junction temperature of 150°C is the constant that must not be exceeded. Therefore, the allowable temperature rise internal within the amplifier when operating at a particular T_A is:

$$T_{Rise} = T_{J\ Max} - T_A = (150 - T_A)°C \qquad T_{Rise} = T_R \tag{1}$$

The amount of allowable power dissipation P_{DA} by the amplifier for this T_A is:

$$P_{DA} = \frac{T_R}{R_{\theta JA}} = \frac{(150 - T_A)}{R_{\theta JA}} \text{ Watts} \tag{2}$$

Since this is the maximum amount of power that is allowed to be dissipated so that the junction temperature will not be exceeded, then the amplifier operating at the selected T_A must be derated to this power from the rated P_D. Therefore, the amount of power derating when operating at the selected T_A is:

$$\text{Rated } P_D - P_{DA} = \text{Amount of Power Derating} \tag{3}$$

This may be expressed as power derating per °C as follows:

$$\frac{P_{DR} - P_{DA}}{T_A - T_{APR}} = \text{Derating Factor W/°C} \tag{4}$$

T_{APR} is the ambient temperature at which the manufacturer has specified the derating factor for the amplifier in question. T_{APR} = 24°C

Determine the derating factor for the power amplifier in question when it is operating at T_A = 125°C.

(b) T_R = ____________

(c) P_{DA} = ____________

(d) Derating Factor ____________

From Equation (4), how much power dissipation is allowed when the T_A = 75°C?

(e) P_{DA} = ____________

The derating factor is given as watts/°C from a particular ambient temperature. When this ambient temperature is exceeded, the amount of power that the power amplifier can dissipate is determined by derating the amplifier using the derating factor times the difference in temperature between the operating T_A and the T_A at which the derating factor is given.

II. POWER AMPLIFICATION

Small-signal operational amplifiers that have insufficient power can be coupled to an amplifier designed to supply large output current in order to boost the capability of the total system to deliver power to the load; such amplifiers must also have the necessary high-voltage breakdown and be heat sinked to be able to supply the required power.

1. Calculating Power Output

Connect the power amplifier A_2, as shown in Fig. 6-20. Make sure all power supply and ground connections are made.

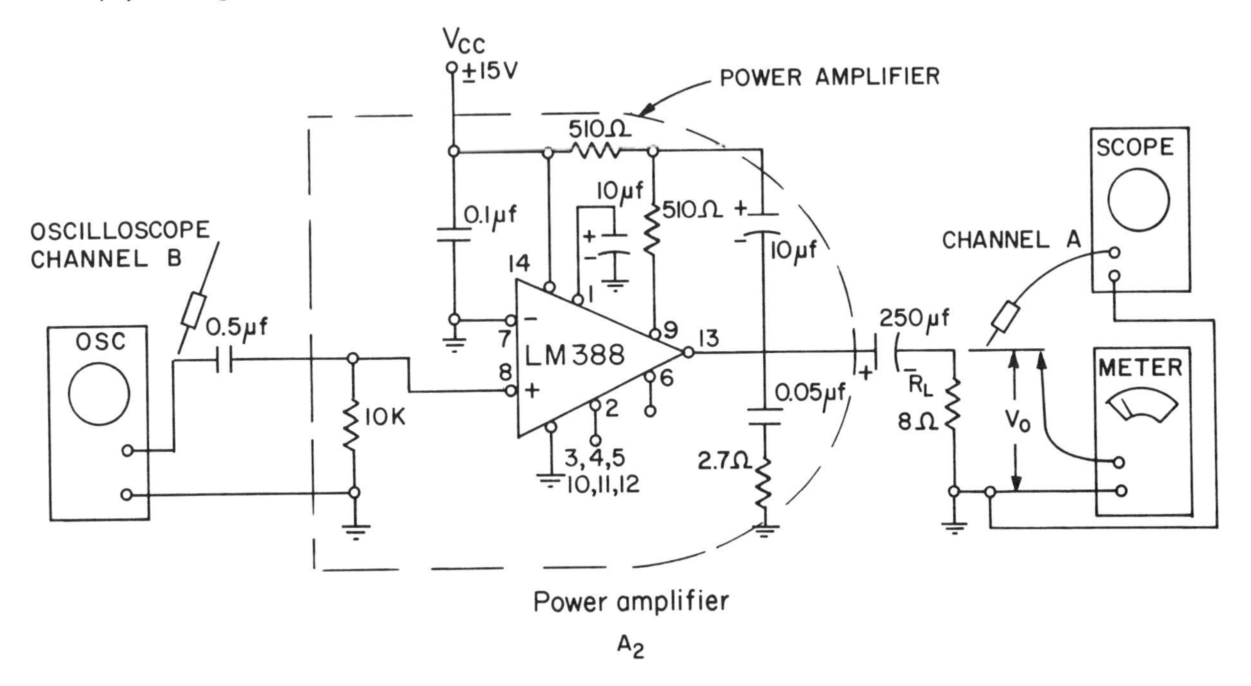

Fig. 6-20. Power Amplifier.

The +15 power supply must now be able to supply 1 amp of current. Make sure the polarity on the capacitor at the output of A_2 is connected correctly. A_2 is assembled so it has adequate heat sinking. A 9.5 μF capacitor is connected at the input so that the operation will be ac amplification. Connect the oscillator to the input.

Adjust the oscillator frequency to 1 KHz and turn the oscillator output control to zero. Connect channel B of the oscilloscope across the oscillator output.

Connect an ac voltmeter and channel A of an oscilloscope across R_L to measure V_o.

2. With R_L = 8 ohms, increase the oscillator output until V_o = 2 volts measured with the ac voltmeter.

Calculate the P_o, the power output of the amplifier by using one of the following:

$$P_o = \frac{E^2}{R}$$

$$P_o = I^2R$$

(a) P_o ______________ Watts

The ac voltmeter measures the RMS (Root-Mean-Square) value of the output voltage V_o. With Channel A of the oscilloscope measure the peak-to-peak voltage $V_{o(pp)}$.

(b) $V_{o(pp)}$ ______________ volts

Using the following, calculate $V_{o(rms)}$ the RMS value of V_o

$$V_{rms} = \frac{V_{o(pp)}}{2} \times 0.707$$

(c) $V_{o(rms)}$ ______________ volts

The dynamic voltage swing of a power amplifier $V_{o(pp)}$ must be 2.8 times V_{rms} to be able to deliver the power output without distortion. The RMS voltage value is the value that will dissipate an equivalent amount of power in a resistor as a dc voltage.

Calculate I_{rms} in the load R_L.

(d) I_{rms} ______________ ma

$$I_{rms} \times R_L = V_{o(rms)}$$

3. Measure V_o with Channel A of the oscilloscope. Increase the oscillator voltage output until a maximum power output of 2 Watts from amplifier A_2 is obtained. Calculate the value of $V_{o(pp)}$ so that the setting can be made correctly. With P_o = 2 Watts, calculate:

   (e) V_{rms} ________________

   and $V_{o(pp)}$

   (f) $V_{o(pp)}$ ________________

   Now measure V_o with the voltmeter.

   (g) $V_{o(rms)}$ ________________

   Amplifiers are rated for maximum power output under various temperature conditions. These specifications can be checked by using the above procedure and making sure the thermal characteristics are adhered to by using the calculations of Part I.

Experiment 6-5 DIGITAL CIRCUITS, LOGIC GATES AND COMBINATORIAL LOGIC CIRCUITS

This set of experiments demonstrates what digital IC's are, and how they can be made to work together. We begin with a study of AND, OR, and NOT logic circuits, which can be coupled together to form what are called "combinatorial logic circuits." Such circuits provide an interface between human information and digital codes. See Fig. 6-20A for cross referenced digital IC's parts information.

Old Part Number	Type IC	New Part Number
SN7402	Quad 2-input NOR	SN74LS02
SN7404	Hex Inverter	SN74LS04
SN7408	Quad 2-input AND	SN74LS08
SN74LS00	Quad 2-input NAND	same
SN7432	Quad 2-input OR	SN74LS32
SN74LS20	Dual 4-input NAND	same
SN74LS30	8-input NAND	same
SN74LS73	J-K F.F.	SN74LS73A
SN74LS175	4-Bit Latch	same

Fig. 6-20 A. In Experiments 6-5 to 6-8, substitute the cross-referenced part number for the old part if it is not available.

PURPOSE

This experiment introduces digital integrated circuits, and discusses how electronic circuits can be designed to make decisions.

DISCUSSION

Linear electronic circuits are circuits that use certain properties of electricity (voltage, current, resistance, inductance, capacitance) to represent analogs of physical quantities that vary in a smooth and continuous fashion. Digital electronic circuits, on the other hand, represent physical quantities by a combination of bits that assume only discrete levels. The respective combination of bits with values at set discrete levels form codes to identify a specific total digital value that can represent a symbol, a character or a number.

SIMPLE DIGITAL CIRCUIT

In Fig. 6-21, the output of the circuit V_0 can only assume two values, either +5 V when the switch is closed, or 0 V when the switch is open.

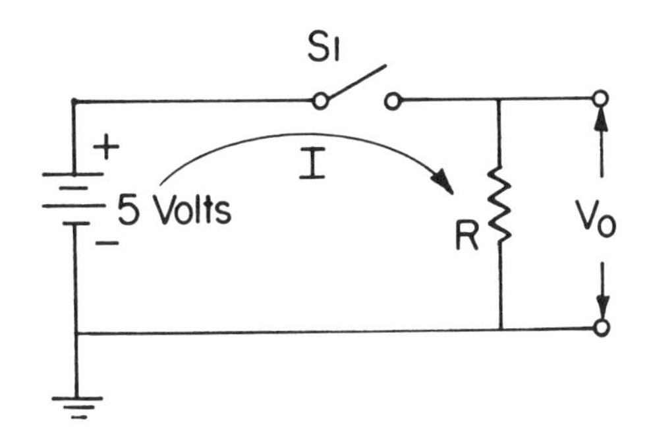

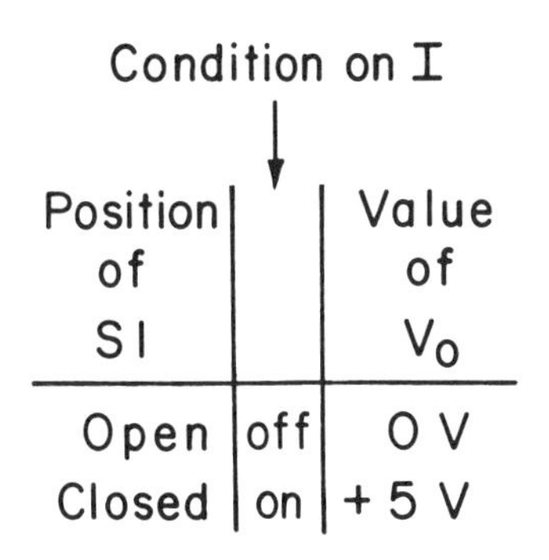

Condition on I

Position of SI		Value of V_o
Open	off	0 V
Closed	on	+ 5 V

Fig. 6-21. Simple Digital Circuit.

This is a simple digital circuit. The switch S1 can be considered an input. It can assume only two values - open or closed. The output can assume only two values, 0 or +5 volts. This digital circuit is called a "binary" circuit (binary meaning two). It represents its bit of information in one of two discrete levels. In like fashion, the current in the circuit I assumes two values: ON when the switch is closed and OFF when the switch is open.

TRUTH TABLES

In order to describe the behavior of a typical digital circuit, the condition of the output for each possible condition of the input is summarized in a table called a "truth" table.

In order that the language of logic circuits could be more universal - and not dependent on a particular voltage or current level - the two discrete values that a binary circuit could have are identified as "1" and "0". Therefore, the truth table for the circuit of Fig. 6-21 is as shown in Fig. 6-22.

SI	V_o
0	0
I	I

Fig. 6-22. Truth Table of Simple Digital Circuit.

DIGITAL CODES

Obviously, in this case there is only one quantity that can assume two values, only one bit of information is represented. To represent more values, more bits must be added to the code. For example, in Fig. 6-23 are shown 2, 3, 4-bit and n-bit codes.

2-Bit Binary Code	3-Bit Binary Code	4-Bit Binary Code	n-Bit Binary Code
00	000	0000	0----------0
01	001	0001	
10	010	0010	
11	011	0011	n-bits
Can represent 4 things	100	0100	
	101	0101	
	110	0110	Can represent 2^n things
	111	0111	
		1000	
	Can represent 8 things	1001	
		1010	
		1011	
		1100	
		1101	
		1110	
		1111	
		Can represent 16 things	

Fig. 6-23. 2, 3, 4, and n-Bit Digital Codes.

The codes shown in Fig. 6-23 can be used to represent numbers, letters, characters, symbols, different conditions of a machine, different locations for storing information. The number of different things or conditions that can be represented by an n-bit binary code is 2n, where every bit in the code assumes a value of 1 or 0.

DECIMAL NUMBER CODES

A decimal number can be represented in a binary code by the equation:

$$\text{Decimal Number} = A_0 2^0 + A_1 2^1 + A_2 2^2 + A_3 2^3 + A_4 2^4 + A_5 2^5 \text{ --- } A_n 2^n$$

where the coefficient A_0, A_1, A_2, A_3 --- A_n can have either a value of 1 or a value of 0. Therefore, the place value of the product term is determined by 2 raised to the power equal to the subscript of the coefficient (the significant position of the coefficient). For example, the place value of the A_0 bit position is $2^0 = 1$; the A_1 position is $2^1 = 2$; the A_2 position is $2^2 = 4$; the A_3 position is $2^3 = 8$, etc. When the coefficient is 1, then the place value is included in the sum of all product terms for the n-bit code. If the coefficient is 0, the product term is not included in the sum.

Using these rules, an equivalent decimal number binary code can be formed as shown in Fig. 6-24. The 4-bit code is just an example. As large a number as required can be represented by adding more significant bits to the code. Obviously, the eight different codes could be used to identify any eight things. Different letters could have been assigned to the codes rather than the decimal numbers, or special symbols to make typewriter characters, or eight different commands to a digital system.

Decimal Number	4-Bit Code *8 4 2 1	Comments	
0	0 0 0 0	Every coefficient is 0	
1	0 0 0 1	$A_0 = 1$	Sum = 1 + 0 + 0 + 0
2	0 0 1 0	$A_1 = 1$	Sum = 0 + 2 + 0 + 0
3	0 0 1 1	$A_0, A_1 = 1$	Sum = 1 + 2 + 0 + 0
4	0 1 0 0	$A_2 = 1$	Sum = 0 + 0 + 4 + 0
5	0 1 0 1	$A_0, A_2 = 1$	Sum = 1 + 0 + 4 + 0
6	0 1 1 0	$A_1, A_2 = 1$	Sum = 0 + 2 + 4 + 0
7	0 1 1 1	$A_0, A_1, A_2 = 1$	Sum = 1 + 2 + 4 + 0
8	1 0 0 0	$A_3 = 1$	Sum = 0 + 0 + 0 + 8

*Weight of Bit Position In Sum

Fig. 6-24. Binary Code for Decimal Numbers.

LOGIC CIRCUITS

Digital logic circuits use digital codes to implement logical statements, make corrections, make decisions, perform arithmetic, etc. There are three basic logic elements that are implemented with digital circuits. The first of these is called the "AND" circuit.

AND Logic Circuit

Look at Fig. 6-25. It is very similar to Fig. 6-21 except it has S2 in the circuit in series with S1.

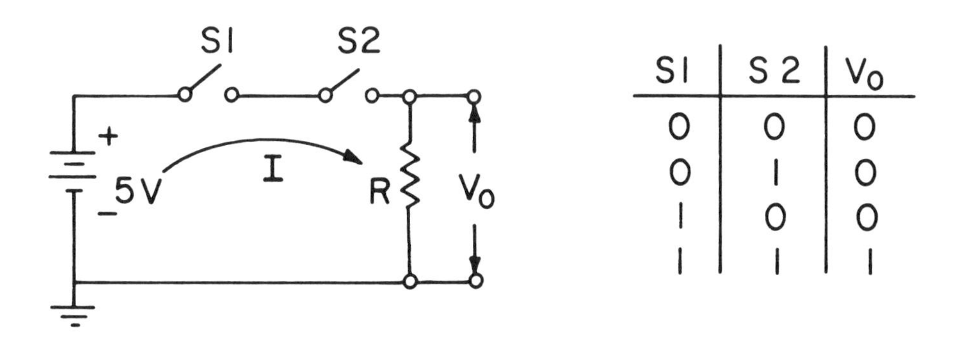

S1	S2	V_o
0	0	0
0	1	0
1	0	0
1	1	1

Fig. 6-25. AND Logic Circuit.

In the truth table, 0 represents an open switch and 1 represents a closed switch for the inputs, and 0 represents zero volts and 1 represents +5 V at the output. Note that a 1 appears at the output only if S1 and S2 are a 1. If S1 is 1 and S2 is 0, V_O will be a 0. The symbolic equation used by logic designers is $S1 \cdot S2 = V_O$. This is spoken as follows: S1 AND S2 equals V_O.

OR Logic Circuit

Next, examine Fig. 6-26. This circuit is very similar to Fig. 6-21 except that now S1 and S2 are in parallel in the circuit.

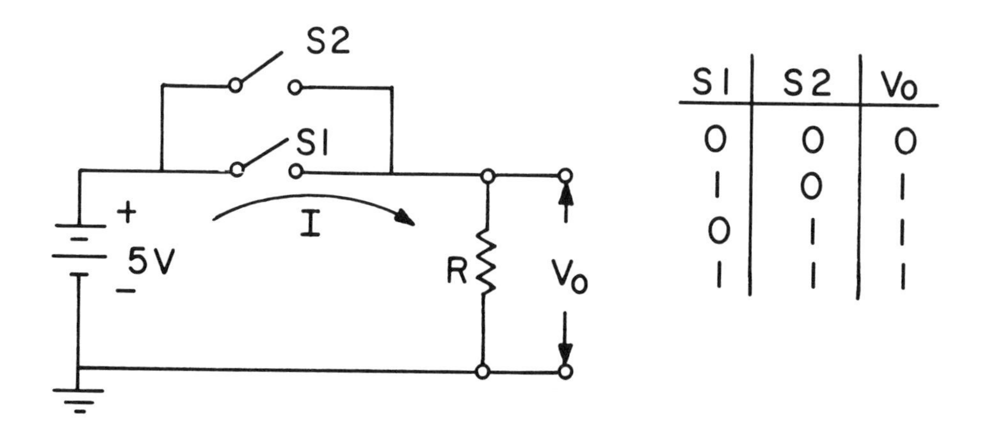

S1	S2	Vo
0	0	0
1	0	1
0	1	1
1	1	1

Fig. 6-26. OR Logic Circuit.

Looking at the truth table, (where again 0 represents an open switch and 1 a closed switch at the input, and 0 represents zero volts and 1 represents +5 V at the output), it is seen that a 1 appears at the output when either S1 or S2 or both are a 1. Only when S1 and S2 are both 0 is the output 0. The symbolic equation used by logic designers is $S1 + S2 = V_O$. This is spoken as follows: S1 OR S2 equals V_O.

NOT Logic Circuit (Inverter)

One other circuit is necessary to complete the assortment required to implement all types of decisions with digital logic circuits. It is called an inverter or a NOT logic circuit. Its truth table is shown in Fig. 6-27.

The output is just the opposite the input (It is NOT the input). It is also very commonly called an inverter because the output is in the inverted state from the input.

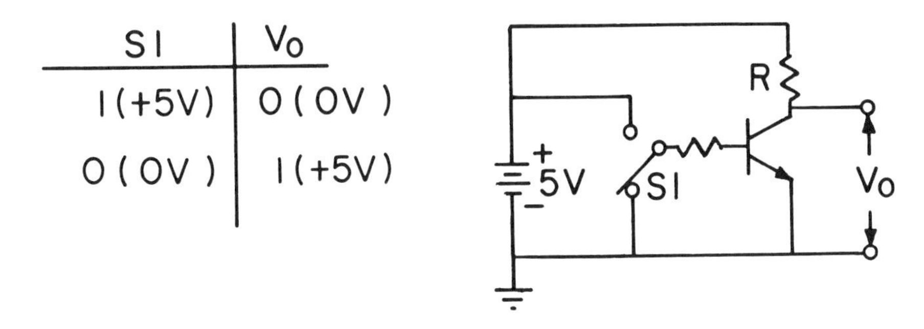

SI	Vo
1(+5V)	0(0V)
0(0V)	1(+5V)

Fig. 6-27. NOT Logic Circuit.

Digital logic circuits built in integrated circuit form implement the truth tables for the AND, OR and NOT logic functions.

EQUIPMENT REQUIRED

Digital Integrated Circuits - (AND, OR, NOT, NOR and NAND)
+5 V Supply - 1 Ampere
Multimeter
Board Circuits with sockets, resistors, capacitors, switches, etc.

I. AND LOGIC CIRCUITS

A digital integrated circuit that has two inputs and one output and performs the AND function is identified with the following symbol:

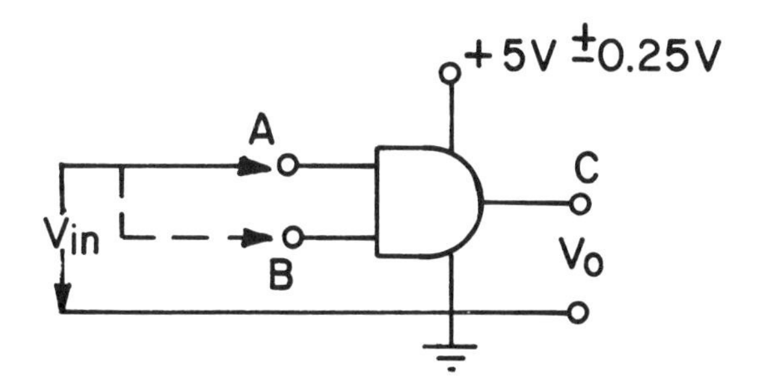

Fig. 6-28. AND Function.

This symbol is commonly referred to as an AND logic gate.

A very common type is a TTL (Transistor - Transistor Logic) circuit which uses a +5 V power supply and ground as the power connections. The 0 and 1 levels are identified by the following voltage levels at the input pins A and B or the output pin C:

Table 6-1. Logic Levels.

Level	Voltage Range
1	+ 2.5 V to + 5 V
0	0 V to + 0.5 V

A standard multimeter may be used to verify the correct voltage at each terminal.

PROCEDURE

1. Connect a SN7408 digital integrated circuit (a 14-pin integrated circuit package that has 4 2-input AND gates) into the circuit of Fig. 6-29.

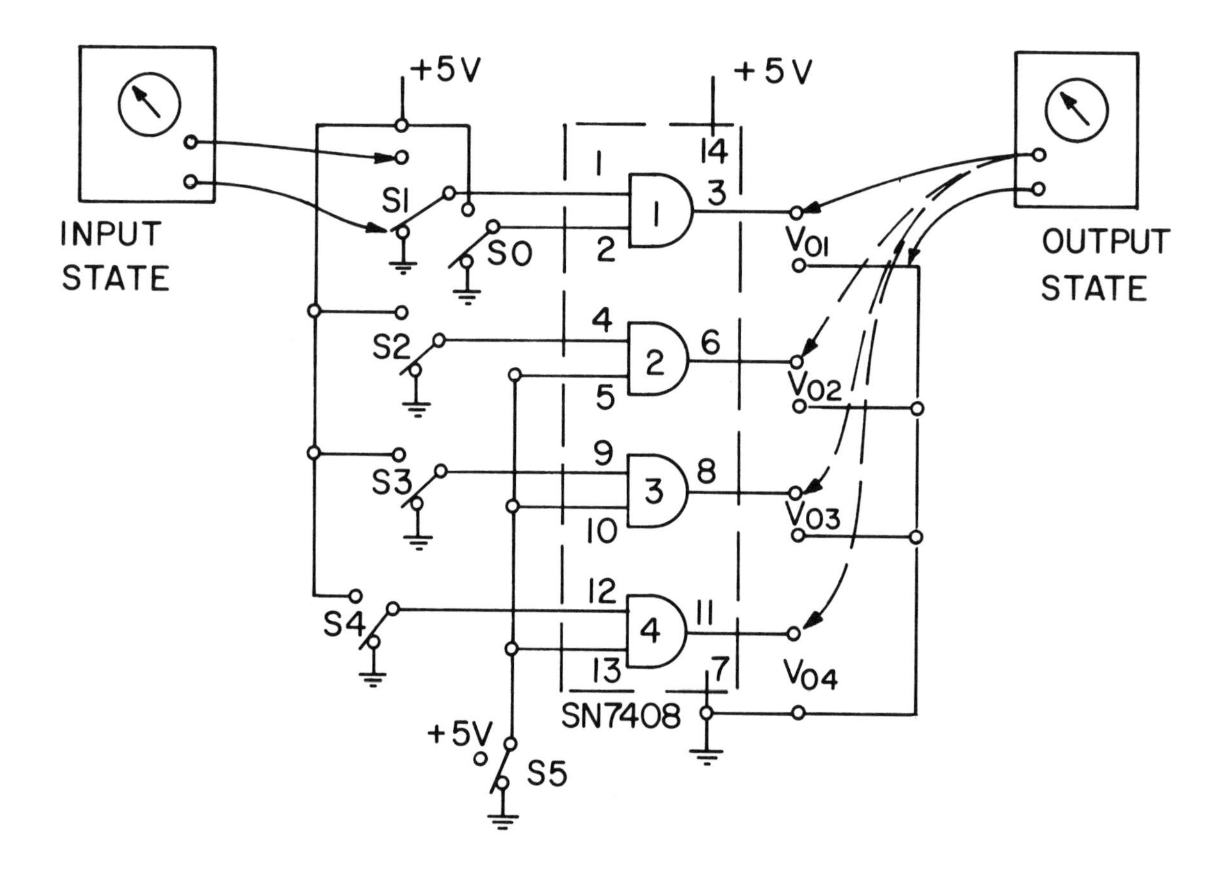

Fig. 6-29. SN7408 and Logic Gate.

The output and input voltage will be measured with the multimeter. If the voltage level is within the ranges shown in Table 6-2 identify the input or output state as a 0 or a 1 accordingly.

2. Set S2, S3, S4 and S5 to ground (zero volts) position. With S0 and S1 and voltage readings, verify the truth table of the 2-input logic gates. Fill in the table with 1's and 0's.

(a)

INPUTS		OUTPUT
1	2	V_0

Table 6-2.

The AND gate should have the same truth table as Fig. 6-25.

3. With S0 at ground (0), change S1 from a 0 to a 1. What happens to the output state?

(b) ____________________

4. With S0 at +5 V (1), change S1 from a 0 to a 1. What happens to the output state?

(c) ____________________

5. Many times an input on an AND gate will be used to simply transfer or "gate through" a signal to the output. To observe how this may be accomplished, try the following:

With the S5 at 0, change the S2, S3, S4 signals from 0 to 1 in any combination and measure the output states of gates 2, 3, and 4. Is there any change?

(d) Yes ________

(e) No ________

(f) Why? ____________________

6. Now set S5 = 1 and change S2, S3 and S4 signals from 0 to 1 in every combination possible and measure the output states of gates 2, 3 and 4. Do the outputs change?

   (g) Yes ________

   (h) No ________

   (i) Why? ____

7. If yes, fill in Table 6-3 for the input and output logic states for all combinations of inputs.

   (j)

INPUTS				OUTPUTS		
S2	S3	S4	S5	VO2	VO3	VO4
			1			
			1			
			1			
			1			
			1			
			1			
			1			
			1			

Table 6-3.

A common input to a network of AND gates can be used to "gate" signals through from input to output. This principle will be used in a later experiment for an encoder circuit.

II. NOT LOGIC CIRCUIT (INVERTER)

A digital integrated circuit that has one input and one output, where the output state is the inverted input state, performs the NOT function and is identified in Fig. 6-30.

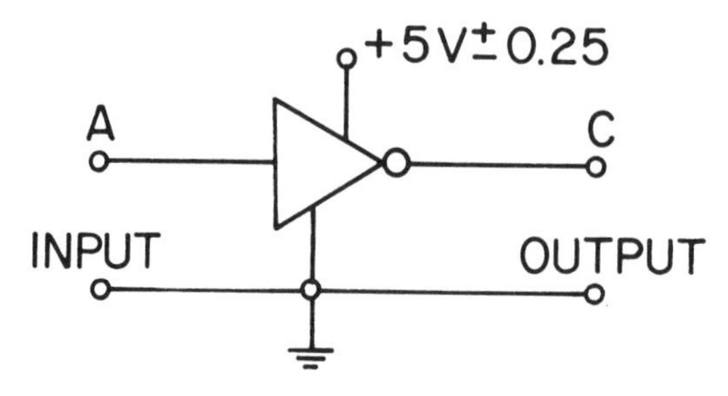

Fig. 6-30. Inverter Logic Circuit.

This symbol is commonly referred to as an INVERTER logic circuit.

TTL circuit voltages will again be used and the logic levels are the same as for the AND gate. As before, measure the voltage with a multimeter.

PROCEDURE

1. Connect a SN7404 digital integrated circuit (a 14-pin package that has 6 inverter circuits) into the circuit of Fig. 6-31. It extends the circuits from Fig. 6-30.

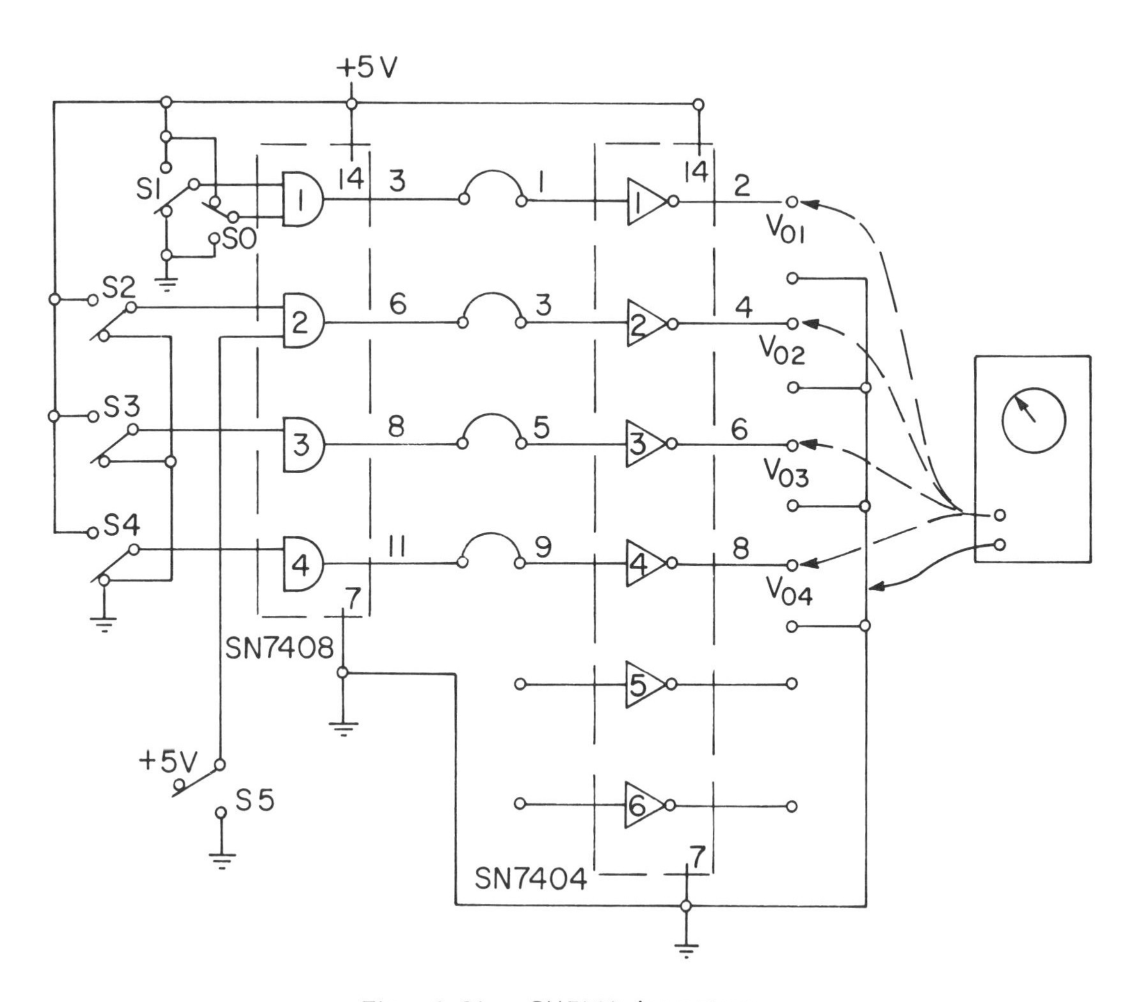

Fig. 6-31. SN7404 Inverters.

2. With S5 = 1 and S0 = 1 (+5 V), change S1, S2, S3 and S4 each separately from 0 to 1, measure the output voltage V_o for inverter 1, 2, 3 and 4 separately as the input is changed. Fill in Table 6-4 with the logic levels at the outputs:

(a)

INPUTS						OUTPUTS			
S0	S1	S2	S3	S4	S5	V_{01}	V_{02}	V_{03}	V_{04}
1	0	0	0	0	1				
1	1	0	0	0	1				
1	0	1	0	0	1				
1	0	0	1	0	1				
1	0	0	0	1	1				

Table 6-4.

How do the output logic levels for V_{02}, V_{03}, V_{04} compare to those filled in for Table 6-3 of Part (j) of the AND gate Experiment 6-5 for the same logic levels on S2, S3, S4 and S5?

(b) ______________________________

3. Switch S5 to a 0 logic level. With voltage readings at S1 and S0 for inputs and V_{01} as the output, repeat the verification of a 2-input gate truth table as in Part 1 of I.

(c)

INPUTS		OUTPUT
1	2	V_{01}
0	0	
0	1	
1	0	
1	1	

Table 6-5.

Describe the difference in the V_{01} output voltage levels you have described in Table 6-5 above from the results you observed in Part 1 of I.

(d) ______________________________

____

____

NAND LOGIC GATE

A logic gate that combines an AND and NOT gate in one gate is called a NAND gate (not AND), and in many cases is much easier to make in integrated circuit form than an AND gate alone. The symbol for a 2-input NAND gate is shown in Fig. 6-32.

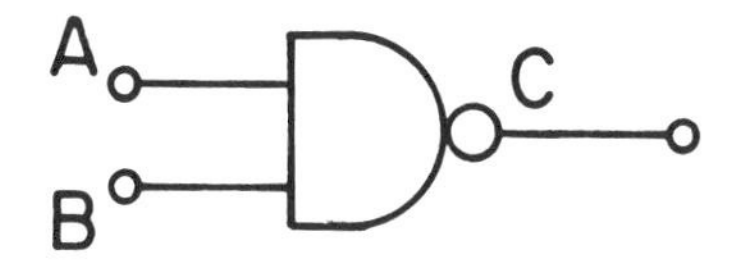

Fig. 6-32. 2-Input NAND Gate.

It is equivalent to Fig. 6-33.

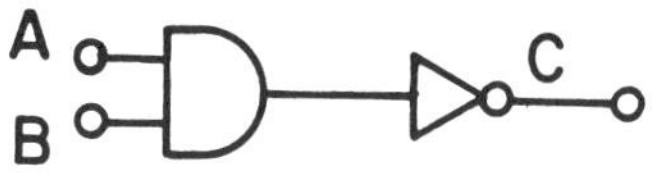

A
B
C

Fig. 6-33.

4. Remove the SN7408 from the circuit of Fig. 6-29 and substitute a SN74LS00, which contains 4 2-input NAND gates rather than 4 2-input AND gates. It is another type of TTL digital integrated circuit called low-power Schottky TTL. It basically is the same as the standard TTL circuit, but will operate at faster speeds with less power dissipation. The SN74LS400 has the same pin connections as the SN7408, each 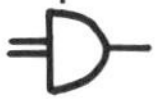 symbol is simply replaced with a  symbol.

   Verify that the 2-input NAND gate of a SN74LS00 has the same truth table as Step 2(c) by switching S5 to a 0 and measuring V_{01} while switching S0 and S1 through every combination. Does it check?

   (e) Yes ________

   (f) No ________

III. OR LOGIC CIRCUITS

A digital integrated circuit that has two inputs and one output and performs the OR function is identified in Fig. 6-34.

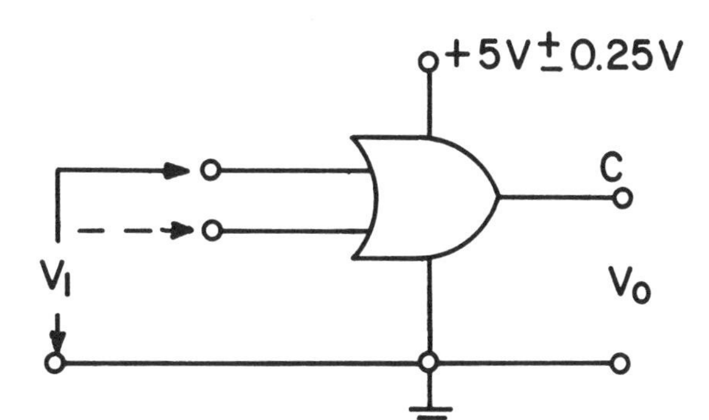

Fig. 6-34. OR Logic Gate.

This symbol is commonly referred to as an "OR" logic gate.

TTL circuit voltages will again be used, and the logic levels are the same as for the AND and NOT circuits. Voltage measurements again are made with a multimeter.

PROCEDURE

1. Into the circuit of Fig. 6-29 plug in a SN7432 (a 14-pin package that has 4 2-input OR gates).

   Set S2, S3, S4 and S5 to 0 logic level. With S0 and S1 and voltage readings verify the truth table of a 2-input OR logic gate. Fill in Table 6-6 with 1's and 0's.

   (a)

INPUTS		OUTPUT
1	2	V_{01}
0	0	
0	1	
1	0	
1	1	

Table 6-6.

   The OR gate should have the same truth table as Fig. 6-26 that we developed earlier.

2. With S0 at a 0 logic level, change S1 from a 0 to a 1. What happens to the output state?

   (b) ______________________________

   With S0 at a 1 logic level, change S1 from a 0 to a 1. What happens to the output state?

   (c) ______________________________

3. Many times an OR gate will be used to gate through data from input to output, or else set a particular bit to a 1. For example, if S5 = 1 in Fig 6-29 with the SN7432 in place of the SN7408, each of the outputs V_{01}, V_{03} and V_{04} are set to the 1 level and whatever appears on the inputs has no effect.

   Now set the S5 = 0 and switch inputs S2, S3 and S4 from 0 to a 1 logic level. What happens to outputs V_{02}, V_{03} and V_{04}?

   (d) ______________________________

   Switch inputs S2, S3 and S4 back to 0. What happens to outputs V_{02}, V_{03}, V_{04}?

   (e) ______________________________

IV. NOR LOGIC GATE

In like fashion to the NAND gate, an OR gate with an inverter on the output is called a NOR (not OR) gate. The symbol is shown in Fig. 6-35.

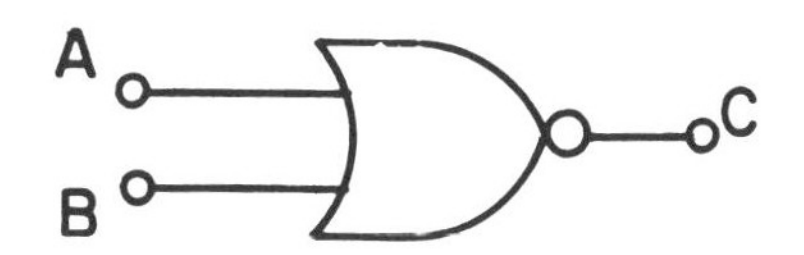

Fig. 6-35. NOR Logic Gate.

which is equivalent to Fig. 6-36.

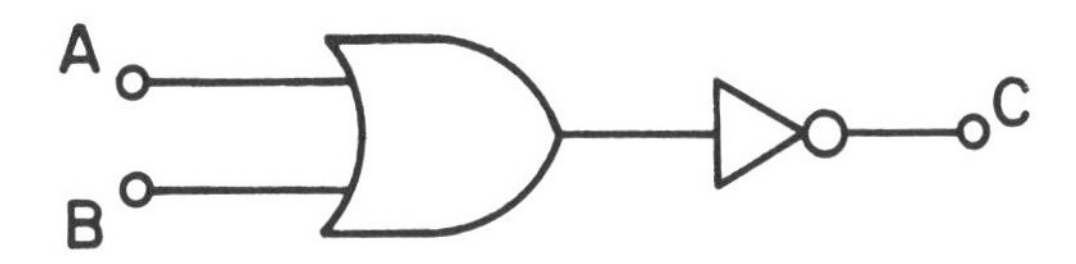

Fig. 6-36.

As an added exercise, students may verify that the above two diagrams are equivalent by measuring voltages on a 2-input NOR gate such as a SN7402. It has the same pin connections as the SN7408 of Fig. 6-29 and this circuit diagram can be used by just plugging a SN7402 into the socket.

Experiment 6-6 COMBINATORIAL LOGIC CIRCUITS: THE ENCODER AND DATA SELECTOR

I. THE ENCODER

Now that the AND, NOT, OR, NAND and NOR logic circuits are understood, circuits which include mixtures of some of these can be formed to make what are known as "combinatorial" circuits. Combinatorial circuits are circuits that always produce the same logic level outputs for a unique combination of logic levels at the inputs.

One such circuit that we will explore in detail is an "encoder" which is used to take human inputs and encode them into digital codes so that electronic circuits can handle the information.

Fig. 6-37 illustrates such an encoder. On the left, there are 10 toggle switches numbered 0 through 9. These are SPDT switches, with the center arm connected at one input of a two-input NAND gate. When a switch is ON the encoder will output the digital code for the respective number when the store button is pushed. Only one switch can be ON at any one time. The STORE push button energizes a common gating line to all the input gates, as shown previously, so that the input switches can be set and will be stable before the data is presented to the encoder.

The encoder output is a 4-bit binary code representing the decimal number of the switch that is ON. Since there are 10 switches, the encoder must output 10 different codes. Four output lines ($2^4 = 16$) must be used to generate the 10 different codes (six of the total of 16 are not used). (Note: At least 4 lines must be used, since 3 lines would provide only 2^3 or 8 outputs.

In summary, the encoder of Fig. 6-37 will output a binary code which is equivalent to the decimal number represented by the switch that is ON. To accomplish this, NAND gates are connected in the matrix shown.

The digital codes will be verified by multi-meter measurements. TTL logic levels identify the bit as a 1 or a 0.

PROCEDURE

1. First, wire the circuit as shown in Fig. 6-37. Make sure the 3 SN74LS00 (Quad 2-input NAND gates), the SN74LS20 (Dual 4-input NAND gates) and the SN74LS30 (8-input NAND gate) are correctly in place. The inputs are numbered from 0 to 9. When the switch is OFF the input level to the encoder is at 0 (zero volts); when the switch is ON the input level to the encoder is at 1 (+5 volts). Only one switch may be on at a time.

   The output code will have 4 bits identified as d_3, d_2, d_1 and d_0. d_3 is the most significant bit (MSB) (i.e., It has the largest equivalent decimal place value weight of 8.) d_2 is the next most significant with a weight of 4. d_1 is next with a weight of 2, and d_0 is the least significant bit (LSB) with an equivalent decimal place value weight of 1. These bits will be identified as having a 0 or 1 level by measuring the voltage level on the outputs as shown in Fig. 6-37.

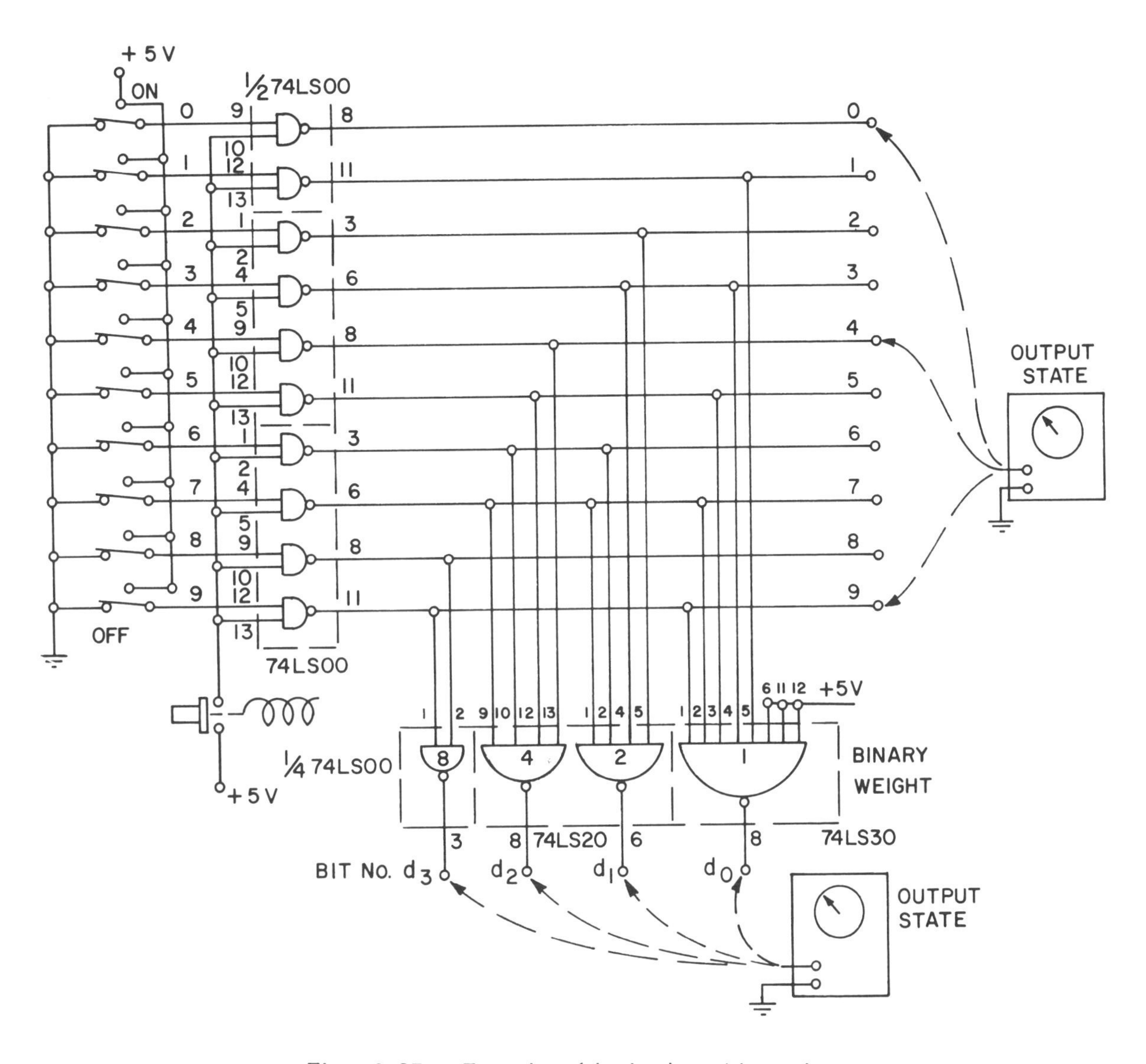

Fig. 6-37. Encoder (decimal to binary).

Note in Fig. 6-37 that the outputs of the input NAND gates are also available to measure. These are not normally available in an encoder but are made available in this experiment to allow further understanding of the operation of the encoder.

2. Refer to Table 6-9A. This is a truth table for the encoder. All of the output conditions are to be plotted for all the possible input conditions. A "1" is plotted for an input condition when the switch is ON; a "0" when the input switch is OFF. The resultant output codes are determined by the voltage level measurement. Make sure the STORE button is pressed when measuring the output voltages.

(a) Start with switch 0 and plot the values of all the inputs. Measure d_3, d_2, d_1 and d_0 and plot the output code. Table 6-7 is an example for switch 0:

Inputs										Outputs			
										d_3	d_2	d_1	d_0
0	1	2	3	4	5	6	7	8	9	(8)	(4)	(2)	(1)
1	0	0	0	0	0	0	0	0	0	0	0	0	0

Table 6-7.

(b) In addition, Table 6-9B is a table for the input NAND gate outputs. Fill this in with the logic levels. Table 6-8 is an example for switch 0:

Input NAND Gate Outputs									
0	1	2	3	4	5	6	7	8	9
0	1	1	1	1	1	1	1	1	1

Table 6-8.

(c) This encoder is called a combinatorial logic circuit. For each input combination there is only one output code. Each time the respective input combination is present, the same unique output code will be present and available just as soon as the encoder is enabled with the STORE button. Each student should verify this for themselves.

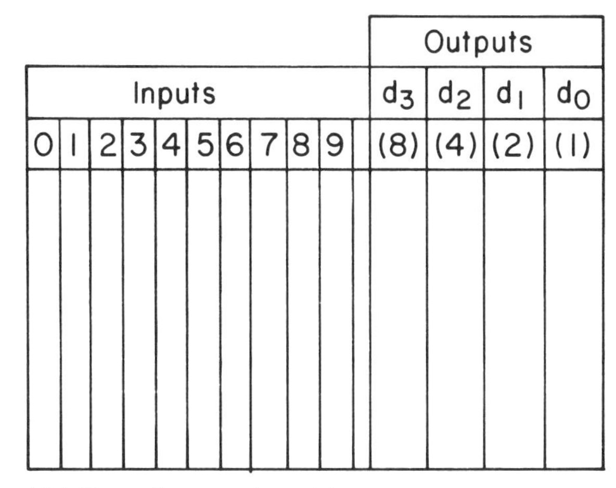

Inputs										Outputs			
										d_3	d_2	d_1	d_0
0	1	2	3	4	5	6	7	8	9	(8)	(4)	(2)	(1)

(A) Encoder truth table

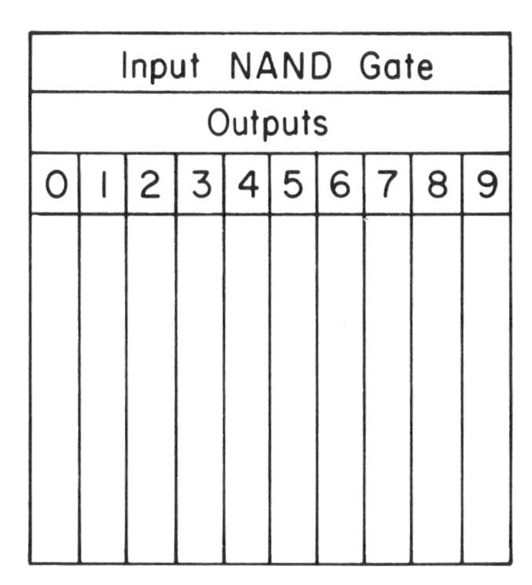

Input NAND Gate Outputs									
0	1	2	3	4	5	6	7	8	9

(B) Table of NAND gate outputs

Table 6-9. Encoder Tables.

3. Using the output code you've recorded in Table 6-9, verify that the code is indeed the decimal equivalent of the input by adding the place values indicated by a 1 in the bit position of the output code. Put zero in the place value if the output code has a 0 on the bit position. Fill in Table 6-10 shown below. (An example is filled in for you.)

INPUT	OUTPUT d_3	d_2	d_1	d_0	DECIMAL EQUIVALENT 8 + 4 + 2 + 1	=	
0	0	0	0	0	0 + 0 + 0 + 0	=	0
1					+ + +	=	
2					+ + +	=	
3					+ + +	=	
4					+ + +	=	
5	0	1	0	1	0 + 4 + 0 + 1	=	5
6					+ + +	=	
7					+ + +	=	
8					+ + +	=	
9					+ + +	=	

Table 6-10.

4. Note in Fig. 6-37 how the outputs of the first set of NAND gates are coupled to the inputs of the NAND gates at the bottom of the picture. This wiring enables the circuit to provide the necessary place value. Each student should verify that the output NAND gate logic interconnections are correct.

5. As an extra added exercise, students may want to verify that the encoder of Fig. 6-37 can also be constructed using input AND gates in place of the input NAND gates and output OR gates in place of the output NAND gates.

II. A DATA SELECTOR

Another combinatorial logic circuit that is quite useful is called a "data selector" or multiplexer (MUX). Such circuits are used to route digital information over selected paths. In digital systems the name "bus" is given to the bundled wires that carry the many bits of a digital code in paralle (at the same time). As shown in Fig. 6-38, data selectors are used to switch digital codes coming in on a common bus to selected output buses. As shown, the output bus selected is determined by the logic level on a control line.

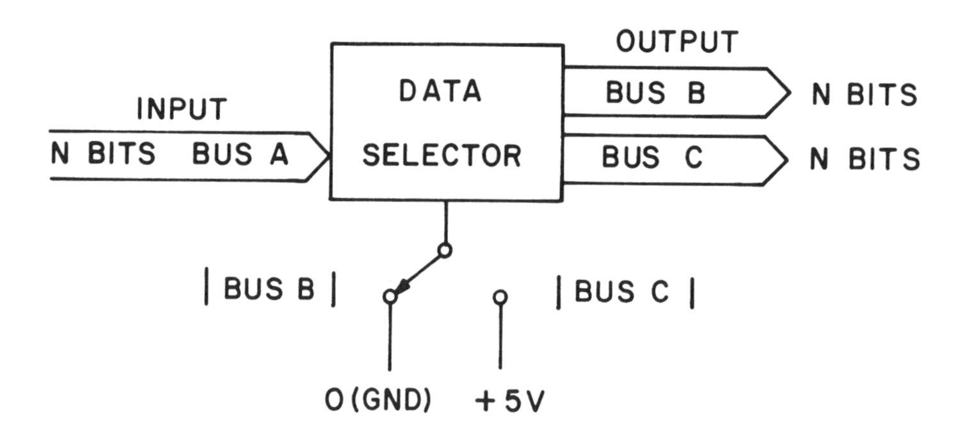

Fig. 6-38. Data Selector.

AND gates used in combination make excellent data selectors. Fig. 6-39 shows such a combinatorial circuit. SN7408 (Quad 2-input AND gates) are used to switch the input logic levels on d_3, d_2, d_1 and d_0 to the corresponding bit line for bus A or bus B. The output bus is determined by a control signal of 1 on the respective control line for the bus.

PROCEDURE

1. Wire the circuit shown in Fig. 6-39, making sure that the 2 SN7408 integrated circuits are in place. Measure the output levels on d_3, d_2, d_1 and d_0 of bus A and bus B for various code combinations of input signals. Set S1 to left so there is a logical 1 on bus A control line. Set S2 to 0 and S3 to 1. Fill in Table 6-11 with output voltage levels for bus A - row "a" on the table.

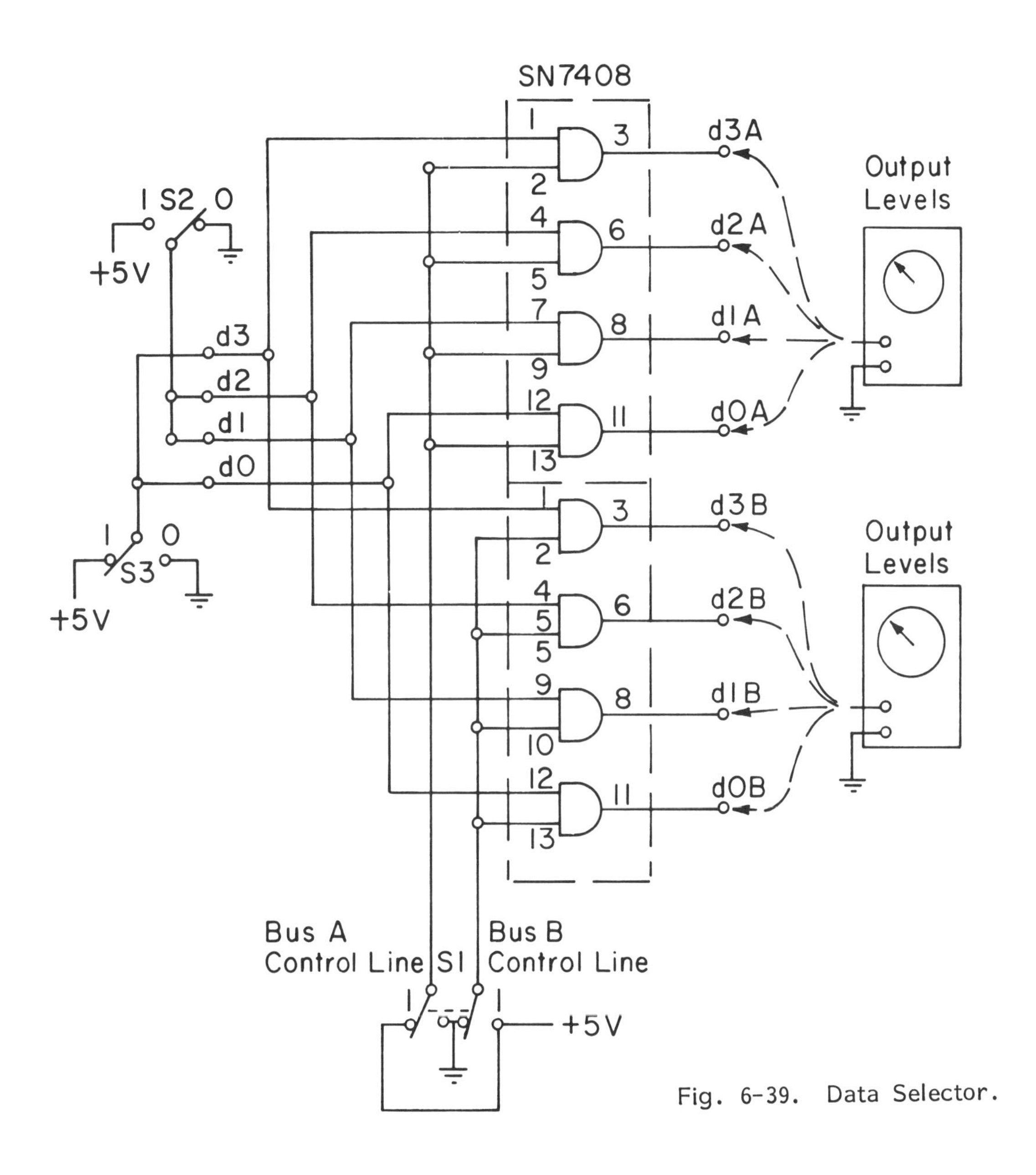

Fig. 6-39. Data Selector.

	S_2 S_3	INPUTS				BUS A				BUS B			
		d_3	d_2	d_1	d_0	d_{3a}	d_{2a}	d_{1a}	d_{0a}	d_{3b}	d_{2b}	d_{1b}	d_{0b}
a.	0 0												
b.	1 0												
c.	1 0												
d.	1 1												

Table 6-11.

2. Set S2 to 1 and S3 to 0. Measure the output voltage levels again. Fill in Table 6-11 with the output levels on line b.

3. Set S1 to right so there is a logical 1 on Bus B control line. S2 is still to 1 and S3 to 0. Measure the output voltage levels again. Fill in Table 6-11 with the output on line c.

4. Set S2 to 1 and S3 to 1. Measure the output voltage levels again. Fill in Table 6-11 with the output levels for d.

Similar kinds of data selectors can be designed that have multiple bus inputs and select the input to be put onto one output bus. The principles are the same as shown for the data selector ypu've worked with here.

Experiment 6-7 DIGITAL CIRCUITS - SEQUENTIAL LOGIC CIRCUITS - FLIP-FLOPS AND LATCHES

Combinatorial logic circuits always produce the same logic level outputs for a unique combination of logic levels on the inputs. Sequential logic circuits differ in that the output logic levels depend, not only on the input logic levels at the present moment, but on what happened previously, on the past history of the circuit. The reason for this is that sequential logic circuits have so-called memory. They remain in states set by past input signals. Sequential circuits are used for flip-flops, latches, registers, random access and sequentially-accessed memories.

PURPOSE

This experiment provides an understanding of the fundamental bistable sequential logic circuit called a "flip-flop". Flip-flops are the basis for other circuits derived from it, such as latches, counters, registers and memory circuits used for storing digital data.

DISCUSSION

When two simple NAND gates are interconnected as shown in Fig. 6-40 they form a special kind of circuit that has two stable states. It is called a bistable logic circuit and has a common name called "Flip-flop". As the name implies, it flips into one of the bistable states upon receipt of an input signal and stays there upon removal of the input signal. Upon receipt of another input signal it "flops" back to the other of the bistable states and stays there upon removal of the input signal.

The circuit shown in Fig. 6-42 is a basic flip-flop called an R-S flip-flop.

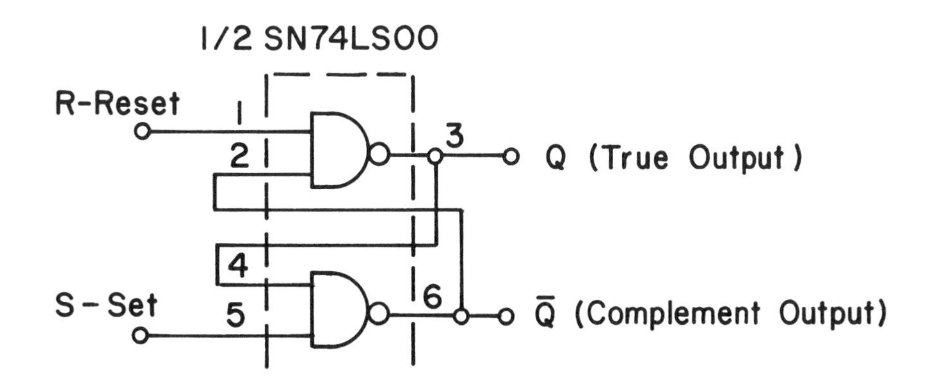

Fig. 6-40. Bistable Logic Circuit Called Flip-Flop.

Using this basic unit, as shown in Fig. 6-41, additional logic circuits can be put at the input or in the input and feedback patch to make various types of flip-flops such as D, T and J-K. Note also that another terminal is added called CLOCK. This is a control line for a timing signal that synchronized when the circuit will change state.

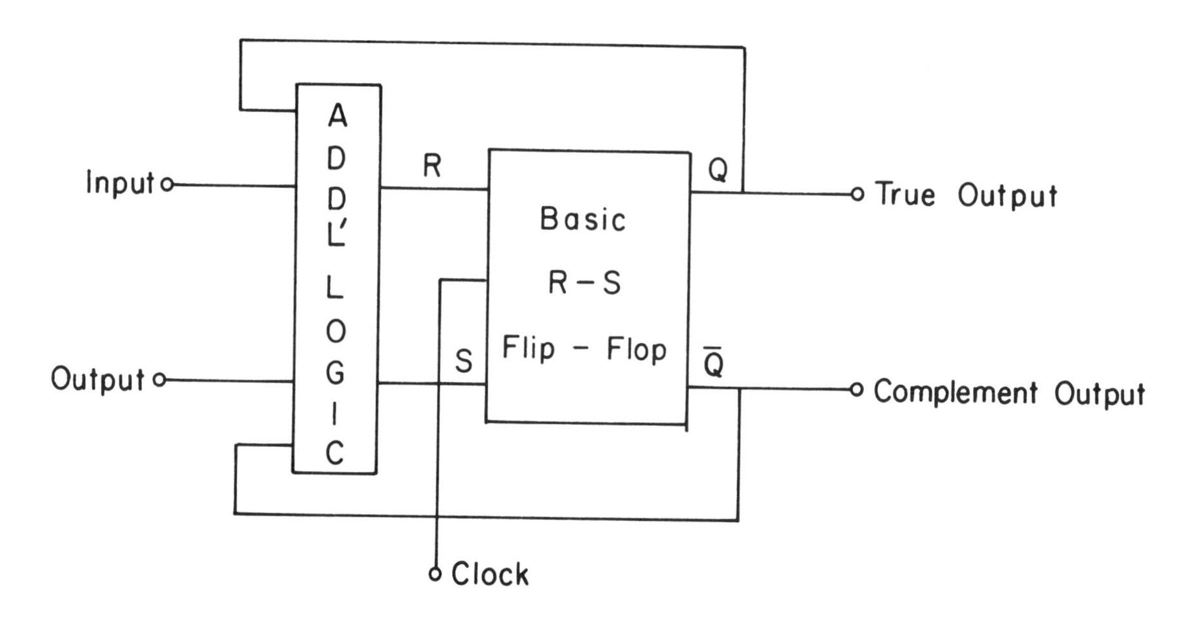

Fig. 6-41. Adding Additional Logic Circuits to R-S Flip-Flop.

Each flip-flop will store one bit of information. When flip-flops are combined in series and/or parallel they form counters or registers, or memories as shown in Fig. 6-42. Inputs for bits o through n may be in parallel (all at the same time) or the digital code may be shifted into the correct bit position by successive clock pulses that move the data right or left until positioned correctly.

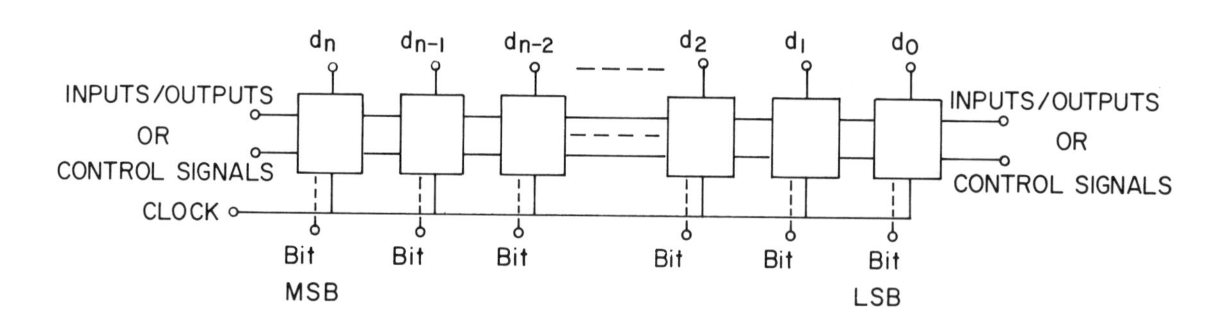

Fig. 6-42. Flip-Flops Coupled Together to Form Counters, Registers, Memories.

Outputs from each bit position may be available in parallel or only selected bit outputs may be available. Almost endless variations can exist. Each flip-flop circuit can have an output Q (called the true output) or an output $\bar{Q}$ (which is called the complement output) or both. $\overline{Q}$ is also called NOT Q.

EQUIPMENT REQUIRED

Digital Integrated Circuits - NAND and Inverter logic circuits, Flip-Flop circuits, clock generator, counter, registers.
+5 V Power Supply - 1 Ampere
Multimeter
Board Circuits with sockets, resistors, capacitors, switches, etc.

I. R-S FLIP-FLOP OR LATCH

Refer to Fig. 6-43 and make sure the SN74LS00 is in place in the socket. This is the same circuit as Fig. 6-40 except two inverter circuits have been added. The reason for this will be explained shortly.

Set S1 = 1, S2 = 0. For this setting of S1 and S2 there will be a 1 on the SET input to the flip-flop and a 0 on the RESET input to the flip-flop. The output of the S inverter is 0; the output of the R inverter is a 1. The S inverter 0 feeds an input of NAND gate A which is the Q NAND gate. A 0 on the Q NAND gate input will make Q a 1 (Refer to a NAND gate truth table). With Q at a 1, this signal feeds one input of NAND gate B, which is the $\overline{Q}$ NAND gate. The output of R inverter is the other input to the $\overline{Q}$ NAND gate. It is a 1.

With both inputs of $\overline{Q}$ NAND gate at a 1, $\overline{Q}$ is 0, which not only is the output but feeds the other input to the Q NAND gate and holds the output Q at 1 even though the S input might change back to 0. Q has been set to a 1 with a 1 on S.

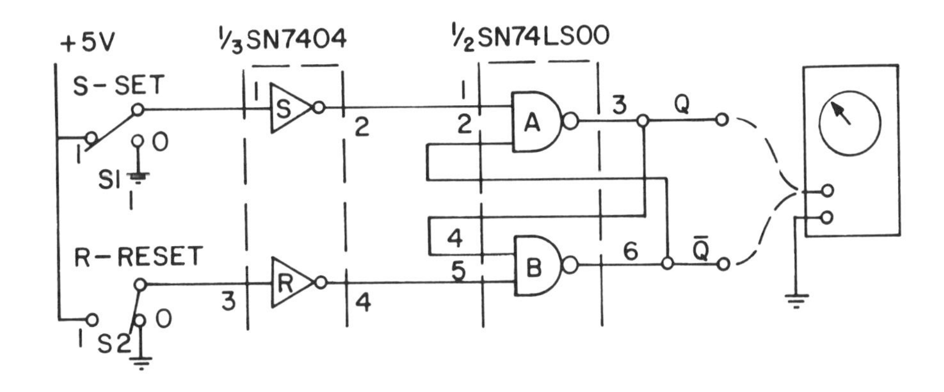

Figure 6-43. R-S Flip-Flop.

PROCEDURE

1. With the multimeter verify that Q = 1 and $\overline{Q}$ = 0 for S1 = 1 and S2 = 0. Fill in entry a in Table 6-12. Next switch S1 back to 0. Verify the state of Q and $\overline{Q}$. Fill in entry b in Table 6-12.

	INPUTS		OUTPUTS	
	R	S	Q	$\bar{Q}$
a.	0	1		
b.				
c.				
d.				

Table 6-12.
R-S Flip-Flop Truth Table.

2. Next set S1 = 0 and S2 = 1 and measure the output states of Q and $\overline{Q}$ and fill in entry c of Table 6-12.

   (e) Has Q been reset to 0?

   Yes ________

   No ________

3. Now set S2 = 0, measure Q and $\overline{Q}$ and fill in entry d in Table 6-12.

   (f) Did $\overline{Q}$ change?

   Yes ________

   No ________

4. Set both S1 and S2 to 0. Now switch both S1 and S2 to a 1 at the same time and measure Q and $\overline{Q}$. Try this a number of times always measuring Q and $\overline{Q}$ after S1 and S2 are switched.

   (g) Were you able to decide on the state for Q and $\overline{Q}$?

   Yes ________

   No ________

The cross coupled NAND gates of Fig 6-43 form a sequential circuit that has two stable states. The output state of Q and $\overline{Q}$ depends on what happened before present inputs were applied. If S = 0 and R = 0, the output states remain in the state they were before the inputs were present. With S = 1, Q is set to a 1; with R = 1, Q is reset to a 0. When S = 1 and R = 1, there is an indeterminant condition, the final reliable state of Q and $\overline{Q}$ are unknown. Therefore, this state is not allowed for an R-S flip-flop.

5. Is Table 6-13 a representative truth table for an R-S flip-flop?

   (h) Yes ________

   No ________

R	S	Q_{n+1}	$\bar{Q}_{n+1}$
0	0	Q_n	$\bar{Q}_n$
0	1	1	0
1	0	0	1
1	1	?	?

Subscripts: n before inputs
n+1 after inputs

Table 6-13.
R-S Flip-Flop Truth Table.

If the inverters had not been added in Fig. 6-43, the indeterminant state would have been S = 0, R = 0 and S = 1, R = 1 would have been the "no change" state. In addition the complementary signals S and R would have had to be maintained at all times because S = 0, R = 0 would be indeterminant.

II. D-TYPE FLIP-FLOP

Input gates can be provided to assure that the indeterminant signal condition is never present on the inputs to an R-S flip-flop. One such unit that results is the D-type flip-flop shown in Fig. 6-44.

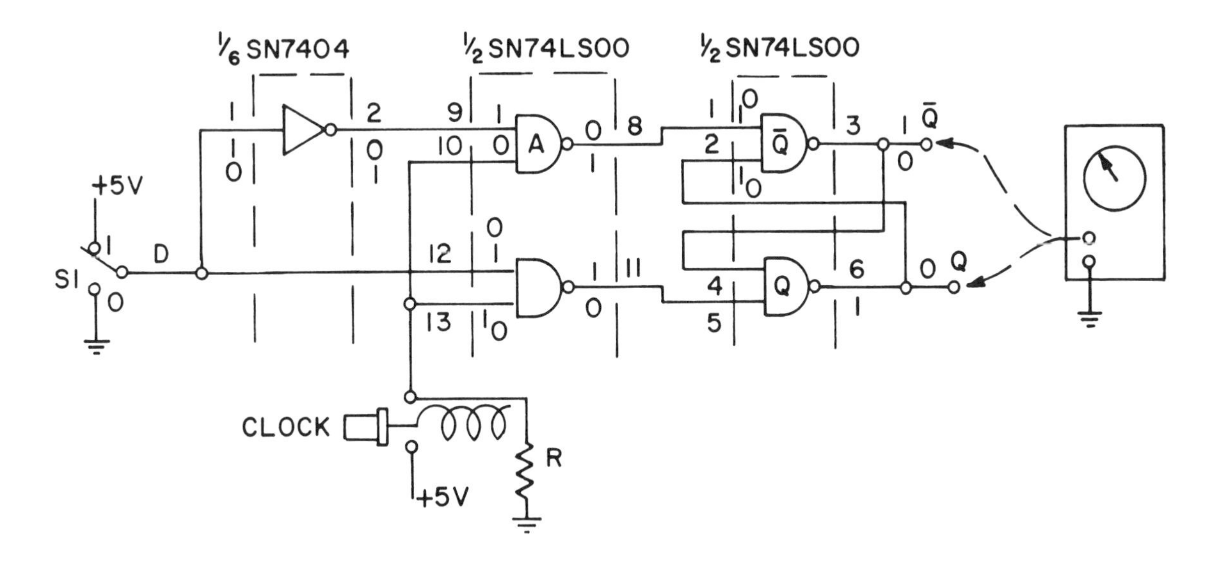

Fig. 6-44. D-Type Flip-Flop.

This is called a D-type flip-flop because the data appearing on D will appear on Q after the data is gated into the flip-flop with the clock. Note that the clock signal must be a 1 in order to do this.

PROCEDURE

1. Set S1 = 1. Press CLOCK and measure Q and $\overline{Q}$. Fill in entry a of Table 6-14.

	INPUT	OUTPUT	
	D	Q	$\overline{Q}$
a.	1		
b.	0		

Table 6-14.
D Flip-Flop.

2. Set S1 = 0. Press CLOCK and measure Q and $\overline{Q}$. Fill in entry b of Table 6-14.

   It might be a good exercise for the student to verify that all the logic levels at the intervening interconnections on Fig. 6-46 give the desired results on Q and $\overline{Q}$. Don't measure with a meter, just verify on paper.

The concept of a CLOCK signal has been introduced for flip-flops. It determines the precise time the signal on the input or control line of a flip-flop is going to be detected and govern the state of the flip-flop. In some flip-flops, it may be a logic level that must be maintained (as a 1 level for the D flip-flop of Fig. 6-45), or it may be a transition from one level to another. Many flip-flops are designed in integrated circuit form whose output state is governed by the truth-table conditions that exist on the inputs when the leading edge or the trailing edge of the clock pulse is transitioning from one logic level to another. This is shown in Fig. 6-47. Flip-flop data sheet specifications must be consulted to determine how they are designed.

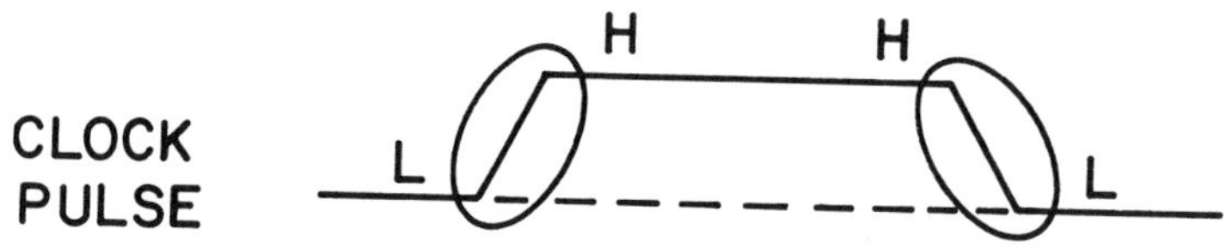

CLOCK PULSE

L H H L

Design A or B
Flip-flop gates in input data power on transition from L to H or from H to L.

Design C or D
Flip-flop gates in input data when it is at logic level H or L.

Fig. 6-45. Flip-Flop Clock Signal Designs.

In the following experiments, a push button clock will provide the clock pulses so a step-by-step clocking procedure can be implemented. The student need not be concerned about which flip-flop design is being used. However, for any electronic circuit design the student may be involved with that uses flip-flops or sequential circuits in general, this is avery important concern of the designer.

III. J-K FLIP-FLOP

There is a flip-flop whose truth table is shown in Table 6-15. In this truth table there is no indeterminant state like the R-S flip-flop. J is the same as SET input and K is the same as a RESET input. If J = 1 and K = 1 when the clock pulse is applied, the flip-flop will "toggle". Toggle means that the output changes to the opposite state. J is sometimes called the true input and K the complement input. As with the R-S flip-flop, if both J and K are 0, the flip-flop remains in the state it was in before the clock pulse arrived.

INPUT		OUTPUT	
t_n		t_{n+1}	
J	K	Q	$\bar{Q}$
0	0	Q_n	$\bar{Q}_n$
1	0	1	0
0	1	0	1
1	1	$\bar{Q}_n$	Q_n

Table 6-15.
J-K Flip-Flop Truth Table.

There is also an input called CLEAR on many flip-flops. Such an input usually does not depend on the clock and all inputs are inhibited while this signal is present.

All of the signals are included in the truth table of the J-K flip-flop, SN74LS73 shown in Table 6-16, including the clock. In this case, the Q_0 means the condition that exists before the clock arrives. The arrow in the clock column indicates the transition of the clock that detects the inputs and governs the state of the outputs after the clock occurs.

INPUTS				OUTPUTS		
CLR	CLK	J	K	Q	$\bar{Q}$	
0	X	X	X	0	1	
1	↓	0	0	Q_0	$\bar{Q}_0$	
1	↓	1	0	1	0	
1	↓	0	1	0	1	
1	↓	1	1	$\bar{Q}_0$	Q_0	(TOGGLE)
1	1	X	X	Q_0	$\bar{Q}_0$	

Table 6-16.
SN74LS73 Flip-Flop Truth Table.

The arrow pointing down means the flip-flop triggers on a transition from a high level to a low level (from a 1 to a 0). X indicates a "don't care" condition. It means the logic level can be a 0 or a 1; it doesn't make any difference.

PROCEDURE

1. Make sure the SN74LS73 is seated in the socket of the circuit of Fig. 6-46. Set S1 = 0 and S2 = 0.

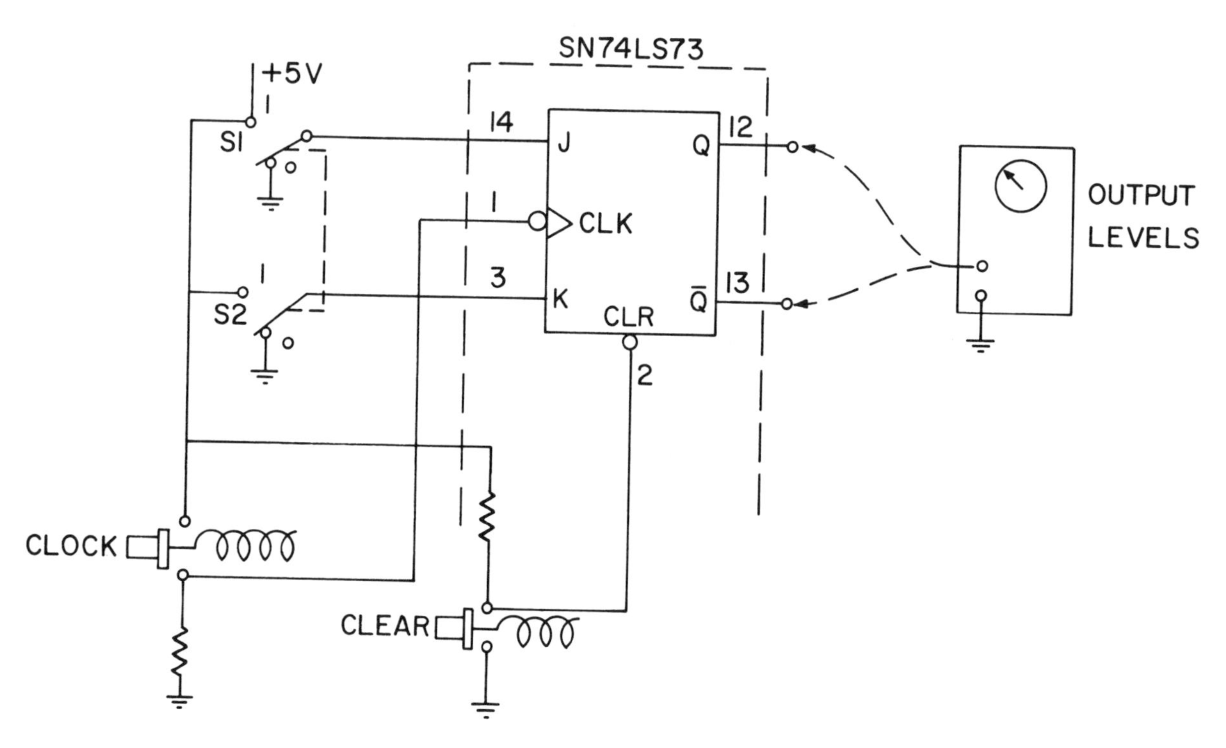

Fig. 6-46. J-K Flip-Flop Test Circuit.

Press CLEAR push-button and release. Measure the output voltage at Q and $\bar{Q}$ and fill in entry a in Table 6-17.

	INPUTS				OUTPUTS	
	CLR	CLK	J	K	Q	$\bar{Q}$
a	0	X	X	X		
b	1	↓	0	0		
c	1	↓	1	0		
d	1	↓	0	1		
e	1	↓	1	1		
f	1	1	X	X		

Table 6-17.
J-K Flip-Flop.

2. Press CLOCK and release and measure output voltage levels of Q and $\overline{Q}$ again. Fill in entry b of Table 6-17.

3. Set S1 = 1 and S2 = 0. Press CLOCK and release. Measure Q and $\overline{Q}$ and record as entry c of Table 6-17.

4. Set S1 = 0 and S2 = 1. Press CLOCK and release. Measure Q and $\overline{Q}$ and record as entry d of Table 6-17.

5. Set S1 = 1 and S2 = 1. Press CLOCK and release. Measure Q and $\overline{Q}$ and record as entry e of Table 6-17.

6. Press CLOCK and hold it. While it is held, switch S1 and S2 back and forth several times. Still holding CLOCK, measure Q and $\overline{Q}$ and record as entry f in Table 6-17.

7. Is Table 6-18 the same as the J-K flip-flop truth Table 6-16?

   Yes ________

   No ________

IV. "T" FLIP-FLOP

Table 6-18 is the truth table for a "T" flip-flop.

INPUT		OUTPUT	
CLK	T	Q	$\overline{Q}$
↓	0	Q_0	$\overline{Q}_0$
↓	1	$\overline{Q}_0$	Q_0

"T" Flip-Flop Truth Table

Table 6-18.

PROCEDURE

1. Use the same J-K flip-flop as in Fig. 6-46. However, S1 and S2 will now be switched together as if they were connected together to make J = K with only one switch to govern the state of J and K as shown in Fig. 6-47.

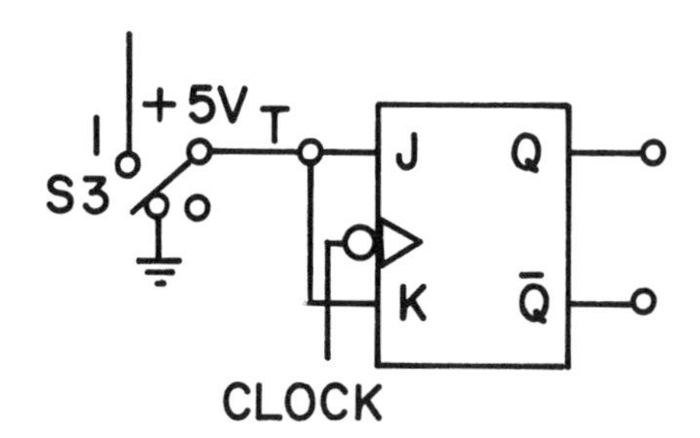

Fig. 6-47.
"T" Flip-Flop.

2. Press CLEAR and release Set S1 = 0 and S2 = 0. (This would represent S3 = 0 for Fig. 6-47. Record the logic levels of Q and $\overline{Q}$ after CLOCK is pressed and released. Enter measurements in (a) of Table 6-19.

	INPUT		OUTPUTS	
	CLK	T	Q	$\overline{Q}$
a	↓	0		
b	↓	1		

Table 6-19.
"T" Flip-Flop.

3. Press CLEAR and release. Set S1 = 1 and S2 = 1. (This would represent S3 = 1 for Fig. 6-47.) Record the logic levels of Q and $\overline{Q}$ after CLOCK is pressed and released. Enter measurements in (b) of Table 6-19.

4. Does this truth table satisfy the truth table for the T flip-flop?

   (c) Yes ________

   No ________

A "T" flip-flop will toggle if it has a 1 on the T input when CLOCK is applied. It will not toggle if there is a 0 on the T input when CLOCK is applied. As shown in Fig. 6-47, a J-K flip-flop is converted to a "T" flip-flop by connecting the J and K inputs together.

V. D TYPE FLIP-FLOP FROM J-K

A D type flip-flop is easily implemented from a J-K flip-flop by adding an inverter in the K input signal path as shown in Fig.6-48a.

PROCEDURE

1. Verify that a D type flip-flop results from Fig. 6-48 by filling in the logic levels a and b of Truth Table 6-20.

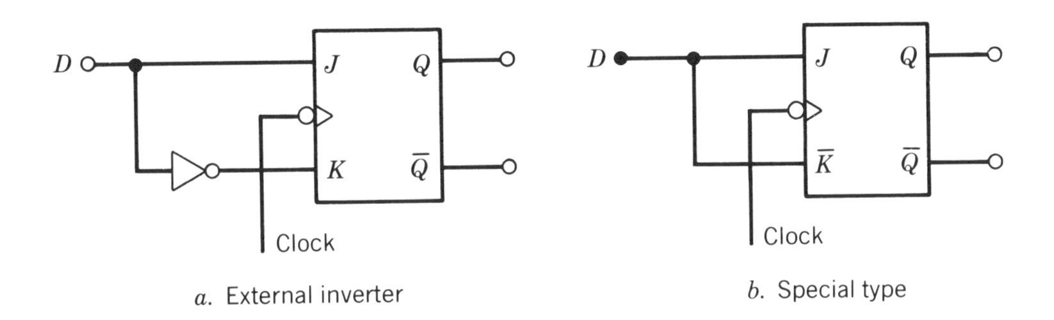

a. External inverter

b. Special type

Fig. 6-48. D-Type from J-K Flip-Flop.

	INPUT				OUTPUT	
	CLK	D	J	K	Q	$\bar{Q}$
a	↓	0				
b	↓	1				
c	1	0				
d	1	1				

Table 6-20. D Flip-Flop from J-K.

There are special types of integrated circuit J-K flip-flops that have a $\overline{K}$ or sometimes called K*) input to make this conversion easy so that a separate inverter gate is not necessary (Fig. 6-48b).

2. Latching

It is important to note that the CLOCK timing signal makes the flip-flop a latched flip-flop. That is, the data on the input D is latched into the flip-flop when the CLOCK signal changes in the transition indicated. Thus, the input signal on D can change before or after the clock transition and it will not affect the outputs. Only the state of D at the time the clock transition occurs (or the level if it is a flip-flop that depends on clock level) will determine the output state.

If the flip-flop depends on a clock transition for latching, then holding the CLOCK input at a high level (or it may be a low level depending on the flip-flop design) will prevent the flip-flop from latching in new data. Its outputs will remain at their last changed condition.

Fill in lines c and d on Truth Table 6-20 based on this information. Use n subscripts to indicate time before inputs arrive and n+1 for time after inputs arrive.

Experiment 6-8 DIGITAL CIRCUITS - SEQUENTIAL LOGIC CIRCUITS - COUNTER, REGISTERS AND MEMORY

This experiment continues the emphasis on sequential logic circuits. It shows many other circuits can result by using a combination of basic flip-flops.

PURPOSE

This experiment will investigate the operation of sequential circuits derived from flip-flops. Three types will be covered: counters, registers and memory circuits.

DISCUSSION

Flip-flop states that are cascaded can be made into temporary storage units called registers, "permanent" storage units called memory, or into circuits that count pulses or events called counters. Let's take counters first. Fig. 6-49 is similar to Fig. 6-46 but has two flip-flops coupled together. The circuit is a two-stage binary counter.

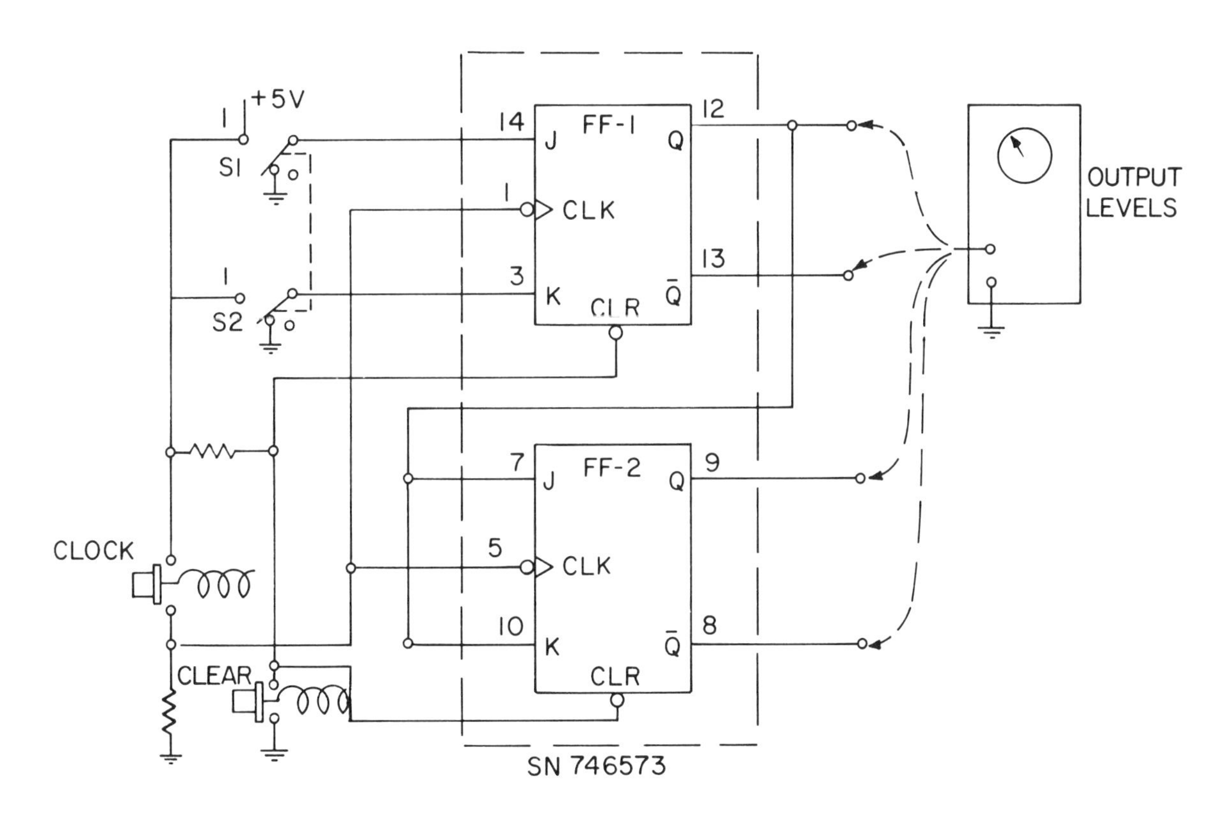

Fig. 6-49. 2-Stage Binary Counter.

EQUIPMENT REQUIRED

Same as Experiment 6-7. Measure logic levels with meter as in previous experiments.

I. COUNTERS

In Fig. 6-49 both flip-flops will be used as T flip-flops. The inputs to FF-1 are S1 and S2, but these will be operated together so that J = K as for the T flip-flop. The Q output of FF-1 is the input to FF-2 connected as a T flip-flop. Each time the CLOCK push-button is pressed it will be considered a clock pulse.

PROCEDURE

1. Press CLEAR push-button and release. Measure Q and $\overline{Q}$ of FF-1 and FF-2 and enter in Table 6-21 in the CLR column.

2. Set S1 = 1 and S2 = 1. Press CLOCK push-button once and release. Measure Q and $\overline{Q}$ of FF-1 and FF-2 and enter in Table 6-21 in the clock pulse #1 column.

3. Press CLOCK push-button again and release and measure again Q and $\overline{Q}$ of FF-1 and FF-2. Enter output levels $\overline{Q}_1$, Q_1, and $\overline{Q}_2$, Q_2 level in clock pulse #2 column.

4. Continue to press the CLOCK push-button and release, each time measuring Q and $\overline{Q}$ of FF-1 and FF-2 and entering data in appropriate columns of Table 6-21.

Clock Pulse No.

	CLR	1	2	3	4	5	6	7	8	9	10	11	12
S1, S2	X	1	1	1	1	1	1	1	1	1	1	1	1
Q_1													
$\overline{Q}_1$													
Q_2													
$\overline{Q}_2$													

Table 6-21. Binary Counter.

5. Plot the levels of Q_1 and Q_2 on the diagram of Fig. 6-50.

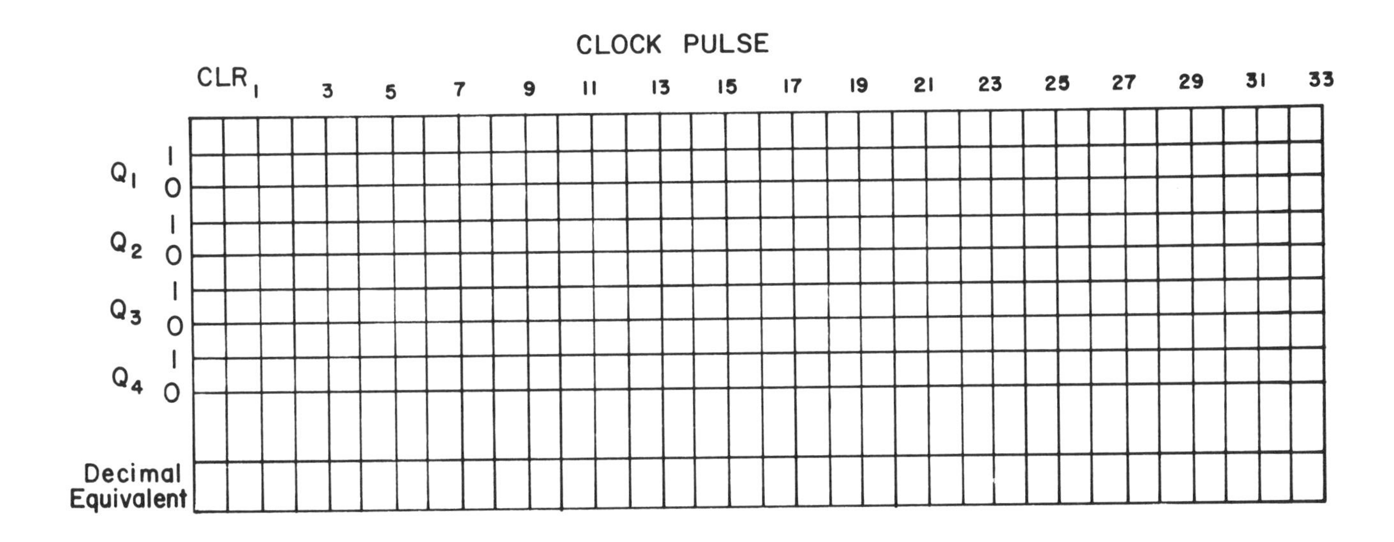

Fig. 6-50. Binary Counter Outputs.

6. Is it apparent that Q_2 changes once for every two transitions of Q_1?

   Yes __________

   No __________

7. Plot on Fig. 6-50 the output levels of Q_3 and Q_4 if two more J-K flip-flops connected as T flip-flops were placed in series with FF-1 and FF-2, as shown in Fig. 6-51.

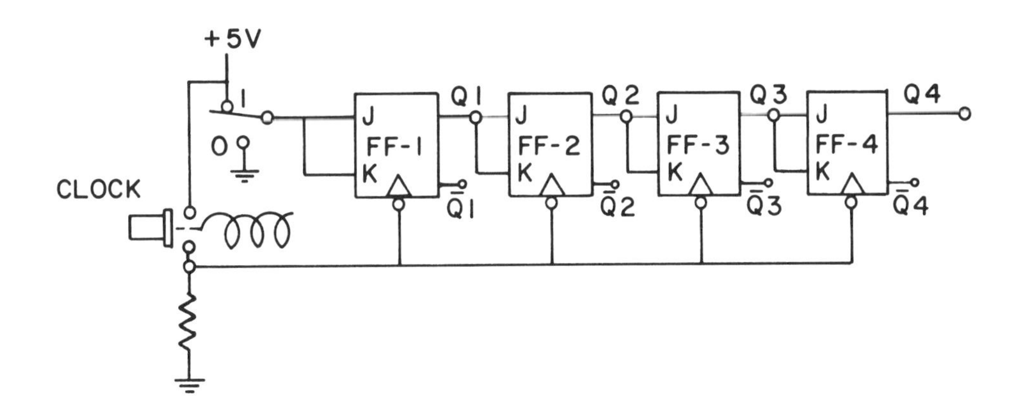

Fig. 6-51. Four-Stage Binary Counter.

How many clock pulses does it take to cause a negative transition (from 1 to 0 level) for:

Q_1? ____________

Q_2? ____________

Q_3? ____________

Q_4? ____________

The binary counter counts the number of clock pulses determined by 2^n stages.

8. As an added exercise, write in the decimal equivalent of the number of clock pulses on the line indicated in Fig. 6-50. The decimal equivalent is obtained by adding together the numbers indicated by the Q output levels. $Q_1 = 1$ is equivalent to 1, $Q_2 = 1$ to 2, $Q_3 = 1$ to 4 and $Q_4 = 1$ to 8. If after the 7th clock pulse $Q_1 = 1$, $Q_2 = 1$, $Q_3 = 1$ and $Q_4 = 0$, then the decimal equivalent is $1 + 2 + 4 + 0 = 7$.

Counters are made by cascading flip-flop circuits. Various types of counters are available and the number of pulses or events that they count before they start over is called the Modulus. The Modulus may be set by the number of initial stages and gating that feeds back signals to make the flip-flops change state at the correct time to accomplish the Modulus. Many varieties are available. Consulting data specifications catalogs from manufacturers will illustrate the variety of types.

II. REGISTERS

As shown previously, a D flip-flop accepts and outputs data on the Q output that is equivalent to the data on the D input when the clock trigger transition or level appeared. Therefore, the D flip-flop is called a gated latch because the data on the D input is "gated" into the flip-flop at a time determined by the clock and is "latched" into the flip-flop and stored until changed (if required) by the next clock pulse.

When a number of flip-flops are ganged together as shown in Fig. 6-52 the combination forms a circuit called a register. The one shown is a 4-bit register.

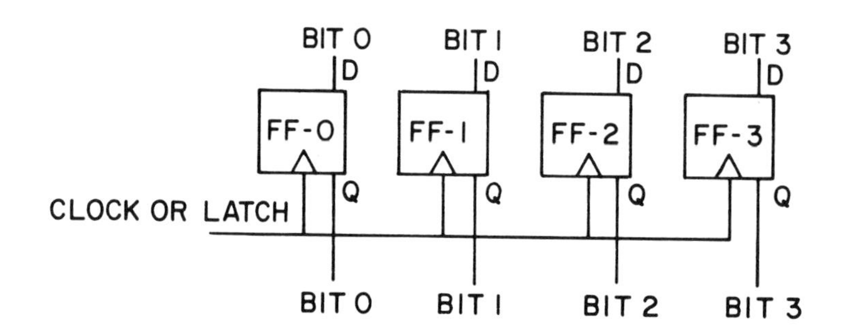

Fig. 6-52. 4-Bit Register (Gated Latch).

Data that is present on each of the D inputs at the time the clock signal latches the flip-flops will be stored in the register until the next clock pulse might change the register data. Registers can be made any length to handle the number of bits in a digital word. Common bit lengths are 4, 8, 16, 32 or 64 bits. A register is called a temporary storage because the data may change at each clock pulse.

PROCEDURE

1. Make sure the 4-bit latch or register integrated circuit SN74LS175 is seated in the socket of the circuit of Fig. 6-53. Press the CLEAR push-button and release. Measure the outputs Bit 0, 1, 2 and 3 and record in (a) of Table 6-22. CLEAR sets all Q outputs to an initial condition of 0.

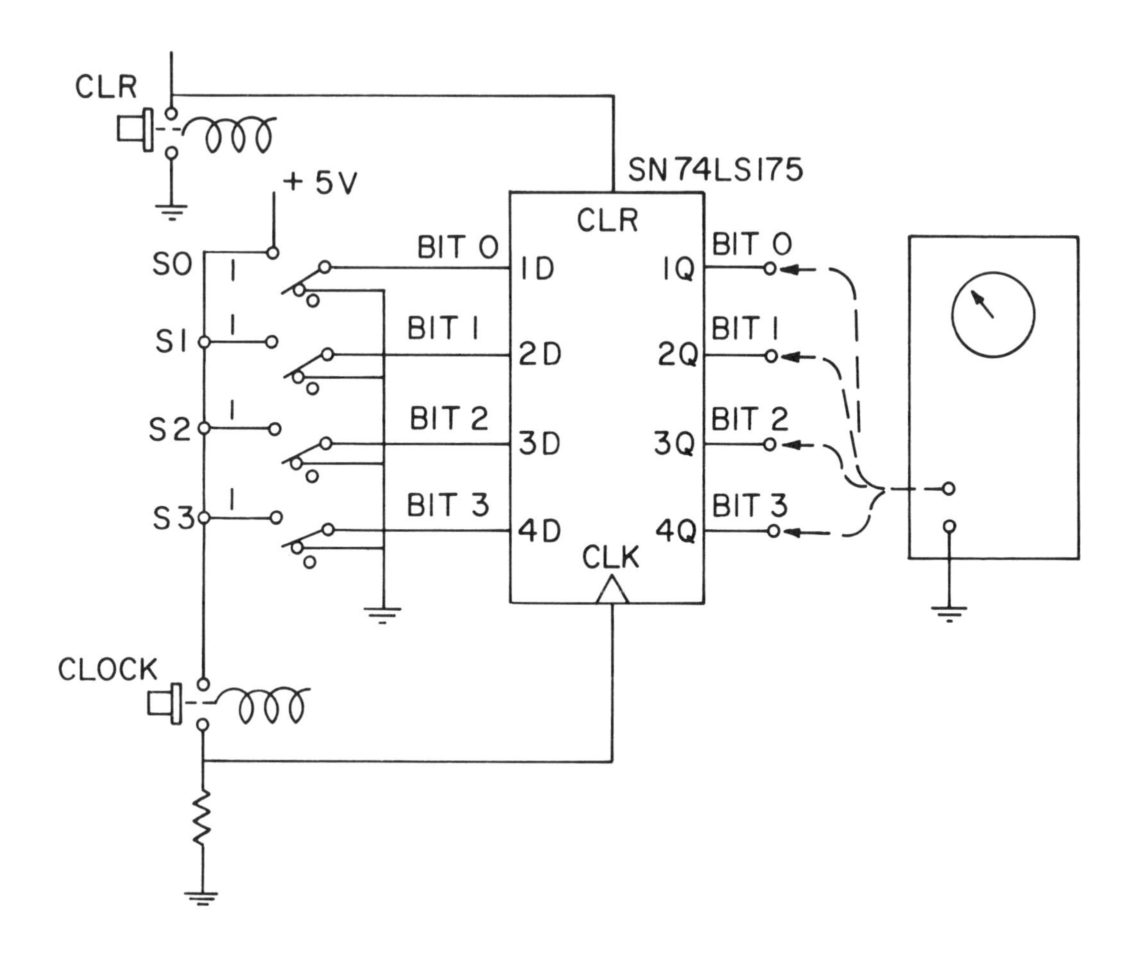

Fig. 6-53. 4-Bit Register.

	INPUTS					OUTPUTS			
						BIT			
	CLR	S0	S1	S2	S3	0	1	2	3
a	0	X	X	X	X				
b	1	1	0	0	0				
c	1	1	1	0	0				
d	1	1	1	1	0				
e	1	1	1	1	1				
f	1	0	0	0	1				

Table 6-22.
4-Bit Register.

2. Set S0 = 1 and S1, S2 and S3 to 0. Press CLOCK and release. Measure Q outputs and record in (b) of Table 6-22.

3. Set S0, S1 = 1 and S2 and S3 to 0. Press CLOCK and release. Measure Q outputs and record in (c) of Table 6-22.

4. Set S0, S1, S2 = 1 and S3 to 0. Press CLOCK and release. Measure Q outputs and record in (d) of Table 6-22.

5. Set S0, S1, S2 and S3 = 1. Press CLOCK and release. Measure Q outputs and record in (e) of Table 6-22.

6. Set S0, S1, S2 = 0 and S3 = 1. Press CLOCK and release. Measure Q outputs and record in (f) of Table 6-22.

III. MEMORY

When a more permanent memory is desired, integrated circuits are designed to have flip-flop type circuits arranged in a matrix as shown in Fig. 6-54. The integrated circuit shown also contains a decoder so that an address code can select a particular word by selecting a particular row.

In Fig. 6-54 there are 4 bits in each word and four words shown. Integrated circuit memories range in all sizes from small memories of 64 bits to very large ones, e.g., 64K bits. These circuits can then be combined to assemble memories with millions of bits storage capacity.

Note that with a control line, either data can be read from memory or else data can be stored (written) in memory. Selection of the word by an address is the same for both operations. Typically a memory of this type is called a random access memory (RAM) because each word can be randomly selected and, when read, the data arrives at the output in approximately the same time from any of the randomly selected locations.

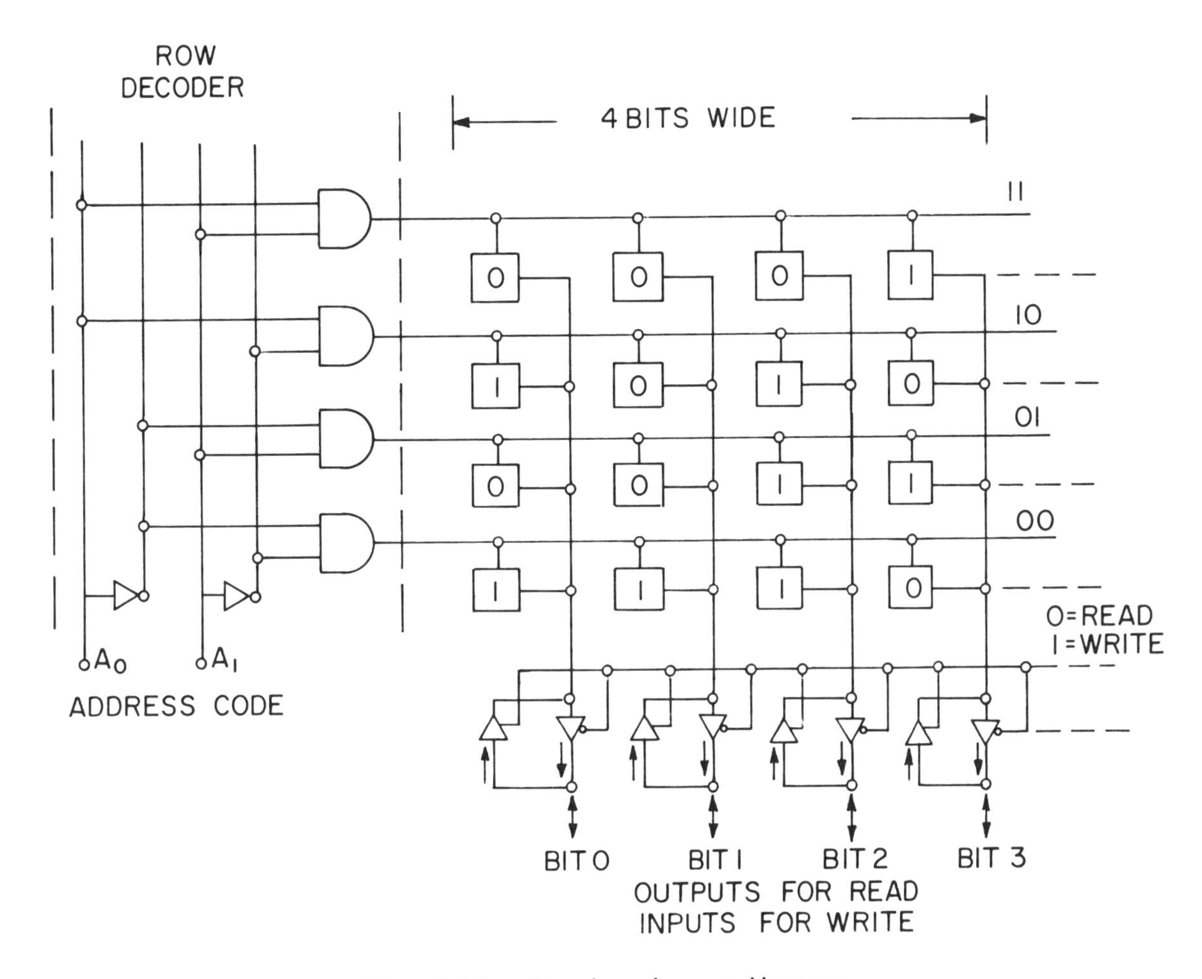

Fig. 6-54. Random Access Memory.

PROCEDURE

In Fig. 6-54 the state of the bits in the memory words are shown. Using the AND and INV gate truth tables determine the data that appears at the output for the address codes applied to A_0 and A_1 and fill in Table 6-23 for a, b, c and d.

	ADDRESS CODE		OUTPUT DATA			
	A_0	A_1	D_0	D_1	D_2	D_3
a.	0	0				
b.	1	1				
c.	0	1				
d.	1	0				

Table 6-23.
Random Access Memory.

ROM - Other memories may have the data fixed in the memory and it cannot be changed once it is programmed. These are called read-only memories (ROM). Some of them are programmable only at the factory when they are made; others can be programmed by the user in the field.

PROM and EPROM - When the programmable read-only memories are programmed by the user they are called PROMS. These are two types: fuse-link and EPROMS. The fuse-link type are manufactured with all 1's in each bit of the word. A 0 is programmed into the bit by passing current through a link and fusing it away. Once programmed the data cannot be changed unless more 1's are made into 0's.

EPROMS are programmed by placing set voltages in the inputs in order to make a 1 or a 0. These programmed bits will remain until erased with an ultraviolet light. Exposing the memory chip to a certain strength of ultraviolet light will erase all the data so that the memory can be programmed with new data.

Read-only memories are very popular in digital systems that require fixed memory.

Experiment 6-9 DIGITAL CIRCUITS - MICROCOMPUTERS

The combinatorial and sequential logic circuits can be combined to form many many varieties of digital systems. One common type of digital system is the microcomputer. It derives its name from the fact that it is small in physical size but yet has full computer capability.

PURPOSE

This experiment will demonstrate the basic functions of a computer by using a board microcomputer, the TM990/U89 University Board.

DISCUSSION

A computer has the basic functions shown in Fig. 6-55. The brains of the system reside in the central processing unit or CPU. It is the nerve center of the system. It does all the computations that are required. It decides what the system should be doing at any particular time as a result of interpreting (decoding) instructions and executing these instructions. Executing the instructions requires that the CPU coordinate what the other parts of the system do so that the task at hand can be accomplished.

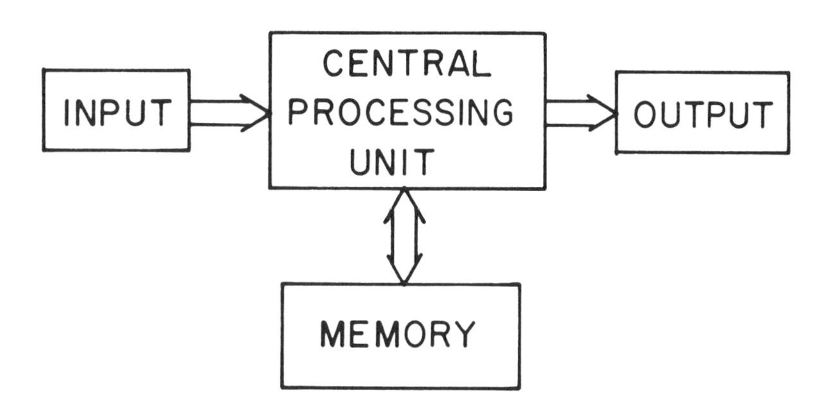

Fig. 6-55. Basic Functions of a Computer.

INSTRUCTIONS

However, the CPU must have instructions to follow. These instructions must be in the correct order and be followed in sequence for the CPU to carry out the operations that result in a particular task getting done. Humans determine what these instructions are and in what order they should be arranged. Such an arrangement is called a program.

PROGRAM AND PROGRAM STORAGE

After the program is decided for a particular system to do a particular task (or it may be while the program is being developed), the instructions are stored so that they are preserved for the CPU to follow. This is called program storage and memory performs that function. We discussed bistable circuits and how they can be used to form memory circuits to store digital information. The instructions are converted from language humans understand into digital form so they can be stored in memory.

DATA STORAGE

While the CPU is performing the operations that are called for by the instructions it must use information. This information is called data and is stored in memory. The data may be original or initial information, it may be interim information that results from operations performed by the program or it may be final information that results from the program being completed. Memory that stores this information is called data storage.

INPUTS

The data stored in memory could also change as a result of something happening external to the system. In other words, input information from external sources provide information (data) to the system initially or while it is operating either under peripheral or direct control of the program. The interface from the computer to the outside world to received information is called input. The input function interfaces signals to the CPU from humans, switch closures, logic signals, timing pulses, push-buttons, keyboards, light sensors and many other types of transducers. Some of the input information goes to memory to be stored; other information is operated on directly by the CPU as it is received.

OUTPUTS

As the operations are performed as a result of instructions, the system may cause outputs to occur. These outputs are in the form of logic signals, codes that occur in serial or in parallel form, contact closures, timing signals, etc. These outputs usually result in some useful work being accomplished. They control motors, relays, provide digital signals to control other digital systems, print out information, show information on a CRT (cathode-ray-tube) screen.

The input and output functions are quite similar as far as the CPU is concerned. Therefore, both functions are usually combined into a term "I/O" and, therefore, the computer is considered to have three functions - CPU, memory and I/O.

TM990/U89 MICROCOMPUTER

The functions just described are contained in the system of Fig. 6-56. This system is a complete microcomputer. It contains a microprocessor as the central processing unit. It contains a calculator type keyboard to input information from humans to the computer. The calculator display is an output. There is also another output via a small ceramic sound disk that can be programmed to make tones as single sounds. It has terminals or connectors on the board (P3, P4 and P5) that are used to input or output external signals to and from the microcomputer. It also contains a connector P2 that is used to input information from a cassette-type recorder or to store information on the tape. Note that additional circuits are present on the board to control these I/O signals. The power to the board is supplied through connector P1. A cable connects power supply TM990/519 to supply +12 V at 250 μA; -12 V at 180 μA; and +5 V at 2A.

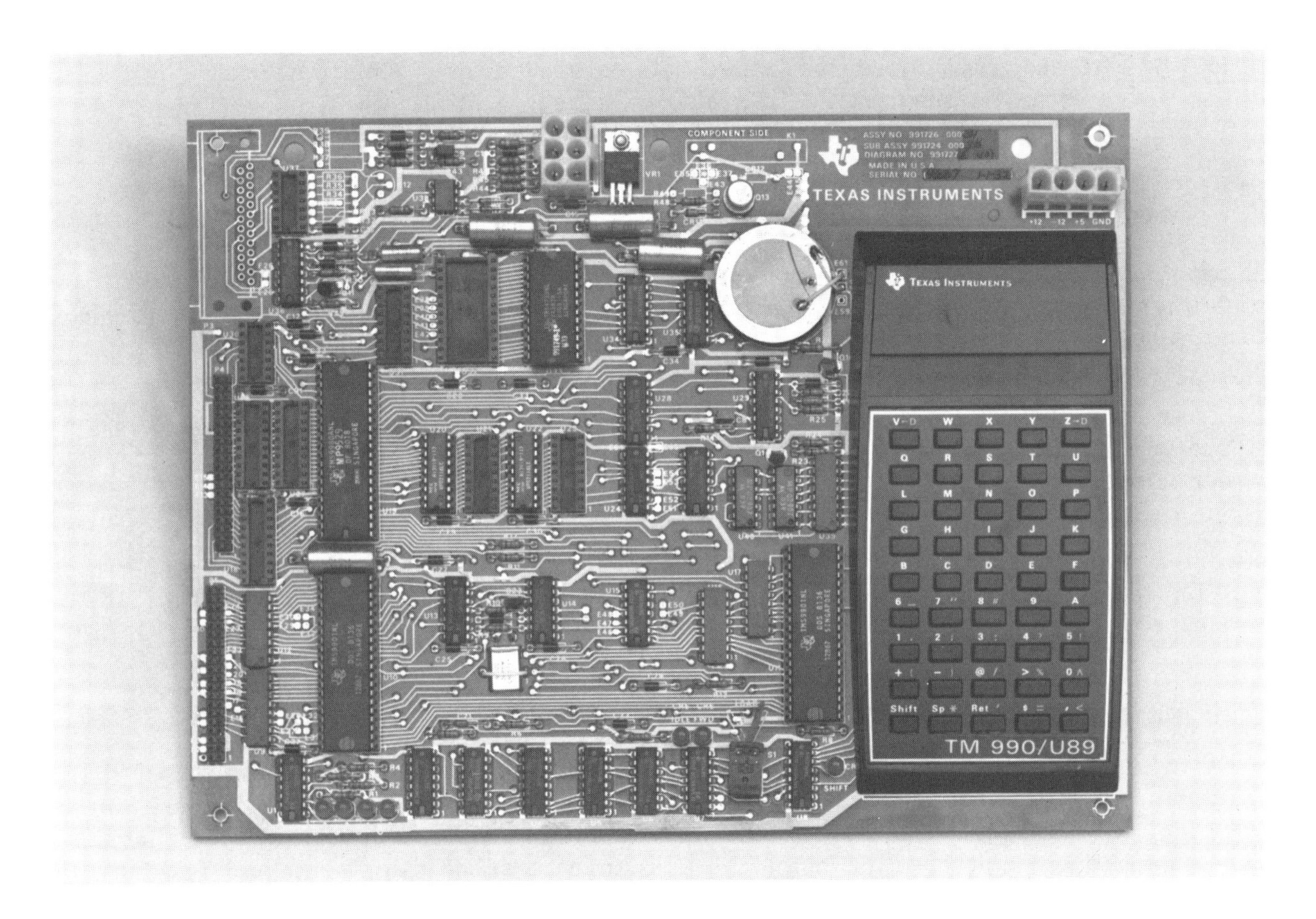

Fig. 6-56. Microcomputer.
(Figure 1-1 of TM990/U89 book page 1-2)

MEMORY - ROM

Two kinds of memory are present on the board. The first is read-only memory or ROM. It is memory in which information is stored "permanently" and is not changed. It is only read. The "permanently" in quotes means that the read only memory can be of the type that can be erased and then programmed again - but only occasionally. In this case the read-only memory is called an EPROM (Erasable Programmable Read-Only Memory). It is shown in Fig. 6-56. Programs that are not going to change but remain the same for all operations of the microcomputer are stored in Read-Only Memory. This is where the instructions that control the inputs from the keyboard are stored. It also has machine-language instructions stored that control all kinds of special functions that the computer or the I/O units connected to the computer must do. In large computer systems, these special program functions are contained in a software (developed program on paper) program called an operating system. In small microcomputers, this software is called a monitor and it is stored in EPROM.

MEMORY - RAM

The second kind of memory shown is random access memory, RAM. It is the memory where the program instructions that are stored are written (are stored) to make the computer do a specific task. Thus, the same hardware can be used for many different tasks by changing the program. This is the significant advantage of computers. Standard hardware can be used to do a variety of tasks just by changing the program. In addition, as the program runs it must have data to operate on. This data that may be required to perform the programmed task also is stored in RAM memory.

BLOCK DIAGRAM

The microcomputer is shown in block diagram form in Fig. 6-57. Auxiliary storage is provided beyond the on-board memory by making the connection through connector P2 that interfaces to a cassette tape recorder. Programs that have been proven can be saved for future use by storing them on cassette tape. When they are to be used again they are read back into memory.

All of the functions shown must occur in steps at a given time. The master timing is provided by a clock oscillator. As previously stated, the power is provided by the power supply connected to P1 and regulators on the board.

The signals that carry the instructions and data addresses for memory, input and output; data; and control commands move between the subsystems of Fig. 6-57 over the multiple line busses. Addresses move over the address bus, data move over the data bus, and control commands move over the control bus. The special CRU bus is for I/O signals.

STORING DATA FROM AN INPUT

Suppose data is to be stored in memory from a particular input. What would the computer do in order to accomplish this? First of all there would be a sequence of instructions to accomplish the task. Each instruction is a particular digital code that can be recognized by logic circuits inside the CPU to be a specific instruction. This is called decoding or interpreting the instruction. After the CPU decodes the instruction it carries out or executes the operation called for by the instruction and then moves on to the next instruction.

For example to store data from the keyboard and place it in memory here are typical instructions that would be followed:

No.	Instruction
000	CLEAR ALL REGISTERS
001	ADDRESS KEYBOARD
002	WAIT FOR KEY TO BE DEPRESSED ON KEYBOARD
003	RECOGNIZE KEY CODE, STORE IT IN REGISTER A
004	GENERATE MEMORY ADDRESS
005	ADDRESS MEMORY
006	STORE DATA FROM REGISTER A IN MEMORY

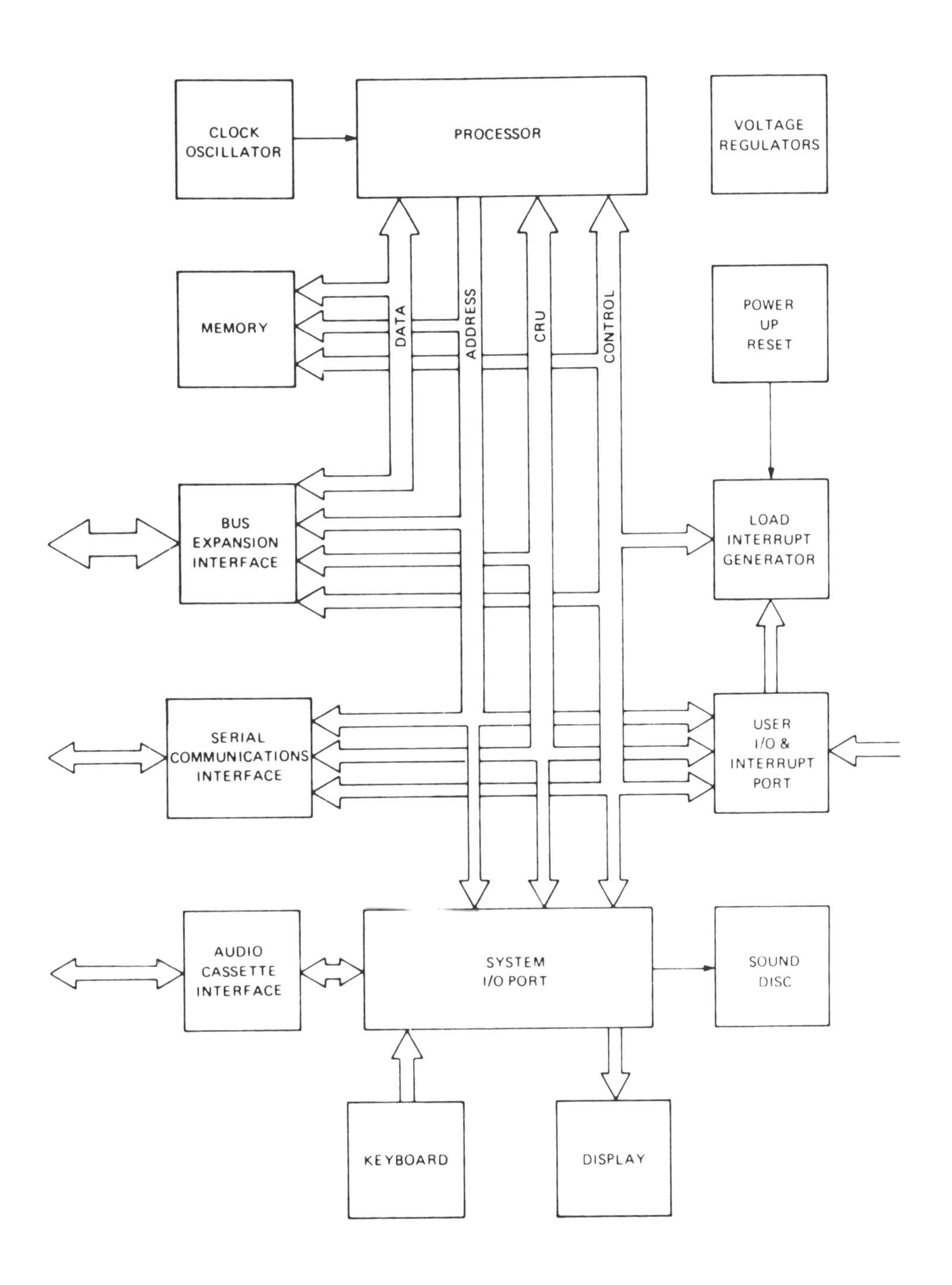

Fig. 6-57. Microcomputer System Block Diagram.
(Figure 1-2 of TM990/189 page 1-3)

When the data is to be stored in memory, a memory address had to be generated (Step 004 on Page 276) and stored in a memory address register in the CPU. At the correct time (Step 005), this memory address is sent over the address bus to address the memory and choose the correct location for storing the information. Previously the information data was stored (Step 003) in a data register in the CPU. At a correct time later (Step 006), a command is given to read the data from the data register in the CPU into memory to complete the task.

Note that there are registers in the CPU that are used as temporary storage places for instructions, addresses, data and results of operations. These registers and memory storage locations are the main items that are available to the programmer to accomplish tasks by computers. The register that keeps track of the next instruction to be executed is called the program counter. It counts in binary code in sequence to generate the address for the next instruction. When this address (a digital code of 1's and 0's) is sent to memory by the CPU, the memory contents (an instruction in digital code) are sent back to the CPU and stored as 1's and 0's in an instruction register. From the instruction register the CPU decodes the instruction and executes the operations called for by the instructions.

PAPER EXAMPLE

To illustrate how a microcomputer is operating and the CPU (microprocessor) is controlling the system in order to make sure that the instructions in the program are carried out properly, you will act as the CPU and provide the required action requested by the program of Fig. 6-59 and observe the interaction between the microprocessor and memory. The program is stored in the memory locations shown in Fig. 6-59.

Acting as the microprocessor, you would have the parts shown in Fig. 6-58: a program counter that contains the address of the next instruction; an instruction register that receives the digital code from program memory so the microprocessor can decide it and decide what the instruction is and what operation needs to be done; three "working" registers that can contain only one decimal digit, registers A, B and C; two registers which can contain three decimal digits, register X and register Y, used to address the data memory; and a status register with two bits. Bit one is set to a 1 to indicate if a carry has been generated on an addition and bit two is set to a 1 to indicate that a comparison has resulted in an equal condition. Bit one of the status register is reset to 0, if used in an addition. All registers are set to 0 initially except the program counter which is set to 001. The CPU automatically increments the program counter by 1 after an instruction is completed. The data memory contains the data shown in Fig. 6-59.

Start with the first instruction in Fig. 6-59 at location 001 and follow the instruction. Fill in the chart of Table 6-24 with the register and memory contents that appear when the instruction is executed. The instruction 001 has been carried out as an example for you.

When you finish, you will have added 2222 to 5973 and placed the sum in data memory within locations 200 to 205. Extend lines to Table 6-24, as needed.

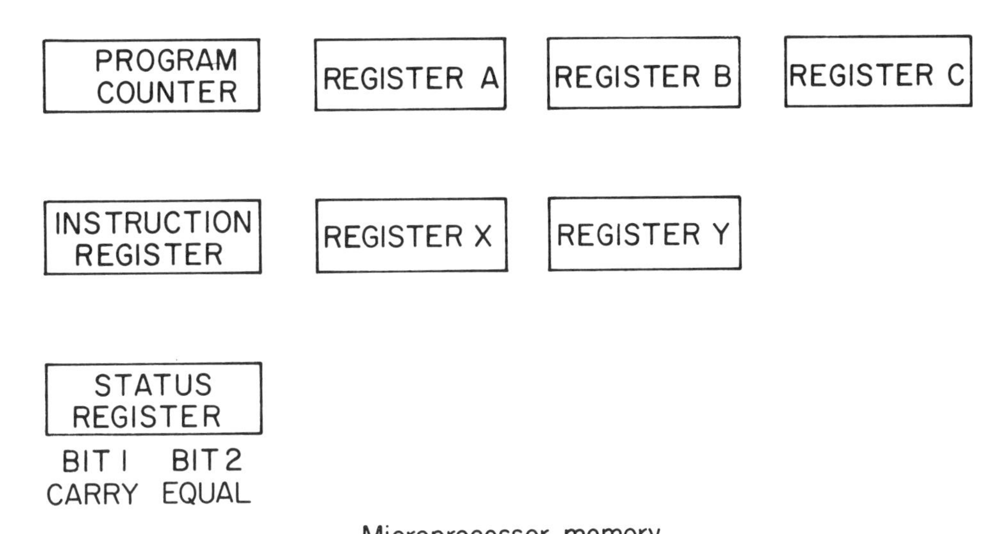

Microprocessor memory

Fig. 6-58

PROGRAM		DATA	
Location	*Instruction	Locatic	Data
001	Clear status register	100	3
002	Load register A with 0	101	7
003	Load register B with 4	102	9
004	Load register C with 2	103	.
005	Load register X with 100	104	0
006	Load register Y with 200	105	0
007	Move data addressed by register X to register A	200	0
008	Add register C to register A and carry (bit 1 of status register) Put sum in register A Set bit 1 if carry is generated	201	0
009	Move data in register A to memory location addressed by register Y	202	0
010	Increment register X and Y	203	0
011	Decrement register B. If register B is 0 set bit 2 to 1	204	0
012	Check status register bit 2. If it is a 1, stop; if it is a 0, load 007 into program counter	205	0

*In digital code form.

Fig. 6-59. Initial Memory Conditions.

PROGRAM COUNTER	DATA REGISTER			DATA ADDRESS REGISTER		STATUS		DATA MEMORY					
	A	B	C	X	Y	1	2	200	201	202	203	204	205
001	0	0	0	0	0	0	0	0	0	0	0	0	0

Table 6-24. Register and Memory Chart.

Chapter 7

MECHANICS

7.1 Introduction

It is difficult to develop experiments that will not only appeal to students but will also satisfy the main aims of a general physics laboratory course. Experiment appeal depends greatly on the attitude of the student. They should be patient and realize that each experiment is designed to satisfy some of the objectives of the laboratory. Laboratory space and available student time act as limiting factors on the type of experiment. Progress in science and engineering is at such a fast pace that the beginner must be immediately introduced to the fundamental methods of analysis and the applicability of theories.

In most of the experiments, students will not blindly substitute data into a formula, but will be directed to use his or her analytical skills in order to reach a conclusion. Constant referral to the introductory chapters must be made in order to master the techniques and concepts demonstrated in this and subsequent chapters. Precision analysis will be required in most of the experiments.

Experiments in this chapter vary in difficulty as well as in length. In some cases the experimental techniques are not simple such as those needed for the Gravitational Balance. In others the experimental procedures are easy but the theory is on an above-average level such as in Scattering by Elliptical Targets or in the Driven Torsional Pendulum. The type of experiment to be performed depends on the kind of laboratory the student is taking, the time available, the degree of freedom he is given to experiment on his own, etc.

Experiments 7-2 through 7-4 are not intended to teach the concepts of simple harmonic motion, but are designed to illustrate the evaluation of data including the methods of graphing. In general, students will perform one of these three experiments. There are also several experiments using inclined planes; the choice of experiments depends on the equipment available and the sophistication desired. Again, it is not expected that students will perform all three of these experiments (7-5, 7-6, and 7-12).

The experiment on the Elongation of an Elastomer (7-16) is an example of a case where the apparatus is very simple - a thin piece of "Koroseal," a meter stick and a timer - but where the results are probably surprising to most students. Experiment 7-17 uses a microprocessor and a CRO to analyze the elastomer.

Experiment 7-13 on Rotational Kinematics and Dynamics is recommended as a review of mechanics; it involves a large number of important concepts.

Three experiments cover the driven damped harmonic oscillator: Experiment 7-18 is rather qualitative and uncomplicated; while the theory is discussed in great detail and all the equations are derived in Experiments 7-19 and 7-20. Once again, the choice between these three experiments depends very much on the emphasis in the course that the student is taking.

The success of Experiment 7-21, the Analysis of Gravitation, depends on the ability to minimize vibrations; this may not be easily possible in some buildings. We feel, however, that this was an important experiment in the history of physics and that it involves a number of significant experimental techniques. For these reasons, students are encouraged to try the experiment, even if their final numerical result is not as close to the expected value as they might hope for.

Further experiments in mechanics are discussed in Chapter 8.

Experiment 7-1 MEASUREMENT OF LENGTH, AREA AND VOLUME

INTRODUCTION

In this experiment you will make measurements with the vernier and micrometer calipers and the polar planimeter discussed in Section 4.2. This section should be studied carefully before using the measuring devices. A detailed error analysis is required; the appropriate sections in Chapter 2 must be studied before proceeding with the calculations.

The following procedure is an example of the type of data that can be taken in order to get some practice in measuring distances, angles, and areas. It explains how to measure the length and diameter of a metal cylinder and the measurement of an area, such as the area of a county or a state on a road map. Check with your instructor about other types of measurements.

In all experiments it is important to properly set up a data table which is concise and clear. For most experiments, this will be left up to the student; in this experiment, however, a sample data table is shown on Page 285 as an example of one way to present the data and results.

PROCEDURE

PART I

(a) Use the English scale on the vernier caliper and measure the length of the sample (metal cylinder) four times. Each time a measurement is taken rotate the cylinder 90° about its longitudinal axis. Record in Table 7-1 at the end of this experiment.

(b) Use the metric scale on the same vernier caliper and duplicate the measurements that you made in (a). Record in the data table.

(c) Determine the random limit of error RE for the measurements of length of the metal cylinder. Add the RE to the instrumental limit of error ILE and get the total (net) limit of error LE.

(d) Use either a metric or an English micrometer caliper and measure the diameter of the sample near one end twice, and twice near the opposite end. Between the measurements turn the sample 90° about its longitudinal axis. Record the four measurements in the data table.

(e) Determine, using the same method as in Part I(c), the limit of error LE for the diameter.

(f) Find the volume of the metal cylinder. Determine the maximum error LE and the relative error in the indirect measurement (volume). Remember "volume" is an indirect measurement because it depends on the length and the diameter of the metal cylinder. Depending on whether a metric or English micrometer caliper was used to measure the <u>diameter</u>, choose the appropriate <u>length</u> measurement.

PART II

Take measurements as requested by the laboratory instructor on the angular vernier. Enter the data in the data table at the end of the report.

PART III

(a) Use the polar planimeter to measure an unknown area. Take nine measurements. Calibrate the planimeter by measuring a known area four times.

(b) Determine the RE and add this to the ILE to get the LE for the tracings N_s around the known area A_s.

(c) Use the given ILE of ±0.02% (for the known area) as the LE for the area and calculate the constant K.

(d) Determine the RE and add this to the ILE to get the LE for the tracings N_x around the unknown area.

(e) Find the area A_x. Consult the instructions on use of the Polar Planimeter, in Section 4.2. Determine the maximum error LE and the relative error % in the indirect measurement of A_x.

QUESTIONS

1. Determine the probable (absolute) limit of error in the indirect measurement (volume) of the cylinder in Part I. Find the relative (%) error of the probable error.

2. Compare the maximum limit of error $L.E._v$ found in (f) of Part I with the probable limit of error and discuss the significance of the variation.

3. What is the "least count" of an instrument?

4. Explain in <u>words</u> the general theory of the polar planimeter.

5. Determine the probable (absolute) limit of error in the indirect measurement of the unknown area A_x in Part III. Find the relative (%) error of the probable limit of error.

6. Would the zero correction of the micrometer be more important in the measurement of the diameter of a hair or the diameter of the brass cylinder?

7. Explain why a vernier is important.

8. List six possible causes of systematic error in the measurement of an area with a polar planimeter.

Part I
Metal Cylinder

	Length (cm)	Deviations	Length (in)	Deviations	Diameter ( )	Deviations
1						
2						
3						
4						
–	M		M		M	
R.E.	–		–		–	
I.L.E.	–		–		–	
L.E.	–		–		–	

Measurements on the Angular Vernier	Vernier Clockwise		Vernier Counterclockwise	
	Degrees (°)	Minutes (')	Degrees (°)	Minutes (')

Part II
Polar Planimeter

Calibration Area A_s				Unknown Area A_x				
After	Before	N_s (rev.)	Deviations	After	Before	N_x (rev)	Dev.	$(Dev.)^2$
–	–	M						
–	–	R.E.						
–	–	I.L.E.						
–	–	L.E.						
Unknown Area found by Direct Measurement A_x								
								Σd^2
				–	–	–	–	$3 \frac{S.D}{\sqrt{n}}$
				–	–	–	–	ILE
Area A_x =				–	–	–	–	LE

Table 7-1

Experiment 7-2 THE VIBRATING SPRING

INTRODUCTION

This experiment is designed so that you can investigate some of the fundamental properties of a vibrating spring. No specific equation is initially given for verification. Instead, the approach to analysis is similar to that followed by any experimentalist who would not expect a certain result. In general, when confronted with a problem involving several variables (period T, mass m or any other combination), the independent variable (in this experiment, m) is varied and the effect on the dependent variable T is investigated.

The independent variable is always plotted along the x-axis and the dependent variable along the y-axis. An equation of this type is expressed as $y = f(x)$. In this experiment $T = f(m)$.

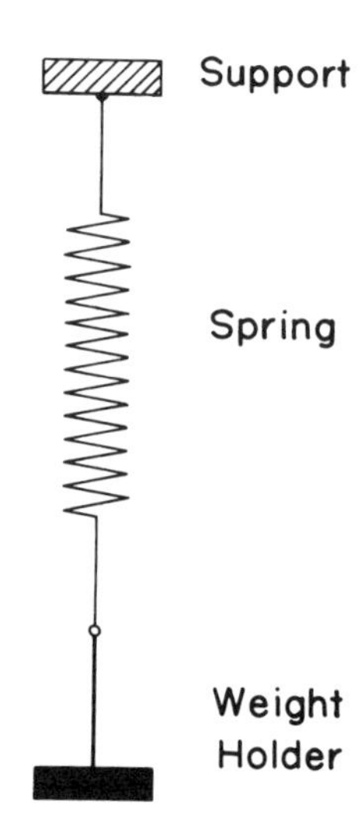

Fig. 7-1.
The Vibrating Spring.

PROCEDURE

1. Detach the spring and determine its mass on the balance scale in the laboratory. Record in Table 7-2.

2. For values of m = 50, 100, 150, 200, 250, and 300 grams, determine the period T of vertical oscillation. The time t (t = the total time for n oscillations should be of the order of 30 to 40 seconds. The time t is limited to a specific range to minimize the effect the ILE of the time clock and the possible fluctuations of the frequency of the input alternating current to the clock might have on the measurement of the period T. The number n of oscillations should be so chosen that $t = nT$ is in this range. Take, record and average four values of nT, and calculate the average value for each load.

   Important: For each load, note the rest (or equilibrium) position. This can be easily found by referring to the meter stick behind the spring. After the spring begins to oscillate, be sure to count only complete oscillations.

ANALYSIS

(Read through several times and note the sample data table at the end of this experiment - use millimeter rectangular coordinate paper for all graphs - even when plotting logs.)

(a) Plot T as a function of m. Note non-linearity of curve. $T = f(m)$ curves down; i.e., T depends on m to a power less than 1.

(b) Since the data yields a curve that is suspected to be of the power law type, plot log T vs. log m. Note the slight curvature which becomes smaller at large values of m. The plotting of the log of the dependent variable versus the log of the independent variable enables the power of the independent variable to be found if a linear plot (no curvature) was obtained. Note - log T is the log base 10 of T, or $\log_{10}$ T.

(c) Derive an equation (see Eq. 3.15, Chapter 3) that will enable you to find the power of T. Show all the work carefully in the computation outline.

(d) Use the linear portion at large values of m of the curve of log T vs. log m and find n, the power of T.

(e) Plot T^2 vs. m. Make sure that both the x and y intercepts can be read from the graph.

(f) Observe the linearity of T^2 vs. m. Find both the intercepts and interpret them. (Hint: Is m the total effective mass of the spring or is it merely the added weights?) Discuss at the conclusion of the report.

(g) Find the force constant k for the spring.

(h) Compare the intercept on the m axis with the measured mass of the spring. Discuss at the conclusion of the report.

QUESTIONS

1. Why is the graph log T vs. log m curved, especially at low values of m?
2. Why is the intercept in (h) above, less than the mass of the spring?
3. At what point in the oscillation (vibration) does the mass have its greatest velocity? Acceleration?
4. Can the mass be increased indefinitely? Why?
5. Explain qualitatively why the spring at certain loadings will oscillate horizontally with a different period than vertically.
6. How much work (energy) is required to stretch the spring 2.0 cm?
7. List three possible sources of systematic error. Explain how they could be eliminated or minimized.
8. Given $T^n = km$, find an expression for the maximum error in T.

Data

Mass of Spring			Force Constant		
mass gm	Equil. position	t sec	T sec	log T	log m

Table 7-2

Experiment 7-3 THE SIMPLE PENDULUM

INTRODUCTION

You will investigate some of the fundamental properties of the simple pendulum. No specific equation is initially given for verification. The analytical approach is similar to that followed by any experimentalist. In general when confronted with a problem involving several variables (period T, length L or any other combination), the independent variable (in this experiment L) is varied and the effect on the dependent variable T is investigated.

The independent variable is always plotted along the x-axis and the dependent variable along the y-axis. An equation of this type is expressed as:

$$y = f(x).$$

In this experiment

$$T = f(L).$$

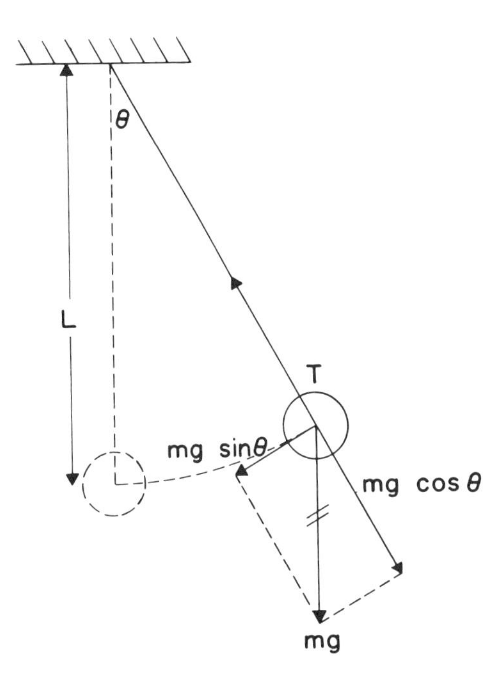

Fig. 7-2. The Simple Pendulum.

PROCEDURE (Read the Procedure and Analysis several times before setting up the data table.)

1. Make certain that the amplitude of swing used for each length setting is always less than one tenth of the length L being used. Why?

2. Choose five lengths. Values of about 20, 30, 40, 60 and 80 cm are satisfactory. It is not necessary to spend a lot of time adjusting the lengths to exactly the values stated, but those used should be carefully measured.

3. For each length value, make three determinations of the time required for 50 complete swings. Read timer to hundredths of a second. Average the three values for the time of 50 swings and compute the period T, the time of one complete swing of the bob.

ANALYSIS (Read Chapter 3, Article 3.4, B-5 before proceeding.)

(a) Use millimeter rectangular coordinate paper and plot T as a function of L. Note the non-linearity of the curve. T = f(L) curves down, i.e., T depends on L to a power less than 1.

(b) Since the data yields a curve that is suspected to be of the power law type, plot log T vs. log L on both millimeter rectangular coordinate paper and log-log paper. Before making a plot on log-log paper, check the data carefully to determine the proper type of cyclic log-log paper.

(c) Derive an equation (see Eq. 3-15, Chapter 3) that will enable you to find the power of T and the proportionality constant between T^n and L. Show all your work carefully in the computation outline.

(d) Find n, the power of T, and k the proportionality constant from both of the log T vs. log L plots. In the conclusion discuss the significance of the variation in your results. Explain which technique is preferable when precision is desired.

(e) Recognizing that the motion of the pendulum bob is simple harmonic (for the small angle used), show in the computation outline that the acceleration of gravity

$$g = \frac{4\pi^2}{k}.$$

Calculate g and compare with the accepted value in your area.

(f) After n and k have been determined, use the final empirical equation and check the curve T = f(L) to see if a smooth fit is obtained. (Read Articles 3.4A, 3.7, Chapter 3). Select values of L and calculate the corresponding values of T. Discuss any discrepancies in your conclusion.

QUESTIONS

1. Why is it necessary that the amplitude of swing used for each length be always less than one-tenth of the length being used?
2. What effect does the mass of the pendulum bob have on the period T?
3. What effect would moving the simple pendulum to an altitude of 4000 miles above the earth's surface have on the period T?
4. What are four possible sources of systematic error?
5. From the necessary condition for simple harmonic motion (F = -kx), derive an expression for the acceleration of the pendulum bob as a function of position.
6. At what point in its swing does the pendulum bob have its greatest velocity? Its greatest acceleration?
7. If the amplitude of swing was much greater than one-tenth of the length being used, what kind of harmonic motion would result? Could the period T be found? How?
8. If the length of the pendulum and period were found by direct measurement, use calculus and find an equation (no numerics) for both the maximum limit of error L.E. and the relative (percent) error in the indirect measurement of g.
9. Find an equation (no numerics) for the probable limit of error and percent probable limit of error in the indirect measurement of g.
10. What are four possible sources of accidental error?

Experiment 7-4 THE VIBRATING RING

INTRODUCTION

The behavior of a series of thin rings of various radii will be investigated. Again no equation is given for verification. Instead the procedure and analysis sections attempt to develop an appreciation of the problems confronted by an experimentalist and to introduce fundamental analytical techniques.

The independent variable (mean diameter, $\bar{D}$) is plotted along the x-axis and the dependent variable (period, T) along the y-axis. An equation of this type is expressed as

$$y = f(x).$$

In this experiment

$$T = f(\bar{D}).$$

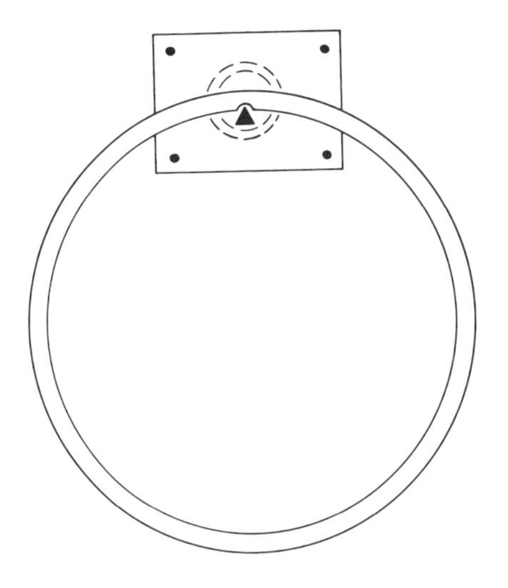

Fig. 7-3. The Vibrating Ring.

PROCEDURE

(Read the Procedure and Analysis several times before setting up the data table.)

1. Measure the outside diameter D_o, the inside diameter D_i and calculate the mean diameter $\bar{D}$, for each of the five rings. Record D_o, D_i and $\bar{D}$ in the data table. (See Fig. 7-3.)

2. Weigh each ring and record its mass in your data table. Use two trip balances to determine the mass of the largest ring. Arrange the ring so that its rim is placed above the knife edge of each trip balance. Make several weighings of each ring and record the average in the data table.

3. Suspend the ring from the knife edge and set it vibrating through a small angle with an amplitude of 5° or less.

4. For each ring make three determinations of the time required for 50 complete swings. Read the timer to hundredths of a second. Average the three values for the time of 50 swings and compute the period T, the time of one complete swing.

ANALYSIS

(Read Chapter 3, Article 3.4, B-5 before proceeding.)

(a) Plot T as a function of $\bar{D}$ on millimeter rectangular coordinate paper. $T = f(\bar{D})$ curves down, i.e., T depends on $\bar{D}$ to a power less than 1.

(b) Since the data yields a curve that is suspected to be of the power law type, plot log T vs. log $\bar{D}$ on millimeter rectangular coordinate paper. (For extra credit plot your data on log-log paper. Check carefully to determine the proper type of cyclic log-log paper.)

(c) Derive an equation (see Eq. 3-15, Chapter 3) that will enable you to find both the power of T and the proportionality constant between T^n and $\bar{D}$. Show all your work carefully in the computation outline.

(d) Find n, the power of T, and the proportionality constant k. If you are working for extra credit, find n, k from both plots. In the conclusion, discuss the significance of the variation in your results. Explain which technique is preferable when precision is desired.

(e) Find by use of the parallel axis theorem, the rotational inertia of a ring about its axis of suspension. Show that:

$$I = \frac{1}{2} M\bar{D}^2, \text{ for each ring.}$$

(I = rotational inertia of the ring about its axis of suspension.)

(f) Recognizing that the motion of the ring (for the small angle used), is simple harmonic, show in your computation outline that the acceleration due to gravity is:

$$g = \frac{4\pi^2}{k}$$

(k = proportionality constant found in (c).)

(g) Derive an equation for the period T of the ring from the constants n, k and the variable $\bar{D}$. Compare with the equation for the period of a physical pendulum.

(h) Select from your data a value for the period T, mass M, and mean radius $\bar{R}$ from any one of the five rings. Calculate from the empirical equation for the period of a ring found in (g) the rotational inertia I of the ring about its axis of suspension. Also calculate from the value for the mean diameter $\bar{D}$ found in (e), the rotational inertia I. Compare the results and discuss any variation in the conclusion to your report.

QUESTIONS

1. See (f) in the analysis and calculate the value of g. Compare your value of g with the accepted value in your area. Discuss any variation in the values.
2. What is the radius of gyration of a ring about its axis?
3. Why is it necessary that the amplitude of swing for a ring in this experiment be 5° or less?
4. What effect would moving the vibrating ring to an altitude 4000 miles above the earth's surface have on the period T?
5. What is the center of oscillation of a ring?

6. If you found an equivalent simple pendulum, would the period T be equal to, greater than, or less than that for a vibrating ring? Explain.

7. Since the vibrating ring is acted on by an elastic restoring torque, derive an expression for its period.

8. What are four possible sources of systematic error?

9. What are four possible sources of accidental error?

10. At what point of its swing does the ring have the greatest velocity? Its greatest acceleration?

11. Would the precision of your results be increased by using the outside diameter D_o and inside diameter D_i instead of the mean diameter $\overline{D}$? Explain.

12. If the diameter of a ring and period were found by direct measurement, use calculus and find an expression (no numerics) for the maximum limit of error (L.E.) and relative (percent) error in the indirect measurement of the acceleration due to gravity.

13. Find an equation for the probable limit of error and percent probable limit of error in the indirect measurement of gravity g.

14. If the amplitude of swing was much greater than 5°, what kind of motion would result? Could the period T be found? How?

Experiment 7-5 ANALYSIS OF RECTILINEAR MOTION

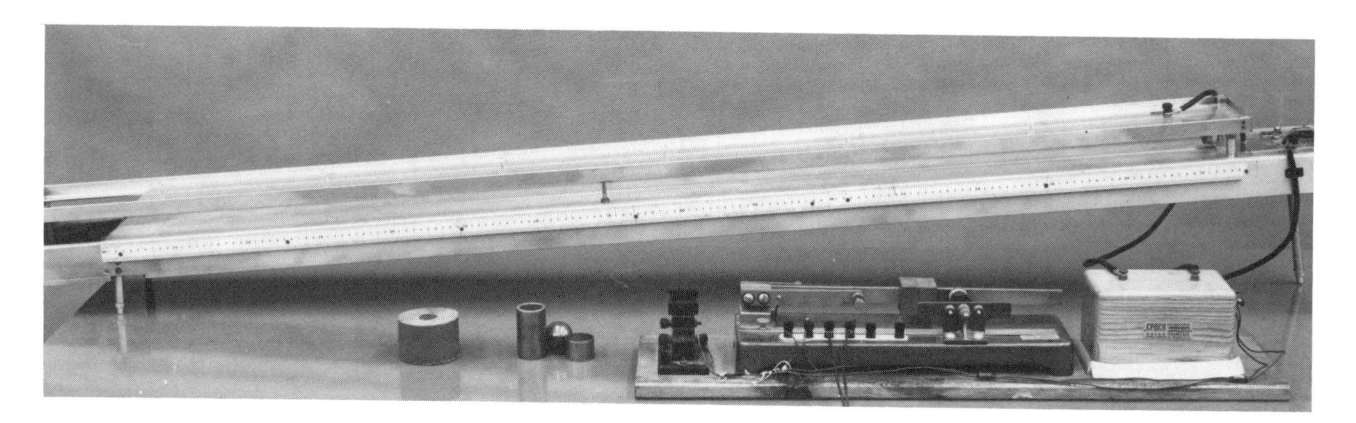

Fig. 7-4. Inclined Plane and Spark Timer.

INTRODUCTION

In this experiment, you will study the rectilinear motion of various objects to obtain a feeling for such quantities as velocity, acceleration displacement and to develop an understanding of vector and scalar quantities.

A sphere, a hollow cylinder, or a solid cylinder (disc) is allowed to roll down an incline. The high voltage terminals of a spark coil are connected to the metal incline and to a wire in the upper plastic track so that sparks, timed by a vibrator, pass from the rolling object to the wire. A length of waxed paper (white side down) placed in the upper track records the sparks and thereby the position of the rolling object for various time intervals. The time between sparks should be 0.1 ± 2% seconds. An impulse counter connected to a spark timer permits the number of sparks per unit time to be checked (descriptions of the spark timer and impulse counter can be found in Chapter 4, Article 4.4).

PROCEDURE

1. Remove the upper half of the plastic track and insert waxed paper with white side down.

2. Inspect the apparatus and try to start it. Do not change the position of the weights on the vibrator arm. Wait until your instructor checks your familiarity with the equipment before attempting to take data. Set up a good data table in your report.

3. Start the vibrator and obtain a spark trace for the sphere (or hollow cylinder or disc) as it rolls down incline. Note that the separation between traces increases with the increased speed of the object. Since you are going to plot the position of a particle against time, <u>from what point should you begin to take measurements?</u>

ANALYSIS

(a) Plot a graph of position Y vs. time t, Y = f(t). Remember that the independent variable t is, in general, plotted along the abscissa (x-axis) and the dependent variable Y along the ordinate (y-axis).

(b) Construct normals to the Y = f(t) curve at half your recorded t values. Once the scale factor is found, determine the actual slope of a tangent line at each selected value of t. Record the speed in your data table. Plot a graph of the magnitude of the instantaneous velocity (instantaneous speed as a function of time, Y' = f(t) or V = f(t).

(c) Inspect your V = f(t) plot. Should your actual slope values be approximately the same for various selected values of time? What is the significance of the actual slope of a tangent line constructed at some point on your graph of V = f(t)?

(d) Plot a graph of the magnitude of the acceleration versus time, Y" = f(t) or a = f(t).

(e) Repeat the procedure and analysis for a different object (use a disc, for example, if a sphere was originally chosen).

(f) Obtain from your plot of speed vs. time, Y' = f(t) or V = f(t), an equation (empirical) which shows how speed increases. Find the numerical values of the constants. Hint - Read Article 3.4B, Chapter 3.

(g) Derive from your plot of speed vs. time, Y' = f(t), an equation that shows that the displacement Y increases quadratically. Find the numerical values of the constants. Can you also determine how far your object rolled before you began to take measurements?

(h) Check the equation found in (g), Y = f(t), by selecting some value for time and calculating Y. The value of the displacement Y should fall near your plot of Y = f(t) within the range of error of your measurement.

QUESTIONS

1. Check with the fundamental dimensions L and T the equations $v = v_o + at$ and $Y = v_o t + \frac{1}{2} at^2$.

2. Does the system of units used have any effect on the magnitude of the velocity or the acceleration?

3. Are your measurements of time and displacement considered to be direct or indirect measurements?

4. Discuss in a qualitative manner the relative accuracy of the direct and indirect measurements made in this experiment.

5. Can you give examples of several types of indirect measurements in this experiment?

6. Explain what is meant by the displacement of an object. Is displacement a scalar quantity? Is it the same as length? Why?

7. Explain what is meant by the speed of an object. Is speed a vector quantity? Can speed be positive or negative?

8. Can the acceleration of an object be negative? Why?

9. Discuss why you do not expect the sphere and cylinder to reach the bottom of the incline with the same speed.

Experiment 7-6 COEFFICIENT OF FRICTION - THE INCLINED PLANE

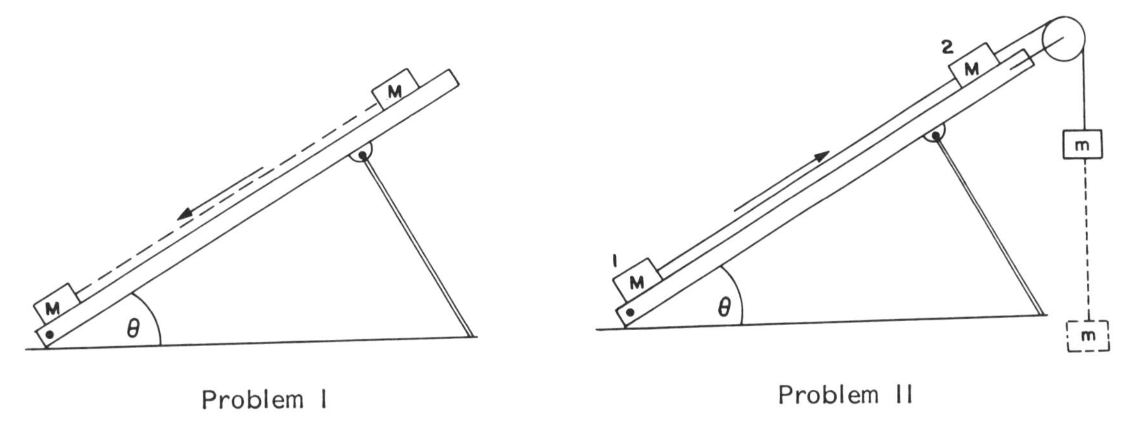

Fig. 7-5. The Inclined Plane.

INTRODUCTION

In this experiment the student is presented with two problems and the instruments necessary for their investigation. The procedure or method of attack is left entirely up to the student.

Treat each problem separately in the computation outline. Draw force diagrams for all bodies. Define every force (or component). Make necessary substitutions of numerics into equations that are used. Record the results of each problem in the "Result Section" of your report and then discuss them in the conclusion.

PROBLEM I

Determine the coefficient of static friction for a wood, a metal, and a rubber block.

PROBLEM II

Find the coefficient of sliding friction for all three blocks. Would it be advantageous to take more than one timed run? The time for each run for a particular block should be fairly constant.

QUESTIONS

1. Why does the coefficient of static friction not remain constant when several measurements in Problem I are taken of the same block?

2. Why does the coefficient of sliding friction not remain constant when several measurements in Problem II are taken of the same block?

3. What are four possible sources of systematic error in Problem I? Problem II?

4. What are four possible sources of accidental error in Problem I? Problem II?

5. What effect does the surface area and the weight have on the coefficient of static friction? Sliding friction?

6. From the direct measurements that were used in Problem I to find the angle θ, determine the maximum limit of error L.E. and relative (percent) error in the indirect measurement μ. Assume a L.E. for each direct measurement.

7. Find an equation for the probable limit of error and percent probable limit of error in the indirect measurement (μ). Use the same assumed limits of errors L.E. for the direct measurements as in Question 6.

8. Explain by use of the concept of resultant torque ($\tau = I\alpha$) how the frictional force acting on the pulley could be found. Draw a force diagram for the pulley and label the torques carefully. Would the tension in the cord of the hanging weight be the same as the tension in the cord attached to the weight on the inclined plane?

9. What effect would the pulley have on the accuracy of the coefficient of sliding friction? Explain your answer.

Experiment 7-7 RADIAL ACCELERATION (CENTRIPETAL FORCE)

INTRODUCTION

Whenever an object moves in a curved path with a constant or variable tangential velocity, it will have radial acceleration. The radial acceleration is the rate of change of velocity due to the change in direction of the tangential velocity. The tangential velocity v_t can be shown by simple calculus to equal $R\omega$, where R is the radius of curvature and ω the angular velocity. Radial acceleration acts perpendicular to the path of the object and directly toward the center of curvature. By calculus a_R can be related to the angular and tangential velocities by:

$$a_R = \omega v_t = R\omega^2 = \frac{v_t^2}{R}$$

In Fig. 7-6, a diagram is shown of the apparatus used for Problem I. A small rubber ball m is rotated in a circular path in such a manner that the angle α does not change. This establishes a steady state condition. The forces on m can be seen in Fig. 7-7.

$\Sigma F_y = 0$, $T_v = w = mg$ (T is resolved into components).

$\Sigma F_x \neq 0$, (because a centripetal force T_c is needed.)

The forces D and P which are equal if friction at the edge E is neglected provide the tension T in the cord (D = P = T). The force D acting on the cord is due to the weight pull of m ($D_w = mg$) and to the centrifugal reaction of m D_c. Therefore, the weight and the centrifugal force are the component forces of D and they act on the cord.

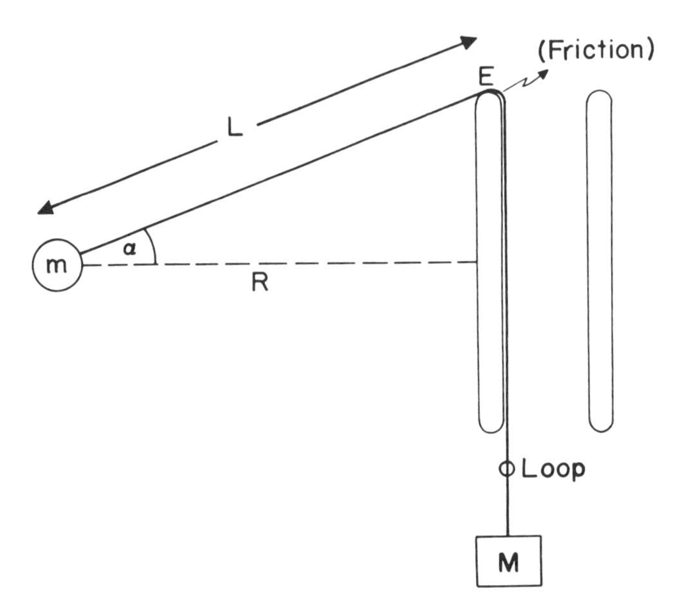

Fig. 7-6. Problem I.

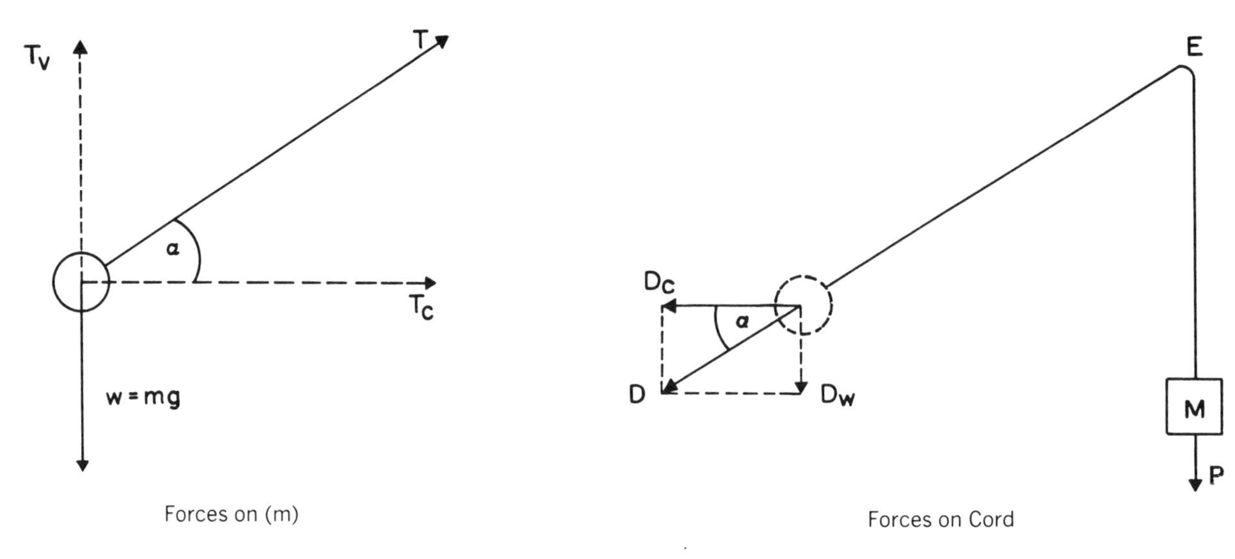

Fig. 7-7.

In summary, the centripetal force provided by the component of the tension in the cord acting directly inward is:

$$T_c = m \frac{v_t^2}{R}$$

The centrifugal force is the reaction to the centripetal force and is represented as D_c, a component of D acting on the cord.

PROCEDURE (This experiment is divided into two parts, Problem I and Problem II. Allot about 1/2 hour to taking of data for each. Use the balance of your laboratory period in writing up the report.)

PROBLEM I (Read first, then set up the data table and record all the data.)

1. Weigh the rubber ball four times. Do not detach cord. The small mass of the cord on the balance pan is negligible. Use the average value of the mass of the ball in your calculations.

2. Slip the cord through the metal cylinder and attach the weight holder. Adjust the mass on the weight holder so that the total mass M including the mass of the weight holder is about ten times the mass of the rubber ball.

3. Slowly rotate the rubber ball. Note that L (see L in Fig. 7-6) varies with the rate of rotation of the ball. Do not swing the ball in a vertical circle. Keep the ball rotating at a constant angular velocity in as close to a horizontal plane as is possible. L can be found by:

a. Holding the cord and the base of the metal cylinder together so that the cord can not slip because of the attached weight holder. Then place on the table.

b. Measuring the distance L from the center of the ball to the center of the top of the metal cylinder.

Find the angular velocity needed for "steady state" for each of 3 different values of L. Use the timer to measure the elapsed time for 25 revolutions.

4. Since ω is small, $\omega^2 L$ should be nearly constant. Compute $\omega^2 L$ for each of the 3 values of L. Discuss the results at the conclusion of your report.

5. Use one value for L and find the angular velocity ω for 3 different values of M. Discuss the results at the conclusion of your report.

PROBLEM II (Requires a qualitative analysis (discussion) only. See Fig. 7-8.)

CAUTION: DO NOT ROTATE THE GLASS GLOBE AT HIGH VELOCITY. HANDLE WITH CARE. BREAKAGE OF THE GLOBE WILL COST $40.00. MAKE SURE THE LOCK SCREW FOR THE GLOBE AT THE TOP OF ROTATOR SHAFT IS TIGHT.

1. Fill the glass globe 1/4 full of water. Slowly increase the angular velocity of the globe by turning the adjustment screw located at the lower end of rotator shaft.

   a. What happened to the water? Why?

   b. Draw a force diagram for a small element of volume of water that is in contact with the glass wall of the globe. Show all the forces acting on the element of volume and all the forces exerted by it. Define every force (or component). Does the element of volume of water have a buoyant force acting on it?

2. Adjust the velocity of the globe so that the liquid surface is about 1-1/2 inches from the top. Drop in the samples one at a time. Wait a few minutes between dropping samples so that they can reach equilibrium. Place a white sheet of paper behind the globe and note the relative positions of the samples.

   a. Record positions of samples.

   b. Draw a force diagram in 3 dimensions for one of the three samples. Assume that it is in contact with the side of the globe. Show all the forces acting on the sample. Hint: When the balls are rotating the buoyant force ($F_B = \rho a V$) has a value for the acceleration which is the resultant of the radial acceleration and gravity. Define every force (or component).

   c. Draw a force diagram in 3 dimensions for one of the three samples when it is not in contact with the side of the globe.

   d. Explain why each sample takes up a different equilibrium position.

   e. Remove the sample balls from the globe. Clean and dry the table.

Fig. 7-8. Rotator and Globe for Problem II.

QUESTIONS

1. Use the values for $\omega^2 L$ found for each of the 3 values of L, (Problem I-4) and calculate the centripetal force. Compare your calculated values with T_c found by using:

   T = constant = W = Mg (W = weight of weight holder)
   T_v = constant = w = mg (w = weight of ball)
   $T_c = \text{constant} = \sqrt{T^2 - T_v^2}$

2. Explain any variations between the results for Question 1.

3. If an error analysis had been performed on all the measurements and the individual limits of error that were found were: ±0.2 for m, ±0.05 rad/sec for ω and ±0.05 cm for L, find the probable limit of error in the indirect measurement for F_c ($F_c = m\omega^2 L$). Express your answer numerically and as a percent (relative) error.

4. What systematic errors in Problem I could affect the validity of your results? Discuss.

Experiment 7-8 INVESTIGATION OF UNIFORM CIRCULAR MOTION

INTRODUCTION

Since the student must derive in the computation outline of this experiment the fundamental relations which will explain the behavior of a body moving with uniform circular motion, the introduction to Experiment 7-7 should be read carefully before proceeding.

PROCEDURE (Error analysis is required - do not forget to show the deviations in your data table.)

Part I (See Fig. 7-9.)

1. Inspect the rotator head H. Note that the value of the mass within the carriage is stamped on the end. Also, movement of the mass will cause the needle to rise or fall - move the mass by hand and see what happens. When data is recorded, the needle must be kept as near as possible to the center of the small disk at the axis of the head. (I.L.E. of mass = ±0.1 gm)

2. Place the rotator head on the rotator. Tighten the lock screw. Note the clutch plate drive with the control screw at the base. Start the rotator and experiment with the apparatus until you become familiar with its operation. By a very slight and gradual adjustment of the screw, the needle can be kept, on the average, opposite the edge of the disk.

3. One student maintains as near as possible constant velocity by keeping the needle opposite the edge of the disk, while the other student engages the revolution counter and times the operation. Record the original reading of the revolution counter before it is engaged and its final reading after a two-minute run. Do this four times. The I.L.E. of the time clock is ±0.02 seconds.

4. Since it is not convenient to measure the radius while the head is rotating, it will be found statically in Part II.

Part II

1. Hang the rotator head from its hook on the stand provided. Attach the mass carrier to the eye in the end of the bob and load with enough mass to produce the same deflection of the needle as occurred during rotation. Record all masses, including that of the carrier and bob in the data table.

2. Since the I.L.E. of the added masses is very small compared to the sensitivity of this measurement, the sensitivity is used as a measure of the precision. When the needle is exactly opposite the edge of the disk, add sufficient mass to just slightly move the needle. This mass will be the I.L.E. of the measurement.

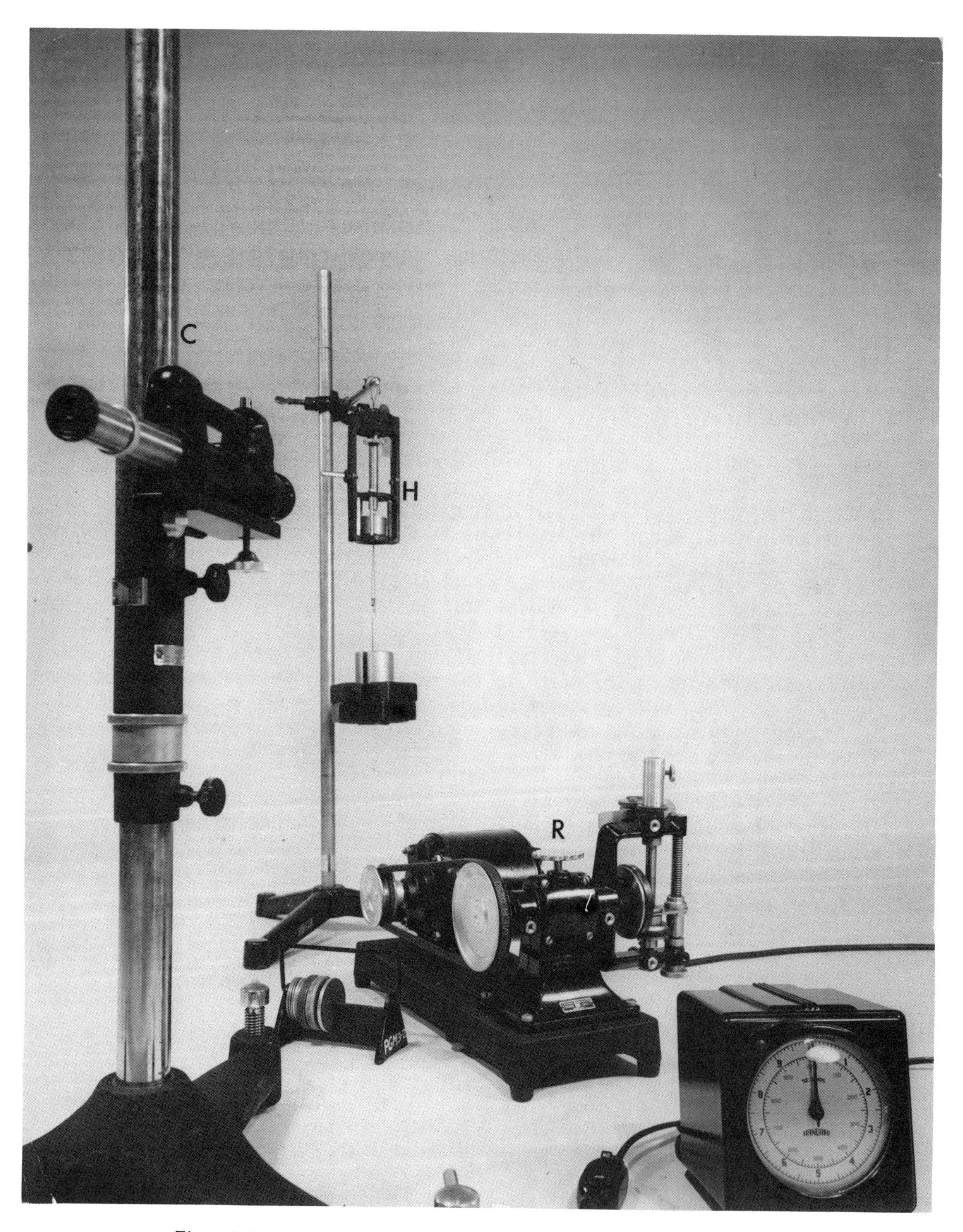

Fig. 7-9. Circular Motion Apparatus.
C = Cathetometer; R = Rotator; H = Rotator Head.

3. Before using the Cathetometer, read Chapter 4, Article 4.8 - Parallax. Since the rotator head's axis is marked with an engraved line and the center of the bod is similarly indicated, use the Cathetometer to measure the distance between the center of mass of the bob and the axis of rotation. Record the initial reading and the final reading. Subtraction of the two readings will give the radius. Make this measurement four times. Use the least count of the scale for the I.L.E. of the Cathetometer. Record in your data table.

ANALYSIS

(a) Derive an expression for the centripetal force of a body moving with constant angular velocity. Illustrate your derivation with a vector diagram. Label the diagram carefully. Number and explain each step in the derivation.

(b) Using the value of the radius from Part II, calculate the angular velocity from the data in Part I and find the value of the centripetal force.

(c) Find the maximum limit of error and the probable limit of error for the indirect measurement of centripetal force. Express your answers as numerical errors, not percent.

(d) From $F = mg$, find the static force necessary to produce the same deflection of the needle as was produced dynamically in Part I.

(e) Find the maximum limit of error and the probable limit of error for the static force.

(f) Compare the dynamic and static values for force in the result section. Discuss any differences in your conclusion.

QUESTIONS

1. Why was the rotator head not rotated around a horizontal axis in this experiment?
2. Why could we assume in Part I that we had constant angular velocity, even though the needle could only be held at an average deviation opposite the disk?
3. If the tension setting were changed in the rotator head, would there be any effect on the value of the centripetal force? Explain.
4. What is an expression for the rotational inertia of the bob in the rotator head?
5. Using your value of angular velocity, find the rotational kinetic energy of the bob.
6. If the tension in the spring were reduced, would the rotational inertia of the bob be greater or smaller? Explain.
7. If the rotator head were rotated in a vertical circle, would the rotational inertia of the bob be greater or smaller at the top of the circle than it would be at the bottom? Explain. What would be the effect on the rotational kinetic energy of the bob?
8. If the head were rotated in a vertical circle, are there any positions were the radial acceleration would be zero? Explain.
9. If the head were rotated in a vertical circle, are there any positions where the tangential acceleration would be present? Explain. What provides the force necessary to produce the tangential acceleration?
10. List four sources of systematic error.

Experiment 7-9 BALLISTIC PENDULUM - PROJECTILE MOTION*

INTRODUCTION

The velocity of a projectile will be found by two methods. The first method requires the application of Newton's Second Law to an object (projectile) in motion. If air resistance is neglected, the only force acting on the moving projectile is that due to the gravitational field of the earth. Remember that the weight (force) of an object is equal to the force of attraction between two masses - that of the earth and the object or:

$$F = W(\text{weight}) = \frac{Gmm'}{r^2} = mg \text{ (or } ma) \qquad 7\text{-}1$$

Eq. 7-1, as shown by Newton's Second Law, is:

$$\Sigma F_y = ma_y \text{ } (a_y = \text{acceleration in the vertical direction - gravity}) \qquad 7\text{-}2$$

$$\Sigma F_x = ma_x = 0 \text{ } (a_x = 0) \qquad 7\text{-}3$$

As a result of Eqs. 7-1, 7-3, the equations of linear motion with constant accelerations (if gravity g is constant) can be applied to the moving projectile. Derive the necessary free flight (or kinematical) equations as the first step in your computational outline.

In the second method, the principles of the conservation of momentum and the conservation of energy must be applied. The conservation of momentum states that the combined momentum of both the projectile m and the ballistic pendulum M is the same before and immediately after impact:

$$mv = (M + m)\,V$$

As a result of the projectile being caught by the catcher at the end of the pendulum, the pendulum M swings about its axis of support and its center of mass rises a distance h above the initial position. v is the velocity of the projectile before impact. V is the common velocity of both the projectile and the pendulum after impact.

The conservation of energy (sum of the initial energies = sum of the final energies, or $P.E._1 + K.E._1 + H_1 = P.E._2 + K.E._2 + H_2$) cannot be easily applied to the projectile before and after impact. The reason is the difficulty in finding the total amount of kinetic energy lost as heat energy and sound energy at the time of impact. Energy is still conserved in the entire process but in order to avoid trying to find the heat energy, this problem is split into two parts:

I. Apply the conservation of momentum.

II. Apply the conservation of energy after impact. After impact there is

*See P.D. Gupta, Blackwood Pendulum Experiment and the Conservation of Linear Momentum, Am. J. Phys. 53(3) 1985.

a small amount of energy loss (friction) at the pivot of the support which can be neglected. Since the height h of swing after impact can be measured, the common velocity V of both projectile and pendulum can be found.

Derive an expression for v, the velocity of the projectile before impact. Your final equation should contain a term which includes h, the height of swing. This derivation should appear as the second step in the computation outline.

In summary this experiment requires an understanding of:

(a) Newton's Universal Law of Gravitation
(b) Newton's Second Law
(c) The Conservation of Momentum
(d) The Conservation of Energy

Fig. 7-10. Apparatus for the Ballistic Pendulum.

PROCEDURE (Read both methods to determine the type of data table that is required.)

First Method (Kinematical)

Error analysis is required for the free flight (kinematical method). You should use for the instrumental limit of error of the meter stick ILE = ±0.2 cm. This ILE is very high because of the crude technique used in taking measurements.

1. Swing the pendulum up out of the way, and then place a piece of carbon paper, face up, under a sheet of plain paper on the floor of the box. Arrange so that one end of the paper is just touching the front of the box. Secure with thumb tacks.

2. Fire the ball four times. Imprints on the paper in the box should form a close cluster. Pick a point that represents the average of the four imprints. Measure the horizontal distance between the center of the ball on the gun in the uncocked position to the average of the four imprints. Do this four times. (Note: The horizontal distance actually should be measured four times to each of the four imprints, but to save time an average point is used. Instead of 16 measurements of the horizontal distance, we only make four.)

3. Measure four times the height of the ball (center of mass) above the average of the imprints on the floor of the box which it lands. Record all the values in the data table.

Second Method (Dynamical)

1. Weigh the ball to the nearest tenth of a gram with a trip balance. Weigh the ball four times and record each measurement. Also record the average of the four measurements. The mass of the pendulum will be found on a card on the apparatus.

2. Shoot the ball into the pendulum case four times. Pull the trigger quickly, but do not yank on the apparatus while doing so. Record the reading of the pawl on the ratchet for each shot. After the four shots, place the pawl of the pendulum on the ratchet at the ratchet reading equal to the average.

3. The position of the center of gravity of the pendulum is indicated by a pointer attached to the pendulum cage. Find the height of this pointer above the base of the apparatus. Take four readings. Next permit the pendulum to hang vertically in its equilibrium position and again find the height above the base of the apparatus. Take four readings. Find the actual displacement of the center of mass of both the pendulum and ball. Record all the measurements and their average in the data table.

QUESTIONS

1. What is the kinetic energy K.E. of the projectile an instant before impact?

2. What is the kinetic energy K.E. of the projectile an instant after impact?

3. Compute the kinetic energy lost during impact - express numerically and as a percent. What becomes of the difference in energies of the ball before and after impact?

4. List some reasons for the differing velocity values of the projectiles in Part I and Part II.

5. If the gun had not been horizontal, how would the result of Part II be changed?

6. Compute the ratio of the mass of the pendulum to the total mass of the ball and pendulum. Compare this with the fractional loss of energy during the collision.

Experiment 7-10 IMPULSE AND MOMENTUM*

INTRODUCTION

When two bodies collide with each other, each will exert a force on the other one. This force F varies with time. We can calculate the impulse

$$J = \int_{t_1}^{t_2} F dt ,$$

where t_1 is a time before the collision and t_2 is a time after the collision. If F is the net force acting on the body, then the impulse is equal to the change in momentum, or $J = \Delta p = p_2 - p_1 = mv_2 - mv_1$. Remember that J, p and v are vector quantities. The apparatus is shown in Fig. 7-11.

In this experiment a small car will collide with a force transducer. A microcomputer measures and records the force every 100 microseconds for about 1/2 second. Also recorded on the same time scale is information which tells us whether or not a photo gate is blocked by a flag mounted on the car. This enables us to calculate v_1 and v_2. You are to find and attempt to explain the discrepancy between J and Δp.

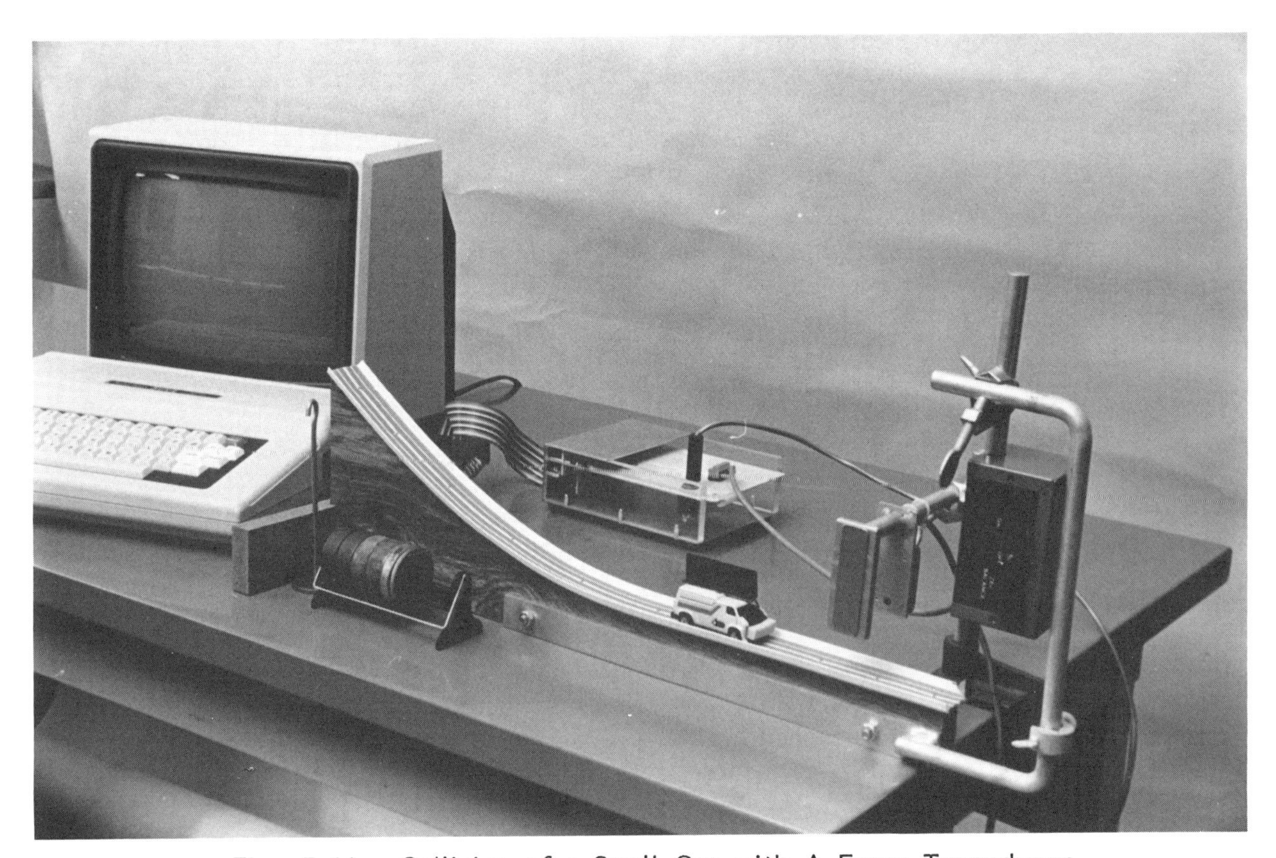

Fig. 7-11. Collision of a Small Car with A Force Transducer.

---

*These directions are written for use with a TRS 80 Color Computer 2 and an interface designed under the supervision of Dr. Thomas Shannon under a grant from the RCA Corporation. A detailed description may be obtained from Professor Walter Eppenstein, Physics Department, Rensselaer Polytechnic Institute, Troy, N.Y. 12180-3590. Note: The Force Transducer and the Impulse Momentum Apparatus can be obtained from Pasco Scientific, 1876 Sabre St., Hayward, CA, 94545

PROCEDURE AND ANALYSIS

It is recommended that you become familiar with the apparatus by making several runs before recording the data. Start the program and choose "calibrate."

Step A - With no masses on the transducer set the "Zero Balance" control to give a reading of 1, then turn the knob to just give a zero reading.

Step B - Add mass to exert a force on the transducer of 1 Newton. Adjust the "Gain Control" to give a reading of 10.0.

Repeat Steps A and B several times until there is no change in the values. This is necessary because the gain and zero controls interact slightly. Next press the space bar. Place the car on the track touching the transducer and position the photo gate so that the beam is just behind the rear of the flag on the car. (Note: The small red light on the gate is on if the beam is clear and off if it is blocked.) Place the car near the top of the ramp. Select "Run." Release the car. (The actual measurements start when the flag first breaks the beam.) At the end of the data collection (about 1/2 second) part of the data is shown on the screen. Approximately 4,000 measurements were made. However, only about 250 can be displayed on the screen at one time. The screen acts as a window with a cursor on the total set of data. The window is initially at the extreme left. (Early time values.) The window can be moved by means of the arrow keys. Holding the shift key down speeds the movement. When you set the point where the force returned to zero after the collision under the cursor, then the value ΣFdt given at the top of the screen is a good approximation to $\int Fdt$, the impulse for this collision.

Measure, using the cursor, the time the flag was in the beam before and after the collision. Measure the length of the flag and calculate v_1 and v_2. Measure the mass of the car and calculate Δp; (Remember p_1 and p_2 are vectors). Compare J and Δp and try to explain the difference.

QUESTIONS

1. What are the units of impulse?

2. Are there ways to find the relation between the force during impact and time without using a microcomputer? Discuss briefly.

3. Is kinetic energy conserved in the collision? If not, why not?

4. Discuss some other applications of a force transducer and a photo gate.

5. When the car hits the force transducer the whole track may move slightly. Is this significant? Verify by measurement.

6. Why is there foam rubber on the car bumper? What would happen if the car runs down backwards? Explain any difference.

7. How many grams are needed to obtain the one Newton force for calibration?

8. Suppose the car stuck (completely) to the transducer. How would you expect the measured impulse to compare with the value when the car bounced back?

Experiment 7-11 SCATTERING

INTRODUCTION

Considerable information about the properties of various nuclei, atoms, and molecules is obtained by bombarding them with uniform beams of photons, electrons, protons, neutrons, or alpha particles and investigating the angles through which the incident particles are deflected. Although the scatterers cannot be seen, quantitative conclusions are obtained from measurements of the relative number of particles scattered at a given angle. We will study scattering with the apparatus in Fig. 7-12.

Fig. 7-12. Scattering Apparatus.
(Courtesy of The Sargent-Welch Scientific Company)

In our analysis, we investigate the relationship between the scattering angle θ in Fig. 7-13 and 7-14 (the angle between the direction of the outgoing particle and the one it would have if it kept moving without any external forces acting upon it) and the "impact parameter" b (the distance between the center of the incident particle and the center of the scatterer measured perpendicular to the line of flight of the incident particle. Although in practice, a uniform beam of particles is ised, (i.e., the number of particles per unit of cross sectional area is constant for the incoming beam) in this model we use single particles.

From measurements of the angle θ, the "impact parameter" b, and the radius r of our particle, and certain assumptions about the nature of the scattering process, we can determine the radius R of the nucleus - in this case the cylinder that is scattering the particles. We can therefore measure the size of our scatterer indirectly by measuring the angular distribution of the scattered particle.

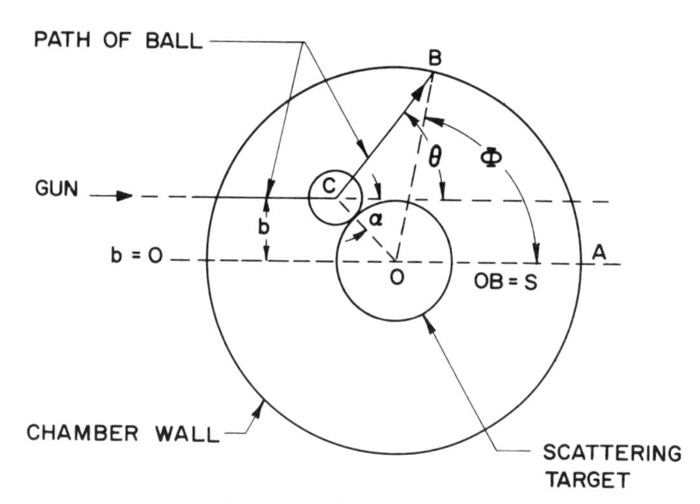

Scattering Apparatus

θ = Scattering angle
b = Impact Parameter
S = OA = OB
$\phi = \frac{\text{arc AB}}{S}$ radians

Fig. 7-13. Scattering Apparatus

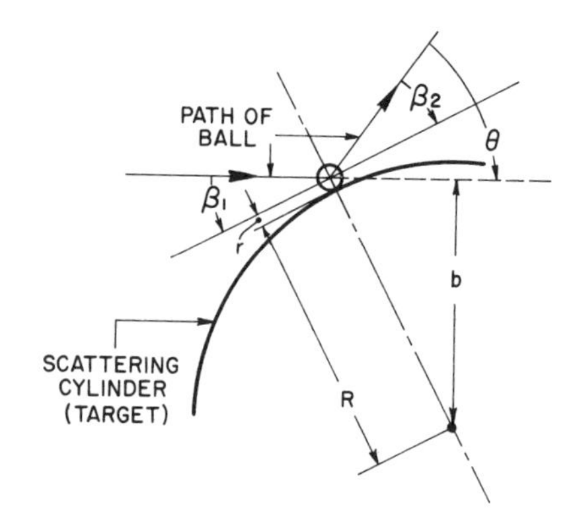

Ball in Contact with Target.

r = Radius of particle
R = Radius of scatterer (cylinder)
b = Impact parameter
θ = Scattering angle

Fig. 7-14. Ball in Contact with Target

If we assume that the incoming particle is reflected by the scatterer (i.e., a perfectly elastic collision), then $\beta_1 = \beta_2$ (see Fig. 7-14). The initial velocity $\bar{v}_i$ and final velocity $\bar{v}_f$ for the particle are, therefore, equal in magnitude. From the conservation of energy we obtain

$$\frac{1}{2} m |\bar{v}_i|^2 = \frac{1}{2} m |\bar{v}_f|^2, \qquad 7\text{-}4$$

where m is the mass of the bombarding particle.

Why can we say that Kinetic Energy is conserved? We also consider the conservation of momentum and the fact that the impulse J acting during a collision is the difference between the momentum before and the momentum after a collision. That is

$$\bar{J} = m\bar{v}_f - m\bar{v}_i \qquad 7\text{-}5$$

The direction of the impulse exerted on the particle must be normal to the surface of the cylinder at the point of contact. Since the interaction forces act only along the line of centers during a collision of this type, the angle β_1 is equal to the angle β_2. The collision reverses the component of momentum perpendicular to the surface of the cylinder and the scattering process can be considered a reflective process. Full understanding of the theory mentioned above is not necessary for purposes of this experiment. Knowing that $\beta_1 = \beta_2$ and also $\theta = \beta_1 + \beta_2$, we find

$$\frac{\theta}{2} = \beta_1 = \beta_2 \quad \text{and} \quad \cos\frac{\theta}{2} = \frac{b}{R + r}. \qquad 7\text{-}6$$

PROCEDURE (<u>WARNING</u>: The gun is dangerous. Do not fire before reading Step 3 carefully.)

1. Attach a strip of waxed paper, white side toward the target, to the chamber wall. Cover about 3/4 of the circumference of the chamber. Locate and mark the zero position on the tape.

2. The value of the impact parameter b is controlled by the distance the gun is displaced from its zero position. It is determined by the pitch of the screw and the number of turns. The screw has 18 threads per inch which, after converting to the metric system, means that the <u>gun moves 0.141 cm per revolution of the screw</u>. Before trying to fire the gun, make certain that you understand how the gun is moved and adjusted.

3. <u>WARNING</u> - BEFORE FIRING GUN MAKE CERTAIN APPARATUS IS COVERED WITH PLASTIC TOP - IF YOU ARE CARELESS, YOU MAY INJURE SOMEONE. DO NOT LOOK THROUGH GUN APERTURE WHEN FIRING THE BALL. IT MAY BE DEFLECTED BACK FROM A HEAD-ON COLLISION WITH TARGET.
   To load the gun, rotate the metal sleeve until the hole in the sleeve lines up with the hole in the barrel. Drop in the ball and rotate the sleeve to close the hole. The apparatus should be so levelled that the barrel of the gun is horizontal, otherwise the ball may roll from the barrel before it can be fired. Best results are obtained when the hose of the aspirator is held straight and the balls are not fired too hard. CHECK - do you know what one complete revolution of the screw does?

4. When taking data, fire the gun five times for each impact parameter (each new screw setting) starting at $b = 0$. Do not forget to replace the top after each shot. After each shot locate the indentation made in the paper by the ball and label it so you can identify it later. When should you stop bombarding the target? Although it would be best to select impact parameters on both sides of $b = 0$, you may not have time to do so during one laboratory period.

5. Remove the tape and measure with a meter stick the distance from the zero position on the tape to each indentation. Average these values for each group and find the angle ϕ corresponding to the arc. Record all the data in tabular form with the corresponding values of b. How do you obtain a value for b in cm for each ϕ observed?

6. Record the average values of the radius R of the <u>target</u>, the diameter and the radius r of the <u>particle</u> (steel ball), and the radius ($OB = S$) of the <u>chamber</u> in your data table. See Fig. 7-13. List all of the multiple measurements with their average group values. Use calipers to measure R and r.

ANALYSIS

<u>Project 1</u>. The students will use two methods to determine R, the radius of the target.

1. Plot $\cos \phi/2$ versus b. Calculate (as a first-order approximation) the value of $R + r$ using the range of b values for which the approximation $\theta = \phi$ is most nearly correct. Since you know the radius r of your bombarding particle, what is R?

2. Compare R found analytically with the directly measured value for R. Discuss the validity of the assumptions used to obtain R + r from the plot of cos $\phi/2$ versus b.

<u>Project 2.</u> Students must complete Project 1 before starting Project 2.

1. Derive an equation which allows the scattering angle θ to be obtained in terms of the following directly measured quantities: the radius S of the chamber, the impact parameter b, and the angle ϕ.

2. Use the equation found in Project 2-1 and calculate values of θ for corresponding values of b and ϕ. Prepare a table of these values, plot cos $\theta/2$ versus b, and find the value of R + r. (Suggestion: consider the slope of the curve.) Compare with R + r measured directly. Discuss the results.

3. Derive an equation for R + r in terms of quantities which can all be measured directly. Substitute your data into this equation and compare these values with directly measured value of R + r. Discuss the results.

QUESTIONS

1. What is the significance of the Y-intercept in the plot of cos $\phi/2$ versus b?

2. What effect does the spin of the ball have on the scattering angle θ?

3. Is surface friction between ball and target the same for all values of b? Explain, remembering which force component affects friction.

4. If the target had a spherical cross-section, what kind of angular distribution would you get? What would be the effect of a hexagonal cross-section?

5. Discuss the validity of the approximation $\theta = \phi$ in terms of:

   A. The radii of the scattering particles.
   B. The radius of the chamber.

6. In a true atomic collision there is no sharp "edge" to the force causing the change of path of the incident particle. Would the meaning of the impact parameter change in this case? Why?

7. For collisions having the same impact parameter, does the velocity of the incident particle affect the scattering angle? Why?

APPENDIX: Calculation of the Scattering Angle and R + r

I. The Self-Consistent Method

Applying the law of sines to the triangle OBC in Fig. 7-13,

$$\frac{R + r}{\sin(\phi - \theta)} = \frac{S}{\sin(\theta + \alpha)}, \quad \text{and}$$

$$\sin(\phi - \theta) = \frac{R + r}{S} \sin(\theta + \alpha), \qquad 7\text{-}7$$

where S = OA = OB in Fig. 7-13.

Using the trigonometric identities,

$$\sin(A + B) = \sin A \cos B + \cos A \sin B$$

and

$$\sin(A - B) = \sin A \cos B - \cos A \sin B$$

the above equation for $\sin(\phi - \theta)$ may be written as

$$\sin\phi \cos\theta - \cos\phi \sin\theta = \frac{R + r}{S}(\sin\theta \cos\alpha + \cos\theta \sin\alpha).$$

Dividing by $\cos\theta$, we get

$$\sin\phi - \cos\phi \tan\theta = \frac{R + r}{S}(\tan\theta \cos\alpha + \sin\alpha).$$

From Figs. 7-13 and 7-14,

$$\sin\alpha = \frac{b}{R + r}, \quad \text{and}$$

$$\sin\phi - \cos\phi \tan\theta = \frac{R + r}{S} \tan\theta \cos\alpha + \left(\frac{R + r}{S}\right) \frac{b}{R + r}.$$

Therefore,

$$\sin\phi - \frac{b}{S} = \tan\theta \left(\frac{R + r}{S} \cos\alpha + \cos\phi\right),$$

and

$$\tan\theta = \frac{\sin\phi - \dfrac{b}{S}}{\cos\phi + \dfrac{R + r}{S} \cos\alpha}, \qquad 7\text{-}8$$

where from the inspection of Fig. 7-13,

$$\cos\alpha = \frac{\sqrt{(R + r)^2 - b^2}}{R + r}.$$

If we examine Eq. 7-8 closely, we see that we must know not only the angles ϕ and α but alos R + r before we can determine θ. Since we want to find out something about the properties of our target (nucleus) from the scattering angles θ, we certainly do not want to measure R + r directly.

We can, however, obtain (R + r) from our experimental data using Eqs. 7-6 and 7-8, if (R + r)/S is small. From Eq. 7-7 it follows that if $(R + r)/S << 1$ the angles θ and ϕ are nearly equal. For this apparatus (R + r)/S is small. Since $\sin(\theta + \alpha)$ can never exceed one, the $\sin(\phi - \theta)$ can never exceed 0.08 and ϕ and θ can not differ by more than 5 degrees. As a first approximation, we assume $\theta = \phi$. Plot $\phi/2$ against b, and from the slope of the resulting curve obtain a first approximation of (R + r). Use this value to compute θ from Eq. 7-8. Plot $\theta/2$ against b and obtain a second-order approximation of R + r. Proceed in this way until the value of (R + r) computed from Eq. 7-6 remains essentially the same from one approximation to the next.

This procedure should yield an accurate value of (R + r) but the calculations are extremely tedious. The suggested procedure is to obtain the first approximation of (R + r) assuming that $\theta = \phi$; bypass the intermediate stages, and use a measured value of (R + r) to compute each value of θ from Eq. 7-8. The extent to which Eq. 7-6 describes our data provides us with an indication of the validity of the assumptions we used to derive it.

II. A Direct Method

The scattering angle θ can also be obtained as follows: From Fig. 7-13,

$$\sin\alpha = \frac{b}{R + r} = \cos\frac{\theta}{2}$$

$$\therefore \cos\alpha = \left(1 - \cos^2\frac{\theta}{2}\right)^{\frac{1}{2}} = \sin\frac{\theta}{2} \ .$$

From the law of sines,

$$\frac{\sin(\phi - \theta)}{R + r} = \frac{\sin(\theta + \alpha)}{S},$$

and

$$\sin(\phi - \theta) = \frac{R + r}{S}(\sin\alpha\cos\theta + \cos\alpha\sin\theta)$$

$$= \frac{R + r}{S}\left(\cos\frac{\theta}{2}\cos\theta + \sin\frac{\theta}{2}\sin\theta\right)$$

$$= \frac{R + r}{S}\cos\left(\theta - \frac{\theta}{2}\right),$$

or

$$\sin(\phi - \theta) = \frac{R + r}{S}\cos\frac{\theta}{2}$$

$$\sin(\phi - \theta) = \frac{b}{S}$$

$$\therefore \theta = \phi - \arcsin\frac{b}{S} \qquad 7\text{-}9$$

Are Eqs. 7-8 and 7-9 equivalent? Explain. Substitute data and see what variation is present.

From Eq. 7-9 it follows that if $b << S$, then the angles θ and ϕ are nearly equal. Eq. 7-9 may be used directly with Eq. 7-6 to find $R + r$.

SUPPLEMENTARY PROJECT

SCATTERING BY AN ELLIPTICAL TARGET*

Elliptical targets may be used in place of the round target to provide an interesting supplementary project. The manner of attaching and using the elliptical targest is the same as with the circular ones. However, the analysis of the data is more complex.

THEORY

The analysis below considers the scattering of a point mass particle by a general (not necessarily circular) cylinder. By examining the angle through which the particle is scattered, the shape and size of the scatterer can be determined.

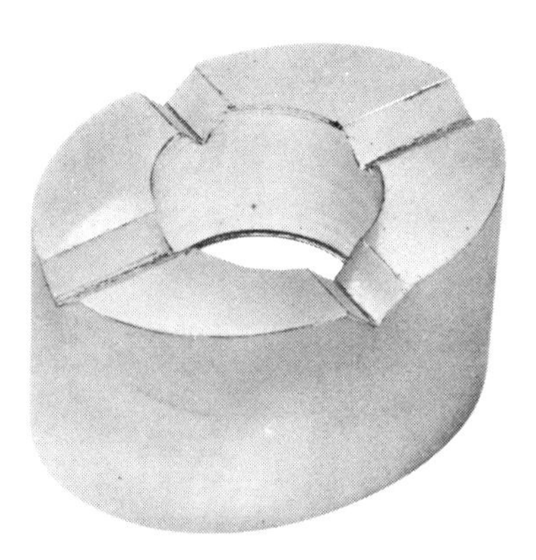

Fig. 7-15. An Elliptical Target for the Scattering Apparatus. (Courtesy of The Sargent-Welch Scientific Company)

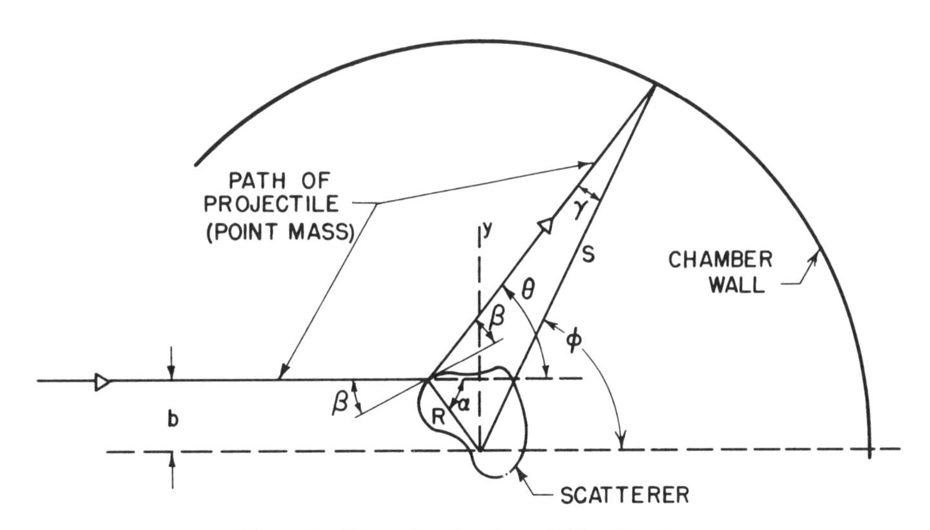

Fig. 7-16. Analysis of Scattering.

*Elliptical targets and the supplementary project were prepared by Prof. Harry F. Meiners.

In Fig. 7-16 the x-axis is the zero scattering parameter line, $b = 0$. The y-axis is the perpendicular at the center of the chamber, as indicated. The chamber radius is S, the scattering angle θ and the angle of incidence of the particle is β.

From geometry it can be see

$$\beta = \frac{\theta}{2}$$

$$\tan \beta = \frac{dy}{dx} \qquad \text{7-10}$$

$$\tan \alpha = -\frac{y}{x}$$

and

$$\phi = 2\beta + \gamma$$

The law of sines yields

$$\frac{R}{\sin \gamma} = \frac{S}{\sin (\theta + \alpha)}$$

or

$$\gamma = \arc \sin [\frac{R}{S} \sin (\theta + \alpha)]$$

In a typical apparatus R/S is less than 0.08. Since the sine function is always less than or equal to one,

$$\gamma \leq \arc \sin (0.08) < 0.08 \text{ radians} < 5^\circ$$

Considering the above, it is reasonable to neglect γ. This approximation is equivalent to

$$\phi \simeq \theta = 2\beta$$

The above analysis can be applied to an elliptic cylinder. The equation of an ellipse is

$$\frac{x^2}{A^2} + \frac{y^2}{B^2} = 1 , \qquad \text{7-11}$$

where A is the semi-major axis and B is the semi-minor axis. An ellipse is characterized by its eccentricity ε which is defined by

$$\varepsilon = [1 - (\frac{B}{A})^2]^{1/2} .$$

The eccentricity determines the shape of the ellipse but not its size. Ellipses with the same eccentricity are "similar." From Eqs. 7-10 and 7-11 it can be seen that

$$\frac{dy}{dx} = -\frac{x}{y} (\frac{B}{A})^2 = \tan \frac{\phi}{2}$$

$$y \tan \frac{\phi}{2} = \frac{B}{A} (B^2 - y^2)^{1/2}$$

$$y^2 \tan^2 \frac{\phi}{2} = \frac{B^4}{A^2} - y^2 \frac{B^2}{A^2} \quad . \tag{7-12}$$

Since both y and ϕ can be determined experimentally, Eq. 7-12 can be used to determine the ratio B/A. If we know B/A, then we can determine the eccentricity ε and values of x. If we know the values of x that correspond to values of y, we can find A and B.

ANALYSIS

The following data table is suggested:

Trial	b=y	$\pi - \phi$ cm	$\pi - \phi$ rad	$\phi/2$	$\tan \frac{\phi}{2}$	$b^2=y^2$	$y \tan \frac{\phi}{2}$	$y^2 \tan^2 \frac{\phi}{2}$	x
1									
2									

Plot y^2 vs. $y^2 \tan^2 \phi/2$ and determine the slope of the curve. From Eq. 7-12, we can show that the slope is equal to $-1/(B/A)^2$. Use this value of $(B/A)^2$ to calculate the eccentricity ε of the elliptical target.

Measure the dimensions of the elliptical target with a micrometer, and calculate the eccentricity of the ellipse. Compare the experimental and measured values of ε and give the percent deviation. A percent deviation of between four and eight percent is normal for this type of experiment.

Plot y vs. x and use the method of selected points to determine A and B in Eq. 7-11. The method of selected points involves taking two values of y and their corresponding x values and substituting them into Eq. 7-11. The two simultaneous quadratic equations are then solved in the usual manner. Use the best experimental values of y and x available, since errors compound very rapidly in this calculation.

Experiment 7-12 ROTATIONAL AND TRANSLATIONAL MOTION

INTRODUCTION

Whenever a body (solid cylinder, hollow cylinder or sphere) rolls down an incline, translational and rotational motions occur simultaneously. We assume rolling, without slipping, and that all bodies are in contact with the plane along a line which is called the instantaneous axis of rotation.

The problem is to determine the linear acceleration of the center of mass (O) and also the linear velocity of the center of mass of each body at the base of an incline. This can be done by three different methods:

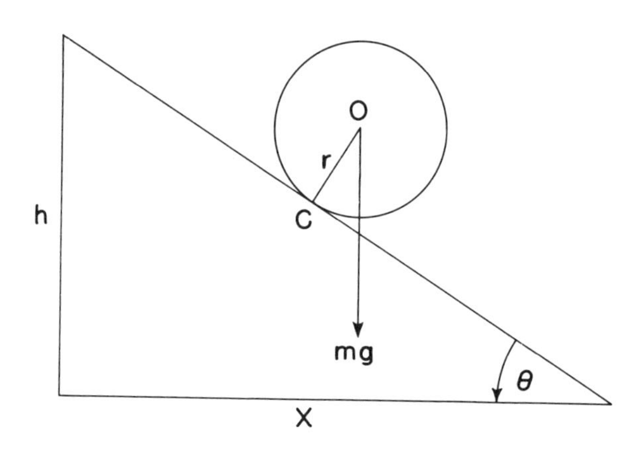

Fig. 7-17

I. Torque. The body in Fig. 7-17 is acted on by the torque due to gravity (τ = mgrsin θ). τ is also equal to the rotational inertia of the body times the angular acceleration ($\tau = I\alpha$) at any instant. Since the rotational inertia of the body with respect to the instantaneous axis of rotation can be found by the parallel axis theorem, the linear or tangential acceleration of the center of mass can be found. After the linear acceleration is calculated, the final velocity of the body at the base of the incline is readily obtained.

II. Energy. The potential energy of the body at the top of the plane is converted into rotational and translational kinetic energy at the base of the plane. Therefore, the final velocity and the linear acceleration can be found.

III. Since both translational and rotational motions occur, we may independently consider the translation of the center of mass and a rotation about the center of mass. Newton's Second Law is applied for translation of the center of mass. Rotation about the center of mass requires the use of $\Sigma\tau = I\alpha = Fr$ where F is the force of friction along the plane. Because the linear (tangential) acceleration is ($a = \alpha r$), the equations for translation and rotation can be solved simultaneously for the linear acceleration of the center of mass.

We assumed rolling without slipping, and that all bodies are in contact with the plane along a line which is the instantaneous axis of rotation. Therefore, friction will convert translational energy into rotational energy.

PROCEDURE AND ANALYSIS (WARNING - DO NOT PUT FINGERS ACROSS WIRE TERMINALS.)

Inspect the apparatus. Plug time clock into A.C. outlet. Note that a trip balance, meter sticks, and a vernier caliper are provided. Each instrument may or may not be needed.

Determine the linear acceleration of the center of mass, and the linear velocity of the center of mass of each body at the base of an incline. Use the timing device, measurement of distance, etc., and compare these results with those calculated by one of the three different methods discussed in the INTRODUCTION to this experiment.

After one of the suggested methods has been chosen, show carefully in the Computation Outline the derivation of an equation for the linear acceleration of the center of mass and the linear velocity of the center of mass for the solid cylinder, hollow cylinder and sphere.

Do not forget to:

1. Record all data.
2. Outline the procedure that was followed.
3. Show derivations with sample substitution of data for all equations in the Computation Outline.
4. Show comparison of results in the Result Section.
5. Discuss variation in answers, possible errors, etc., in the Conclusion.

SUMMARY

In this experiment you are presented with a problem and the instruments necessary for its investigation. The method of attack is left entirely up to you. Before attempting to perform this experiment, read the references and review the laws of motion for rotation, the conservation of angular momentum, and the parallel axis theorem.

Note: Do not use the entire period making measurements. Use the first thirty minutes in finding the linear acceleration and linear velocity of the center of mass. Then choose one of the suggested methods, derive the proper equations, and again make any measurements. If you cannot get started after trying for half an hour, ask your instructor for help. Do you have time in this experiment to make multiple measurements of every quantity?

QUESTIONS

1. Does the mass or the radius of any of the bodies enter into the answers for linear acceleration and linear velocity? Explain.

2. What is the rotational inertia of the solid cylinder, hollow cylinder and sphere with respect to their instantaneous axis of rotation? (See Fig. 7-17 - C.)

3. What are four possible sources of systematic error?

4. What are four possible sources of accidental error?

5. Given: mass of a cylinder = 100.0 ± 0.1 gm, radius = 1.50 ± .01 cm. Assume it was placed on an incline of angle 5.0 ± .1°. What is the maximum limit of error L.E. and the relative (percent) error in the torque due to gravity? Acceleration due to gravity = 980.3 ± .1 cm/sec^2.

6. Find an expression for the probable limit of error and percent probable limit of error in the torque due to gravity. Use the values and limits of error given in Question 5.

Experiment 7-13 ROTATIONAL KINEMATICS AND DYNAMICS

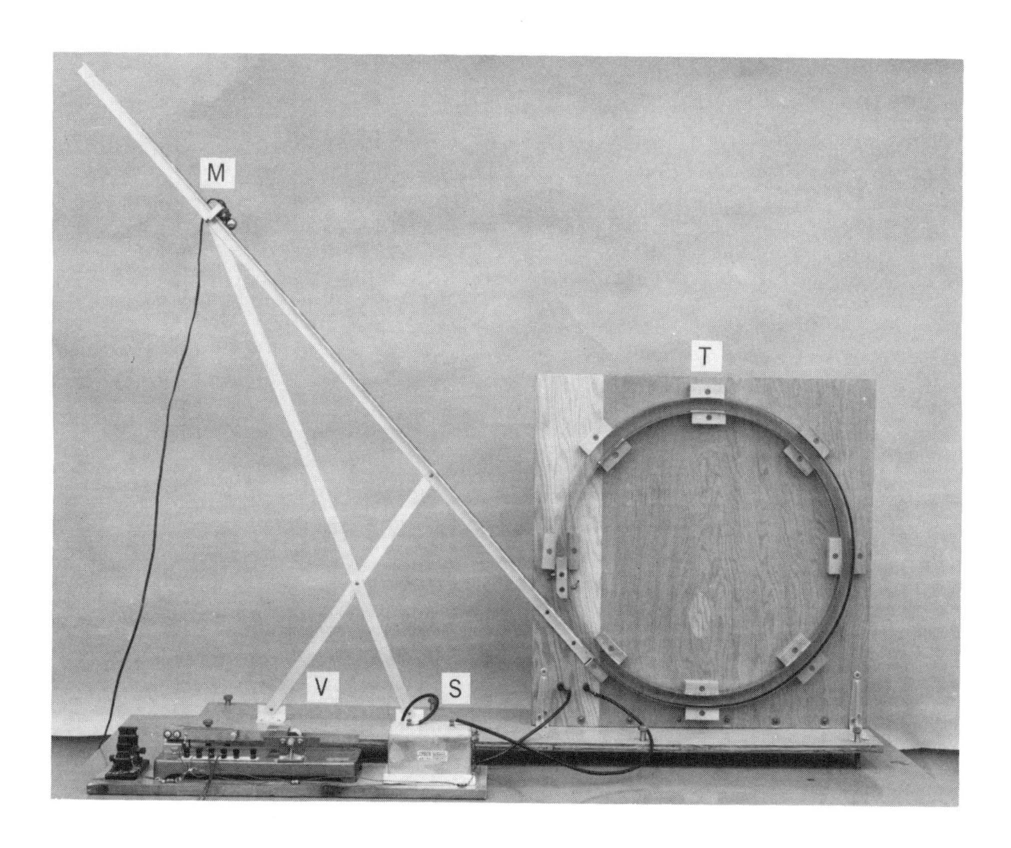

Fig. 7-18. Apparatus for the Analysis of Rotational Motion.* M = electromagnetic release. T = circular spark track. V = vibrator. S = spark coil.

*Developed by Professor Harry F. Meiners under a grant from the General Electric Co. For complete construction details, see Robert G. Marcley, Apparatus Drawings Project (Plenum Press, New York, 1962, p. 205); Robert G. Marcley, Am. J. Phys. 30, 336(1962).

INTRODUCTION

In this experiment the motion of a steel ball traveling around a vertical circle will be analyzed to develop an understanding of the relationships between translational and rotational motion. You will also be introduced to a physical application of the principle of conservation of energy.

Since we are interested in investigating the behavior of the center of mass of the ball, the data obtained from the spark trace of the surface of the sphere around the inner circle must be transformed to record the displacement of the center of mass. How could this be done?

The high voltage terminals of a spark coil are connected across the two circular metal tracks A and B. The sparks, timed by a vibrator, pass between the ball as it moves on the inner surface of the outer track A and the inner track B.

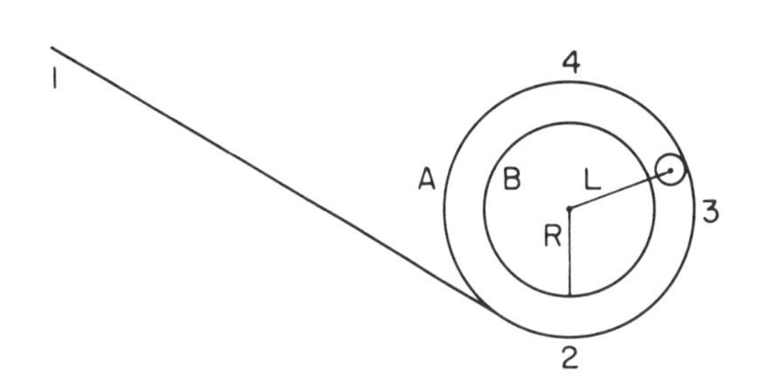

Fig. 7-19. Vertical Circle Diagram.

PROCEDURE

1. A piece of waxed paper, red surface towards the center of the circle, is placed on the inner track B. After starting the vibrator, the sphere is released by opening the electromagnet switch so that the ball's position can be recorded on the waxed paper by the high voltage applied across the tracks.

2. Check the elapsed time ΔT between sparks with the electric timer. The time should be 0.05 sec. Mark the tape at positions 2, 3, and 4 as shown in Fig. 7-19 before removing it from the apparatus.

   Draw a line across the waxed paper at the beginning of the spark trace, (after the trace has been taken). Measure to the nearest millimeter the distance of each spark hole from the line (origin). Record the increasing distances S and the increasing times T in your data table.

3. R is 12 inches. It is the distance from the center of the inner track to the surface of the inner track. The distance L from the center of both tracks to the center of mass of the moving ball is 12.625 inches. Calculate the increasing distances S for the center of mass of the ball. Record in data table. (Be sure to show the derivation of this correction factor in your computation outline.)

ANALYSIS

To save time, half of each team may work on Steps (a) through (e) of the analysis while the other(s) may work on the remainder. At the end of the period, every student must understand all the steps.

(a) Plot a graph of the distances traveled by the center of mass of the ball versus t, the elapsed time [S = f(t)]. Remember that the independent variable t is always plotted along the abscissa and the dependent variable S along the ordinate.

(b) Construct normals to the S = f(t) curve at all recorded t values and find the corresponding actual slopes. Do not lose too much time in drawing normals and calculating slopes, but be careful. Record the speed S' in the data table.

(c) Plot a graph of the magnitude of the instantaneous velocity (instantaneous speed) versus time [S' = f(t)]. On the graph mark points that correspond to the top, middle and bottom of the Vertical Circle. (See points 4, 3 and 2 in Fig. 7-19.)

(d) Draw normals to the instantaneous speed versus time curve [S' = f(t)] at all recorded t values and find the corresponding actual slopes. Record the magnitude of the acceleration in the data table.

(e) Plot a graph of the magnitude of the acceleration versus time [a = S'' = f(t)]. On the graph mark the points that correspond to the top, middle and bottom of the Vertical Circle.

(f) Weigh the ball, and record its mass in the data table.

(g) The potential energy of the steel ball at point 1 is converted into kinetic energy of rotation and kinetic energy of translation at point 2 if we assume rolling, without slipping or other losses, and that the ball is in contact with the surface along a line which is its instantaneous axis of rotation. Measure and record the change in position of the center of mass of the ball between 1 and 2, (height h). Determine from Step (c) the speed of the center of mass at point 2 and record it in the data table. Show carefully in your computation outline whether or not the potential energy of the ball at point 1 is completely converted at point 2 into kinetic energy of rotation plus kinetic energy of translation. Calculate the magnitude and percent of any variation.

(h) Find from the graph plotted in Step (c) the speed of the center of mass at point 3. Record the value in your data table. What is the value of the centripetal force at point 3? What is the normal force? From Step (e) find the acceleration of the center of mass at point 3. What is the resultant acceleration of the center of mass; the resultant force?

(i) Duplicate Step (h) at point 4, the top of the vertical circle.

(j) Optional - to be done if time permits. Proceed as in Step (g) and determine if the kinetic energy of rotation and translation at point 2 equals the kinetic energy of rotation and translation plus the increase in potential energy at point 4. Discuss any variation.

QUESTIONS

1. Draw a complete force diagram for the steel ball at the position shown in Fig. 7-19. Is there a net force acting on the ball? Why?

2. Is the acceleration plotted in Step (d) in the Analysis radial or tangential? Why?

3. Discuss the causes for the variation between the theoretical and experimentally determined values of the rotational and translational kinetic energy at point 2.

4. Discuss the effects, if any, of using a steel cylinder of the same mass as the sphere. Would the initial energy at point 2 be the same? Would the rotational energy at point 2 be the same?

5. Is the minimum initial height necessary for the ball to make a complete revolution greater than, equal to, or less than the diameter of the circle? Why?

6. Which parameter(s) affect the value of the centripetal force on the ball at each of the positions 2, 3, and 4? Which affect the normal force? The total force?

Experiment 7-14 ROTATIONAL INERTIA*

INTRODUCTION

The rotational inertia (often referred to as the moment of inertia) of a body depends on the axis about which the body is rotating as well as on the shape of the body and the manner in which the mass is distributed. In SI units the rotational inertia I is expressed in $kg \cdot m^2$. The rotational inertia of a collection of particles, each of mass m and a distance r from the axis of rotation, is given by

$$I = \Sigma mr^2 .$$

In this experiment, the rotational inertia of a rigid body will be measured by two different methods and the results will be compared. First, we calculate the rotational inertia of the body by breaking it into parts and finding the rotational inertia of each by knowing the mass and shape of each. The total rotational inertia is simply the sum of the rotational inertias of the individual parts.

In the second method, a torque τ will be applied to the rigid body so that the rotational inertia can be found from the equation

$$\tau = I\alpha ,$$

where α is the angular acceleration.

The apparatus consists of a 9" aluminum disk and a hub of smaller radius, both mounted on an axle with ball bearings. We also have a notched disk which allows us to measure the rotation of the body.

PROCEDURE AND ANALYSIS

Part A

To find the rotational inertia from the mass and shape of the individual parts, use the UNASSEMBLED parts found next to the balance - DO NOT DISASSEMBLE THE APPARATUS TO BE USED IN PART B. Measure the mass and all relevant dimensions and compute the resulting rotational inertia. Estimate the uncertainty for each of the parts and the overall uncertainty of the calculated value of the rotational inertia.

Show all of these measurements, calculations and uncertainties in the form of a table.

*These directions are written for use with a TRS 80 Color Computer 2 and an interface designed under the supervision of Dr. Thomas Shannon under a grant from the RCA Corporation. A detailed description may be obtained from Professor Walter Eppenstein, Physics Department, Rensselaer Polytechnic Institute, Troy, N.Y. 12180-3590.

Part B

We are using a microcomputer to measure the angular velocity of the rotating body. The micro will count the slots in the notched wheel. There are 90 slots, but there will be 180 counts per revolution since we have two photo gates. The micro will count, record and display the number of pulses each 0.1 seconds. Thus, the micro measures the angular velocity as a function of time. The angular acceleration α is the derivative of the angular velocity with respect to time, or the slope of the plot of angular velocity as a function of time.

Fig. 7-20. The Unassembled Disks (Part A) to the Left and the Assembled Rigid Body to the Right of the Microcomputer.

The micro is turned on by pressing the button on its back towards the left side. Turn on the monitor and wrap the string (with the 100 gm on the end) onto the larger radius hub. Release the mass and immediately press any key on the micro. The micro will record the angular velocity of the rotating body. At the end of the run the velocity vs. time will be plotted on the screen. Using the arrow keys to position the cursor on the screen, select two representative points and calculate the slope which is the angular acceleration α. Show this calculation in your notebook.

Now you can use the micro to calculate the slope for you. Position the cursor near one end of a straight line portion of the curve. Press the space bar. A second cursor will appear which can be moved with the arrow keys. Position it near the other end of the line. Pressing the space bar will cause the micro to perform a "least squares" fit to the data between the cursors and report the slope. The process can be repeated by pressing the space bar. The micro is reset for a new run by pressing the "Break" key.

We can now calculate the value of the rotational inertia I by making use of the following relations:

$$\tau = I\alpha, \qquad \tau = rT, \qquad mg - T = ma, \qquad \text{and} \qquad a = r\alpha,$$

where r is the radius, T the tension and a the linear acceleration.

Part C

Repeat Parts A and B after adding the two disks. Calculate the new rotational inertia by considering the added masses to be point masses and using the parallel axis theorem. Note the percent difference between the two calculations.

Part D

Replace the disks with the two rectangles. First, align them with their long axis along the radii of the disk, then with their long axis perpendicular to the radii. Would you expect different results in each case? Run the micro program for each to verify your expectation.

Part E

The bearings are not frictionless. Using the basic rigid body, spin it and run the micro program, obtain a value for α and using $\tau = I\alpha$, find the frictional torque.

QUESTIONS

1. Does the frictional torque account for the sign of the difference of I from Part A and I from Part B?

2. Can you legitimately use your derived value of I to find the frictional torque? Explain.

3. The string is wound such that the mass is to the left of the axle as you are looking at it. The mass is released, the body spins, the string falls free and the body slows to a stop. Describe the direction of the following vectors during the motion of the body:
   (a) displacement
   (b) angular velocity
   (c) angular acceleration
   (d) torque due to the hanging mass
   (e) frictional torque

4. The actual string has a thickness. Does this significantly affect your results in Part B? Explain.

Experiment 7-15 INVESTIGATION OF VARIABLE ACCELERATION

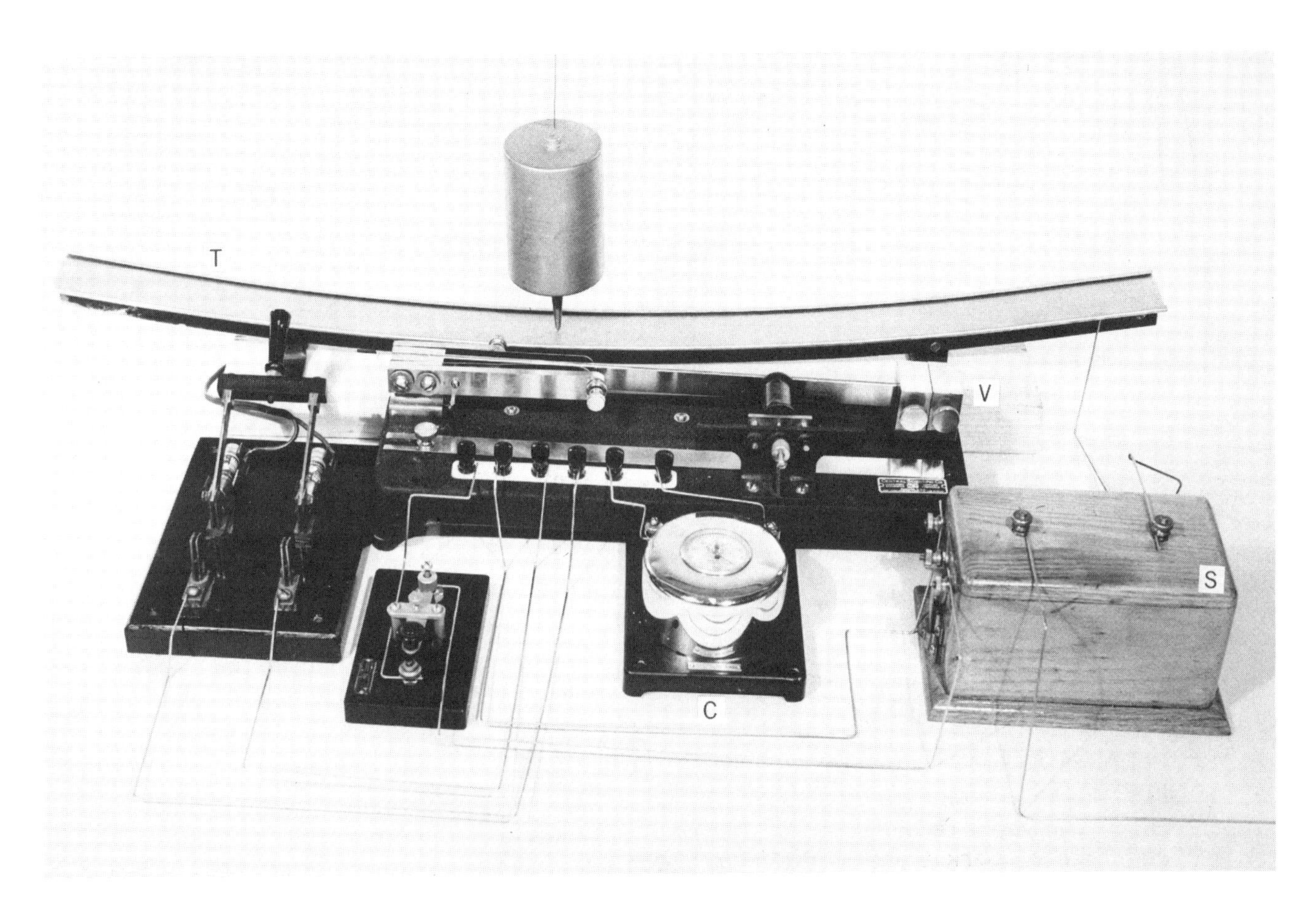

Fig. 7-21. Variable Acceleration Apparatus.
T = spark track. S = spark coil. V = vibrator. C = electric timer.

INTRODUCTION

When a simple pendulum consisting of a massive bob m at the end of a fine wire oscillates about a knife-edge support (Fig. 7-21), it moves with variable velocity and variable acceleration. You will investigate the behavior of the center of gravity of the bob m.

The high voltage terminals S of a spark coil (Fig. 7-21) are connected to K (Fig. 7-22) and to a curved metal track T so that sparks, timed by a vibrator V pass between the point of the bob and the track. A piece of waxed paper placed on the track T records the sparks and thereby the position of the bob at various times.

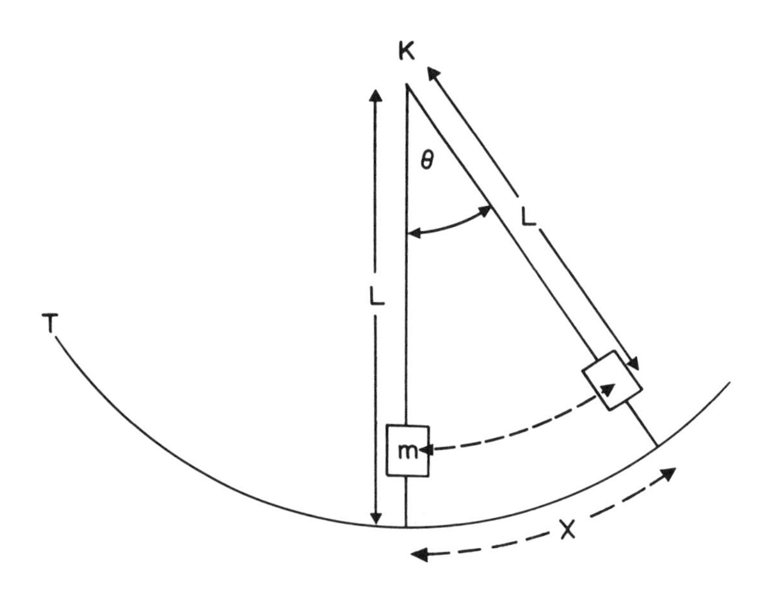

Fig. 7-22

PROCEDURE (Read Procedure and Analysis several times before setting up the data table.)

1. Cut the waxed paper to fit track. Draw a line across the paper at the center (equilibrium or rest position of bob). Place paper on the track and label the right side + and the left side -.

   PLEASE DO NOT TOUCH THE ADJUSTMENT SCREWS ON THE VIBRATOR. The instructor will take the necessary spark trace for you in order to save time. After you have recorded all of the data, duplicate the instructor's procedure and take a spark trace in order to become familiar with the equipment's operation. Discard the trace that you take.

2. The time between sparks has been adjusted to 0.1 seconds ± 2%. Measure the displacement X to the nearest millimeter of each spark hole from the center. In order to avoid fractional multiples of time increments, assign t = 0 to the <u>first</u> spark hole made on the trace. Therefore, X will not equal 0 at t = 0. The next spark hole is at t = 0.1 sec. Time increases throughout the measurement, but the displacement of the bob from the equilibrium position varies. Make a table and record the time values and the corresponding X values.

   Note that these measurements are for the <u>tip</u> of the bob and not for its center of gravity. The motion of the tip of the bob is similar to that of its center of gravity.

ANALYSIS

(a) Plot a graph of displacement X as a function of time t. $X = f(t)$. Remember that the independent variable t is always plotted along the abscissa and the dependent variable X along the ordinate.

(b) Draw tangent lines to the $X = f(t)$ curve at all recorded t values. Find the actual slopes (see Art. 3.2, A8) of the tangent lines. Do not lose too much time in drawing tangents and calculating slopes but be careful. Record values in the data table.

(c) Plot a graph of velocity, $X' = f(t)$ or $V = f(t)$, versus time.

(d) Draw tangent lines to the $X' = f(t)$ curve at all recorded t values. Determine slopes and enter values in data table.

(e) Plot a graph of acceleration, $X'' = f(t)$ or $a = f(t)$, versus time.

(f) Label the amplitude and period on $X = f(t)$. Give the values of frequency, period and amplitude in the computation outline.

QUESTIONS

1. At what point of the swing does the pendulum have its greatest velocity? Acceleration?

2. How does the mass of the pendulum affect the period?

3. How would the curves be affected if the amplitude of the pendulum were increased?

4. How could you secure a graph of displacement versus time for the center of gravity of the pendulum, $X = f(t)$? See Fig. 7-22.

5. Explain how the maximum velocity of the pendulum found from the graph could be checked if it were possible to calculate the height of rise of the center of gravity above the equilibrium position. Be sure to show how h could be found from the geometry of Fig. 7-22.

6. If $x = 30 \sin 2t$, how long a time is required for the bob to move from its equilibrium position to a point 10.0 cm away? What is the amplitude, frequency and period?

7. If direct measurements were made of the amplitude A, period T and the instantaneous time t as follows:

   $A = 30.0 \pm 0.1$ cm $T = 3.00 \pm 0.05$ seconds $t = 0.30 \pm 0.03$ seconds,

   and the position equation of the bob is $X = A \sin \omega t$, what is the maximum numerical error in the indirect measurement X?

8. Find for Problem 7, the probable (numerical and percent) limit of error in the indirect measurement X.

Experiment 7-16 ELONGATION OF AN ELASTOMER

INTRODUCTION

The elastomer (Fig. 7-23 and Fig. 7-24) is made of 1/8" diameter Koroseal* cord and possesses properties which are somewhat different from most metals under stress. Metals are subject usually to two types of deformation, elastic and plastic. In an elastic deformation, any change in the stress causes an instantaneous deformation that obeys Hooke's law. The body under tension will stretch until its elastic limit is reached; the elongation will be proportional to the stretching force. If the force (load) is removed, the body will return to its original length. Therefore, an elastic deformation is a reversible process.

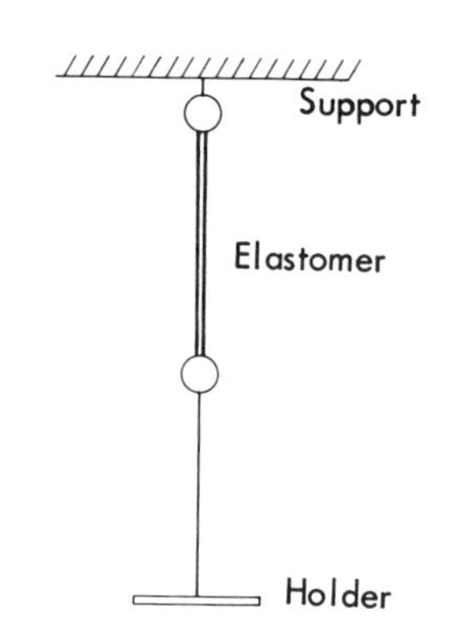

Fig. 7-23. Apparatus.

In a plastic deformation the body is stressed beyond its elastic limit and will not return to its original length. The body retains a permanent set (strain), and the process is irreversible.

If the stress is increased beyond the elastic limit, there is usually a large increase in strain for slowly increasing stress. Eventually, if the stress is continued, the specimen will break (fracture).

The term "creep" is applied to the process of plastic deformation. In a plot of strain at constant load versus time (Fig. 7-25), the first stage is called primary creep - a region where the material adjusts itself to the load. In the second stage, strain is almost linear with time and elongation is at a constant rate. If the stress is high enough, a third nonlinear region associated with the failure of the material is obtained.

In addition to understanding Hooke's law, stress, strain, elastic limit, and the several types of deformation, review WORK (Energy) and the process of INTEGRATION.

This experiment has two important objectives:

(a) It allows you to discover the actual significance of integration by analyzing a physical system.

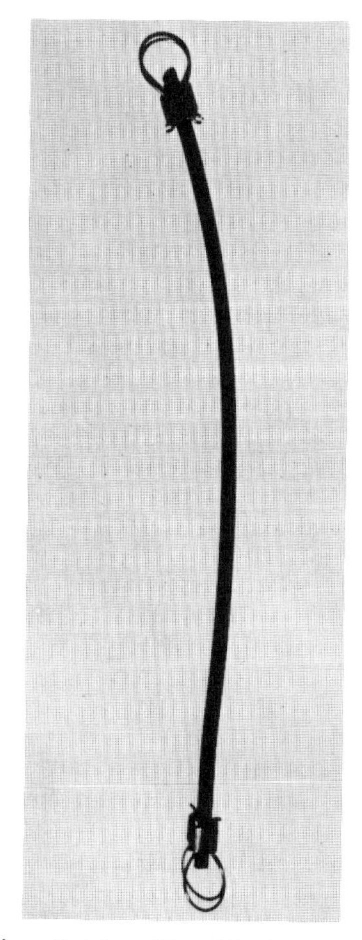

Fig. 7-24. The Elastomer.

*Koroseal can be obtained from the B.F. Goodrich Company, Marietta, Ohio. The chemical composition is 53% Polyvinyl Chloride, 38% Tricresyl Phosphate and 9% Stabilizers, Lubricants, etc. (By weight).

(b) It emphasizes that WORK can be done by a variable force that is a function of position.

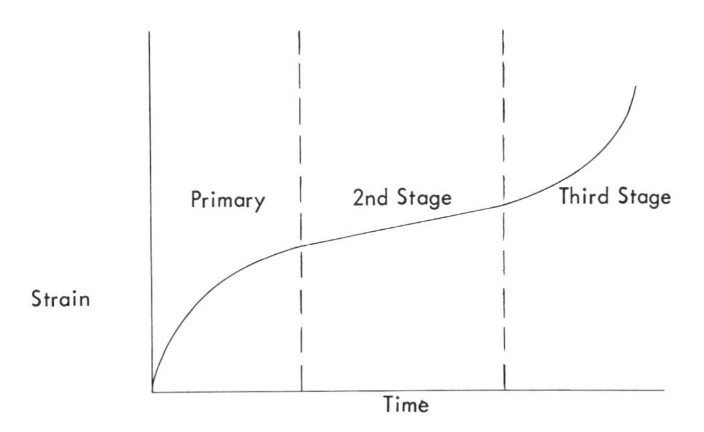

Fig 7-25. Creep. Plot of Strain at Constant Load versus Time. The elastomer is of interest because creep takes place very slowly and hence is observable. This is not true for most metals.

PROCEDURE

1. Observe and record the position of the elastomer with no load on it.

2. Use equal increments of 200 grams, and determine the increase in length Y for loads ranging from 0 to 1000 grams. Note that the holder is 50 grams. After each loading wait five minutes before adding another load. Do not at any time attempt to change the combination of masses after they have been placed on the holder. RECORD ALL THE VALUES.

3. Wait five minutes after the 1000-gram loading has been reached and then remove the masses in the same order as they were originally added. Again wait five minutes before removing another load. RECORD the decrease in elongation or contraction Y.

ANALYSIS (Study operation of the polar planimeter in Section 4.2.)

(a) Plot M vs. Y. (M = f(Y)).

(b) Choose a square of graph paper four blocks by four blocks. (10 mm. per block) Label using the same scale for ordinate and abscissa as in your plot of M = f(Y). The area of this square corresponds to the known area of the calibration disk or the area obtained by use of a calibrated radius arm (Fig. 4.4, D; p. 69). Use the known area to find the constant of the polar planimeter (Eq. 4.2, p. 69). Should you trace more than once around the known area? Why? RECORD DATA.

(c) Use the polar planimeter to find the area between the curves of your M = f(Y) plot. Should you trace more than once around the unknown area? Why?

(d) After the area between the two curves has been obtained in terms of grams and centimeters, convert to ergs. Why?

QUESTIONS

1. Was hysteresis observed? Could you say that the effect was elastic? Explain.

2. Did the elastomer undergo a reversible or an irreversible deformation? Explain. Was the deformation elastic or plastic? Explain.

3. Did the sample exhibit primary creep? Did it show the second or third stage of creep? Explain.

4. Did the sample have an elastic limit? Was Hooke's law obeyed? Explain.

5. Why doesn't a plot of mass (force) versus elongation give you a straight line?

6. What type of energy is retained by the elastomer? Since the elastomer returns to its original length in 24 hours, how is the energy released?

7. Define the scalar product of two vectors? Is the dot product the same as the scalar product? Explain.

8. If one vector corresponds to a force F and the other vector represents the displacement dy of the point of application of the force, what is their scalar product called?

9. During an infinitesimal displacement dy does the force F approach a nearly constant value? Why?

10. Express mathematically the total work done in displacing the elastomer from y_1 to y_2.

11. When the constant of the polar planimeter was obtained, a square with 40 millimeter sides was used. What is the "least count" of this square? What is the ILE for the known area?

12. What is the ILE for the polar planimeter? Determine the RE and add this to the ILE for the number of tracings made around the known area of the calibration square. Be careful - the ILE for the polar planimeter is not the same as that found in Question 11. Remember that ILE + RE = LE (LE of N_s).

13. Use the ILE found in Question 11 as the LE for the known area A_s and the LE for the number of tracings around the known area N_s to obtain the maximum LE of the constant K. What is the relative error RE?

14. From the tracings around the unknown area (the area between the curves of your M = f(Y) plot), find the RE for N_x. What is the LE for N_x?

15. What is the maximum limit of error LE for the indirect measurement A_x?

16. What is the probable limit of error for the indirect measurement A_x? Compare the probable limit of error with the maximum limit of error for A_x and discuss the variation.

17. If the mass (including the mass of holder) placed on the holder = 200.00 ± .05 gm, g = 980.3 ± .1 cm/sec^2, and the displacement = 1.00 ± .02 cm, what is the maximum limit of error LE and the relative (percent) error in the indirect measurement of the work (energy)?

18. In Question 17 what is the probable limit of error and the relative (percent) error in the indirect measurement of the work (energy)?

Suggested Data Table:

A_s =		
M (grams)	Elongation (Y) in cm.	Contraction (Y) in cm.
0		
200		
400		
600		
800		
1000		

Experiment 7-17 INVESTIGATION OF THE ELONGATION OF AN ELASTOMER WITH A MICROCOMPUTER*

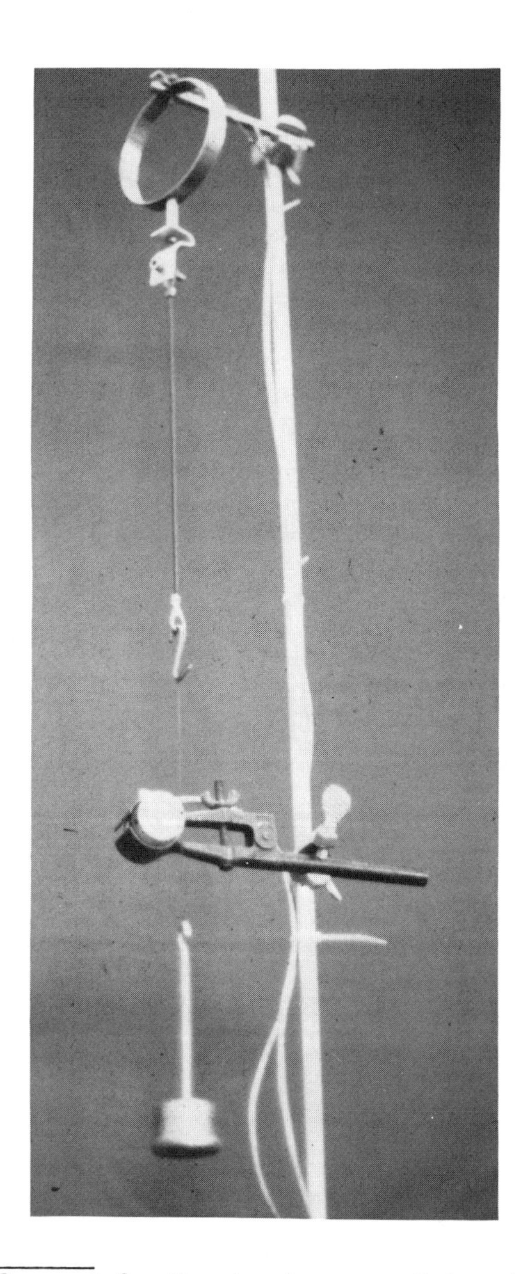

Fig. 7-26.
Apparatus Used to Investigate Elastic Hysteresis. Force applied to the elastomer is measured by two strain gauges mounted on opposite sides of a copper ring. The displacement of the elastomer is measured with a potentiometer.

*Support for the development of the microprocessor experiments was provided in part by grants from Erik and Margaret Jonsson of the Texas Instruments Corporation. The following students prepared the routines for the microprocessor: Chris D. Delise, Gary Polz, Sam Fortin and Philip H. Grove, among others. The development project was supervised by Harry F. Meiners. Circuit diagrams and routines can be obtained from Prof. Meiners. These details will enable any school to duplicate the experiments and the microprocessor.

INTRODUCTION

We have linked a microcomputer to ordinary physics laboratory equipment to provide a comprehensive, interactive instrument for teaching purposes; the system described is used to investigate the "elastic hysteresis" phenomena of some materials. The student is able to operate the experiment directly and therefore becomes familiar with several different disciplines - mechanics, electronics, and microcomputers.

THEORETICAL CONSIDERATIONS

The application of a force to an elastic material, such as a string, can lead to many different results. For example, the system could oscillate, undergo elastic deformation, demonstrate inelastic (plastic) deformation, or even result in a complete breakdown of the material. The effects encountered depend not only on the material, but also on the method of application, the magnitude of the force, and the speed with which changes are carried out. In the region where deformations are below the "elastic limit" of the substance being analyzed, we may describe the results by Hooke's Law as:

$$F = -Ky \qquad 7\text{-}13$$

where x represents the displacement. This is called the proportional region and corresponds to the straight-line curve on a Force vs. Displacement plot. The potential energy U from which this conservative force can be derived can be written as:

$$U = \frac{1}{2} Ky^2 \qquad 7\text{-}14$$

Therefore, for small displacements, where Hooke's Law is valid, simple harmonic motion may result. However, if we exceed the elastic limit of the sample, then the force can no longer be simply expressed. It will depend on other conditions such as the speed of deformation and the previous history of the material.

For the conservative case, the work required to stretch an elastic string by an amount (b-a) Is given by:

$$U = -\int_a^b -Ky\ dy \qquad 7\text{-}15$$

If the displacement occurs infinitesimally, this energy is conserved. However, if the elastic limit of the string is exceeded while the breaking point is not exceeded, there will be a different assignment of energy; the force is not conservative and energy will be used in breaking bonds, heat, etc. This means that the linear force-displacement relationship no longer applies. This feature is often described as elastic hysteresis and is quite similar to the electromagnetic hysteresis curves.

THE EXPERIMENTAL APPARATUS

The apparatus in Fig. 7-26 must allow for the measurement of both the force applied and the displacement observed of the elastic cord made of Koroseal. Force is being measured by the use of two strain gauges which are mounted on opposite surfaces of one side of a copper ring; see Fig. 7-27. The gauges are connected in a differential configuration which reduces the effects of external noise and temperature variations and permits the measurement of the force applied through the ring. The output from the gauges is then amplified approximately 5000 times before being measured by the computer. The displacement of the elastomer is measured with a potentiometer whose shaft is connected to a pulley. As the displacement varies, the pulley is rotated, and the voltage taken from the wiper of the potentiometer is sent to the computer.

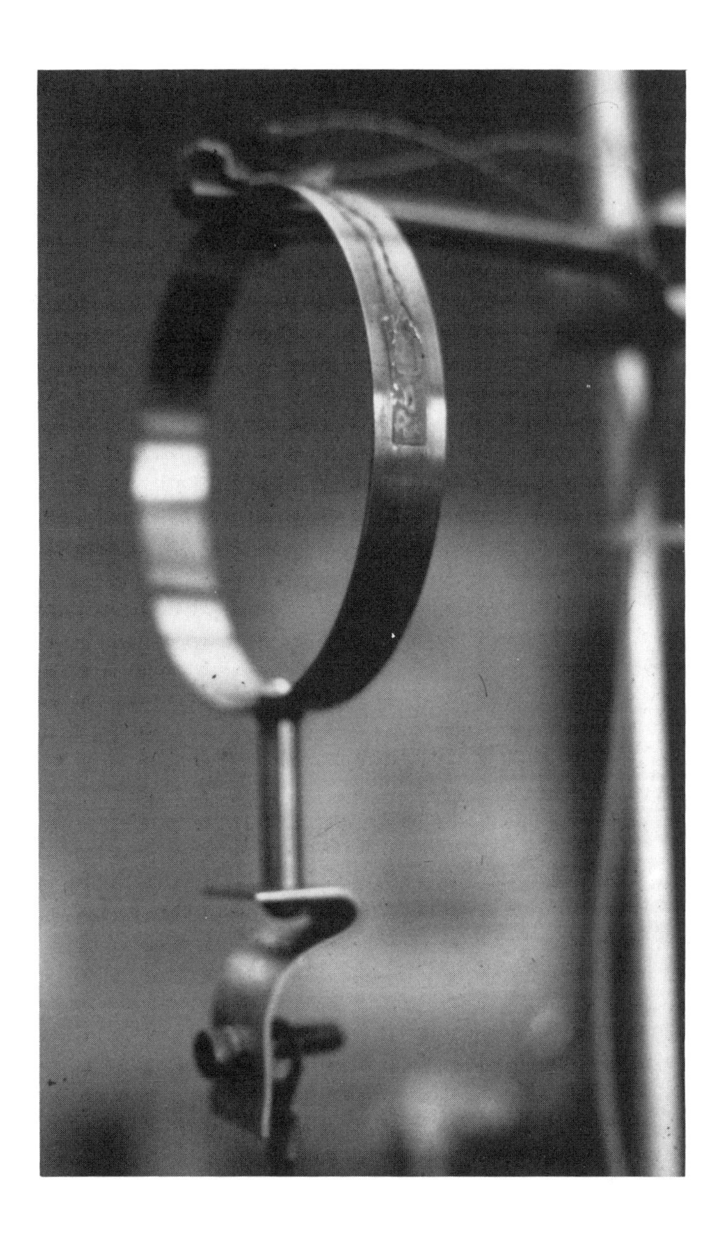

Fig. 7-27.
Two Strain Gauges Are Mounted on Opposite Surfaces of One Side of a Copper Ring.

PROCEDURE

1. After the apparatus is completely connected and the computer is given power, press the reset key to start the monitor. Choose the option to execute the experiment. The options displayed next will be those for the experiment itself. At initial entry, choose the option to initialize the experiment. When asked, specify

(in seconds) the time interval between measurements. This will allow the computer to prompt for action at a programmable rate. Intervals of five minutes are typical for the actual laboratory.

2. Press the "Take data" key. The screen will show the instantaneous values of the force and displacement. When ready, press any key to start taking data. Add a 100-g mass to the hanging mass holder, and watch the screen. As values change, the computer will draw a line from the previous force-displacement position to the new one. After the preset interval, a tone will sound and it is time to press "F" to freeze the point in memory. Afterwards, add another weight as described above. Do not remove weights after they have been added; such action would invalidate the experimental data.

3. After a total of 1.0 Kg (1 Kg x 9.8 m/sec 2) has been added to the holder, wait for the tone to sound and then press the "B" key to start the reverse portion of the experiment. Remove one weight, and observe the effect on the screen. When another tone is heard, again press the "Freeze" key and remove another weight. Repeat this procedure until no weights remain on the holder. After a last tone is heard, press "Freeze" again. The data is now complete. Press "E" to return to the menu. In Fig. 7-28 is shown a typical plot of Force vs. Displacement - F vs. x for Koroseal.

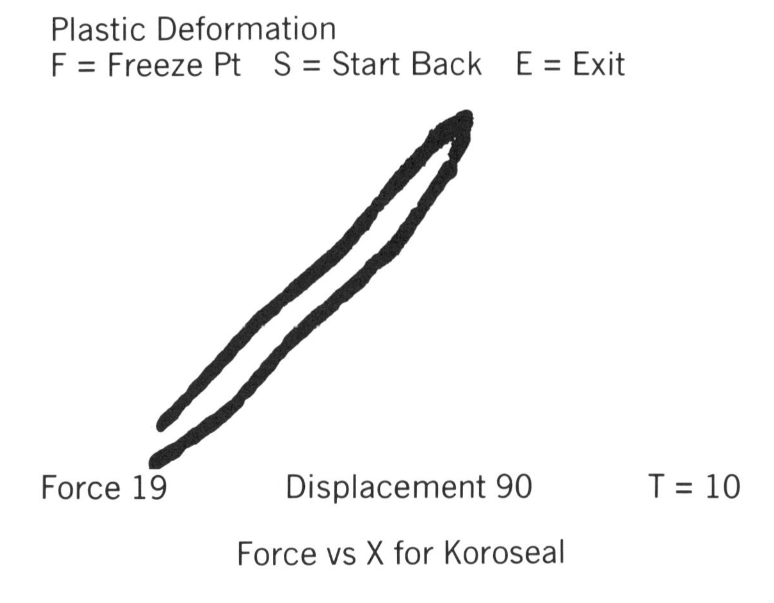

Fig. 7-28. Actual Image Fine Line on CRO.

ANALYSIS (Review ANALYSIS for Experiment 7-16.)

(a) Place a sheet of mm graph paper against the CRO screen and trace as carefully as possible your hysteresis curve. Add appropriate scales.

(b) Use a polar planimeter to find the area between the curves of your M = f(y) plot. Should you trace more than once around the unknown area? Why? Note - see

Paragraph (b) in ANALYSIS of Experiment 7-16. You must calibrate the Polar Planimeter.

(c) After the area between the two curves has been obtained in terms of grams and centimeters, convert to ergs. Why? Record data.

OTHER EQUIPMENT REQUIRED

A moderate quality oscilloscope with 0-5 volt Z-axis control and triggered sweep. We use a Tektronix T922.

SOURCES OF ERROR

1. Since the A/D interface has a resolution of only 8 bits, a limited 2.5 significant digits may be used in analysis.
2. The CRO and/or its driving interface may not be properly calibrated and an error may easily occur in measurement using the polar planimeter on a replica of the CRO trace.
3. The vectors drawn between measured points may not yield as accurate areas as hand-interpolated curves. Still, they permit more accuracy than would be possible with just the data points themselves.

CONCLUSIONS

Benefits to User

1. Real-time interaction with the experiment
2. Ease of recording and analyzing large amount of data
3. Hands-on experience with a small computer, as used in research

Experiment 7-18 DAMPED DRIVEN LINEAR OSCILLATOR*

INTRODUCTION

The behavior of a driven one-dimensional damped spring system is investigated. Mechanical resonance, the relationship between frequency and amplitude, the relationship between phase and amplitude, and the effect of damping are studied. From the data obtained, the Q of the resonant system can be determined.

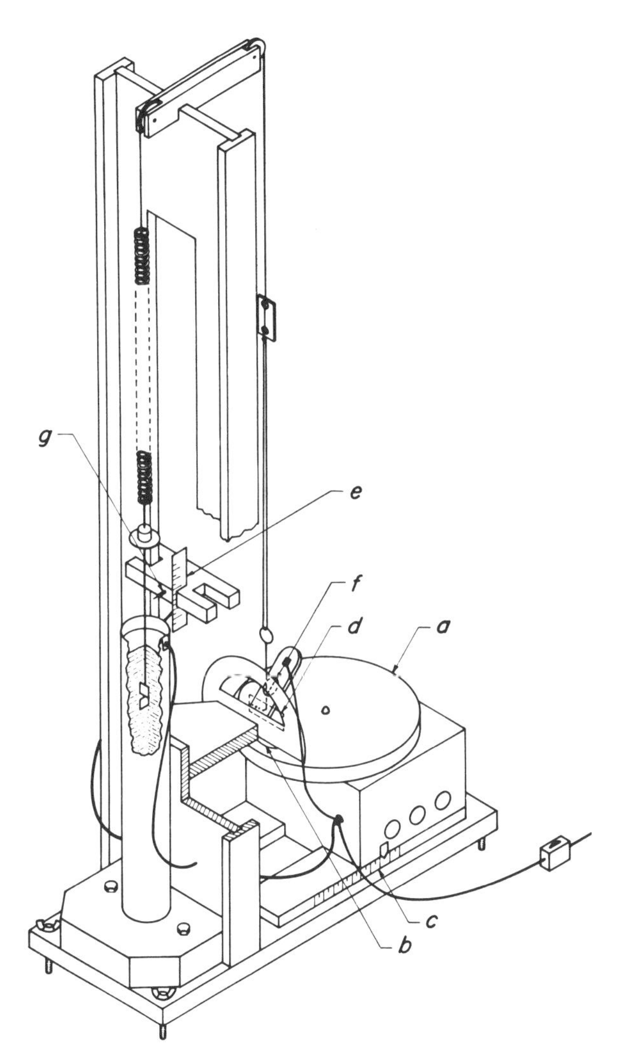

Fig. 7-29. Detail of the Damped Driven Linear Oscillator.

The apparatus consists of a mass and spring system driven at various frequencies by means of a modified phono turntable as shown in Fig. 7-29. The 4-speed turntable a is moved inward and outward beneath a drive-wheel b thereby providing a range of driving frequencies varying between 10 rpm to nearly 150 rpm. A scale c on the side of the turntable box is used for reference when driving frequencies are selected. Attached to the face of the drive wheel is a rectangular plate d containing four holes at various distances from the wheel's center of rotation. The driven mass, which can be varied by adding extra washers, is suspended from a spring. The top end of the spring is given periodic impulses by a thread that runs up over the top of the upright and down to the drive wheel at the turntable. By inserting a hook into the various holes on the drive wheel, the amplitude of the impulse can be varied. The maximum displacement (amplitude) of the driven mass is measured by sighting against the ruler e.

Since the mass cannot keep up with the driving force, it will lag behind. If the force has a maximum displacement (amplitude) at t_1, the mass will reach its maximum displacement at some later t_2. The mass, therefore, lags the driving force by

*See Experiment 5-6, Simple Harmonic Oscillations, Case c, Damped, Driven Oscillations; Case d, Undamped, Driven Oscillations.

a phase angle

$$\theta = \omega_d(t_2 - t_1) ,$$

where ω_d is the angular frequency of the driver. To obtain the phase angle θ, the reference pointer f is moved until a spark jumps from the point in the red wire g to the wide washer that fits over the mass. The position of the reference pointer is changed until the spark is as nearly horizontal as possible and the angle recorded. A new driver angular frequency ω_d is then chosen by moving the turntable to a different scale setting.

Energy is inserted into the spring mass system by a periodic force. As the angular frequency of the driver ω_d approaches the undamped natural angular frequency ω_o of the spring mass system, the amplitude of oscillation (observed with the aid of the ruler) increases. When the angular frequency of the driver and the natural angular frequency of the spring-mass system are nearly equal ($\omega_d \cong \omega_o$), the amplitude of oscillation of the driven mass becomes very large. This phenomenon is called resonance. At the resonant angular frequency ω_r, the driving force is doing a maximum amount of work on the system during each successive oscillation. Thus, the amplitude increases in each cycle and would continue to do so if there were no damping. When the amplitude reaches a steady value, the work done in each cycle just equals the energy loss per cycle because of damping. It is also important to understand that even in an undamped system (oscillations in air) there is always some inherent damping due to the properties of the system and air resistance. Therefore, at resonance the maximum amplitude does not become infinite.

At angular frequencies of the driver other than at resonance (where $\omega_d \neq \omega_o$), the rate of doing work by the driving force on the system is less than that at resonance. The oscillator accepts energy from the driving system during a part of the cycle and then transfers it back by doing work on the driver during the remainder of the cycle. Hence, when the driver and the oscillator are not "in phase" the net energy transfer is less than that of the "in phase" resonant case. Therefore, the amplitude of the driven mass is less at all angular frequencies below and above the resonant frequency.

In mechanical and electrical systems, the energy initially stored in the system divided by the energy lost per cycle is often called the "Q" of the system. Sometimes, Q is referred to as the quality or quality factor. It also is used as a measure of the "sharpness" of the response of a physical system to the resonant driving force.

Systems with small energy losses (in our case caused by viscous damping) are characterized by high Q values. Three methods of measuring Q are as follows:

1. For a damped, undriven, oscillating system, Q is equal to π times the number of oscillations required for the mass motion amplitude (one-half of the total excursion of the washer at the lower end of the spring) to decrease to $1/\varepsilon$ of its initial value. The mass is given an initial displacement and released from rest.

2. The amplitude quotient method can be used to obtain Q when the damped, oscillating system is driven at the resonant frequency ω_R. In this case, Q is equal to the ratio of one-half of the total excursion of the washer at the lower end of the spring (amplitude of mass motion) to one-half of the total excursion of the upper-end of the spring (amplitude of the driving motion) at resonance. Extra care must be used in measuring the amplitude of the driving motion at the upper end of the spring because the total excursion is quite small and difficult to estimate by sighting against a ruler.

3. We can also determine Q from a resonance curve by selecting two points where the amplitude is $1/\sqrt{2}$ times the maximum amplitude at resonance. The frequency difference $(\omega_2 - \omega_1)$ at these two half power points is commonly called the bandwidth of the system. Low loss (large Q) systems are characterized by small bandwidths and sharply-peaked resonance curves.

PROCEDURE

1. If necessary, add liquid to the glass tube until the level is about one cm below the end of the spark terminal. Adjust the height of the block supporting the ruler and spark wire until the point in the wire is directly opposite the wide washer on the plunger. At this position, the shaft of the plunger should be about half immersed in the liquid. Level the glass tube so that the plunger can oscillate freely without rubbing against the glass. While the system is being driven, do not allow the mass and washer to enter the fluid.

2. To obtain the phase angle θ at a scale setting of the turntable, move the reference pointer f over the protractor until the spark appears to be horizontal and then record the angle. Since it is difficult to determine exactly when the spark is horizontal, make four observations of the spark and record the average of your readings of the phase angle. The effect of the spark curving upward and downward as the washer moves past the tip of the wire is then minimized.

3. Do not touch the high voltage connections of the spark coil or the 110-v a.c. connections to the transformer.

4. The experiments use water as a damping fluid. If possible, repeat each using (a) 50% glycerin, 50% water and (b) 75% glycerin, 25% water. Plot the results on the graph for water. For example, the resonance curves of mass amplitude versus driver angular frequency ω_d for the three damping fluids can be plotted on the same graph.

5. Before starting to take data, discuss the experiments you want to perform with your instructor.

Experiment 1 (Undriven System)

1. With the system not being driven, release the mass from rest (use the washer for reference) from an initial displacement of 5 cm. Plot a graph of amplitude (cm) versus number of oscillations for about 10 oscillations. Record in a table the amplitude (cm), the number of oscillations, and the period (seconds) of each oscillation.

2. Determine Q.

Experiment 2 (Driven System)

1. Insert the drive hook into the second hole of the rectangular plate d. Set the turntable control at 45 rpm and move the turntable box until the pointer on its side is at 2.0 cm on the plastic scale. Record the readings from the turntable scale for values between 2.0 and 8.0 cm. When the amplitude of mass motion begins to increase or decrease quickly, take readings every 0.2 cm. At each scale setting allow the system to reach equilibrium before proceeding.

2. At each turntable scale setting, record the phase angle θ (degrees) and the corresponding amplitude (cm) of oscillation of the driven mass. When resonance is obtained, record as carefully as possible the maximum amplitude of motion of the driver and the maximum amplitude of excursion of the upper end of the driving spring.

3. Plot a resonance curve of the amplitude of oscillation (cm) versus the scale reading in cm. The scale reading (S.R.) in cm is proportional to the angular frequency of the driver ω_d. Why? Plot also on the same graph a curve of phase angle ($0° - 180°$) versus scale reading (S.R.).

Experiment 7-19 HARMONIC MOTION ANALYZER*

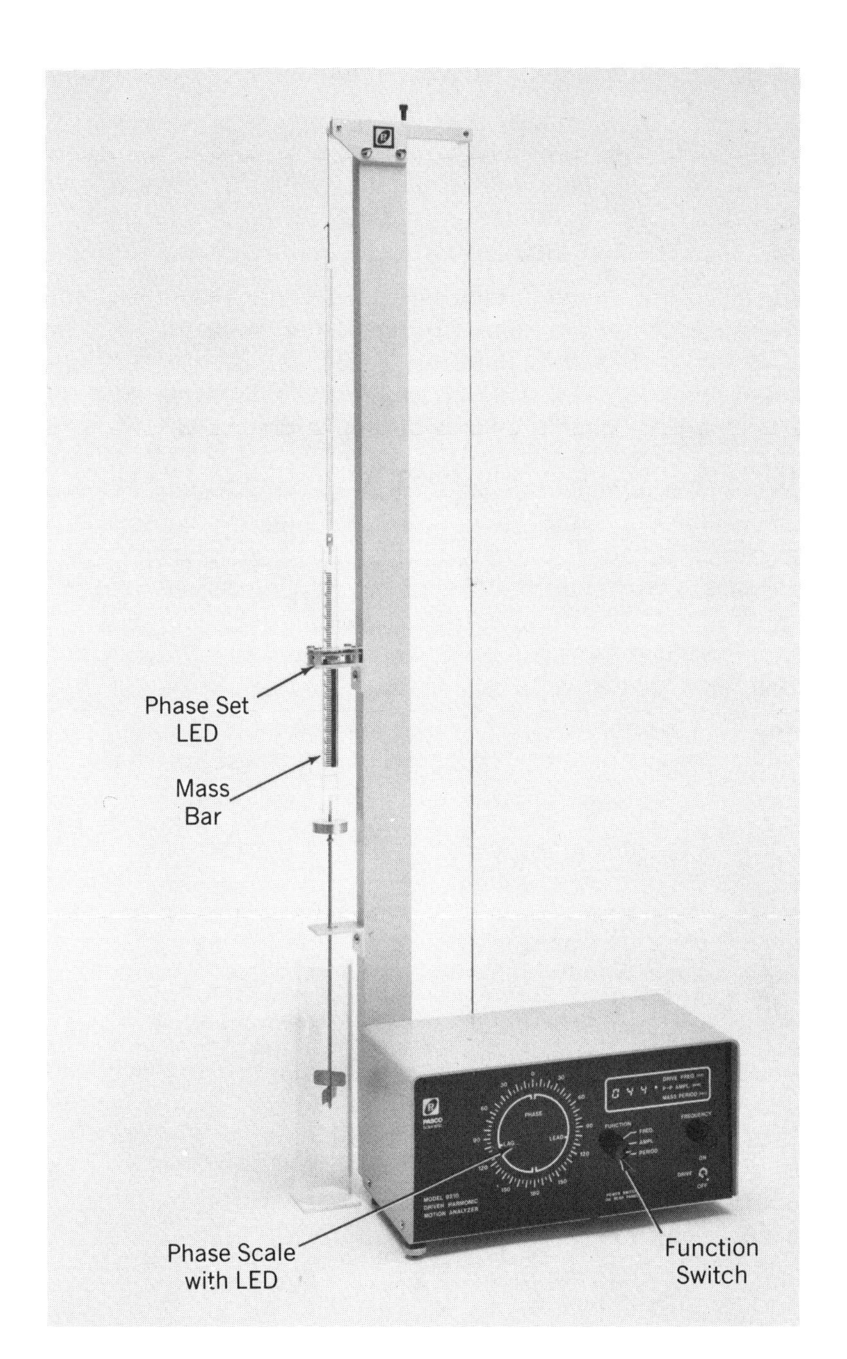

Fig. 7-30. Harmonic Motion Analyzer. (Photo Courtesy of The Pasco Scientific Co.)

*See Experiment 5-6, Simple Harmonic Oscillations, Case c, Damped, Driven Oscillations; Case d, Undamped, Driven Oscillations.

INTRODUCTION

The driven harmonic motion apparatus makes possible the quantitative study of harmonic motion (free, driven and damped). This can be accomplished with a minimum of difficulty in taking measurements and good accuracy.

As can be seen in Fig. 7-30, the apparatus uses a vertical spring/mass system which may be driven over a wide frequency range (0.5 Hz to 3.5 Hz). A servo controlled motor that uses an optical encoder on the drive shaft and a phase locked loop network provide the sinusoidal driver frequency.

The system can be damped by allowing the lower end of the plunger with the vanes to move vertically up and down within a fluid dash pot. This can be either a glass or metal container. Water or some convenient proportion of water and glycerin, for example, 75% water and 25% glycerin, can be used in the dash pot.

In mechanical and electrical systems, the energy initially stored in the system divided by the energy lost per cycle is of ten referred to as the Q of the system. It is sometimes called the quality or quality factor. It is also used as a measure of the "sharpness" of the response of a physical system to the resonant driving force. Systems with small energy losses, for example, like ours with vixcous damping have high Q values.

In summary, the harmonic motion analyzer gives you complete flexibility in changing the following variables in the experiment:

(1) spring constant (using different springs);
(2) mass - by adding additional mass to the top of the plunger;
(3) driving frequency - change the controls on the front panel;
(4) driving amplitude - variable from 0 to 15 mm; and
(5) damping - change the proportions of water to glycerin.

Of equal importance to changing variables is their measurement and obtaining results. The unique design of the harmonic motion analyzer enables you to measure such important factors as:

(a) Driving frequency. It is measured once a second and displayed digitally with an accuracy of ±1%.

(b) The peak-to-peak excursion of the mass once each cycle is displayed digitally to eliminate the guess work involved in trying to measure the displacement of a rapidly moving object.

(c) Period. The period of each oscillation is measured with a resolution of 0.01S and an accuracy of ±1%. As the Mass Rod moves up and down in the guide, the LED goes on and off.

(d) Phase. A flashing LED on a 360° scale on the front panel displays the phase between the sinusoidal driver and the motion of the mass. A phase measurement is made every cycle.

IMPORTANT NOTES

1. The apparatus should not be set in direct sunlight, or even brightly reflected sunlight since this might affect the operation of the oprical detectors. Normal room light, however, should not cause any problems.

2. The peak-to-peak amplitude of the drive excursion is twice the radius. A scale located on the driving wheel indicates the radius of the eccentric.

3. The plastic Mass Bar and Damping Rod (with approximately 1/3 of the mass of the spring) have a mass of 50 g.

4. If the driver amplitude, mass, or spring are changed, the Phase measuring system must be calibrated again.

5. Phase: (Check with your laboratory instructor.)

   (a) As the Mass Rod moves up and down in the Mass Guide, the Phase Set LED on the upper guide goes on and off. At this point, the Mass Rod Scale will read about 7.5. Adjust the cord length so that this takes place.

   (b) While the Phase Set LED is flashing on and off, look at the Phase scale on the front panel. Somewhere on the scale, another light should be flashing. If not, set the mass oscillating again. Now turn the Drive Wheel on the back panel until the flashing light on the Phase scale is at the top of the scale.

   (c) Again adjust the drive cord so that the Phase Set LED just goes on and off. On the top of the support column is a fine adjustment screw for cord length. Use the screw to adjust the Mass Rod position so that a very small displacement of the Mass causes the LED to flash on and off.

6. When measuring amplitude near resonance, observe a few beats and use the average of the highest and lowest amplitudes measured for your reading.

7. Measuring phase in an undamped system is very difficult since the phase change primarily occurs close to resonance. At resonance the driving amplitude is small and, therefore, the resolution for measuring phase is coarse.

PROCEDURE

1. Select a spring, and determine its spring constant k. Adjust the drive cord length so that with only the mass bar (mass = 50 g) attached and no other weights, the bar scale reads 15 cm. Then add 50 g and measure and record the new displacement. Repeat this procedure two more times and record your data. Calculate an average value for the spring constant k.

2. To find the system's period T, adjust again the cord length so that the phase set LED on the upper mass guide is just at the point of going on. Set the phase scale on the front panel at zero. Set the function switch on period, and start the Mass Bar in motion in air with an initial displacement of about 5 cm below its equilibrium position. Repeat your measurements. Record the periods T; calculate the frequencies f_0. Obtain some other fluid mixtures selected from the list of damping fluids that follow and again calculate the frequencies f_0: (1) water; (2) 75% water and 25% glycerin; (3) 50% water and 50% glycerin; (4) 25% water and 75% glycerin; and (5) glycerin.

3. Use air as the damping fluid. Set the system into motion by giving it an initial displacement of 5 cm; then record the amplitude (mm) for 25 oscillations. Record the data. Repeat the experiment first using water as a damping fluid and then again with a mixture of water and glycerin (see your instructor). Plot a graph of amplitude vs. number of oscillations for the undamped and damped cases.

4. Set the driver amplitude to about 3 mm. Vary the driver frequency from 0.5 Hz to 2.5 Hz - taking amplitude readings every 0.2 Hz, more frequently near resonance. Do this for 5 different damping fluids. Record the data. Plot a resonance graph of amplitude (mm) vs. driver frequency (Hz). Label each curve plotted with the damping fluid used.

5. Set the driver amplitude at 4 mm. Again vary the driver frequency in 0.2 Hz steps from 0.5 Hz to 2.5 Hz. Record the corresponding phase angle (degrees) for each driver frequency selected. The phase should be close to zero, with 5 to 10 degrees of lag. Obtain data for three damping mixtures, including water. Plot phase angle θ (degrees) vs. driver frequency. Mark on your plot the resonance position.

6. For a damped, undriven oscillating system, we can obtain the quality factor Q by multiplying by π the number of oscillations required for the mass motion amplitude to decrease to 1/e of its initial value. Calculate Q from the data you obtained in Step 3 when you plotted amplitude vs. oscillations.

7. The quality factor Q can be obtained from the mass motion amplitude divided by the driver amplitude. The mass motion can be read off the front panel when the function switch is set to "amplitude." The driver amplitude can be read from the scale attached to the rear of the apparatus. Set this amplitude to about 2 mm which is nearly the maximum amplitude that will not result in the mass hitting the table at resonance when no damping is used. The driver amplitude is twice the scale reading. Calculate Q for air and again for water. Discuss your results.

QUESTIONS

1. In Step 2 of the procedure, the period T of the oscillations in air and in other damping fluids is obtained. From the expression $T = 2\pi\sqrt{m/k}$, determine T and discuss the accuracy of your results. Do the results compare within the limit of error L.E. of the measurements?

2. Compare the values of the quality factor Q found in Steps 7 and 8. Discuss possible explanations for their variation.

3. Does the driver motion lead or "lag" the mass motion? Explain.

4. Does the driver motion lead or "lag" the mass motion at resonance? Explain.

7.2 Analysis of a Linear Oscillator

In this analysis, we are concerned primarily with the driven motion of a one-dimensional damped spring-mass system. As an introduction we will first consider the following two cases: Undamped, Undriven Case and the Damped, Undriven Case.

I. UNDAMPED, UNDRIVEN CASE (Free Oscillations)

Assume that the mass is freely suspended from a liner (Hooke's law) spring. Take the upward x direction as positive, let x be zero when the mass is at its equilibrium position, and assume the spring is massless. (If x is measured from the equilibrium position, the weight mg does not appear in the analysis.) The resultant vertical force on the mass is just the restorative force exerted on it by the spring, and is given by F = -kx (the minus sign arises from the fact that F and x are in opposite directions). From Newton's second law, we have

$$\Sigma F_x = ma_x$$

$$-kx = ma = m\frac{d^2x}{dt^2},$$

$$m\frac{d^2x}{dt^2} + kx = 0. \qquad 7\text{-}16$$

Eq. 7-16 is a linear, second-order, total differential equation with constant coefficients whose solution is given by

$$x = A\cos\omega t + B\sin\omega t, \quad \text{where } \omega^2 = \frac{k}{m}. \qquad 7\text{-}17^*$$

The constants A and B can be evaluated from the initial conditions. The most general initial conditions are $x = x_0$, $v = dx/dt = v_0$, when t = 0. From Eq. 7-17, when t = 0,

$$x_0 = A\cos(0) + B\sin(0)$$

$$\therefore \quad A = x_0.$$

Also, from Eq. 7-17, $v = \frac{dx}{dt} = -\omega A\sin\omega t + \omega B\cos\omega t$; and when t = 0 we obtain:

$$v_0 = -\omega A\sin(0) + \omega B\cos(0)$$

$$\therefore \quad B = \frac{v_0}{\omega}.$$

*Note that $\omega = \sqrt{k/m} = \omega_0$, where ω_0 is called the natural angular frequency of the system. ω_0 is determined by the constants of the system k and m. In this case the mass undergoes simple harmonic motion in the vertical direction, with amplitude a.

Inserting the values of A and B into Eq. 7-17 yields

$$x = x_o \cos \omega t + \frac{v_o}{\omega} \sin \omega t \; . \qquad 7\text{-}18$$

With the aid of the trigonometric relation sin (A + B) = sin A cos B + cos A sin B, and if we associate a sin A with x_o, a cos A with v_o/ω, A with ϕ, and B with ωt, Eq. 7-18 can be re-written as:

$$x = a \sin (\omega t + \phi) \; , \qquad 7\text{-}19$$

where[1]

$$a = \sqrt{x_o^{\,2} + \frac{v_o^{\,2}}{\omega^2}}$$

is the amplitude of the motion, and the phase angle ϕ is given by[2]

$$\phi = \tan^{-1} \frac{\omega x_o}{v_o}$$

II. DAMPED, UNDRIVEN CASE

Suppose now a damping vane is attached to the mass and immersed in the liquid. As a result of the viscosity of the liquid, a damping force will also be exerted on the mass in a direction opposite to that of the motion (i.e., velocity) of the mass at any instant. To a very good approximation, this damping force will be directly proportional to the velocity of the mass, and can therefore be written as

$$-bv = -b \frac{dx}{dt}$$

where b is a positive constant. Let us neglect the mass of the damping vane. From

---

[1]The value of a is obtained from the fact that

$(a \sin \phi)^2 + (a \cos \phi)^2 = (x_o)^2 + (\frac{v_o}{\omega})^2$; $\quad a^2 \sin^2 \phi + a^2 \cos^2 \phi = x_o^{\,2} + \frac{v_o^{\,2}}{\omega^2}$;

$a^2(\sin^2 \phi + \cos^2 \phi) = x_o^{\,2} + \frac{v_o^{\,2}}{\omega^2}$; $\quad a^2 = x_o^{\,2} + \frac{v_o^{\,2}}{\omega^2}$;

$\therefore \quad a = \sqrt{x_o^{\,2} + \frac{v_o^{\,2}}{\omega^2}}$

[2]Since $a \sin \phi = x_o$, $a \cos \phi = \frac{v_o}{\omega}$, then $\frac{a \sin \phi}{a \cos \phi} = \frac{x_o}{v_o/\omega}$; $\tan \phi = \frac{\omega x_o}{v_o}$

$\therefore \quad \phi = \tan^{-1} \frac{\omega x_o}{v_o}$

Newton's second law $\Sigma F_x = ma_x$, we obtain for motion in either direction

$$-kx - b\frac{dx}{dt} = m\frac{d^2x}{dt^2}$$

so that the differential equation of damped harmonic motion is given by

$$m\frac{d^2x}{dt^2} + b\frac{dx}{dt} + kx = 0 \; . \qquad 7\text{-}20$$

The nature of the solution to Eq. 7-20 depends upon the amount of damping present in the system (i.e., upon the relative magnitude of the damping factor b). There are three possible cases.

1. Underdamped Case; $b < \sqrt{4mk}$.

The general solution to this case is given by the equation

$$x = Ae^{-\frac{b}{2m}t}\sin(\omega' t + \delta) \; , \qquad 7\text{-}21$$

where $\omega' = \sqrt{\frac{k}{m} - \frac{b^2}{4m^2}}$. In this case the phase angle δ is:

$$\delta = \tan^{-1}\frac{x_o}{(v_o + \frac{bx_o}{2m})/\sqrt{\frac{k}{m} - \frac{b^2}{4m^2}}}$$

$$\delta = \tan^{-1}\frac{\omega' x_o}{v_o + \frac{bx_o}{2m}} \; ,$$

and

$$A = \left[x_o^2 + \frac{(v_o + \frac{bx_o}{2m})^2}{(\frac{k}{m} - \frac{b^2}{4m^2})}\right]^{\frac{1}{2}} \; .$$

x_o and v_o are the initial displacement and initial velocity, respectively.

Eq. 7-21 is an equation of simple harmonic motion where the amplitude decreases exponentially with time; the greater the damping, the faster the oscillations die out. Note also that the effect of damping is to increase the period of the motion, since the value of ω' is less than that of $\omega_o = \sqrt{k/m}$, the undamped natural angular frequency. Fig. 7-31 illustrates the variation of x with t for the case where $x = x_o$ at $t = 0$.

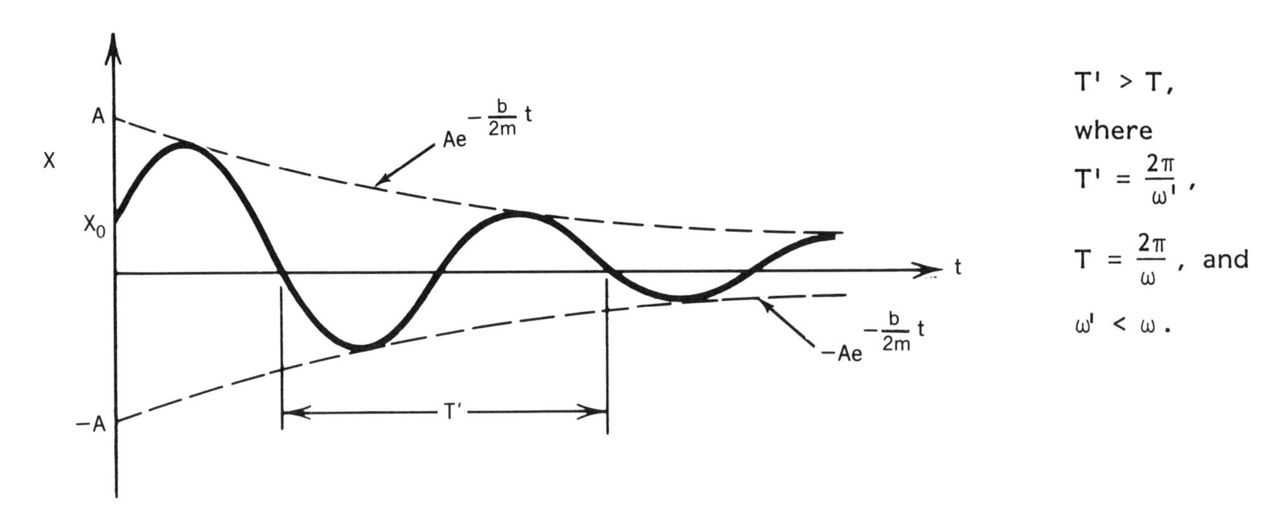

Fig. 7-31

Note that if $b = 0$, corresponding to the case of no damping, Eq. 7-21 reduces to Eq. 7-19 for the case of free oscillation (as it should).

2. <u>Critically Damped Case;</u> $b = \sqrt{4\,mk}$.

In this case, the general solution is given by

$$x = (B_1 + B_2 t)\, e^{-\frac{bt}{2m}} \qquad 7\text{-}22$$

where the constants B_1 and B_2 can be determined when the initial conditions are specified. The motion in this case is no longer oscillatory, but it again dies out as time progresses, as shown in Fig. 7-32.

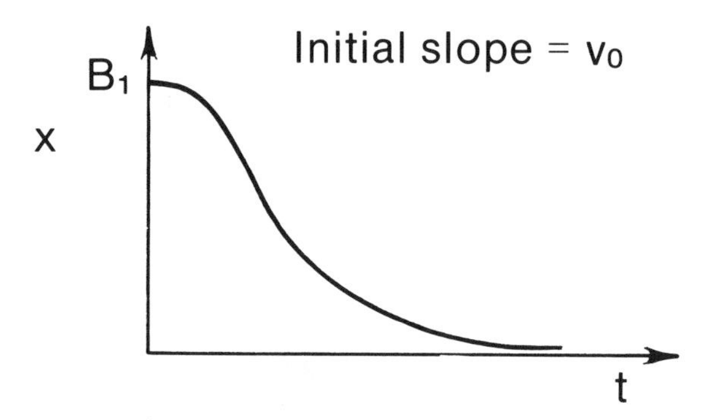

Critically Damped Harmonic Motion

Fig. 7-32

You should observe that this provides several methods of measuring the value of b, the damping factor. If either the value of T' or the envelope ($Ae^{-bt/2m}$) of the graph of x vs. t can be determined experimentally, the value of b can be calculated.

Note: If $x = x_o$, $v = v_o$, when $t = 0$, then $B_1 = x_o$, and $B_2 = v_o + \frac{x_o b}{2m}$.

These follow from the fact that when

$$x = (B_1 + B_2 t)\, e^{-\frac{bt}{2m}},$$

then $t = 0$, $x = x_o$ and $x_o = (B_1 + 0)\, e^{-0}$.

$$\therefore\ B_1 = x_o\,.$$

To determine B_2, we first calculate $v = \frac{dx}{dt}$.

$$v = \frac{dx}{dt} = (B_1 + B_2 t)\, e^{-\frac{bt}{2m}} \left(-\frac{b}{2m}\right) + e^{-\frac{bt}{2m}} (0 + B_2)$$

$$= -\frac{b}{2m}(B_1 + B_2 t) e^{-\frac{bt}{2m}} + B_2 e^{-\frac{bt}{2m}}\,.$$

At $t = 0$, $v = v_o$:

$$v_o = -\frac{b}{2m}(B_1 + 0)e^{-0} + B_2 e^{-0}$$

$$= -\frac{bB_1}{2m} + B_2$$

$$B_2 = v_o + \frac{bB_1}{2m}\,.$$

The value of B_1 found above is $B_1 = x_o$. Therefore,

$$B_2 = x_o + \frac{bx_o}{2m}\,.$$

3. Overdamped Case; $b > \sqrt{4\ mk}$.

For this case, the general solution has the form

$$x = C_1 e^{\gamma_1 t} + C_2 e^{\gamma_2 t}, \qquad \text{7-23}$$

where

$$\gamma_1 = -\frac{b}{2m} + \sqrt{\frac{b^2}{4m^2} - \frac{k}{m}}\,,$$

and

$$\gamma_2 = -\frac{b}{2m} - \sqrt{\frac{b^2}{4m^2} - \frac{k}{m}} \; .$$

The constants C_1 and C_2 are again evaluated from the initial conditions. The motion in this case is also non-oscillatory, and dies out as time progresses (although more slowly than in Case 2), as shown in Fig. 7-33.

Since $b^2 > 4\ mk$,

$$\sqrt{\frac{b^2}{4m^2} - \frac{k}{m}} > 0 \; .$$

However,

$$\sqrt{\frac{b^2}{4m^2} - \frac{k}{m}} < \frac{b}{2m} \; ,$$

so that both γ_1 and γ_2 in Eq. 7-23 are negative. Therefore, x approaches zero as t increases indefinitely.

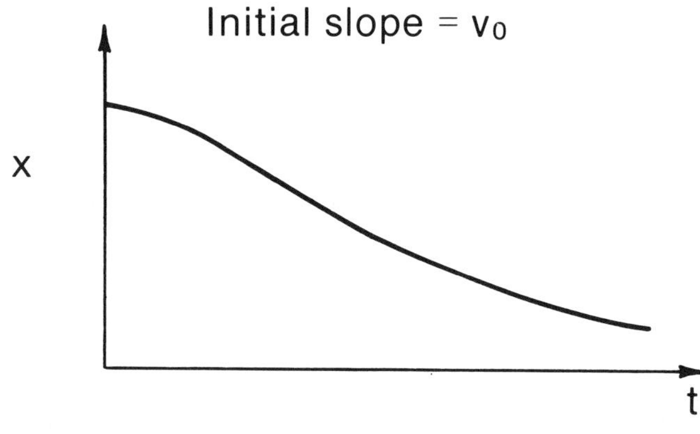

Fig. 7-33

Over Damped Harmonic Motion

If $x = x_o$ and $v = v_o$ when $t = 0$, then

$$C_1 = \frac{v_o - \gamma_2 x_o}{\gamma_1 - \gamma_2} \; , \qquad \text{and} \qquad C_2 = \frac{\gamma_1 x_o - v_o}{\gamma_1 - \gamma_2} \; .$$

From $x = C_1 e^{\gamma_1 t} + C_2 e^{\gamma_2 t}$, at $t = 0$, we find $x = x_o$ and

$$x_o = C_1 + C_2 \; .$$

Since $v = \frac{dx}{dt} = \gamma_1 C_1 e^{\gamma_1 t} + \gamma_2 C_2 e^{\gamma_2 t}$, then at $t = 0$, $v = v_o$ and $v_o = \gamma_1 C_1 + \gamma_2 C_2$.

Now substitute C_2 into Eq. 7-24 and

$$v_o = \gamma_1 C_1 + \gamma_2(x_o - C_1) = \gamma_2 x_o + C_1(\gamma_1 - \gamma_2) \; . \qquad 7\text{-}25$$

$$\therefore \; C_1 = \frac{v_o - \gamma_2 x_o}{\gamma_1 - \gamma_2} \; .$$

Then

$$c_2 = x_o - C_1 = x_o - \frac{v_o - \gamma_2 x_o}{\gamma_1 - \gamma_2} \; , \qquad \text{and}$$

$$\therefore \; C_2 = \frac{\gamma_1 x_o - v_o}{\gamma_1 - \gamma_2} \; .$$

Thus

$$x = \left(\frac{v_o - \gamma_2 x_o}{\gamma_1 - \gamma_2}\right) e^{\gamma_1 t} + \left(\frac{\gamma_1 x_o - v_o}{\gamma_1 - \gamma_2}\right) e^{\gamma_2 t} \; . \qquad 7\text{-}26$$

Note that $\gamma_1 - \gamma_2 = 2\sqrt{\frac{b^2}{4m^2} - \frac{k}{m}}$ is greater than zero.

For those who are familiar with the hyperbolic functions, Eq. 7-23 can be re-written as

$$x = e^{-\frac{bt}{2m}} (C_3 \cosh \gamma t + C_4 \sinh \gamma t) \; , \qquad 7\text{-}27$$

where C_3, C_4 are constants and $\gamma = \sqrt{\frac{b^2}{4m^2} - \frac{k}{m}}$.

III. DAMPED, DRIVEN CASE (Forced Oscillations)

Eqs. 7-21, 7-22, and 7-23 represent the behavior of a one-dimensional damped spring-mass system that has been given an initial disturbance (displacement, velocity, or both) and then left to itself. They govern (within the limitations of their derivation) the natural, or free, motion of the system. The equation that applies to a given physical system depends, as we have seen, upon the amount of damping present. In all cases, these effects die out exponentially with time so that the system eventually comes to rest at its equilibrium position. For this reason, the equations above are known as the transient solutions.

If now an additional external driving force is applied to the system we would expect that the subsequent motion should be a combination of such a transient effect together with that produced by the external force itself. If we wait long enough, the transient motion will die away so that the response of the system will be governed by the driving force.

In this experiment we consider only the effect of a constant-amplitude oscillatory driving force given by $F_0 \sin \omega_d t$, where F_0 is the amplitude of the force and ω_d is the angular frequency; the period of the driving force is then $T = 2\pi/\omega_d$. By considering all of the forces acting on the mass, we obtain from Newton's second law, $\Sigma F_x = ma_x$.

$$- kx - b\frac{dx}{dt} + F_o \sin \omega_d t = ma_x = m\frac{d^2x}{dt^2},$$

or

$$\frac{d^2x}{dt^2} + \frac{b}{m}\frac{dx}{dt} + \frac{k}{m}x = \frac{F_o}{m}\sin \omega_d t . \tag{7-28}$$

Eq. 7-28 is the differential equation of motion governing the forced oscillations of the system. It contains Eq. 7-20. Therefore, the general solution of Eq. 7-28 will contain a transient term of the type discussed earlier, plus a steady-state response that persists after the transient effect has died away (actually, this is a characteristic possessed by any linear differential equation - of which Eq. 7-28 is an example). It is this steady-state solution that we now wish to determine.

From physical and mathematical considerations, we expect a steady-state oscillation with an angular frequency that is the same as that of the driving force. This proposed solution can be written in the form

$$x = A \sin \omega_d t + B \cos \omega_d t , \tag{7-29}$$

where the constants A and B must be determined so that the differential equation Eq. 7-28 is satisfied.

Differentiation of Eq. 7-29 yields

$$\frac{dx}{dt} = \omega_d A \cos \omega_d t - \omega_d B \sin \omega_d t ; \tag{7-30}$$

$$\frac{d^2x}{dt^2} = -\omega_d^2 A \sin \omega_d t - \omega_d^2 B \cos \omega_d t . \tag{7-31}$$

Substituting Eq. 7-29, 7-30 and 7-31 into the equation of motion, Eq 7-28, gives

$$-\omega_d^2 A \sin \omega_d t - \omega_d^2 B \cos \omega_d t + \frac{b}{m}\omega_d A \cos \omega_d t - \frac{b}{m}\omega_d B \sin \omega_d t + \frac{k}{m} A \sin \omega_d t + \frac{k}{m} B \cos \omega_d t$$

$$= \frac{F_o}{m} \sin \omega_d t \, .$$

Collecting like terms, we have

$$[-\omega_d^2 A - \frac{b}{m}\omega_d B + \frac{k}{m} A - \frac{F_o}{m}] \sin \omega_d t + [-\omega_d^2 B + \frac{b}{m}\omega_d A + \frac{k}{m} B] \cos \omega_d t = 0.$$

For non-trivial solutions, this equation is satisfied if and only if the coefficients in the brackets are both zero; thus, changing the sign of all the terms in the brackets,

$$\omega_d^2 A + \frac{b}{m}\omega_d B - \frac{k}{m} A + \frac{F_o}{m} = 0, \qquad \text{and } \omega_d^2 B - \frac{b}{m}\omega_d A - \frac{k}{m} B = 0.$$

Recalling that $\omega_o = \sqrt{k/m}$, we have

$$\begin{cases} (\omega_d^2 - \omega_0^2)A + \dfrac{b}{m}\omega_d B = -\dfrac{F_o}{m} \\ -\dfrac{b}{m}\omega_d A + (\omega_d^2 - \omega_o^2)B = 0 \, . \end{cases} \qquad 7\text{-}32$$

The unknown coefficients A and B are determined by solving Eqs. 7-32 simultaneously, either by means of straightforward algebra or by use of determinants. The coefficients are, as follows,

$$A = \frac{F_o(\omega_o^2 - \omega_d^2)}{m\,[(\omega_o^2 - \omega_d^2)^2 + \frac{b^2}{m^2}\omega_d^2]} \qquad \text{and} \qquad B = \frac{-b\, F_o \omega_d}{m^2[(\omega_o^2 - \omega_d^2)^2 + \frac{b^2}{m^2}\omega_d^2]}$$

Then, from Eq. 7-29 together with these values of A and B, we obtain

$$x = \frac{F_o}{m\,[(\omega_o^2 - \omega_d^2)^2 + \frac{b^2}{m^2}\omega_d^2]} \left\{ (\omega_o^2 - \omega_d^2) \sin \omega_d t - \frac{b}{m}\omega_d \cos \omega_d t \right\}. \qquad 7\text{-}33$$

This equation can be re-written as

$$x = \frac{F_o}{m\sqrt{(\omega_o^2 - \omega_d^2)^2 + \frac{b^2}{m^2}\omega_d^2}} \left\{ (\sin \omega_d t) \cdot \frac{(\omega_o^2 - \omega_d^2)}{\sqrt{(\omega_o^2 - \omega_d^2)^2 + \frac{b^2}{m^2}\omega_d^2}} - (\cos \omega_d t) \cdot \frac{\left(\frac{b\omega_d}{m}\right)}{\sqrt{(\omega_o^2 - \omega_d^2)^2 + \frac{b^2}{m^2}\omega_d^2}} \right\} \quad 7\text{-}34$$

The terms within the brace, however, are just of the form

$$\sin \omega_d t \cos \theta - \cos \omega_d t \sin \theta = \sin (\omega_d t - \theta). \quad 7\text{-}35$$

We therefore define the phase angle θ by means of the relations

$$\sin \theta = \frac{\left(\frac{b\omega_d}{m}\right)}{G} \qquad \cos \theta = \frac{\omega_o^2 - \omega_d^2}{G}, \quad 7\text{-}36$$

where

$$G = \sqrt{(\omega_o^2 - \omega_d^2)^2 + \frac{b^2}{m^2}\omega_d^2}.$$

Fig. 7-34 illustrates the trigonometry associated with these relations.

Then

$$\tan \theta = \frac{\frac{b\omega_d}{m}}{\omega_o^2 - \omega_d^2}, \quad \text{so that}$$

$$\theta = \tan^{-1}\left[\frac{\frac{b\omega_d}{m}}{\omega_o^2 - \omega_d^2}\right] \quad 7\text{-}37$$

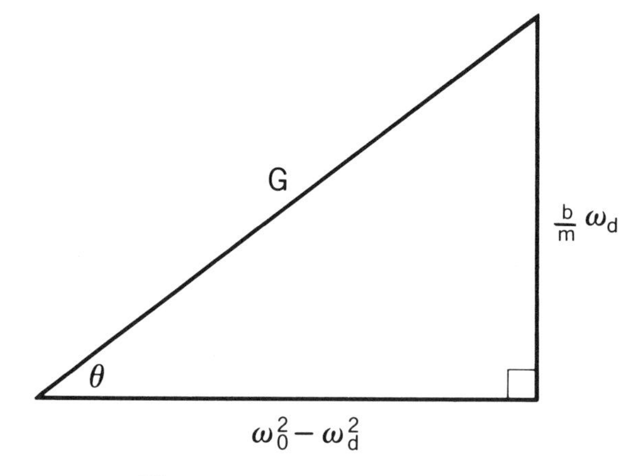

Phase Angle θ Diagram.

Fig. 7-34

Hence the steady-state solution to the sinusoidally-driven damped linear oscillator is given by

$$x = \frac{F_o}{m\sqrt{(\omega_o^2 - \omega_d^2)^2 + \frac{b^2}{m^2}\omega_d^2}} \sin(\omega_d t - \theta) , \qquad 7\text{-}38$$

where

$$\theta = \tan^{-1}\left[\frac{\frac{b\omega_d}{m}}{\omega_o^2 - \omega_d^2}\right] .$$

For a given system (fixed values of b, m and ω_o), θ depends upon the value of ω_d.

Exercise: Verify that Eq. 7-38 is the solution to the differential equation of motion by computing $\frac{dx}{dt}$, $\frac{d^2x}{dt^2}$, and substituting it into Eq. 7-28.

Eq. 7-38 gives the displacement of the mass from its equilibrium position as a function of time (and also of ω_d). The mass oscillates with an undamped harmonic motion at the same angular frequency as the driving force, but its oscillation is out of phase with (lags behind) that of the driving force by the angle θ.

As $\omega_d \longrightarrow 0$, note that $x \longrightarrow \frac{F_o}{k}$ (as it should). If ω_d becomes large, re-arrangement of Eq. 7-38 shows that $x \sim \frac{F_o}{m\omega_d^2}$, so that $x \longrightarrow 0$ as $\omega_d \longrightarrow \infty$.

For a given system with fixed values of ω_o and b, the displacement x will increase as $\omega_d \longrightarrow \omega_o$, and will become very large if the damping factor b is small when $\omega_d \approx \omega_o$. The amplitude will be a maximum when the quantity $(\omega_o^2 - \omega_d^2)^2 + \frac{b^2}{m^2}\omega_d^2$ is a minimum. From calculus, the value of ω_d that maximizes x is obtained by setting the first derivative of the last expression with respect to ω_d equal to zero:

$$\frac{d}{d\omega_d}\left[(\omega_o^2 - \omega_d^2)^2 + \frac{b^2}{m^2}\omega_d^2\right] = 0 .$$

$$2(\omega_o^2 - \omega_d^2)(-2\omega_d) + 2\frac{b^2}{m^2}\omega_d = 0 ;$$

$$\therefore \omega_d = \sqrt{\omega_o^2 - \frac{b^2}{m^2}} .$$

(We reject the result $\omega_d = 0$ as unrealistic, since it would correspond to the case of no driving force at all.) The "resonant" frequency ω_R is defined as that (driving) angular frequency at which the amplitude is a maximum; thus

Note: You should check to see that this does maximize x by examining the second derivative.

$$\omega_R = \sqrt{\omega_o^2 - \frac{b^2}{m^2}} \,. \qquad 7\text{-}39$$

We see that $\omega_R < \omega_o$, and that $\omega_R \longrightarrow \omega_o$ as $b \longrightarrow 0$. It should be observed, therefore, that the maximum amplitude does not occur at the angular frequency $\omega_d = \omega_o$ (except in the case of no damping). The difference between ω_R and ω_o increases as b increases.

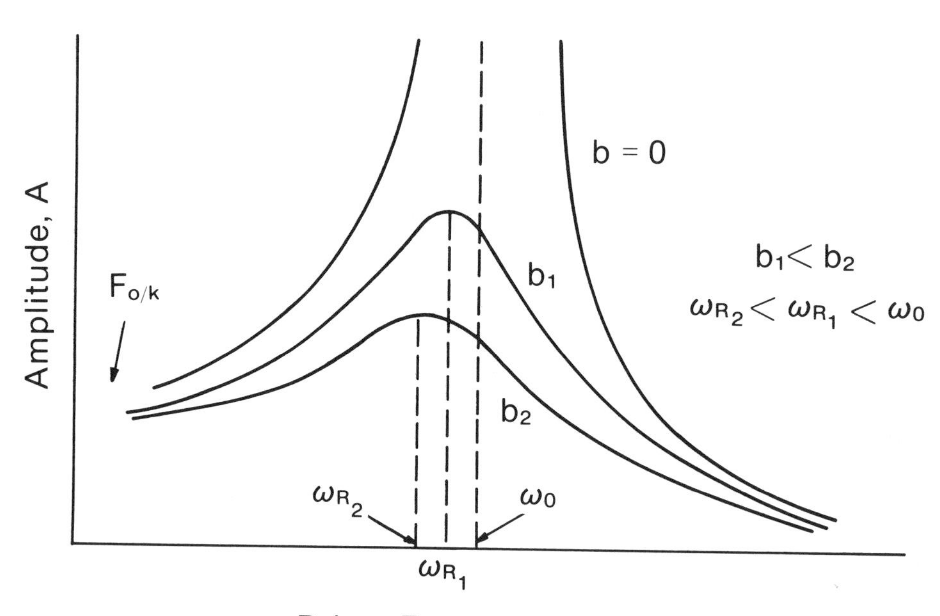

A versus ω_d for $b_1 < b_2$. For the special case where there is no damping ($b = 0$), the amplitude of the motion would become infinite. This is a mathematical result only; no physical system would every attain infinite amplitude.

Fig. 7-35

For the special case in which there is no damping (i.e., $b = 0$), the denominator of Eq. 7-39 becomes zero and the amplitude of the motion becomes infinite when the driving angular frequency and the natural angular frequency of the system are equal; i.e., when $\omega_d = \omega_o = \sqrt{k/m}$. This is a mathematical result only; no physical system would ever attain infinite amplitude!

Fig. 7-35 illustrates the general behavior of a resonant system as a function of both b and ω_d. Whenever damping is present the amplitude of the motion passes through a maximum value as the driving angular frequency ω_d is varied. The driving angular frequency corresponding to each amplitude maximum is the resonant angular frequency (for that particular value of b); the resonant angular frequency, therefore, depends on the amount of damping present in the system. Furthermore, the phenomenon of resonance does not occur at a single angular frequency but, rather, is exhibited over a range of angular frequencies. It is not necessary that the driving angular frequency coincide with the resonant angular frequency in order that the amplitude of the motion become large. If, however, the damping is small, large amplitudes are obtained only within a small angular frequency range near the resonant angular frequency; in such a case the system is said to have a "sharp" response to the driving force (i.e., a high "Q"). As the amount of damping increases, the amplitude enhancement becomes smaller and takes place over a wider range of angular frequencies; and then the system's response is said to be "broad" (i.e., low "Q").

Behavior of the Phase Angle

θ is the phase difference between the driver and the mass, where $\theta = \omega_d(t_2 - t_1)$, and t_2 and t_1 are the successive times at which the driving force and the displacement of the mass take on their maximum values, respectively (during the same portion of a cycle of the motion); the driver, of course, has to lead the mass. From Eq. 7-37

$$\theta = \tan^{-1}\left[\frac{b\omega_d/m}{\omega_o^2 - \omega_d^2}\right] .$$

$$\text{As } \omega_d \longrightarrow \omega_o,\ (\omega_o^2 - \omega_d^2) \longrightarrow 0, \text{ and } \frac{b\omega_d/m}{\omega_o^2 - \omega_d^2} \longrightarrow \infty ;$$

Thus

$$\theta = \tan^{-1}\left[\frac{b\omega_d/m}{\omega_o^2 - \omega_d^2}\right] = \tan^{-1}\infty = \frac{\pi}{2}$$

Since θ approaches $\pi/2$ through positive angles, the driver leads the mass by θ degrees.

Similarly, as $\omega_d \longrightarrow 0$, $\theta \longrightarrow 0$, and the driving force and displacement are then very nearly in phase. Finally, as $\omega_d \longrightarrow \infty$, $\theta \longrightarrow \pi$, so that the displacement and driving force are nearly 180° out of phase.

Since ω_R and ω_o are not equal, in general, then θ will not be exactly $\pi/2$ when $\omega_d = \omega_R$ (i.e., at resonance). From Eq. 7-39,

$$\omega_o^2 - \omega_R^2 = \frac{b^2}{m^2} ,$$

so that at resonance the phase angle θ_R is:

$$\theta_R = \tan^{-1}\left(\frac{m\omega_R}{b}\right) = \tan^{-1}\left[\frac{m}{b}\sqrt{\omega_o^2 - \frac{b^2}{2m^2}}\right] < \frac{\pi}{2} ,$$

and will be $\sim \pi/2$ if b is small.

The phase angle θ_R at resonance is then generally somewhat smaller than $\pi/2$; the difference between its actual value and $\pi/2$ depends on the value of b. The variation of phase angle with the angular driver frequency is shown in Fig. 7-36.

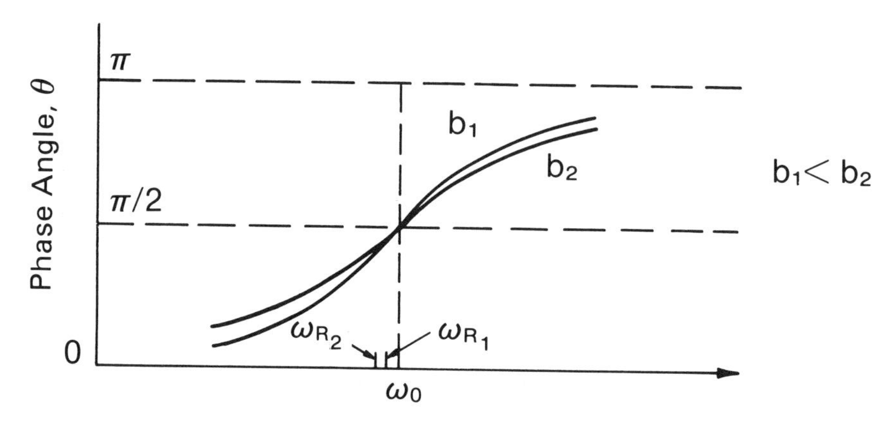

"Driver Angular Frequency"

Difference in phase θ between the driver and the mass versus the angular frequency ω_d of the driver. The variation between the actual value of the phase angle θ_R at resonance and $\pi/2$ depends on the value of b.

Fig. 7-36

Experiment 7-20 ANALYSIS OF RESONANCE WITH A DRIVEN TORSIONAL PENDULUM*

INTRODUCTION

As the angular frequency μ of the driver approaches the undamped natural angular frequency ω_0 of the torsional pendulum (copper disc), the angular amplitude increases (pointer attached to the top of the copper disc swings through a greater angle). When the angular frequency of the driver and the natural angular frequency are nearly equal, the angular amplitude of swing of the pointer reaches a maximum value. This phenomenon is called resonance. Resonance of a driven system is that condition reached when the angular frequency of the driver is such that maximum response is obtained from the system being driven. It is important to understand that even in our so-called undamped system (no electromagnetic damping) there is always some inherent damping. Therefore, at resonance, the angular amplitude does not become infinite. The greater the degree of electromagnetic damping, the larger the transfer ratio G (See Eq. 7-58) and the smaller the maximum angular amplitude reached by the pointer at the top of the copper disc.

You are expected to review carefully the procedure and analysis sections so that a neat complete data table can be made. All measurements must be recorded and units indicated at the top of the data columns.

PROCEDURE

1. Check to see that the D.C. to the electromagnet is off. Start the D.C. motor and adjust the rheostat (Fig 7-37, R) until resonance of the undamped (no external damping) copper disc is obtained. The angular amplitude should be approximately 15 units to the right or left of equilibrium at resonance. If the angular amplitude is greater than 15 units, the amplitude of the driver can be decreased by changing slightly the position of the driver arm in the adjustment sleeve on the D.C. motor (Fig. 7-37, J). DO NOT CHANGE THE DRIVER AMPLITUDE DURING THE EXPERIMENT. Shut off the D.C. motor.

2. Deflect the copper disc (pendulum) about 15 units BY HAND and find the average period T_0 for ten oscillations. Record T_0 and the proper angular frequency ω_0 of the undriven system.

3. The damping ratio of two successive angular amplitudes in the same direction (on the same side) is called K. To obtain K_1 (without electromagnetic damping), deflect the copper disc BY HAND to a reading of 15 units. Record successive same-side amplitude values until the oscillations die out. Find the average K_1. For currents with electromagnetic damping of about two and four amperes, again deflect

*See Experiment 5-6, Simple Harmonic Oscillations, Case c, Damped, Driven Oscillations; Case d, Undamped, Driven Oscillations. Torsional Pendulum can be obtained from Leybold-Heraeus GMBH Co., KG, 5000 Köln 51, Germany.

Fig. 7-37. Apparatus Used in the Analysis of Resonance. M = D.C. motor used to drive arm which is attached to top of coiled spring. R = rheostat used to regulate speed of D.C. motor. D = copper disk which is the driven torsional pendulum. P = variable D.C. for electromagnet. A = D.C. ammeter. C = time clocks used to determine the difference in phase between the maximum amplitudes of the pendulum disc pointer and the pointer to which the coiled spring is attached. E = electromagnet. J = Adjustment sleeve for driver arm. V = driver arm from D.C. motor.

the copper disc BY HAND and determine the average K_2 and K_3. Record all data, including the average period T, of K_2 and K_3.

4. Without electromagnetic damping, start motor (exciter) which causes the forced oscillations. The angular frequency of the exciter ($\mu = 2\pi/T$) can be found from the average period T_o for ten oscillations at each rheostat setting. Record steps (a) and (b) carefully before taking data.

   (a) Find and record nine different values for the angular frequency μ of the exciter distributed from well below the proper angular frequency ω_o to as far above as possible. For each μ also record the maximum disc amplitude θ_o.

   (b) For each value of the angular frequency μ of the exciter, find β, the difference in phase between the maximum amplitudes of the copper disc and the exciter. The difference in phase is obtained from observation of the relative position of the pointer attached to the center of the copper disc and the pointer to which the coiled spring is attached. Since $\beta = \mu\Delta t$, where Δt is the time difference between the pointer attached to the coiled spring (driven by the exciter or motor) reaching a particular maximum angular amplitude and the pointer on the disc reaching the same point, we must obtain the difference in time as accurately as possible. Two time clocks are used. As the pointer to which the coiled spring is attached passes the midpoint of the scale, one observer gets ready to start his time clock. The clock is started when the driver pointer reaches maximum amplitude. The second observer starts his time clock when the pointer on the copper disc reaches its maximum amplitude on the same side as that reached by the driver pointer. Both clocks are stopped by one observer at the same time. Record all data.

   Note - β for each different μ is obtained after the average period T for a particular rheostat setting is found. Therefore, one time clock is used in getting T and consequently μ. Two time clocks are used to find β.

5. With electromagnetic damping (currents of about two and four amperes), repeat (a) and (b) of Step 4. Record all data.

ANALYSIS The student is permitted to choose from the following options, the analysis he prefers. Each additional option requires more technique or mathematical ability than the one preceding.

Option 1

Use the data obtained from Steps (1), (2), (3), (4-a-b), and (5-a-b). Plot a graph of the angular amplitude θ_o of the disc versus the ratio of the angular frequency of the exciter to the natural angular frequency of the disc (μ/ω_o). Do this for each of the various damping factors K on the same graph. Label each plot with the correct numerical K. Discuss the properties of resonance in the conclusion of your report. What is the effect of electromagnetic damping?

Option 2

Use the data obtained from Steps (1), (2), (3), (4-a-b), and (5-a-b). Plot a graph of β versus the ratio of the angular frequency of the exciter to the natural angular frequency of the disc (μ/ω_o). Do this for each of the various damping factors K on the same graph. Label each plot with the correct numerical K. Discuss the properties of resonance in the conclusion of your report. What is the effect of electromagnetic damping?

Option 3

Plot the graphs required in Options 1 and 2. In addition, plot the angular amplitude θ of the free vibrations versus time for K_1, K_2, and K_3. After reading the theory, Equation of Motion of a Forced Oscillator, show that Eq. 7-44 is a solution to the underdamped condition of the system analyzed. Hint - how is Eq. 7-49 obtained? Discuss the properties of resonance in the conclusion of your report.

Option 4

Plot the graphs required in Options 1 and 2. After reading the theory, Equation of Motion of a Forced Oscillator, find, for a fixed K, the value of (μ/ω_o) which will give resonance. Plot a graph of (μ/ω_o) [at resonance] versus K for both your derived and experimental results. Compare and discuss the results in the conclusion of your report.

Hint - how is Eq. 7-63 obtained? Can you derive

$$(\mu/\omega_o)_{reson.} = \left[1 - \left(\frac{\ln K}{2\pi}\right)^2\right]^{\frac{1}{2}} \qquad \text{from Eq. 7-63?}$$

$$\text{"} \qquad \cong 1 - \frac{\ln K}{2\pi}$$

THEORY - Equation of Motion of a Forced Oscillator

Because the problem of forced oscillations is quite general, the development of a solution to the equation of motion for a driven torsional pendulum can be easily extended to acoustical systems, alternating current circuits and to atomic physics. Since the mathematics required is quite advanced, most steps are shown in detail. Values of the arbitrary constants are determined from the properties of the driven system.

If the copper disc (Fig. 7-37, D) is displaced by an amount θ, the axle to which the coiled spring is rigidly attached is displaced through a similar angle. The restoring torque $\tau_D = n\theta$, where n = the constant of proportionality, is proportional to the angular displacement θ. The driver arm V is attached to a lever pivoted at the axle and connected to the top of the coiled spring. When the driver arm displaces the top of the coiled spring through an angle Φ a torque $\tau_S = n\Phi$ is produced. The constant of proportionality n is the same for the driver torque τ_S and the restoring torque τ_D. Why?

In addition to the restoring torque of the copper disc τ_D and the driver torque τ_S, there is a natural damping torque τ_M which is aided by the electromagnetic damping. The damping torque $\tau_M = b\dot{\theta}$, where b = the constant of proportionality, is proportional to the angular velocity $\dot{\theta}$.

The driver torque τ_S drives the copper disc while the restoring torque τ_D and the damping torque τ_M oppose the motion. Since the resultant of all the torques $(\Sigma\tau = I\alpha)$ acting upon the system is equal to the rotational inertia of the system times the angular acceleration, the equation of motion becomes:

$$\tau_S - \tau_D - \tau_M = I\alpha \, . \qquad \text{7-39}$$

In terms of angular variables, Eq. 7-39 becomes

$$n\Phi - n\theta - b\dot{\theta} = I\ddot{\theta}. \quad \text{7-40}$$

When the driver arm V is detached, the equation of motion for the un-driven system is

$$I\ddot{\theta} + b\dot{\theta} + n\theta = 0 \quad \text{7-41}$$

Eq. 7-41 has three types of solutions which depend on the magnitudes of the proportionality constants b, n and the moment of inertia (or rotational inertia) of the system. The solutions correspond to critically damped, overdamped and underdamped conditions of the un-driven system (see Fig. 7-38).

1. Critically damped $\frac{b}{2I} = \sqrt{\frac{n}{I}}$ 7-42

2. Overdamped $\frac{b}{2I} > \sqrt{\frac{n}{I}}$ 7-43

3. Underdamped $\frac{b}{2I} < \sqrt{\frac{n}{I}}$ 7-44

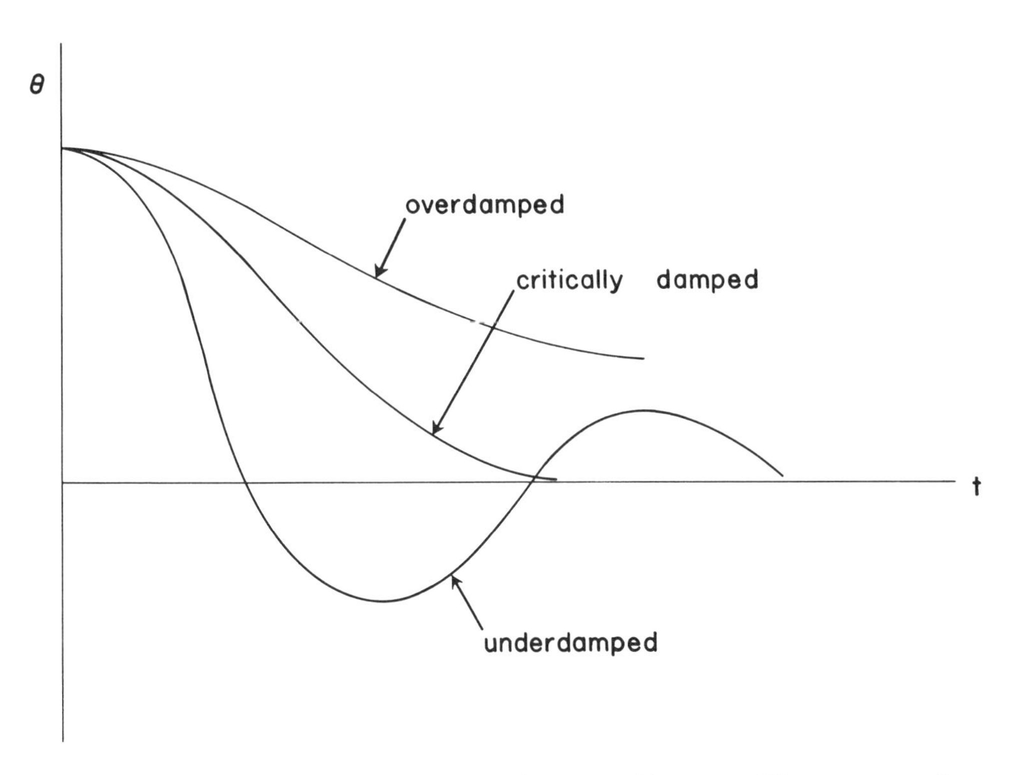

Fig. 7-38. Graphs of Damped Harmonic Motion. The copper disc is given an initial positive angular displacement θ_0 and released from rest.

Since our torsional pendulum (copper disc) oscillates after being released from an initial positive angular displacement θ_o, the system represents Case 3, the underdamped condition.

We know from the experiment that the angular amplitude θ decreases with time (a plot of θ vs. t shows an exponential decrease in amplitude). Therefore, a possible solution to the equation of motion for the un-driven system (Eq. 7-43) could consist of the product of an oscillatory term ($\cos \omega t$ or $\sin \omega t$) and an exponential term corresponding to the decay or decrease in amplitude of the oscillations (e^{-pt}), or

$$\theta = Ce^{-pt} \cos \omega t \ , \qquad 7\text{-}45$$

where C = the maximum angular amplitude θ_o of the un-driven copper disc and p is a constant to be evaluated.

If Eq. 7-45 is differentiated first with respect to time $\dot{\theta}$ and then again with respect to time $\ddot{\theta}$ we can substitute $\dot{\theta}$ and $\ddot{\theta}$ into Eq 7-41 and see if a solution is obtained.

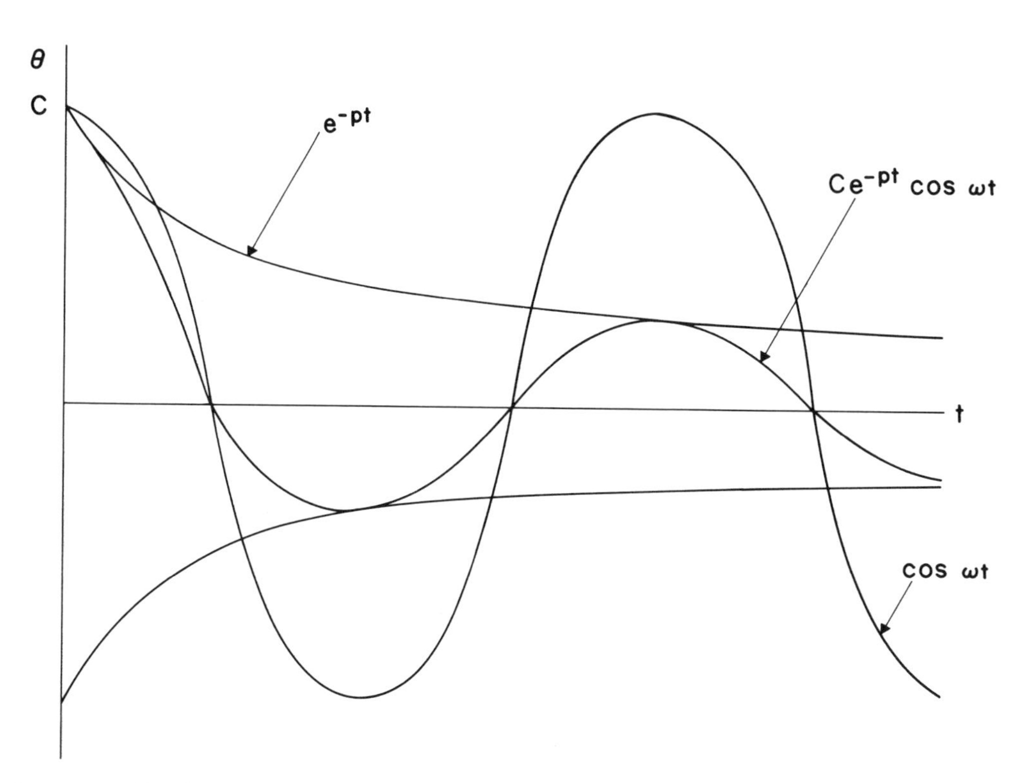

Angular displacement θ versus time t. An oscillatory term ($\cos \omega t$), a damping term e^{-pt}, and the product of both terms [$Ce^{-pt} \cos \omega t$] represent the damped harmonic motion of the un-driven system. (To simplify construction, C was chosen equal to one.)

Fig. 7-39

Therefore,

$$\dot{\theta} = C(-p)e^{-pt} \cos \omega t + Ce^{-pt} (+\omega)(-\sin \omega t),$$

and

$$\ddot{\theta} = Cp^2e^{-pt} \cos \omega t + C(-p)e^{-pt} (+\omega)(-\sin \omega t) + Ce^{-pt}(+\omega^2)(-\cos \omega t) + Ce^{-pt}(-p)(+\omega)(-\sin \omega t).$$

For convenience, Eq. 7-41 is expressed as

$$\ddot{\theta} + \frac{b}{I}\dot{\theta} + \frac{n}{I}\theta = 0 , \qquad 7\text{-}46$$

which becomes when $\dot{\theta}$ and $\ddot{\theta}$ are substituted,

$$Cp^2e^{-pt} \cos \omega t + p\omega Ce^{-pt} \sin \omega t - C\omega^2e^{-pt} \cos \omega t + p\omega Ce^{-pt} \sin \omega t$$

$$+ \frac{b}{I}(-Cpe^{-pt}\cos \omega t - C\omega e^{-pt} \sin \omega t) + \frac{n}{I}Ce^{-pt} \cos \omega t = 0 .$$

Collecting terms and canceling C and e^{-pt}, we obtain

$$\cos \omega t \left[p^2 - \omega^2 - p\frac{b}{I} + \frac{n}{I}\right] + \sin \omega t \left[2p\omega - \frac{b}{I}\omega\right] = 0. \qquad 7\text{-}47$$

Since Eq. 7-47 must be true for all t (if Eq. 7-45 is a solution), the bracketed terms must be identically zero. The value of p can be obtained from the bracketed part of the sine term.

Therefore,

$$2p\omega = \frac{b}{I}\omega,$$

and

$$p = \frac{b}{2I} \qquad 7\text{-}48$$

When p is substituted into the bracketed part of the cosine term, ω is found. Hence,

$$\frac{b^2}{4I^2} - \omega^2 - \frac{b^2}{2I^2} + \frac{n}{I} = 0$$

$$\omega^2 = \frac{n}{I} - \frac{b^2}{4I^2},$$

or $$\omega = \sqrt{\frac{n}{I} - \frac{b^2}{4I^2}} . \qquad 7\text{-}49$$ [ω is real because of the magnitudes of n, I and K]

Eq. 7-45 with p substituted from Eq. 7-48 and ω from Eq. 7-49 is a solution to our particular un-driven system ($C = \theta_0$).

The ratio K of two successive same-side amplitudes in our experiment was nearly constant and greater than one (K > 1). Therefore, by using Eq. 7-45, an expression for b, the constant of proportionality for natural or aided electromagnetic damping is obtained.

Since

$$\theta \text{ (at } t = 0) = Ce^{-p0} \cos 0 = C,$$

then

$$\theta = \text{(at } t = T\text{, the period of 1 osc.)} = Ce^{-pt} \cos \frac{2\pi}{T} T = Ce^{-pt},$$

and the ratio

$$\frac{\theta_{t} = 0}{\theta_{t} = T} = \frac{C}{Ce^{-pt}} = K \qquad [K \text{ must be} > 1].$$

Therefore,

$$e^{+pT} = K, \qquad \text{(See Eq. 3-6 - Chapter 2.)}$$

$$pT = \ell n\, K,$$

and

$$p = \frac{\ell n\, K}{T} \qquad 7\text{-}50$$

If we equate Eq. 7-48 equal to Eq. 7-50,

$$b = \frac{2I}{T} \ell n\ K \qquad 7\text{-}51$$

And when b is substituted into Eq. 7-49,

$$n = \left(\frac{4\pi^2 + \ell n^2 K}{T^2} \right) I, \qquad 7\text{-}52$$

where K = ratio of successive angular amplitudes, I = rotational inertia of the copper disc, T = period. Note - the constant of proportionality n is the same for the un-driven and driven systems. Why?

In summary, we have now obtained b, p, and n, expressed in terms of quantities that can be measured. These parameters (b,p,n) can also be used in the solution to the equation of motion for the forced or driven oscillator (Eq. 7-40). Why?

From the theory of differential equations, it is known that the equation of motion for the forced or driven oscillator (Eq. 7-40) has two terms in the general solution. A transient term analagous to the un-driven or free damped oscillatory term (e^{-pt}) and a steady state term characterized by the frequency μ of the driver. The transient term will "die out" if we wait for equilibrium to be obtained because of the term (e^{-pt}). This actually means that all of our measurements are to be made after such a length of time that $e^{-pt} = e^{-(\ell n\, K/T)t} \cong 0$. We then can attempt to find a steady state solution to Eq. 7-40 for our driven system.

Since all of our measurements begin when θ has reached its maximum value (position of maximum angular amplitude) for any particular driver frequency μ, and since the driver motion is to be simple harmonic, a possible steady state solution is,

$$\theta = \theta_o \cos \mu t.^* \tag{7-53}$$

If Eq. 7-40 is expressed as

$$\ddot{\theta} + \frac{b}{I}\dot{\theta} + \frac{n}{I}\theta = \frac{n}{I}\Phi , \tag{7-54}$$

we can proceed to verify our assumption that Eq. 7-53 is a possible solution by taking $\dot{\theta}$ and $\ddot{\theta}$ and substituting into Eq. 7-54.

Therefore,

$$\dot{\theta} = -\mu\theta_o \sin \mu t,$$

and

$$\ddot{\theta} = -\mu^2\theta_o \cos \mu t.$$

Then

$$-\mu^2\theta_o \cos \mu t - \frac{b}{I}\mu\theta_o \sin \mu t + \frac{n}{I}\theta_o \cos \mu t = \frac{n}{I}\Phi ,$$

or

$$\theta_o\left[\left(-\mu^2 + \frac{n}{I}\right)\cos \mu t - \frac{b}{I}\mu \sin \mu t\right] = \frac{n}{I}\Phi,$$

and when multiplied by $\frac{I}{n}$,

$$\theta_o\left[\frac{I}{n}\left(\frac{n}{I} - \mu^2\right)\cos \mu t - \frac{b}{n}\mu \sin \mu t\right] = \Phi . \tag{7-55}$$

If

$$\frac{I}{n}\left(\frac{n}{I} - \mu^2\right) = G \cos \beta , \tag{7-56}$$

and

$$\frac{b}{n}\mu = G \sin \beta , \tag{7-57}$$

where Eq. 7-56 and Eq. 7-57 are components of G, then

---

*A cosine function is used because of the initial condition that $\theta = \theta_{max.}$ when t = 0. We have avoided the introduction of boundary conditions, etc., at this time because of the limited mathematical background of the student. Other oscillating functions such as cos (μt + β), sin μt + cos μt, etc., could be used.

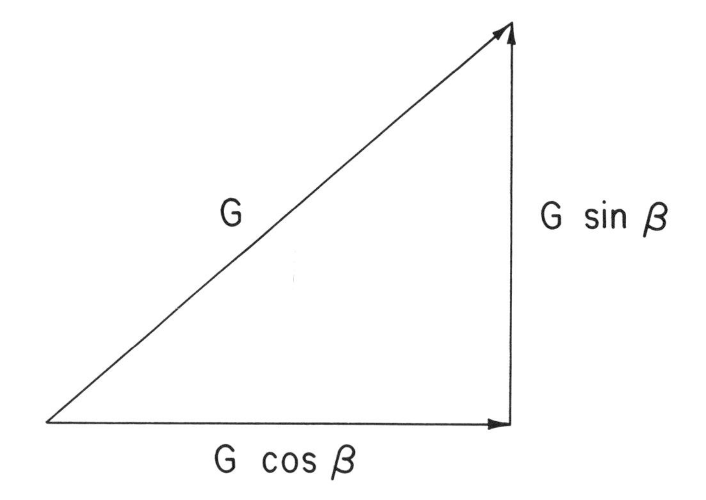

and

$$G = \sqrt{\left(1 - \frac{I}{n}\mu^2\right)^2 + \left(\frac{b}{n}\mu\right)^2}\,. \qquad 7\text{-}58$$

If Eq. 7-56 and 7-57 are substituted into Eq. 7-58, then

$$G\theta_o[\cos\beta\ \cos\mu t - \sin\beta\ \sin\mu t] = \Phi\ , \qquad 7\text{-}59$$

A trigonometric identity is substituted for the bracketed part of Eq. 7-59 and we obtain,

$$\Phi = \theta_o\ G\cos\ (\mu t + \beta), \qquad 7\text{-}60$$

or

$$\Phi = \Phi_o\ \cos\ (\mu t + \beta).$$

Physically G is a transfer ratio (Φ_o/θ_o), because Φ is equal to Φ_o times an oscillating function. Φ_o is the maximum angular amplitude reached by the top of the coiled spring as it is displaced by the driver arm V. θ_o is the maximum angular amplitude reached by the resonator or copper disc. β is the phase angle between the driver and oscillating disc amplitudes. In the experiment you observed for very small frequencies that the pointer on the copper disc and the pointer to which the coiled spring was attached both moved in the same direction and their difference in phase β was zero. With an increase in μ, the exciter angular frequency, the maximum angular amplitude of the driver "leads" the maximum angular amplitude of the copper disc or resonator.

We will now discuss the significance of G, the transfer ratio, and analyze the system as the ratio of the angular frequency of the driver to the natural frequency of the copper disc approaches unity, $\mu/\omega \longrightarrow 1$, the condition for resonance.

The product of Eq. 7-52,

$$\frac{I}{n} = \frac{T^2}{4\pi^2 + (\ln K)^2}, \qquad 7\text{-}61$$

and Eq. 7-51

$$\frac{b}{I} = \frac{2\ \ln K}{T}\ ,$$

is

$$\frac{b}{n} = \frac{2T \ln K}{4\pi^2 + (\ln K)^2} \qquad 7\text{-}62$$

If we substitute I/n and b/n into Eq 7-58 and let $T = 2\pi/\omega$, then

$$G^2 = \left(1 - \frac{4\pi^2 (\mu/\omega)^2}{4\pi^2 + (\ln K)^2}\right)^2 + \left(\frac{4\pi \ln K\ (\mu/\omega)}{4\pi^2 + (\ln K)^2}\right)^2 . \qquad 7\text{-}63$$

Since K is very small when compared to $4\pi^2$, $G^2 \rightarrow (\ln K/\pi)^2$ as $\mu/\omega \rightarrow 1$, the condition for resonance. A steady state solution for our driven system or forced oscillator at resonance is obtained by substituting into Eq. 7-53, $\theta_o = \Phi_o/G$ or $\theta_o \rightarrow \pm\Phi_o/\ln K$ as $\mu/\omega \rightarrow 1$. Therefore,

$$\theta = \frac{\Phi_o}{G} \cos \mu t, \quad \text{which at resonance becomes} \qquad 7\text{-}64$$

$$\theta = \pm \frac{\pi \Phi_o}{\ln K} \cos \mu t.$$

When $\mu/\omega \longrightarrow 1$, θ_o becomes very large because the $\ln K$ is quite small (in the experiment $\ln K_1 \cong \ln 1.07 \cong .07$).

In Eq. 7-64 as $K \longrightarrow 1$, the $\ln K \rightarrow 0$ and $\theta_o \rightarrow \infty$ as $\mu \rightarrow \omega$. In Eq. 7-63, when $\mu << \omega$, then $G \geqq 1$ and $\theta_o \longrightarrow \Phi_o$ as $\mu/\omega \rightarrow 0$. If $\mu >> \omega$, then G becomes very large and $\theta_o \rightarrow 0$. Our resonance curve is, therefore, somewhat lopsided (see Fig. 7-40).

Eq. 7-63 shows that we do not have "clear" resonance. Although the $(\ln K)^2$ is small, it is still present. As $\mu \rightarrow \omega$, there is a value for $\mu \lesseqgtr \omega$ such that the first squared bracket of G^2 becomes zero and G attains its smallest value. Therefore, the resonance peak broadens and decreases in height as K increases (see Fig. 7-40).

We will now investigate the behavior of the phase angle β as $\mu \rightarrow \omega$. From Equations 7-56 and 7-57, we obtain

$$\tan \beta = \frac{\frac{b}{n}\mu}{1 - \frac{I}{n}\mu^2} \qquad 7\text{-}65$$

When I/n from Eq. 7-61 and b/n from Eq. 7-62 are substituted into the tan β ($T = 2\pi/\omega$),

$$\tan \beta = \frac{4\pi \ln K\ (\mu/\omega)}{4\pi^2 + (\ln K)^2} \quad \frac{1}{1 - \dfrac{4\pi^2 (\mu/\omega)^2}{4\pi^2 + (\ln K)^2}} = \frac{4\pi (\ln K)(\mu/\omega)}{4\pi^2 (1 - \mu^2/\omega^2) + (\ln K)^2} \qquad 7\text{-}66$$

If K is very small (K for natural damping ratio), then $(\ln K)^2$ can be omitted. If, in addition, $\mu << \omega$, then

$$\tan \beta \cong \frac{\ell n\ K}{\pi}\left(\frac{\mu}{\omega}\right) \qquad 7\text{-}67$$

We would expect the $\tan \beta \cong 0$ when $\mu \ll \omega$ for all values of K. However, as K becomes greater, the $\tan \beta$, and hence β, approaches zero more slowly. (See Fig. 7-41.) In all cases for $\mu = 0$, $\tan \beta = 0$ and therefore $\beta = 0$. As $(\mu/\omega) \longrightarrow 1$ and with $(\ell n\ K)^2 \cong 0$, $(1 - \mu^2/\omega^2)$ in Eq. 7-66 also approaches zero. Therefore, $\tan \beta \longrightarrow \infty$, and $\beta = \pi/2$. Careful examination shows that as K increases, μ/ω must be a value greater than one for β to be 90°. A plot of β versus μ/ω actually would not have all the graphs for different values of K crossing $\beta = 90°$ at $\mu/\omega = 1$. Because the variation is quite small, experimental errors may conceal any difference so that all graphs for different values of K apparently cross $\beta = 90°$ at $\mu/\omega = 1$.

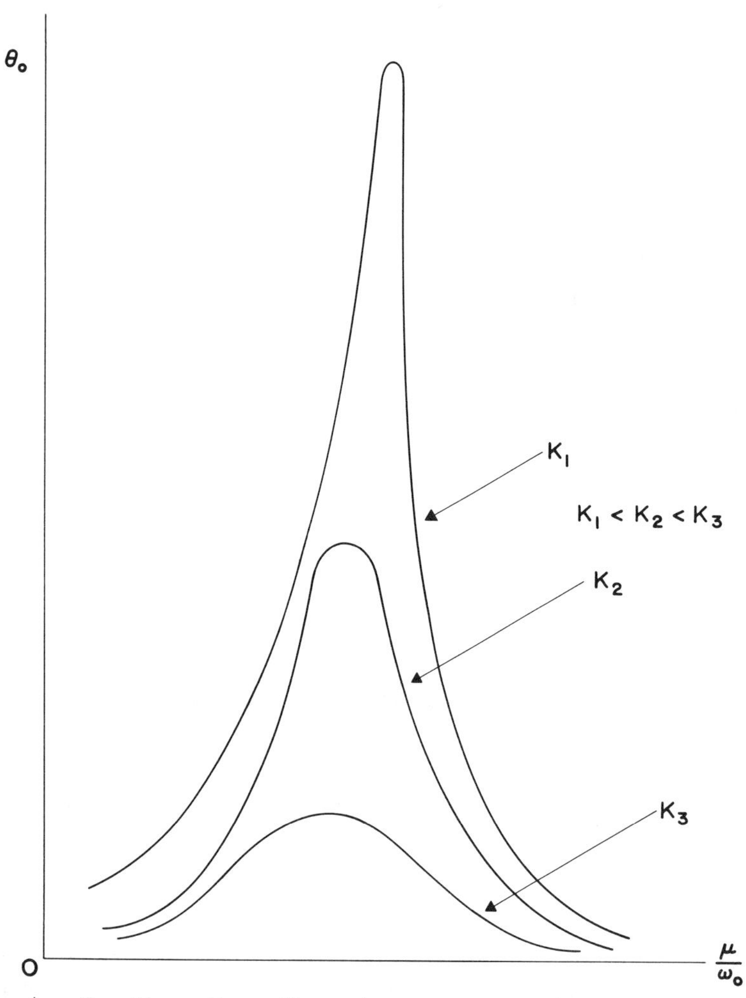

θ versus μ/ω_o for $K_1 < K_2 < K_3$. At resonance the angular amplitude of the copper disc (resonator) becomes very large. The peak of the plot increases and the width decreases as K approaches the value of the natural damping factor. Top curve-natural damping only.

Fig. 7-40

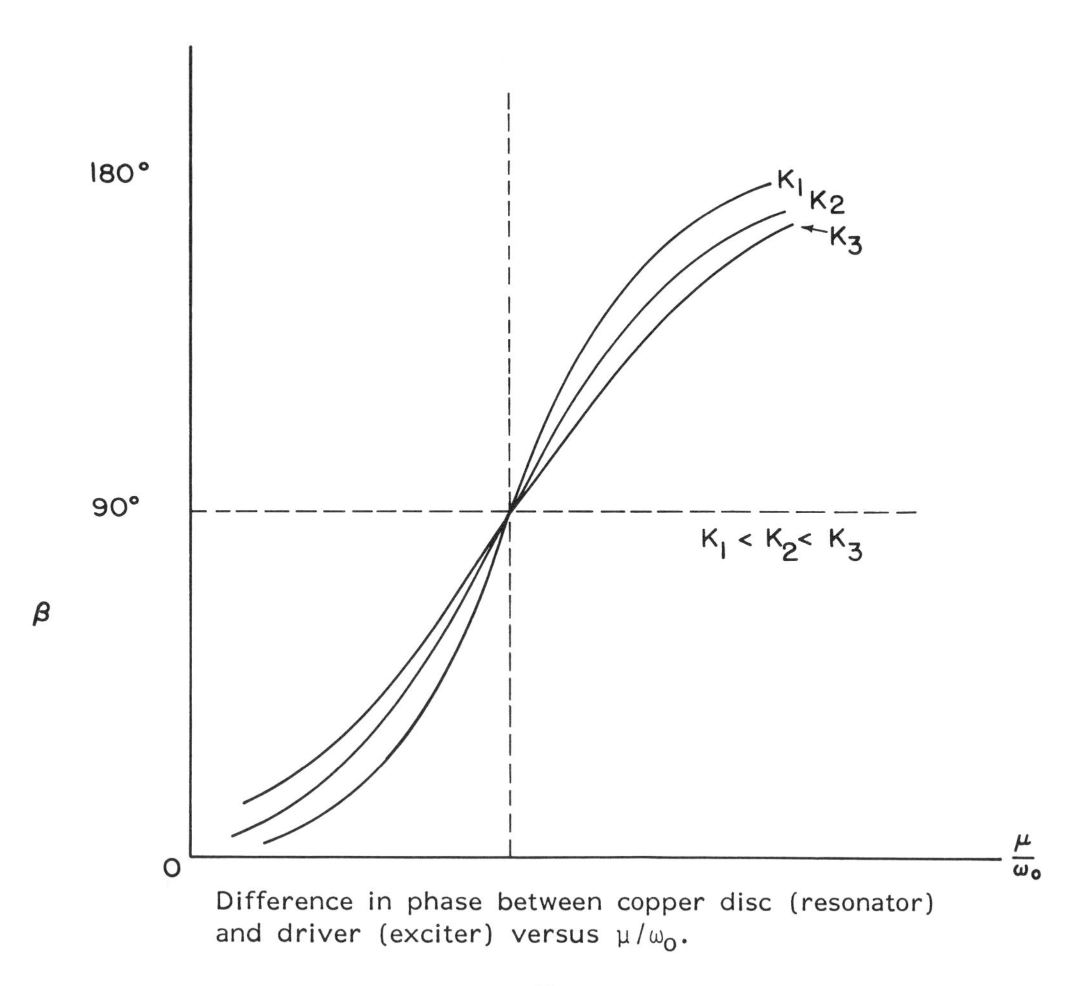

Difference in phase between copper disc (resonator) and driver (exciter) versus μ/ω_0.

Fig. 7-41

When $\mu >> \omega$, the tan β also approaches zero through negative values because $(1 - \mu^2/\omega^2)$ becomes more negative. Therefore, $\beta \longrightarrow \pi$ because it reverses sign immediately after resonance. As K becomes larger with $\mu >> \omega$, β approaches π more slowly.

In Equations 7-45 to 7-67 the angular frequency of the copper disc (pendulum or resonator) is ω. This allows a general approach to be followed in the development of a solution to our driven or forced oscillator. The system analyzed is real and therefore has natural damping. The natural damping (K_1) or the higher damped conditions (K_2, K_3, etc.) can be used. In the Procedure and Figures 7-40 and 7-41, ω_0 refers to the natural (or proper) angular frequency of the un-driven copper disc (given an initial positive angular displacement θ_0 and released from rest). The natural angular frequency ω_0 is obtained for the system without external electromagnetic damping but includes natural damping. Therefore, at resonance, the angular amplitude can not become infinite (Fig. 7-40).

In summary, at resonance, the difference of phase between the copper disc (resonator) and the driver (exciter) is 90° (see Fig. 7-41). The end of the spring is moving to the right through its equilibrium position at the instant the copper disc is turning back at its position of maximum amplitude toward the left. As the angular frequency of the driver (exciter) increases, the phase difference increases to 180°. The pointer on the copper disc and the pointer at the end of the coiled spring then pass through their position of equilibrium at the same instant but in opposite directions. At resonance, energy is continually being supplied to the copper disc throughout its entire motion because the accelerating couple is leading by 90°. At this point the amplitude would increase without limit if it were not for the energy losses caused by damping.

QUESTIONS

1. Derive an expression for the fractional decrease in energy per cycle for the successive same-side decrease in angular amplitude obtained in Step 3 of the Procedure for K_1 (natural damping). Hint - when the angular displacement is a maximum, the energy is all potential.
2. From Question 1, obtain the "Q" of the system.
3. The rate at which oscillations are damped out is often called the <u>logarithmic decrement</u> δ, which is defined as the natural logarithm of the ratio of the amplitudes of two consecutive oscillations. What is the logarithmic decrement δ for the condition of natural damping? For K_2? For K_3?
4. When does the maximum energy input to the resonator take place?
5. At some frequencies the energy input from the driver greatly exceeds the energy obtained by the disc. Where does the excess energy go? Can the energy acquired by the disc ever exceed that of the driver input? Why?
6. From your values of K and T obtain a reasonable time to wait for steady state conditions. Show how you arrive at the estimate.
7. What is meant by "the driver motion leads (lags) the disc motion?"
8. Does the driver motion lead or lag the disc at resonance? Explain.
9. At what value of β does resonance occur? If $K > 0$, does resonance occur at the same, a larger, or smaller value of β?
10. If the radius of the copper disc is 10.0 cm and the mass = 200 gm, find the values of b, n, p, and ω for the condition of natural damping K_1. Hint - review Equations 7-48 to 7-52.
11. What is the meaning of G, the transfer ratio? What happens to G if the exciter angular frequency μ is much greater than the natural angular frequency ω_o? How does G behave if $\mu \ll \omega_o$?
12. Under what conditions does $\beta \longrightarrow \pi$?
13. What are your values for G, the transfer ratio, at resonance for the damping conditions K_1, K_2 and K_3?
14. Use the data given in Question 10 and find a solution to the equation of motion for the un-driven system.
15. What are your steady state solutions at resonance to the equation of motion of your forced or driven oscillator (copper disc) for all damping conditions (K_1, K_2 and K_3)?

EQUIPMENT: Distributor - J. Klinger, 82-87 160 St., Jamaica 32, N.Y.
Caution: For optimum results use well regulated D.C. Do not run more than one unit of apparatus from each outlet. Apparatus highly sensitive to line fluctuations.

Experiment 7-21 ANALYSIS OF GRAVITATION

INTRODUCTION

The fact that every object is attracted to the earth is known by everyone. Few people are aware that bodies mutually attract each other because the attractive forces are so small that they can not be observed without very delicate instruments. The torsion balance was first constructed by Coulomb in 1784 for use in electrostatic measurements. Cavendish modified the apparatus in 1798 and used it for the first accurate measurement of the gravitational constant G.

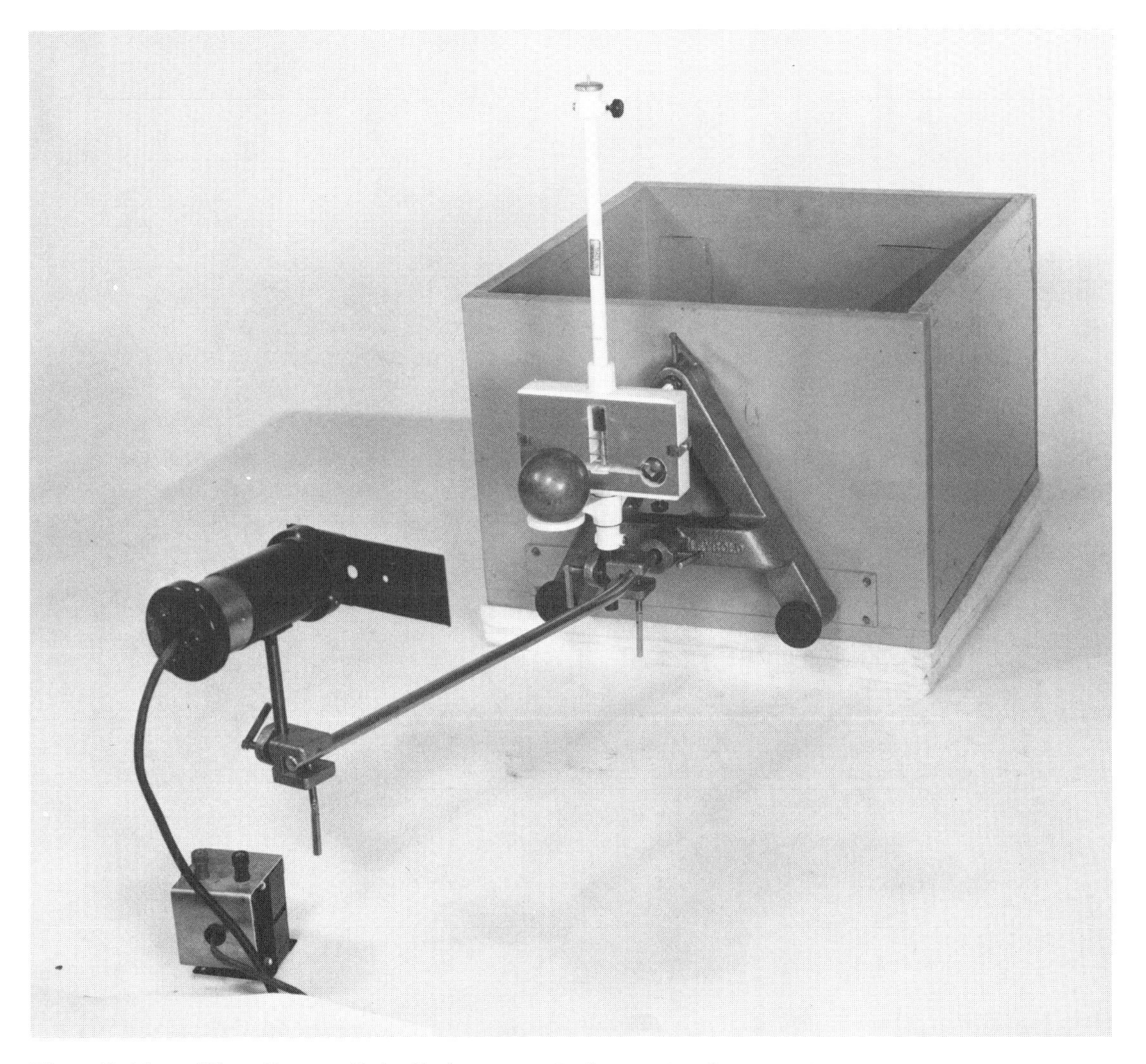

Fig. 7-42. The Cavendish Balance. Balance is fastened to a heavy stand which is attached at its vertex by a bolt to a sand-filled wood box. In order to minimize the transmission of room or building vibration, the box is supported by thick layers of sponge rubber. Although not shown in the photograph, the complete unit of balance, box, and layers of sponge rubber is held by a steel support countersunk into the outer brick wall of the building. L = light source. B = large lead ball on rotary support. Other large lead ball is on the same rotary support behind the metal-glass casing. M = Metal glass casing. V = Adjustment screws for stand. The balance can be obtained from Leybold-Heraeus EMBH Co., KG, 5000, Köln, Germany.

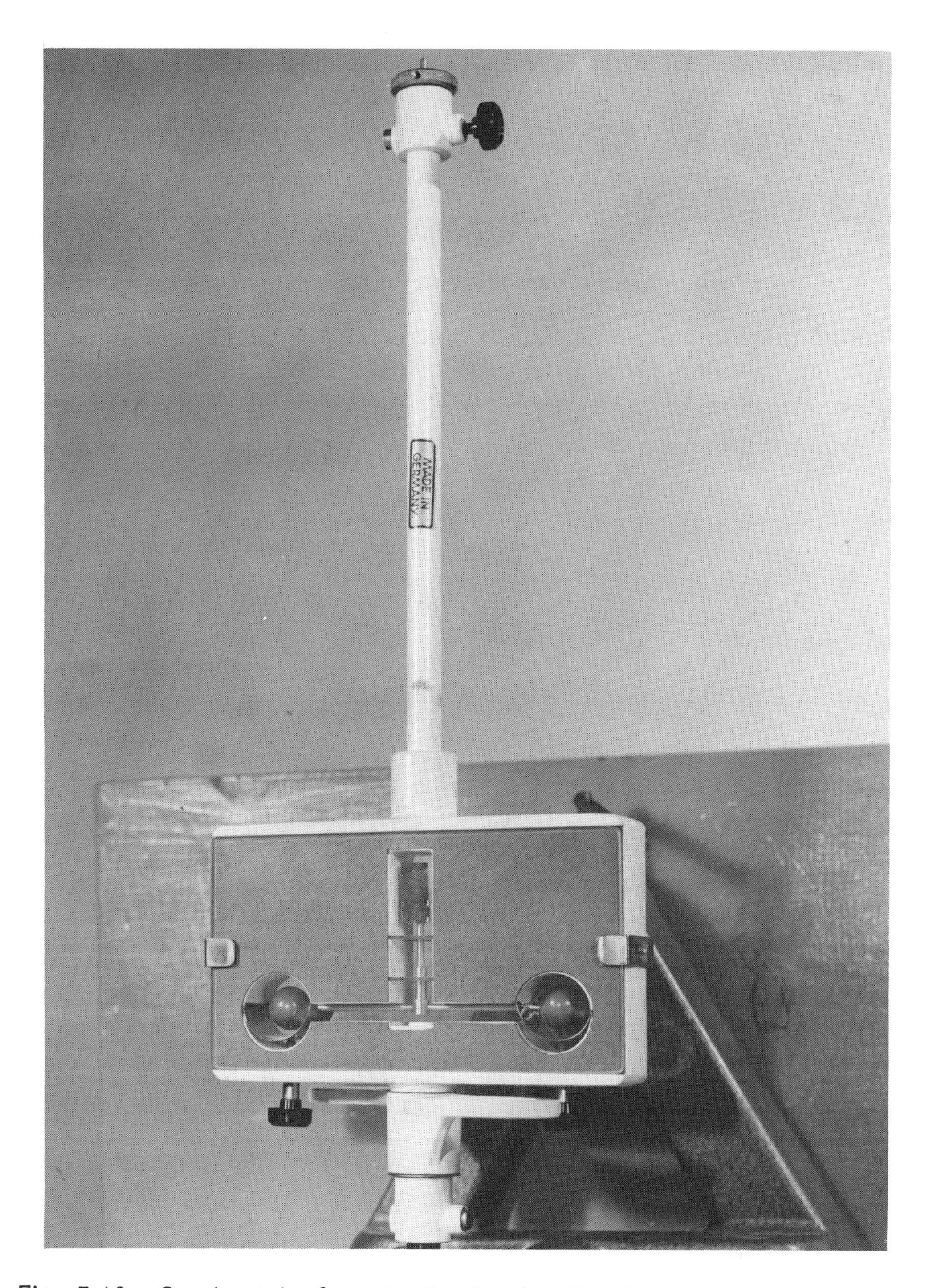

Fig. 7-43. Carrier tube for a torsion band and casing containing suspension with two small lead balls. Large lead balls removed to show inner detail. C = Carrier tube. (1) = Lock screw for torsion band. (2) = Adjustment screw for tension in torsion band. W = Screws to lock perpendicular suspension with two small balls. R = Rotary support for large balls. O = Mirror. P = Perpendicular suspension. A = Adjustment ring for rotary support.

The apparatus in Fig. 7-42 consists of a carrier tube (Fig. 7-43-C) for a torsion band which is made of bronze with a cross-section of 0.01 mm x 0.15 mm. The band is connected to a small concave mirror O of 30 cm focal length. The mirror is attached to a perpendicular suspension P which supports two 15-gm balls each 5 cm from the axis of rotation. The mirror and the suspension are housed in a rectangular metal casing which is covered by glass plates so that the movable system can be seen. The space for the suspension within the casing has been kept as small as possible to avoid air currents and is subdivided by plastic panes.

The perpendicular suspension is locked by two knurled-head screws W beneath the metal-glass casing. Beneath the casing is also a rotary holder for two approximately 1.5-k masses (Are they 1.5 k?). The rotary support rests upon a ring A which permits centering of the small and large balls. They must be on the same level!

On the side of the carrier tube near the top is a knurled screw (1) which has to be untightened before the top screw (2) is turned. The top screw when turned slightly changes the torque in the torsion band and, therefore, the position of the perpendicular suspension supporting the two small lead balls in equilibrium.

A lamp with condenser (convex lens) and image slide (Fig. 7-42-L) is attached to a support rod beneath the casing. The slide has a thin wire in the center of a large diaphragm. An image of the thin wire is reflected to a wall (over 5 meters away) by the concave mirror. The lamp housing is about 30 cm away from the glass front of the casing. The housing is adjusted until the wire image is sharply focussed on the opposite wall.

The two adjustment screws (Fig. 7-42-V) beneath the heavy stand rest on a heavy metal plate and are used to make certain that the apparatus is exactly vertical.

Important objectives of this experiment are:

(a) Show that bodies mutually attract each other.
(b) Determine the gravitational constant G.
(c) Develop an appreciation of the difficulties encountered in experimentation when delicate instruments are used.

PROCEDURE (Record all data in a neat table. Read ANALYSIS to find out which project to select.)

1. Adjust the position of the light source until the image of the wire in the large diaphragm is sharply focussed on the opposite wall. Are the centers of the small and large lead balls at the same level? Place the rotary support perpendicular to the glass of the casing.

2. Use a level and adjust the carrier tube position until it is as near vertical as possible by use of the two adjustment screws at the lower end of the heavy stand. Next, release very slowly and uniformly the lock screws of the perpendicular suspension. If the torsion band holding the suspension should rub the two horizontal plastic panes, then center the band by again slowly turning the proper screw at the lower end of the heavy stand.

3. The instant the suspension is released, the system will oscillate. Use extreme care in release so as to minimize the amount of motional energy (rotational and translational kinetic energy) that will cause the small balls to strike against the

inner glass walls. Since lead has diamagnetic properties, a quick dampening of the swinging system can be obtained with a small magnet which repels lead. Place the magnet on the glass for a short time until the system moves back. The balance can be brought to rest by using the magnet alternately at the front and rear glass plates.

4. While the system is oscillating, the image of the wire on the opposite wall will move backward and forward between two extreme values. When the system is at rest the image of the wire should be at the center of the two extreme reversal points. If the rest position deviates greatly from the center of the two extreme reversal points, the knurled screw on the top of the carrier tube should be slowly loosened and the suspension turned through a small angle toward the desired zero point. Don't forget to tighten the lock screw (1) after the suspension's position has been changed.

[Note - Steps 1 to 4 are required adjustments before data can be taken. While adjustments are made and during the complete experiment, students must avoid disturbing the apparatus and must not walk heavily on the floors. Any unusual vibration will be transmitted to the balance.]

5. Move the rotary holder from its position perpendicular to the glass of the casing to a point where both large lead balls just touch the glass. (See Fig. 7-44.) The two large balls M attract the two small ones m within the casing and the system will begin to oscillate. Equilibrium will take place in one to two hours between the gravitational force of attraction and the force due to the twisting of the torsion band. Record the equilibrium position of the wire image on the opposite wall.

6. ALTERNATE I. ACCELERATION METHOD. Swing the large balls on the rotary holder so that they change their position as shown in Fig. 7-44. BE VERY CAREFUL in the reversal process. Balls should be brought to rest very gently so as to avoid transmitting energy to the balance. If the suspension begins to swing violently, you must start at Step 5 and begin again. After reversal record at 15-second intervals until 90 seconds are reached, the position S of the wire image on the opposite wall. Record the data. The ILE of the metric scale, because of the properties of the optical system, is ± 0.5%. What is the ILE of the time clock?

7. ALTERNATE II. DEFLECTION METHOD. (Most accurate but requires much more time!) Proceed as in Step 6, but record the position S at 15-second intervals up to 2 minutes, and then at intervals of one minute until the system comes to rest. About one hour is required to take data. Record the data.

[Note - After the data for either Alternate I or Alternate II has been obtained, use screws W and lock the perpendicular suspension. Check the side adjustment screw (1) near the top of the carrier tube and make certain it is tight. DO NOT TOUCH TOP ADJUSTMENT SCREW (2).]

8. Use a vernier caliper (look up the ILE) and measure the thickness of the casing. Record one-half the casing thickness.

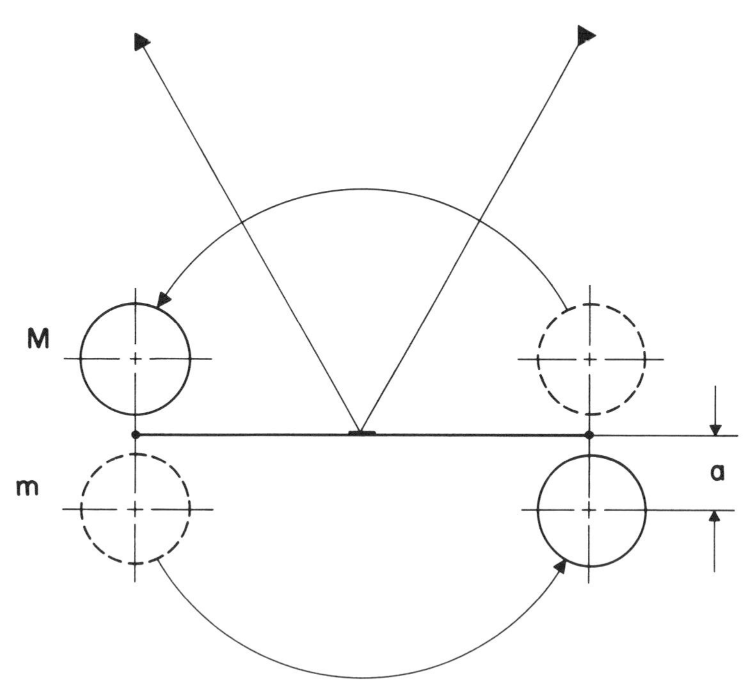

Equilibrium position of large M and small m balls.
a = Distance between the centers of mass.

Fig. 7-44

9. Measure the diameter of a large ball M. Record the radius and the ILE in your data table. Find (a) which is equal to one-half the thickness of the casing plus the radius of the large ball M.

10. The mass of the small balls m = 15.0 ± 0.1 gm. Find the mass M of the large balls. What is the ILE of the scale?

11. Use a tape measure (ILE = 0.5%) and find the distance from the surface of the glass to the opposite wall. Add to this measurement one-half the thickness of the casing and obtain L (Fig. 7-45). Is the ILE for one-half the thickness of the casing (Step 8) significant when compared to the ILE of the tape measure?

ANALYSIS (The student is given the choice of several projects. Read and decide which project you would like to work on. If you select the DEFLECTION METHOD, you will have to accommodate your instructor and make arrangements to put in additional time outside your laboratory period. Naturally, extra credit will be given.)

Project I. ACCELERATION METHOD (No error analysis)

In Alternates I and II of the procedure, the torsion band remains for a moment twisted in a direction opposite to the direction of reversal because the suspension (small lead balls m plus mirror) has a long period of oscillation (about ten minutes). This long period is due to the large rotational inertia and small angle through which the system rotates. When the large balls are in their reversed positions and touching the glass, the mutual attraction between both sets of balls provides the accelerating force. The force due to the twisted band acts in the same direction and is of the same magnitude as the gravitational force at the start of the motion. Each small ball is, therefore, accelerated by force:

$$F = \frac{2GmM}{a^2} \quad \text{(Can you explain the factor of two?)} \tag{7-68}$$

As the small balls are accelerated, the tension in the torsion band will be at the same time relieved and then twisted in the opposite direction. The system will oscillate as a result. If we neglect the variation of the twist of the torsion band, the small balls are accelerated almost uniformly for a time which is about 1/10 of the period of oscillation (with an accuracy of about 5%.

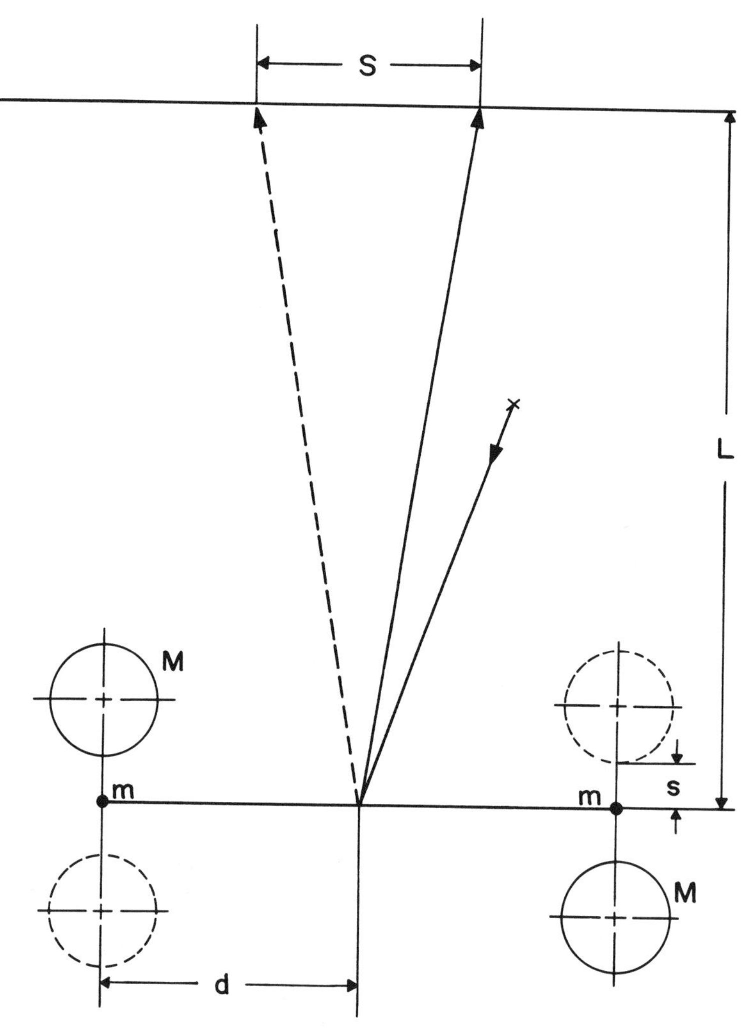

Reflection of light by convex mirror. s = Path of small balls within casing. d = Distance of small balls from the axis of rotation (d = 5.00 ± .02 cm). S = Path of wire image on wall. L = Distance of wall from mirror.

Fig. 7-45

1. Inspect Fig. 7-45, and derive (in terms of L, S and d) an expression for s, the path of the small balls within the casing. Careful - what happens to the optical path if the angle of the incident light is doubled by reflection? Construct diagram and show the angles, etc.

2. Find for each value of S, the corresponding value of s. USE THE DATA OF ALTERNATE I in the PROCEDURE.

3. Plot s, the path of the small balls within the casing, versus time (s vs. t).

4. Plot v, the velocity of the small balls within the casing, versus time (v or s' vs. t).

5. Plot (A or s" vs. t). Is the acceleration A constant as we have assumed over the time interval chosen?

6. From Eq. 7-68 and the average of A, the acceleration of the small balls within the casing, find G, the gravitational constant. In the conclusion of your report compare your value of G with that in your text and discuss any variation.

Project II. ACCELERATION METHOD WITH ERROR ANALYSIS

1. After reading Project I carefully do Steps 1 to 6.

2. Use the ILE of each measurement as the final LE and find the maximum LE in the indirect measurement G.

3. What is the maximum LE in G expressed as a relative % error?

4. We have not considered that the small spheres are also attracted by the more distant large spheres. The value of G is affected by this SYSTEMATIC ERROR. Since SYSTEMATIC ERRORS must be eliminated or minimized, G must be increased by a correction factor. Inspect Fig. 7-46 and derive an expression for F_0. Find k which reduces F, the force due to the mutual attraction between the large and small spheres. Knowing k and F, find the correction factor β.

5. Use β and obtain G, corrected for systematic error. Although we will not carry the error analysis further, it is important to point out that a maximum LE, etc., should be obtained for β because it depends on independently measured quantities. β not only will increase the magnitude of the numerical value of G, but will also cause the final LE of G to be increased. Remember LE and PI are the same!

Project III. DEFLECTION METHOD (No error analysis). Read discussion in Project I carefully before proceeding.

When the large balls are reversed, a turning moment τ acts on the system. The moment (or torque) is due to the gravitational forces exerted by the large balls on the small ones (see Eq. 7-68). The turning moment reaches equilibrium by twisting the torsion band through an angle of $\theta/2$, where θ (in radians) is measured from the wire image's zero position to the point of maximum deflection. How is $\theta/2$ obtained?

If the couple per unit angle of twist of the torsion band is K, the torque $\tau = K\frac{\theta}{2}$.

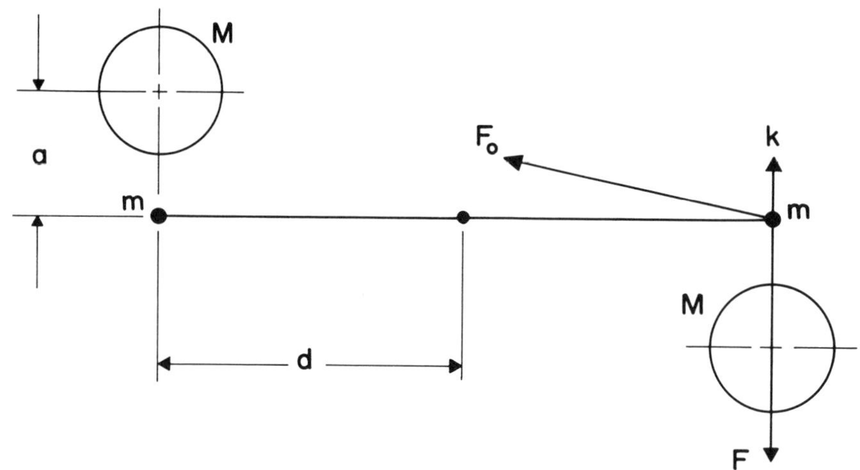

Force of Attraction between M and m.

Fig. 7-46

1. Inspect Fig. 7-45 and derive (in terms of S and L) an expression for θ. Careful - what happens to the optical path if the angle of the incident light is doubled by reflection?

2. Obtain an expression for the torsional constant K (in terms of I and T). If the rotational inertia of the suspension and the mirror is negligible, what is I for the two small balls?

3. From Eq. 7-68 and Fig. 7-45, obtain an additional expression for the torque.

4. Find an expression for G, the gravitational constant. Check your result with your instructor before proceeding.

5. Use data from Alternate II in the Procedure and plot S vs. t. Is the motion damped harmonic? Find an average value for the period T of the oscillations.

6. Find a numerical value for G, the gravitational constant. In the conclusion of your report compare your value of G with that in your text and discuss any variation.

Project IV. DEFLECTION METHOD WITH ERROR ANALYSIS

Review preceding projects and use your own judgement.

QUESTIONS

1. Why is the value of G, the gravitational constant, increased when the SYSTEMATIC ERROR correction is made?

2. If the small lead balls were replaced with more dense ones, would the measured value of G, the gravitational constant, be larger, smaller or the same? Why?

3. Draw a complete force diagram for a small ball at
   (a) Equilibrium.
   (b) After reversal of the large spheres.

4. If the casing containing the suspension was not quite parallel to the wall, would the value of G, the gravitational constant, be affected significantly. Explain by use of a diagram showing the optical system.

5. Of what significance is the horizontal component of the force F_0 in Fig. 7-46?

EQUIPMENT: Distributor - J. L. Klinger, 82-87 160 St., Jamaica 32, N. Y.
Caution: For best results support unit so as to minimize effect of structure vibrations.

Chapter 8

LOW-FRICTION DEVICES

8.1 Introduction

In many of the experiments in mechanics, frictional forces must be taken into account. To demonstrate some physical phenomena, it is desirable to try to reduce friction as much as possible. In this chapter a number of commercially available pieces of equipment are described which may be used to perform experiments not practical without a considerable decrease in friction. This is accomplished by introducing a thin layer of gas between the two surfaces involved, usually air or carbon dioxide.

The low-friction devices may be used for qualitative observations as well as quantitative data. The apparatus is simple in design and use and the principles involved are easily understood. Therefore, this equipment does lend itself to the performance of experiments conceived by students. We want to encourage "free" experiments without directions as to procedure and analysis. Here is an opportunity for you to be more innovative!

Four different low-friction devices are discussed:

(1) The linear air track - no experiments are described in this section.

(2) Low-friction pucks using dry ice or compressed air - in this section details for eight different experiments (8-1 through 8-8) are given.

(3) The air table - here the air is supplied through the table surface and not through the pucks. Again, no specific experiments are described, but Experiments 8-1 through 8-8 can easily be adapted.

(4) The air bearing rotational apparatus - one typical experiment is described, but many others are possible.

8.2 The Linear Air Track[1]

A linear air track with its gliders as shown in Fig. 8-1 is an almost friction free system. It can be used for many experiments such as those involving velocity, acceleration, momentum, oscillatory motion, etc. Air from a blower[2] - an industrial vacuum cleaner - is forced through small holes in a track consisting of a hard aluminum - alloy extrusion. Gliders are supported by the air jets on each side of the vertex. Tracks are usually equipped with leveling screws, scales, spring bumpers at the ends, etc.

Fig. 8-1. Linear Air Track.
(Courtesy of the Pasco Scientific Corporation)

[1]Linear air tracks with instructions may be obtained from a number of manufacturers, including Pasco Scientific Corporation, Daedalon Corporation, and the Sargent-Welch Scientific Company, among others. Each manufacturer publishes a manual describing the track and its accessories and suggests a number of experiments.

[2]Several air tracks can be supplied by one blower.

Many qualitative observations are possible with this almost frictionless system. Quantitative data may be obtained in four different ways:

(1) Spark recording - produces a continuous full scale record of position as a function of time.

(2) Strobe photography - produces a continuous reduced scale record of complex interactions.

(3) Photocell gates - the fastest way to measure velocities at one or two points.

(4) Stopwatch and counting - for studying periods of oscillations or other similar events.

Gliders of various sizes are available for all tracks. A number of accessories, such as the liquid accelerometer shown in Fig. 8-2, may also be used. Colored water in a narrow chamber clearly shows the acceleration and deceleration by a change in the water level. Small magnets may be attached to the gliders for interesting variations in experiments. The track may be slightly curved for a study of harmonic oscillations.

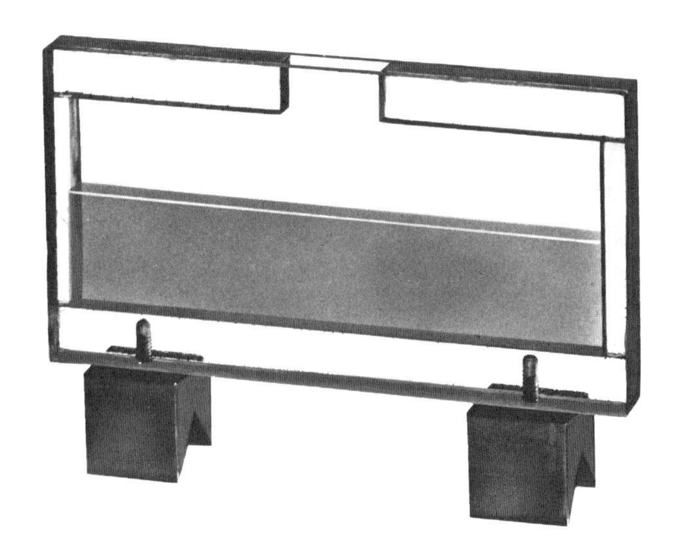

Fig. 8-2. Liquid Accelerometer for the Air Track. (Courtesy of The Sargent-Welch Scientific Company)

The linear air track offers an excellent opportunity to the students to devise his own experiments. A large number of concepts from mechanics can be studied, including kinematic relationships, Newton's Laws of Motion, elastic and inelastic collisions and simple harmonic motion.

8.3 Low-Friction Puck Experiments*

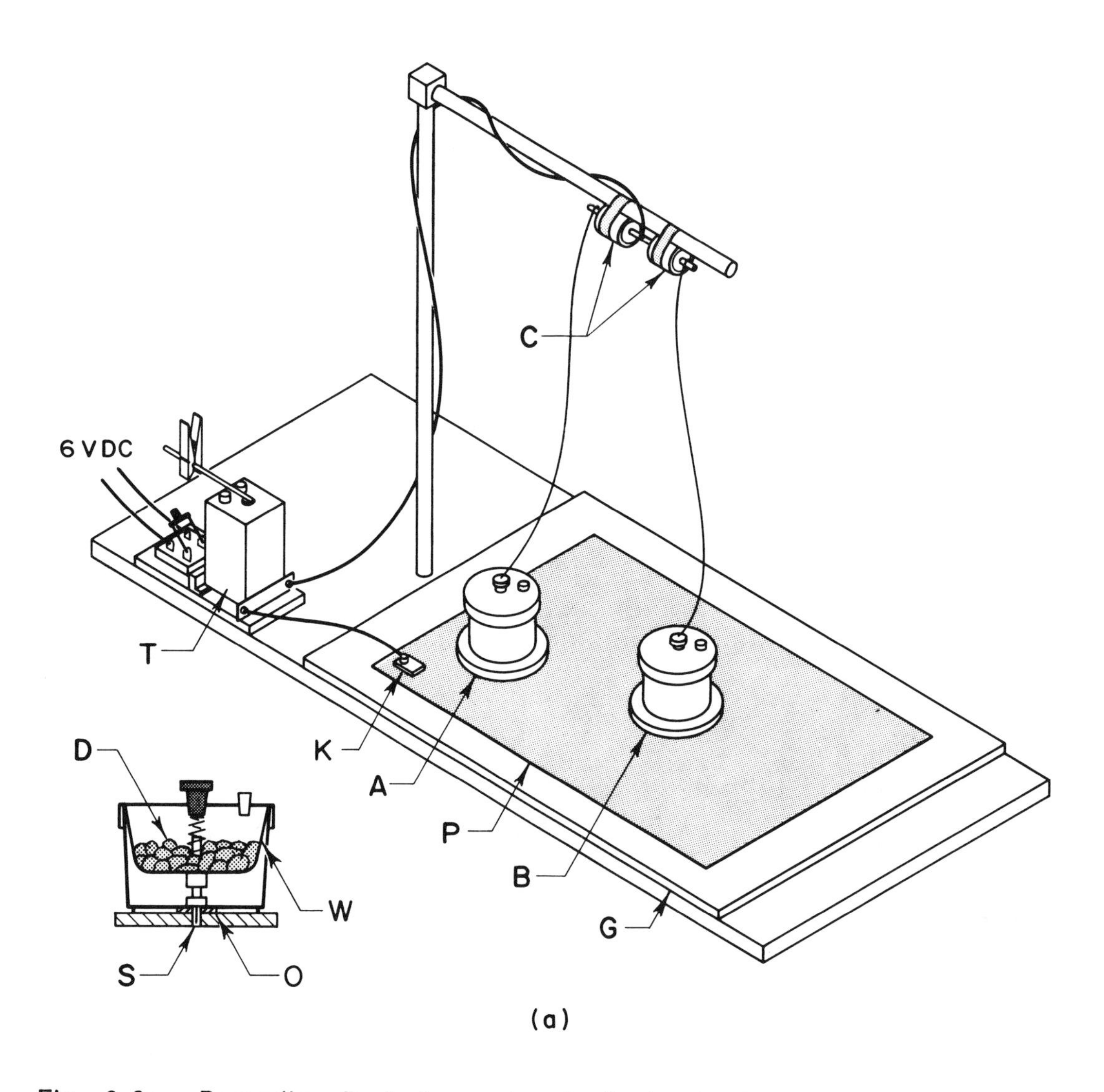

Fig. 8-3a. Recording Puck Apparatus.* Pucks A, B are each connected to a 500 μμf capacitor C. Dry ice D is supported by wire screen W. Gasket O isolates plastic container from base so that it will not become too cold and adhere to conducting paper P placed on carefully leveled glass plate G. Electrode S provides spark from variable spark timer T which is connected to capacitors and clip K.

---

*Developed by Professor Harry F. Meiners. The Recording Puck Apparatus using dry ice pucks or pucks fed by compressed air through surgical tubes was first sold by the Macalaster Scientific Corporation, then later by the Raytheon Corporation. See also Harry F. Meiners, Ed. Physics Demonstration Experiments, Vol. I (Robert E. Krieger Publishing Co., Malabar, FL 32950, 1985, p. 228).

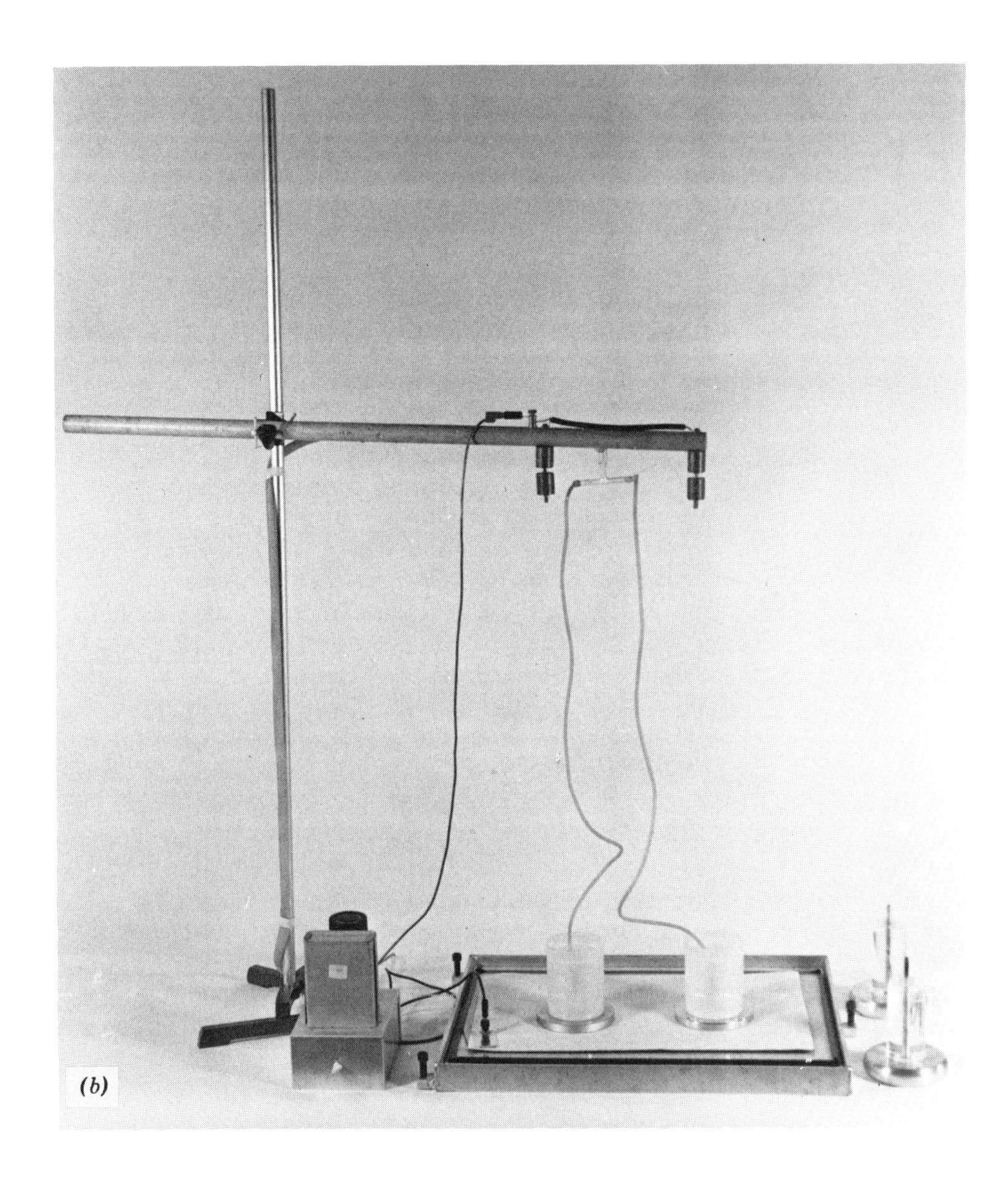

Fig. 8-3b. Recording Puck Apparatus. Compressed air is supplied by surgical tubing to the pucks. Wires pass through the surgical tubes and connect to electrodes in the center of the pucks.

APPARATUS

The recording pucks shown in Fig. 8-3a can be used to study the causes of motion and the way in which bodies influence each other's motion. They also can be used to investigate two-body, or multi-body collisions by applying the principles of the conservation of energy.

Pucks are supported on a nearly friction free film (μ_k < 0.0005) of gas generated by sublimating nuggets of dry ice. Compressed air from the building, a pressure tank, a fish tank aerator, or a pressure bulb can also be used. See Fig. 8-3b . The gas film acts as a cushion between the base of the puck and any very smooth surface, for example, a ¼-in. plate glass resting on jack screws. To insure that the base of the puck does not get too cold and thereby tend to stick to the paper, a wire mesh screen (Fig. 8-3a) is used to isolate the dry ice from the base.

The motion of the pucks is recorded on electrical recording paper by a high voltage, pulse-timed spark which jumps to the paper from an electrode in the center of the gas opening in the puck. Only one spark timer and one induction coil are used for the timing system.* As shown in Fig. 8-4, two pucks are connected in parallel across the terminals of the secondary of the spark coil. A 20 kv, 500 pf capacitor is placed in series with each puck. If a hotter spark is required, a 2 µf 600-volt capacitor can be connected across the vibrator points at the terminals on top of the coil. Also, if desired, another 20 kv, 500 pf capacitor can be added in series to the other to obtain a stronger spark. The resistance R represents normal leakage; no resistors are actually used, and

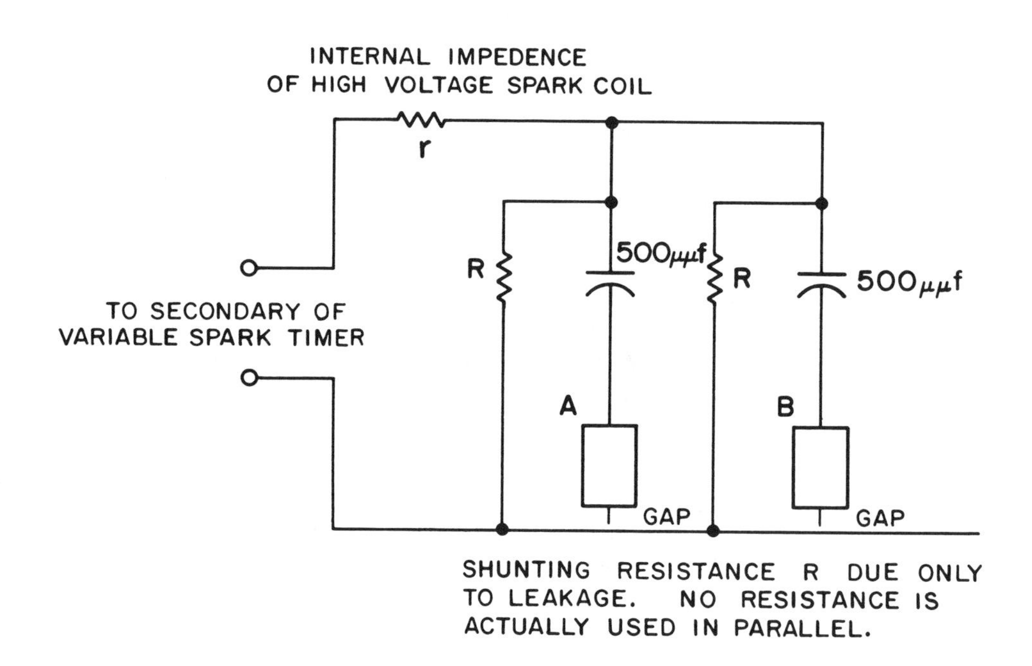

Fig. 8-4. Wiring Diagram.

*The Variable Spark Timer, Cat. No. 4500, was initially sold by the Macalaster Scientific Company.

r is the internal resistance of the high voltage spark coil. Since the time constant of each of the puck circuits is very short compared with the duration of the high-voltage pulse, which is essentially unidirectional, the spark gap of each puck will fire on the same pulse, almost simultaneously. An incoming voltage pulse builds up the voltage on the capacitors until one of them discharges across the spark gap, which reduces the voltage across the corresponding gap until that capacitor ceases to fire. After firing, a given capacitor re-charges rapidly bringing the voltage to a maximum so that the second capacitor can fire. Because of the short time constants, both firings occur during the same voltage pulse from the spark coil. When the high-voltage pulse is broken by the mechanical make-and-break of the induction coil, both capacitors discharge through the shunting leakage resistance. With the next pulse, the process is repeated. Because of the very short time constants, the interval between sparks can be considered constant to within one per cent.

ADJUSTMENTS AND CALIBRATION

1. Before taking data, level the glass plate carefully by using a puck; when the plate is level, the puck should remain at rest at the center of the plate. Make certain the outer plastic surfaces of each puck are dry before using. Do not use pucks if a film of ice is on the outer surface. If dry ice is used within the pucks, do not turn down the plastic tops too tightly, or otherwise the plastic catch strips will break, and air leaks will be caused by the broken bits of plastic caught in the grooves inside the cover. Inspect the grooves before using the pucks.*

2. Do not set the pucks directly upon the glass plate unless you are certain that they are working properly. Scratched glass will impede the motion of pucks. Place the pucks upon a paper towel, when they are not being used.

3. The variable spark timer can be calibrated accurately by using a triggering cathode-ray oscilloscope equipped with a calibrated sweep time adjustment, for example, a Tektronix 504. The output of the spark timer is applied to the vertical plates of the CRO; to prevent damage from the coil's high voltage secondary, the connection is best made by placing the coil's output line parallel to the vertical plate input lead. The induced current will then appear on the screen.

   When making time measurements, the oscilloscope is set to internal triggering and the sweep time adjustment knob is used to place two well-spaced pulses on the screen (all sweep time knobs remain in the calibrated position!). After the pattern of pulses is locked in place on the screen, the distance between pulses is read from the scale marked on the face of the screen. The product of the pulses read from the screen and the setting of the sweep time (in units of time per length) is the time between pulses. For this group of experiments, a time interval between sparks of $0.05 \pm 1\%$ seconds is used.

* If leaks still exist, disassemble the pucks and use vacuum wax on the gasket between the metal base of the puck and the plastic top; also use a small amount on the rubber seal in the cover and on the rubber stoppers.

4. Motions of the pucks are measured by spark records on electrically conducting paper (Teledeltos or Electrosensitive*). The coated paper is placed, conducting side down, upon the glass plate, and the pucks are moved over it. If a metal surface is used, instead of glass, wax-coated paper can be substituted for the conducting paper.

5. It is very important that you check the time interval between spark marks immediately before taking data for any experiment. Use a stroboscope, find the frequency of oscillation of the vibrator, and obtain the period. The time interval between spark marks can also be checked by moving a puck for a short distance over a sheet of recording paper and using a time clock to find the average time interval between spark deposits.

Experiment **8-1** MOTION IN ONE DIMENSION

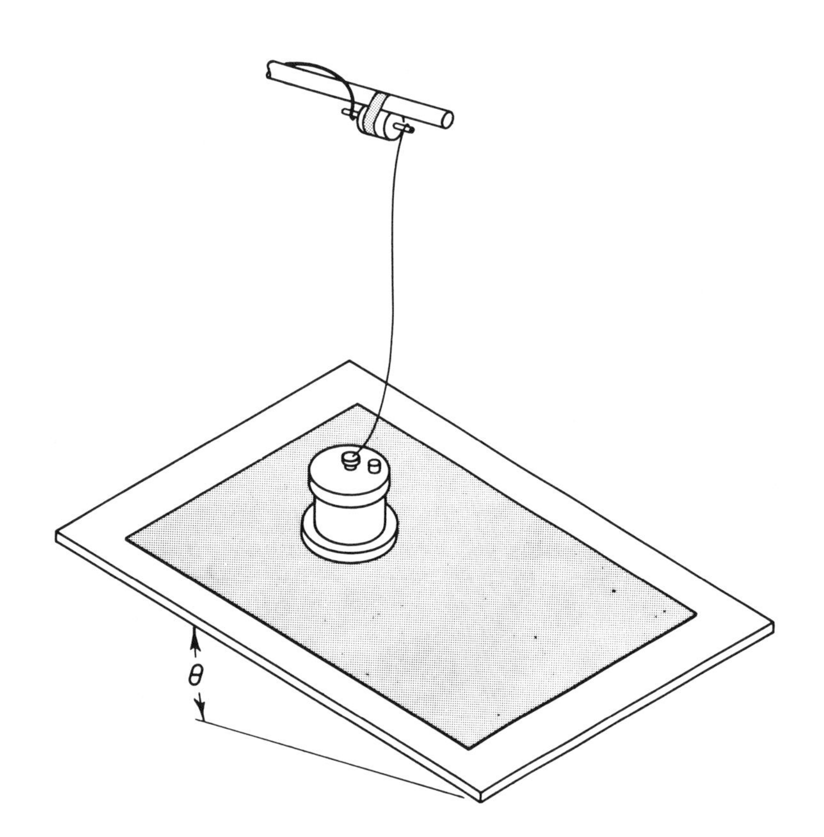

Fig. 8-5.
Motion in One Dimension.

---

*Since various widths and lengths of conducting paper can be used in performing the experiments in this section, paper in roll or sheet stock can be obtained by special order from the following companies: Teledeltos paper - Western Union Telegraph Co., Development and Research Dept., 60 Hudson Street, New York, New York; Electrosensitive paper (Timemark 14) - Communications Papers, Inc., Box 1106, Scranton, Pa.

1. Support the glass plate, as shown in Fig. 8-5, at a slight angle θ to the horizontal. Before taking data, make several trial runs without using the spark timer.

2. It is instructive to be able to compare two values for the speed found independently, for example, by the use of the conservation of energy and by graphical techniques. Therefore, use energy relationships to find the final speed of the puck after it accelerates a distance down the plane. Mark the point on the trace where you find the speed by energy methods so that you can compare it to that obtained graphically.

3. Plot the following graphs: (1) position x vs. time t, or $x = f(t)$; (2) $\dot{x} = f(t)$, or $v = f(t)$; (3) $\ddot{x} = f(t)$, or $a = f(t)$.

4. From your plot of $v = f(t)$, obtain an algebraic equation which shows how the speed increases with time; also find an expression which shows how the distance traveled increases quadratically.

5. Discuss the variation between the two values found independently for the final speed of the puck.

Experiment 8-2 CONCEPT OF MASS: NEWTON'S SECOND LAW OF MOTION

1. To determine if the ratio of mass m_1 to mass m_2 is the same regardless of the force used, we can define mass in terms of the acceleration produced by a force exerted by a spring. When the spring attached to a puck is elongated beyond its unstretched length to a particular reference position, the puck will be accelerated. If the spring is stretched further, the puck will be given a greater acceleration. We define the ratio of the masses to be

$$\frac{m_1}{m_2} = \frac{a_2}{a_1}, \quad \text{(given Force)}$$

   where m_1 is the mass of a puck containing a small amount of dry ice; m_2 is the mass of another puck with a much greater amount of dry ice.

2. In Fig. 8-6, a puck held by a light cord and a spring is shown on a flat glass plate. When the cord is released, the puck is accelerated by the spring. We, therefore, can measure mass quantitatively.

   Springs of suitable lengths with reasonable k can be made from No. 22 piano wire wound about 92 turns on a $\frac{1}{2}$-in. mandrel. Cord and spring are connected to hooks tapped into the puck base. Hooks should be arranged so that both the spring and cord can be removed quickly from the puck.

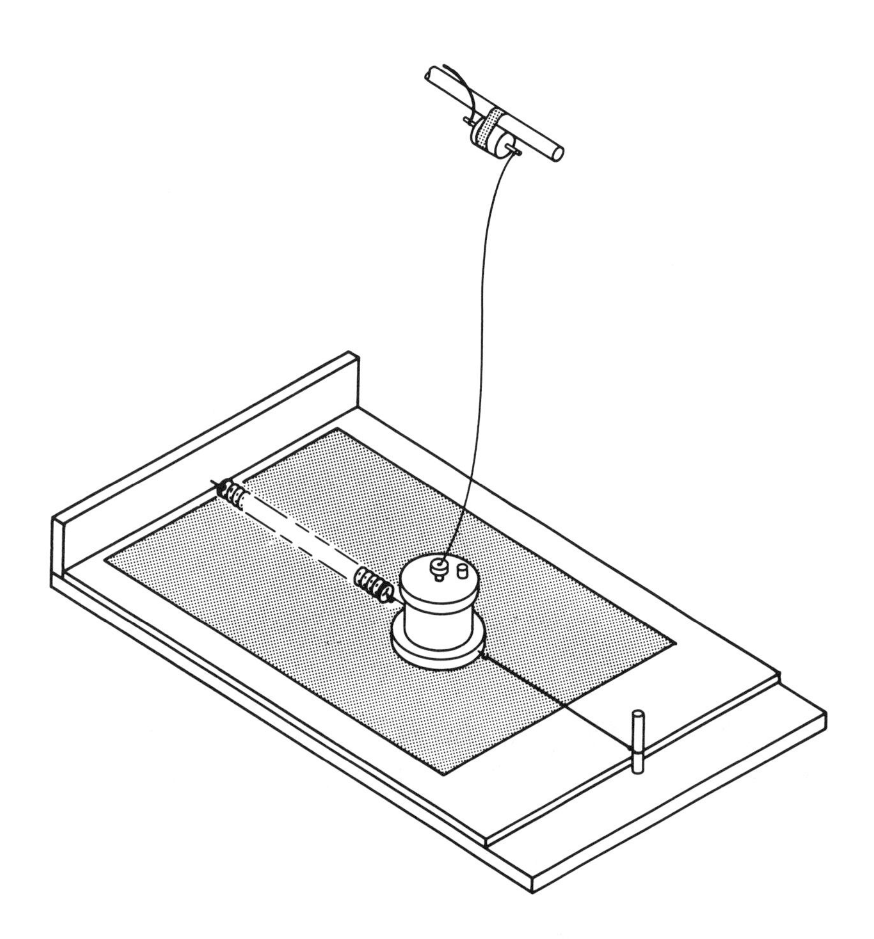

Fig. 8-6.
Concept of Mass;
Newton's Second Law
of Motion.

3. To perform the experiment, two cords (one longer than the other) are provided. Quickly obtain spark tracks for mass m_1, since you want to minimize the loss of the mass of dry ice by sublimation. After obtaining the first spark trace with the long cord, move the conducting paper slightly so as to obtain a second trace for m_1 with the short cord. Repeat the procedure for m_2 (second puck).

4. Show that $\frac{m_1}{m_2} = \frac{a_2}{a_1} = \frac{a_{2'}}{a_{1'}}$ within experimental limits of error. Repeat the experiment, if possible, in order to improve the accuracy. Discuss your results.

5. Assign an arbitrary mass unit, say one "unit" to m_1, and obtain the mass m_2. Discuss the meaning of Newton's second law, in terms of your measured quantities.

Experiment **8-3** CENTRIPETAL FORCE

1. One of the concepts in physics which is quite difficult to understand is centripetal force. The apparatus shown in Fig. 8-7 can be used to investigate the fact that when a particle is constrained to move in a circular path at constant speed, the acceleration at any instant is directed toward the center of the circle. The force acting inward which produces the circular motion is called centripetal force.

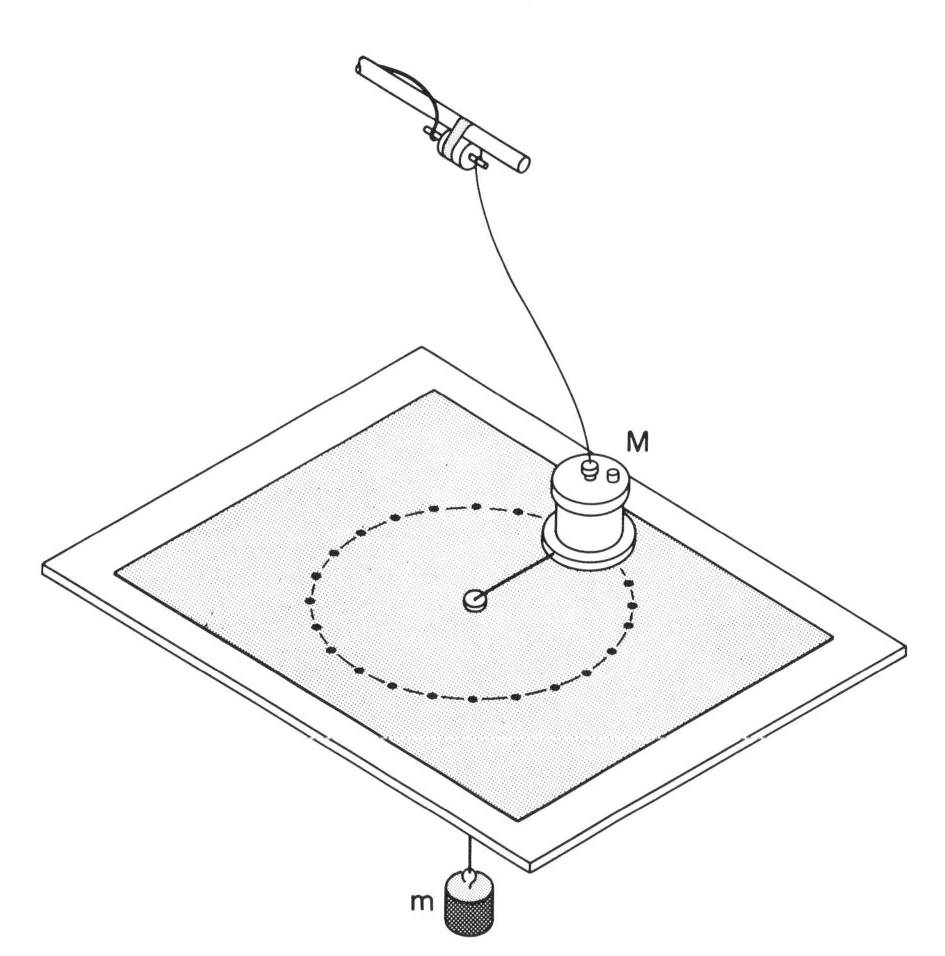

Fig. 8-7.
Centripetal Force.

2. For best results, a roller bearing or a short Teflon tube should be used for an insert in the hole in the glass plate; energy loss by the moving cord is thereby minimized. Try several combinations of mass m on the holder hanging below the plate in order to obtain the best radius R for a particular puck mass M. Make several trial runs before taking data with the spark timer. When taking data, allow the puck to complete only one orbit. Determine the mass M of the puck immediately after the orbit is completed.

3. From the spark trace find the "radial" acceleration of the puck. Derive an expression for the "centripetal" acceleration using only the values of m, M, and g. Discuss the reasons for the variation between your results.

4. What is the magnitude of the centripetal force? On what body does it act? What is the magnitude of the centrifugal force? On what does it act?

Experiment 8-4 LINEAR OSCILLATOR

1. Connect two springs (see Experiment 8-2, Part 2, p. 397) with about the same spring constants ($k_1 = k_2$) to each side of a puck, as shown in Fig. 8-8. Use a meter stick and carefully mark the equilibrium position O near the edge of the conducting paper; also mark a reference line at a distance of 10 cm (amplitude = ±10 cm) on each side of the equilibrium position. Before taking data, record the mass of each spring.

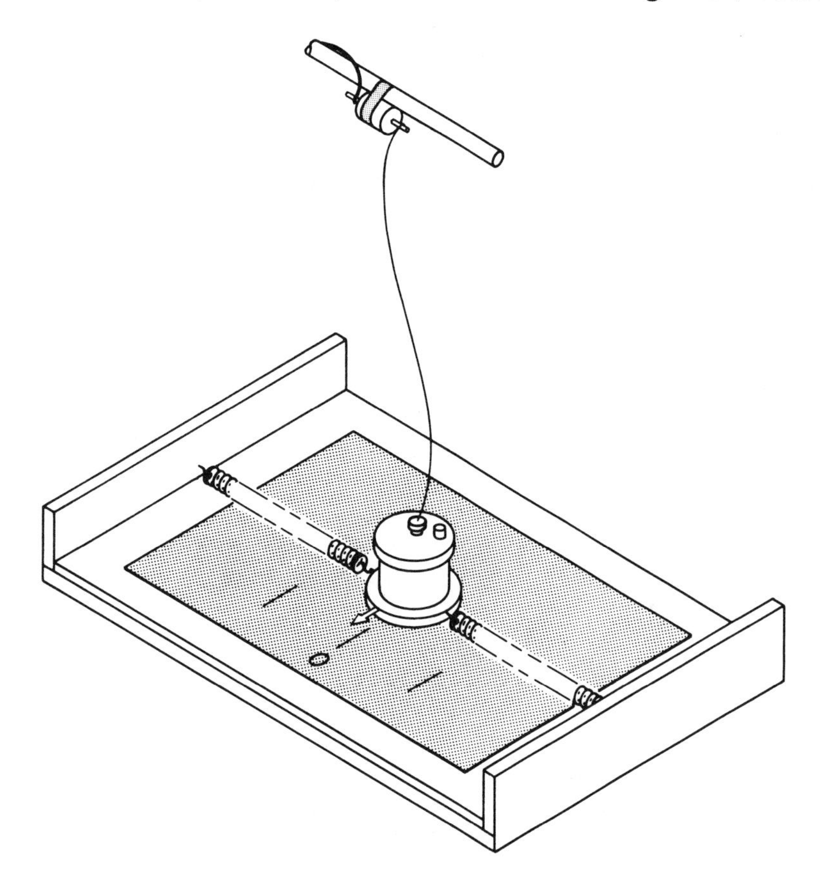

Fig. 8-8.
Linear Oscillator.

2. Move puck to an amplitude of +10 cm. Turn on the spark timer or spark source and release the puck. Stop it at -10 cm. After the spark trace is obtained, immediately find the mass of the puck and then shift the paper to a new position.

3. With the timer operating, again release the puck from +10 cm. While the puck is oscillating, slowly slide the paper along beneath it. What kind of curve is traced out?

4. Determine if the total mechanical energy of the system of springs and puck is conserved (KE + PE = Const.) within experimental limits of error. Hint: PE(springs) $\cong 1/2\,(k_1 + k_2)\,x^2$. Discuss your results.

5. In your calculations for Step 4, did you allow for the fact that each spring also oscillates? Compute a correction term that can be added to the mass of the puck which will consider the equivalent mass of both springs. Repeat the calculations in Step 4 and discuss the results.

6. If time permits, take sufficient data to enable you to plot a potential energy vs. distance curve [$V = f(x)$]. Draw a horizontal line at a height equal to the total energy E.

Experiment 8-5 ONE-DIMENSIONAL COLLISION

1. As indicated in Fig. 8-9, obtain a spark trace of a collision in one dimension between two pucks of the same mass; one puck is at rest. Before taking data, you should practice with the apparatus using a reasonable amount of dry ice in each puck.

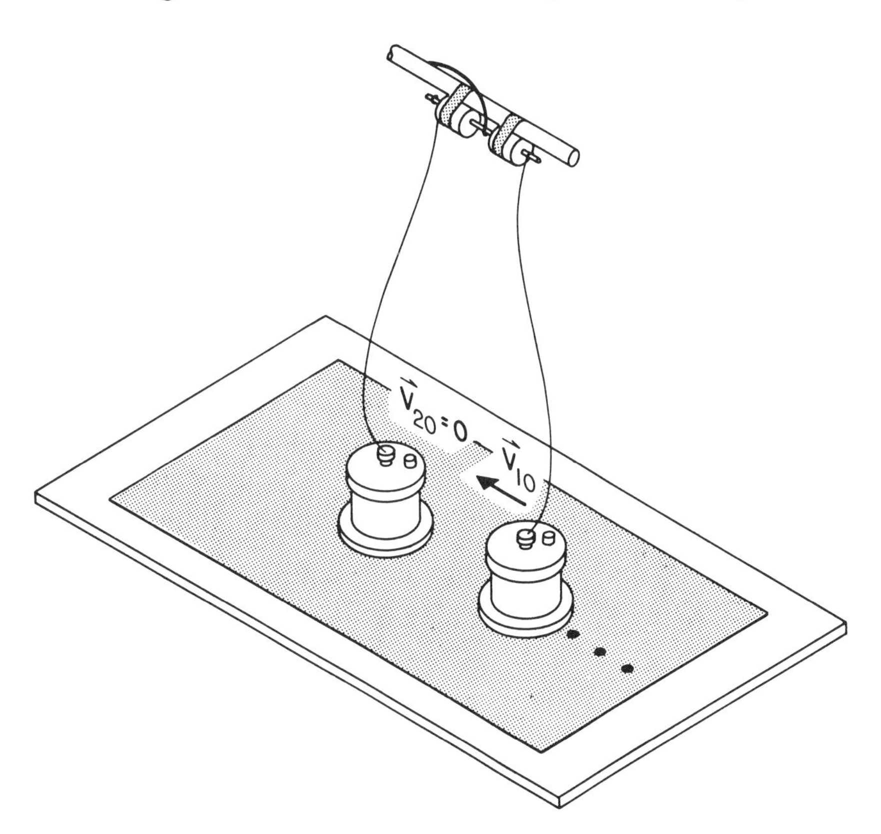

Fig. 8-9.
One-Dimensional
Collision.

2. Apply the conservation or momentum principle and show within experimental limits of error that the moving puck stops and the second puck moves off with the velocity the first puck originally had.

3. Was the collision elastic, or inelastic? Explain.

4. Was the total energy of the system conserved? Explain.

5. What percentage of the initial kinetic energy was lost? Since pucks were supported on a gas film, how was the energy lost?

Experiment 8-6 CENTER OF MASS MOTION

1. Connect two pucks together with a long rubber band or rubber string (Fig. 8-10). If possible, use a glass plate at least 3 ft long of some convenient width; a similar sheet of conducting paper is also needed. Fill one puck with a reasonable quantity of dry ice; use a much smaller amount in the other puck; experiment with the system before starting the spark timer. Pucks should not move too quickly toward each other.

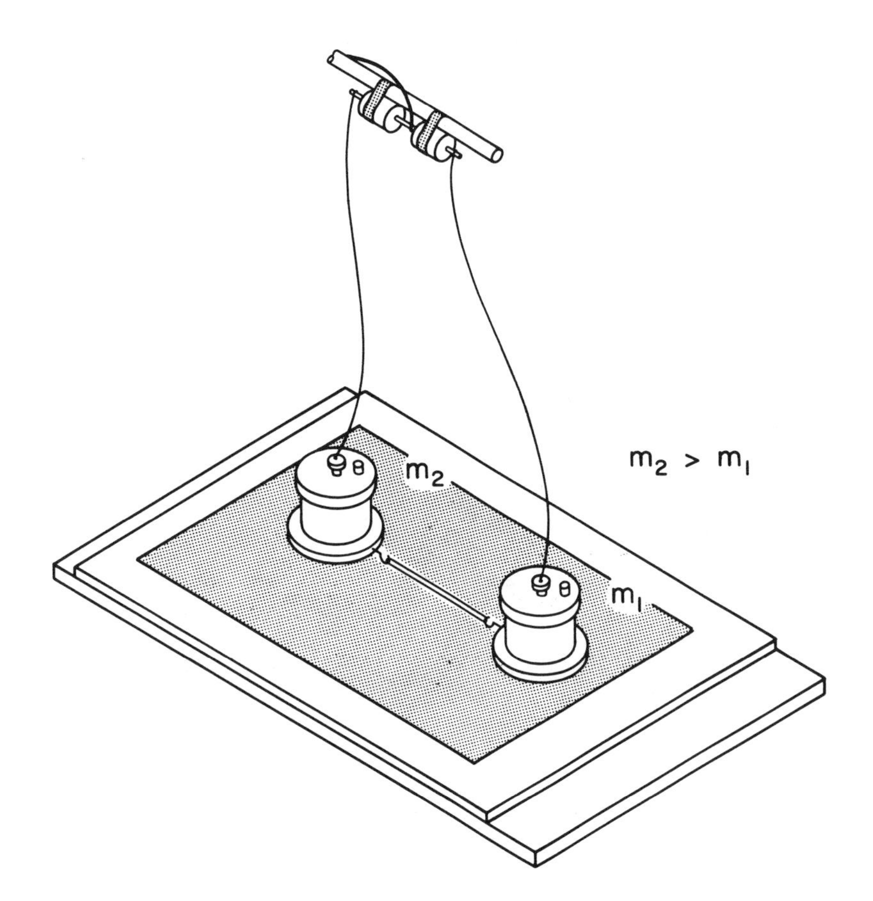

Fig. 8-10.
Center of Mass Motion.

Obtain a spark record of the motion only as pucks travel toward each other. Immediately after obtaining spark trace, use a trip balance and find the mass of each puck.

2. Plot on the same graph the x coordinates of each puck as functions of time [x = f(t)]. Also show on the graph the location of the center of mass.

3. On another graph, plot the velocities of each puck as functions of time [V = f(t)]. Show on the graph the center of mass velocity.

4. Plot the accelerations of each puck as functions of time [a = f(t)]. What is the acceleration of the center of mass? What is the total momentum of the system? Explain.

5. Are the kinetic energies of the pucks inversely proportional to their respective masses within experimental errors? Use data from your graphs in your calculations and discuss your results.

Experiment **8-7** LINEAR MOMENTUM

1. Insert a small spring, as shown in Fig. 8-11, between two pucks and tie the system together with a light cord. Before attempting to take data, make several sample runs in order to become familiar with the apparatus. Start the variable spark timer; check to see if both spark gaps are firing and then burn the cord. As soon as the pucks stop moving, carefully measure their masses on a trip balance; also calculate velocities of pucks, etc.

2. Apply the conservation of linear momentum to the motion and find out if the total momentum of the system is zero within the experimental limits of error. Explain your results.

3. Determine if the kinetic energy of the system is conserved within the experimental limits of error. Explain your results.

4. Determine if the kinetic energies of the pucks are inversely proportional to their respective masses. Explain your results.

5. What is the motion of the center of mass of the two-puck system?

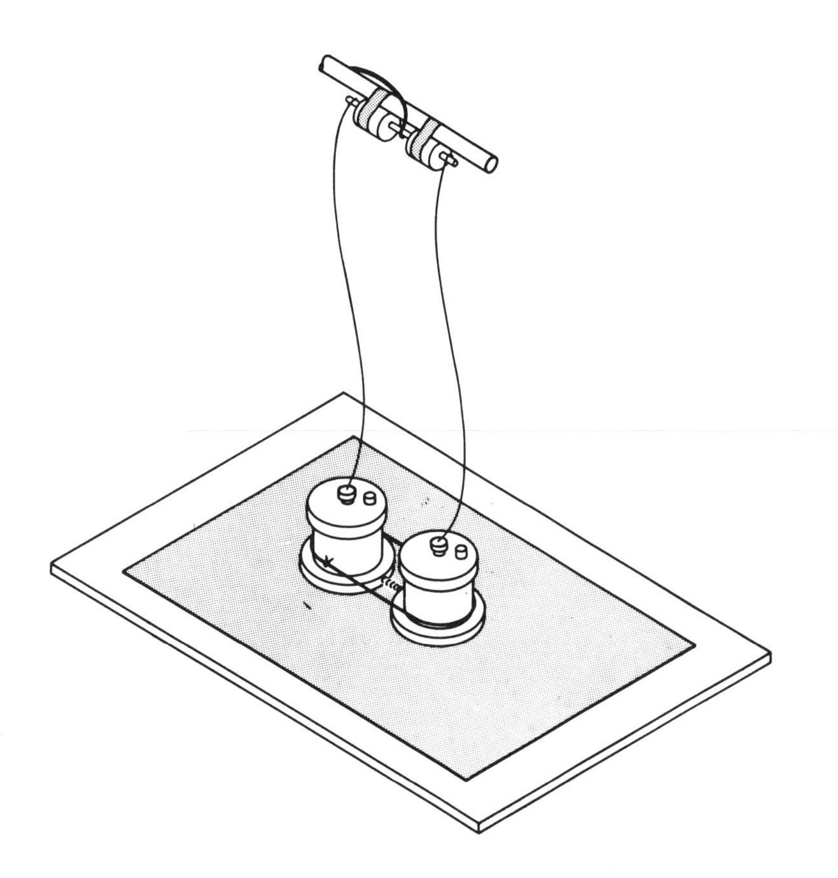

Fig. 8-11. Linear Momentum. After a small spring is inserted between two pucks, they are tied together with a cord. When the cord is burned, the pucks move apart and the sum of their momenta remains zero within experimental limits of error.

Experiment **8-8** TWO-DIMENSIONAL COLLISIONS

1. Fill both of the recording pucks with dry ice (or use thin-walled surgical rubber tubing, and connect the pucks to a source of compressed air). Set one body A at the center of the conducting paper. Start the spark timer. With both gaps firing, push the other body B gently toward A. Bodies must collide and then scatter. Try to obtain a decent scattering angle ($\theta > 20°$). After the bodies come to rest, stop the spark timer.

2. Use a trip balance and determine quickly the mass of each object. Before a puck is placed on the scale, cover the pan with a paper towel so that the puck will not slip off.

3. Connect the spark points with straight lines and label the paths of m_A and m_B, as shown in Fig. 8-12. The large spark mark was made by m_A, when stationary. It is not necessary to obtain a spark mark at the intersection (collision point) of the lines connecting m_B points - before and after collision. The distance from the collision point to the large spark mark, made by puck A, is equal to the diameter of a puck.

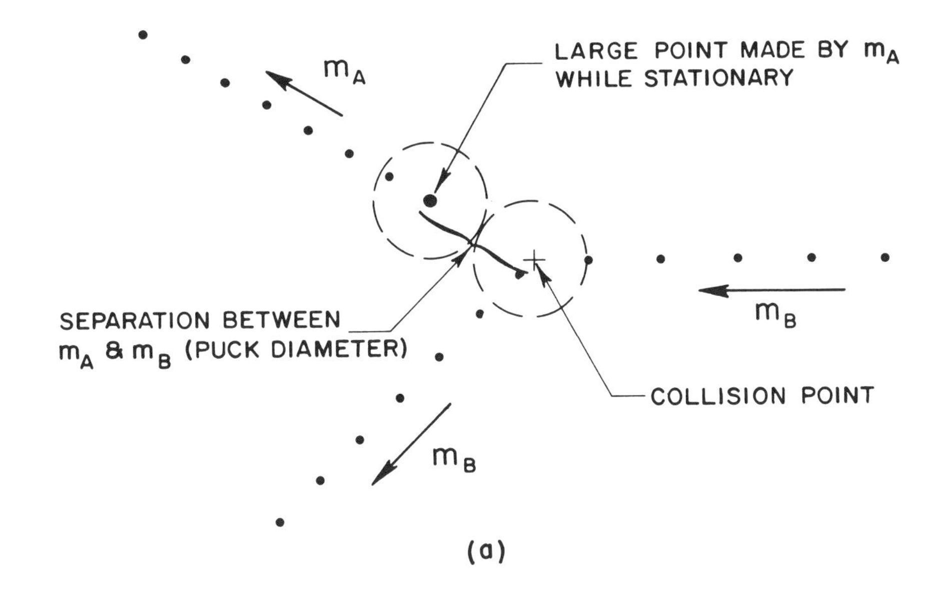

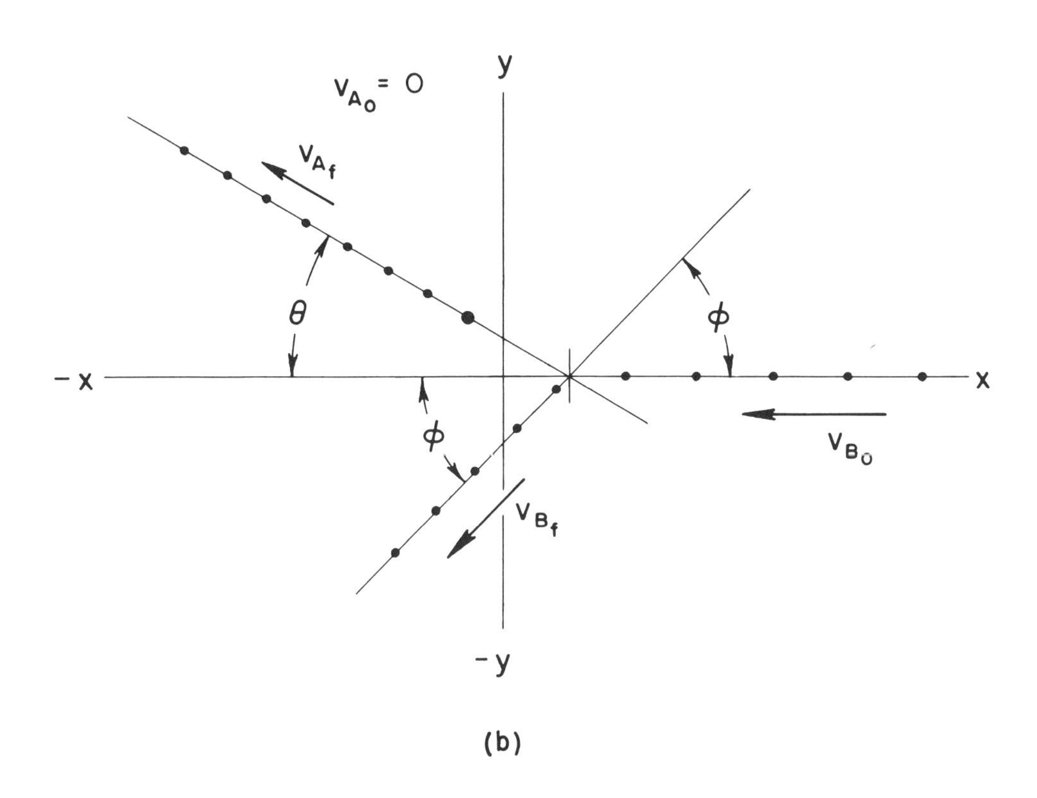

Fig. 8-12. Sample Trace of a Typical Collision (on top) and Coordinate System Constructed Through Collision Point (at the bottom).

4. Construct a coordinate system through the COLLISION POINT. Label axes and angles and record their values in a data table.

5. Since the separation between spark points before and after collision is nearly constant (Why?), the velocity of each body is easily obtained. Select about ten spark traces near the collision point along each path. Measure the distance, record it in your data table, and find the velocity of each body before and after collision. Avoid using the first few spark traces near the collision point. (Why?) Record the velocities of each body before and after collision, and their x and y components in a data table.

6. Determine if linear momentum is conserved within the experimental limits of error for the two-dimensional collision. For the analysis, use the following limits of error; mass (gms) of a puck ±1% (as obtained quickly on a trip balance); distance (cm) measured with a meter stick, ±0.5%; time (sec), ±1%; angle (degrees), ±1°.

7. Determine if the kinetic energy of the system is conserved during the two-dimensional collision. Explain your results.

8. What linear impulse did body B exert on body A? What linear impulse did body A exert on body B? Do your results compare within the experimental limits of error?

9. Repeat the above experiment, but instead of setting one puck at the center of the conducting paper and then pushing the second puck toward the first so that they collide, give each body a different initial velocity before the collision. Again, be sure that you obtain a decent scattering angle ($\theta > 20°$) before analyzing the results using the methods discussed above.

8.4 The Air Table[1]

The experiments 8-1 through 8-8 may also be performed on an air table instead of using dry ice pucks or compressed air supplied to the pucks. The air table consists of a rigid box with a flat surface on top. Air is forced into the box and leaves through a large number of small holes in the flat surface. A flat object, such as a puck, will float on a thin film of air when placed onto the surface. The air is usually forced into the box by an industrial vacuum cleaner. In the laboratory several air tables may be supplied by a single blower unit.

A large number of pucks of various shapes and sizes are available, including magnetic pucks which may be used for elastic scattering and collision experiments.

Data may be obtained by either one of three methods[2]!

[1]Air tables of various sizes are available from Educational Instruments, Inc., Watertown, MA., Fisher Scientific, Sargent-Welch Scientific Co. and the Central Scientific Co. Tables are also available from other companies.

[2]Details of the three methods of obtaining data as well as a large number of suggested experiments are discussed in the 103-page pamphlet "Experiments on an Air Table," by George Marousek and T. Walley William, III, published by the Ealing Corporation.

(1) Spark recorders may be used - similar to those discussed previously.

(2) Multiexposure photographs with a Polaroid camera give a compact permanent record. Multiexposures of moving objects are obtained by using a rotating disk in front of the camera or by modulating the illumination using a strobe lamp with a variable flash rate.

(3) The air table may be equipped with photocell gates which start and stop clocks or counters.

The air table as well as the puck experiments described previously, give the student an excellent opportunity to devise experiments of his own. A large number of topics from mechanics can be illustrated by means of these almost frictionless surfaces.

8.5 Air Bearing Rotational Apparatus[1]

The air bearing rotational apparatus consists principally of metal disks rotating on a layer of air resulting in an almost frictionless surface. Air bearings are used to further reduce friction. Data may be taken by either observing the rotation of the disk or by a recording spark timer.

Since two disks may be used, both rotating freely, a number of different experiments may be performed involving concepts such as angular momentum, rotational inertia, torque, energy, and harmonic motion. The student is presented with an opportunity to devise his own experiment. One possible experiment, dealing with the conservation of angular momentum, is described below.

Experiment 8-9 CONSERVATION OF ANGULAR MOMENTUM

In this experiment, the angular momentum of a steel ball before impact is compared with the angular momentum of an aluminum disc, ball-catch, and ball after impact. The equipment is shown in Fig. 8-13.

To determine the velocity of the ball as it leaves the ramp, attach the ball ramp to the base so the ball will roll off the ramp and follow a parabolic path to the floor. To determine precisely the point at which the ball hits the floor, place a piece of carbon paper, carbon side up, in the approximate area where the ball will hit the floor and cover with a piece of onion skin paper. When the ball strikes the paper a trace will be made on the paper by the carbon paper. Starting from the same height, roll the ball a number of times down the ramp and number the impact points. Measure the vertical distance d through which the ball has dropped and determine the time of free fall from the relationship $t = \sqrt{2d/a}$. Measure the horizontal distance from the point at which the ball leaves

[1]The Pasco Rotational Dynamics Apparatus Models 9270 and 9279 are available from the Pasco Scientific Corporation, 1876 Sabre St., Hayward, CA., 94545. A laboratory guide is also available; it describes the details of the apparatus and ten experiments.

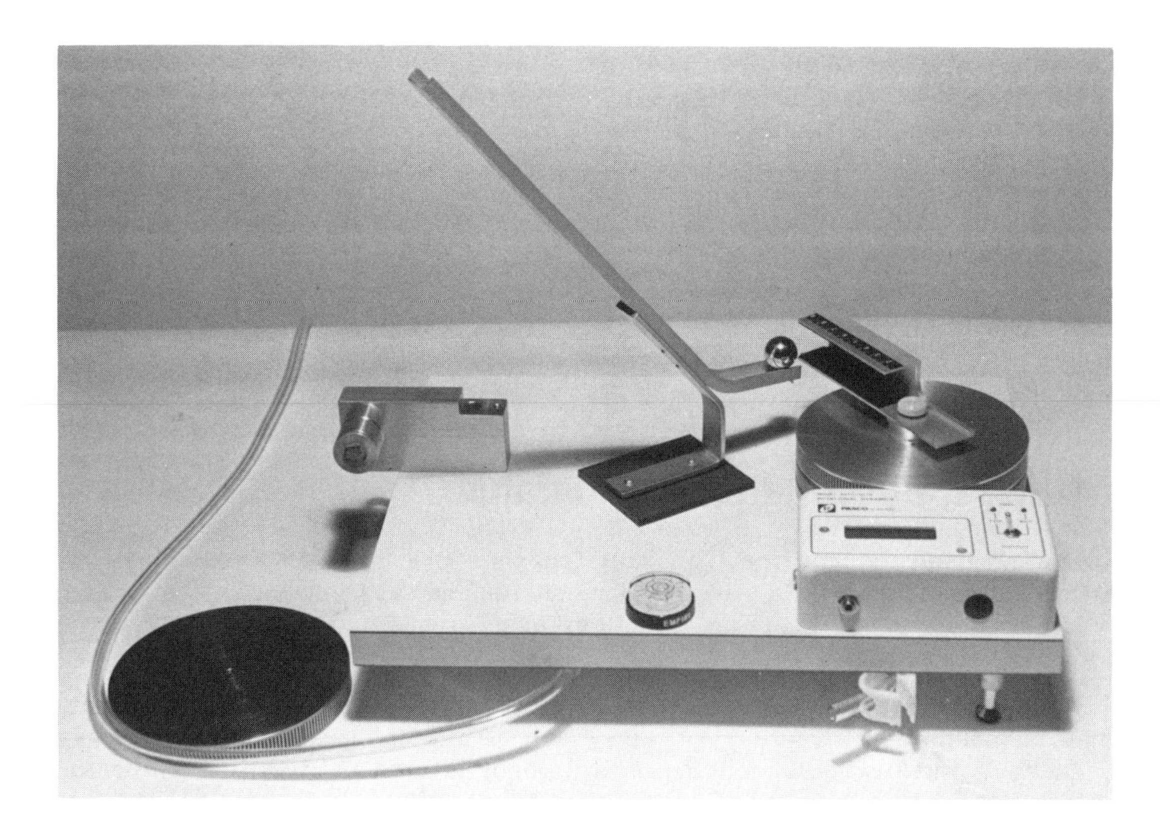

Fig. 8-13. Pasco Rotational Dynamics Apparatus Used for the Conservation of Angular Momentum Experiment.

the ramp to the point where it hits the paper and calculate the horizontal velocity of the ball.

After these preliminary experiments, point the ramp so that the ball will be caught in the ball catch. Determine the angular velocity after the ball has been caught by either measuring the time for several revolutions (you must turn the ramp out of way of the moving system), or by using a calibrated sparktimer. You can vary the impact parameter (i.e., the perpendicular distance between the line of motion of the ball before impact and the axis of rotation of the system) by moving the ball ramp to any one of the different holes in the base. For the holes close to the disc, the impact should be perpendicular to the ball catch. For the holes placed on a circle, the ramp should be rotated so that the ball is caught each time near the end of the ball catch.

Repeat the experiment with different ball velocities. Attach the ball catch to the steel disc and repeat with this system which has a much larger rotational inertia.

Use this apparatus to do a separate experiment on the Conservation of Energy and Momentum. Calculate the velocity with which the ball should leave the ramp from the conservation of energy principle, compare this calculated value with the value obtained at the beginning of the experiment. If your calculated value differs significantly from the measured one, you may not have taken into account all the details of the particular motion of the ball down the ramp.

Chapter 9

HEAT

9.1 Calorimetry

Every good college physics textbook describes the use of the fluid calorimeter for determining such items as specific heat capacities, latent heats, heats of combustion and reaction, mechanical or electrical losses to heat, etc. Theoretically, such procedure should be simplicity itself. An insulated can or cup, made of metal whose specific heat capacity is known or can be readily measured, is filled with a measured amount of fluid, usually a liquid, and most commonly water. When the calorimeter and its contents are in thermal equilibrium (all at the same temperature), a hot body or cold body is added, fuel is burned and mechanical or electrical energy is converted into heat – in the fluid of the calorimeter.

The result of the heat transfer within the calorimeter is a measurable rise or fall in the temperature. If the process is ADIABATIC (no gain or loss of heat, to or from the surroundings - an insulated process), heat loss can be equated to heat gain and the appropriate heat balance equations can be written. These can be solved for such items as the specific or latent heats already mentioned. For example:

(a) If a sample of metal of mass m_s and specific heat capacity c_s is heated to a temperature t_s and is then dropped into a water calorimeter which is at temperature t_1:

Heat (energy) lost by metal = Heat (energy) gained by calorimeter and contents.

$$m_s c_s (t_s - t_2) = m_w c_w (t_2 - t_1) + m_c c_c (t_2 - t_1) \qquad 9\text{-}1$$

where t_2 = final temperature of calorimeter and contents

m_w, m_c = mass of water and calorimeter

c_w, c_c = specific heat capacities of water (fluid) and calorimeter

(b) If an electric heater (such as a current-carrying resistance wire) immersed in a calorimeter supplied heat (energy) which was converted from electrical energy (I^2RT) or (EIT) where:

I = current
E = potential drop
R = resistance
T = time
J = conversion factor

then the heat balance equation becomes:

$$J(EIT) = m_w c_w (t_2 - t_1) + m_c c_c (t_2 - t_1) \qquad 9\text{-}2$$

If mechanical work W is supplied, the left hand term in Eq. 9-2 is simply J(W).

(c) As a final example: if a piece of ice of mass M at its normal equilibrium temperature 0° C is dropped into a calorimeter and melts, the temperature of the calorimeter decreases from t_1 to a new and lower value t_3. Heat (energy) is absorbed by the ice both in melting and in warming, as water, to t_3. Heat (energy) is given up by the calorimeter and its fluid.

$$ML + Mc_w(t_3 - t_0) = m_w c_w (t_1 - t_3) + m_c c_c (t_1 - t_3) \qquad 9\text{-}3$$

L = latent heat of fusion of ice
t_0 = melting temperature of ice, essentially 0° C
$t_1 - t_3$ = corrected change in temp. (See Fig. 9-3.)

Similar equations can be written for other like processes.

In practice, the operation is not quite as simple as it sounds - in fact, the attainment of really high precision in calorimetric measurements requires fine equipment, great care, much experience and correction or compensation for a number of systematic errors. Some of the points which must be watched in order to avoid gross errors are :

1. Thorough stirring is necessary since temperature should be as near as possible identical throughout the calorimeter. WHEN DONE BY HAND IT SHOULD BE CONTINUED FROM THE MOMENT THE EXPERIMENT IS STARTED UNTIL THE END OF THE PROCESS. Stirring should not be so vigorous as to convert undue amounts of mechanical energy into heat. Use complete, smooth strokes.
2. Losses of calorimeter fluid by splash, careless stirring or by evaporation must be avoided.
3. Insulation should be as complete as possible. This is accomplished by use of polished metallic surfaces for the calorimeter can, dead-air spaces, vacuum flasks, porous insulating material and other methods. (See Fig. 9-1.)

With reasonably decent equipment operated with care by competent observers, the main remanent sources of error are as follows:

4. Heat energy is added by the essential stirring, during which the unavoidable converted mechanical energy tends to produce an unwanted temperature rise.

5. Heat transfer to (or from) the calorimeter vessel, when the surroundings are at a different temperature, due to the inevitable lack of perfect insulation.

Heat transfer to (or from) the calorimeter vessel can be minimized by adjusting your approach so that the differences in temperature between the calorimeter vessel and the surroundings are as small as possible. This is necessary because the rate of heat transfer from a hot body to a cooler one is proportional for small temperature differences to the temperature difference. Some modern precision methods involve the control of the temperature of the surroundings (usually a water jacket surrounding the calorimeter vessel) so that the jacket temperature follows that of the calorimeter. Losses or gains can thus be greatly reduced - but the cost, both in cash and convenience, is considerable. The use of a water jacket is desirable even when not temperature controlled, since it provides reasonably constant surrounding temperature due to its large thermal capacity. (See Fig. 9-1.)

A method of "compensated loss" (or gain) is often used. In this case the temperature-time relationship is important. The temperature of the calorimeter is measured and recorded at regular intervals before, during and after the "mixture" process. If the specific case of dropping ice into a water calorimeter is considered, then a time-temperature run is divided into three intervals (all part of a continuous series). These are usually referred to as the FORE INTERVAL, the TEST INTERVAL, and the AFTER INTERVAL. The fore-temperature is adjusted to be about as far above (or below) the temperature of the surroundings as the after-temperature is below (or above). This ordinarily means a "dry run" to investigate the general behavior of the equipment.

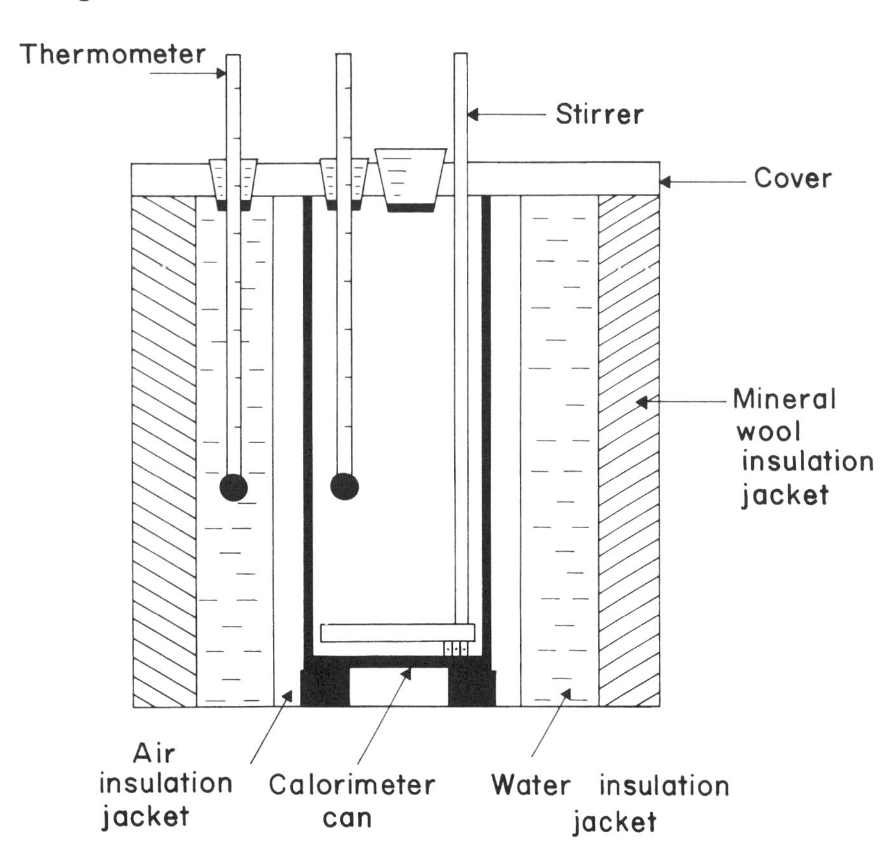

Fig. 9-1. Calorimeter.

CORRECTION CURVES FOR THE LATENT HEAT OF FUSION

The calorimeter vessel is weighed, filled about 3/4 with water and weighed again. The water should be warmer than the jacket or other surroundings. The calorimeter vessel is placed in whatever jacketing arrangement is available (see Fig. 9-1) and stirring is begun. Temperature is recorded at half-minute intervals for about 5 minutes or until it is dropping uniformly. A piece of ice wiped dry to avoid the transfer of water is carefully dropped into the vessel at one of the reading times. Stirring and temperature readings are continued while the ice melts and for approximately five minutes after the ice has melted or until the temperature is rising slowly and steadily. Another weighing at the conclusion of the experiment gives the weight of ice added. The temperature of the surroundings is recorded every few minutes to see if it remains fairly constant. A sample TEMPERATURE vs. TIME graph is shown in Fig. 9-2.

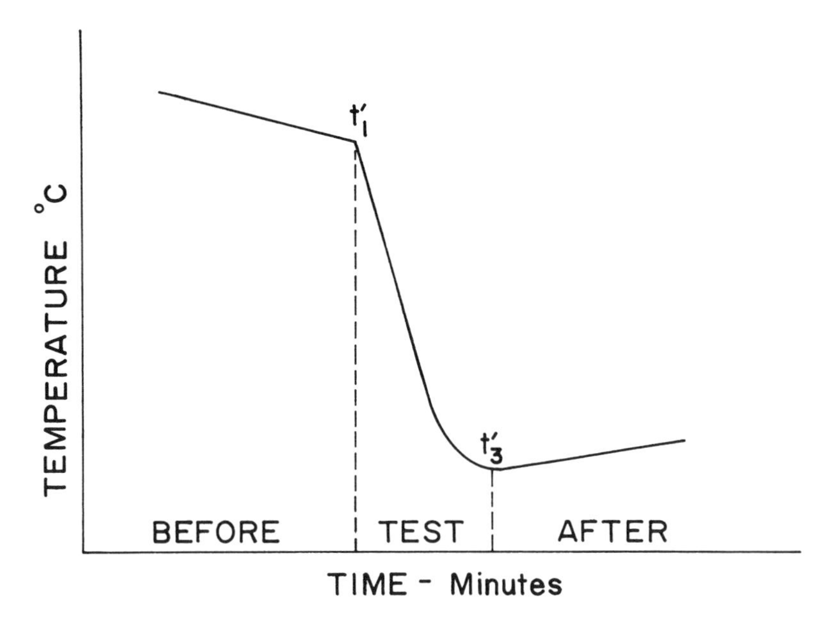

Fig. 9-2.
Uncorrected Temperature-Time Curve for the Latent Heat of Fusion.

The drop in temperature which would have occurred WITHOUT LOSSES t_1-t_2 and the rise in temperature of the water which resulted from the melting of the ice t_3-t_o are desired. The compensated drop in temperature is obviously not equal to t_1'-t_3' but to a corrected Δt as shown in Fig. 9-3. Determining the best corrected value of t_1-t_3 can be a complicated and tedious process.

As a first approximation which is not strictly correct, the partially corrected values of t_1 and t_3 can be obtained by projecting the temperature-time lines for the fore and after intervals. A vertical line (t_1,t_3) should be drawn so that the area A_1 is about equal to A_2. (See Fig. 9-3.) The corrected drop in calorimeter temperature t_1-t_3 and the corrected ice-water temperature rise t_3-t_o will be as accurate as the precision of the thermometers warrants. The dotted line labeled surrounding temperature is approximately the effective temperature of the surroundings and is at the intersection of the vertical line (t_1, t_3) with the actual data curve t_1',t_3'.

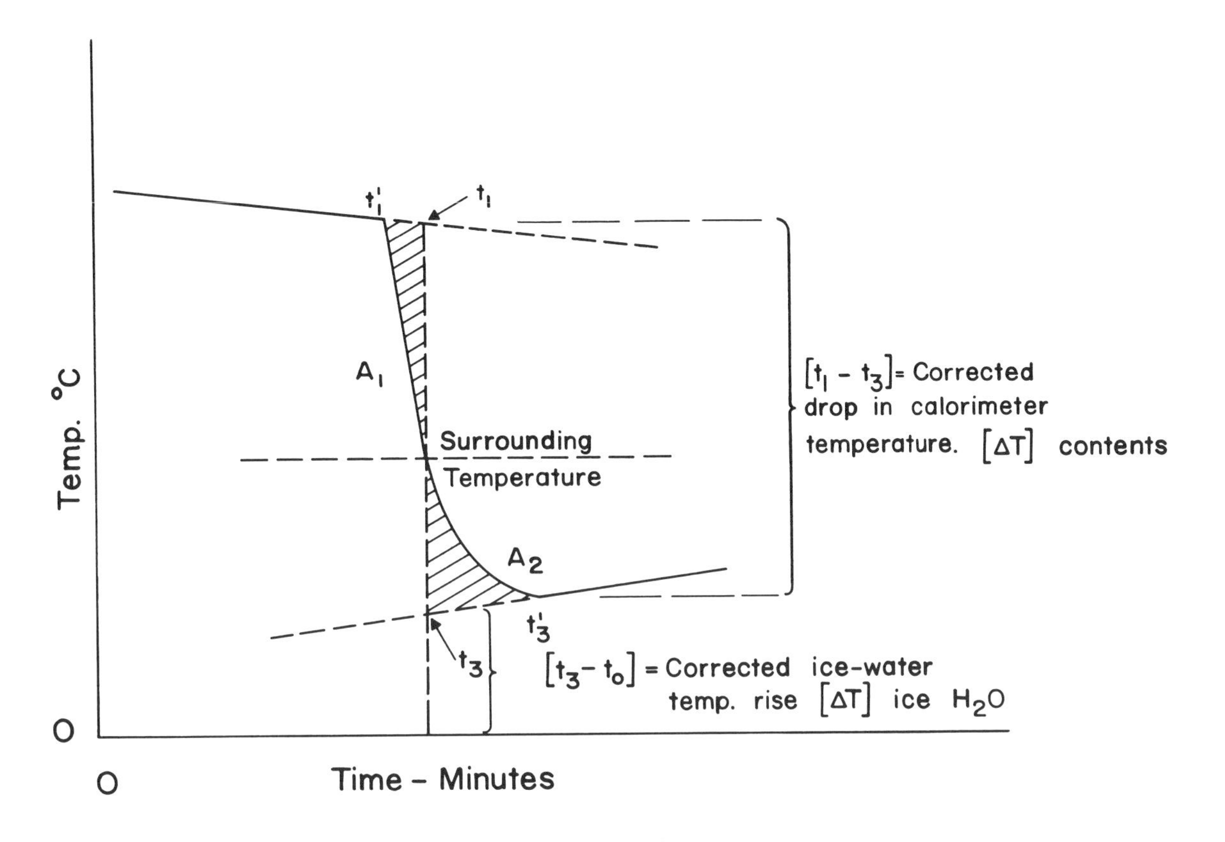

Fig. 9-3. Corrected Temperature-Time Curve for Latent Heat of Fusion.

Experiment 9-1 CALORIMENTRY - SPECIFIC HEAT AND LATENT HEAT OF FUSION

Fig. 9-4. Calorimetry Apparatus.

INTRODUCTION

Before reading the experiment, fill the steam boiler A 3/4 full of water. Connect the rubber hose to the top of the steam jacket C and turn the heating unit on to high. Move the calorimeter B beneath the chute in the steam jacket. Make several test runs with the sample metal specimen down the chute into the can. Remember that the longer it takes to get the specimen into the can, the more energy will be lost to the surroundings. Specimen must be quickly though gently dropped into the can without splashing. After the test runs have been made, weigh the sample to the nearest decigram and suspend it half way down the chute. Place a 0 - 110° C thermometer carefully within the central hole.

This experiment presents two problems:

I. Find the Latent Heat of Fusion of water.
II. Find the Specific Heat of an unknown metal specimen.

The objectives in addition to determining several unknown quantities are to:

(a) Develop an appreciation of the difficulties present in an experimental science.

(b) Acquaint the student with the fact that apparent small errors in stirring technique, weighing of masses, reading of thermometers, spilling of water, absorption of water by corks, etc., can vitally affect the final results.

(c) Attempt to introduce student to the feeling of satisfaction that results from performing a series of rather complex processes during which care was exercised in the taking of data.

The heat balance equations discussed require additional energy (heat) gain or loss expressions for the thermometer 0° - 50° and the stirrer which are placed directly in the mixture in the calorimeter can. (See Fig. 9-1.)

Energy (heat) gain or loss of thermometer = $m_t c_t \, (\Delta T_{corrected})$

$m_t c_t$ = water equivalent of the thermometer. It is unity (1) when using the c.g.s. system.

Energy (heat) gain or loss of stirrer = $m_r c_r \, (\Delta T_{corrected})$

PROCEDURE (PROBLEM I) (Record all the data in a neat table.)

1. Inspect carefully the calorimeter can and the stirrer to find out the type of metal used. Check your textbook for the proper specific heat c. WARNING: The stirrer and the calorimeter may be of different metals. Read 0 - 50° thermometers to the nearest hundredth and the 0 - 110° to the nearest tenth.

2. Weigh the calorimeter can and the stirrer separately and record their masses in the data table. Be sure that they are dry.

3. Fill the calorimeter can 3/4 full of water. Use the extra heating unit that is in the room to heat the water about 10° C above the temperature of the water jacket. After heating, weigh the calorimeter, stirrer and water.

4. Replace the calorimeter can in the jacket, cover, and insert the thermometers. FROM THE BEGINNING TO THE END OF THE EXPERIMENT, STIR CONSTANTLY WITH GENTLY COMPLETE STROKES. Read the calorimeter temperature every half minute and the jacket temperature every minute for five minutes or more. The rubber plug should be kept in the hole in the cover except when adding ice, metal specimen, etc. in order to avoid excess loss of heat.

5. The ice should be on the table for at least ten minutes before being used so that it will reach an equilibrium temperature of 0° C. Five minutes after stirring has begun, add the ice. Before attempting to put the ice through hole in the calorimeter cover, check to make certain that the pieces of ice are not too large. About 70 to 80 grams of ice should be sufficient (this value is not used in the heat balance equations). BE VERY CAREFUL TO WIPE ALL WATER FROM THE SURFACES OF THE ICE AND DO NOT SPLASH ANY WATER OUT OF THE CALORIMETER WHILE ADDING THE ICE.

6. After insertion of the ice, the calorimeter temperature will decrease. Note carefully the time at which it reaches its lowest value. Continue temperature readings for at least five minutes after this time to observe the slow rise in temperature due to the absorption of heat from the surroundings. Remember not to stop stirring - continue throughout the experiments.

7. Remove the calorimeter and reweigh to determine the mass of ice that has melted.

PROCEDURE (PROBLEM II) (Record all data in a neat table.)

1. Spill out about 1/4 of the water in the calorimeter can. The can should now be about 3/4 full. Reweigh to the nearest decigram. Replace the calorimeter can in the jacket, cover, and insert the thermometers. Move so that tip of chute is directly over the hole in cover.

2. Check the temperature of sample metal specimen which has been (during Problem I) suspended in the steam jacket. When the temperature is about 95° and changing very slowly, you can proceed.

3. Before lowering the specimen carefully but quickly into the calorimeter can, read the water temperature and the jacket temperature at half-minute intervals for five minutes. Do not forget to stir gently and uniformly.

4. As the rise in temperature in the water due to the addition of the specimen is about three to four degrees, the water in the calorimeter can should be very near the jacket temperature when the specimen is added.

5. After insertion of the specimen, the temperature will increase. Continue stirring and taking readings on the jacket and calorimeter temperatures at half-minute intervals for five minutes after the calorimeter has reached its highest temperature.

6. Empty and dry apparatus. Clean up the table and make sure the equipment is prepared for the next group.

ANALYSIS (PROBLEM I)

1. Plot a temperature-time curve. Refer to Fig. 9-3 and then determine the corrected change in temperature ΔT for both the drop in the calorimeter temperature and the ice water temperature rise.

2. Set up the proper heat balance equations. Check the specific heats and the masses of all quantities before solving for the Latent Heat of Fusion.

3. The discussion at the conclusion of this report should briefly indicate your ability to locate sources of error. Try to criticize your work and outline exactly how, if the process had been repeated, the errors could have been avoided.

ANALYSIS (PROBLEM II)

1. Plot a temperature-time curve. If room permits, continue on the same graph used in Problem I. Determine the corrected change in temperature for the metal specimen and for the temperature rise of the calorimeter.

2. Set up the proper heat balance equations. Check the specific heats and masses of all quantities before solving for the unknown specific heat of the specimen metal. What kind of metal was used in the specimen?

3. See Paragraph 3 above in Problem I.

QUESTIONS

1. Why should the water in Problem I be above the temperature of the surroundings before the ice is added?

2. Why should the water in Problem II be below the temperature of the surroundings before the hot metal specimen is dropped into the calorimeter can?

3. Does it make any difference in Problem I whether the water is weighed before or after heating? Why?

4. Why should the ice be at an equilibrium temperature of 0° C before being added?

5. What is the per cent difference between your calculated value for the latent heat of fusion and the value indicated in your text?

6. Can you suggest a method that would minimize heat loss to the surroundings when the metal specimen is dropped from steam jacket to calorimeter?

7. If the ice were initially at -5° C, write the heat balance equations necessary to find the latent heat of fusion.

8. What is the per cent difference between your calculated value for the specific heat and the value for the same type of metal in your text?

Experiment 9-2 CALORIMENTRY - MECHANICAL EQUIVALENT OF HEAT

Fig. 9-5. Mechanical (Electrical) Equivalent of Heat Apparatus.

INTRODUCTION

The calorie and the BTU were invented as logical experimental units for the measurement of the quantity of heat long before anyone knew just what heat was. As the concept of energy developed, units such as the erg, the joule and the foot pound, based on mechanical concepts, were set up and used for most forms of energy other than heat. As time passed, it was discovered that practically all forms of energy - mechanical, electrical, sound and light are finally dissipated into the random thermal motion of the molecules of matter, the effect being the same as the addition of heat. Any kind of energy can be measured by dissipating it in a calorimeter and measuring the heating effect. This is not often the most desirable method for measuring energy, since calorimetric methods are subject to many errors and precision is difficult unless great care is taken.

If electrical energy is dissipated in a calorimeter, the heating of the calorimeter and contents will permit the calculation of the amount of heat addition (in calories) which would produce the same result.

PROCEDURE (Review the objectives listed in the introduction to Experiment 9-1. Do not forget to record all the data in a neat table. IMPORTANT: Make certain that the coil is covered by the water before throwing switch.)

1. Connect the heating coil, the fixed resistance, and the A.C. ammeter in series and attach them to the terminals of the power supply switch. Include a circuit diagram of this arrangement in your report.

2. Since the stirrer and resistance with support arms are part of the same unit, their mass is given on the apparatus. Record the mass and resistance in your data table.

3. Inspect carefully the calorimeter can and the stirrer-resistance unit to find out the type of metal used. Check textbook for the proper specific heat c. WARNING - the stirrer and calorimeter may be of different metals.

4. Weigh the calorimeter can and then put enough water into the can so that the resistance coil is completely covered by water. Reweigh the can and water and record the mass of the can and water in your data table. Is it advisable to make more than one weighing? Why? The temperature of the water in the can should be about 15°C below the temperature of the water jacket. Note - the temperature of the water in the calorimeter can be lowered before weighing by the insertion of small pieces of ice.

5. Replace the calorimeter can in the jacket, cover and insert the thermometers. FROM THE BEGINNING TO THE END OF THE EXPERIMENT, STIR CONSTANTLY WITH GENTLE, COMPLETE STROKES. Read the calorimeter temperature every half minute and the jacket temperature every minute for five minutes or more. Read the 0 - 50° thermometer that indicates the temperature of the water in the can to the nearest hundredth and the 0 - 110° thermometer in the water jacket to the nearest tenth. CAUTION - The 0 - 50° thermometer must not touch the resistance coil. Why?

6. After recording the temperatures for five minutes, close the power supply switch. Continue the temperature readings until the calorimeter temperature has risen about 30°C. Read the ammeter every half minute during this period. Record carefully the total elapsed time in seconds from the instant the switch is closed until opened.

7. After opening the switch, continue the temperature readings for 5 minutes. Why?

8. After the current was turned off, why did the temperature of the water continue to rise for a short while before falling?

ANALYSIS

1. Plot a temperature-time curve. Refer to Fig. 9-3 and then determine the corrected change in temperature ΔT for the temperature rise of the contents of the calorimeter.

2. Compute the electrical energy delivered to the coil. Are the units for electrical energy different than those for heat energy?

3. Find the quantity of heat energy required to produce the corrected change in temperature of the contents of the calorimeter. Review the introduction to Experiment 9-1 and do not forget to include the water equivalent of the part of the 0 - 50° C thermometer immersed in the water.

4. Determine the relation between the calorie and the joule, both energy units. What is this relation called?

5. The discussion at the conclusion of this report should briefly indicate your ability to locate sources of error. Try to criticize your work and outline exactly how, if the process had been repeated, the errors could have been avoided.

QUESTIONS

1. Was the process that took place in the calorimeter completely adiabatic? Explain your answer.

2. Could the process in the calorimeter be considered isothermal? Explain.

3. If the resistance is considered as the system being investigated, is there a flow of heat into it? Is there a flow of heat into the water? Is work done on the resistance? If the state of the resistance is considered unchanged, apply the First Law of Thermodynamics to the process. Set up the proper equations but do not solve.

4. While the electrical energy was being supplied to the calorimeter, was there any loss of energy by radiation? By convection? By conduction? Explain.

5. If the instrumental limits of error for the ammeter, resistance and time clock are:

   ±1% of any scale reading on the ammeter,
   ±.02 seconds for the time clock,
   ±.2 ohms for the resistance,

   what is the maximum limit of error L.E. in the indirect measurement for the electrical energy supplied to the resistance coil? Express answer numerically and as a percent (relative) error.

6. What is the probably limit of error L.E. in the indirect measurement for the electrical energy supplied to the resistance coil? Express answer numerically and as a percent (relative) error.

7. What is the percent difference between your calculated value for the mechanical equivalent of heat and the value indicated in your textbook?

8. Why was the initial temperature chosen so that it was approximately as far below the temperature of the surroundings as the final temperature was above?

9. List four possible sources of random and systematic errors.

10. Does the water equivalent of the thermometer affect appreciably the heat gain equation? Explain.

Experiment 9-3 LINEAR EXPANSION

Fig. 9-6. Expansion Apparatus.

INTRODUCTION

The expansion of a pure metal over a limited temperature range is proportional to the change in temperature ΔT and also to the original length L_0. The coefficient of proportionality α is called the coefficient of linear expansion. The change in cross section is very small and can be neglected. The apparatus is shown in Fig. 9-6.

The coefficient α varies for metals from a high value for zinc and aluminum to a very low value for invar. It is important to specify the temperatures between which ΔL is measured because α varies somewhat with temperature. In Fig. 9-7 a typical $L = f(T)$ plot for a metal is shown. Note at point A the change in the slope of the curve. In this experiment, α will be measured for a common metal in the range from room temperature to about 100° C. As the first step in your computation outline, find an expression for α. Define all symbols.

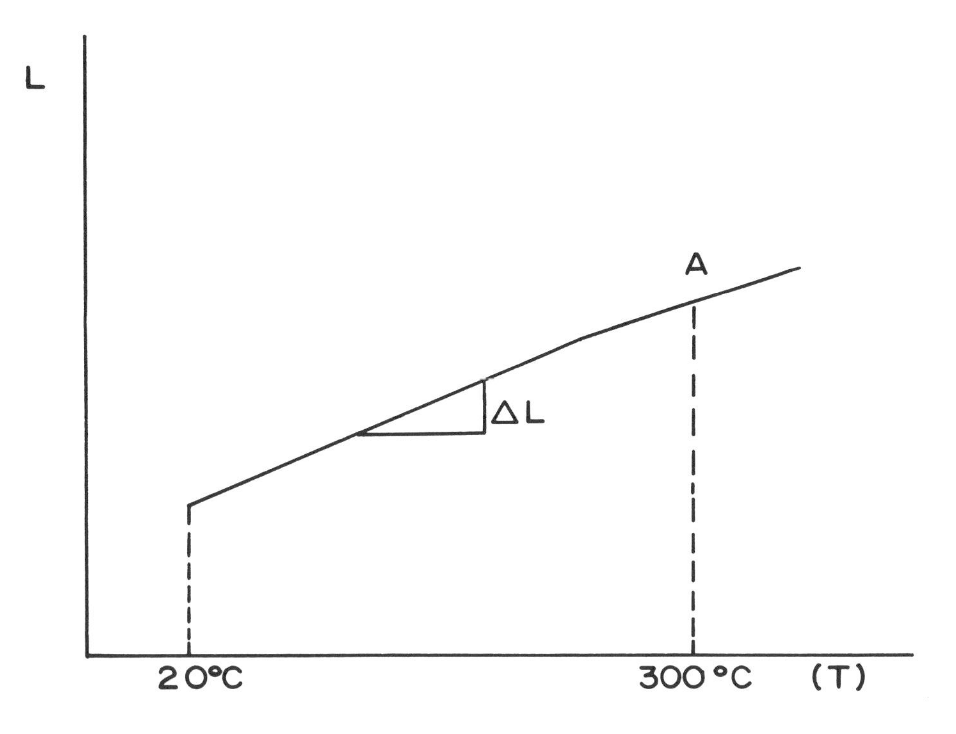

Fig. 9-7.

PROCEDURE (ERROR ANALYSIS IS REQUIRED)

1. Turn the heating unit to high. Fill the boiler 3/4 full of water and start heating. DO NOT CONNECT TO THE STEAM JACKET SURROUNDING THE SAMPLE. Place the heating unit with the boiler as far as possible from the expansion apparatus while getting measurements at room temperature. Why is this necessary?

2. Find the length L_o of the specimen that is enclosed in the steam jacket. Use the meter stick with attached probes. DO NOT REMOVE THE SPECIMEN FROM THE STEAM JACKET. Measure the specimen four times and record the data in the data table. I.L.E. of meter stick = ±0.02 cm.

3. The left end of the expansion apparatus is equipped with an anvil with which the end of the specimen makes contact. The right end of the apparatus contains an adjustable screw micrometer that completes an electric circuit when contact is made with the sample. I.L.E. of the micrometer = ±0.0005 cm. The micrometer permits a ΔL measurement because the left anvil of the apparatus remains fixed throughout the experiment.

   At room temperature take four readings of the temperature of the sample t_o and four micrometer readings of the specimen L_R. A good method is to take readings every minute. I.L.E. of thermometer = ±0.2° C. Record the data in the data table.

4. BACK OFF MICROMETER from the right end of the specimen in order to permit expansion without buckling. Connect the steam generator, heat until the thermometer goes as high as it will, then wait five or ten minutes longer until the sample has reached equilibrium with the steam jacket. Take a new series of thermometer and micrometer readings (four). If the readings show a gradual rise, the sample has not reached equilibrium. If necessary, continue until equilibrium is reached. Call the new micrometer reading L_t and the temperature t_t. Record the data in data table.

   The two series of measurements (at room temperature and heated to equilibrium) allow $L_t - L_R$ and $t_t - t_o$ to be found.

5. Determine α. Use calculus and find the maximum limit of error and the relative error (%) for the indirect measurement α. Remember to show carefully in the COMPUTATION OUTLINE exactly how the maximum limit of error and the relative error for α are found.

QUESTIONS

1. Find the probable limit of error in the indirect measurement α. What is its relative (%) error?
2. List four sources of possible systematic and random errors. Discuss them briefly.
3. Why is it sufficiently accurate to measure the initial length of the sample rod with a meter stick, while the elongation is measured by means of a micrometer?
4. Can you assume from the results of this experiment that the coefficient of linear expansion is the same for all temperatures? Explain.
5. Does the rod expand in more than one dimension? If so, how would this affect the results of this experiment?
6. What are random errors? Are they determinate or indeterminate?
7. Refer to Paragraph 2 in the Procedure section. Compare your R.E. (Random Error) for the four measurements of the length L_o of the specimen with the I.L.E. of the meter stick. Are more observations necessary? Why or why not?
8. What is the relation between linear, area and volume expansion?
9. How does the expansion of water differ from metals? What is the cause of this peculiarity?
10. A clock which runs correctly when the temperature is 25° C has a pendulum made of a slender steel rod with a heavy ball at the lower end. The temperature drops to 10°C. Does the clock gain or lose? How many seconds per day?

Experiment 9-4 THERMAL CONDUCTIVITY

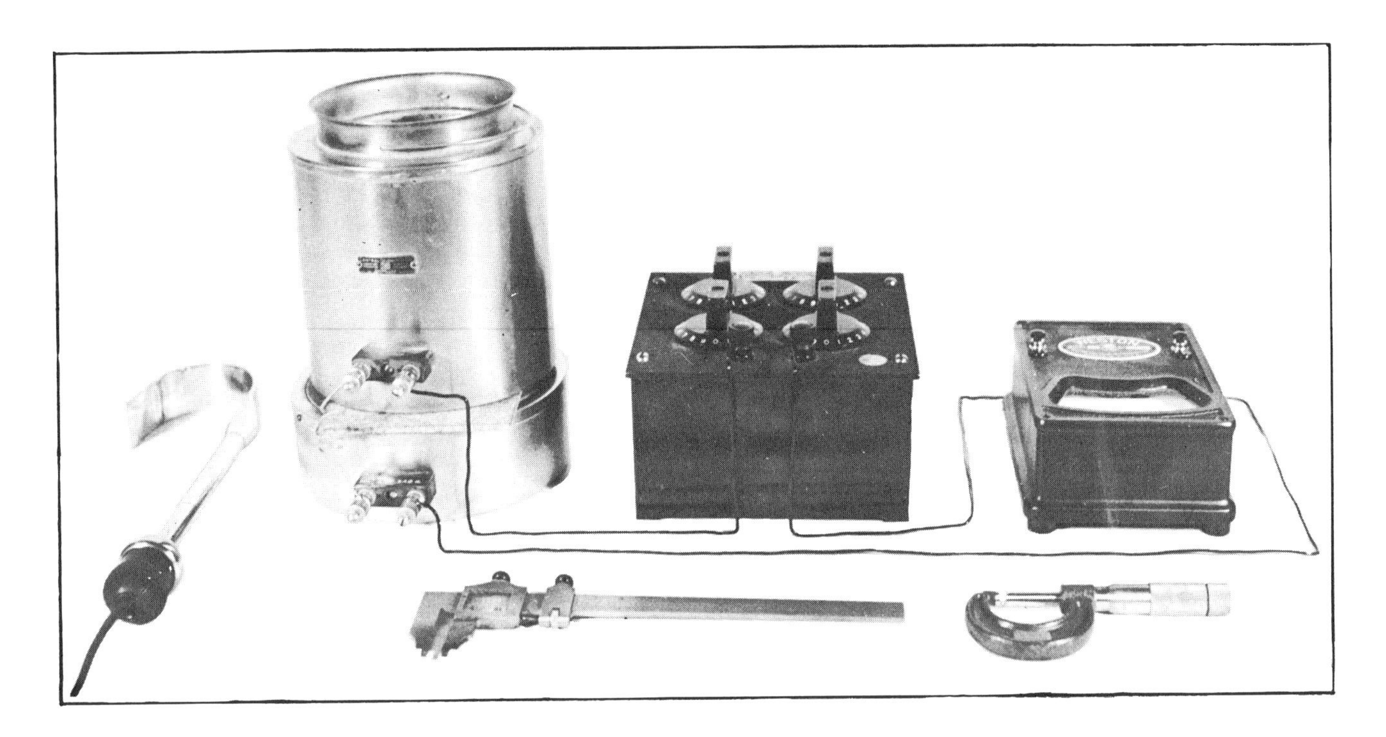

Fig. 9-8. Thermal Conductivity Apparatus.

INTRODUCTION (Symbols used in the equations are defined in your theory text.)

The steady state flow of heat (energy) through a slab of material can be expressed as:

$$H = \frac{KA(T_2 - T_1)}{L} \qquad 9\text{-}4$$

If a slab of material is placed between a constant temperature source (A, in Fig. 9-9) and a receiving unit consisting of a thermally insulated cylindrical block of known thermal capacity B, then the amount of heat conducted through the slab per unit time is equal to that received by the copper block (if no heat escapes). The upper face of the sample slab in contact with the constant temperature source A is at a constant temperature T_2, while the temperature of the lower face T_1 changes slowly. The heat (energy) that must be supplied to a body of mass m and specific heat c to increase its temperature through an interval ΔT is:

$$dQ = mc\Delta T \qquad 9\text{-}5$$

This must be differentiated with respect to time in order to consider the rate of increase in the temperature due to the change in T_1. Equation 9-5 then becomes:

$$H = mc\frac{dT}{dt} \qquad (\text{where } H = \frac{dQ}{dt}) \qquad 9\text{-}6$$

If no heat escapes from the copper block, the heat conducted through the slab per unit time is equal to that received by the block per unit time or:

$$H = \frac{KA(T_2 - T_1)}{L} = mc\frac{dT}{dt} \qquad 9\text{-}7$$

The temperature differences are measured by two copper-constantan thermocouples. One is placed in the source and the other in the receiving unit. In Fig. 9-9, the "hot" junction of the thermocouple is embedded in the heavy copper base of the source A. The "cold" junction is embedded in the receiver B which is a face-ground, nickel-plated, heat insulated copper plug. The leads of both are brought to outside binding posts.

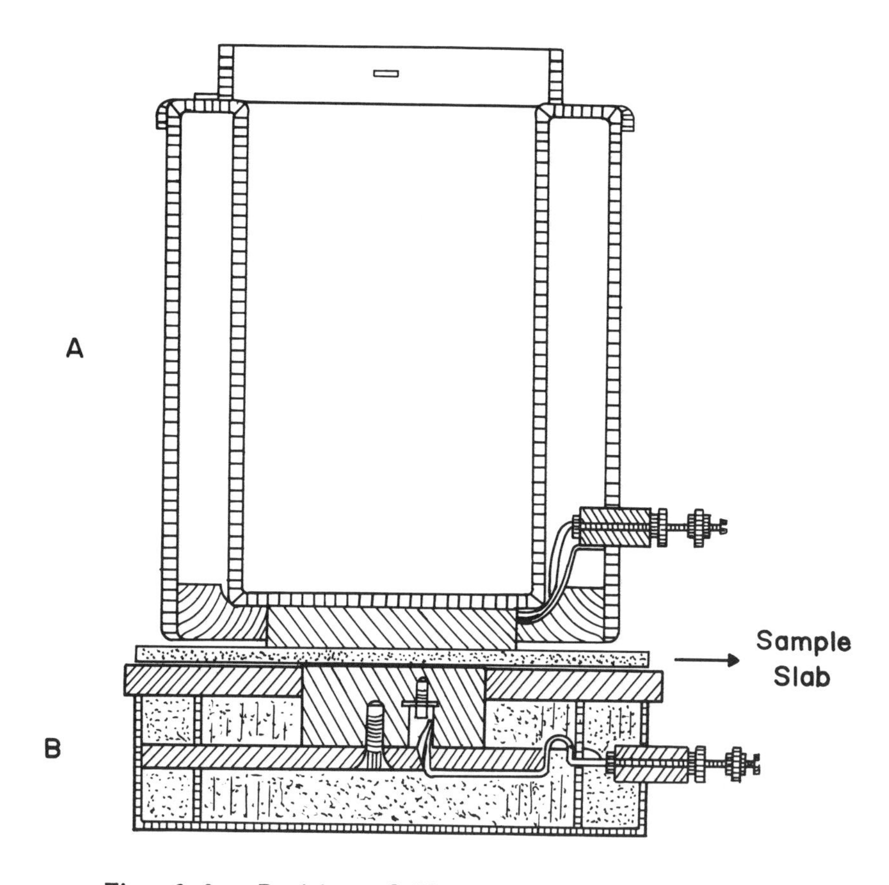

Fig. 9-9. Position of Thermocouples.
(Courtesy of the Central Scientific Co.)

When two metals are joined to form a closed circuit (Diagram II in Fig. 9-10) and the junctions are kept at different temperature ($T_2 > T_1$), a detectable current passes through the closed circuit - in this case from copper to constantan at the hot junction. The current is due to a Seebeck potential difference. (See text on Electricity and Magnetism for more detail.)

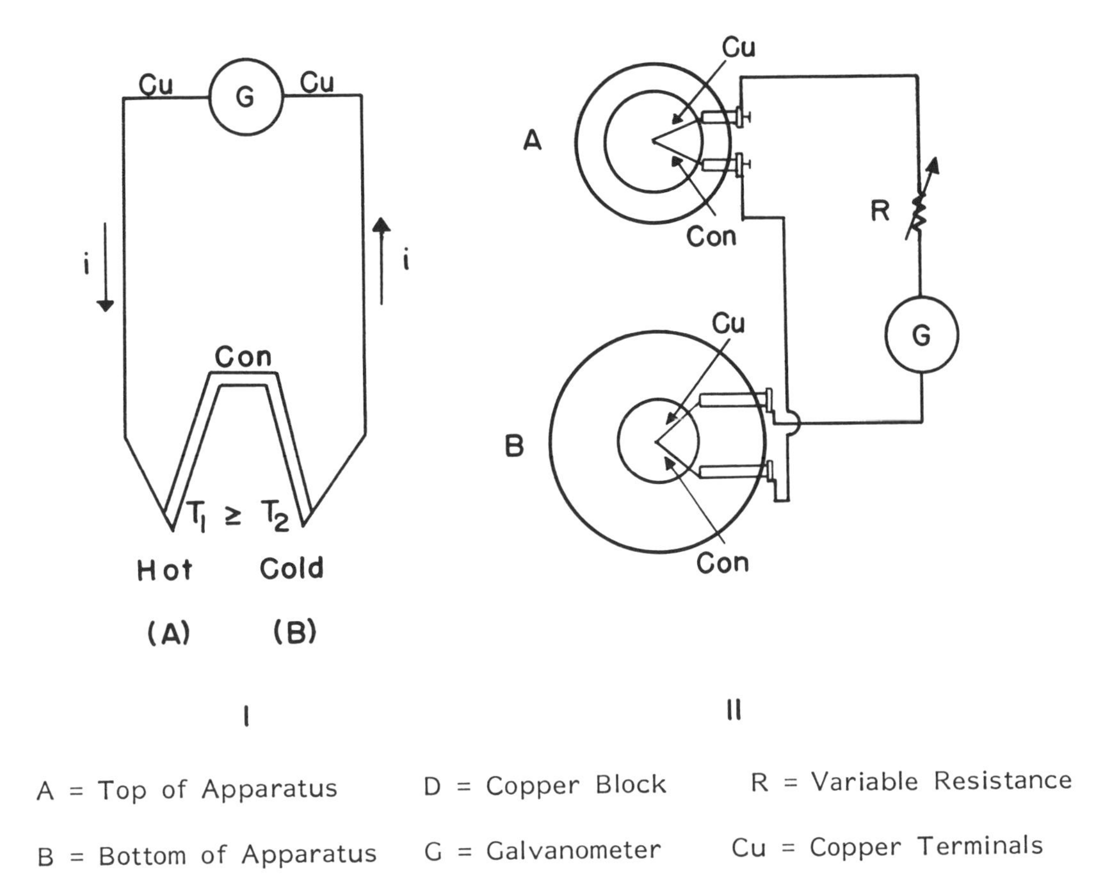

A = Top of Apparatus D = Copper Block R = Variable Resistance

B = Bottom of Apparatus G = Galvanometer Cu = Copper Terminals

Con = Constantan Terminals

Fig. 9-10. Thermocouple Wiring Diagram.

When the temperature difference is small, the potential difference produced by the thermocouples is proportional to the temperature difference. The current caused by the potential difference is also proportional to the temperature difference of the two junctions $(T_2 - T_1)$, or:

$$I = C(T_2 - T_1), \tag{9-8}$$

where "C" is a constant of proportionality which corresponds to the current produced when junctions (T_2, T_1) are at unit temperature difference.

Since the current i is measured by a galvanometer with a linear scale, the deflections are proportional to the current. Deflections of the galvanometer are substituted for currents in Eq. 9-8. Because T_2 remains constant and T_1 changes, any change in T_1 causes a change in current. Thus, the instantaneous change in the current i that corresponds to an instantaneous change in the temperature block $\frac{dT}{dt}$ can be found by differentiating Eq. 9-8.

$$\frac{di}{dt} = -C\frac{dT}{dt}, \qquad 9\text{-}9$$

where the (-) sign is required because of the flow of heat (energy) from a high temperature reservoir to one of low temperature.

When Equations 9-8 and 9-9 are substituted into Eq. 9-7,

$$\frac{KAi}{L} = -mc\frac{di}{dt}. \qquad 9\text{-}10$$

If the variables are separated and the initial conditions $i = i_o$ at time $t = 0$ are used, then: (See Eq. 3-5.)

$$\ln_\varepsilon i = \ln_\varepsilon i_o - \frac{KAt}{Lmc}, \qquad 9\text{-}11$$

which expressed in $\log_{10}$ becomes

$$\log_{10} i = \log_{10} i_o - \frac{KAt}{2.3\ Lmc}. \qquad 9\text{-}12$$

The actual slope of the graph of $\log_{10} i$ versus t is:

$$a = -\frac{KA}{2.3\ Lmc}. \qquad 9\text{-}13$$

The objectives of this experiment are to:

1. Demonstrate the difficulties present in finding the thermal conductivities of commercial insulators.
2. Introduce the use of the thermocouple as a relative temperature measuring device.
3. Demonstrate that small errors in technique in measuring the thickness of the sample, the area of the plug, improper galvanometer reading, presence of dirt on sample, air film between sample and source of receiver, etc., will affect the final results.
4. Determine the thermal conductivity K of the sample.

PROCEDURE (Read several times before setting up data table. Refer to Chapter III and read the sections on galvanometers and resistance plug boxes.)

1. Place the immersion heater into the source. Fill the source with enough water to completely cover the base of the heater. Plug in the heater. WARNING - HEATER MUST BE COVERED WITH WATER BEFORE PLUGGING IT INTO THE A.C. LINE OR IT WILL BURST.

2. Set the source with the heater on a flat board. The source must be heated as high as possible before it is placed on the receiving unit (sink).

3. Measure the thickness L of the sample slab four times with a micrometer. Record the four measurements and their average in the data table.

   Measure the diameter of the cylindrical plug in the receiving unit (sink) four times with the vernier calipers. Record the four measurements and their average in the data table. This measurement permits the calculation of the area A of the plug. The area of the plug is equal to the effective area of the sample slab through which the heat (energy) is flowing.

   Record in the data table the mass of the cylindrical plug in the receiving unit. The mass is given on the insulator for the thermocouple terminals. Look up the specific heat of copper in your theory text.

4. Check the terminals of the thermocouples to make sure that proper connections are made: copper washers - copper wire; constantan washers (silver) - constantan wire.

5. Do not touch the terminals after the experiment has begun - it will vary the galvanometer reading.

6. The sample slab, the bottom of the source and the top of the receiving unit (sink) must be clean and dry. We want to measure the conductivity of the sample and not that of dirt and water.

7. After the source is heated, place the source with the immersed heater on the receiving unit (sink). Heater should remain in the source during the entire experiment. Be sure that water covers the heater.

8. Place about 4 kilograms on top of the source so that no air pockets between sample and metal faces of source and sink exist.

9. Galvanometer: HANDLE GALVANOMETER WITH CARE.

   a. Always push No. 1 button first and lock and then the No. 2. Pushing the No. 2 button makes the galvanometer more sensitive and less current is required for a deflection. When No. 1 is pushed, an internal protective shunt resistance protects the galvanometer from current overload.

   b. Button No. 1 should not be pushed until after the heater is unplugged and the source with the enclosed heater is placed on the receiving unit.

   c. Just after the heated source is placed on the top of the sink, the galvanometer reading may still increase. This is due to the unheated sample. It is absolutely necessary that the readings of the galvanometer are decreasing (toward zero - either from a + or - direction).

   d. Galvanometer needle can be set by careful variation of the resistance in series with the meter. When the needle drops to full-scale deflection, time runs begin and readings are taken every minute for ten minutes.

10. Record the time and the corresponding galvanometer deflection every minute for ten minutes in the data table. Convert to seconds.

11. After taking data, empty the water out of the source and make sure the apparatus is dry and clean for the next group.

ANALYSIS

1. Plot $\log_{10}$ i vs. t on millimeter rectangular coordinate paper.

2. Plot i vs. t on semi-log paper. Check the data carefully in order to choose the proper type of cyclic semi-log paper.

3. Determine, from the actual slope of each graph, the thermal conductivity K of the sample.

4. Discuss at the end of your report the significance of the variation in your values for K. Explain which technique (Step 1 or 2) is preferable when precision is desired. Do not forget to discuss sources of error and how they might have been minimized.

QUESTIONS

1. Substitute Equations 9-8 and 9-9 in the INTRODUCTION into Eq. 9-7 and show that

$$\frac{KAi}{L} = -mc\frac{di}{dt}.$$

2. Separate the variables in Eq. 9-10, use the initial boundary conditions ($i = i_o$ at time $t = 0$) and show that:

$$\ln_\varepsilon i = \ln_\varepsilon i_o - \frac{KAt}{Lmc}.$$

3. Why is the area of the copper plug in the receiver B in Fig. 9-10 less than the area of the copper base of the source A?

4. Draw a thermocouple circuit that could be used to measure the temperature of a lighted match. How could you calibrate the thermocouple to read in centigrade degrees C°?

5. How could you determine the thermal conductivity of a metal?

6. What is the purpose of the variable resistance in series with the galvanometer?

7. Is the loss of heat (energy) due to convection and radiation in this experiment negligible? Explain your answer.

8. How could you determine the energy loss due to convection?

9. Express K in terms of Btu-in/hr-ft^2 - F°.

10. If there was excess energy loss during the process, would the plot of $\log_{10}$ i vs. t be a straight line? Explain.

11. Would an air film between the sample and the source or the sample and the receiving unit cause the value for K to be higher or lower?

9.2 Introduction to Microprocessor Heat Experiments

A microcomputer system* that can also be used with many different physics experiments including the heat experiments previously described is now introduced. It is designed to improve the general quality of the physics laboratories by providing a source of motivation and valuable experience with equipment similar to that found in research and industry. In addition, the microcomputer allows the scientific observation of phenomena that occurs beyond the normal limits of our perception.

This microprocessor is quite unique since it can be used directly with the average quality oscilloscopes found in most laboratories. As a result, other devices, such as a monitor and additional interfacing circuitry are not needed.

By using a hardware design with 2000 bytes of memory RAM and 4000 bytes of program ROM, a low cost multipurpose microprocessor possesses many functions. It is very adaptable and simple to use by the laboratory student. It has two analog inputs, one analog output, and 12 bits of digital input or output. It will display graphs, word text, and/or commands on a CRO. It will also accept input from a keypad, print results on hard copy and perform calculations via the CPU. The system contains an easily replaceable 'experiment chip' (Eprom) that contains an experiment ready for execution. When this program chip and its corresponding interface are changed, a new experiment can be performed.

The microcomputer chosen was the KIM-1; a single board machine with a 6502 central processor chip. This computer contains a keyboard and display, terminal interface, tape storage facilities, and 1000 bytes of read-write memory (RAM). It is supported by a small monitor located in an onboard read-only memory (ROM). The board by itself did not have enough I/O ports to support the planned uses so it was decided to expand the KIM's capabilities by adding external interface ports. Also added were an additional 1000 bytes of RAM and 4000 bytes of program ROM.

We soon realized that the expanded system lacked one major ability in its performance as a laboratory device - the only method of communication with the user was by means of a six-digit display. Considering that we already had designed the graphic output to be used with a moderate quality laboratory oscilloscope, we decided to design an alphanumeric display generator to produce text on the CRO along with the graphic information. These functions together with a two-channel analog input port make up the majority of the custom circuitry added to the basic KIM-1 computer.

With the extensive amount of hardware design came a great need for a set of general purpose sub-programs which could be called upon to perform specific functions. These functions include reading in analog data, driving the CRO to generate graphs and text, and performing calculations. As a result of this need, a software package was de-

*Support for the development of the microprocessor experiments was provided in part by grants from Erik and Margaret Jonsson of the Texas Instruments Corporation. The following students prepared the routines for the microprocessor: Chris D. Delise, Gary Polcz, Sam Fortin and Philip H. Grove, among others. The development project was supervised by Harry F. Meiners. Circuit diagrams and monitor routines can be obtained from Professor Meiners. These details will enable any school to duplicate the experiments and the microprocessor.

veloped that contains on a single memory chip the following:

(1) all of the I/O functions for a particular experiment;

(2) small video editor;

(3) several double-precision mathematical routines: and

(4) a flexible monitor program which displays menus of program options for student selection.

The computer system includes one additional ROM socket for the installation of an 'experiment chip' (Eprom) which may contain several experiments ready for execution. By providing this rather large set of system subroutines, individual experiment programs can be quite small, and programming time is decreased. The microprocessor is shown in Fig. 9-11.

Fig. 9-11. Microprocessor.

Experiment 9-5 THERMAL CONDUCTIVITY WITH MICROPROCESSOR

INTRODUCTION

These instructions are a supplement to the laboratory Experiment 9-4 on Thermal Conductivity. They explain how the experiment is conducted when it is interfaced with a microprocessor. Since the method of data acquisition is slightly different, the following describes the procedure necessary to successfully complete the experiment. Fig. 9-12 shows the Thermal Conductivity apparatus including the Microprocessor.

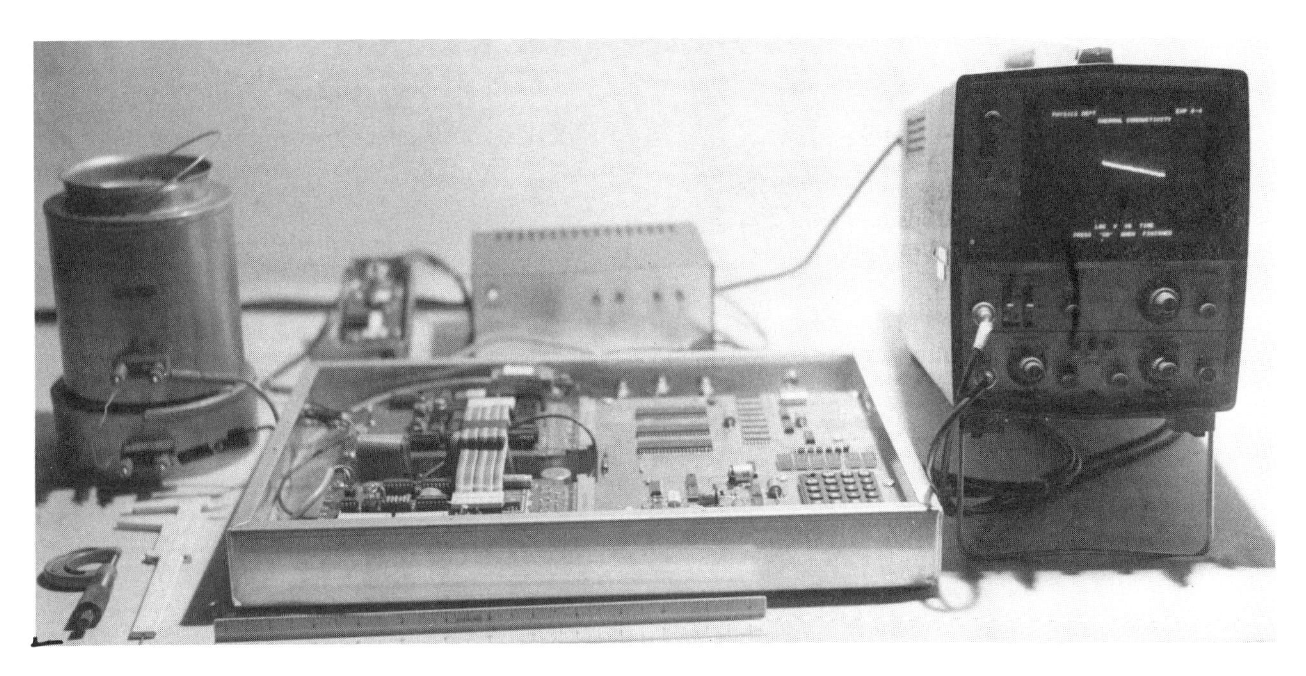

Fig. 9-12. Thermal Conductivity Apparatus.

In order for the microprocessor to be interfaced with the constantan-copper thermocouple, the induced current that is proportional to the temperature difference between the two junctions must be converted into a proportional voltage. Fig. 9-13 shows how the connection is made between the thermocouple and the microprocessor. Details of the amplifier are given in Fig. 9-14.

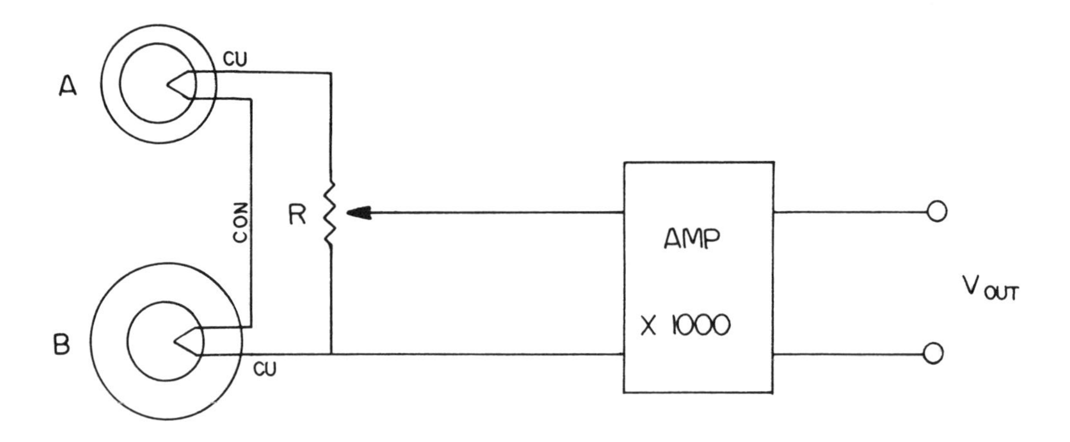

A = Top of Apparatus
B = Bottom of Apparatus

Cu = Copper material
Con = Constantan Material

Fig. 9-13. Thermocouple Wiring Diagram.

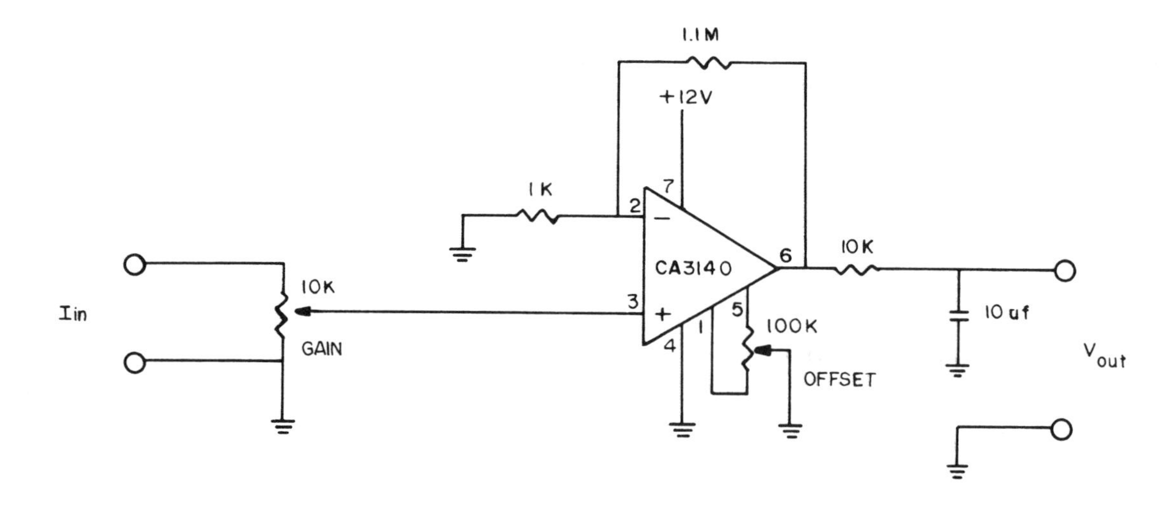

Fig. 9-14. Amplifier Circuit.

PROCEDURE

1. Heat the source with an immersion heater while stirring frequently. The heater must be covered with water before it is plugged in. Heat the source as high as possible before placing it on the receiving unit (sink). Heating the source should take about twenty minutes.

2. Measure the thickness L of the sample slab four times with a micrometer. Record the four measurements and their average in your data table.

3. Measure the diameter of the cylindrical plug in the receiving unit (sink) four times with a vernier caliper. Record the four measurements and their average. Calculate the area of the plug. The area of the plug is equal to the effective area of the sample slab through which the heat (energy) is flowing. Record in data table.

4. Record the mass of the cylindrical plug in the receiving unit (sink). The mass is given on the insulator for the thermocouple terminals.

5. Look up the specific heat of copper in your theory text.

6. Do not touch any of the apparatus or wires during the experiment, because it will affect the readings. The thermocouple amplifier is very sensitive.

7. The sample slab, the bottom of the source, and the top of the receiving unit (sink) must be clean and dry.

8. After the source is heated, leave the immersion heater in the water. Stir the water so that it is at the same temperature at both the top and bottom. Leave the stirrer in the water and place the source on the receiving unit (sink).

9. Press key 'O' on the KIM to leave the system monitor and run "RUN", which is the program which runs the experiment.

10. If the screen goes blank when "RUN" begins, the voltage is too high for the KIM to read. Wait for the screen to come up again before continuing the experiment.

    As the current passes through the constant resistance R, a voltage drop across the resistor is created. This voltage drop is proportional to the current i by:

    $$V = iR \qquad 9\text{-}14$$

    If we combine C from Eq. 9.8 with R from A-1 and let C' = RC be our new constant of proportionality, we obtain:

    $$V = C'(T_2 - T_1). \qquad 9\text{-}15$$

    Because T_2 remains constant and T_1 changes, any change in T_1 causes a change in the voltage. Thus, the instantaneous change in the voltage V that corresponds to an instantaneous change in the temperature block dT/dt can be found by differentiating Eq. 9-15.

    $$dV/dt = -C'\ dT/dt, \qquad 9\text{-}16$$

    where the (-) sign is required because of the flow of heat (energy) from a high temperature reservoir to one of low temperature. When Equations 9-15 and 9-16 are substituted into Eq. 9.7,

$$\frac{KAV}{L} = -mc\,\frac{dV}{dt}\,, \tag{9-17}$$

where K is the constant of proportionality called the thermal conductivity. If the variables are separated and the initial conditions ($V = V_o$ at time $t = 0$) are used, then:

$$\ln V = \ln V_o - \frac{KAt}{Lmc}\,, \tag{9-18}$$

which expressed in log (base 10)

$$\log V = \log V_o - \frac{KAt}{2.3\ Lmc}\,. \tag{9-19}$$

And we find that the actual slope of the graph of log V versus t is:

$$a = -\frac{KA}{2.3\ Lmc}\,. \tag{9-20}$$

11. When the voltage reading on the screen begins to decrease, hit a key to begin taking data. For best results, the voltage should start above 1.10 volts. The experiment runs for ten minutes, taking a reading every six seconds. As the experiment progresses, a plot of log V vs. T is displayed. When the experiment is over, press "GO" to return to the monitor.

12. To look at graphs of Voltage vs. Time and Log of Voltage vs. Time, press key '1' to run "ANALYSE". "ANALYSE" will also calculate the slope of the Log vs. Time graph, although you should do it yourself. Using Eq. 9-20, the thermal conductivity K = -2.3*a*L*M*c/A. Calculate the thermal conductivity constant K for your sample slab.

13. If your KIM is connected to a 300 baud printer, you may run "PRINT", which will print your data. Be sure to turn the printer on. Press key '3' to "PRINT" the Time, Voltage, and Log of Voltage at the end of each 6 seconds.

14. If you need to stop a program at any time, press 'RS' which will return you to the monitor.

15. For your analysis, follow the instructions in your laboratory book, but substitute voltage for current.

ANALYSIS

The accuracy of this experiment depends partly on the materials being used. For example, when porous cork is selected, the results are greatly affected by the techniques that are applied to measure the thermal conductivity. If cork is compressed by a large mass, it becomes more dense and the thermal conductivity changes. Therefore, published values for the thermal conductivity of cork vary greatly. This is not the case, as our results demonstrate for dense substances such as Teflon, Plexiglass, Phenolic Paper, and Glass. The thermal conductivity of these materials checks very well with known values.

Since this equipment is run by a microprocessor, numerous temperature readings are made with ease. The readings are then averaged over a six-second period before the data is given to the experimenter. This type of averaging allows the data to be very smooth and illustrates the running experiment more clearly, as can be seen in Fig. 9-A, Voltage vs. Time. This graph was obtained directly from the screen of the CRO. Notice that the graphs of the raw data are also very smooth showing a curve that corresponds to the theory in predicting an exponential curve. Once the data is plotted in semi-log fashion, the graphs are also very linear. Thus, the data corresponds to the theory of a logarithmic relation between volts and time. In Fig. 9-B is shown a graph of the Log Voltage vs. Time obtained from the screen of the CRO.

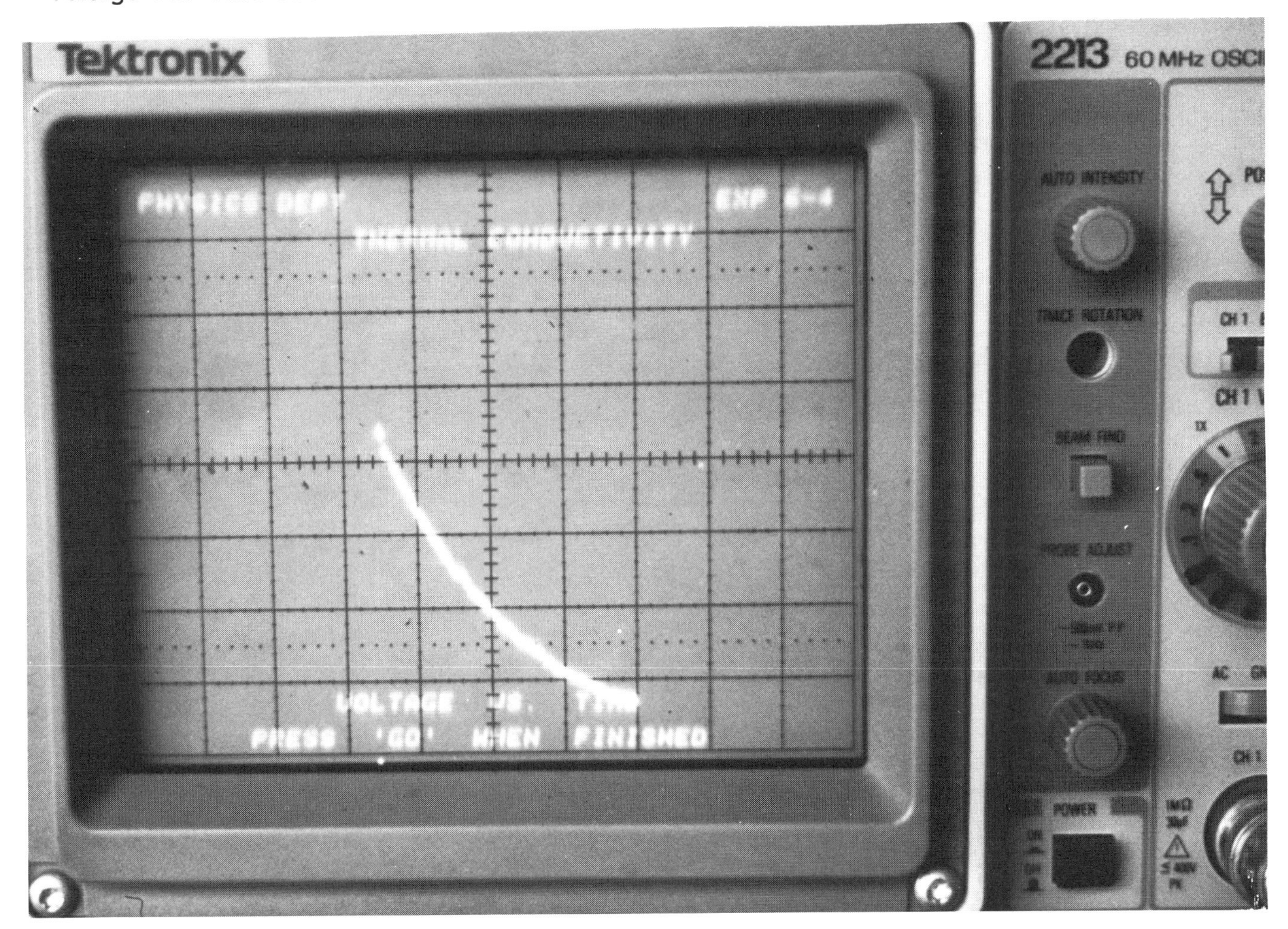

Fig. 9A. Voltage vs. Time.

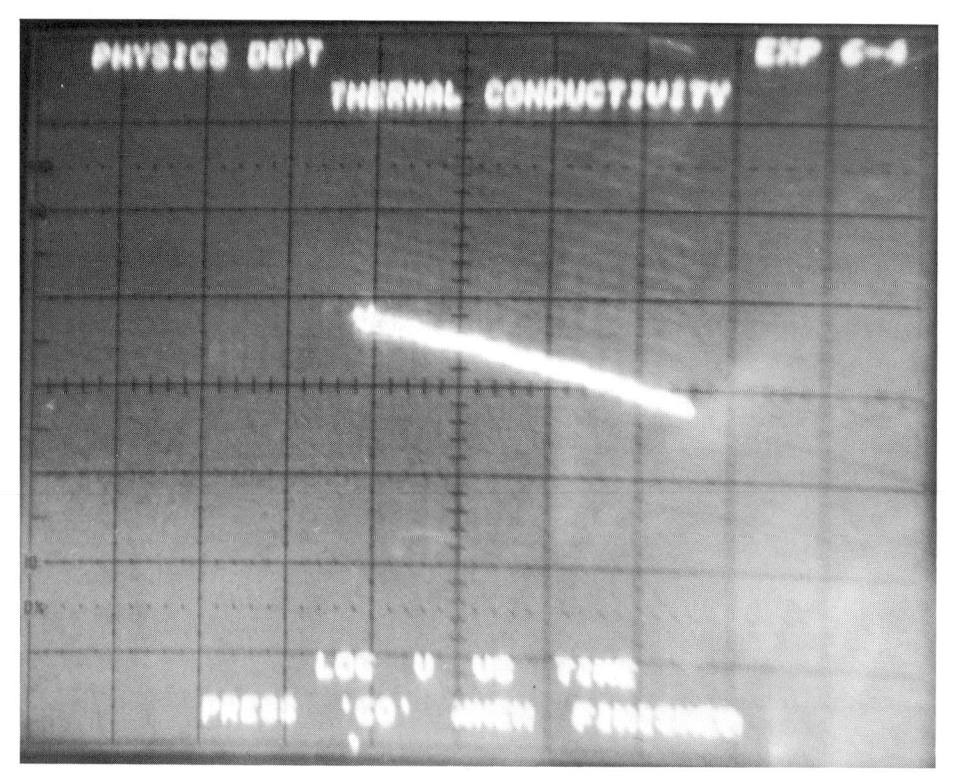

Fig. 9B. Log Voltage vs. Time.

In summary, the accuracy also depends to a limited extent on the apparatus used for the following reasons: the computer is limited to 8-bit resolution; the heat source and receiving unit do dissipate a very small amount of heat that is not accounted for in the experiment; and the input amplifier to the system is very sensitive since it is amplifying the thermocouple signal by 1000 times. Not withstanding these limitations, this microprocessor system gives very good results with a high degree of repeatability.

Experiment 9-6 CALORIMETRY EXPERIMENTS

9-6-1 Latent Heat of Fusion

9-6-2 Specific Heat of Metal

9-6-3 Mechanical (Electrical) Equivalent of Heat

INTRODUCTION

Essentially, the Calorimetry Experiments are performed by the laboratory student, as outlined in this chapter pp. 409-421. Therefore, study carefully the instructions presented previously that apply to the heat experiment that you will do with the microcomputer. Assisting in the data acquisition portions and the analysis sections of each experiment is the microcomputer. Designed with the necessary equipment to change the temperature measurements into a form that the computer can understand, the microcomputer's main duties include taking temperature measurements at regular timed intervals and generating a "real-time" display of these measurements in the form of a graph on a conventional oscilloscope. These instructions are a supplement to the laboratory Experiments 9-1, Calorimetry-Specific Heat and Latent Heat of Fusion and 9-2, Calorimetry-Mechanical Equivalent of Heat.

After the student is finished with a particular data run, the choice will be given of saving the results away in memory for later examination and continuing on with other trial runs, or the ability of "stepping" through the graph to determine such things as initial starting temperature, final temperature, the change in temperature, and the duration of time during which electrical energy was being supplied to the fluid within the calorimetry apparatus.

Advantages of this system versus conventional laboratory techniques include better data measurements (less error and more reliability), and more motivation on the part of the student. The student will see how a rather simple machine can be interfaced to a "real world environment" and how it interacts successfully with it. These experiments are not meant to be "black box machines" whose purpose is to do all the work, but rather to assist the student using 'state of the art' equipment. This is probably the most important aspect of what these systems intend to accomplish. They will provide the student with more accurate results, obtained in a less tedious fashion, and will afford the student with more time to concentrate on the theory and error analysis portions of the experiments.

While calorimetry is straightforward in its procedure - the student has at his disposal a machine capable of recording temperature readings hundreds of times each second. Obviously a student cannot accomplish such a task; yet to the microcomputer, this is relatively simple. One of the more interesting aspects of the program that will accomplish this is a routine to record successive measurements over a 5 to 7 second interval and average these values to obtain a "mean" temperature. This will tend to limit the influence of systematic errors such as A.C. interferance within the fluid medium, temperature discontinuities within the fluid due to convection heat fluxes, and perhaps some of the irregularities of stirring on the part of the student.

The computer will also provide the student with a real time graph in the form of a Temperature vs. Time plot on an oscilloscope. On the face of the CRO - the student will see this graph grow instantaneously with time, as shown in Fig. 9-17. The temperature within the calorimetry apparatus is observed in the form of a floating decimal number, calibrated in degrees Celsius. The student will not be required to think in terms of "Hexadecimal Notation" provided by the limited LED display. Instead, all the interaction between the student and the computer will be displayed on the CRO.

MONITOR

On the CRO, you will see a listing of the options available to you. The first 3 options, numbered 1, 2, and 3 correspond to the 3 heat experiments. See Fig. 9-15. One will be performed during a laboratory period. Below these are the options listed by the letters A, B and C. Letter A refers to the option titled: "Analyze." You will use this option later on during the experiment. B refers to the option titled "Printout." Essentially, this option will be used after you are finished with the experiment run. It is used to printout the temperature values of your Temperature vs. Time plot. C refers to "Calibrate." This is an option intended primarily for adjusting the oscilloscopes so that both character and graphical information can be seen by the student. THIS OPTION SHOULD BE USED ONLY ONCE - BEFORE THE EXPERIMENT IS STARTED. AFTER THAT, THE OPTION SHOULD NOT BE SELECTED AGAIN. If small adjustments in vertical or horizontal positioning are necessary, make them with assistance of a teaching assistant.

KEYBOARD

As you might have noticed, the keyboard is unlike most others that you are accustomed to. The keyboard has keys numbered 0-9, A-F, and special keys marked: "RS", "ST", "GO", "+", "AD", and "DA". Only two of these special keys are of interest to you. The RS key, when hit, will cease operation of the experiment program and return you to the monitor. As mentioned before, the monitor lists several experiment options available to you. While in the monitor, pressing the key corresponding to a desired option will begin executing that option. In other words, if you were in the Monitor and you selected option B (Printout), then by hitting the letter B on the keyboard, a printout of your results would come out on a printer - providing, of course, that a printer was connected to your system and turned on. The other key of importance is labeled GO. The program will tell you when to use this key. See Fig. 9-16 for a diagram of the temperature interface.

PROCEDURE

1. Select the experiment you wish to perform (Hit 1, 2, or 3 on the keyboard). (See Figs. 9-17, 9-18, 9-19.)

2. You should now see the experiment name and number at the top of the oscilloscope screen. You are asked to input the calorimeter jacket temperature to the nearest 0.1 degrees Celsius. Since the computer has no decimal point, you should enter the temperature assuming an "invisible" decimal point after the second digit entered. For example, if the jacket temperature was 32.6 degrees Celsius, then you would enter the three digits: 3, 2, and 6. After this you would hit "GO". If you make a mistake and hit "GO" too soon, hit "RS" and start over again at Step 1

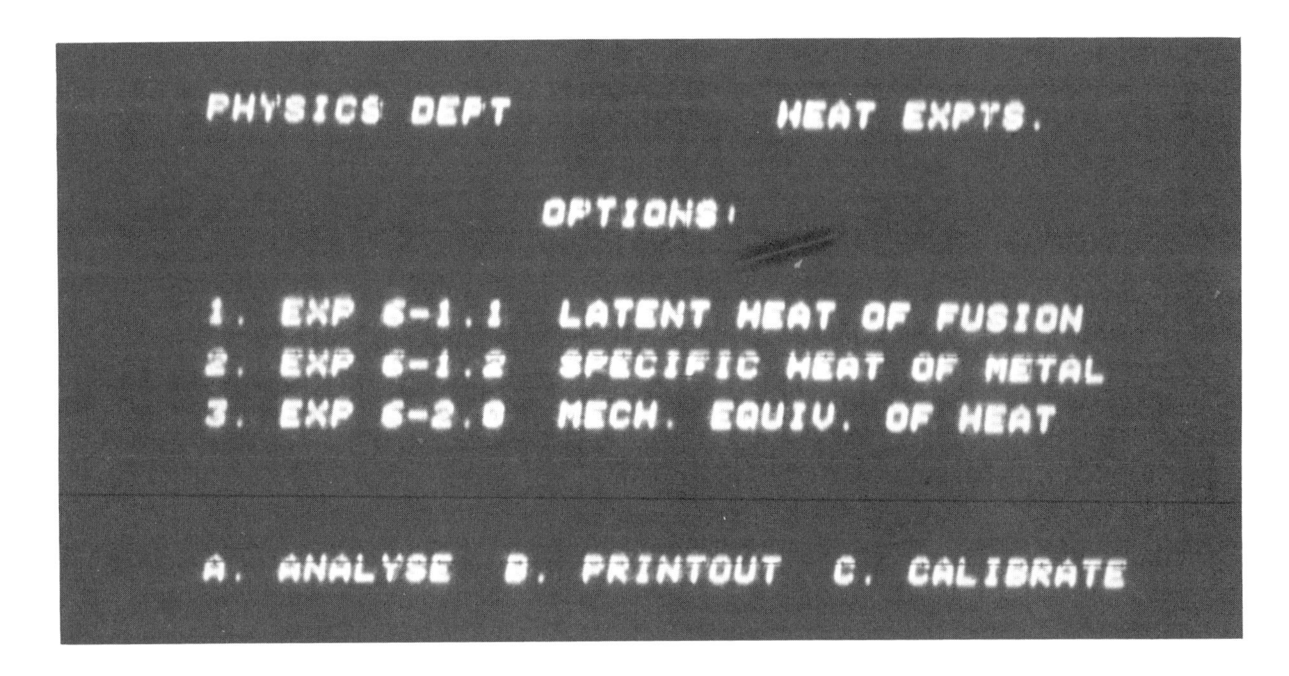

Fig. 9-15. List of Options for Heat Experiments. Photo was taken of CRO screen. Actual image on screen is sharp and bright.

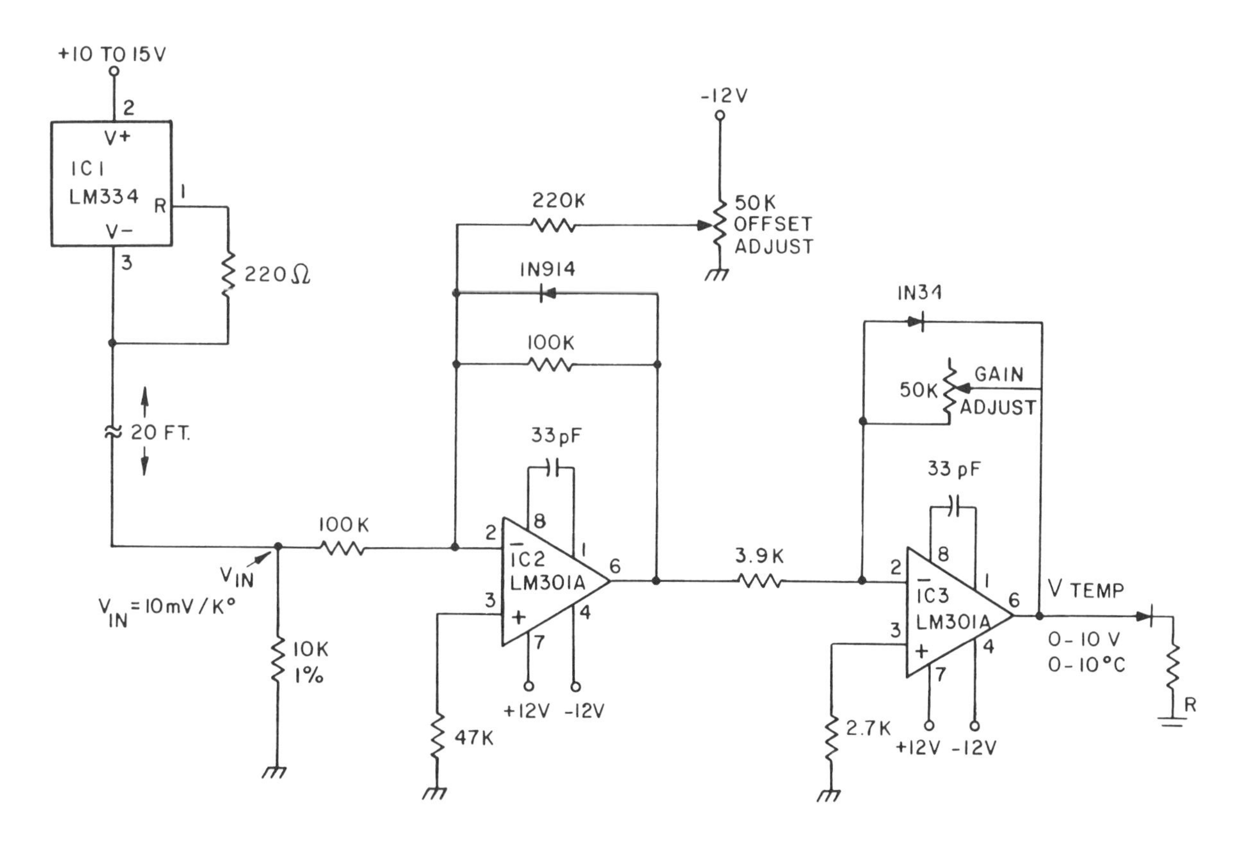

Fig. 9-16. Diagram of the Temperature Interface.

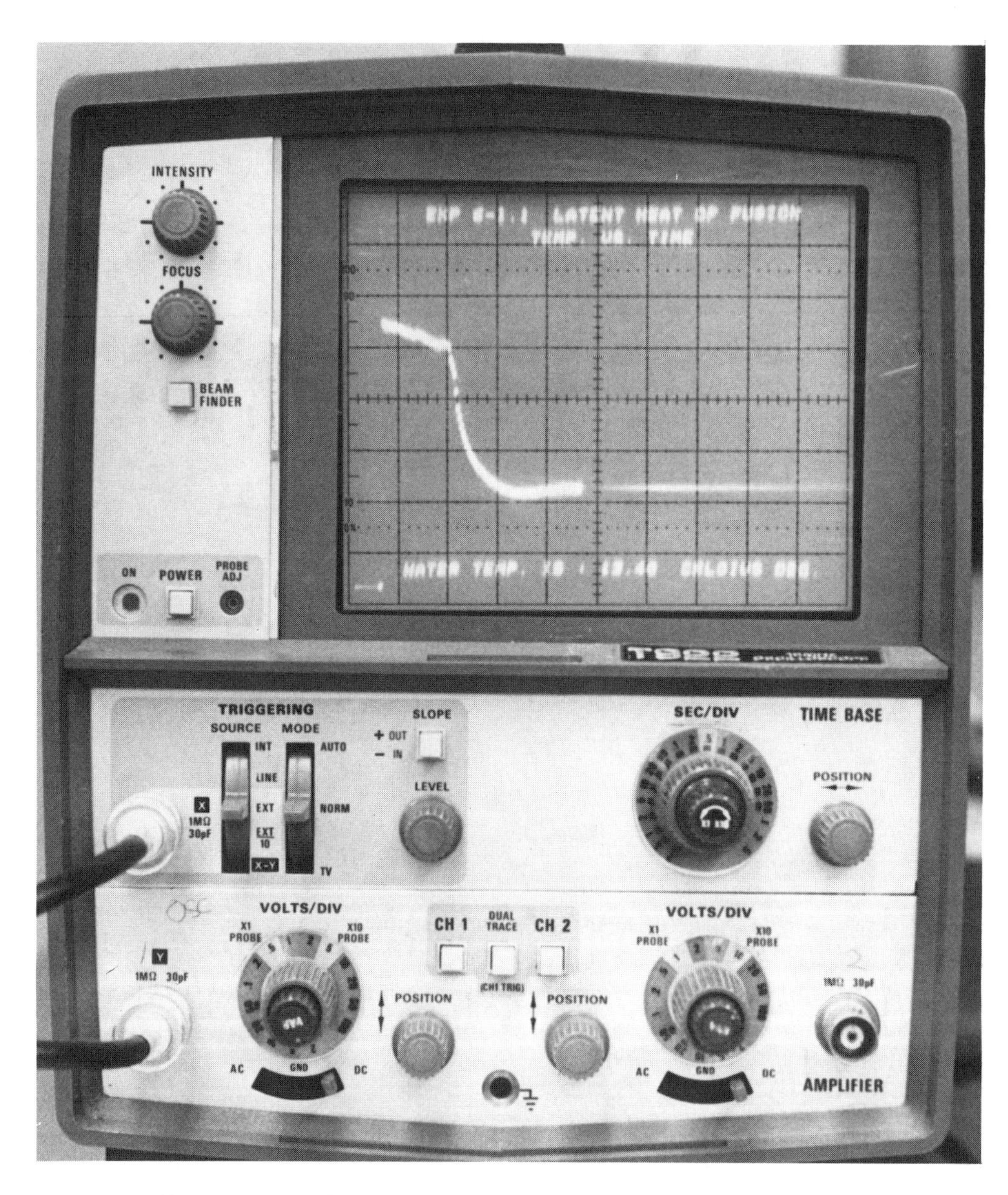

Fig. 9-17. Temperature vs. Time Plot for the Experiment on the Latent Heat of Fusion.

Fig. 9-18. Specific Heat Apparatus.

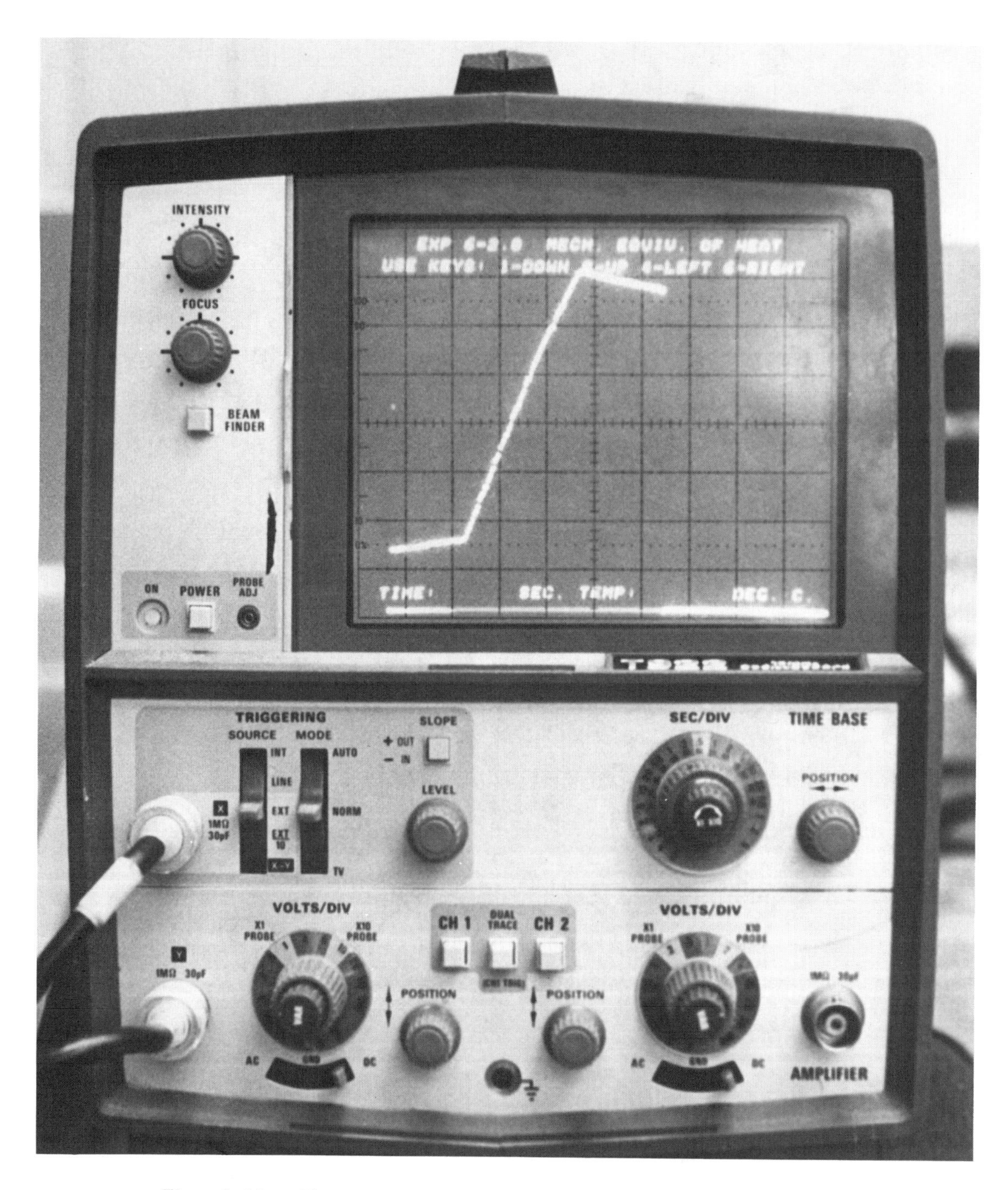

Fig. 9-19. Temperature vs. Time Plot for the Experiment on the Electrical (Mechanical) Equivalent of Heat.

on Page 440. If you hit a wrong number but haven't hit "GO" yet, press "D" to delete the incrorrect numbers entered, then re-enter the appropriate ones.

3. Now you should see a continuously changing water temperature readout at the bottom of the oscilloscope screen. This section of the program allows you to heat up or cool down the calorimeter apparatus before you start taking data. See Note.*

4. When you hit GO, the computer will start taking temperature measurements. A plot of TEMP. VS. TIME will grow in real time as data is added to the graph. IT CANNOT BE OVER-EMPHASIZED THAT DURING THE TIME THE COMPUTER STARTS RECORDING DATA UNTIL THE TIME IT FINISHES, YOU MUST STIR CONSISTENTLY AND UNIFORMLY. For best results, hold the stirrer between thumb and middle finger. The computer will "blip" every time a point is added to graph. It should be noted that a new point is added to the graph every 7.56 seconds and that a complete graph consists of 256 points.

5. When the computer has finished taking 256 points, a short "razz-like" tone will be heard from the computer indicating the graph is completed and will be displayed on the oscilloscope.

6. You should now select option A for "Analyze." The oscilloscope should display one blinking dot. This dot is the jacket temperature indicator. At the bottom of the CRO screen you will see a time/temperature indicator. By using the direction keys: 1, 9, 4, and 6, (as displayed on the CRO), you can move another dot about the screen. This is a "cursor" and by holding down the proper key, this "cursor" will move about the CRO face. Place your finger on key 9 (up). Alternately press key 6 (right). You should note that this dot will move up and to the right. Try these keys yourself to become comfortable with their operation. Key E or key RS will return you to the Monitor. As the point moves about the screen, the corresponding time/temperature information will be displayed at the bottom of the screen. So, if you wanted to know the temperature of the calorimeter at a particular point on the graph - you would move the "cursor" (using keys: 1, 9, 4 and 6) on top of the graph to the point of interest. Here, you could read the temperature and the time that the point added to the graph directly from the screen.

---

*Note: (The following information applies only to the Mechanical Equivalent of Heat Experiment.) It is advisable to preset the heating element current supply to 4 amps. Your teaching assistant should have done this for you. When the computer turns on the heater element, a red light will turn on (located on the experiment interface) indicating that current is being supplied to the heating element. You should start your timer then, and stop it when the light goes out. The computer turns on the heater at 18 degrees Celsius below your entered jacket temperature and turns it back off at the jacket temperature plus 18 degrees (T+18). SO BEFORE YOU PRESS GO (TO START TAKING DATA), THE WATER TEMPERATURE OF THE CALORIMETER CAN SHOULD BE 19-20 DEGREES CELSIUS BELOW JACKET TEMPERATURE AND BEGINNING TO INCREASE SLOWLY. USE ICE TO COOL THE WATER IN THE CALORIMETER CAN DOWN TO THIS POINT. IMPORTANT: INSERT THE AC PLUG OF ELECTRICAL (OR MECHANICAL) EQUIVALENT OF HEAT FROM THE INTERFACE INTO THE TOP OF THE VARIAC.

ANALYSIS

There are two methods of obtaining the corrected change in temperature of the calorimeter apparatus. Please use both methods.

Method 1

While in the "Analyze" mode, place a clear sheet of acetate over the oscilloscope screen. Use a small ruler to extend or to project the slopes of the "before" and "after" intervals (see this Chapter, p. 413) using a water soluble marker. Move the cursor over to the blinking dot indicating the jacket temperature. If you are at the same temperature as the dot (the vertical position will be identical between the blinking dot and the cursor), then use key 6 to move your cursor over and on top of the graph. Once there use key 9 to move the cursor up until it lies on top of the projected line made with the soluble marker. Read the temperature at the bottom of the screen and record it as T_1. Move down using key 1 until you intercept the other projection. In the same fashion, read the temperature and record it as T_3. The formula: $T_1 - T_3$ = CORRECTED CHANGE IN THE CALORIMETER TEMPERATURE.

Method 2

Use option "B" to "Printout" the temperature and time values on a printer. Plot these points on a piece of graph paper from the back of your laboratory book. Refer to p. 412 for an explanation of corrected ΔT using the extrapolation of two slopes.

Experiment 9-7 DETERMINATION OF A THERMODYNAMIC CONSTANT

INTRODUCTION

A length of precision bore glass tubing is inserted with the aid of a rubber stopper A into a 10-liter aspirator. If a steel ball that fits closely within the tubing is released, damped harmonic motion results.

When the ball is in equilibrium (y = 0), the equilibrium pressure (see Fig. 9-20) due to the slight compression of air is (neglecting friction),

$$P = P_o + \frac{mq}{A} \quad . \qquad 9\text{-}21$$

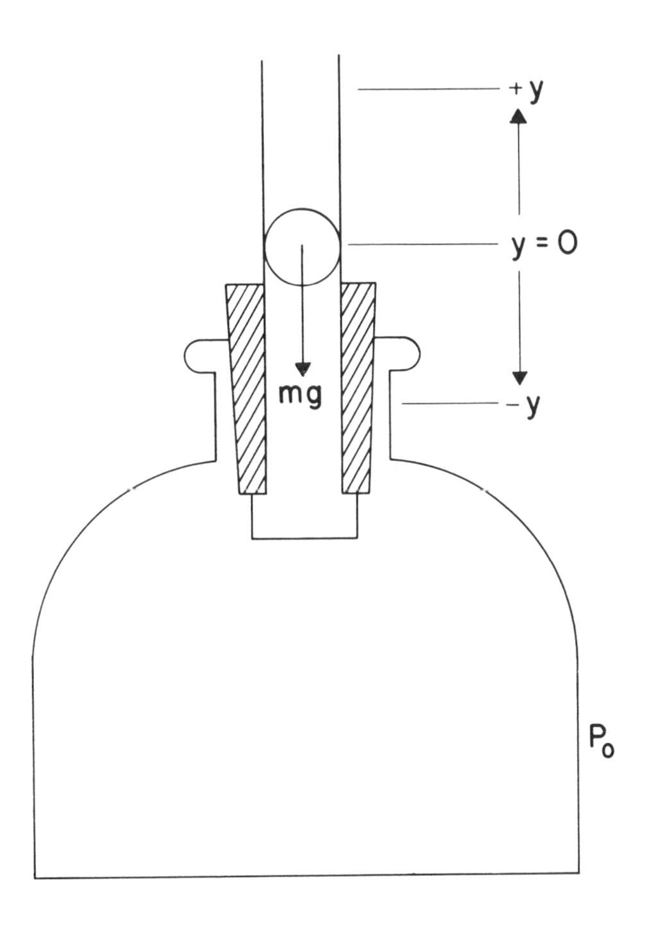

Fig. 9-20. Steel Ball in Equilibrium.

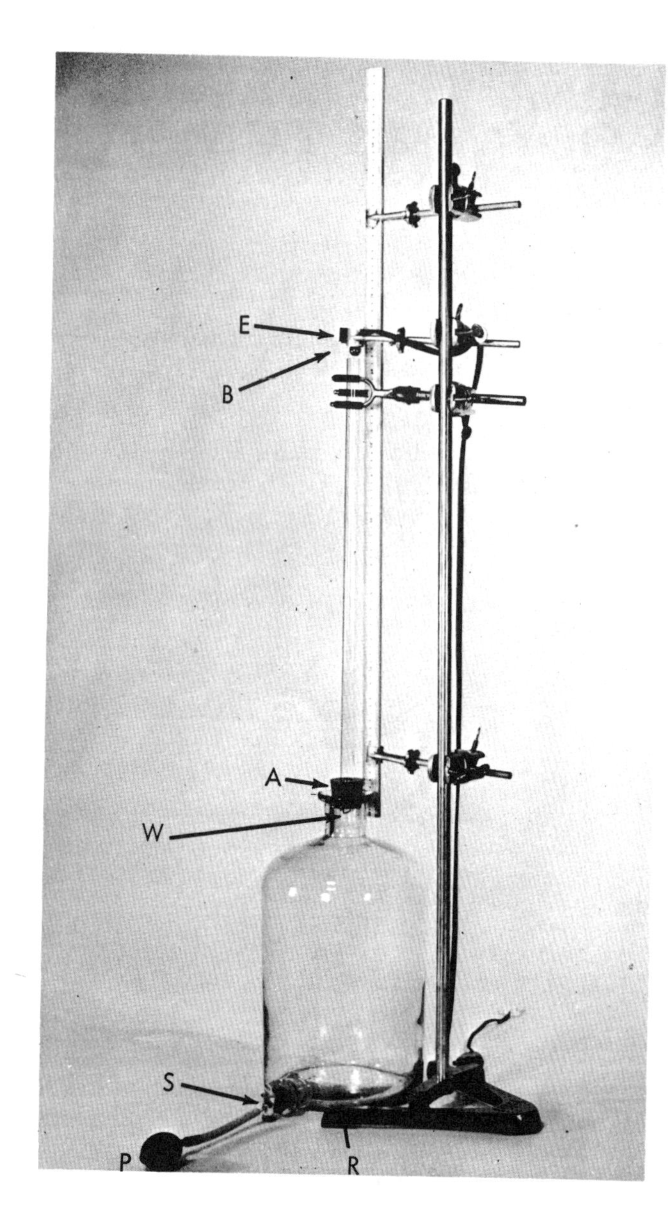

Fig. 9-21.
Apparatus Used to Determine $\gamma = C_p/C_v$.
E = Electromagnet - 110 V. A.C.
B = Steel ball. A = Rubber stopper.
S = Stop-cock. P = Pressure bulb.
W = Wire stop. R = Rubber pad.

A positive upward displacement +y of the ball above equilibrium causes a small increase in volume dV with respect to the equilibrium volume V,

$$dV = A\,dy. \qquad 9\text{-}22$$

The small positive displacement +y also causes a small <u>decrease</u> in pressure,

$$dP = \frac{dF}{A}. \qquad 9\text{-}23$$

For a positive +y displacement, the <u>resultant</u> force dF on the ball is negative and corresponds to a restoring force. Since the ball oscillates somewhat rapidly, the changes in P and V can be considered as a quasi-static <u>adiabatic</u> process. Actually, because of the small fluctuations in P and V, the gas passes through various states of equilibrium and the process is represented by

$$PV^{\gamma} = C. \qquad 9\text{-}24$$

PROCEDURE (Use care in handling glass. Do not twist the precision glass tube in the rubber stopper. Students will pay for any glass breakage due to carelessness. Maximum charge $35.00. Select either Method I or II.)

Method I (Electromagnet and pressure bulb used.)

1. Inspect the system and find out how it operates. Don't put hands across the 110 V. A.C.; use a dry chamois to handle the ball. Use level and make certain the tube is vertical. If the steel ball sticks in the tube, do not force the ball upward by using the pressure bulb. If the ball scratches inner wall of precision tube, you will pay for another! When the ball sticks, the tube is either dirty or not vertical. If, after checking with the level, the ball still sticks, notify our instructor and move to another apparatus. The tube will have to be cleaned.

2. If the ball hits the wire stop on the first drop, the system is not air tight. Check the seal of the rubber stopper A and stop-cock S. Use care while adjusting tube in the stopper. Twisting will break glass.

3. Find the average period T of ten (or more) oscillations. Record.

4. Repeat Step 3 and record, as accurately as possible, the maximum and minimum positions of the ball for each of ten oscillations.

5. Record the atmospheric pressure, P_0.

Method II (Electromagnet and pressure bulb not used.)

1. Use the equipment provided and assemble the apparatus. Precision glass tube must be carefully cleaned before using. Inside of tube must be free of dirt, or grease. Use strong soap solution and wash carefully with a chamois plunger. Rinse and then dry with chamois. Run, finally, silicon eye glass cleaner paper through the tube. Before using steel ball, wash and dry. Use dry chamois to handle ball.

2. Use level and make certain that tube is vertical. Read, again, the INTRODUCTION and ANALYSIS and determine what you want to do. Ball should oscillate about 25 times if system has been cleaned properly! [Note - Method II requires twice as much time as Method I - getting the system to operate without breaking glass is quite a challenge. Method II will require use of the laboratory at times when classes are not scheduled. See your instructor and arrange for permission.]

ANALYSIS

1. Differentiate Eq. 9-24 and substitute from Eqs. 9-22 and 9-23, dV and dP. Solve for dF. Check the result with your instructor before proceeding. What does the equation imply about the force? Does the motion approximate S.H.M.? Explain.

2. Find an expression for the period T in terms of the parameters of the system.

3. Given mass of ball m = 1.67×10^{-2} kgm, cross-sectional area of the tube A = 2.01×10^{-4} m^2, and the volume of the aspirator V = 1.09×10^{-2} m^3, find γ, the thermodynamic constant. What is the meaning of γ?

4. Express γ in terms of the fundamental dimensions M.L.T. Do the fundamental dimensions cancel?

5. From Step 4, Method I of the PROCEDURE, find the total distance that the ball fell the moment it was released (before rising to complete the first oscillation). The ball was released from a position where the pressure was exactly atomospheric P_0. If y is measured from this initial position, the loss of potential energy is equal to the work done in compressing the gas,

$$\int_0^L F\,dy = +mg \int_0^L dy \qquad 9\text{-}25$$

[Note: Steps 1 to 4 use Rüchardt's method. Step 5 is Rinkel's method.]

6. Find γ by Rinkel's method.

7. Compare γ found by Rüchardt's and Rinkel's methods. Discuss any variation. Would you expect Rinkel's method to be more accurate than Rüchardt's? Explain.

8. [Extra Credit] Plot amplitude of oscillation versus time (A vs. t) for ten oscillations.

   (a) Why does the equilibrium position shift?

   (b) Neglect the shift in the equilibrium position and try to obtain an empirical equation that will fit the graph. Hint - See Fig. 7-39 and Eq. 7-45 in Experiment 7-20.

   (c) If the shift in equilibrium was not neglected, suggest how to fit the graph.

9. In damped harmonic motion is the period T constant? Explain.

Experiment 9-8 KINETIC THEORY MODEL*

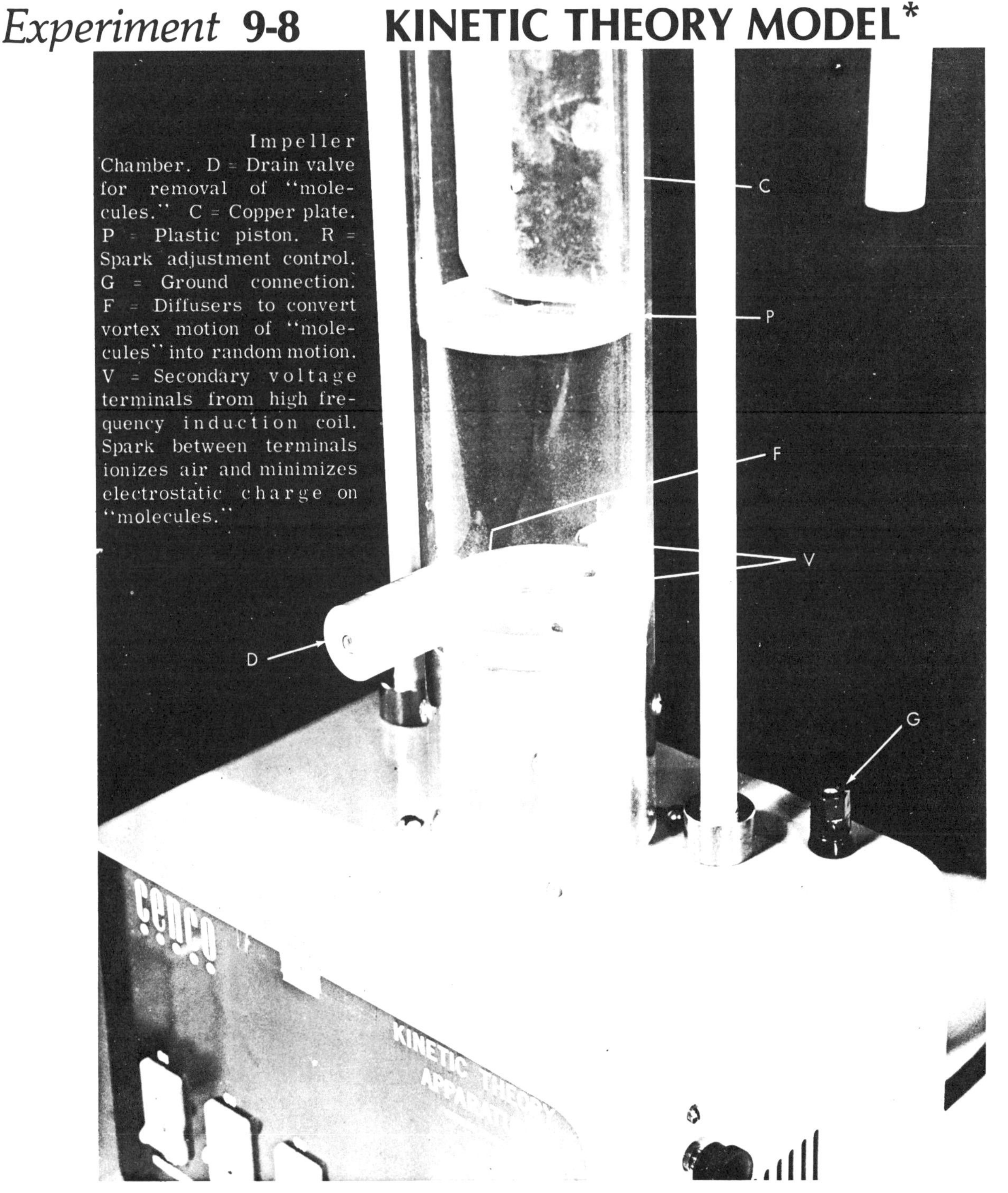

Fig. 9-22. Kinetic Theory Apparatus. M = Permanent Magnet. C = Copper plate. D = Drain valve for removal of "molecules." W = mass holder on top of piston. R = Spark adjustment control. G = Ground connection. B = Counter balance weight. O = Window for stroboscope calibration. T = Drain container for collecting "molecules" from impeller chamber. S = Stroboscope.

*Adapted from apparatus developed by Harry F. Meiners, Professor of Physics, Rensselaer Polytechnic Institute. Kinetic Theory Apparatus can be obtained from the Fisher Scientific Corp., Educational Materials Division.

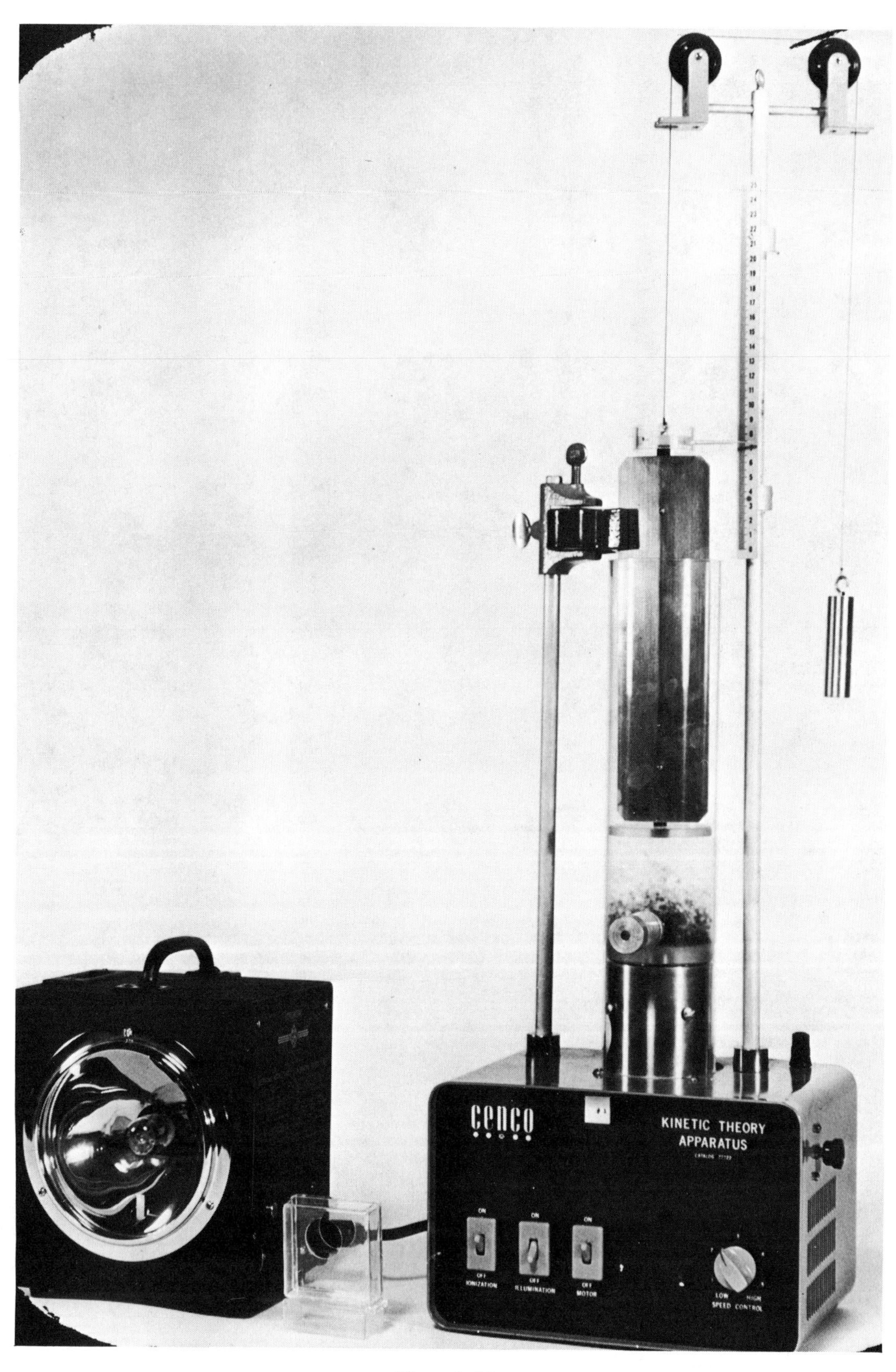
cenco
KINETIC THEORY
APPARATUS
ON
OFF
IONIZATION
ON
OFF
ILLUMINATION
ON
OFF
MOTOR
LOW
HIGH
SPEED CONTROL

Fig. 9-23

INTRODUCTION

A gas of tiny plastic balls or "molecules" exerts a pressure against the piston. The pressure is the average effect of many minute blows caused by the "molecules" striking the piston. The piston oscillates somewhat because of the not quite random molecular impacts.

We will assume, when the piston oscillates about some position H above the impeller which drives the "molecules," that the kinetic energy input from the impeller is almost equal to the total random kinetic energy of all the "molecules."

The finite size of the "molecules" and the effects of gravity and electrostatic forces require the ideal gas equation to be modified. Because van der Waals' equation of state contains a correction for volume b and another correction for the pressure (a/V^2), we will try to apply a similar expression to our model. Obviously, our gas is not real and the assumptions apply only to our system.

Since the volume V_b of a tiny plastic ball (see Fig. 9-24), is

$$V_b = \frac{4}{3}\pi\left(\frac{Db}{2}\right)^3, \qquad 9\text{-}26$$

and the centers of two "molecules" can approach only to a distance equal to the diameter D_b of one "molecule" (Fig. 9-24) each "molecule" has an effective volume,

$$\frac{4}{3}\pi D_b^{\,3} = 8\,V_b, \text{ where } V_b \text{ is the volume of one molecule.} \qquad 9\text{-}27$$

The probability that a "molecule" can be found in a volume V (impeller chamber, for example) is proportional to V. If a second "molecule" is placed into the volume, the space remaining is $V - 8\,V_b$; for a third, $V - 2 \times 8V_b$, etc. The probability of finding n_o "molecules" in V is proportional to,

$$V(V - 8V_b)(V - 2 \times 8V_b)\ldots\ldots[V - (n_o - 1)\,8V_b],^* \qquad 9\text{-}28$$

where n_o = the total number of "molecules," and V can be replaced by the $n_o{}^{th}$ root of the product. Since we know V_b is very small, and $n_o 8V_b$ is still small when compared to V, the product in Eq. 9-28 can be replaced approximately by,

$$V^{n_o} - 8V^{n_o - 1}\,V_b(1 + 2 + \ldots\ldots\ldots(n_o - 1) \cong V^{n_o}(1 - \frac{8V_b n_o^{\,2}}{V\;2}\;. \qquad 9\text{-}29$$

If the $n_o{}^{th}$ root is taken, we find,

$$V\left(1 - \frac{n_o 4V_b}{V}\right) = V - b_1, \qquad 9\text{-}30$$

where b_1 is the unavailable volume (or excluded volume) due to the volume occupied by the individual "molecules." Therefore,

$$b_1 = n_o 4V_b\;. \qquad 9\text{-}31$$

*See Born, Atomic Physics, G.E. Stechert & Co, N.Y., (1936 Edition), pp. 263-264.

Inspection of Fig. 9-25 shows that an additional correction term b_2 is required because the "molecules" can approach only to a distance $D_o/2$ of the plastic wall. Hence,

$$b_2 = \frac{D_b S n_o}{2n} , \qquad 9\text{-}32$$

where D_b = diameter of a molecule, S = total internal surface area of the impeller chamber, n_o = total number of molecules and n = effective number of molecules at height H. The effective concentration of "molecules" n at height H is obtained by considering the amount of work required to lift a "molecule" through a distance H. The decrease in "molecules" with H is,

$$n = n_o e^{\frac{-MgH}{C\omega^2}} \qquad 9\text{-}33$$

where MgH is the gravitational potential energy of each "molecule," and $C\omega^2$ (see Eq. 9-41) is a reference energy factor.

From Equations 9-31 and 9-32, the total unavailable volume b is,

$$b = b_1 + b_2 \quad \text{or} \qquad 9\text{-}34$$

$$b = 4n_o V_b + \frac{n_o D_b S}{2n} \qquad 9\text{-}35$$

For convenience, Eq. 9-35 can also be written as,

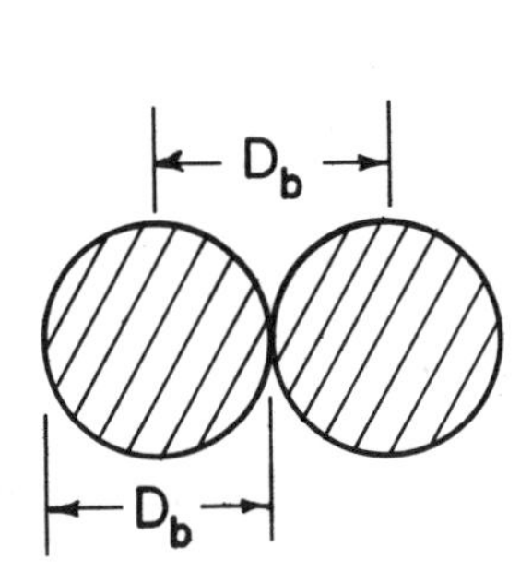

Fig. 9-24. The centers of two molecules can approach only to a distance equal to the diameter of one molecule.

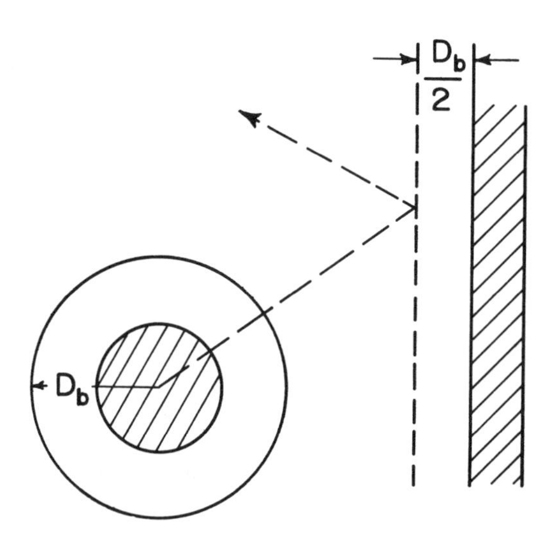

Fig. 9-25. Molecules can approach only to a distance from the wall equal to one-half of their diameter.

$$D_b^{\ 3} + D_b\left(\frac{2V}{D} + A\right)\frac{3}{2n\pi} - \frac{3b}{2\pi n_o} = 0\ , \qquad 9\text{-}36$$

where A = area of the piston and D = diameter of the piston.

Inspection of the "gas" shows that the plastic balls, moving at high speed, become electrostatically charged due to friction when hitting against each other or against the walls of the impeller chamber. "Molecules" cohere to the walls of the impeller chamber, bottom of piston and to the impeller. Even though the charge is minimized by the ionization of the air in the chamber, we can perhaps interpret the cohesion as having the general effect of reducing the pressure. We will also assume, because of the cohesive forces, that one element of volume acts on another of equal size with a force perpendicular to N^2, where N is the number of molecules per cm^3. If the total number of molecules n_o in the volume V of the chamber is $n_o = NV$, then the pressure reduction can be written as,

$$k\left(\frac{n_o}{V}\right)^2 \qquad 9\text{-}37$$

where k is a factor depending on the <u>special properties of our system</u>. Since $a = kn_o^{\ 2}$, Eq. 9-37 becomes,

$$\frac{a}{V^2}\ . \qquad 9\text{-}38$$

When the correction terms a and b are used for P and V, we obtain

$$(P + \frac{a}{V^2})(V - b) = K. \qquad 9\text{-}39$$

Since the energy <u>input</u> is obtained from the impeller, we will assume that the total random kinetic energy K of all the molecules is,

$$K \cong \frac{1}{2}\, n_o M\, \overline{v^2}^{\,av.}\ , \qquad 9\text{-}40$$

where m = mass of one "molecule." Eq. 9-40 can also be expressed as,

$$K \cong \frac{1}{2}\, n_o M\ \overline{(\omega r)^2}^{\,av.}, \text{ or}$$

$$K \cong n_o C\omega^2, \qquad 9\text{-}41$$

where ω is the angular velocity and r is the effective radius of the impeller.

Eq. 9-41 now becomes,

$$(P + \frac{a}{V^2})(V - b) \cong n_o C\omega^2. \qquad 9\text{-}42$$

In summary, we have obtained an expression (Eq. 9-42) which is similar to, but not the same as van der Waals' equation. We will attempt in our investigation to determine if our assumptions can be used.

It is important to realize that in the actual van der Waals' equation for a real gas, the constants a and b do not allow quantitative agreement with actual experimental curves. A qualitative check only is obtained! (See references.) We will see if our approach allows reasonable agreement with curves obtained from our particular system. We will also find the diameter D_b of our tiny plastic "molecules" from our graphs.

General Operating Instructions

1. Check carefully the pulley system and make certain that the piston moves freely in chamber. Copper plate must not catch in permanent magnet. Pointer on weight holder should not catch on adjustable scale. Adjust counter balance so that piston will move downward with about 2 gm in the holder. Make sure that unit is grounded. Check the adjustable scale and determine how to find the piston height H for any rpm.

2. Strobotac Operation. Connect to a.c. line. Set toggle switch on back to direct. Set rotary panel selector switch to Strobotac Low. Allow Strobotac to warm up for at least five minutes before using. Set the illuminated calibration dial at 600 rpm (the lowest possible frequency). Shine flash through window at impeller drive shaft. A black painted groove with a center mark permits observation of the optical effect of slowing down or stopping motion when viewed under the flashing light of the Strobotac. When the impeller shaft is rotating, the illuminated calibration dial (RPM Scale) is adjusted until a single stationary image on the shaft is observed. The RPM correspond within ± 1% to the angular velocity of the shaft. Before actually taking data, check with instructor and make certain that you can read instrument properly.

   Since the tube life of the Strobotac is much greater at low speeds than at high, the LOW scale should be used as much as possible. When measuring speeds above 3600 rpm, adjust illuminated dial with the rotary panel selector at STROBOTAC HIGH, then turn switch to Strobotac Low. The pattern observed will still be stationary because of a 4:1 relationship between the High and Low scale.

3. Damping. A permanent magnet is used to minimize statistical fluctuations which affect the piston. Use magnet at all times when obtaining data.

4. Loading and Unloading of Molecules. To load a desired number of molecules into the impeller chamber, shut the motor off, lift the piston, and slip it away from the magnet. NEVER LOAD MOLECULES WHILE THE IMPELLER IS ROTATING. To remove molecules, slip the drain container over the drain valve. Turn motor on, set speed control knob to LOW and operate the ionization. Open drain valve. Always close the drain valve after draining the impeller chamber. Remove drain container and empty into storage box for reuse of molecules.

5. Since the pressure of the gas within the impeller chamber is a function of the mass upon the holder, place masses symmetrically about the center of the piston in order to prevent unbalance of the piston assembly.

6. When the system is in operation (before recording data), check equilibrium position of pointer by carefully depressing and lifting by hand the piston above and below its equilibrium position.

PROCEDURE AND ANALYSIS [Choose one or more of the following projects. Notify instructor before taking data. Don't forget to check the sensitivity of the piston system before taking data. Piston should move downward when a load of from 2 to 4 grams is placed upon the holder. ALL PROJECTS REQUIRE ACTUAL PRESSURE (dynes/cm^2) AND ACTUAL VOLUME (cm^3) TO BE CALCULATED.]

Project 1

1. Place 100 molecules in the impeller chamber (n_o = 100). Place 4 grams on the holder ($m = m_o$ = 4 grams). The unbalance caused by the 4 grams can be interpreted as constant pressure. Raise the motor speed until 3000 rpm is obtained (N = 3000 rpm). Find H, the height of the piston above the surface of the impeller (H is not the height above the diffusers). Record all data.

2. For n_o = 200, m_o = 4 grams, and N = 3000 rpm, find H. Check motor control carefully with increase in the number of molecules. Why? Record all data. Next, add mass until the H recorded in Step 1 is reached. Record m ($m = m_o$ + added mass).

3. For n_o = 300, m_o = 4 grams, and N = 3000 rpm, again find H. Again, add mass until the H recorded in Step 1 is reached. Record all data. Proceed for n_o = 400 and n_o = 500.

4. Plot V vs. n_o (P = constant, ω^2 = constant).

5. Plot P vs. n_o (V = constant, ω^2 = constant).

6. Show that ω^2 corresponds to temperature T. Discuss any departure of the curves obtained in Steps 4 and 5 from linearity. Explain the significance of the curves in terms of kinetic theory. Does the ideal gas law apply? Explain. Do the "molecules" behave as a real gas?

Project 2

1. For n_o = 500, m_o = 4 grams, and N varying in increments of 500 rpm from N = 2000 rpm to the maximum motor speed, obtain data for a plot of volume V vs. ω^2 at constant pressure. Record all data.

2. Plot V vs. ω^2, (P = constant).

3. For n_o = 500, H = constant (the volume V, or H, is held constant at the value reached when N = 2000 rpm), and N varying in increments of 500 rpm from N = 2000 rpm to the maximum motor speed, obtain data for a plot of pressure P vs. ω^2 at constant volume. Start with P_o = 4 grams. Record all data.

4. Plot P vs. ω^2, (V or H = constant).

5. Show that ω^2 corresponds to temperature T. Discuss any departure of the curves obtained in Steps 2 and 4 from linearity. Explain the significance of the curves in terms of kinetic theory. Does the ideal gas law apply? Explain. Do the "molecules" behave as a real gas?

Project 3

1. For n_o = 500 and N = 5000 rpm, obtain data for a plot of pressure P vs. volume V at constant temperature (ω^2 = constant). Let m = 3, 5, 7, 10, 15 +....90 grams. Record H for each different mass m placed on the holder.

2. Plot P vs. V, (T or ω^2 = constant).

3. Compare your curve with the isotherms of a real gas. Discuss. How would your curve compare with an isotherm of an ideal gas? Discuss.

Project 4 [To be done after Projects 1 and 2 are completed]

1. Extrapolate your V vs. ω^2 curve to $\omega^2 = 0$. The V-intercept is b, the volume which is not available to the "molecules." What is the significance of $\omega^2 = 0$?

2. From b obtained in Step 1, calculate the diameter D_b of the "molecules" (see Eq. 9-36). Assume that the effective concentration of "molecules" n at height H equals the total number of molecules n_o. The cubic equation can be solved quickly by using Newton's method of approximations. Any standard text on the theory of equations or college algebra can be used for reference, e.g. Thomas, Calculus and Analytic Geometry, Addison-Wesley, Reading, Mass.

3. Measure the diameters of 25 plastic balls. Compare the average diameter of the balls with the diameter D_b found in Step 2 and discuss the variation.

Project 5 [To be done after Projects 1 and 2 are completed]

1. Do Step 1 of Project 4.

2. Show, using Eq. 9-42 that $P = -\frac{a}{V^2}$ when either $n_o = 0$ or $\omega^2 = 0$.

3. Extrapolate your P vs. n_o curve to $n_o = 0$ and your P vs. ω^2 curve to $\omega^2 = 0$. Find an average value for (a) from the intercepts.

4. From the P = 0 intercepts of the extrapolated curves in Step 3, find an average value for K, the total random kinetic energy of all the molecules. (See Equations 9-41 and 9-42). Hint - use (b) from Step 1 and (a) from Step 3.

5. Since b is actually not a constant but varies with n, the effective concentration of molecules at height H, should the kinetic energy K of all the molecules be higher or lower than that obtained in Step 4? Discuss.

<u>Project 6</u> [To be done after Project 3 is completed]

1. In Eq. 9-39 a and b depend on P and V. K is a constant. Over small ranges of P and V, a and b can be considered constant. Select three sets of values of P and V, separated by about 50 dynes/cm^2, from the high pressure region of your plot of P vs. V, (T or ω^2 = constant). Find a, b, and K. Hint - see Question 1 at the end of this experiment. Record all data.

2. Find from K, $C\omega^2$, the reference energy factor (see Eqs. 9-33 and 9-41).

3. Find the average mass M of 25 plastic balls.

4. Average your three values of the chamber volume V and obtain H, the height of the piston above the impeller.

5. Find n, the effective concentration of "molecules" at height H.

6. Review Step 2 of Project 4. Find the diameter D_b of the "molecules" (see Eq. 9-36). This technique is similar to that actually used to predict the diameter of molecules in real gas. [Note - you can't assume n_o = n; b obtained in Step 1 must be used.]

7. If you have additional time, repeat Steps 1, 2, 4, and 6 at other points on your plot of P vs. V.

8. Measure the diameters of 25 plastic balls. Compare the average diameter of the balls with the diameter D_b found in Step 6 and discuss any variation.

9. Recognizing that the system does not contain a real gas, discuss the assumptions made in the introduction to this experiment.

QUESTIONS

1. Derive from $\left(P_1 + \frac{a}{V_1^2}\right)(V_1 - b) = \left(P_2 + \frac{a}{V_2^2}\right)(V_2 - b) = \left(P_3 + \frac{a}{V_3^2}\right)(V_3 - b)$,

$$a = \frac{V_1V_2}{V_2 - V_1}\left[b(P_1 - P_2) - (P_1V_1 - P_2V_2)\right]\left[1 - b\,\frac{V_1 + V_2}{V_1V_2}\right]^{-1},$$

and

$$b^2\left\{\frac{V_3(V_2 - V_1)}{V_1(V_3 - V_2)}(P_2 - P_3)\frac{V_1 + V_2}{V_1V_2} - (P_1 - P_2)\frac{V_2 + V_3}{V_2V_3}\right\}$$

$$- b\left\{\frac{V_3(V_2 - V_1)}{V_1(V_3 - V_2)}\left[(P_2 - P_3) + \frac{V_1 + V_2}{V_1V_2}(P_2V_2 - P_3V_3)\right]\right.$$

$$\left.- \left[(P_1 - P_2) + \frac{V_2 + V_3}{V_2V_3}(P_1V_1 - P_2V_2)\right]\right\}$$

$$+\left\{\frac{V_3(V_2 - V_1)}{V_1(V_3 - V_2)}(P_2V_2 - P_3V_3) - (P_1V_1 - P_2V_2)\right\} = 0.$$

2. Express a, b and K in Question 1 in terms of the fundamental dimensions M, L, T.

3. What are the units for a, b and K in Question 1 in the c.g.s. system?

4. Find the mean free path L for several different values of volume. Do your calculated values compare reasonably with your visual observations? Discuss. Compare your values with the mean free path at standard conditions of an actual molecule in a gas.

5. Show that the root-mean square speed of a "molecule" in our system is,

$$V_{rms} = \sqrt{\frac{2C\omega^2}{M}}, \quad \text{where } M = \text{mass of one "molecule."}$$

6. Use the root-mean square speed from Question 5 and find the collision frequency Z for the various values of volume used in Question 4. Compare your values with the collision frequency at standard conditions of an actual molecule in a gas. Discuss.

7. Show that $4/3\ \pi D_b^3 = 8V_b$, where V_b is the volume of one molecule.

8. Derive Eq. 9-36 from Eq. 9-35.

9. What are the units of C in Eq. 9-41 in the c.g.s. system? Of $C\omega^2$?

10. From your data in Project 6 calculate the work done in an isothermal expansion (ω^2 = constant) from a volume $V_1 = 200\ cm^3$ to $V_2 = 400\ cm^3$.

SYMBOLS USED IN PROJECTS 1 - 6

Symbol	Meaning
D_b (cm)	Diameter of a "molecule."
n_o	Number of "molecules" placed in impeller chamber.
n	Effective concentration of "molecules" at height H.
m (grams)	Mass placed on holder.
M (grams)	Mass of one "molecule."
D (cm)	Diameter of impeller chamber.
H (cm)	Actual height of piston above impeller.
$V = AH$ (cm^3)	Volume of impeller chamber.
A (cm^2)	Piston or impeller area.
$P = \frac{mg}{A}$ (dynes/cm^2)	Pressure.
$S = 2A + \pi DH = 2A + \frac{4V}{D}$ (cm^2)	Entire internal surface area of the impeller chamber.
N (rpm)	Number of revolutions per minute of the impeller drive shaft.
ω (rad./sec)	Angular velocity of the impeller drive shaft.

Suggested Data Table for Projects 1-3 (Add units).

A = ____________ D = __________

Settings and Recorded Data				Calculated Values				Notes
n_0	N	H	m	ω	ω^2	V	P	

KINETIC THEORY MODEL

Suggested Data Tables for Projects 4 - 6 (Add units).

Average D_b (measured) = ____________ Average mass (M) of "molecules" ____________

Set	P_1	P_2	P_3	V_1	V_2	V_3	n	Notes
A								
B								
C								

Set	P_1V_1	P_2V_2	$P_1V_1 - P_2V_2$	P_3V_3	$P_2V_2 - P_3V_3$	$P_1 - P_2$	$P_2 - P_3$	Notes
A								
B								
C								

Set	$\frac{V_3(V_2 - V_1)}{V_1(V_3 - V_2)}$	$\frac{V_1 + V_2}{V_1 V_2}$	$\frac{V_2 + V_3}{V_2 V_3}$	Notes
A				
B				
C				

EQUIPMENT: Manufacturer-Central Scientific Company, 1700 Irving Park Road, Chicago 13, Illinois. Apparatus adapted from original model designed by Prof. H. F. Meiners.

Chapter 10
ELECTRICITY

Experiment 10-1 ELECTRIC FIELDS

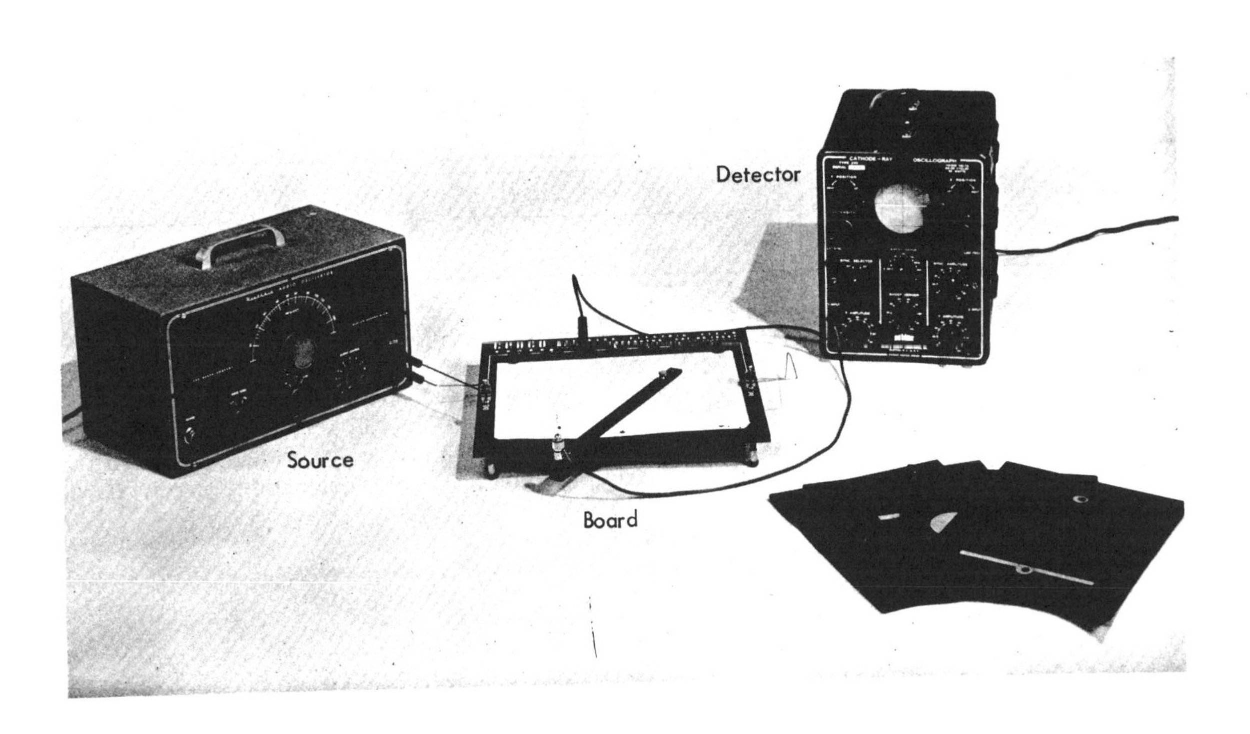

Fig. 10-1. Field Mapping Board with Potential Source (Audio Oscillator) and Detector (Cathode Ray Oscilloscope).

INTRODUCTION

By plotting the equipotential lines of an electric field, the electric lines of force can be determined. It is possible to find a large number of points in any electric field, all of which are at the same potential. If a line or a surface is drawn that includes all such points, this line or surface is known as an equipotential line or surface. The electric line of force, on the other hand, is the path which a free positively charged test charge would follow in traversing an electric field. This path is tangent to the direction of the electric field intensity. The electric lines of force must always be perpendicular to the equipotential lines or surfaces.

The experiment may be performed by using field mapping boards or by using a shallow water tray.

PART A - Field Mapping Boards

The field mapping boards include sheets of high-resistance paper supporting metallic terminals at both sides fastened beneath the board. An electric field is produced by connecting the terminals to a source of potential. Points of equal potential in this field are found by a movable probe connected to a null-point detector. The eight resistors connected across the two terminals make it possible to draw seven equipotential lines. Five different field plates for differently shaped electric fields are provided.

Carefully place one of the five field plates on the underside of the field-mapping board and secure it to the conducting bar using the two thumb screws provided. Connect the audio oscillator to the two binding posts on the upperside marked "Bat. or Osc." Fasten a sheet of paper $8\frac{1}{2} \times 11$ to the board by depressing the board from either side and slipping the paper under the four rubber bumpers provided. Select the template which carries the field plate design you have chosen. Place the template on the template guides just above the upper edge of the paper and trace the design onto the paper to correspond with that of the field plate on the underside.

Connect the cathode ray oscilloscope (or other null-point detector such as ear phones) to the U-shaped probe using the binding post provided and to a fixed point on the resistors using the banana jack E1. The U-shaped probe is now moved over the paper to find a zero reading (or no sound position). The circular hole in the top arm of the probe is directly above the contact point which touches the graphite coated paper, thus the location of the equipotential point may be directly transferred to the paper. Move the probe to another null-point position and record it. Continue in order to find a series of these points across the paper. Connect these equipotential points with a smooth curve to show equipotential lines.

Connect the detector to a new position and plot its equipotential line. Repeat until all points E1 through E7 have their equipotential lines plotted. Since the potential difference across each similar resistor is the same, the equipotential lines obtained will be spaced to show an equal drop of potential between successive lines.

Upon completion, select a different field plate and repeat the above procedure until all five electric fields are drawn.

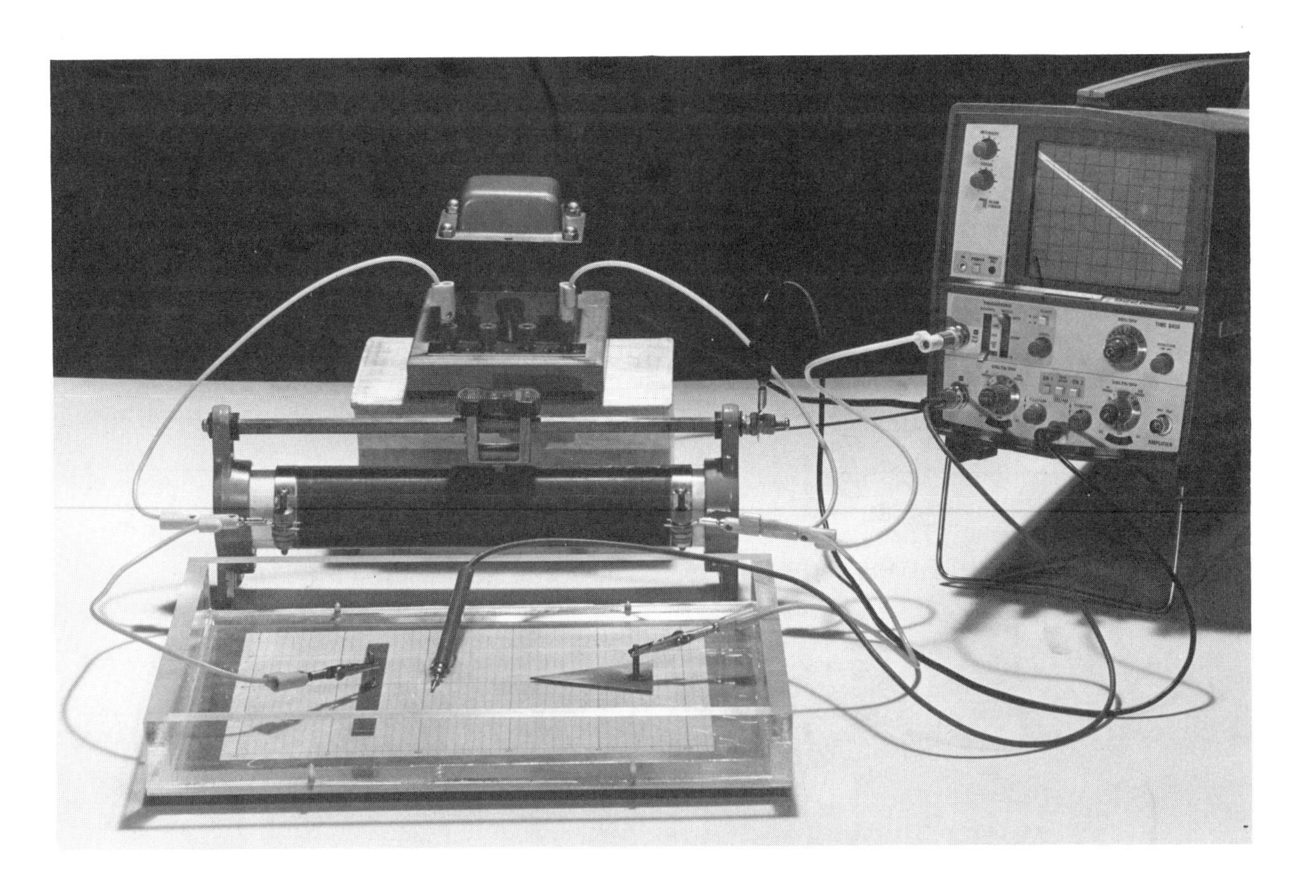

Fig. 10-2. Shallow Water Tray Used to Find Equipotential Surfaces.

PART B - Water Tray

A shallow tray should be filled with tap water to a depth of about 6 mm (1/4 in.). Two electrodes (circular plates, parallel blocks, etc.) are placed in the water in such a way as to allow the equipotential lines or surfaces between them to be measured. The electrodes are connected to an audio-frequency oscillator set at 1000 Hz. Points of equal potential are found by using a movable probe connected to a null-point detector, either earphones or an oscilloscope. The probe is used to locate an equipotential line with respect to a percentage of the total voltage drop between the two electrodes.

Plot the equipotential lines for each set of electrodes that you are given. Carefully construct the electric lines of force between the electrodes.

QUESTIONS

1. Do two different equipotential lines or two lines of force ever cross each other? Explain.

2. How much work is done in transferring a charge of one microcoulomb from one end of an equipotential line to the other end? How much work is done in transferring that same charge from one terminal to the other terminal?

3. Explain why the lines of force are always perpendicular to the equipotential surfaces.

4. Could this experiment be performed with a direct current source? Explain any changes that would have to be made if a D.C. supply were used.

5. Name other types of detectors that could be used in the experiment.

6. Discuss the electric field between two parallel plates. Use Gauss's Law to show the variation of the field with distance.

7. How does the electric field vary with distance near an isolated point charge? Near a spherical charge? Near a cylindrical charge?

8. Sketch the equipotential lines and the lines of force for (a) two isolated negatively charged particles, (b) an isolated positive charge Q close to a negative charge 2Q.

9. Prove that the electric field strength is numerically equal to the potential gradient.

Experiment 10-2 THE ELECTROSTATIC BALANCE*

INTRODUCTION

In this experiment electrostatic properties will be investigated by measuring the forces between charged spheres. Because these forces are relatively small, measurements are made by the use of a light beam reflected from a small mirror. Charges are produced by means of a belt or Van de Graaff generator.

APPARATUS

1. The Electrostatic Balance

The principle of the electrostatic balance - or torsional balance - is similar to the gravitational balance described in Experiment 7-21. The weight S_1 of the sphere suspended on a tight wire between the bottom and top of the instrument is balanced by the damping vane D . Above the sphere is a mirror M which deflects a beam of light onto a distant screen as a force acts on the sphere. Although this force cannot be measured directly, its variation with the distance between two charged spheres and with the charge on the spheres can be investigated.

* Another electrostatic balance is described in conjunction with Experiment 11-2 on the current balance. The electrostatic balance used in this experiment can be obtained from Leybold-Heraeus EMBH & Co., 5000 Köln 51, Germany.

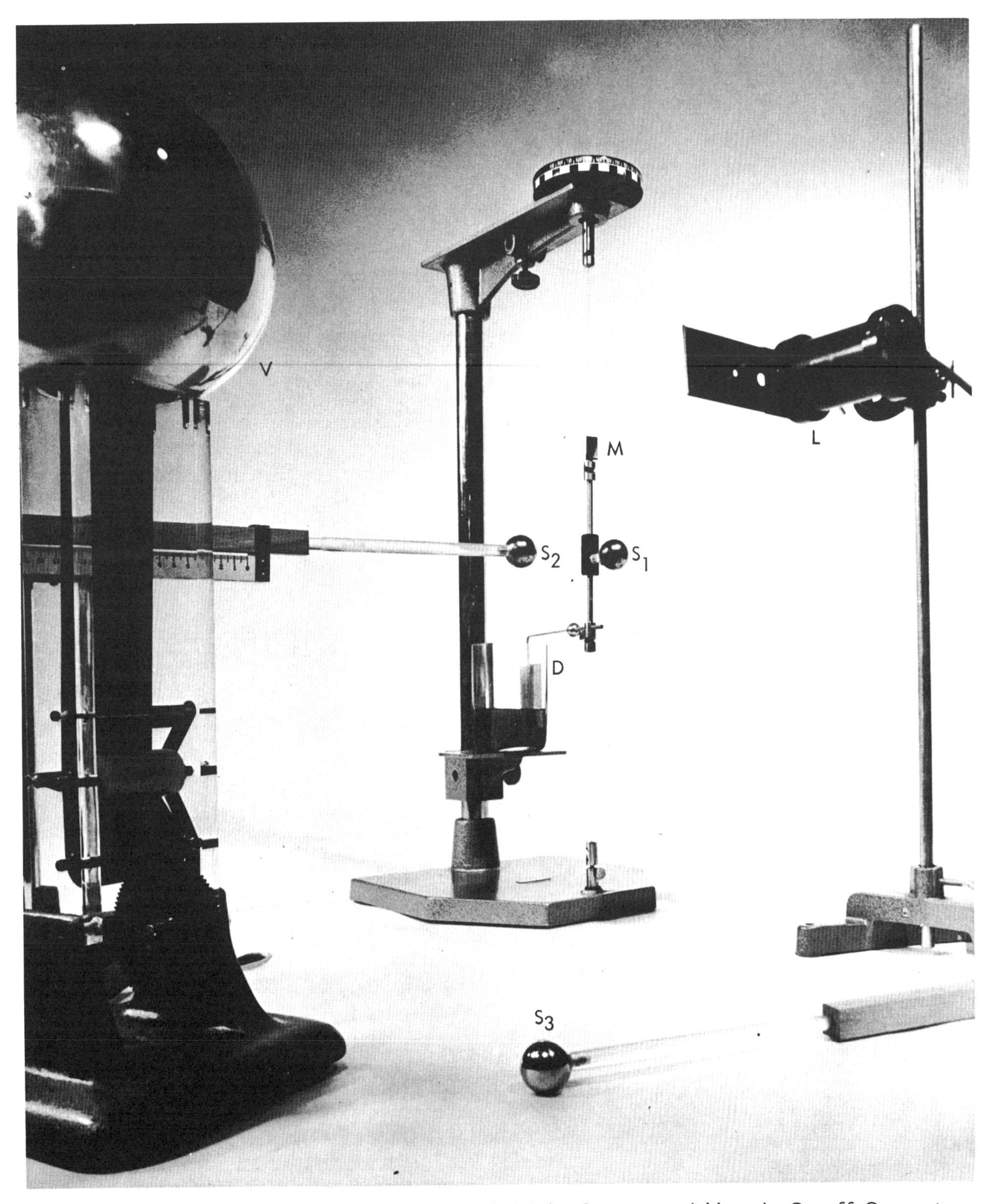

Fig. 10-3. The Electrostatic Balance with Light Source and Van de Graaff Generator.

Unless the torsional constant is given, the balance has to be calibrated. This can be done by finding the period of the torsional oscillation of an object of known rotational inertia. A rod is inserted into the torsional balance; its rotational inertia is given by $I = 1/12\ m\ell^2$, where m is the mass and ℓ its length. The rotational inertia of the torsional pendulum is negligibly small compared to that of the rod inserted.

For small twists, the restoring torque is given by $\tau = -\kappa\theta$, where κ is the torsional constant to be determined. It is computed by measuring the period of oscillation expressed as

$$\tau = 2\pi\sqrt{I/\kappa}\ .$$

2. The Van de Graaff Generator

The generator is an electrostatic high voltage machine based on the principle of self-excitation that supplies a continuous no-load voltage of up to several hundred KV. The charges are so small that the generator may safely be touched.

At its upper end, the generator is provided with a hollow metal sphere resting on a plastic insulator. An endless rubber belt passes into the metal sphere through a hole in the bottom of the sphere. The belt runs over a plastic roller (4) and three metal rollers (1,2,3). Three metal knife-edges are used to give the belt the required load.

Electric charges generated on the belt, primarily through friction between the belt and cylinders, are deposited on the spherical electrode. The charge on the surface of the sphere increases to some limited value, determined by losses through sparks and leakage. In general the belt will be loaded with negative charges and the charging cylindrical roller will accumulate positive charges.

Because the charging roller (4) is situated opposite blade C, a negative charge is induced on C. Because the blade is sharp edged, the negative charge will spark from it to the rubber belt, thus increasing the charge on the belt. The belt bearing these charges travels upward into the spherical electrode K. Shortly after delivery of the charge to the sphere's surface through cylinders (2) and (3), it passes blade A which is in contact with the ball surface. A positive charge is therefore induced upon blade A and imparted to the descending belt. This, in turn, increases the negative charge of the ball. The positive charge traveling downward on the belt increases the charge on roller (4), thus increasing its efficiency. Blade B located above cylinder (4) serves to increase the efficiency of blade A by inducing a positive charge on the belt.

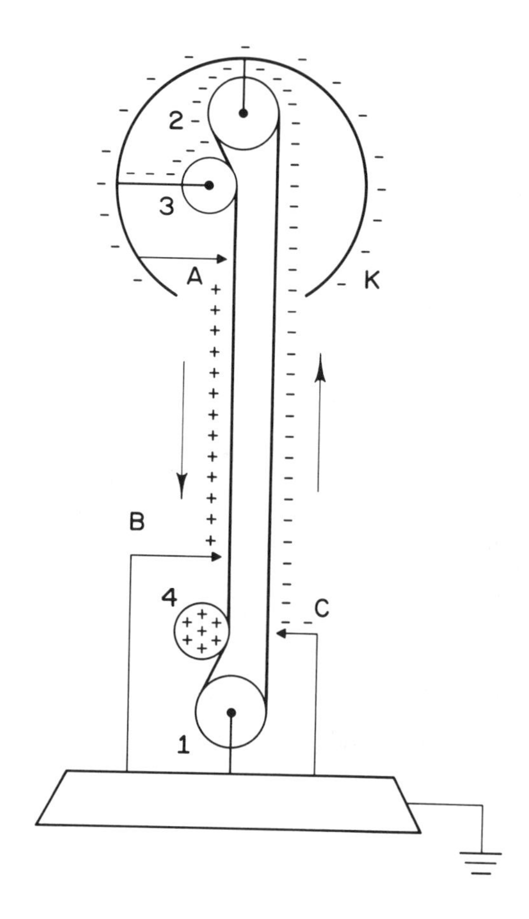

Fig. 10-4. Sketch of the Van de Graaf Generator.

PRECAUTIONS

1. Do not handle spheres or the insulating rods of the electrostatic balance. If it is necessary to stop the moving sphere, use the damping vane as a handle.
2. Keep apparatus free of drafts or vibrations.
3. Make certain damping vane is free to move in oil container.
4. Prior to beginning the experiment, discharge the spheres by touching them with a ground wire.

PROCEDURE

With spheres uncharged, set the pointer on the moving sphere (S_2) scale at 3.1 cm. Then move the entire stand until the distance between spheres is approximately 1 mm. The scale will now read the center to center distance between the spheres. The diameter of the spheres is 3 cm. The small distance that the suspended sphere moves during the experiment may be neglected in calculating distances between spheres.

Turn on the light source L and direct it on the mirror M mounted above the suspended sphere S_1 so that its reflection will fall on a meter stick placed about 4 meters from the mirror. Focus the hair-line on the meter stick. Record this point as a reference point for uncharged spheres.

Part A

Set the distance between S_1 and S_2 at 10 cm and charge spheres inductively, i.e., charge the Van de Graaff generator and place its sphere near S_1 and S_2. Touch the back sides of S_1 and S_2 momentarily with a ground wire. This allows the free negative electrons on S_1 and S_2 to escape to ground, leaving S_1 and S_2 with an excess of positive charge. Remove and discharge the Van de Graaff generator. Record the reading on the meter stick. Take and record deflection readings at 10 cm, 14.1 cm and 20 cm.

Compare the results with calculated values of the deflections for 14.1 and 20 cm using the deflection at 10 cm as a reference. Discuss results.

Part B

Recharge S_1 and S_2. Record the S_1 deflection when d = 10 cm. Reduce the charge on S_2 by one-half by touching it with the uncharged third sphere S_3. Record the deflection and explain the results.

Part C

Take the data needed to plot the force as a function of $(1/d)^2$. Discuss the significance of the slope and the intercept.

Part D

How can the electrostatic balance be used to measure the potential of spheres? If time permits, measure the potential of a sphere. For the calculation, the torque constant of the balance must be known or found (see Item 1, APPARATUS).

QUESTIONS

1. Show that the rotational inertia of the calibrating rod is $1/12\ m\ell^2$.

2. Show that the period of oscillation obtained in the calibration of the torsional balance is given by $\tau = 2\pi\sqrt{I/\kappa}$.

3. What other means could be used to determine the torsion constant?

4. When a hollow conducting sphere is given a charge q, what is the electric intensity inside and outside the sphere? What is the potential inside and outside the sphere? Does it matter if the conducting sphere is solid? Why?

5. Discuss all possible sources of error in this experiment.

6. What is the function of the belt in the Van de Graaff generator?

7. What limits the voltage that can be produced by means of the Van de Graaff generator?

8. Devise a means by which you could determine the sign of the charge on the sphere of the Van de Graaff.

9. Compare the electrostatic attraction or repulsion between the two small spheres with their gravitational attraction.

10. Can the potential at a point be zero when the electric field at that point is not zero? Explain.

11. Can the electric field be zero at a point where the potential is not zero? Explain.

12. Can both the electric field and the potential be zero for some charge distribution? Explain.

13. The gravitational field near the surface of the earth is nearly uniform. Give possible reasons why this is not true for the electrostatic field near the charged spheres.

14. Why must the resultant electrostatic field be perpendicular to the surface of the conductor? What would result if this were not true?

15. Consider a weightless spring with a small spherical mass M attached to its end. If the sphere is charged and the spring is allowed to vibrate above a grounded conducting plane, do you expect its period of oscillation for small displacements to be greater or less than when uncharged? Explain.

Experiment 10-3 ELECTRICAL RESISTANCE

PART A - Resistance and Power

The electrical resistance R of a device is defined as the potential difference across it divided by the current through it, $R = V/I$. An ohmic device obeying Ohm's Law is one in which the ratio V/I is a constant for different values of current. A commercial resistor is essentially ohmic within its power rating, i.e., a 1/4-watt resistor has a nearly constant resistance R provided the power dissipated by the resistor is less than 1/4 watt. A light bulb is an example of an non-ohmic device. Both, ohmic and non-ohmic resistors will be used in this experiment.

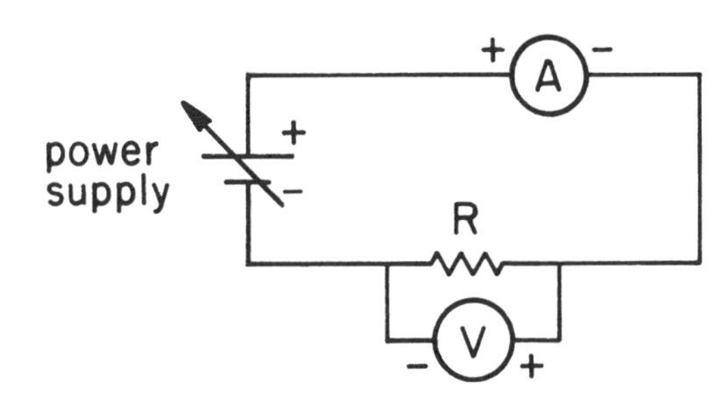

Fig. 10-5.
Ammeter-Voltmeter Connection.

In this part of the experiment a variable D.C. power supply (0-20 volts is a typical range) is connected across various resistors, one at a time. The voltage may be measured either directly by a meter built into the power supply or an external voltmeter of appropriate range, connected across the resistor. The current is measured (usually in milliamperes) by an ammeter in series with the rest of the circuit. If the ammeter has different ranges, always try the highest range first. Make certain to observe the correct polarities in connecting the voltmeter and the ammeter so that all readings are possible.

This experiment is designed to heat, indeed overheat, some of the resistors. Therefore, it is suggested that you start with a low voltage setting allowing the current to come to an essentially equilibrium value for each setting. Caution: for high power settings the resistors will become hot! If the resistance of the ammeter is small, the voltage drop across it may be neglected. (Explain why.)

Read the currents and voltages for the light bulb and the resistors supplied. Compute the resistance R and the power P (i^2R) for each setting. For each of the resistors, plot the resistance R vs. the power P. Discuss the significance of these plots, especially any difference between the curves for the different resistors used.

PART B - Resistivity

The resistance R of a wire depends on the material and the geometrical character (usually the length and diameter of the wire). In this experiment the relationship between the resistance and the length, the area and the material will be investigated.

Connect a fixed-voltage power supply (such as a battery) in weries with a known resistance R_S, and a resistance R_E which limits the current drawn, and a number of wires of different diameters and materials as shown in Fig. 10-6. A voltmeter is used to measure the various potential drops across the wires and the resistor R_S, one at a time. Measure the length and diameter of all wires. From the data taken determine how the resistance varies with length, area and material. Compute the resistivity of the various wires. Use the appropriate instrumental limits of error to express the experimental uncertainties (limits of error) for the resistivities calculated. Results must be expressed in the appropriate number of significant figures.

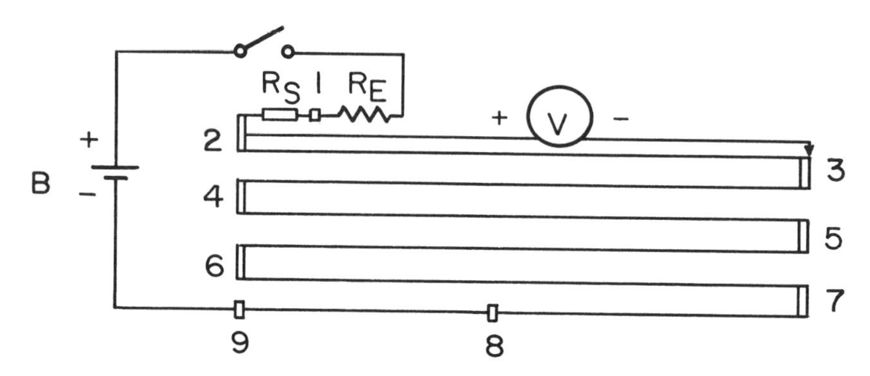

Fig. 10-6. Wiring Diagram for Resistance Board.

QUESTIONS

1. What do the curves obtained in Part A indicate about the relation between the resistance and the temperature in the case of each resistor?
2. Discuss the difficulties in determining whether the filament of a light bulb obeys Ohm's Law.
3. Discuss the significance of the slope of the graphs obtained in Part A.
4. What is meant by "linear" and "non-linear" circuit elements?
5. When does Ohm's Law hold and under what conditions can Ohm's Law not be used?
6. How does the resistance of a wire depend upon the length, area, diameter, and temperature?
7. Discuss the difference between resistivity and conductivity.
8. Compare electrical conductivity with thermal conductivity.
9. Which of the materials used in the experiments would be of greatest value in the construction of a power line? Why?
10. How does the resistivity change with temperature?
11. Assume that an error of 1 mil (±0.001 inch) was made in the measurement of the diameter of the wire 3-4 (see Fig. 10-6). What is the Precision Index (PI) in the resistivity computed?
12. A standard resistance of one ohm is made by using constantan wire of 1 mm diameter. What length of wire is required?

Experiment 10-4 THE R-C CIRCUIT

INTRODUCTION

We wish to determine the response of an RC circuit (Fig. 10-7) as the switch S is thrown back and forth between a and b. In the actual experiment we will substitute a pulse generator for the switch. When the switch is in position a, the equation governing the circuit is

$$V = R\frac{dq}{dt} + \frac{q}{C}, \qquad 10\text{-}1$$

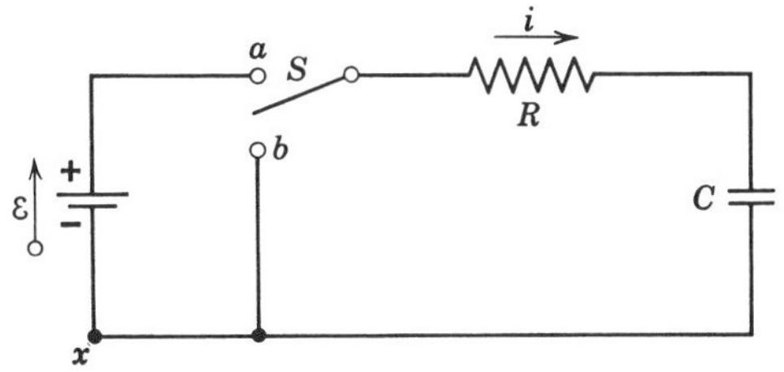

Fig. 10-7. An RC Circuit.

where q is the charge on the capacitor. The solution to this equation, as can be verified by direct substitution, is

$$q = CV(1 - e^{-t/RC}). \qquad 10\text{-}2$$

Thus the charge on the capacitor builds up exponentially in time. If we differentiate Eq. 10-2, we obtain for the current in the circuit

$$i = \frac{dq}{dt} = \frac{V}{R}e^{-t/RC}. \qquad 10\text{-}3$$

When the switch has been thrown to b, the equation governing the circuit is

$$R\frac{dq}{dt} + \frac{q}{C} = 0 \qquad 10\text{-}4$$

The solution to this equation is (assuming the capacitor carried a charge $q_o = CV$ before the switch was thrown to b)

$$q = CV\,e^{-t/RC}. \qquad 10\text{-}5$$

Differentiating Eq. 10-5 we obtain for the current

$$i = -\frac{V}{R}e^{-t/RC}. \qquad 10\text{-}6$$

The potential difference V_C across the capacitor and the potential difference V_R across the resistor can be easily written down from the above equations, realizing that $V_C = q/C$ and $V_R = iR$. Note that the sum of V_R and V_C must always be V, the voltage of the battery, as long as the switch is in position a. When the switch is in position b, V_R plus V_C must be equal to zero at any time.

The time constant of the R-C circuit is given by $\tau = RC$. It is the time at which the charge has reached a value of $q = CV - 1/e\ CV = .63\ CV$ or 63% of the final charge for a charging capacitor where τ is in seconds, R in ohms and C in farads. The time constant $\tau = RC$ for a discharging capacitor is the time at which the charge has been reduced to 1/e (37%) of the original charge.

PART A - Time Constant - Measurement of Resistance

In this experiment, a capacitor is charged to a known voltage and then discharged through a conventional voltmeter. By measuring the time constant and knowing the value of the capacitor, the internal resistance of the voltmeter may be determined. The charging voltage is obtained from a control box which is connected to the wall outlet. The control box will allow you to charge the capacitor to any voltage from 0-30 volts by means of coarse and fine adjustment knobs.

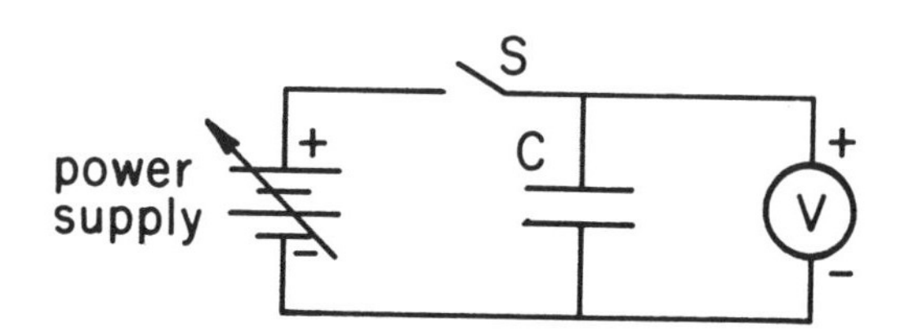

Fig. 10-8. Measuring the Resistance of a Voltmeter.

Set up the wiring as shown in Fig. 10-8. Connect the capacitor to the control box and charge the capacitor to a convenient voltage (e.g. 30 V). Then open the switch to disconnect the control box and start the timer. When the voltmeter indicates a value of 1/e of the original voltage (about 37%), stop the timer. The time interval measured is one time constant, equal to the product RC where R is the internal resistance of the meter and C is the capacitance of the capacitor which must be given. Compute the internal resistance of the voltmeter used. Repeat the procedure for a different initial charging voltage. Note that this procedure can be used to measure the resistance of any large resistor.

PART B - Time Constant - Measurement of Capacitance

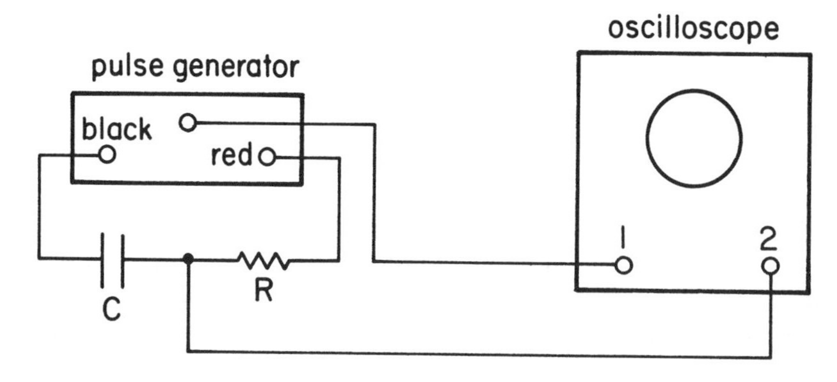

Fig. 10-9. Measurement of Capacitance.

In this experiment, the voltage is supplied by a pulse generator, a device which alternately provides a positive output and a zero output. If this voltage supply is connected in series with a resistor and capacitor, as shown in Fig. 10-9, the capacitor will alternately charge and discharge through the resistor. This process can be observed on a dual channel oscilloscope.

The pulse generator has one output brought to two different terminals. The top jack is connected to channel 1 of the scope so that you can monitor the generator output. The two banana jacks are used to drive the capacitor resistor combination.

The knob on top of the pulse generator varies the duty cycle of the pulse, the percentage of time that the output is high vs. the time it is low. With the knob in the center of its range, the duty cycle is approximately 50%, i.e., half the time the output is high ($\sim$10 volts) and half the time it is low ($\sim$0).

Connect the 10-K (10,000-ohm) resistor and the unknown capacitor in series, as shown in the diagram. Channel 2 on the oscilloscope will now display the voltage across the capacitor as a function of time. Adjust the duty cycle of the pulse generator and the time base of the scope to measure the time constant (conceptually Part B is the same as Part A except the times involved are very different; $\sim 10^2$ vs. $\sim 10^{-3}$ sec). Check that the red knob of the time base control on the scope is fully counterclockwise (x1). Calculate the value of the capacitor.

The value of the resistor is accurate within 1%. The scope has a 3% limit of error. Find reasonable limits of error for your measurement of C.

Interchange the resistor R and the capacitor C. Channel 2 now will display the voltage across R which is also proportional to the current through R. Note the wave shapes for various settings of the duty cycle.

PART C - Time Constant Using a Microcomputer[1]

The microcomputer has been set up to automatically charge or discharge a 1 μF capacitor through a 1 MΩ resistor and display a plot of the voltage across the capacitor vs. time. Run the experiment making appropriate choices of the menus displayed. Store each run in a different file (0-9).

1. Verify that the time constant is the same for charging and discharging.
2. Compare the measured time constant with the product of R and C.
3. From one of the displays, find the percent charged or discharged after 3 and 5 time constants.

[1] These directions are written for use with the microprocessor and its interface designed under the supervision of Dr. Thomas Shannon under a grant from the RCA Corporation. A detailed description may be obtained from Professor Walter Eppenstein, Physics Department, Rensselaer Polytechnic Institute, Troy, N.Y. 12180-3590.

Fig. 10-10. Using a Microcomputer to Study R-C Circuits. The monitor shows the Logo students see when they turn it on - directions for use of the micro appear on the screen.

Comments on the Microcomputer

Power switch on the microcomputer controls both the microcomputer and the monitor (leave monitor switch on). In making choices from the menus, you can change them any time before pressing "GO" or "ENT". The "ESC" (ESCape) key aborts what is running and sends the computer to the first menu. This can be used to abort a useless run; it is also the method to terminate the display. In the mode there is a cursor to identify any point on the curve. It is moved by the "H" (High, right) and "L" (Low, left) keys.

QUESTIONS

1. Derive Eq. 10-1 from the principle of the conservation of energy.
2. Show by direct substitution that Eq. 10-2 is a solution of Eq. 10-1.
3. Solve Eq. 10-4 by rearranging and integrating.
4. Prove that the product of the units of resistance and capacitance can be expressed in seconds.
5. If, in Fig. 10-7, the switch was closed in Position a for a long time and is then opened (without closing it in Position b), what happens to the charge? to the energy?
6. In the circuit of Fig. 10-7 the charge on the capacitor is zero an instant after the switch is initially thrown to Position a; however a current exists in the circuit at this instant. How can this be?

Experiment 10-5 THERMOELECTRICITY

WARNING: The Frigistor should be handled carefully to avoid mechanical damage.

INTRODUCTION

Thermoelectricity deals with the direct transformation of thermal energy into electrical energy and the reverse process of producing or removing heat directly by electrical energy. There are three well known thermoelectric effects: the Thomson effect, the Peltier effect and the Seeback effect.

The Thomson effect refers to the establishment of an electromotive force in a conductor when a temperature gradient exists in the conductor. The temperature gradient results in the diffusion of the charge carriers through the conductor, which in turn leads to the establishment of the emf. One finds that when a current is passed through a conductor in which a Thomson emf has been established, heat is converted to electrical energy if the current is in the same direction as the emf; electrical energy is converted to heat if the current and Thomson emf are in the opposite direction. The rate of heat absorption (or production) per unit length of conductor is given by:

$$W = \tau I(dT/dX), \qquad 10\text{-}7$$

where

τ = Thomson coefficient
I = current
dT/dX = temperature gradient

The Peltier effect refers to the establishment of an emf at the junction of two dissimilar conductors. The emf can be thought of as due to the diffusion of charge carriers from one conductor to the other. As with the Thomson effect, one finds that if a current is passed through the junction, heat is converted to electrical energy or vice versa depending on the relation between the current and emf. The heat in watts produced (or absorbed) at a junction of two dissimilar conductors, A and B, is given by:

$$W = \pi_{AB}(T)I, \qquad 10\text{-}8$$

where $\pi_{AB}(T)$ is in volts and I, the current through the junction, is in amperes. $\pi_{AB}(T)$, which is called the Peltier emf, depends on the nature of the two conductors, A and B, as well as on the temperature of the junction.

A cooling device based on the Peltier effect is shown schematically in Fig. 10-11. When a current I is passed through the circuit, heat is converted to electrical energy at one junction and electrical energy is converted to heat at the other junction. Thus the device acts as a refrigerator.

Since the junctions will be at two different temperatures T_1 and T_2, a temperature gradient will exist in both materials, which results in a Thomson emf setup in each. Also, since the junctions are at different temperatures the Peltier emf at the junctions are unequal. In general, the net emf is not zero. This net emf is called a Seeback emf. This Seeback voltage exists in addition to the IR drop in the circuit. For currents near the maximum useful values for any junction, the IR drop is usually 5 to 10 times the back Seeback voltage.

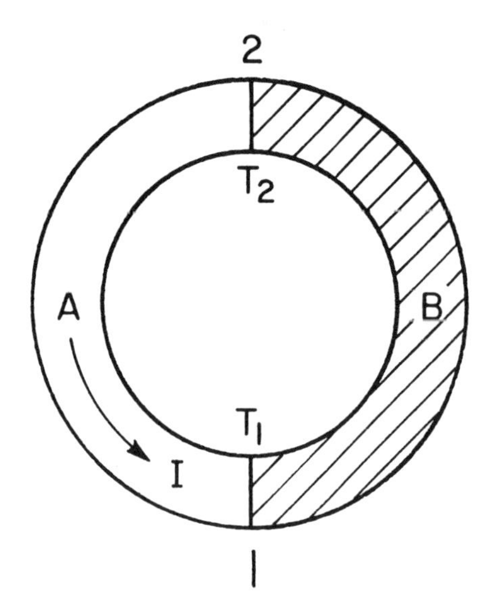

Fig. 10-11.
Peltier Effect.

APPARATUS

The cooling device in our experiment, which is called a Frigistor (Fig. 10-12), is made up of "couples," one of which is shown in Fig. 10-13. It is made of two types of semi-conducting material, known as N-type and P-type. N-type material has free negative carriers, or electrons, and will, when heated at one end, show a negative potential at its cold end with respect to its heated end (Thomson effect). P-type material has positive carriers, or "holes" and will when heated at one end, show a positive potential at its cold end with respect to its heated end. Connecting the P and N type material by copper connectors is a convenient way to combine these materials to obtain maximum effects. The construction of the Frigistor is indicated by the detailed view in Fig. 10-14. It shows the arrangement of the N and P legs of the semi-conducting material and indicates the hot and cold junctions when the unit is connected to a battery with polarity as indicated.

Connected to the hot side of the frigistor is the heat sink which serves to dissipate heat from the Frigistor. The heat sink is anodized, which makes its surface electrically insulating. Do not scratch the surface.

The cold block is made of an insulating material and encloses the cavity which is in contact with the cold side of the Frigistor. The cold block is also anodized.

PROCEDURE

PART A - Qualitative Investigation of the Peltier Effect

Remove the Frigistor from the heat sink. Connect the Frigistor, carbon rheostat, and ammeter in series with the battery. The red lead of the Frigistor is connected to the positive terminal of the battery. Adjust the current to a about 5 amperes. After a short time, the side of the Frigistor with the four connectors will become cold while the side with five connectors will become hot. Reverse the direction of the current and note the results.

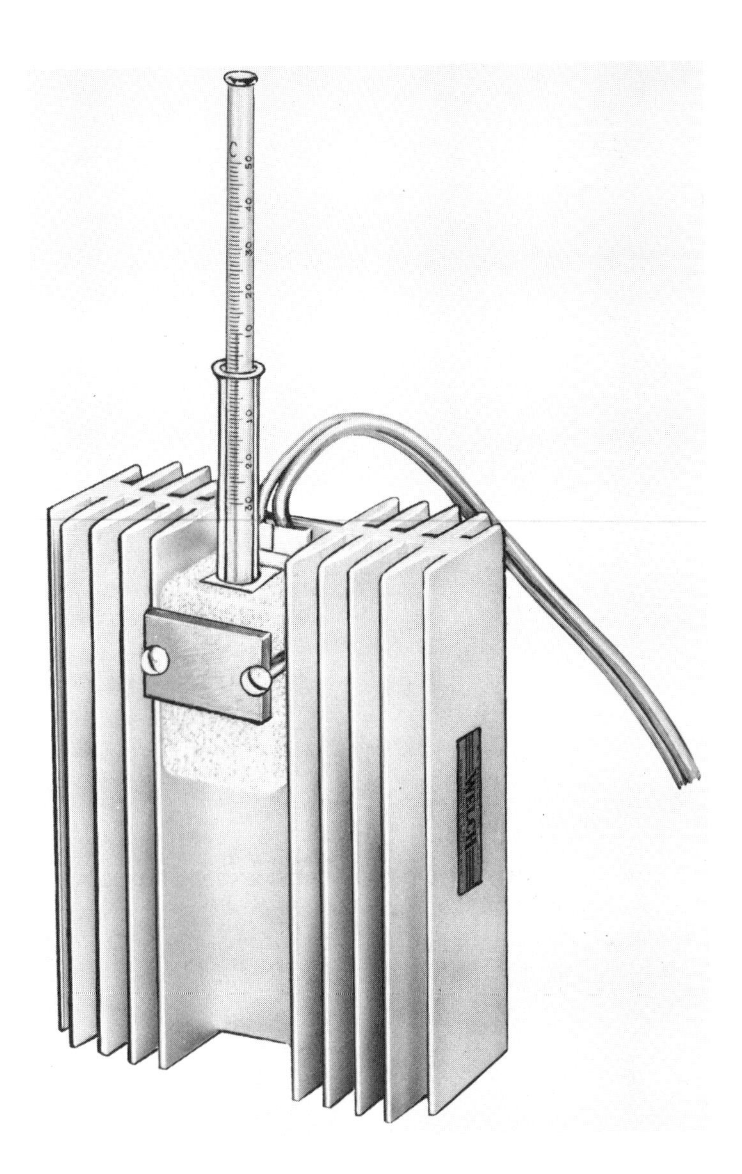

Fig. 10-12.
Peltier Apparatus - Frigistor Type.
(Courtesy of The Sargent-Welch Scientific Company)

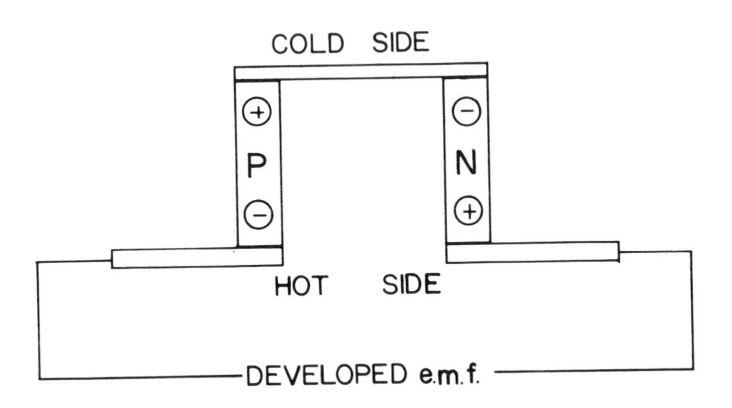

Fig. 10-13.
One Couple of Frigistor.

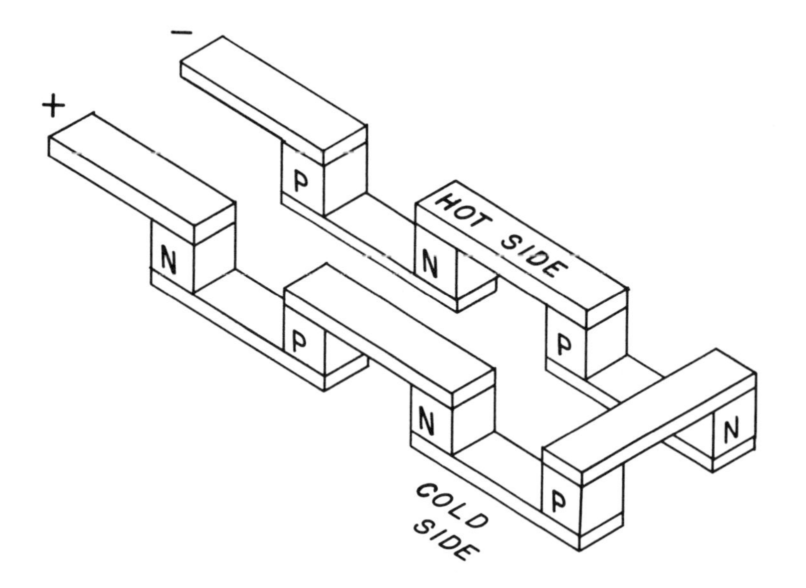

Fig. 10-14. Construction of Frigistor.

PART B - Qualitative Investigation of the Seeback Effect

Assemble the Frigistor to the heat sink with the leads extending downward and use the anodized aluminum strip in place of the cold block (see Fig. 10-15). Connect the leads of the Frigistor to a galvanometer having a sensitivity of 50 millivolts or less for full scale deflection. Heat the anodized aluminum strip with a match and note the deflection of the galvanometer.

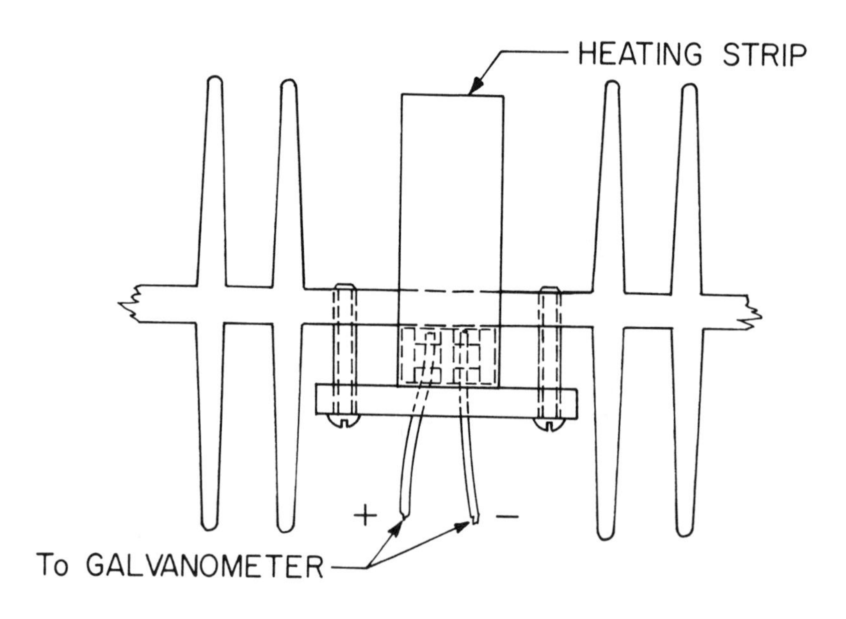

Fig. 10-15. Arrangement for Seeback Effect.

PART C - Investigation of the Performance Curves of Frigistor with Load

Assemble the Frigistor and the cold block to the heat sink. The side of the Frigistor with the five copper sections should be against the heat sink. (See Fig. 10-16.)

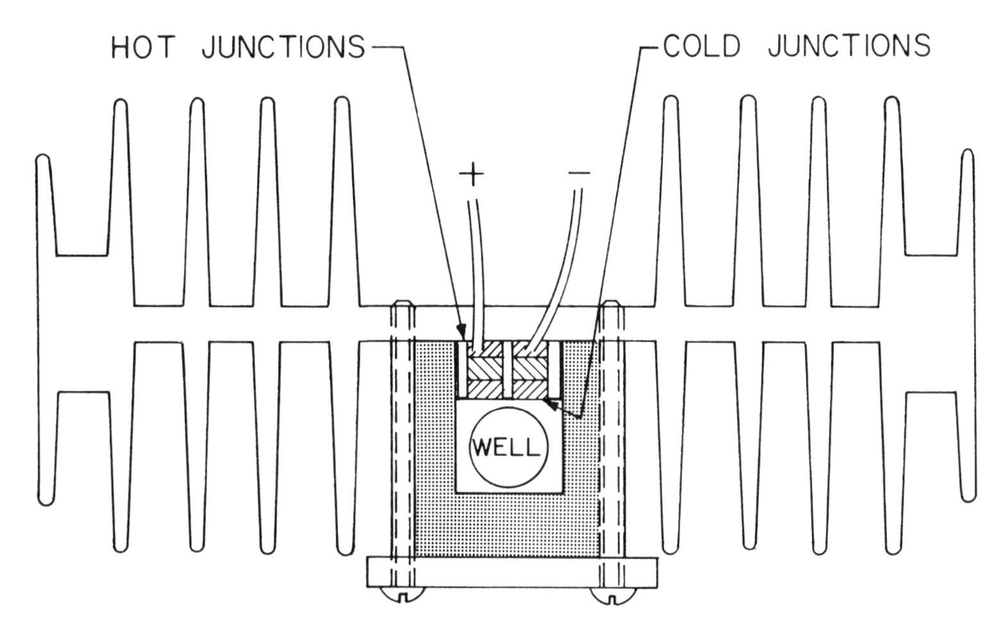

Fig. 10-16. Connections of Frigistor for Performance Curves.

Connect the Frigistor, ammeter, and carbon rheostat in series with the battery, connecting the red lead of the Frigistor to the positive terminal of the heat sink. Place the mercury thermometer (representing the load) in the cavity and pour in enough mercury to cover the thermometer bulb. Attach another mercury thermometer to the heat sink, securing it in place with scotch tape, which gives adequate thermal contact for this purpose. Even though the heat sink has a large radiating surface, its temperature may rise during the experiment. A compressed air jet may be used to keep the heat sink temperature constant.

Adjust the current through the Frigistor to two amperes. Note the drop in temperature of the mercury in the cavity. Allow sufficient time for temperature stabilization. Several minutes are usually required for the material in the cavity to reach the lowest temperature for the current being used. For each reading record the current and corresponding temperatures of mercury in the cavity and of the heat sink. Increase the current in steps of two amperes to a maximum value of 18 amperes. For each value of current, allow the same time for stabilization and record values of current and temperatures.

Plot the following two curves on the same graph:

(a) T_C, the temperature of the material in the cavity against the corresponding current, and

(b) ΔT, the temperature difference between the heat sink and cavity, against the current.

These are characteristic performance curves for the Frigistor under the existing load. Discuss the significance of the curves obtained.

PART D - Freezing of Liquids

Use the same experimental setup as that employed in Part C. Operate the Frigistor with about 12 amperes for several minutes to lower the temperature of the cavity below 0°C. Place the mercury thermometer in the test tube and fill it about 1/3 full of distilled water. Place the test tube in the cavity, starting the timer at the same time. Record time and corresponding temperature at regular, short intervals as the temperature drops. If the water in the test tube is not disturbed, the temperature of the water will fall several degrees below 0°C without freezing. This is supercooling. Freezing can be started by a slight movement of the thermometer. At that time the temperature will immediately rise to zero and remain there until all the liquid is frozen before resuming its downward trend. Record temperature and corresponding time at regular intervals during the entire process.

Plot temperature against time without disturbing the thermometer when each reading is taken. Discuss the significance of each region of this plot.

QUESTIONS

1. What are some of the differences between Peltier and Joule heating?
2. What is the reason for reversing the current in Part A?
3. Discuss some applications of the Peltier effect.
4. Discuss some applications of the Seeback effect.
5. How do the Seeback and Peltier coefficients differ from the Thomson coefficient?
6. How can you measure the coefficients in Question 5?
7. Why does the curve of T_c vs. current in Part C approach a minimum and then begin to increase?
8. Explain the maximum in the curve of ΔT vs. current in Part C.
9. Are the Seeback, Peltier and Thomson effects reversible?
10. Discuss both the advantages and disadvantages of thermoelectric refrigeration compared with conventional methods of refrigeration.

Experiment 10-6 PLASMA PHYSICS

The purposes of this experiment are:

(1) to study the phenomena accompanying the passage of an electric current through a gas at different pressures and

(2) to become familiar with several components of a high vacuum system.

DISCHARGE IN GASES

When an electric current passes through a gas, many phenomena are observed which do not exist for metallic conduction. When a gas becomes highly ionized, it is a good conductor and charged particles in such a gas can interact with electromagnetic fields. To these ionized gases, Langmuir[1] gave the name "plasma."

In a metal generally free electrons serve to carry the current, whereas both positive and negative charges are free to move in a gas. These charged particles drift under an applied voltage, some rather slowly, others rapidly. New charges are created at various points along the gaseous path, while others recombine with charges of opposite sign and are lost as electrical carriers.

A wire emits light only when the current is comparatively large; the emission of light in the case of a gas discharge tube, on the other hand, is a familiar phenomenon. As a result of the multiplicity of phenomena occurring simultaneously in gas conduction, the process is much more complicated than normal metallic conduction.

One of the simplest methods to produce a glow discharge is to apply an increasing potential across a pair of parallel plates (electrodes) immersed in a gas.

For the discharge to occur, a minimum voltage - called the <u>striking potential</u> V_S - must be applied across the tube. The value of V_S is a function of the geometry of the tube, the type of gas and its pressure. At high pressure the existing ions, because of their high collision frequency or short mean free path, cannot acquire sufficient energy to ionize the neutral gas atoms. At low pressures there are so few gas atoms that it is difficult to initiate the discharge.

After a discharge has started, the applied potential may be lowered below the striking potential, to the <u>extinction potential</u> V_X before the discharge ceases. This gives rise to a hysteresis phenomenon used in the construction of "relaxation" oscillators. Other applications of glow discharge are for rectification, amplification, and voltage regulation.

At reduced pressure, the glow discharge is a spectacular display. It has been classified into various regions which have particular physical characteristics. A typical glow is shown in Fig. 10-17.

---

[1] Langmuir, Physical Rev. 33, 954 (1929).

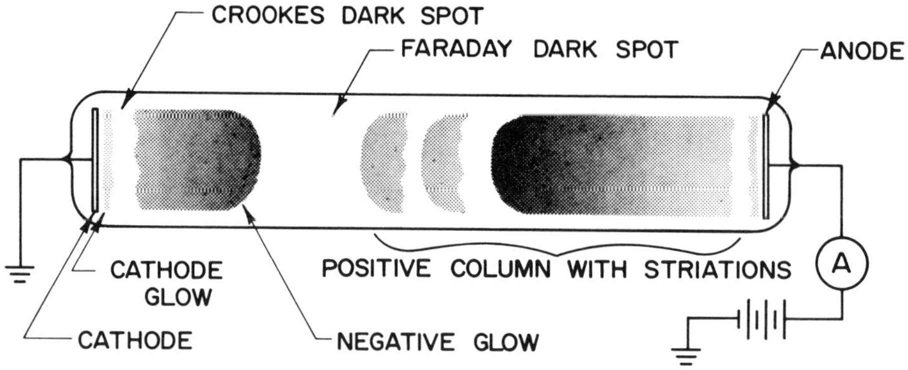

CROOKES DARK SPOT
FARADAY DARK SPOT
ANODE
CATHODE GLOW
POSITIVE COLUMN WITH STRIATIONS
A
CATHODE
NEGATIVE GLOW

Fig. 10-17. A Glow Discharge at Reduced Pressure.

VACUUM SYSTEM

The vacuum system consists of a forepump A connected through stopcocks B and C to the discharge tube D and the thermocouple gauge E. The control unit F supplies the current to the thermocouple gauge and measures its output, while the power supply G provides the high voltage required to produce a discharge.

DISCHARGES IN GASES

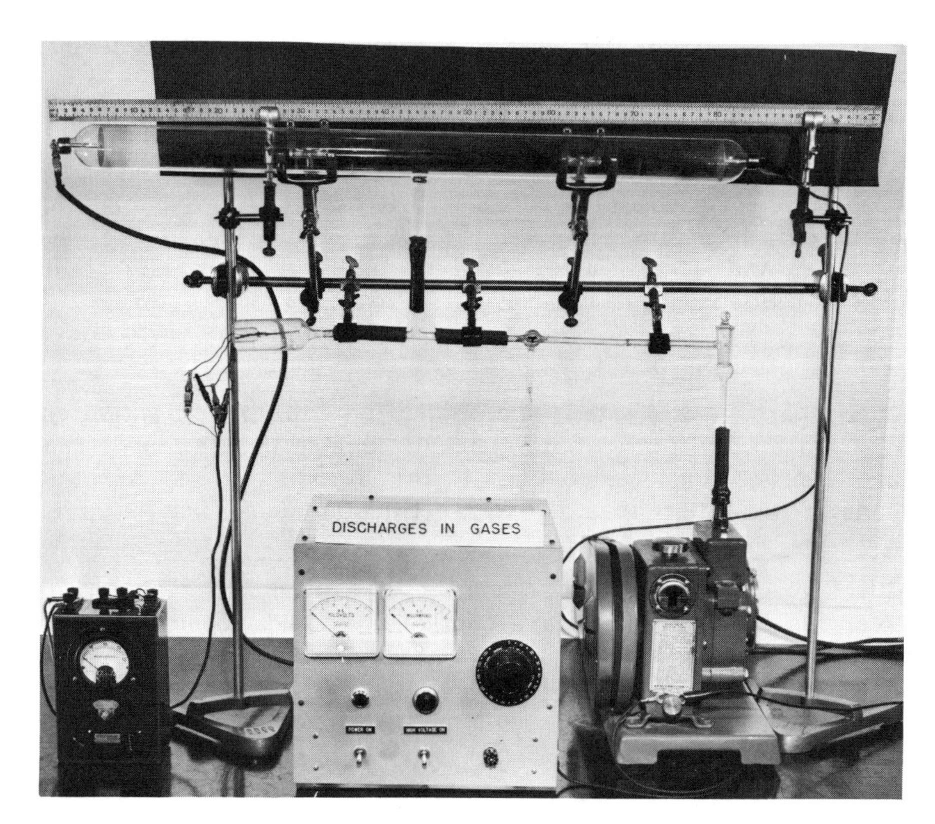

Fig. 10-18. System for Studying the Discharge of Gases at Reduced Pressure.

To operate the system, the pump should be turned on first; then the stopcock B is opened. TURN ALL STOPCOCKS SLOWLY. Close stopcock B BEFORE TURNING OFF THE PUMP. Because oil will be drawn out of the pump into the vacuum system, the stopcock should never be open without having the forepump running.

The three-way stopcock C is used to shut off the discharge tube from the rest of the system, and also to let air into the setup. Make certain that this stopcock C is never open to the atmosphere while stopcock B is open.

Inside the thermocouple gauge E a junction is made at which four wires are spot-welded together. Two of these wires act as a filament for producing heat and the other two are made of dissimilar materials acting as a thermocouple. A current is passed through the pair serving as the heater, thus raising the temperature of the junction. The potential due to the Peltier effect developed by the junction of the two dissimilar wires may be measured on a sensitive device, such as a microammeter. Since the temperature attained by the junction for a fixed current through the heater is determined by the number of collisions which the molecules of the surrounding gas make with the heater wires, the reading of the microammeter may be directly calibrated in terms of pressure.

For the operation of the thermocouple control unit F the manufacturer's instructions should be followed - different models are available. In general the current is supplied by 1.5-volt dry cells. The meter may be calibrated in microns or mm of mercury. If the scale reads microamperes, a calibration curve is required.

PROCEDURE

To obtain a discharge, the power switch of the power supply should be turned on first, then the high voltage switch. After the supply has warmed up, the voltage should slowly be increased until a discharge appears. When a discharge is produced, the voltage will suddenly decrease and the current will suddenly increase. Use this result to record the striking voltage of the tube. DO NOT TOUCH THE HIGH VOLTAGE LEADS (5000 VOLTS). The power supply should always be grounded.

Take data necessary to plot the striking potential V_s as a function of pressure in the tube. Also record the color, intensity, and length of the various regions of the glow discharge for five different pressures over as wide a range as possible.

A few regions in the glow may be considered separately to determine some of their physical characteristics:

(1) Crook's Dark Space. This is the most important part of the glow discharge. The glow can exist independent of the other regions; however, it cannot exist without the Crooke's dark space.

Plot the length of Crooke's dark space vs. the pressure in the tube. Discuss the merits of using the length of the Crooke's dark space as a vacuum gauge. Also plot the distance from the cathode to the end of the Crooke's dark space spot vs. pressure. How is the Crooke's dark space related to the mechanism of a "self-maintaining" discharge tube?

(2) The Positive Column. This is probably the most spectacular part of the glow; however, it is the least important. Most of the current in this part of the glow is carried by electrons due to their high mobilities.

In many cases, the column will be broken up into a series of striations which are alternately light and dark. Note that they are always convex toward the cathode. These striations often give rise to what are known as "plasma" oscillations. Plot distance between striations as a function of pressure.

QUESTIONS

1. Explain briefly the operation of the forepump.

2. Explain the operation of the thermocouple gauge.

3. What does the distribution of the potential and the electric field look like inside the tube before breakdown?

4. At a given temperature and pressure how does the mean free path of an electron compare with a nitrogen atom? Explain physically why the length of the Crooke's dark space should vary with pressure as it does.

5. When a discharge is produced why does the voltage suddenly drop and the current suddenly rise?

6. Explain why the striations in the positive column are all convex toward the cathode.

7. List all properties that effect the color of the discharge.

8. Does the current in a gas differ essentially from that in a copper wire? In a semiconductor? Explain.

9. Convert the lowest pressure obtained in the experiment from mm of mercury to (a) dynes/cm^2, (b) lb/in^2, (c) atmospheres, (d) millibars, (e) pascals. How does this pressure compare with atmospheric pressure?

10. How does a fluorescent light compare to a discharge tube?

Experiment 10-7 GALVANOMETERS

PART A - Measuring Current with a d'Arsonval Galvanometer

INTRODUCTION

The data necessary for the constants of a d'Arsonval galvanometer can be obtained from the deflections of the galvanometer. A constant voltage is applied to the series combination of galvanometer and the various added resistances.

For this purpose, a low voltage is necessary because of the sensitivity of the meter. A voltage divider, R_A-R_B connected across a cell of known E.M.F. gives the required voltage. The resistance R_N may be smaller than the galvanometer so that the voltage E_A is affected by a negligible amount. Therefore, E_A may be considered a constant.

From Fig. 10-19

$$\frac{E_A}{R_A} = \frac{E}{R_A + R_B} \qquad 10\text{-}9$$

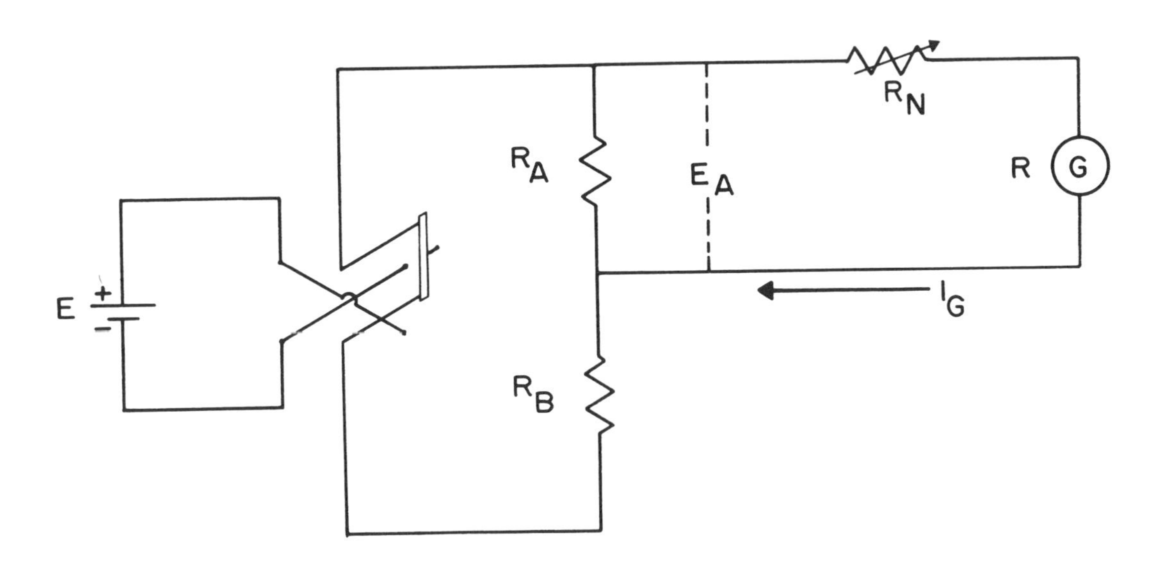

Fig. 10-19. Wiring Diagram for the d'Arsonval Galvanometer.

A simple equation for determining the galvanometer resistance may be derived as follows. If d_1 and d_2 are the galvanometer deflections when R_N is equal to R_1 and R_2, then I_1 and I_2 are the galvanometer currents. Then

$$I_1 = E_A/(R + R_1)$$
$$I_2 = E_A/(R + R_2) \qquad 10\text{-}10$$

and

$$\frac{I_1}{I_2} = \frac{R + R_2}{R + R_1}. \qquad 10\text{-}11$$

Hence

$$R = \frac{I_2R_2 - I_1R_1}{I_1 - I_2}. \qquad 10\text{-}12$$

Assuming that the galvanometer sensitivity does not change to any extent between I_1 and I_2 we have, using k as the current sensitivity of the galvanometer,

$$I_1 = kd_1$$
$$I_2 = kd_2 \qquad 10\text{-}13$$

and R becomes

$$R = \frac{d_2R_2 - d_1R_1}{d_1 - d_2}. \qquad 10\text{-}14$$

Any two values of R_N may be used in making this computation.

If k varied appreciably, it would be necessary to use closely spaced values of R_N or an entirely different method. Variation in k results from a non-linear relationship between galvanometer readings. This plot is obtained during the experiment and will show whether our assumption in regard to the constancy of k is justified.

PROCEDURE

After hooking up the apparatus (leaving one pole of the cell disconnected) and then having it checked by the instructor, record the given values of E, R_A, R_B and compute E_A.

After setting the galvanometer scale at zero for zero current, observe and record left and right values of the galvanometer deflections for all available values of R_N.* The zero setting is obtained by slowly turning the knob above the telescope on the scale.

(a) The galvanometer resistance R is indicated on a small card inside the window of the instrument. Compute the value of I_G in microamperes (10^{-6} amp) from

$$I_G = E_A/(R + R_N). \qquad 10\text{-}15$$

(b) Plot I_G as a function of the deflection d. The slopes of this line will give the working microampere sensitivity for the scale distance used. Or,

$$k_1 = I_G/d \quad [k_1 = \frac{\Delta I_G}{\Delta d}] \qquad 10\text{-}16$$

*If you have a plug box resistance for R_N, use 0, 200, 500, 1000, 2000 and 5000 ohms.

(c) Compute the working megohm sensitivity from

$$k_2 = \text{one volt}/k_1 \tag{10-17}$$

(d) Compute the working voltage sensitivity from

$$k_3 = k_1 R \tag{10-18}$$

PART B - Measuring Capacitance with a Ballistic Galvanometer

INTRODUCTION

The charge q on a capacitor is a function of the potential V and the capacitance C, (q = CV). By discharging capacitors through a ballistic galvanometer and observing the corresponding deflections, we may determine K, the galvanometer constant. K may be found from a graph of Q vs. d.

The galvanometer may be used to measure the capacitance of unknown capacitors and their values in series and parallel. In fact, small charges (pulses of current) resulting from a variety of phenomena may be readily measured.

PROCEDURE

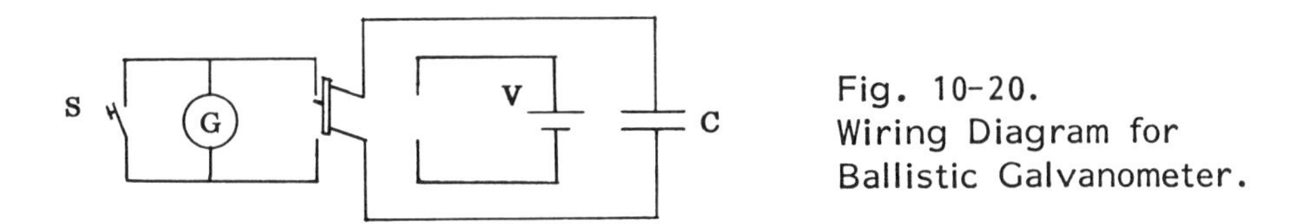

Fig. 10-20. Wiring Diagram for Ballistic Galvanometer.

Connect the circuit elements as shown in the diagram and have the circuit checked by your instructor.

(a) Using the standard capacitor (C in diagram), observe seven deflections obtained from the discharge of capacitances between 0.05 and 1 microfarad. Compute the values of q. Plot q vs. d, and determine K, where K is the slope of the graph.

(b) Replace the standard capacitor C with the unknown capacitors and measure the deflection produced by their discharge.

(c) Calculate by use of K and the deflection of the unknown capacitors the value of their capacitance.

(d) Determine the capacitance of the same two unknown capacitors in series and in parallel by use of the methods in (b) and (c).

(e) Find the capacitance of the two unknown capacitors by the use of the laws for parallel and series capacitors. Compare the calculated values with those found experimentally in above steps (b) through (d).

QUESTIONS

1. Compute the galvanometer resistance from: $R = \dfrac{d_2R_2 - d_1R_1}{d_1 - d_2}$ and compare the result with the resistance used. (See the card mounted inside the window of the instrument.)

2. What current flows through R_A, R_B, and R_N when $R_N = 0$?

3. Discuss briefly the physical significance of:
   (a) current sensitivity k_1 (b) megohm sensitivity k_2 (c) voltage sensitivity k_3

4. What is a voltage divider and when is it used?

5. How could the current sensitivity be improved? What factors limit the current sensitivity?

6. Why are "moving coil" galvanometers used for most current measuring work?

7. How could a galvanometer be converted to read voltage instead of current?

8. How could the galvanometer be converted into an ammeter with a full scale reading of one ampere?

9. How could the galvanometer be converted into a voltmeter with a full scale reading of one volt?

10. Determine the Precision Index (PI) for K. The I.L.E. of the standard capacitor is ±0.25% and the I.L.E. for the graph paper is ±1%.

11. Find the Precision Index (PI) for the two unknown capacitors.

12. Find the Precision Index (PI) for the same two capacitors in series and in parallel. Compare the experimental results for the series and parallel combinations with those found directly by calculation. Do the results check within the limits of error?

13. Determine the amount of energy in joules present in the system of each of the two capacitors after they have been charged. What happens to this energy when the capacitors are discharged through the galvanometer?

14. What are the main differences in construction and use between a ballistic d'Arsonval galvanometer and a current d'Arsonval galvanometer?

15. What are approximately the smallest and largest capacitors that could be measured with the equipment provided?

16. Assuming that you wanted an unknown capacitance, but found that the ballistic galvanometer provided too small a deflection, how would you get it?

17. If the total resistance of the dry cell and the connecting wires is one ohm, how long will it take to charge the one microfarad capacitor to 63% of its maximum charge? What is the time constant of the circuit?

Experiment 10-8 TEMPERATURE COEFFICIENT OF RESISTORS AND THERMISTORS

INTRODUCTION

The resistance of most substances is a function of temperature. The temperature coefficient of resistance α is defined as the ratio of the change of resistance in a metal due to a change in temperature of 1°C, to its resistance at 0°C. This is given by the expression

$$\alpha = \frac{R_t - R_o}{t \times R_o} \qquad 10\text{-}19$$

where R_t and R_o are the values of the resistance of the metallic conductor of t°C and 0°C, respectively. Most metals have a small positive temperature coefficient of resistance.

Equation 10-19 may be written as:

$$R = R_o + \alpha R_o t \qquad 10\text{-}20$$

If we obtain a series of values of R_t for corresponding temperatures, t, and plot these values, the resulting curve is a straight line. The intercept of this line with the resistance axis is R_o and the slope divided by R_o is the value of α.

In contrast with metals, which have small positive temperature coefficients of resistance, thermistors have large negative temperature coefficients. Thermistors are thermally sensitive resistors made of semi-conducting materials. Oxides of manganese, nickel and cobalt are mixed in the desired proportion with a binder and pressed or extruded into shape. They are then sintered under carefully controlled atmospheric and temperature conditions to produce a hard ceramic-like material.

They are very sensitive to small temperature changes and the change of resistance with change of temperature is non-linear. They have been found to have good stability.

The resistance of a thermistor is given by the expression

$$R = R_a\, e^{\beta(1/t - 1/t_a)} \qquad 10\text{-}21$$

where
- R is the resistance in ohms at temperature t(°K)
- R_a is the resistance in ohms at temperature t_a(°K)
- t_a is the initial temperature,

and
- β is a property of the material (in degrees Kelvin) and is constant over a limited temperature range.

Equation 10-21 may be written as

$$\ln \frac{R}{R_a} = \beta x, \qquad 10\text{-}22$$

where $x = 1/t - 1/t_a$.

When the resistance is measured at various temperatures and $\ln R/R_a$ is plotted against x, a straight line results. The slope of this line is β, which is thereby demonstrated to be constant for a limited temperature range.

APPARATUS

The two principal parts supplied are a thermistor assembly and a copper wire assembly. The thermistor assembly consists of a disk-type thermistor mounted on a bakelite cover provided with binding posts for electrical connections in making resistance measurements. The copper wire assembly is essentially the same but with a copper wire coil wound on a brass cylinder replacing the thermistor unit.

A metal vessel with a permanently mounted immersion heater for operation on 115 volts at 250 watts is also supplied. In use it is filled with water, either assembly immersed in the water, and the temperature changed by means of the electric immersion heater.

In this experiment a Wheatstone bridge will be used to measure the resistance. The Wheatstone bridge is a null instrument (zero current for balance). The galvanometer indicates a condition of balance - that is, zero - when the shunt arm is at the most sensitive position. Hence the differences of potential across its terminals must be zero. Therefore:

$$R_1 I_1 = RI$$

$$R_2 I_1 = XI$$

so that

$$X = \frac{R_2}{R_1} R \qquad 10\text{-}23$$

Fig. 10-21. Apparatus for Measuring the Temperature Coefficient of Resistors (on left) and Thermistors (on right). (Courtesy of The Sargent-Welch Scientific Company)

The Wheatstone bridge approaches its greatest sensitivity when R_1 and R, and when R_2 and X are as nearly equal as possible.

A Wheatstone bridge may consist of a number of plug resistance boxes connected as shown in Fig. 10-22. A commercially available Wheatstone bridge may also be used; in this case the manufacturer's directions must be followed carefully.

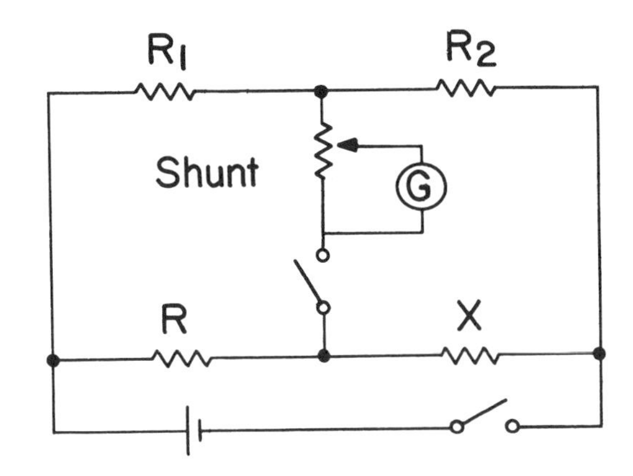

Fig. 10-22. Wiring Diagram for the Wheatstone Bridge.

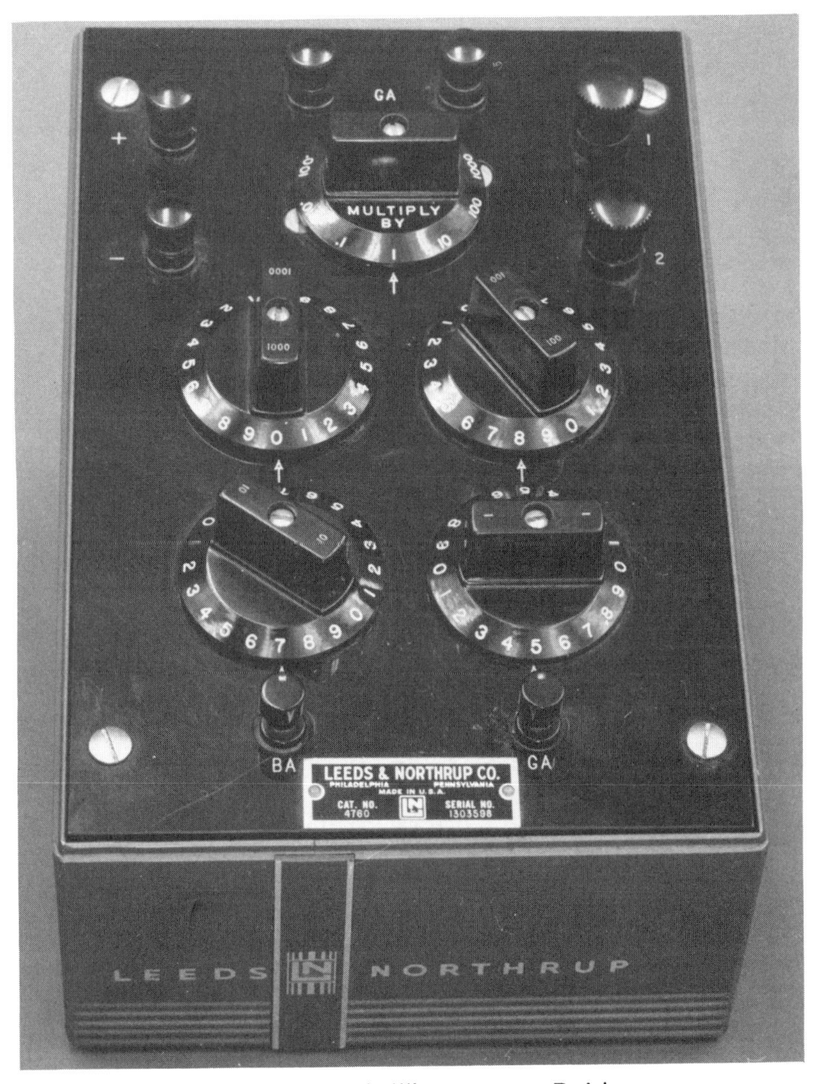

Fig. 10-23. A Wheatstone Bridge.

PROCEDURE

Set up the tripod support. Fill the vessel with water to within an inch of the top. Support the vessel on the fiber ring on the tripod support. Using the one-hole rubber stopper mount the thermometer through the hole of the bakelite cover of the unit to be tested. Place this unit in the vessel with the thermistor or copper wire completely immersed in water. Tighten the two clamps to secure the unit and vessel in place.

Attach the connecting cord to the prongs of the heating unit and with the line switch in the "off" position plug into the 115-volt power supply. Connect the binding posts on the unit to the proper terminals on the Wheatstone Bridge.

Measure the resistance of the unit at room temperature. Record this value and the corresponding temperature. Record resistance values to a tenth of an ohm and temperature readings to a tenth of a degree. Increase the temperature of the water about 10 degrees by turning on the heater. Turn off the heater one or two degrees before the desired temperature is reached, stirring well during the heating. After turning off the heater continue stirring until the maximum temperature is reached and measure the resistance. Care must be taken to insure that the temperature is constant while the resistance is being measured. Repeat the measurements using steps of about 10 degrees until the temperature of 80°C is reached.

PRECAUTIONS

1. Position the thermometer so that its bulb is as close to the thermistor as possible or is centrally located in the copper coil.

2. The temperature of the water will continue to rise for several degrees after the heater is turned off. It seems best to let the temperature reach a maximum before making a resistance measurement.

3. Keep the water well stirred especially just before and during the time a resistance measurement is being made.

ANALYSIS

Plot the measured quantities and determine the constants R_o and α for copper and β for the thermistor.

QUESTIONS

1. Explain how a resistor could be used as a thermometer. What type of substance must be used?

2. In what temperature range would a resistance thermometer be desirable? Explain.

3. What are some commercial applications of thermistors?

4. Sketch the dependence of the resistance of a superconductor on the temperature.

5. If a Wheatstone bridge is balanced, will interchanging the battery and galvanometer affect this balance? Explain. If a Wheatstone bridge is balanced, what would be the effect of removing the battery completely from the circuit?
6. Why is the bridge method not adapted to measuring the resistance of an electric lamp?
7. Discuss the advantages of a "null" statement.
8. Discuss the accuracy and the sensitivity of the Wheatstone bridge as the ratio R_2/R_1 is increased.
9. Find the current in R_1, R_2, R and X when the bridge is balanced assuming that the emf of a dry cell is 1.5 volts.
10. What other methods are available for measuring resistance? Name some advantages and disadvantages.

Experiment 10-9 THE EMF OF A SOLAR CELL

INTRODUCTION

A potentiometer measures an unknown emf E_x, or an unknown potential difference V_x, by comparison with a standard cell E_s. This method is used in precision calibration of meters and resistors. If used carefully, the settings of a simple slide-wire potentiometer can be relied upon to ±0.2% or ±0.5 mm, whichever is the larger.

In the circuit in Fig. 10-24 the length L_s in the potentiometer bridge is proportional to the voltage drop required to balance the E_s of the standard cell (principle of opposing emf's for null current).

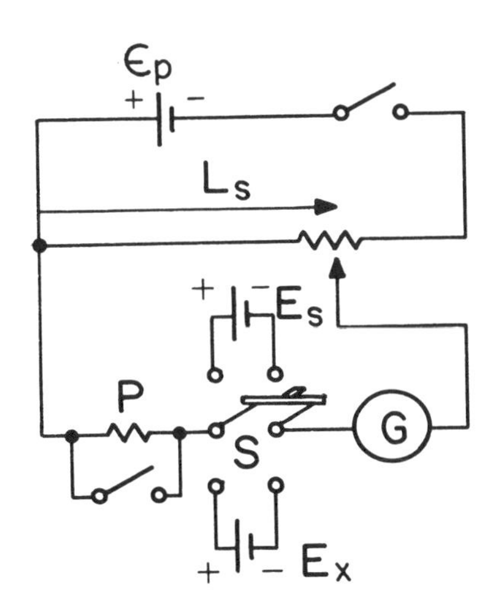

Fig. 10-24. Wiring Diagram for the Potentiometer. ε_p - power supply; E_s - standard cell; E_x - unknown E.M.F.; P - protective resistance.

After balance has been reached for the standard cell, the switch S can be thrown in the reverse direction and a corresponding balance secured for an unknown emf(E_x or the potential drop). Therefore, the length on the potentiometer bridge L_x is now proportional to the voltage drop necessary to balance the unknown E_x. Hence

$$E_x = I_x R_x$$

Dividing,

$$E_x = \frac{I_x R_x}{I_x R_s} E_s$$

$$E_s = I_x R_s$$

Since R_x is proportional to L_x, and R_s is proportional to L_s, we can express E_x as

$$E_x = \frac{L_x}{L_s} E_s. \qquad 10\text{-}24$$

In this experiment the potentiometer will be used to measure an unknown emf, such as the emf of a solar cell.[1] With the rapid advancement of solid state physics and the increasing demand on space exploration, the solar cell has become an integral part of modern technology.

PROCEDURE

Set up the apparatus as illustrated in the diagram (Fig. 10-25). Be sure to check polarity. With the standard cell in the circuit, balance the potentiometer for a null reading. When the current is close to zero, short-circuit the protective resistance in order to obtain a more sensitive balance. Take a reading on the meter stick. Repeat with the unknown emf in the circuit. Take four readings on the meter stick for the standard and the unknown. Compute the unknown emf and its limit of error.

If the unknown is a solar cell, mount it in a lens holder on an optical bench with a light source also mounted on the bench. (Are you using a point source?) Measure the emf developed by the solar cell as a function of light intensity. The light intensity may be varied by changing the distance from source to cell or by the use of two sheets of Polaroid. Show the results in the form of appropriate graphs and discuss these. Are the results what you would expect? Does the emf of the solar cell vary linearly with light intensity?

If a more accurate potentiometer is available (such as a Leeds and Northrup Student Potentiometer), repeat the measurements with this instrument and compare the results (emf with its limit of error) with the results obtained with the slide wire arrangement. Consult the manufacturer's direction for the operation of the potentiometer.

QUESTIONS

1. Why do we use a potentiometer rather than a voltmeter for the measurement of an emf?
2. Do temperature changes in the slide wire affect the results obtained from the potentiometer?

---

[1] The S-1A Silicon solar cell consists of a large area silicon p-n junction, designed for the conversion of solar energy into electrical energy.

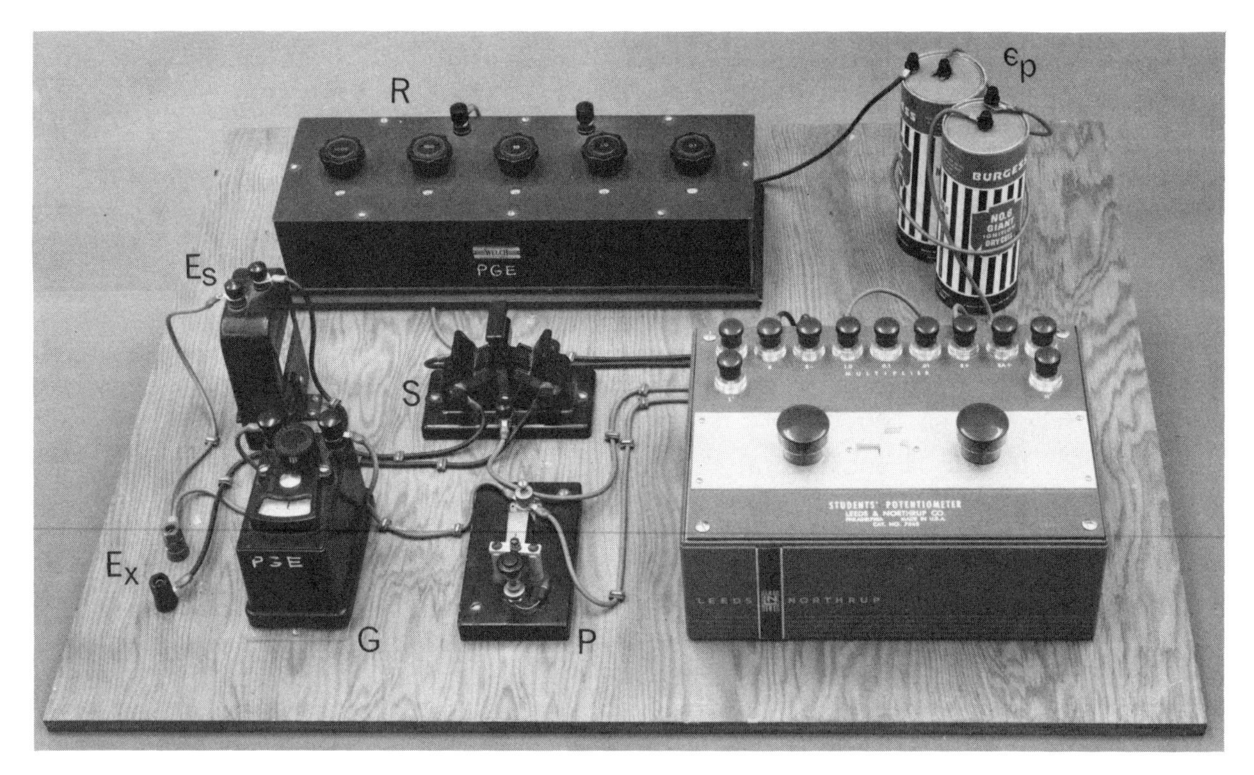

Fig. 10-25. A Leeds and Northrup Student's Potentiometer Connected to a Power Supply ε_p, Resistance R, Standard Cell E_S, Double Pole Double Throw Switch S, Protective Resistance P and Galvanometer G. The unknown potential to be measured is connected across the terminals E_X.

3. What is a standard cell? Describe briefly its construction and use.
4. How do the area and the material of the slide wire affect results obtained by a potentiometer? Explain.
5. Draw a circuit diagram and show in which parts the current is flowing when the potentiometer is balanced.
6. What effect would a slide wire ten times the length have on the accuracy of the measurements?
7. What is the purpose of the protective resistance?
8. Suppose the resistance of the slide wire is 4 ohms and the power supply has 2 volts. What is the current in the main circuit when the potentiometer is balanced?
9. How can a potentiometer be made to read directly in volts?
10. Does the emf of the power supply affect the results obtained by the potentiometer? Should the power supply be larger, smaller or the same as the unknown emf?
11. Discuss the various uses of the solar cell as well as some of its properties.
12. Theoretically how do you expect the intensity of light to vary with distance from a point source? Do you expect this to hold true in this experiment? Explain.
13. If the I.L.E. for the standard cell is ±.1% and that of the slide wire is ±.05%, find the Precision Index PI for one unknown emf.

Experiment 10-10 THE CATHODE RAY OSCILLOSCOPE*

PART A - Introduction to the Oscilloscope

Familiarize yourself with the oscilloscope. Understand the fundamental physical principle involved (not necessarily the electronic circuits used) and know the purpose of the control knobs on the front panel.

Turn on the scope and look at the pattern with the sweep circuit off and then with the sweep circuit turned on. Make sure that you never allow a bright spot to appear on the screen for any length of time because it will burn the screen. Investigate the effects of the various controls, including that for intensity, vertical and horizontal centering, etc.

Connect the vertical plates of the scope across a voltage divider connecting it, in turn, to a 110-V a.c. source. Sketch the pattern obtained with the sweep frequency adjusted to about 30, 60 and 120 Hz. The pattern obtained could be used to calibrate the frequency dials. Look at the effect of the vertical and horizontal gain on the pattern.

Disconnect the 60-Hz voltage and connect the vertical plates of the scope to the variable audio oscillator. Observe the effects of changing the frequency of the input signal. If the oscillator generates square waves as well as sinusoidal waves, observe both wave forms on the scope.

QUESTIONS

1. Describe briefly the primary function of the cathode-ray tube.

2. Sketch the pattern you would expect on the scope if a 60-Hz voltage is applied across the vertical plates and the sweep frequency is (a) 20 Hz, (b) 30 Hz, (c) 60 Hz, (d) 120 Hz.

3. Under what conditions is it advantageous to use external synchronization?

PART B - Observation of the Wave Forms of Sounds

Connect the vertical plates of the oscilloscope to the amplifier and microphone, and observe the wave forms for various sounds of different pitch, loudness and quality. Describe all patterns observed.

If you have a dual trace oscilloscope, compare the frequency of a tuning fork with the known oscillator frequency.

---

*Before starting on this experiment, the description of the cathode ray oscilloscope in Section 4-7, should be studied. This experiment gives a few examples for the use of the oscilloscope - other applications are included in various experiments throughout this book.

QUESTIONS

1. Briefly discuss the physical significance of the pitch, the loudness, and the quality of a musical sound.

2. The frequency of a "300-Hz" tuning fork is to be checked against the frequency of the alternating current wall outlet. Draw a block diagram showing all the necessary equipment. Discuss the procedure to be followed and the accuracy expected.

3. Consider two waves $\psi_1 = A \cos \omega_1 t$ and $\psi_2 = A \cos \omega_2 t$, where ω_2 is nearly equal to ω_1. Derive an expression for the resultant disturbance.

4. Design a setup so that one uses the phenomena of beats to determine an unknown frequency. Describe what will be observed on the scope.

PART C - Lissajous Figures*

Connect the horizontal plates of the scope across the voltage divider supplying 60 Hz ac off the 110-V line. Connect the vertical plates of the scope to the output of the audio oscillator. Turn the sweep control to "off." Now you are putting two simple harmonic motions at right angles into the scope; the pattern on the screen gives the resultant motion, or Lissajous Figures. Sketch the patterns observed for oscillator frequencies of 60 Hz and multiples of 60 Hz. Obtain as many patterns as you can.

For the frequency ratios 1:1, 2:1 and 3:2, construct Lissajous figures for phase angle difference of 0, $\frac{\pi}{2}$, and π. Check the resulting figures with the sketches obtained on the scope, for the corresponding frequency ratios.

By observation of the orientation and shape of a 1 to 1 ratio pattern, an idea of the phase difference between two signals can be obtained.

* A more detailed analysis of Lissajous figures is given in Experiment 12-4.

In Fig. 10-26, the distance A represents the intersection of the ellipse with the Y axis. By drawing a line parallel to the X axis tangent to the top of the ellipse, the distance B is determined by the interception of this tangent with the Y axis. The phase difference θ is calculated from $\sin\theta = A/B$.

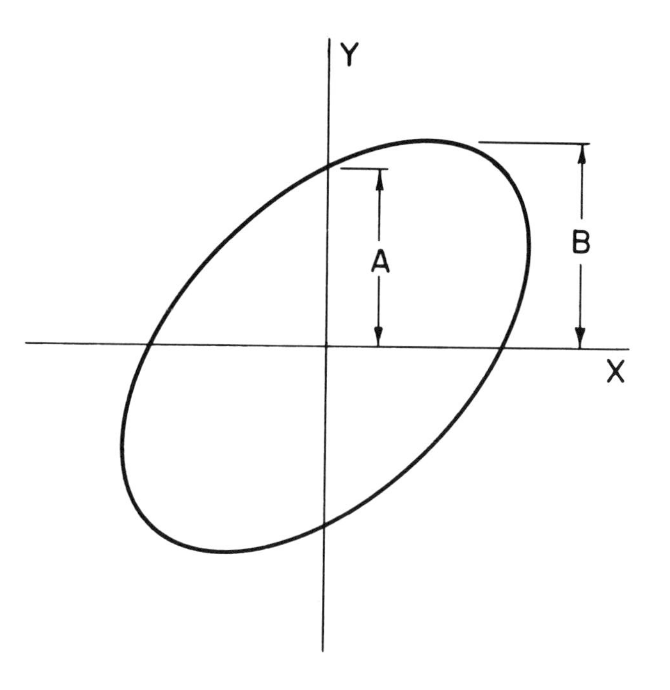

Fig. 10-26.
Phase Measurement with Lissajous Figures.

Another point to note is that if the spot traces out the curve in the clockwise direction, then the phase angle is between 0 and 180°. If the spot traverses the curve in the counterclockwise direction, the phase angle is between 180° and 360°. When the spot traces out the pattern in the counterclockwise direction, the phase is usually taken as

$$\theta = 360^\circ - \sin^{-1}\left(\frac{A}{B}\right)$$

Fig. 10-27 shows an example. In (a) $A/B = 0$; therefore $\theta = 0$ or $\theta = 360^\circ$. In (b) of Fig. 10-27 $A/B = 0.5$: therefore $\theta = 30^\circ$ or $\theta = 330^\circ$.

Note: Without any external signals, center the spot on the scope. When the signals are applied the intercepts of the figure can then be read directly off the Y axis. This method centers the figure on the screen.

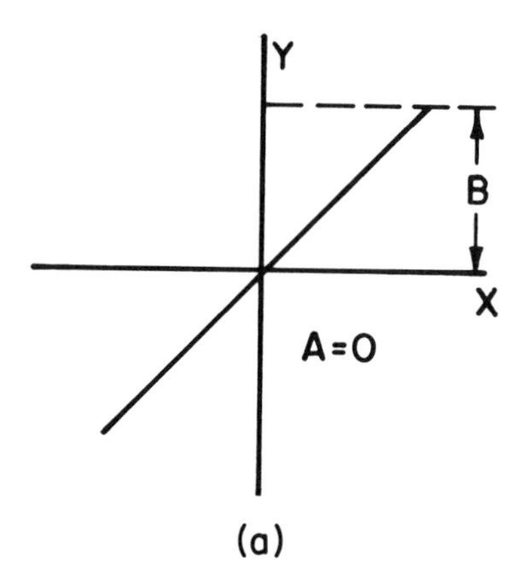

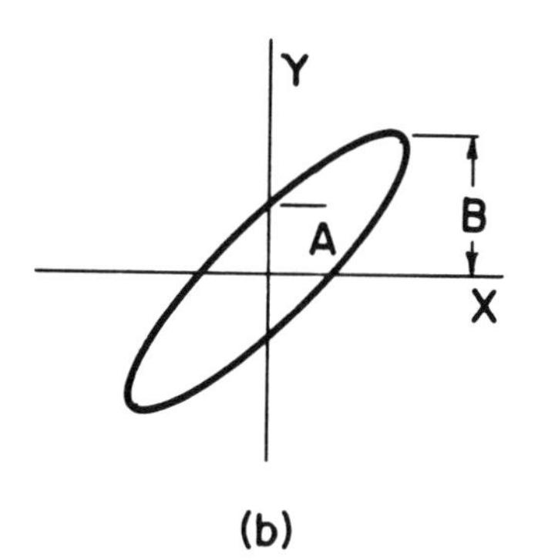

Fig. 10-27.
(a) phase angle = 0 or 360°
(b) phase angle = 30 or 330°

QUESTIONS

1. Construct Lissajous figures for phase differences of $\pi/4$ and $3\pi/4$, and frequency ratios of 1:1, 2:1, and 3:1.

2. Discuss briefly one other method of obtaining Lissajous figures besides the oscilloscope.

3. When making a phase measurement could the X axis also be used as a measuring reference? Explain.

4. Using a frequency ratio of 1:1, draw the possible Lissajous patterns if the A/B ratio is: (a) A/B = 1, (b) A/B = 0. What are the respective phase angles? Repeat for a frequency ratio of 2:1.

5. Discuss how Lissajous patterns may be used to determine unknown frequencies.

PART D - Simple Circuits

Use the oscilloscope to investigate any simple circuit that may be available. Sketch and explain the patterns observed.

Experiment 10-11 A.C. SERIES CIRCUITS

INTRODUCTION

This experiment is divided into three independent parts, all of which illustrate the behavior of a.c. circuits. Before starting the experiment you should read through all three parts.

PART A - Resistance and Capacitance in Series

For the R-C circuit in Fig. 10-28, we define the following qualities:

- R resistance of the resistor
- C capacitance of the capacitor
- X_C capacitive reactance
- I current through the series circuit
- V_R voltage drop across the resistor
- V_C voltage drop across the capacitor

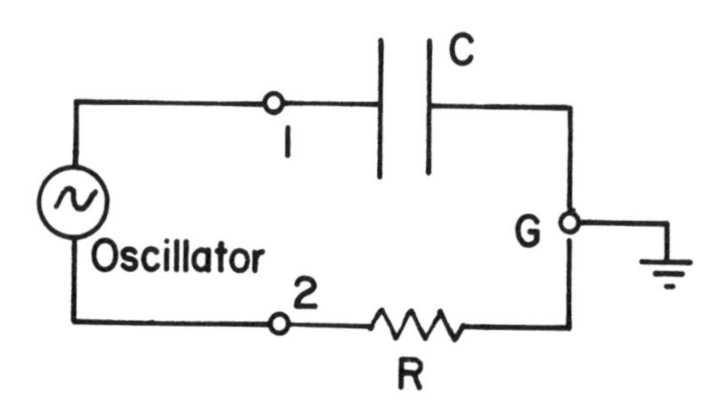

Fig. 10-28. R-C Series Circuit.

The current through the series circuit is

$$I = \frac{V_C}{X_C} = \frac{V_R}{R} \qquad 10\text{-}25$$

and therefore the capacitive reactance is

$$X_C = \frac{V_C}{V_R} R \qquad 10\text{-}26$$

Connect point G on the diagram to the oscilloscope ground. To measure V_C, connect point 1 to the vertical input on the oscilloscope. To measure V_R, connect point 2 to the vertical input of the oscilloscope.

(a) V_C/V_R

Measure V_C and V_R for frequencies between 0.5 and 4.5 kHz (500-4500 Hz) in steps of 1 kHz. For convenience, all voltages are measured off the scope as peak to peak, twice the amplitude of the sine wave. Plot the ratio V_C/V_R vs. frequency.

(b) Capacitive Reactance

Verify for one of the points that $X_C = 1/\omega C = \frac{1}{2\pi fC}$; i.e., calculate $\omega X_C = 2\pi fR \frac{V_C}{V_R}$ and show that it is approximately equal to 1/C.

(c) Phase Angle

Note the difference in phase (phase angle ϕ) between the two curves on the oscilloscope. Estimate the phase angle in degrees (or radians). On a dual-trace oscilloscope the phase angle can be observed directly. If a dual-trace oscilloscope is not available, the phase angle can be estimated by using Lissajous figures (see Experiment 10-10, Part C). How does this phase angle change with a change in frequency?

(d) $V_C + V_R$

For one of the frequencies used in Part (a) measure V_o, the voltage drop across the resistor and capacitor in series, by connecting the oscilloscope across the oscillator (points 1 and 2). Note that the total voltage V_o is not the algebraic sum of V_C and V_R. Why not? Consider the results of Part (c).

PART B - Addition of Voltages in a R-C-L Series Circuit (See Fig. 10-29.)

(a) Use the digital a.c. voltmeter (make certain you use the correct scale) to measure the voltages across the resistor V_R, across the inductor V_L, across the capacitor V_C and across all three V_{LCR} at approximately 1000 Hz. Is V_{LCR} the sum of V_R, V_C and V_L? Explain. Remember that a.c. meters generally measure RMS values.

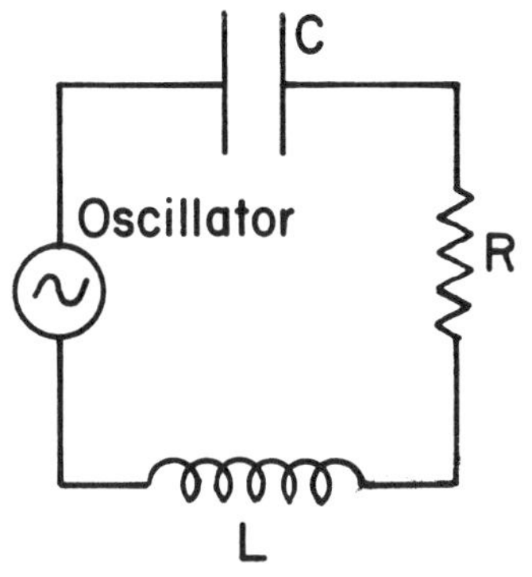

(b) Compute the impedance Z of the circuit and the current I.

(c) Measure the current I (using the digital a.c. ammeter) and compare with the computed value.

Fig. 10-29. R-C-L Series Circuit

(d) Compute the power factor and the average power dissipated in this circuit.

PART C - Resonance in a Series R-C-L Circuit

When a capacitor is connected in series with an inductor, resonance will be exhibited when the inductive reactance X_L is equal to the capacitance reactance X_C. We shall investigate the amplitude and the phase of the response signal relative to the driving signal.

(a) Voltages Near Resonance
Connect the resistance R, the inductance L and the capacitance C in series across the oscillator as shown in Fig. 10-30. Use Channel 1 of the dual-trace oscilloscope to measure the voltage V_o across the oscillator and Channel 2 the voltage across the inductor - capacitor combination V_{LC}. Vary the frequency to find resonance. Measure V_{LC} and V_o below resonance, at resonance and above resonance. Also sketch the ratio V_{LC}/V_o vs. frequency.

(b) Phase Angle Near Resonance
Interchange scope connections at Channel 1 and ground so that Channel 1 shows the voltage V_o across LCR and the signal on Channel 2 measures the voltage across the resistor which is proportional to the current through the circuit. For each of the frequencies used, measure the phase angle ϕ and plot phase angle vs. frequency.

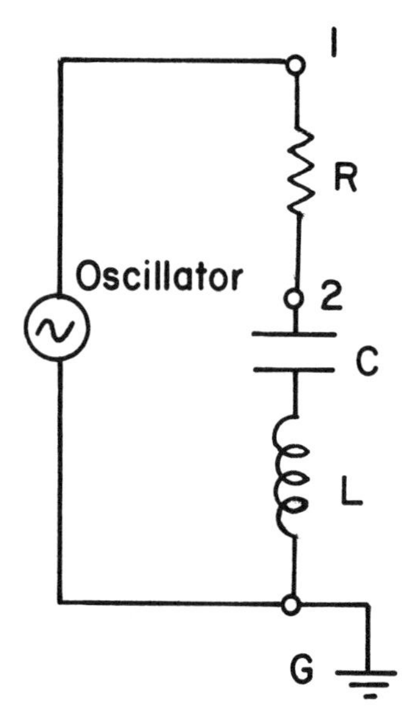

Fig. 10-30. R-C-L Series Circuit.

QUESTIONS

1. Explain the difference between the following voltages in an a.c. circuit: instantaneous, maximum, peak to peak, average, and root-mean square. Which of these can be measured by an oscilloscope? a voltmeter?

2. What is the advantage of plotting a ratio of two voltages vs. frequency in Parts A and C of the experiment over plotting just a voltage vs. frequency?

3. Could an oscilloscope be used in Part B to measure the voltages, instead of a voltmeter? Could an oscilloscope be used to measure the current?

4. Describe two different ways to measure phase angles on the oscilloscope.

5. If henrys and farads are the units of a capacitance and inductance respectively, show that the inductive and capacitive reactances both have the units of ohms.

6. What is the phase relation between the instantaneous potential difference across the inductance and capacitance at resonance? What would a voltmeter read if it were placed across an inductance and capacitance connected in series at resonance? Why?

7. If a fixed inductance of 400 microhenrys is used to form a tuned circuit with a variable tuning condenser, what are the maximum and minimum values of capacitance needed for a tuning range over the broadcast band from 200 to 600 meters?

Chapter 11

MAGNETISM

Experiment 11-1 THE EARTH'S MAGNETIC FIELD

INTRODUCTION

The tangent galvanometer consists of a current carrying coil producing a magnetic field perpendicular to the horizontal component of the earth's magnetic field, as indicated by a compass needle. A pointer shows the direction of the resultant magnetic field. The current in the coil will produce a magnetic field along the axis of the coil with a magnitude given by:

$$B_c = \frac{\mu_o N i}{2a}, \qquad 11\text{-}1$$

where

B_c = magnetic field (Tesla)
N = number of turns
μ_o = permeability H/m of free space
i = current A
a = radius m of coil.

When the pointer has come to rest, it points in the direction of the resultant of the field produced by the current in the coil and the horizontal component of the earth's magnetic field. If the angle that the needle makes with the axis of the coil is θ, then

$$\tan \theta = \frac{B_c}{B_e}, \qquad 11\text{-}2$$

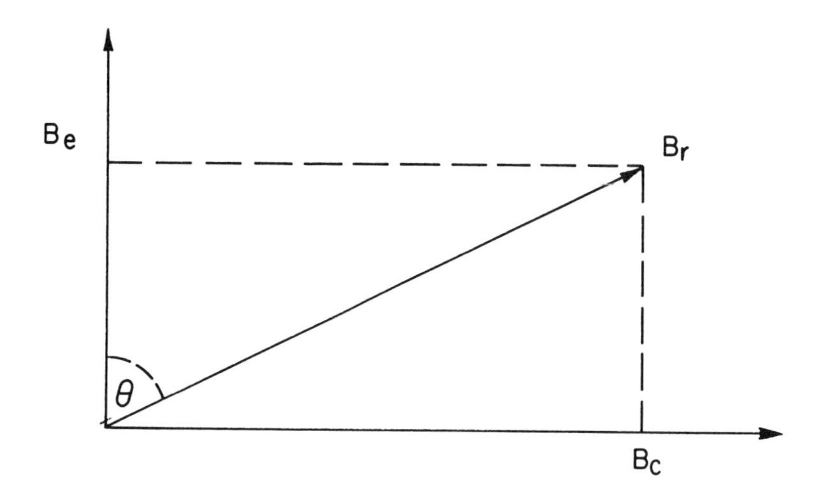

Fig. 11-1.
Vector Diagram.

$$\text{or } B_e = \frac{B_c}{\tan\theta} = B_c \cot\theta, \text{ where} \tag{11-3}$$

B_e = magnetic field due to the earth
B_r = resultant magnetic field
B_c = magnetic field produced by current in the coil

PROCEDURE

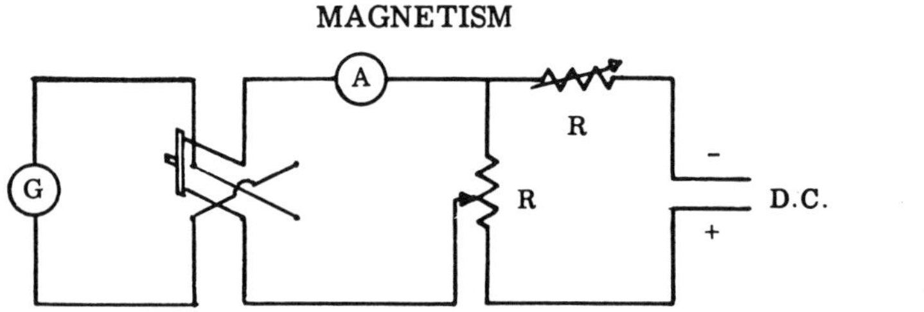

Fig. 11-2.
Wiring Diagram for the Tangent Galvanometer G.

(a) The tangent galvanometer is first leveled so that the pointer can move as freely as possible. Test by observing the swing. The plane of the coil should make an angle of 90° with the pointer. The magnetic needle points in the north-south direction and should lie in the plane of the coil.

(b) Adjust the resistances R to allow a current to flow through five turns of the coil so that the pointer deflects about 45°. Because the pointer has friction in the bearings, tap the case <u>slightly</u>. <u>Remove all magnetic materials such as metal pencils, watches, current carrying wires, etc., as far from the galvanometer as possible</u>.

(c) Record the angular readings θ from both ends (a and b) of the pointer and then reverse the current in the coil and record the two new values of θ, a' and b'.

(d) Repeat the above procedure for ten and fifteen turns.

(e) Measure the diameter of the coil.

(f) Compute B_e from the data for 5, 10 and 15 turns and then find the average of the three results.

(g) Compare your value with that given in a handbook or textbook, and discuss the reasons for the discrepancies in the results.

QUESTIONS

1. If the coil is initially a few degrees off the north-south direction, how would this affect the results? What effect, if 30°?

2. Does the compass needle point toward the North Pole? Explain.

3. What is the total magnetic field B_r when there is a current in the 15-turn coil?

4. If the angle of dip (the angle that the earth's magnetic field makes with the horizontal) is 73°, what is the magnitude of the vertical component of the earth's magnetic field? What is the total earth's magnetic field?

5. Is the magnetic field of the earth symmetrical and constant? Explain.

6. Find the Precision Index (PI) for the horizontal component of the earth's magnetic field B_e. Assume that a meter stick was used to measure the diameter of the coil, and that the I.L.E. for the angle is ±0.5°.

7. Why should the tangent galvanometer be kept as far as possible from the rest of the equipment? Why should the leads through which the current flows be close together?

8. If the current through the coil were doubled, would the angle of deflection become twice as large? Explain.

9. Do the angles measured at both ends of the magnetic needle agree? If not, why not?

10. How could the sensitivity of the tangent galvanometer be increased?

Experiment 11-2 THE CURRENT BALANCE

INTRODUCTION

If a current I flows through an infinitely long straight wire, a magnetic field is set up which can be found from Ampere's law to be

$$B = \frac{\mu_o I}{2\pi d} , \qquad 11\text{-}4$$

where d is the distance from the center of the wire to the point where the field is found and μ_o is the permeability of free space ($4\pi \times 10^{-7}$ H/m). If another wire carrying the same current I is placed parallel to the first wire, it will experience a force given by:

$$F = BIL = \frac{\mu_o I}{2\pi d} ; \quad IL = \frac{\mu_o}{2\pi} \frac{L}{d} I^2 , \qquad 11\text{-}5$$

where L is the length of the second wire and d the distance between the wires. If the two currents are in opposite directions, the conductors repel each other with the force given above.

The expression for the force is, strictly speaking, valid only for very long wires, but we will assume it to be sufficiently accurate for this experiment.

The above expression for the force between two wires is used to define the ampere in the MKS system: "One ampere is the current which, if present in each of two parallel conductors of infinite length and one meter apart in a vacuum, causes each conductor to experience a force of exactly 2×10^{-7} newton per meter of length."

PART A - Use of Telescope

In this experiment, the current I is passed in opposite directions through two parallel horizontal bars which are connected in series. The lower bar is fixed; the upper one is balanced a few millimeters above it by adjusting a counter-poise. The upper bar supports a small pan onto which analytical weights are placed, thereby causing the upper bar to drop down toward the lower one.

When the current is turned on and increased sufficiently, repulsion between the two bars causes the upper bar to return to its initial equilibrium position. The position of the bar is observed by means of a mirror, a telescope and scale. With this experimental setup, the relationship between the force on either conductor and the current passing through the conductors can be obtained.

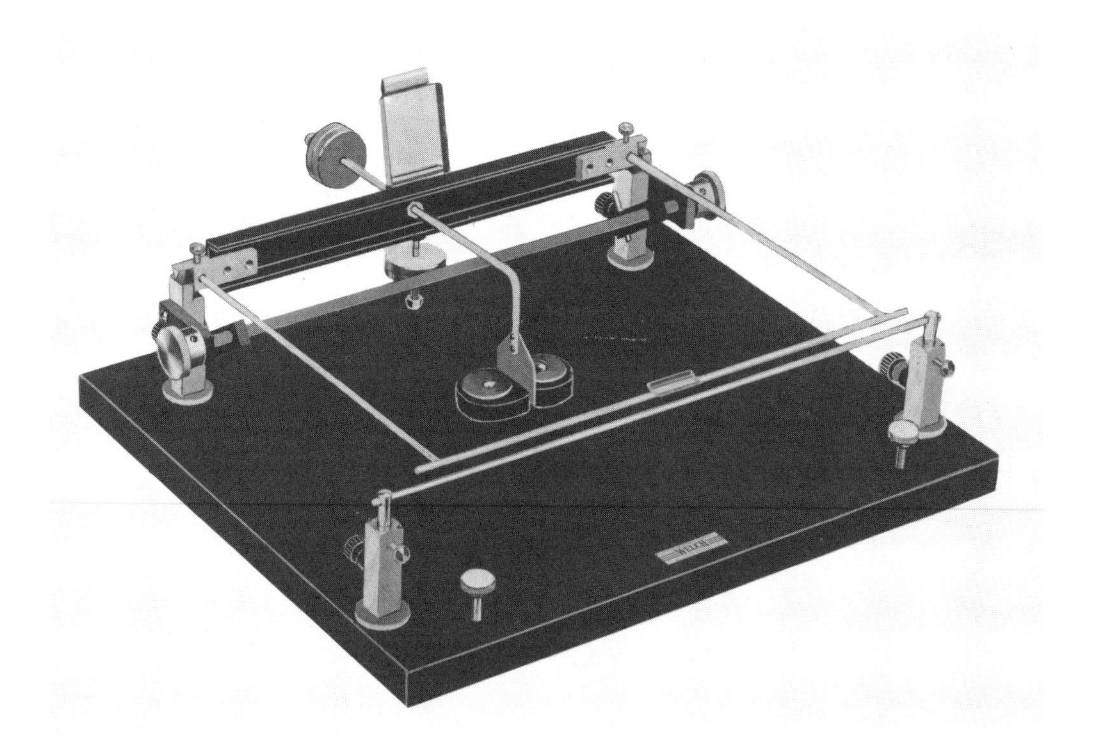

Fig. 11-3. The Current Balance.
(Courtesy of the Sargent-Welch Scientific Company)

PROCEDURE AND ANALYSIS

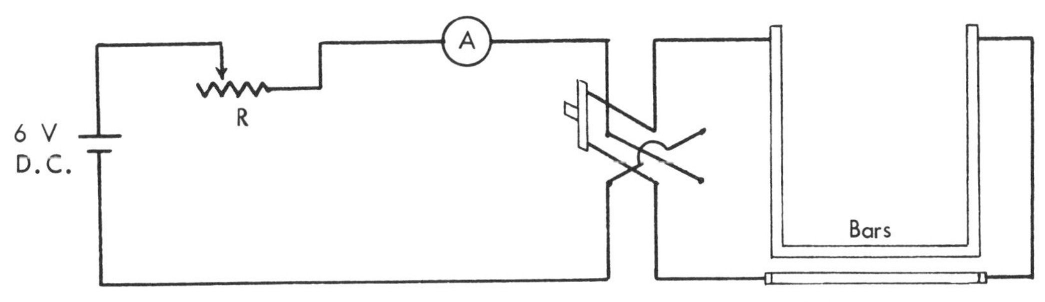

Fig. 11-4. Wiring Diagram for the Current Balance.

1. Level the base of the current balance.

2. Align the two bars. By careful adjustment the two bars should be aligned as accurately as can be determined by the unaided eye when viewed from the front and top.

3. Connect apparatus as indicated in the diagram. Wires should leave binding posts at right angles. Why?

4. Adjust the counter-poise until the upper bar is a few millimeters above the lower bar.

5. Set up the telescope and scale 1 to 3 meters from the mirror. Adjust the telescope until you can see the scale clearly. Record the equilibrium point indicated by the cross hairs on the telescope scale.

6. In increments of 5 mg, place weights in the pan. Adjust the current until the scaling reading returns to its equilibrium value. Record the current. Reverse the current and repeat. Find the average current.

7. Measure the length of the upper bar. The length L is the distance between the supporting bars.

8. In order to verify Eq. 11-5, we also need to know d, the center-to-center distance between the bars at equilibrium. This is determined as follows: measure the lever arm a or the distance from the knife-edge to the center of the front bar at each side and take the average. Rerecord the scale reading at equilibrium. Depress the upper bar by placing a small coin on the scale pan. Note the new scale reading. The distance d_o between the two bars is given by:

$$d_o = \frac{Da}{2b}, \tag{11-6}$$

where D is the difference in scale readings,
a is the mean distance from the knife edge to the bar, and
b is the distance from the mirror to the scale.

The center-to-center distance d is obtained by adding the diameter of either rod to d_o.

9. Using the data obtained in Part 6, plot the force F as a function of I^2 and determine the slope of the resultant curve. From the slope, determine the value of the permeability of free space μ_o.

PART B - Use of Microcomputer *

Part of the purpose of this experiment is to give an example of how a microprocessor can be used to control and collect data from a piece of apparatus. The experiment is essentially the same as Experiment 11-2, Part A above, except that the equilibrium position is determined by a photo-detector instead of an optical lever and the micro controls the current needed to reestablish equilibrium.

* These directions are written for use with the microprocessor and its interface designed under the supervision of Dr. Thomas Shannon under a grant from the RCA Corporation. A detailed description may be obtained from Professor Walter Eppenstein, Physics Department, Rensselaer Polytechnic Institute, Troy, N.Y. 12180-3590.

Fig. 11-5. The Current Balance with Photo Detector, Microprocessor, Monitor and Auxiliary Equipment.

The photo detector is an IR LED and photo-transistor pair with a flag which is connected to the moveable bar. The spacing between the bars is established by means of a piece of wire (diameter 1.31 ± .03 mm). The flag then should block approximately half of the beam of the IR source-detector pair. The spacer is removed and the balance adjusted to about the same equilibrium position as indicated by the reading taken by the micro. This step is not critical. When mass is placed on the balance and the run started the micro will check the position of the balance and increment the current until the balance is back to the spacing as set by the wire. The value of this current is stored. (You have the option of selecting an automatic repeat to get an average value for each mass.) Once you have your data points they can be transferred to the Apple which will calculate a straight line least squares fit to the data and send back the value of the slope and intercept of the line. From the slope, spacing between the bars and the length of the top bar a value of the permeability μ_0 may be obtained.

The procedure is essentially to follow the instructions given on the monitor screen and make appropriate choices. Measure the length of the top bar as the center-to-center distance of the two bars which are perpendicular to the balance bar. The diameter of each of the bars is 1/8". Calculate μ_0.

Suggestion: Make a trial run; do not worry about establishing equilibrium of the balance; select an average of one, using masses 10, 30, 50, 100 mgs. Then reset the micro and perform the experiment carefully using an average of four with masses 10-100 mgs in 10-mg steps. The masses may be added or removed in any order.

Comments on the Microcomputer

- "Reset" clears the memory, i.e., all previous data is lost.

- "Esc." jumps to display of data.

- Additional runs of any mass may be made; only the last value is stored.

- Transfer to the Apple (calculate) takes about 15 sec; longer if someone else is already using the line.

- Data points which are far off the line can be deleted by putting the cursor on it then pressing shift then "D". All points will be restored by shift "B".

PART C - Coulomb Balance Attachment

The current balance described in this experiment can be converted to a Coulomb balance measuring the electrostatic force between two oppositely charged plates. The two wires are replaced by two plates as shown in Fig. 11-6. A potential difference is applied across the two plates - up to 200 volts D.C. A megohm resistor should be in one branch of the circuit to limit the current in case the two plates accidentally touch - otherwise, the power supply may be damaged.

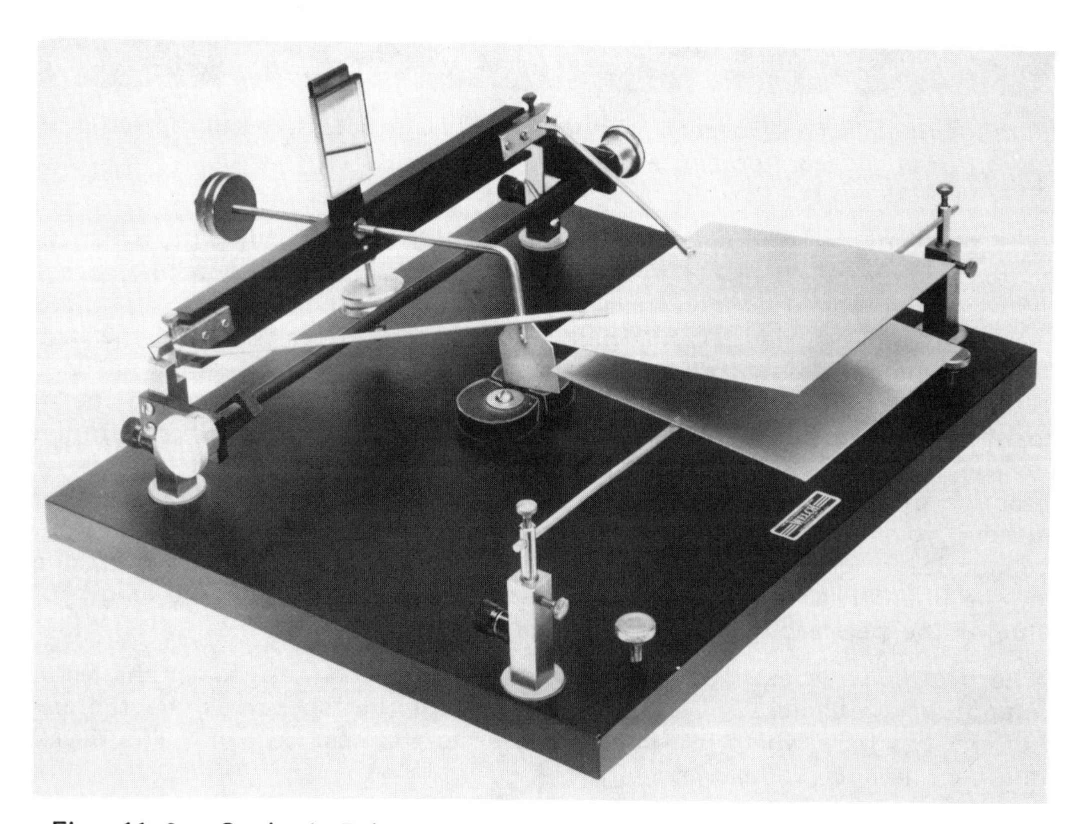

Fig. 11-6. Coulomb Balance Attachment Mounted on Current Balance.

Measure the force between the plates for different voltages. Also measure the area of the plates and their separation (Eq. 11-6).

From the data obtained compute the value of ε_0, the permittivity of free space (derive necessary equations). Use the value of ε_0 and the μ_0 obtained from the data of the current balance to compute the speed of light. Give the percentage error and discuss possible sources of error.

QUESTIONS

1. Derive Eqs. 11-4 and 11-5.

2. What is being assumed when you use Eqs. 11-4 and 11-5 in this experiment?

3. If you want to calculate the magnetic field set up by a finite straight wire, can you use Ampere's law or must you use the Biot-Savart law?

4. Is μ_0 (the permeability of free space) an assigned quantity or a measured quantity? Is ε_0 (the permittivity constant) an assigned quantity or a measured quantity?

5. Why do you not have to worry about the weight of the upper bar when you calculate the force F between the two bars?

6. Explain in detail the reasons for the repulsive force between the two wires when they carry currents in opposite directions. What would happen if the currents in the two wires were in the same direction?

7. Could the force between the conducting bars be determined in this manner if an alternating current were used?

8. What is the purpose of reversing the current and using the mean current in plotting F vs. I^2?

9. Why is it important that the wires leave the binding posts at right angles?

10. Derive Eq. 11-6. State clearly the assumptions you make!

Experiment 11-3 DETERMINATION OF e/m*

INTRODUCTION

In this experiment the effect of electric and magnetic fields on charged particles will be investigated and the ratio of the charge to the mass of an electron will be measured. Electrons - or cathode rays - are ejected from a cathode, heated indirectly. They are accelerated by a voltage applied to the anode of a cathode ray tube. The velocity of the electrons in the emitted beam is determined by the anode potential applied, or $\frac{1}{2}mv^2 = Ve$, where m and e are the mass and charge of the electron respectively, v is their velocity and V is the anode potential.

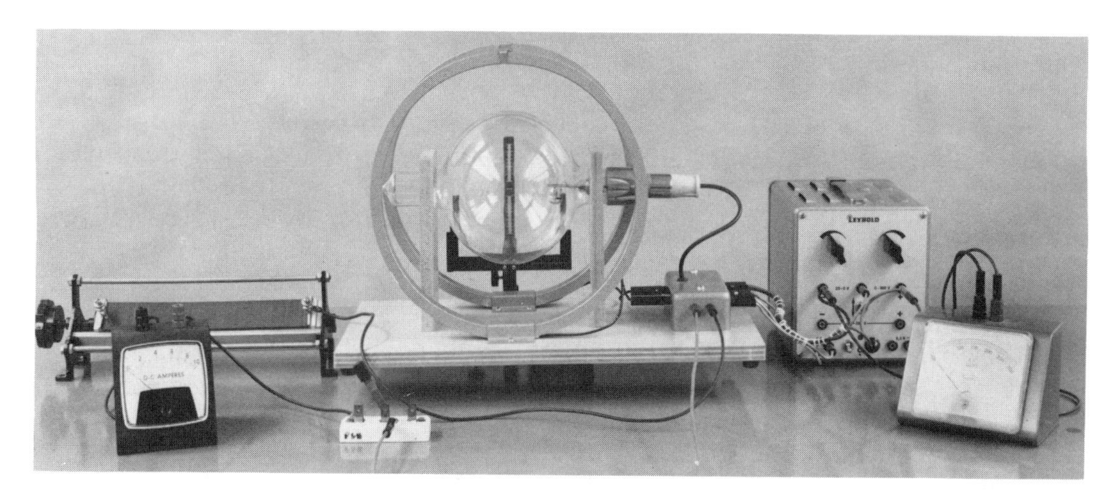

Fig. 11-7. The e/m Apparatus. The carbon rheostat and ammeter on the left control and measure the current through the Helmholtz coils at the center. The power supply on the right controls the filament voltage and the potential difference through which electrons are accelerated.

The cathode ray tube used is filled with hydrogen gas at a pressure of about 10^{-2} mm Hg. Due to the ionization of the hydrogen by the electrons, the electron beam is made visible as a luminous streamer in a completely darkened room, as shown in Fig. 11-8.

A homogeneous magnetic field is produced in the region of the cathode ray tube by a current through two circular coils. Whenever a charged particle - such as an electron - is ejected into a magnetic field, a force $\vec{F} = q\vec{v} \times \vec{B}$ acts on the particle and will deflect it. If the particle moves perpendicular to the magnetic field, the force exerted is at right angles to the velocity and will not affect the magnitude of the velocity, but will merely

* The directions in this experiment refer to the apparatus manufactured by Leybold; equipment from other manufacturers is available and the general principles are the same.

alter its direction. The particle, therefore, moves under the influence of a force whose magnitude is constant, but whose direction is always at right angles to the velocity of the particle. Therefore, the orbit of the particle is a circle and the radial acceleration gives rise to a force F, the centripetal force.

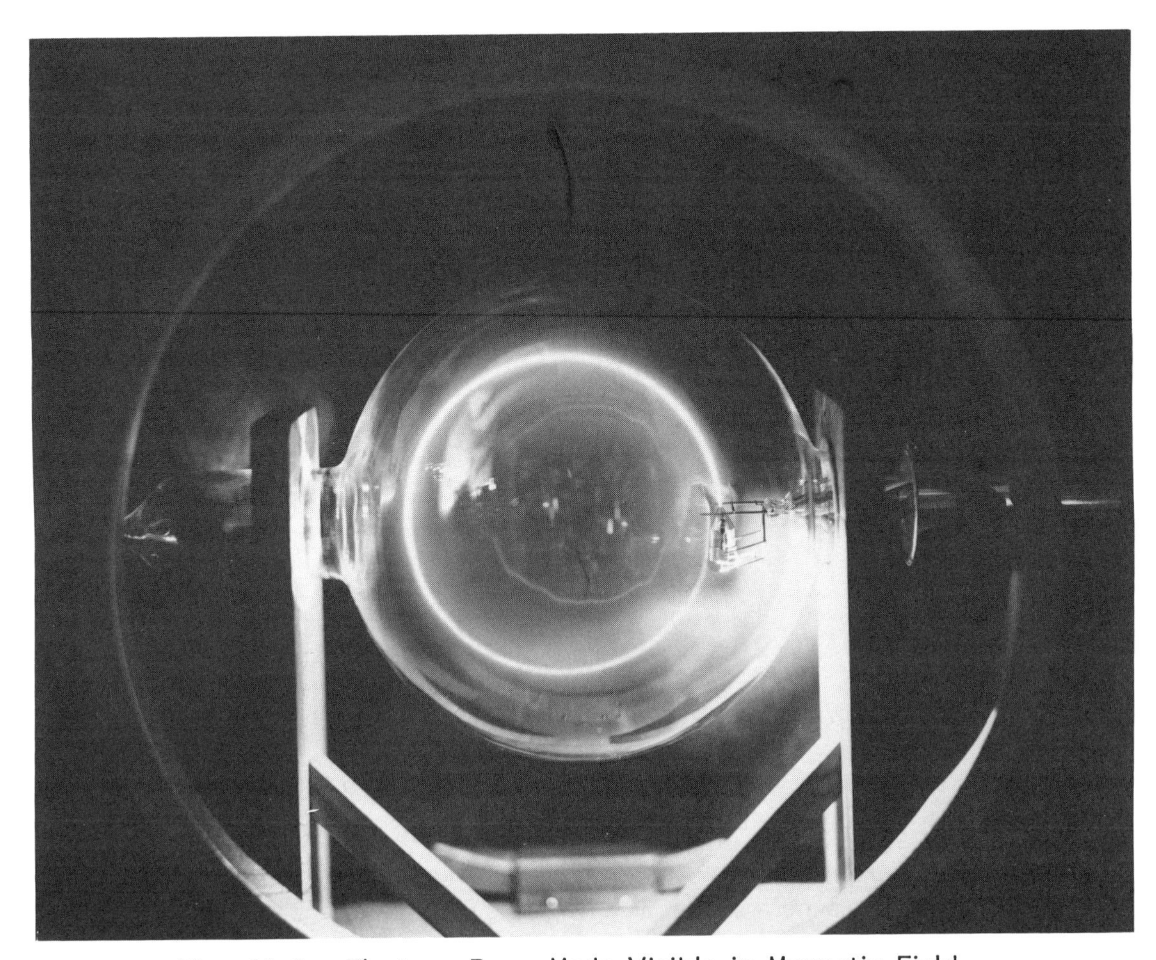

Fig. 11-8. Electron Beam Made Visible in Magnetic Field.

The diameter of the circular electron path can be measured by means of a marked mirror behind the tube or an internal scale projector as shown in Fig. 11-7. This device projects a virtual image of an external millimeter scale into the plane of the electron beam. This procedure eliminates errors due to parallex present when a marked mirror is mounted behind the tube.

PROCEDURE

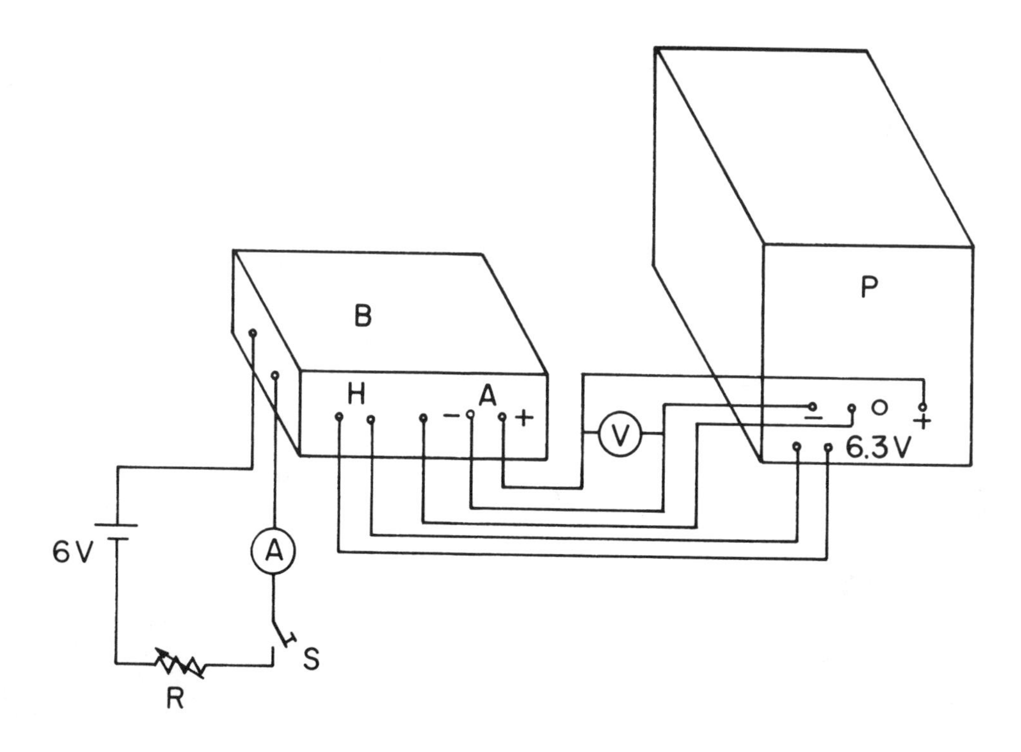

Fig. 11-9. Connections for Setting Up the e/m Apparatus.

Check the wiring and ask the instructor for help if there are any questions about the operation of the equipment. The filament voltage of the hot cathode should be about 6 volts and is supplied by the power unit. The anode voltage should be between 150 and 250 volts, and is also supplied by the power supply. As soon as the cathode starts to glow, step up the anode voltage. Adjust voltages until the electron beam is as sharp as possible.

A 6- or 12-volt battery is used to produce the magnetic field. Depending on the size of the circular electron beam desired, the current should be between 0.5 and 2 amps. Measure the diameter of this beam for at least four different settings of anode potential and magnetic field current. Record the anode potential, the current through the coil and the diameter of the electron beam.

Gently turn the tube and observe the electron beam when the velocity of the electrons is not perpendicular to the magnetic field. Also observe the path of the electrons under the influence of a permanent magnet.

Apply a potential across the small parallel plates inside the tube and observe the effect on the electron beam at various voltages. Study the path of the electrons under crossed electric and magnetic fields.

ANALYSIS

The magnetic field on the axis of a circular coil due to the current in the coil is given by

$$B = \frac{\mu_o}{2} \frac{ia^2}{(a^2 + b^2)^{3/2}} \qquad \text{11-7}$$

For the special case of two coils with N turns each, the magnetic field at the midpoint between the coils is given by

$$B = \mu_o \left(\frac{4}{5}\right)^{3/2} \frac{Ni}{a}, \qquad \text{11-8}$$

where the distance between the two coils is equal to the radius of the coils. The number of turns on each coil is 130 and the distance between the coils is 15 cm. Compute the magnetic field for each reading taken.

The electrons are accelerated through a potential difference equal to the voltage on the anode of the tube. By setting the potential energy of the electrons equal to its kinetic energy, write down an expression for the velocity in terms of e and m.

Now consider the forces acting on the electron due to the magnetic field and the circular motion of the electron. Use the expression for velocity found above and derive an expression for the ratio e/m in terms of measurable quantities. Compute the ratio e/m for each reading taken during the experiment. Compare the average value with the correct value and discuss any discrepancy.

QUESTIONS

1. Use the Biot-Savart law to derive Equations 11-7 and 11-8.

2. Consider a Helmholtz coil system where the radius of the coils is "a" and the distance between the coils is "s". Show that if s = a, then the magnetic field along the axis is nearly constant for points which are not too far from the midpoint between the coils.

3. Find the fractional change in B as you move a distance s away from the midpoint between the coils. Does this result justify the assumed homogenuity of the magnetic field between the coils? (Hint - Expand B in a Taylor series about the midpoint, carry out to 4th order.)

4. In your derivation of an expression for e/m, can you justify why it is only necessary to consider a single electron in the magnetic field, rather than the many electrons which are actually present? Explain.

5. Discuss the effects of the earth's magnetic field on the result of this experiment. How could you correct for the effect of the horizontal and vertical component of the earth's magnetic field?

6. Compute the velocity of the electrons for all anode voltages used.

7. Compute the Precision Index (PI) of the measurement of e/m. Does the result check with the true value within the PI?

8. How does the diameter of the electron beam vary with the magnetic field at a constant anode potential?

9. Consider an electron rotation in a uniform magnetic field B. How much must the field be changed if a pi-meson is to rotate in exactly the same orbit as was occupied by the electron?

10. How could a similar experiment be used to determine the relation between the mass of an electron and its velocity? What is this relation?

11. In the experiment you are asked to turn the tube, thus causing the velocity of the electrons to no longer remain perpendicular to the magnetic field. What path will the electrons now follow? Physically, why must this be the case?

12. Explain the principle difference between this method used to measure e/m and the method used by J. J. Thomson.

Experiment 11-4 THE MAGNETIC FIELD OF A CIRCULAR COIL*

INTRODUCTION

A current carrying conductor produces a magnetic field $\vec{B}$. The magnitude and direction of the field depend on the geometry of the conductor, the current flowing through the conductor, and the location of the point at which the field is determined.

In certain cases it is quite easy to calculate the magnetic field using Ampere's law or the Biot-Savart law. Ampere's law is useful only when dealing with geometrics which possess a high degree of symmetry. The Biot-Savart law can be used in all cases; however, it is often impossible to perform the necessary integrations and we must be content with writing the fields in terms of integrals or infinite series. For this reason it is often necessary to determine the field experimentally.

The simplest procedure for measuring a magnetic field is through the use of a "search coil" (a small coil of wire which can be placed at various positions in the field). We pass a low frequency alternating current through the conductor and measure the emf induced in the search coil. (Low frequencies must be used; otherwise the field set up will be significantly different from the static field which one is trying to determine.) The emf induced in the search coil depends on the magnetic field at the position of the coil,

---

* Construction details of the equipment are given in the Am. J. Phys. 28, 147 (1960). Also see Robert E. Marclay, AAPT Apparatus Drawing Project, Plenum Press, New York, 1962, p. 5.

the construction of the coil, and the orientation of the coil relative to the field. The dependence of the emf on the above can be derived as follows: The flux through the coil is approximately

$$\phi = BA \cos \theta \qquad 11\text{-}9$$

where A is the area of the coil and θ is the angle between the normal to the plane of the coil and the direction of the field. From Faraday's law the emf induced in the search coil is

$$\text{emf} = -N \frac{d\phi}{dt} = -NA \cos \theta \frac{dB}{dt}. \qquad 11\text{-}10$$

Since the current setting up the field varies sinusoidally with time, the field itself has a sinusoidal time variation. We can assume

$$B = B_o \cos \omega t$$

where ω is the angular frequency of the alternating current which is setting up the field. Thus we can write

$$\text{emf} = \omega NA \cos \theta \, B_o \sin \omega t.$$

In practice we usually measure the r.m.s. voltage which can be written as

$$\text{emf}_{r.m.s.} = \omega NA \cos \theta \, B_{r.m.s.}$$

Since the induced emf's are quite small it is usually necessary to amplify them by a certain amount, k. Thus the final relationship between the measured voltage and the magnetic field is

$$V_{measured} = k\omega NA \cos \theta \, B_{r.m.s.} \qquad 11\text{-}11$$

In principle we could determine the magnetic field from this relation. However, in practice it is simpler to calibrate the experimental setup by measuring the amplified induced emf established when the search coil is placed at a position where the magnetic field is known.

For a circular coil a convenient position to use for the calibration is a point on the axis of the coil. At such a point the field is given by

$$B = \frac{N\mu_o i a^2}{2(a^2 + x^2)^{3/2}} \qquad 11\text{-}12$$

where
- i = current through the coil
- $\mu_o = 4\pi \times 10^{-7}$ H/m
- a = radius of the coil
- x = distance from the center of the coil along the axis
- N = numbers of turns on the coil

To determine the field for a point on the axis of the coil, you need to know the value of the alternating current through the coil.* You can determine the current from a knowledge of the voltage across the coil and the "impedance" of the coil. The current amplitude is given by

$$I = \frac{V}{Z} \tag{11-13}$$

where V is the voltage amplitude and Z is the impedance of the coil. For a coil Z is given by

$$Z = \sqrt{R^2 + (\omega L)^2}\ , \tag{11-14}$$

where

R = resistance of the coil
ω = angular frequency of the oscillator
L = inductance of the coil

APPARATUS

A schematic drawing of the apparatus is shown in Fig. 11-10. The search coil is free to ride in a track cut in the lucite arm, its position being indicated by a scale along the track. The angular orientation of the search coil relative to the track is indicated by a protractor. The lucite arm is free to turn about a point at the center of the large coil. Protractors are used to indicate the angular orientation of the lucite arm.

The large coil is driven by an oscillator set at about 2000-3000 Hz. The emf induced in the search coil is measured by a voltmeter used in conjunction with an amplifier.

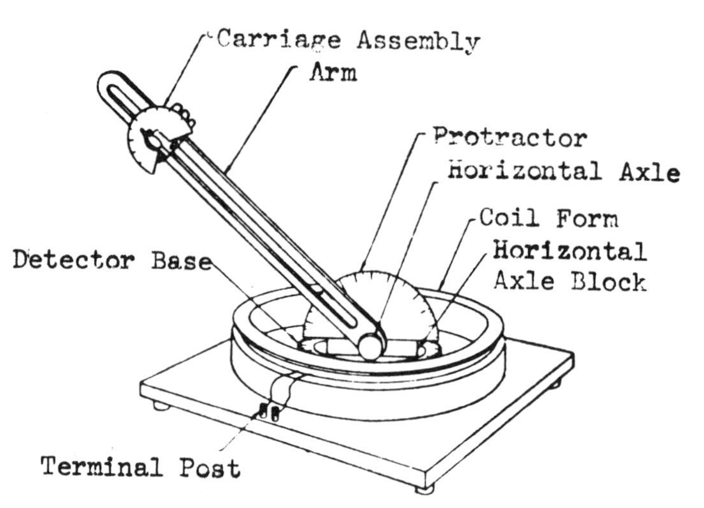

Fig. 11-10. Schematic of Apparatus.

PRECAUTIONS

1. Always set the "voltage range switch" on the voltmeter to a higher range than needed until ready to read the voltage; then adjust this switch to obtain as nearly a full scale reading as possible.

2. Allow all equipment a few minutes to warm up.

3. Do not change the frequency or the gain of the oscillator during the experiment.

---

* See Experiment 10-11 for details regarding alternating currents.

PROCEDURE

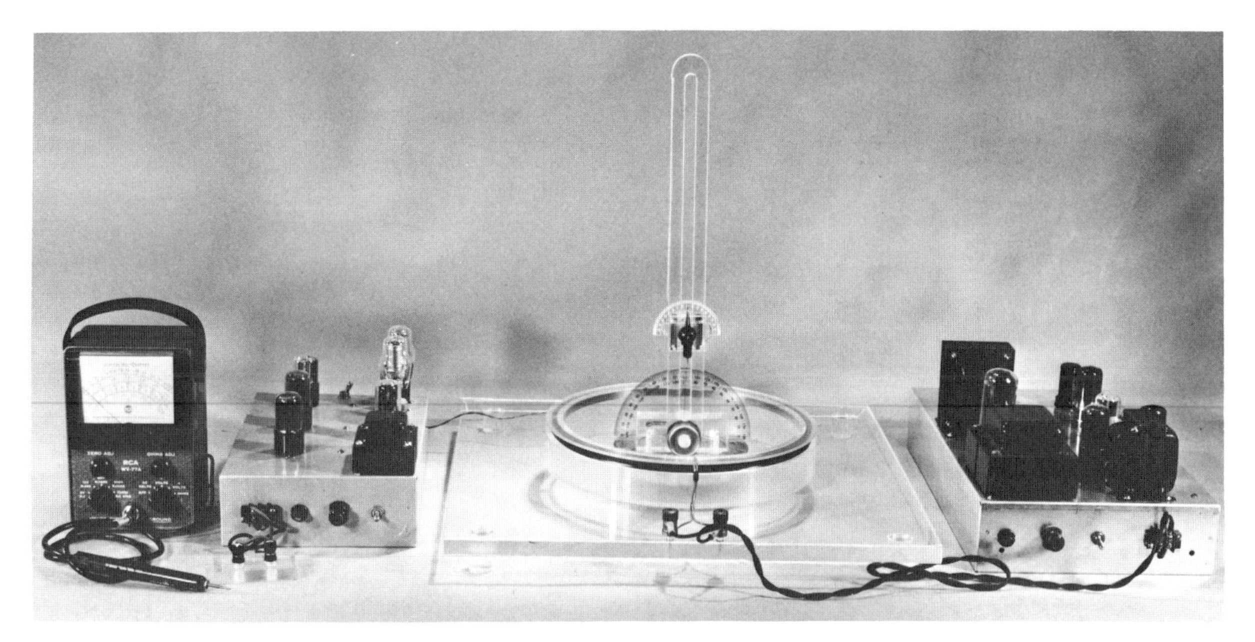

Fig. 11-11. From Right to Left: Oscillator (a regular audio oscillator may be used), Coil with Search Coil, Amplifier, Voltmeter.

1. Calibration of the Equipment

   (a) Record the following quantities: the resistance and inductance of the coil, the diameter and number of turns of the coil, and the frequency at which the oscillator is set (Remember that the angular frequency ω is equal to 2π times the oscillator frequency in Hz.)

   (b) Compute the impedance of the coil from Eq. 11-14.

   (c) Measure the r.m.s. voltage across the coil. Determine the r.m.s. current from Eq. 11-13, $I = V/Z$.

   (d) Calculate the magnetic field B of the circular coil for a point on the axis of the coil and at a distance of 15 cm from its plane, using Eq. 11-12.

   (e) Disconnect the voltmeter from the coil and connect it to the amplifier. Place the search coil on the axis of the large coil at a distance of 15 cm along the arm, making sure the arm is normal to the plane of the large coil. Rotate the search coil until the voltmeter reading is a maximum. Record this voltage.

   (f) Since we know the field at this point from the calculation in Part (d) and have measured the corresponding voltage in Part (a), we can now compute the calibration constant A for the equipment defined as

$$A = \frac{\text{calculated } B \text{ (in teslas)}}{\text{measured voltage (in volts)}} \tag{11-15}$$

The calibration constant depends on the geometry of the search coil, the gain of the amplifier, and the frequency of the oscillator.

2. Determination of the Magnetic Field. The magnetic field B can now be determined systematically in terms of the independent variables r, θ, ϕ defined as shown in Fig. 11-12.

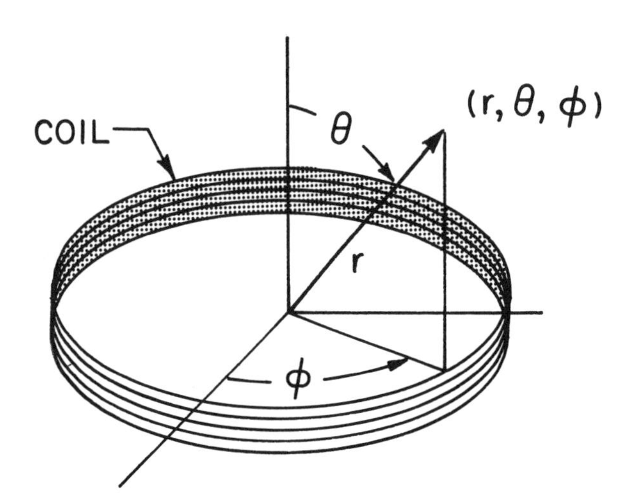

Fig. 11-12.
Geometry of the Apparatus.

(a) At a convenient point in the field determine the voltage as a function of the orientation of the search coil relative to the field. Take readings at about every 10°. You should observe the measured voltage varying from a maximum to a minimum (ideally zero) as the search coil's orientation changes by 90°.

(b) The magnitude of B at any point (r, θ, ϕ) is determined by rotating the search coil about its own axis until a maximum voltage is obtained.

(c) The direction of the field B is found by rotating the search coil about its axis until a minimum or zero voltage is maintained. Under these conditions the plane of the search coil is parallel to the magnetic field. The direction of the field relative to the arm can be determined from the protractor attached to the arm. Be sure to reduce all data so that the direction is specified with reference to the axis of the circular coil.

ANALYSIS

(a) Plot the results of 2(a) and compare with Eq. 11-11.

(b) The results of 2(b) and (c) can most conveniently be presented by plotting them on polar graph paper. Decide upon a suitable scale factor for plotting the vectors determined in the experiments. The following plots should be made:

Plot B as a function of r with θ and ϕ constant. This could be done for several angles θ such as 0, 25, 50, and 75 degrees. Let the graph paper be the plane of constant ϕ.

(c) Plot B as a function of θ for constant r and ϕ. This could be done for several distances r.

QUESTIONS

1. Derive Eq. 11-12.

2. Why do we determine the direction of the magnetic field from the orientation of the search coil that corresponds to a minimum in the induced emf?

3. How would a plot of B vs. ϕ look if r and θ are kept constant?

4. Would the plot suggested in Question 3 differ substantially if a square coil was used instead of a circular coil? Explain.

5. The construction of the equipment assumed that B at all points lies in a plane including the axis of the circular coil. What modifications in the apparatus would have to be made if the above assumption were not true?

6. Could the search coil technique for the measurement of the magnetic fields used in this experiment be applied to fields due to direct current? Explain.

7. Could a search coil be used at all in the measurement of constant magnetic fields? If so, what equipment would be necessary?

Experiment 11-5 THE HALL EFFECT

INTRODUCTION

The Hall effect, discovered in 1879,* may be used to determine the predominate type of charge carrier in a conductor or semi-conductor by the sideways deflection of these moving carriers in an external magnetic field. When a conductor carrying a current I_c (see Fig. 11-13) is placed in a transverse magnetic field B, a force is exerted in the charge carriers. As a result of this force, the charge carriers are displaced in a direction perpendicular to both the direction of the magnetic field and the original direction of motion of the charge carriers. Hence the carriers have now become segregated, causing transverse potential difference between (x) and (y) known as the Hall voltage.

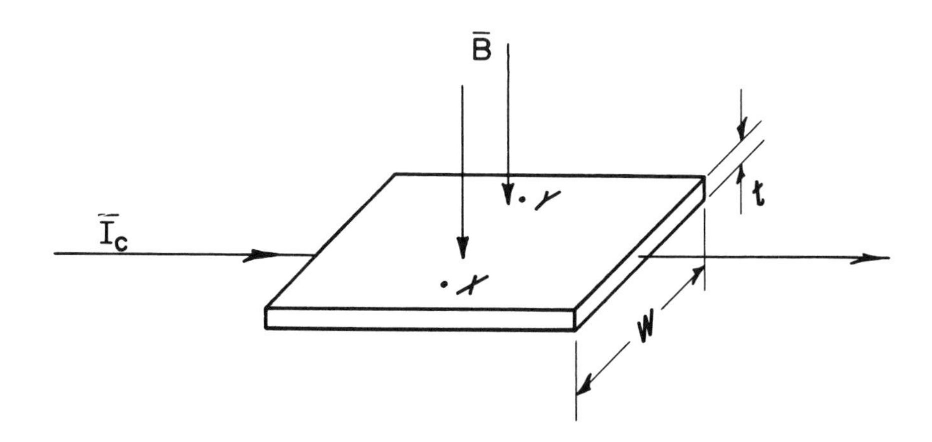

Fig. 11-13. Current Flowing through Conductor.

The force acting on the charge carriers can be written as

$$F = qv_dB = qE = q\frac{V_H}{w}, \qquad 11\text{-}16$$

where V_H is the Hall voltage (in Volts), B the magnetic field strength (in tesla) and w the width of the sample (in meters). By measuring or knowing V_H, w and B, we can calculate the drift velocity v_d of the charge carriers.

Once the drift velocity v_d is known and the current I_c is measured, the number of charge carries per unit volume, n can be found since

$$V_d = \frac{I}{nAe}, \qquad 11\text{-}17$$

---

* The original paper of E.H. Hall can be found in the American Journal of Mathematics 2, 287 (1879).

where the area A is the width w times the thickness t of the sample and e the charge of an electron.

Combine Equations 11-16 and 11-17 to find an expression of the Hall voltage V_H as a function of the current I.

The Hall effect has become of great importance with the development of semiconductor physics. In particular, along with conductivity measurements, it is used in determining current carrier mobilities in metals and in semiconductors. As readily suggested by this experiment, the Hall effect is also used for magnetic field measurements, as in the Hall generator. Using a semiconductor, the Hall generator is essentially a solid state multiplying device that provides a voltage output proportional to the product of two electrical quantities, i.e., the current passing through it and the magnetic field perpendicular to it. Hall generators can be used for many types of instrument applications as well as analog computer elements.

PART A - The Hall Effect in Bismuth

APPARATUS

The Hall effect in a bismuth strip will be investigated by means of the circuit shown schematically in Fig. 11-14.

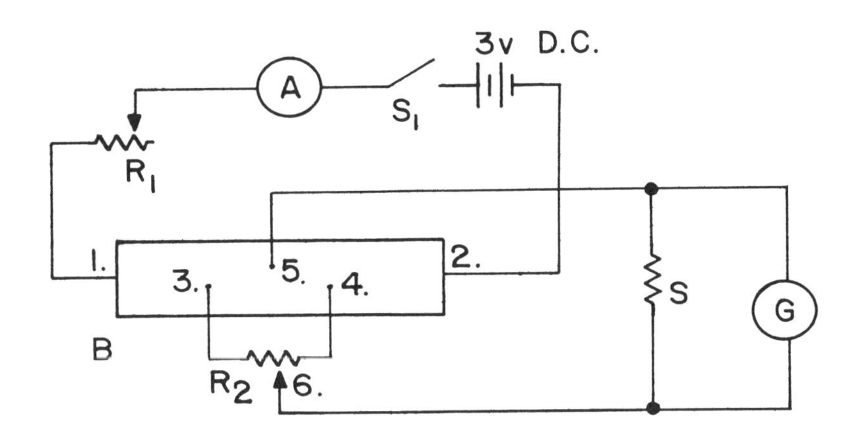

Fig. 11-14. Circuit for Measuring the Hall Voltage Bismuth.

The ammeter A is used to indicate the control current I_c passing through the bismuth strip B at points (1) and (2). The magnetic field is applied in a direction perpendicular to the bismuth strip by a permanent magnet (not shown in Fig. 11-14). The magnitude of the magnetic field is varied by means of soft iron pole pieces. Since a potential gradient due to I_c exists along the bismuth strip, it is necessary to adjust R_2 until the potential difference between points (5) and (6) is zero. The voltage indicated by the arrangement of the spotlight galvanometer G and the Aryton shunt S is then the Hall voltage V_H.

PROCEDURE

Take the necessary data for a plot of V_H vs. I_c for a constant magnetic field. Be sure the use of the Aryton shunt is understood so that the galvanometer is never allowed to go off scale. Remove the magnet from the vicinity of the bismuth strip. Close the switch S_1 and adjust I_c to about .2 amperes by varying R_1. Adjust R_2 until the galvanometer is zeroed on its most sensitive scale. Keep the magnetic field constant and place the magnet so that the bismuth strip is in the center of the magnet gap. If the galvanometer deflection is negative, reverse the direction of the magnetic field by turning the magnet around. The galvanometer deflection obtained is proportional to V_H. Repeat, varying I_c to 1 ampere. If I_c is held near 1 ampere for any length of time, excessive drain on the dry cells will result. The thickness of the bismuth strip is given. Measure the width of the strip and the average distance between probes (3) and (4) and probe (5). Record information from the galvanometer to obtain its voltage sensitivity in volts/cm.

The result of changing the magnetic field is to be investigated. Change the number of pole pieces in the magnet gap. Consult the chart provided to obtain the corresponding values of the magnetic field. Plot V_H vs. I_c for as many values of the magnetic field as time permits.

NOTE: The contact between the probes and the bismuth is very sensitive to movement of the apparatus. If the mount is moved, it may be necessary to repeat an entire run. Also, you should move the magnet, not the mount, when adjusting the magnetic field. Never handle the probes.

ANALYSIS

1. By observing the polarities of V_H, I and B, draw a conclusion about the sign of the charge carriers in the material used - are they positive or negative?

2. From the slope of the plot of V_H vs. I calculate the approximate number of charge carriers per unit volume.

3. From your data calculate the drift velocity of the charge carriers.

NOTE: If you were not able to vary the magnetic field and therefore could not plot V_H vs. I, do item (1) above and then calculate the drift velocity (item 3) and finally the number of charge carriers per unit volume (item 2).

PART B - The Hall Effect in Germanium

APPARATUS

The Hall effect in the semiconductor germanium will be investigated by means of the circuit shown schematically in Fig. 11-15.

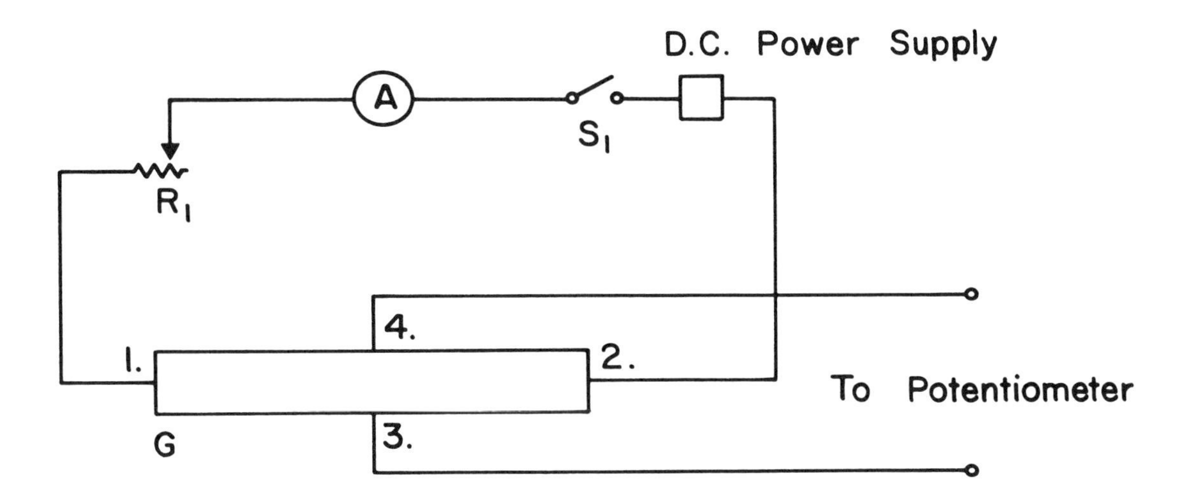

Fig. 11-15. Circuit for Measuring the Hall Voltage in Germanium.

PROCEDURE

Take the necessary data for a plot of V_H vs. I_c for a constant magnetic field. Read the instructions provided with the potentiometer and understand the operation of each component of the circuit. Keeping the magnetic field constant, place the magnet so that the germanium strip is in the center of the magnet gap. Close the switch S_1 and adjust I_c to 2 mA by varying R_1 and the output of the power supply (6 or 12 volts). Record the voltage on the potentiometer. Reverse the direction of the magnetic field by turning the magnet around. Again record the voltage on the potentiometer. Repeat, varying I_c up to 15 mA. Record pertinent information indicated on the mount of the germanium specimen.

The result of changing the magnetic field is to be investigated. Change the number of pole pieces in the magnet gap. Consult the chart provided to obtain the corresponding value of the magnetic field. Determine V_H vs. I_c for as many values of the magnetic field as time permits.

ANALYSIS

1. By observing the polarities of V_H, I and B, draw a conclusion about the sign of the charge carriers in the material used - are they positive or negative? Did you use p-type (positive) or n-type (negative) germanium?

2. From the slope of the plot of V_H vs. I calculate the approximate number of charge carriers per unit volume.

3. From your data calculate the drift velocity of the charge carriers.

NOTE: If you were not able to vary the magnetic field and therefore could not plot V_H vs. I, do item (1) above and then calculate the drift velocity (item 3) and finally the number of charge carriers per unit volume (item 2).

QUESTIONS

1. Consider a free electron model of a metal. Discuss why the measurement of the Hall coefficient does reveal the type of conduction, whereas the electrical conductivity does not.

2. How would you expect a plot of power dissipation vs. Hall voltage to look?

3. Using the equation for Hall voltage derive an expression for power dissipation in terms of V_H, geometrical parameters, and the conductivity of the substance.

4. Qualitatively discuss what is meant by "p-type" and "n-type" germanium.

5. What is the order of magnitude of the drift velocity of the electrons in a typical copper wire for a current of a few amperes?

6. How does the value of the drift velocity of charge carriers compare to the speed with which electrical energy is propagated?

Experiment 11-6 MAGNETIZATION AND HYSTERESIS

PART A - The Rowland Ring Method

INTRODUCTION

This experiment is designed to illustrate some of the fundamental principles of ferromagnetism. Ferromagnetic substances display two important characteristics, the S-shaped magnetization curve and a hysteresis (or memory effect). Since the normal magnetization curve is just the locus of maxima of hysteresis loops for a given sample, the same type of circuit may be used to obtain information for both curves.

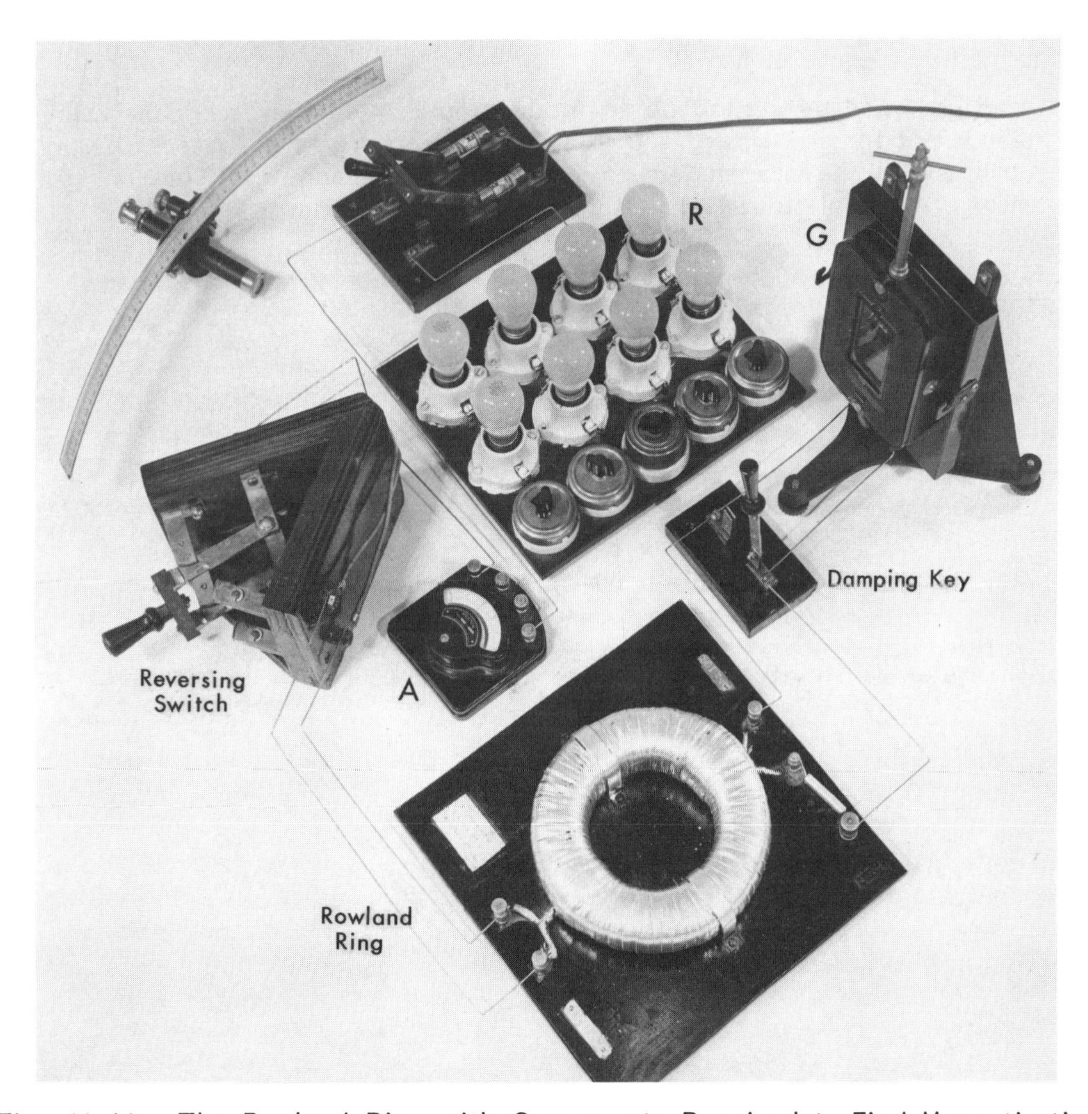

Fig. 11-16. The Rowland Ring with Components Required to Find Magnetization.

The Rowland Ring Method is used in this experiment. See Fig. 11-16. The specimen is made into a toroidal ring which is wound with many turns of wire for the primary and on top of these are wound several turns for the secondary. The specimen to be investigated is placed inside the toroidal winding.

If the primary contains n_p turns per unit length carrying a current i_p, the magnetic field inside the toroid without an iron core is

$$B_o = \mu_o n_p i_p.$$

The actual magnetic field inside the toroid with an iron core in place is greater than B_o by the contribution of the iron core B_M to the total magnetic field B or

$$B = B_o + B_M.$$

B_M is associated with the alignment of elementary atomic dipoles in the iron

B_o can be found by measuring the current i_p through the primary and by knowing the number of turns per unit length n.

B can be found by setting up an induced current in the secondary coil by reversing the current in the primary and measuring the charge that passes through the secondary coil during the time that the magnetic field is changing. The charge is measured by means of a ballistic galvanometer. From Faraday's law,

$$\text{emf} = -N_s \frac{d\phi}{dt} = NA \frac{dB}{dt} = i_s R_g \tag{11-18}$$

or

$$-N_s A \int_B^{-B} dB = R_g \int i_s \, dt \tag{11-19*}$$

$$2N_s AB = R_g q \tag{11-20}$$

and

$$B = \frac{R_g q}{2N_s A} \tag{11-21}$$

If the galvanometer sensitivity is given by

$$K = \frac{q}{d} R_g , \tag{11-22}$$

then

$$B = \frac{Kd}{2N_s A}. \tag{11-23}$$

* $\int_B^{-B}$ indicates an integration from one point on the hysteresis curve to the corresponding point in the opposite direction - i.e., from +B to -B on the diagram.

In the preceding equations,

N_s = turns in secondary
i_s = current in secondary
R_g = resistance of galvanometer circuit
A = cross sectional area of Rowland Ring
K = galvanometer sensitivity in teslas/mm
d = galvanometer deflection in mm

PROCEDURE

1. Turn on all the lamps. This permits maximum current to pass through the primary.

2. The sample must be demagnetized since it does have a small amount of magnetism. To remove any small amount of magnetism, the reversing switch is thrown from one side to the other at the rate of 20 times per minute. See Fig. 11-17. During the reversal process increase the resistance in the circuit (increasing the resistance reduces the current in the primary) by turning off the lamps one group at a time. Allow about six reversals of the switch between groups. The sample should be demagnetized after all lamps are turned off.

3. One group of lamps R is turned on and the reversing switch is thrown back and forth at about the rate of 20 times per minute. The switch is left on the right side before proceeding to take measurements. Leave the switch on the same side for all measurements. Open the damping key and note the zero galvanometer G reading.

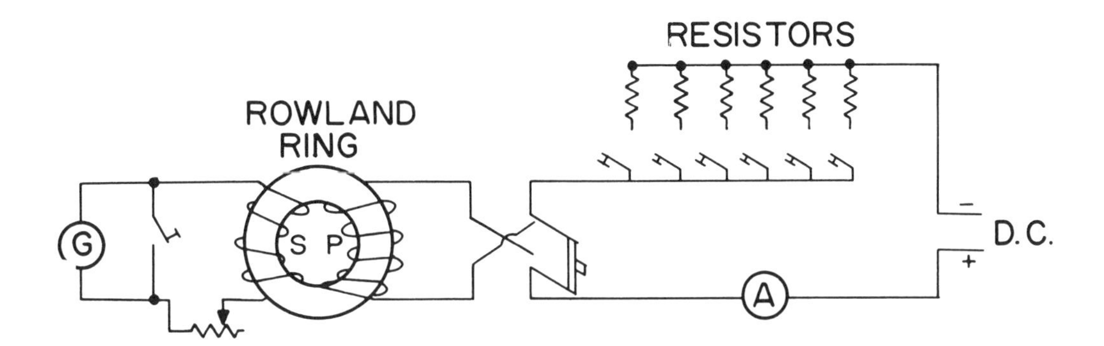

Fig. 11-17. Wiring Diagram for the Rowland Ring.

4. Quickly throw the reversing switch from the right side to the left. This induces an E.M.F. in the secondary which causes the galvanometer to deflect. The E.M.F. is produced in the secondary by the linkage of the magnetic field caused by the current in the primary. When the reversing switch is thrown from the right side to the left, the magnetic field collapses across the secondary. An E.M.F. is set up as indicated by the deflection of the galvanometer. Therefore, the point A on the B vs. B_0 graph in Fig. 11-18 can readily be determined. Record the ammeter reading. It should be constant for this group of lamps.

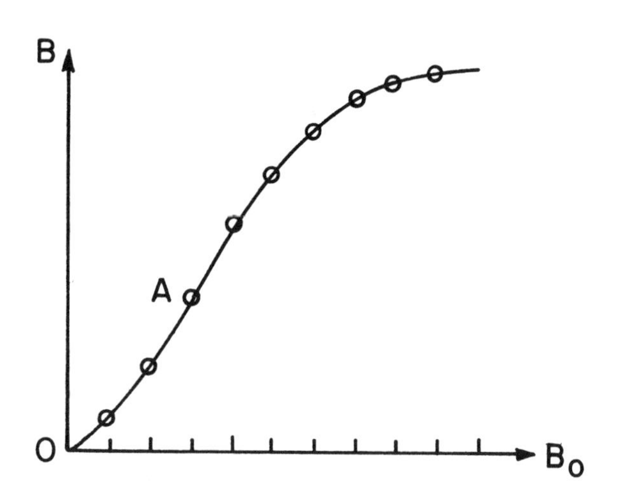

Fig. 11-18. Magnetization Curve.

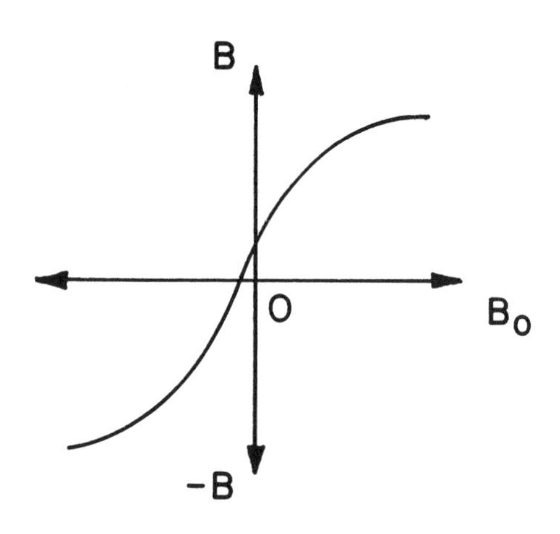

Fig. 11-19. First Half of Hysteresis Curve.

5. Turn on the next group of lamps (leave on the first group). This reduces the resistance of the primary circuit and increases the current. Repeat Steps 3 and 4 - i.e., close damping switch, reverse switch 20 times per minute, stop switch on right side, opening damping switch and note zero reading, and finally throw switch from right to left and read the galvanometer deflection. Again record the meter reading.

   Continue these steps until all lamps are on.

   WARNING: DO NOT ATTEMPT TO RETAKE A READING, OTHERWISE THE ENTIRE EXPERIMENT MUST BE REPEATED.

6. Calculate B_o and B and plot the magnetization curve (B vs. B_o). See Fig. 11-19.

TYPICAL FORM OF DATA:

A__________ n_p__________ K__________ N_s__________

i_p (amp)	Galvanometer Readings Zero (mm)	Throw (mm)	Defl (mm)	$B_o = \mu_o n_p i_p$	$B = \frac{Kd}{2ANs}$

PART B - The Oscilloscope Method

INTRODUCTION

One possible circuit which may be used to obtain the hysteresis loop on the oscilloscope is shown in Fig. 11-20. The vertical input of the scope V_y is proportional to the magnetic field B, while the horizontal input V_x, is proportional to the magnetic field B_0. This is shown by the following derivation.

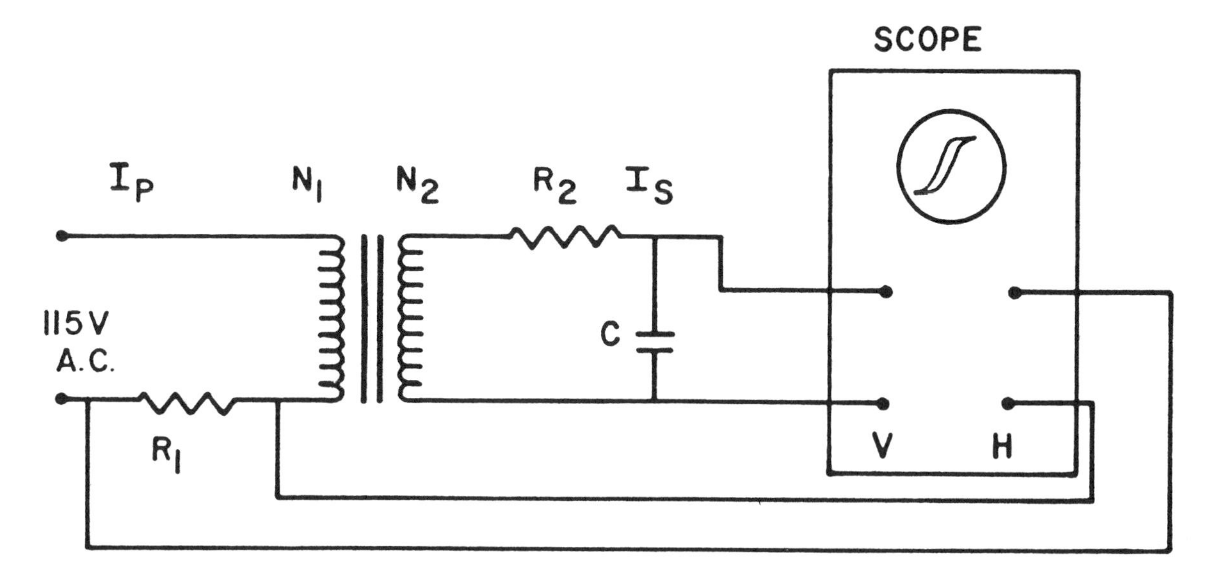

Fig. 11-20. Wiring Diagram for Displaying Hysteresis Curve on the Cathode Ray Oscilloscope.

The voltage V_s across the secondary coil of the transformer is given by

$$V_s = N_2 \frac{d\phi}{dt} = N_2 A \frac{dB}{dt}, \qquad 11\text{-}24$$

where N_2 = number of turns of the secondary coil
ϕ = flux through the secondary coil
A = cross sectional area of the transformer core
B = magnetic field

We can also write (from Fig. 11-20)

$$V_s = V_c + I_s R_2 , \qquad 11\text{-}25$$

The charge on the capacitor C is given by

$$Q = CV_c = CV_y .$$

Using the fact that $I_s = dQ/dt$ we can rewrite Eq. 11-25 as

$$V_s = \frac{Q}{C} + R_2C\frac{dV_y}{dt} . \qquad 11\text{-}26$$

Since C is very large (1 μf) we can write approximately

$$V_s = R_2C\frac{dV_y}{dt} . \qquad 11\text{-}27$$

Combining this with Eq. 11-24 we obtain

$$V_s = R_2C\frac{dV_y}{dt} = N_2A\frac{dB}{dt} . \qquad 11\text{-}28$$

Integrating we obtain

$$B = \frac{R_2C}{N_2A}V_y . \qquad 11\text{-}29$$

This shows that the vertical input of the scope is proportional to B. From Ampere's law we know that the magnetic field strength B_o is given by

$$B_o = \mu_o\frac{N_1I_p}{\ell} , \qquad 11\text{-}30$$

where N_1 = number of turns of the primary coil
I_p = current through the primary coil and
ℓ = mean circumference of the transformer core.

Since $I_p = V_x/R_1$ (see Fig. 11-20), we can write

$$B_o = \mu_o\frac{N_1}{\ell R_1}V_x . \qquad 11\text{-}31$$

Thus the horizontal input of the scope is proportional to B_o. The energy loss per cycle per unit volume U is proportional to the area under the hysteresis curve or

$$U = \frac{1}{\mu_o}\int B_o\, dB . \qquad 11\text{-}32$$

Using Equations 11-29 and 11-31 we can write

$$U = \frac{R_2CN_1}{N_2R_1A\ell}\int V_y\, dV_x . \qquad 11\text{-}33$$

Since $A\ell$ is the volume of the transformer core, the total energy loss per cycle is given by

$$U = \frac{R_2N_1}{R_1N_2} C \int V_y \, dV_x \ . \qquad 11\text{-}34$$

The integral in Eq. 11-34 is the area enclosed by the loop displayed on the oscilloscope.

PROCEDURE

Set up the circuit as shown in Fig. 11-20.

Calibrate the oscilloscope.

If a camera is available, photograph the hysteresis loop; otherwise trace the loop on transparent paper.

From the area of the hysteresis loop determine the energy lost per cycle by the transformer. Show that U in Eq. 11-32 has the units of energy.

QUESTIONS

1. Find the Precision Index (PI) for B. The error in K = ±1%, the error in d = ±0.5 mm and the error in A = ±5%.
2. Find the Precision Index (PI) in B_o. The error in i_p = ±2%, and the error in L (mean circumference of toroid) = ±2%.
3. Why is it necessary to demagnetize the sample before measurements are made?
4. What takes place in the core of the Rowland Ring when the reversing switch is thrown from right to left?
5. Prove that the galvanometer sensitivity K in teslas /mm = the galvanometer sensitivity k in coulombs/mm times the resistance R_g of the galvanometer.
$$K = k \times R_g$$
6. Locate the point of maximum permeability on the normal magnetization curve. Discuss its significance.
7. What is the energy density of the Rowland Ring when the flux density B is maximum?
8. Show that the area of the hysteresis loop is the energy dissipated in the cycle per unit volume.
9. Explain each part of the hysteresis loop.
10. How can a hysteresis loop be used to demagnetize an object?
11. Is the magnetization process a reversible or an irreversible process? Explain in terms of the hysteresis loop.

12. What is a ferroelectric substance?

13. What is the relation between the hysteresis observed in magnetic materials and the hysteresis effect in an elastomer (Experiment 6-14)?

14. Refer to Eq. 11-32 and show that the units of $\frac{1}{\mu_o} \int B_o \, dB$ give the units of energy per unit volume.

15. Refer to Eq. 11-34 and show that the units of $C \int V_y \, dV_x$ give the units of energy.

16. Use Eqs. 11-24, 11-30 and the definition of energy dW = (emf)i dt to prove that the energy lost per cycle per unit volume in a hysteresis loop is given by

$$U = \frac{1}{\mu_o} \int B_o \, dB$$

Chapter 12

WAVE MOTION

Experiment 12-1
TRANSVERSE STANDING WAVES

PART A - STANDING WAVES IN A STRING

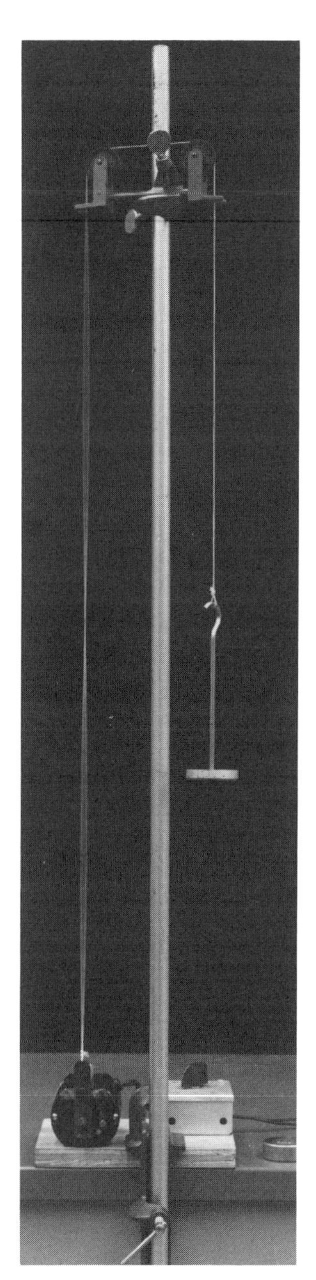

Fig. 12-1. A Variable Speed Motor Used to Set-Up Vertical Standing Waves. Frequency is measured by strobotac.

INTRODUCTION

The general equation for the velocity of propagation of waves in a medium of continuously distributed stiffness and inertia is given by

$$v = \sqrt{\frac{\text{elasticity or stiffness factor}}{\text{density or inertial factor}}} \qquad 12\text{-}1$$

In the case of transverse waves propagated along a flexible string, the elasticity or stiffness factor is due to the longitudinal tension in the string and the inertial factor is the linear density of the string. The equation for the velocity of propagation of a transverse wave therefore becomes

$$V = \sqrt{\frac{T}{\mu}}\,, \qquad 12\text{-}2$$

where T = tension in the string and μ = linear density of the. string. If the tension T and the linear density μ are measured, the velocity can be calculated. If T is in Newtons (kgm-m/sec^2) and μ is in kgm/m in the SI system, the velocity (v) will be in m/sec.

The velocity can also be determined by setting up standing waves and using the basic relation

$$v = f\lambda\,, \qquad 12\text{-}3$$

where f = frequency and λ = wave length. It should be noted that the wave length is twice the distance between successive nodes.

Fig. 12-2. A Vibrator Sets Up Horizontal Standing Waves in A String.

In the experimental setup a vibrator, operated by a rotator equipped with a revolution counter, sends sinusoidal pulses of measurable frequency down a flexible string whose length is fixed (Fig. 12-2). The tension may be adjusted by weights attached past a pulley of negligible friction. We can therefore change the frequency f by adjusting the speed of the rotator, the tension T by changing the weights, and the linear density μ by changing the string being used.

PROCEDURE

1. Measure the exact length of the string (use an extra piece of string about one meter long) by stretching it along a meter stick. Then weigh it to the nearest milligram, and calculate its linear density (mass per unit length).

2. With a small tension T_1, adjust the rotator so that resonance occurs with the cord vibrating as a whole or with its fundamental frequency. The frequency is determined by measuring the number of revolutions per minute of the rotator. The distance between nodes should be measured, from which the wave length is then computed. The nodes will not be exactly at the end points of the string because the end points are not fixed.

3. Keep the tension T_1 and the linear density μ_1 constant and raise the frequency to resonance at the second harmonic. Again measure the frequency and the distance between nodes. Two such distances can now be measured and these may be averaged.

4. Repeat the same measurements for the third harmonic. For this case, and for higher harmonics, the distances between nodes can be measured on segments that do not have nodes at the ends of the string. This reduces the errors in the measurements.

5. Determine the ratio of each frequency found to the fundamental frequency.

6. Compute the velocity ($v = f\lambda$) for each resonant frequency and compare the values.

7. Compute the velocity from the values obtained for the tension and the linear density and compare the result with the velocity found in Step 6 of the Procedure.

8. Keeping the linear density μ_1 the same, increase the tension by 10-15 grams and determine the new frequency for the third harmonic. Compare the result with theoretical predictions.

9. Use a string with a different linear density μ_2. Keep the tension the same as it was before and adjust the rotator to the resonance frequency for the third harmonic. Check the relation that is predicted from theory between frequency and linear density.

PART B - STANDING WAVES IN A VIBRATING WIRE

INTRODUCTION

The apparatus consists of a wire inserted between the pole faces of a magnet. The wire is under tension produced by weights hung from one end. A 5-10 volt A.C. signal from an audio-oscillator is supplied to the wire which is part of a simple circuit. The frequency is measured with an audio-oscillator.

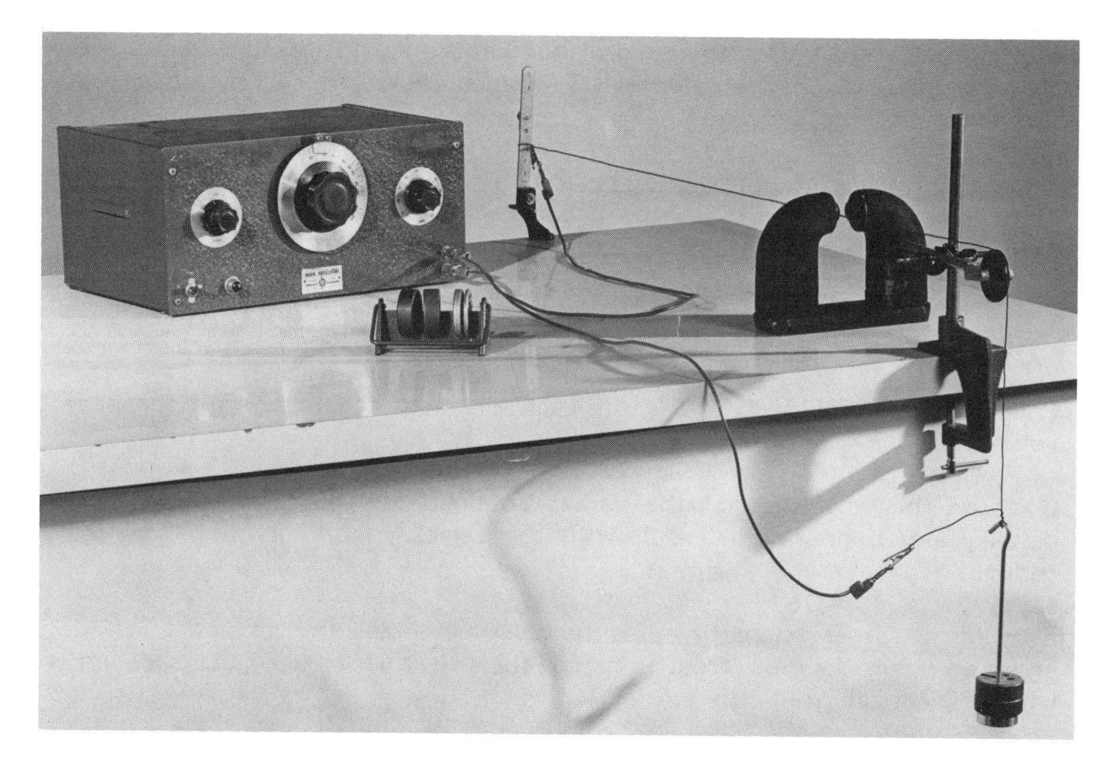

Fig. 12-3. Standing Waves on a Current-Carrying Wire.

PROCEDURE

1. Measure the length of the wire. For a specific mass (such as 150 g), find the fundamental frequency. Find the wavelength by measuring the distance between the nodes. Repeat the procedure for the second and third harmonics by changing the frequency. How many harmonics can you obtain? Repeat for a different mass (600 g). Record your data carefully in a table.

2. Compute the velocity ($v = f\lambda$) for each resonant frequency. Compare the values of the velocities.

3. Calculate the linear mass density of the wire for each of the tensions used. Did the wire stretch?

PART C - STANDING WAVES WITH MICROWAVES

If the necessary equipment is available, set up standing waves with microwaves as shown in Fig. 12-4.* Determine the wavelength of the microwaves by measuring distances between nodes. For details on the microwave equipment see Page 106. Compute

---

*For details see Project I, Experiment 14-4 (Page 605).

the frequency of the microwave generator used.

Fig. 12-4. Measuring the Wavelength of Microwaves by Setting Up Standing Waves. (See also Fig. 14-15, Page 609).

QUESTIONS

1. What is meant by a standing wave? A progressive or traveling wave? Can one type of wave be obtained from the other? Explain.

2. Distinguish between transverse and logitudinal waves. Give examples of each.

3. Consider a string of uniform density and tension. Derive an expression for the displacement of a standing wave at any point of the string at anytime. Write down the corresponding expression for a traveling wave.

4. Distinguish clearly between the velocity of propagation of a transverse wave and the particle velocity of a small segment of the string or wire at any given time. Write an equation for each of these two velocities.

5. Are nodes formed at both ends of the vibrating string or wire used in this experiment? Explain.

6. The velocity of propagation was found by two methods. Discuss the relative accuracies of these two calculations and the reasons for any discrepancies.

7. Show that the units of $\sqrt{T/\mu}$ are the units of velocity.

8. Is the total mechanical energy of the vibrating string or wire conserved? Explain.

9. When the string or wire is in resonance, why does the amplitude of vibration remain finite?

10. If the tension and the linear density are fixed, how does increasing the length of the string or wire effect the resonance frequencies? How does decreasing the length or wire effect the resonance frequencies?

11. Find the maximum PE and the probable PE in the calculation of the velocity if the weights used have an instrumental limit of error of ±2% and the linear density was ±1%. Express both PE's as numerical and relative (percent) errors.

12. What is the velocity of the microwaves? How does it compare with the wave velocity of the string?

13. Give as many other examples of standing waves that you can think of in addition to stretched strings, the vibrating wire and microwaves.

Experiment 12-2 VELOCITY OF SOUND IN AIR

INTRODUCTION

One of the simplest and most useful methods for determining the propagation velocity of waves in any substance consists of making use of the fundamental relation between the velocity, the frequency of the wave, and the wavelength:

$$v = f\lambda , \qquad 12\text{-}4$$

where v = velocity of propagation, f = frequency, and λ = wave length.

In order to determine the velocity, we have to set up a vibration in the substance, determine the frequency and wavelength, and calculate the velocity of propagation from the above equation.

The general equation for the velocity of propagation of waves in a medium of continuously distributed stiffness and inertia is given by:

$$v = \sqrt{\frac{\text{elasticity or stiffness factor}}{\text{density or inertial factor}}} . \qquad 12\text{-}5$$

In the case of longitudinal (sound) waves in a gas this equation takes the form of

$$v = \sqrt{\frac{\gamma p_o}{\rho}} , \qquad 12\text{-}6$$

where γ = a constant depending on the gas, p_o = the static pressure of the gas, and ρ = volume density of the gas. In this experiment, a tuning fork of known frequency will be used to produce vibrations in a resonating air column. By setting up standing waves in a tube, the position of the nodes can be located and, from the distance between nodes, the wavelength computed.

As the length of the air column is changed by raising or lowering the water level in the tube, there are certain lengths at which the volume of sound radiated into the room is noticeably greater than at all other lengths. This increase in loudness occurs when the air column is in resonance. Two such resonance conditions are shown in Fig. 12-5. The first one occurs when the tube is approximately one-quarter wavelength long, the second when it is approximately three-quarters of a wavelength, and the third (not shown) for five-quarters of a wavelength. A correction must be applied to the measured length of the tube because the loop is not located exactly at the open end of the tube but a little beyond. This additional distance, which should be added to the length of the tube measured, is about 0.7 the radius of the tube.

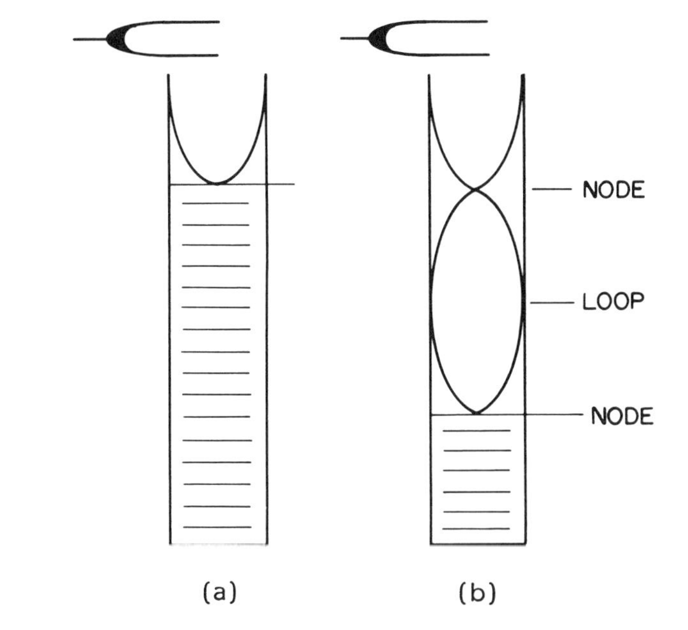

Fig. 12-5. Resonating Air Column.
(a) Tube length is 1/4 wavelength.
(b) Tube length is 3/4 wavelength.

PROCEDURE

1. Take a reading of the room temperature and of the barometric pressure at the beginning of the experiment and at the end. The value of γ, the ratio of the specific heat at constant pressure to the specific heat at constant volume for air is 1.402 ± .002. The density of air depends on the temperature and pressure and may be found from density tables (see Appendix 2). These values of pressure, γ, and density should not be used to calculate the velocity of sound in air. Care should be taken that all values are given in appropriate units.

2. Excite the tuning fork by striking it gently with the rubber hammer, and determine the three consecutive points of resonance by adjusting the water level with the funnel. If there is too much water in the tube, pour some into the jug provided. In determining each position of resonance, make eight careful readings: four when

the water level is passing up through the point, and four when it is passing down through the point.

3. Measure the diameter of the glass tube and add the necessary correction factor to the lengths measured. Calculate the velocity of sound in air from the average of the readings taken. Determine the limits of error for the value obtained for the velocity of sound in air. Do complete error analysis. Find also the maximum and probable limits of error. Express both limits of error as numerical and relative errors.

4. Repeat Steps 2 and 3 with a tuning fork of a different frequency. Compare the wavelength obtained by the two tuning forks.

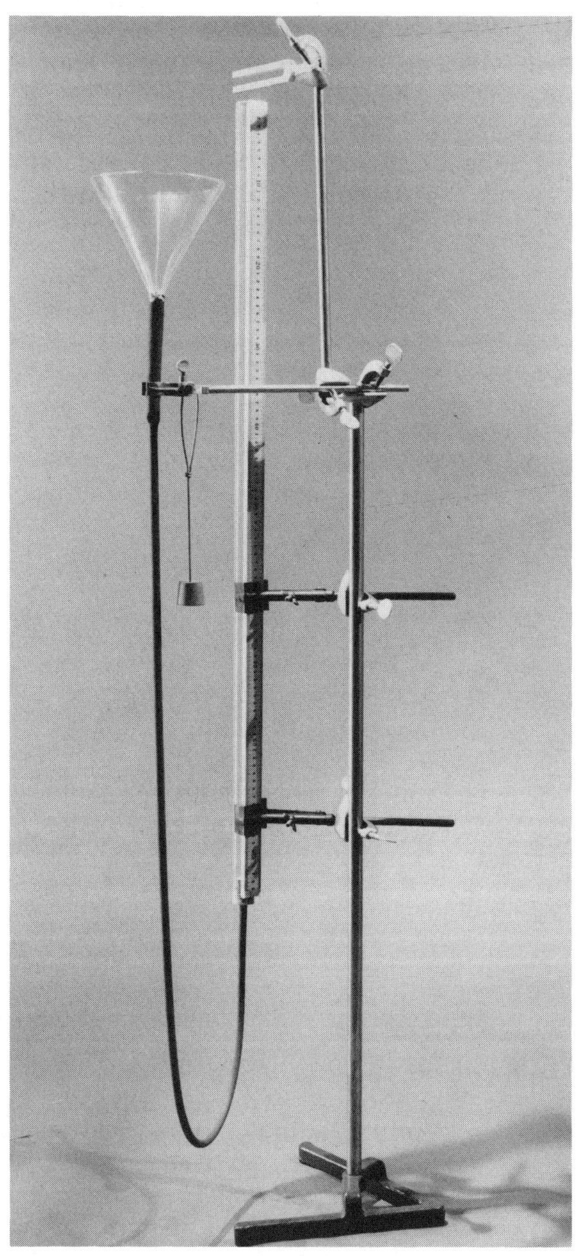

Fig. 12-6. Resonance Apparatus.

QUESTIONS

1. How does the velocity of sound in air vary with temperature?

2. Draw diagrams of the standing waves produced in the air column at each resonance length.

3. Draw diagrams of the standing waves produced in the air column at three resonance lengths for an open tube.

4. How would the results of the experiment be changed if the tube were open at both ends?

5. Which of the two methods used in determining the velocity of sound in air is more accurate? More reliable?

6. Neglecting the end correction, use the data obtained in the experiment to calculate the velocity of sound in air. By what per cent is this value off?

7. Define transverse and longitudinal waves. Are sound waves transverse or longitudinal?

8. Define resonance.

Experiment 12-3 VELOCITY OF SOUND IN METALS

PART A - KUNDT'S TUBE

INTRODUCTION

As mentioned in the introduction to Experiment 12-2, the simplest method for determining the velocity of propagation of waves through any medium is obtained from the relation between velocity, frequency, and wavelength. In the case of the velocity of sound in air, the length of the air column was changed until resonance was reached. However, for the velocity of sound in a metal, it is not conveniently possible to change the length of the metal rod. We therefore use a definite length, determine the wavelength in the metal from this fixed length, and find the frequency of vibration by some other means.

One method of determining this frequency is the use of the Kundt's tube. We let a vibrating metal rod set up vibrations in a glass tube containing some light dust (such as cork dust) which is used as an indicator. If the length of the glass tube is properly adjusted, standing waves will be set up. The positions of the nodes and loops of these standing waves are revealed by the pattern assumed by the cork dust. The length of the waves in air can be obtained by measuring the distance between two consecutive nodes. If the velocity of sound in air is known, the frequency of the air waves may be calculated. Since these waves are caused by the vibrations in the metal bar, their frequency must be the same as that of the air vibrations. Knowing both the frequency and wavelength in the metal bar, the velocity of sound in the metal can be calculated.

PROCEDURE

1. Take a reading of the room temperature at the beginning and at the end of the experiment.

2. Form the cork dust into a narrow line at the bottom of the glass tube by tapping, and then rotate the tube slightly. This will make the positions of nodes very apparent when the dust is disturbed by the air waves, since it will fall down in

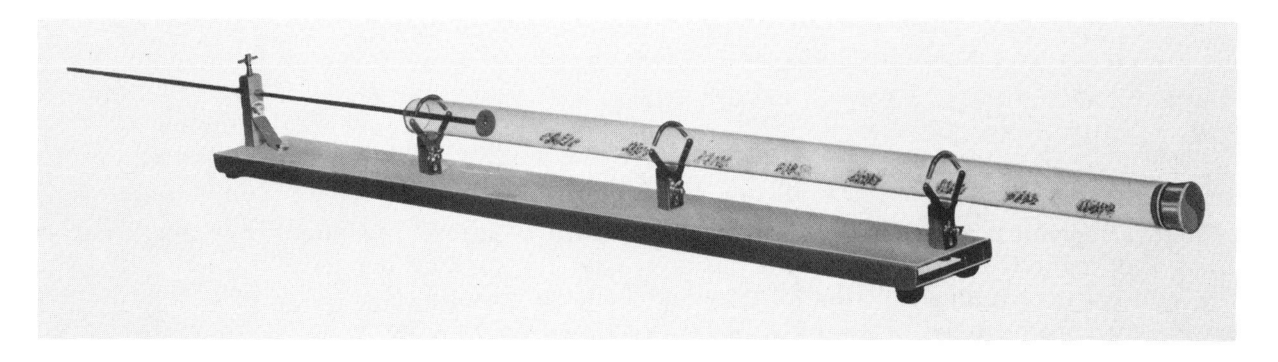

Fig. 12-7. Kundt's Tube. (Courtesy of The Sargent-Welch Scientific Company)

festoons and remain undisturbed only at the nodes. Next, stroke the metal rod with a piece of rosined leather or cloth until the rod rings out clearly. Moderate pressure and a slow, smooth stroke produce the best results. While causing the bar to "sing," adjust the position of the glass tube until a well-defined pattern of the cork dust is obtained.

3. Place a meter stick up against the tube and read the positions of six nodal points. Then subtract the reading of the first point from that of the fourth, the second from the fifth, and the third from the sixth. Taken the average of these three measurements of $3/2\ \lambda$ and compute the wavelength. Since the value of $3/2\ \lambda$ may be determined with only one-third of the relative error involved in measuring $\lambda/2$, this method of determining the wavelength is more accurate than measuring half the wavelength six times and averaging.

4. Since the metal rod is clamped at its mid-point, there must be a node at that point, and since both ends are free to move, loops will be found there. Therefore, the bar must be half a wavelength long. Measure the length of the metal rod and compute the wavelength.

Fig. 12-8. Standing Waves in a Kundt's Tube.

5. Compute the velocity of sound in air from the relation

$$v = 331.7 + 0.6t, \tag{12-7}$$

where t is the temperature in degrees Celsius. Use the data taken to find the velocity of sound in the metal. If the metal is known, look up the density and Young's modulus and check the value for the velocity of sound by using

$$v = \sqrt{\frac{Y}{\rho}} \tag{12-8}$$

where Y is Young's modulus and ρ the density of the metal.

PART B - VIBRATING RODS

An aluminum rod (between 1.0 and 1.5 cm in diameter and about 2 m long) is cleaned and sanded with coarse emery cloth so that its surface provides friction when the rod is rubbed longitudinally. Rolling the rod in rosin powder is also recommended. Grooves have been cut all around the rod with a saw or on a lathe at distances that are 1/2, 1/4, 1/6, ... of its length from each end. What do the grooves correspond to?

Holding the rod between two fingers of one hand at a groove and rubbing it lengthwise with the other hand produces a ringing of surprisingly long duration. Six or more longitudinal harmonics can be produced separately. The lowest mode has a frequency of about 1750 Hz. Some practice is needed to produce these results carefully.

To measure the frequency of the ringing on an oscilloscope, place a microphone in a convenient spot so that it will pick up the harmonics that you set up in the rod. Calculate the velocity of sound v for the fundamental and the first overtone for the aluminum rod. Record in your data table the wavelength λ, frequency f, and velocity v for each distance selected on the rod corresponding to the fundamental and first overtone.

PART C - ULTRASONIC WAVES

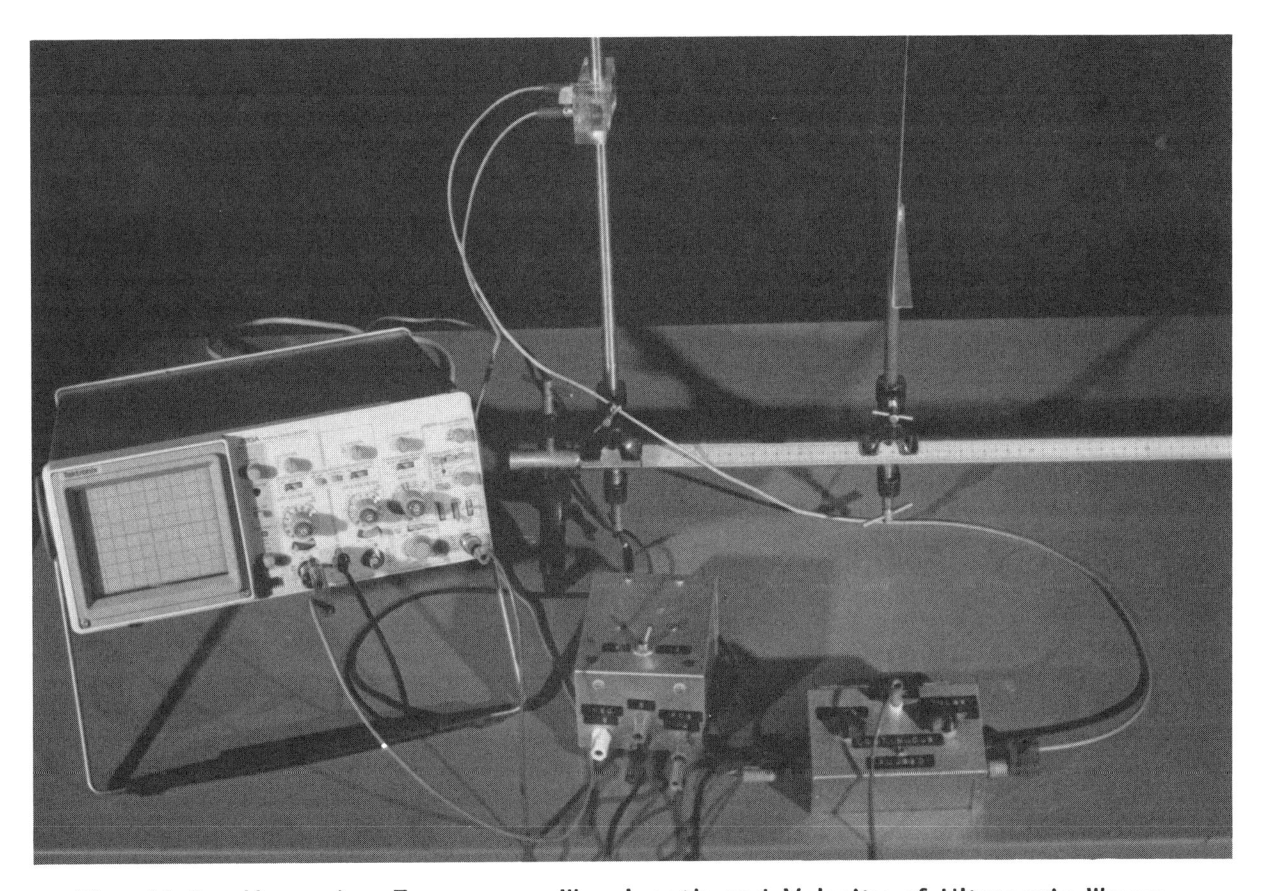

Fig. 12-9. Measuring Frequency, Wavelength and Velocity of Ultrasonic Waves.

INTRODUCTION

In this experiment, we will measure three properties of a sound wave in air:

(1) its frequency,
(2) its wavelength and
(3) its velocity.

The measurements will be carried out at ultrasonic frequencies (frequencies above the audible range). The apparatus consists of a transmitter driven by a tone burst generator to create the sound and a receiver, connected to an oscilloscope, to detect the sound. The tone burst generator is a variable frequency oscillator (about 5 KHZ to 50 KHZ) which can be pulsed on and off periodically. The "on" time is variable (about 0.1 ms to 1 ms). The "off" time is fixed (about 20 ms). A switch is provided to allow the oscillator to run continuously. This is useful for measuring the frequency and wavelength. The "sync" output is a pulse to trigger the oscilloscope at the beginning of each tone burst. Thus, in the pulsed mode the scope is placed on external trigger; in the continuous mode the internal trigger of the scope must be used. The transmitter and receiver are identical piezoelectric transducers. You may observe that they will create and detect sound over a wide frequency range. However, they are most effective (by several orders of magnitude) at their resonant frequency which is about 40 KHZ.

PROCEDURE

1. The velocity of sound in air is measured by measuring on the oscilloscope the time required for a pulse to leave the transmitter and return to the receiver after being reflected. This time, combined with a measurement of the path length of the sound waves gives the velocity of this sound in air. To see the pulses most clearly turn on the equipment and set the frequency near its maximum (clockwise); set switch to "pulsed." The setting of the pulse width is not critical. Now place a reflector about 1 meter away. If the oscilloscope is properly adjusted, a burst of oscillations should appear about 5 ms from the left end of the trace. The frequency should now be adjusted to maximize the size of the return pulse. Measure the time between the start of the pulse and its echo, and the path length of the sound waves. Calculate the velocity of sound.

2. The frequency is measured by measuring the period of the oscillator on the oscilloscope. This is most easily done if the switch on the tone burst generator is thrown to "continuous." Note that you will have to change the time base rate of the scope.

3. The wavelength is measured by noting the interference between the "direct" and reflected sound. With the oscillator in the "continuous" mode you may note that as the reflector is moved relative to the transmitter and receiver the amplitude displayed on the scope will vary periodically. The distance moved from one minimum (or maximum) to another is one-half of a wavelength. Thus, if you move the reflector through say N minima (or maxima) and divide the distance traveled by N/2, you will have the wavelength of the sound wave.

ANALYSIS

Compute the velocity of the sound waves from the frequency (Part 2) and the wavelength (Part 3) and compare the result with the velocity found in Part 1 and with,

$$v = v_o + .61t ,$$

where v_o is the velocity in meters/sec at 0°C, and t is the temperature of the air in °C. Use 331 meters/sec for v_o.

Discuss some of the reasons for any discrepancies you have in your results.

QUESTIONS

1. What is the precision index PI for the value of the velocity of sound in the metal rod?
2. Describe how a Kundt's tube may be used to measure the velocity of sound in gases.
3. Show why the method used for finding the wavelength in air is more accurate than simply measuring the distance between nodes and averaging.
4. Explain in detail why the metal rod is clamped at its mid-point.
5. What is the approximate frequency range of audible sound? What are the corresponding wavelengths?
6. What is the approximate frequency range of ultrasonic waves? What are the corresponding wavelengths?

Experiment 12-4 INVESTIGATION OF LONGITUDINAL WAVES

INTRODUCTION

Wavelength, phase, the shape of wave fronts, and the decrease of intensity with increase of distance from the source are important properties of not only sound waves but also are characteristic of water waves and electromagnetic waves. The results obtained in this experiment are quite general and apply to waves of all kinds. Mastery of the concepts emphasized will provide an excellent foundation for future experimental and theoretical treatment of diffraction and interference.

APPARATUS

Fig. 12-10. Aluminum tube is lined with Kimsul which absorbs sound. Any similar soft material should perform in the same manner. The dynamic type microphone has cardiod directional characteristics that reduce pickup of extraneous sound due to wall reflections.

The experiment uses a sound tube; this consists of a 2-meter long aluminum tube, about 55 cm in diameter and lined with sound absorbing material. One end of the tube is closed with a speaker at the center. An 8-inch diameter speaker with two separate voice coils and cones and a frequency response of 50 to 15 000 Hz is suggested. At the open end of the tube a microphone with a frequency response from 40 to 15 000 Hz is mounted on a boom. See Fig. 12-10. The microphone can be moved along the boom and the boom can be rotated in a horizontal plane. A scale from -90° to 0° to $+90^\circ$ is used to indicate the position of the microphone boom.

In addition a 20-watt amplifier with a pre-amplifier section is used, as well as an audio oscillator and an oscilloscope.

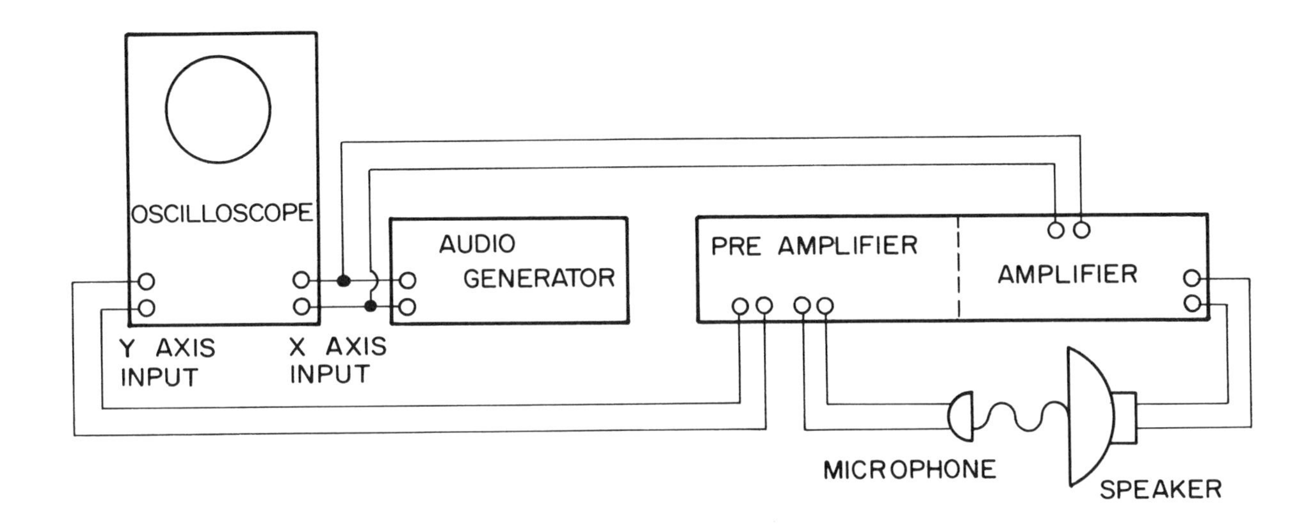

Fig. 12-10A. Block Diagram of Equipment.

As shown in Fig. 12-10A, the power amplifier increases the small signal amplitude (about 1 volt) output of the audio oscillator to a signal amplitude (about 13 volts) high enough to efficiently operate the speaker. The same small signal amplitude (about 1 volt) from the audio oscillator is applied to the horizontal or x-plates, of the cathode ray oscilloscope CRO . The voltage which drives the speaker ($V = V_1 \sin \omega t$) is therefore in phase with the voltage which appears across the horizontal plates of the CRO. The very small (millivolts) output of the microphone which is proportional to the pressure amplitude of the incident sound waves is amplified by the pre-amplifier to a signal amplitude (about 1 volt) that is high enough to operate the CRO. This small signal $V = V_2 \sin(\omega t + \phi)$ is connected across the vertical, or y-plates of the CRO. The resulting pattern that appears on the CRO screen is a closed curve with the x-coordinate, $x = V_1 \sin \omega t$ and the y-coordinate, $y = V_2 \sin(\omega t + \phi)$. Note that the maximum voltage amplitudes are not identical ($V_1 \neq V_2$). The closed curve is called a Lissajous figure and the phase angle ϕ can be found from the relative orientation and shape of the patterns. (See Experiment 10-10, Part III.)

As indicated in Fig. 12-10A, the signal supplied to the y-input of the CRO from the output of the pre-amplifier has the same frequency (ω) as that supplied directly to the x-input of the CRO by the audio oscillator. The phase of the y-input signal, $y = V_2 \sin(\omega t + \phi)$ will be different from that of the x-input signal, $x = V_1 \sin \omega t$, because ϕ is a function of the distance R between speaker and microphone and also depends upon the constant phase variation ϕ_o introduced by the instruments. Therefore, $\phi = \phi_o + 2\pi R/\lambda$. Because the frequency ω is the same, the shape of the Lissajous figures will depend on the amplitudes V_1 and V_2 and the phase angle ϕ.

Four cases will be illustrated:

1. If $\frac{2\pi R}{\lambda} = -\phi_o$; i.e., $\phi = 0$,
   then $x = V_1 \sin \omega t$ and $y = V_2 \sin \omega t$, and, therefore, $y = \frac{V_2}{V_1} x$.
   This is a straight line through the origin with slope $\frac{V_2}{V_1}$ (see Fig. 12-11a).

2. If $\frac{2\pi R}{\lambda} = -\phi_o + 90°$; i.e., $\phi = 90°$,
   then $x = V_1 \sin \omega t$ and $y = V_2 \sin(\omega t + 90°) = V_2 \cos \omega t$ and
   $\frac{x^2}{V_1^2} + \frac{y^2}{V_2^2} = 1$.
   This is an ellipse with semi-axes V_1, V_2 (see Fig. 12-11b).

3. If $\frac{2\pi R}{\lambda} = -\phi_o + 180°$; i.e., $\phi = 180°$,
   then $y = -\frac{V_2}{V_1} x$. (See Fig. 12-10c.)

4. If $\frac{2\pi R}{\lambda} = -\phi_o + 270°$; i.e., $\phi = 270°$, then $\frac{x^2}{V_1^2} + \frac{y^2}{V_2^2} = 1$.
   Again we obtain an ellipse as in Case 2.
   What is the difference between this ellipse and the one described in Case 2?

At any other values of ϕ the Lissajous figures will simply be distortions of the basic ellipse.

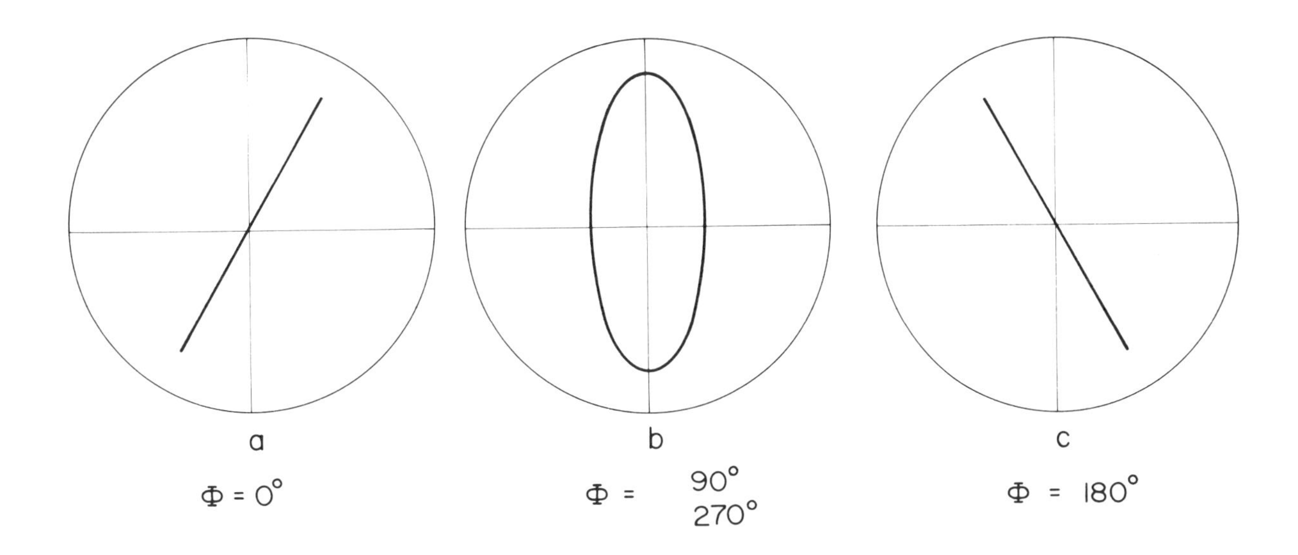

Fig. 12-11. Lissajous Figures for Three Different Phase Angles.

In the special case where $V_1 = V_2$, the straight lines are oriented at 45° and the ellipses degenerate into circles of radius V_1. Can you show this?

It is important to notice here that if the frequencies of the two simple harmonic motions impressed along the x and y axes are not the same, then very complicated patterns may result. These may be traced by geometrical and other means which are treated in many standard texts.

PROCEDURE

1. Remove any aperture or covering from the end of the aluminum tube. Move the microphone to the 0 position directly in front of the tube opening. Turn on the amplifier power switch - DO NOT TOUCH THE MICROPHONE VOLUME CONTROL! This must be left at zero to effectively isolate the preamplifier section in the circuit. If it is not, a mixture of the audio oscillator output and the microphone output will constitute the amplifier output to the speaker. No conclusive results will be obtained under such circumstances. Set the controls so that the red circle is up. When the red circle is up, the amplification is uniform over the range of frequencies. Adjust the volume control.

2. Set the amplitude control of the audio oscillator to zero and the frequency controls to 20,000 Hz. Turn the power switch on.

3. Turn on the CRO using the intensity control or power switch. Adjust the intensity control so that a spot appears on the screen. Vary the horizontal and vertical

centering controls and focus control if necessary until a sharply defined spot appears at the center of the screen. Make sure that the bright spot does not remain on the screen for any length of time or that area of the tube will be burned. Set the selector switch to amplified horizontal input.

4. Turn up the fine output control of the audio oscillator, which was set previously in Step 2 at a minimum, until the resultant closed Lissajous figure fills the screen. Adjust the frequency controls until as good a figure as possible appears at as high a frequency as possible. This will be a resonant point of the crystal microphone.

5. Move the microphone outward in the 0° (perpendicular to the tube opening) direction until a straight line slanted at approximately 45° appears. The angle is not exactly 45° because $V_1 \neq V_2$. To which value of ϕ, approximately, does this line correspond?

   NOTE: When making any adjustments, always stand behind the microphone. Why?

6. Record the distance at which this figure appears, reading the measurement on the boom-mounted meter stick at the indicator.

7. Move the microphone directly away from the opening until another straight line appears which is slanted again approximately 45° from the vertical but in a direction opposite to that of the first. What is the approximate ϕ value here? Measure and record.

   Again move the microphone until a straight line appears in a position exactly the same as in Step 6. Record. Finally, adjust the microphone so that a straight line appears in same position as Part (a) of Step 7. Record.

8. Use the -90° to 0° to +90° scale and move the microphone boom to 15°. Slide the microphone into the tube opening. Make certain that the first reading is taken with the straight line slanted in the same direction as in Step 5. Repeat Steps 5 to 7 with the boom at 15° position. One student should hold the boom end in order to insure that there is no drifting while another student changes the microphone position. Record all distances. Repeat for 30° and 45° on the same side and then for 15°, 30° and 45° on the other side of the 0° position.

9. Reset the audio-oscillator frequency control for an output of 10,000 Hz. Move the clamp and rod supporting the microphone out along the boom until the front of the microphone is approximately 75 cm from the tube opening. Repeat Steps 5 to 8 for the new signal. Record all data.

10. Return the boom to the zero position and move the microphone to about 75 cm from tube opening. (Audio oscillator is still set for an output of 10,000 Hz.) Remove the x-axis input from the oscillator to the scope. Set y-axis amplifier so that peak to peak height of signal on screen is about two inches and centered on reference grid lines. This initial signal is used for reference. Move the microphone out along the boom. Record the position and the amplitude of the signal on the CRO screen for about six different microphone positions 10 cm apart. Select values between x = 75 and x = 150 cm.

ANALYSIS

1. From the 0° microphone position data in Steps 5 to 7 and 9 determine the average wavelength (λ) for both frequencies.

   (a) Calculate the velocity of sound in air for both frequencies. How do your values compare with standard values? Did you check room temperature?

   (b) What is the phase angle ϕ at the recorded positions?

2. Plot on polar paper the microphone distance x along the radii corresponding to the various angular positions recorded in Steps 5 to 9. Use separate sheets of polar graph paper, one for each frequency.

   (a) How can you determine the profile of the wave?

   (b) Locate the centers of the most nearly plane portion of the wave fronts. What should be the direction of the perpendicular to the wave front at this point? Do all the perpendiculars lie along the same line? If not, why not?

   (c) Using your wavelength as a standard, for what distance did the wave fronts begin to appear most plane? Is this distance the same for both frequencies?

3. Plot from the data in Step 10, the reciprocal of the relative amplitude (1/A) versus x. From theory, the average intensity of the sound waves measured at a distance r from a point source varies inversely with the square of the distance. This is also true for an extended sound source like ours if intensity measurements are made far enough away from the source. To analyze properly the fall-off of intensity for such extended sources, the distance r must be replaced with (R - K) where K is a constant which depends on the properties of the source. K, the effective source center, is the position at which a point source should be located to give our observed variation of intensity.

   (a) Why does the plot of 1/A vs. R show that the average sound intensity is inversely proportional to the distance squared?

   (b) From the wave profile plot in Part 2 of the analysis, estimate the effective source center.

   (c) Could the effective source center be found from the plot of 1/A vs. R? Explain.

QUESTIONS

1. What type of wave is the sound wave, transverse or longitudinal? Why?

2. With what, physically, do we associate the "amplitude" of the wave?

3. Are the Lissajous figures which we have obtained here equally valid for the superposition of light waves or electric current variations? Why?

4. Discuss some simple Lissajous figures obtained when the frequencies of the superimposed waves are not the same. Consider the cases where they have the rations 2:1, 3:1, or 3:2.

5. How is Huygen's principle related to the waves we have studied in this experiment?

6. What effect does the tube filled with sound absorbing material have on the shape of the wavefront?

Chapter 13

OPTICS

Experiment 13-1 LASER RAY TRACING

LASER SAFETY

NEVER LOOK INTO THE LASER WINDOW OR STARE INTO THE BEAM. Make certain that the laser beam is not pointed towards someone else's eyes. If not in use, turn off power to laser. Do not completely darken the room. Remove highly reflective objects from the beam's path such as rings, watches, tools, etc. The laser used has a power of less than 1 mw visible output which places it within the class II category, as defined by the Bureau of Radiological Health Standards for laser safety. Its use is completely safe as long as you do not look into the beam.

PROCEDURE

Ray diagrams are commonly used to illustrate light paths. This experiment allows you to see actual paths created by real light beams. To produce the beams on the table, the narrow parallel laser beam is diverged in a vertical plane by a thin glass rod mounted horizontally outside the laser window. A sheet of white paper on the table enables you to observe and trace the rays. Use a ruler and protractor to make measurements on the paper after the laser is turned off.

A. Reflection and Refraction

Use the semicircular body as shown in Fig. 13-1 and measure the angle between the incident and reflected rays. Find the angle of incidence, θ_1.

Measure the angle of refraction θ_2 and use Snell's Law to compute the index of refraction of the semicircular body. The incident ray must enter the semicircular body at its center.

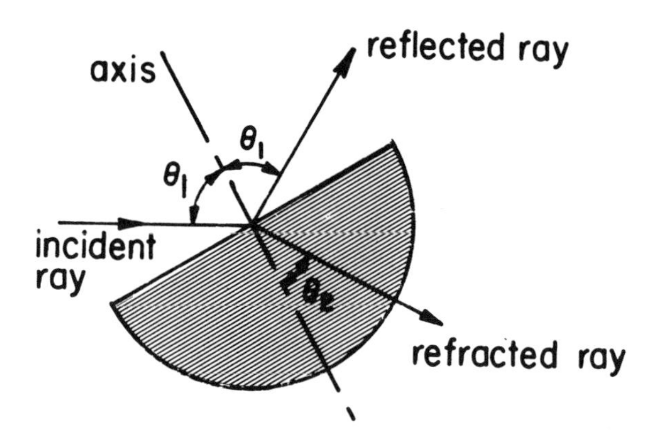

Fig. 13-1. Reflection and Refraction.

B. Total Internal Reflection

Use the semicircular body as shown in Fig. 13-2 and observe the intensity of the reflected and the refracted rays as the cylindrical surface is rotated around its center. The critical angle is the angle θ_c at which the refracted ray points along the surface of the semicircular body. Rotate the body until you reach the critical angle. Measure the critical angle and compute the index of refraction. Compare with the results obtained in Part A.

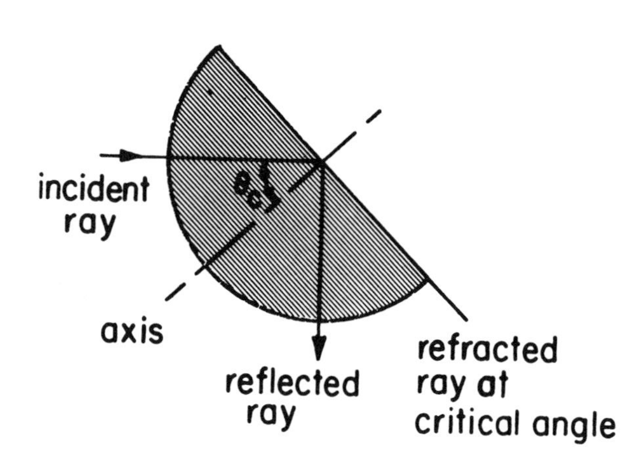

Fig. 13-2. Total Internal Reflection.

C. Index of Refraction of a Rectangular Transparent Block

Use the rectangular block as shown in Fig. 13-3 to proof that a ray of light incident on the surface emerges from the opposite face parallel to its initial direction but displaced sideways, by a distance x.

Measure the angle of incidence, θ, when this angle is small. Also measure the thickness of the block, t, and the distance x. Calculate the index of refraction n from the relation

$x = t\theta \frac{n-1}{n}$ (See Question 11.)

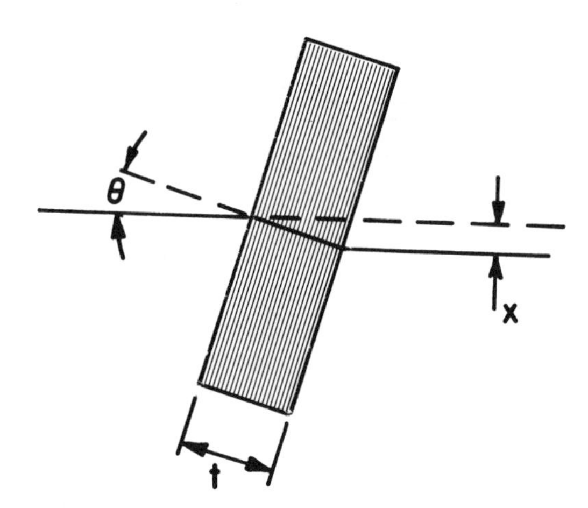

Fig. 13-3. Sideways Displacement of Light Passing Through Rectangular Block.

Take data for a large angle of incidence and show that the preceding equation does not hold.

D. Index of Refraction of Transparent Liquids

Repeat Part C replacing the rectangular block by a hollow glass cell with flat, parallel walls, containing a liquid.

E. Convex and Concave Mirrors

Use the beam splitter shown in Fig. 13-4 which produces parallel rays by multiple reflections. Observe the ray patterns and measure the focal lengths of the concave and the convex mirrors.

Observe the patterns for paraxial rays and for non-paraxial rays. Rays which lie close to the mirror axis are paraxial rays. For paraxial rays, the rays diverging from the object make small angles with the axis of the mirror.

Note the aberrations. These "defects" arise because the assumption that all rays are paraxial is never completely justified.

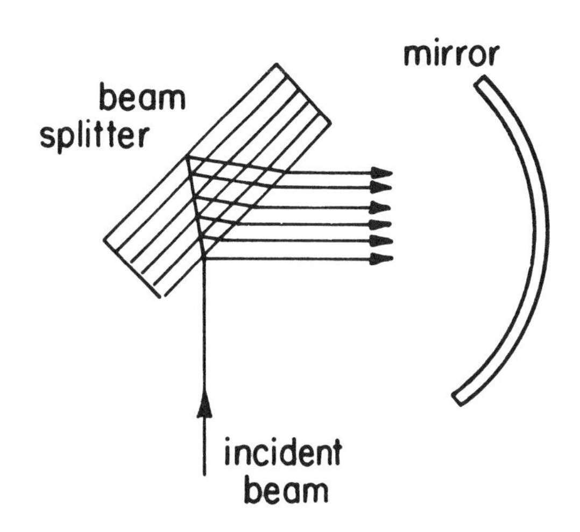

Fig. 13-4. Beam Splitter Used to Measure Focal Length of Mirror.

QUESTIONS

1. Explain in detail the purpose of the thin glass rod in front of the laser beam. Could this experiment be performed without this glass rod?

2. How can you proof that the angle of reflection is always equal to the angle of incidence?

3. Why should the incident light in Part A enter the semicircular body at its center? Draw a ray diagram for the case in which the light does not enter at the center 0.

4. What happens to the light rays at the semicircular surfaces in Parts A and B? Is it refracted?

5. What is the velocity, wavelength and frequency of the laser light inside the semicircular body used in Parts A and B?

6. Derive the relationship between the critical angle (Part B) and the index of refraction.

7. Would you expect the index of refraction found in Parts A, B and C to be larger, smaller or the same if blue light was used instead of the red laser light? Explain.

8. How would you go about measuring the index of refraction of a transparent liquid (water, for instance)?

9. How would you expect the index of refraction of a liquid to change with its density? How could you show this change experimentally?

10. Prove that a ray of light incident on the surface of a sheet of plate glass of thickness t emerges from the opposite face parallel to its initial direction but displaced sideways.

11. Show that, for small angles of incidence, θ, the sideways displacement mentioned in Question 10 is given by $x = t\theta \frac{n-1}{n}$, where n is the index of refraction and θ is measured in radians.

12. Derive an equation for the sideway displacement mentioned in Question 10 if the angle of incidence is not small.

Experiment **13-2** LENSES

PROCEDURE

PART A - The Convex Lens

Use the calibrated optical bench for object and image distances, and the small rule for object and image sizes. The ground glass side of the focusing screen should face the lens. The object slides are to be placed in the holder on the outside of the lamp-house, and the color filters (when used) inside. When focusing, always observe the image by transmitted light.

Use the line object at the lamp-house. For various object distances, as indicated below, obtained by setting the lens at the required distance from the lamp-house and object, focus the image on the ground glass screen by moving the latter. Using a small celluloid scale, measure and record object and image sizes (O and I). Also record object and image distances (S and S') as read from the scale on the optical bench.

Compute the magnification $m = \frac{I}{O}$, also the value of $\frac{S'}{S}$, which should be the same in value within the precision of the experiment and the focal length.

Draw a complete ray diagram to scale for one of these measurements.

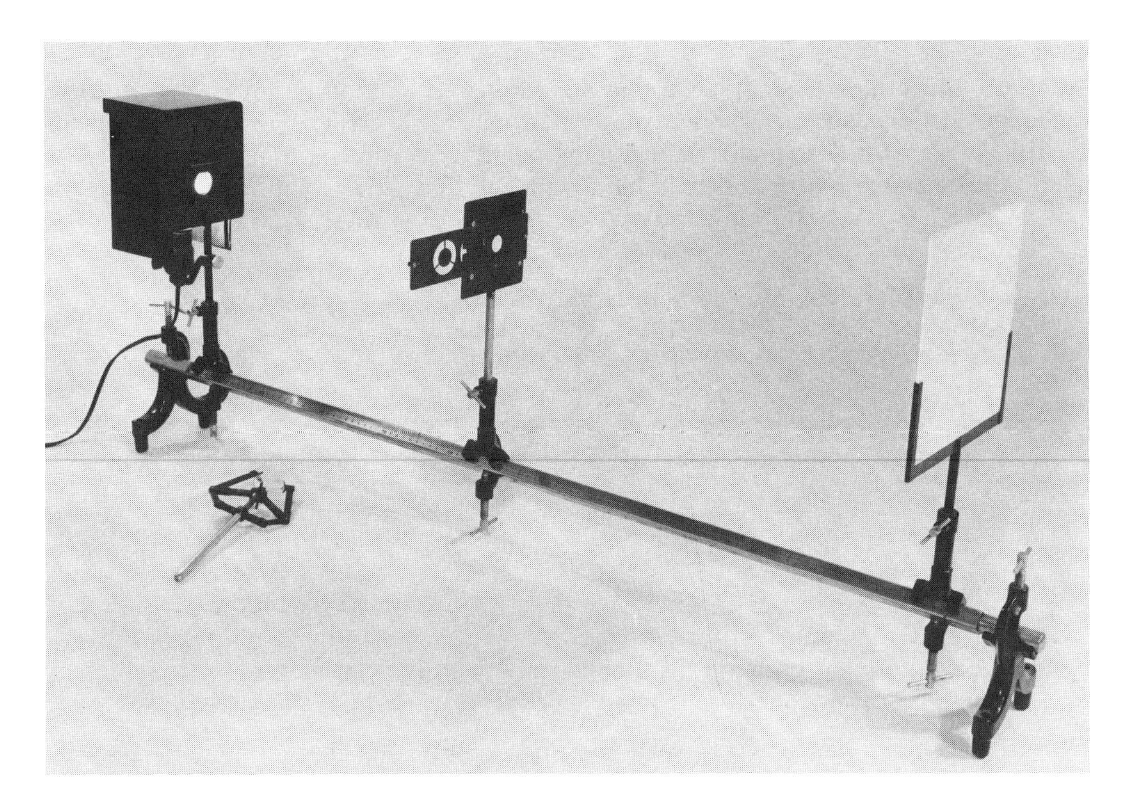

Fig. 13-5. The Optical Bench.

PART B - Aberrations

Many types of aberration occur when a simple lens is used. The two simplest and most noticeable are chromatic and spherical aberrations. Both of these may be corrected by the use of combinations of lenses. Chromatic aberration occurs because the index of refraction of a glass (and therefore the focal length of a lens made from that glass) varies for different wave lengths. Spherical aberration occurs because the focal length of a simple lens is not the same for all parts of the lens. The General Lens Equation does not fit all parts of the lens at the same time.

1. Chromatic aberration (Use wire mesh object.)
   Insert a blue filter at the lamp-house. Adjust the image distance carefully for a set object distance. Measure S and S' and calculate f (blue). Repeat, using a red filter, and determine f (red).

2. Spherical aberration (Use wire mesh object.)
   Block the outer or peripheral part of the lens with one of the stops provided, so that only the central half of the lens is used. Set object distance to 15 cm. Adjust the image distance carefully. Measure S and S' and calculate the value of f (central). Repeat, using the other stop to block the central part of the lens so that only the outer or peripheral portion of the lens area is used. Determine f (peripheral).

PART C - The Concave Lens

Set up the object and place a convex lens in front of it so as to produce an image on the screen. Interpose the concave lens between the convex lens and the image and withdraw the screen until the image is again formed. Measure the distances from the two positions of the screen to the concave lens and calculate the focal length from the lens equation. Remember that the object distance is negative when light is converging toward a "virtual object."

Draw a complete ray diagram to scale for this measurement.

PART D - The Fresnel Lens

Investigate the characteristics of a fresnel lens, if one should be available.

PART E - Use of Laser

If a laser is available, study the characteristics of lenses by the use of the laser. This may include an investigation of spherical aberrations as well as coma and astigmatism. Why is there no chromatic aberration when a laser is used?

The laser beam has a very small divergence; it therefore may be necessary to collimate and expand the beam. This can be accomplished by using two lenses, one of short focal length and the other one of long focal length. The details are left up to the student.

NOTE: NEVER LOOK DIRECTLY INTO THE LASER BEAM!

QUESTIONS

1. Explain what is meant by real and virtual images; by real and virtual objects. How can a virtual object be obtained?

2. What is meant by "aberration"? Explain the cause of spherical and chromatic aberration.

3. How can aberrations be corrected in optical instruments?

4. The "power" of a lens is measured in diopters. Define a diopter.

5. Plot a curve of the image distance as a function of the object distance for a convex lens of given focal length. Let the object distance vary from minus infinity to plus infinity.

6. Repeat Problem 5 for a concave lens.

Experiment 13-3 PRISM SPECTROMETER

INTRODUCTION

If a ray of light of a given wavelength passes from air through a glass prism, the ray will change in direction by an angle of deviation θ . The minimum angle of deviation D is the smallest value of θ for varying values of incidence. The index of refraction n of the prism is given by the equation

$$n = \frac{\sin \frac{A + D}{2}}{\sin \frac{A}{2}} \qquad 13\text{-}1$$

where A is the prism angle and D is the angle of minimum deviation.

Different colors (i.e., different wavelengths) have different values of θ for a given prism orientation. This is the phenomenon of dispersion. The angle of minimum deviation, therefore, depends on the wavelength of the incident light, and the index of refraction varies with different wavelengths.

A standard prism material illustrates what is known as normal dispersion and shows an increase in refractive index with decreasing wavelength. For the usual optical materials in the visible region, an appropriate expression for the dispersion is given by the Cauchy formula $n \cong A + B/\lambda^2$ where A and B are constants for the particular substance under consideration.* The constants A and B can be found by measuring the index of refraction for two different wavelengths and solving the two simultaneous equations. By using the spectrometer and determining the angles of minimum deviation for different wavelengths, the index of refraction for each wavelength may be obtained.

PROCEDURE AND ANALYSIS

The spectrometer should not be used until it is thoroughly understood. The instrument section (Pg. 102) must be consulted for details - the use of all set screws must be known before they can be properly used. The telescope must be focused on the slit and the crosshairs must be in sharp focus. Parallax must be eliminated and the width of the slit adjusted.

The angle of the prism is first determined. Use a light bulb as a source in this part of the experiment. Mount the prism as indicated in Fig. 13-6. Move the telescope until an image of the slit, reflected from one surface of the prism, is in the field of view and then turn it until the reflection from the other face can be seen. Read the vernier settings at the two positions. The prism angle A is one half the difference between the two readings. Turn the prism table so as to change the readings, and make another determination of the prism angle.

---

*Cauchy developed this approximate formula for the visible region in 1836. His theoretical reasoning turned out to be false, but the formula checks with experimental results and is therefore considered empirical.

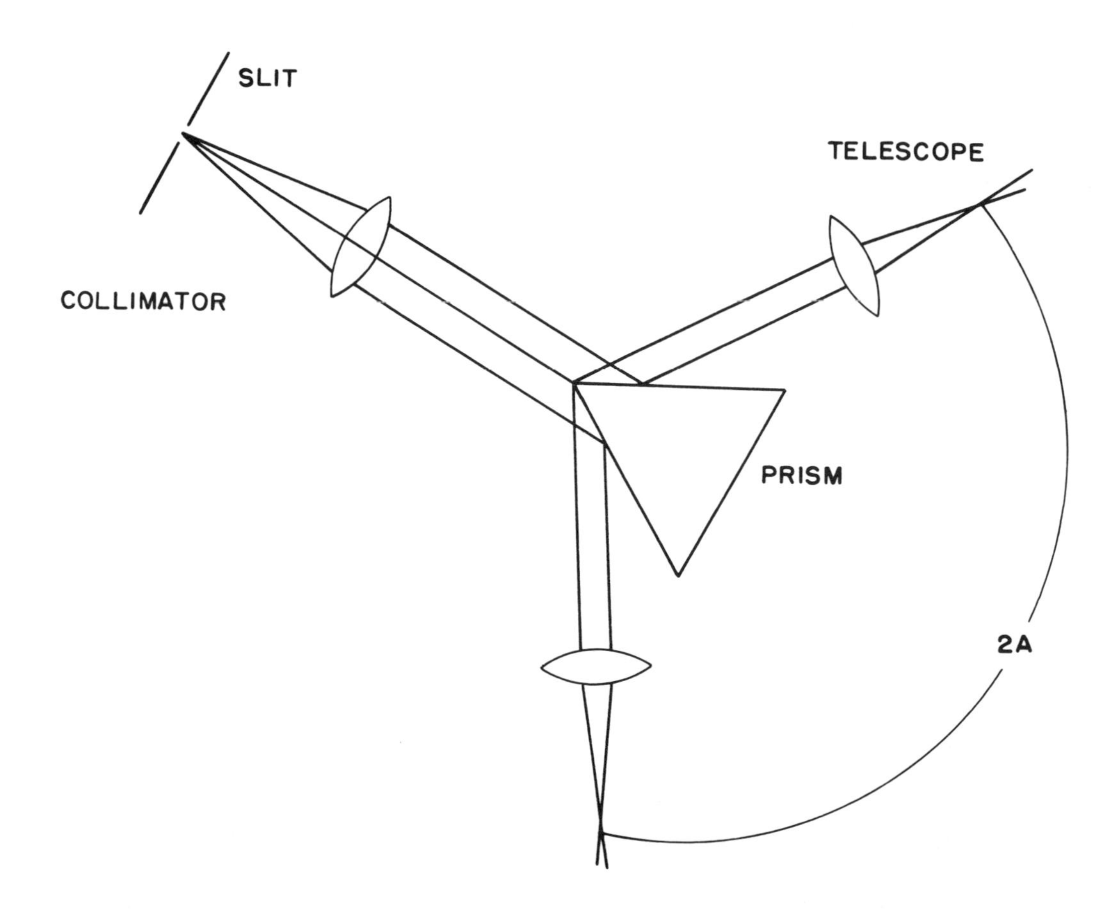

Fig. 13-6. Measurement of Prism Angle.

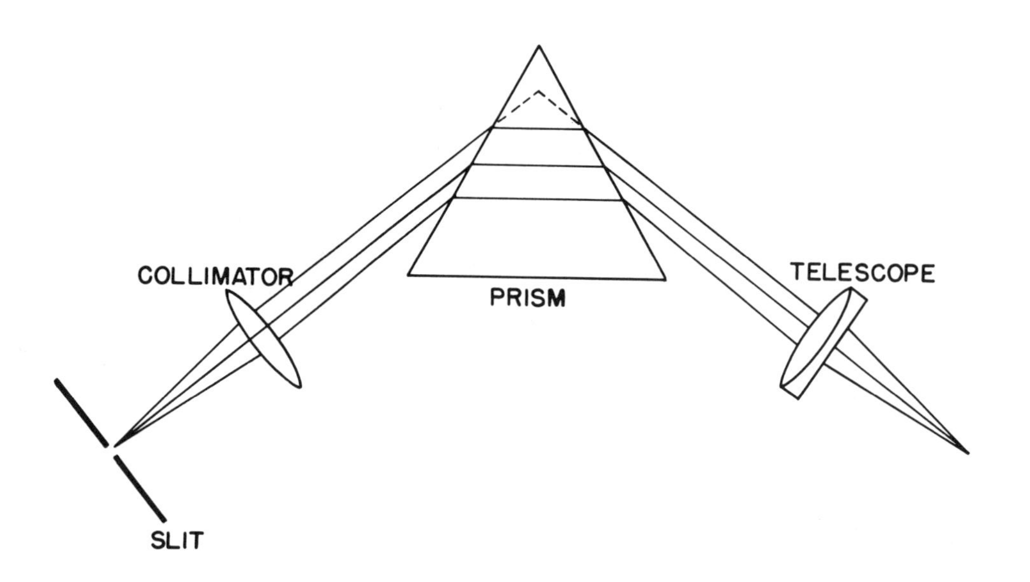

Fig. 13-7. Measurement of the Angle of Minimum Deviation.

In measuring the angle of minimum deviation, a helium discharge tube may be used as a light source. Two angles are read on the circular scale, one corresponding to the direction of the undeviated beam of light, and the other corresponding to the direction of the light deviated by the prism. The first of these angles is read off the scale before mounting the prism. The prism is then mounted as indicated in Fig. 13-7. The telescope is rotated in the proper direction until the slit image is found - it will appear as a line spectrum of different colors. The angle through which the telescope is turned gives the angle of deviation, but not necessarily the angle of minimum deviation. To find this angle, the prism table is rotated until the image of the slit comes to a standstill, and then reverses its motion as rotation of the prism table is continued. At this position where the image comes to a standstill, $\theta = D$. This angle of minimum deviation should be found for as many lines as possible of the helium spectrum by taking the difference in reading from the undeviated light and the deviated light. Using the prism angle (A) found previously, calculate the index of refraction of the prism for the various colors.

The wavelength of each line can be determined by a diffraction grating (Experiment 13-5) or can be looked up in charts or tables. Plot the index of refraction vs. $(1/\lambda)^2$ where λ is the wavelength of each line. Discuss the significance of the plot.

From the data obtain the coefficients A and B in Cauchy's formula.

If a hollow prism is available, determine the index of refraction of water or other liquids by the same method used for a glass prism.

Use microwave equipment to determine the index of refraction of a paraffin or plastic prism for microwaves. How does the value compare with the index of glass in the visible region?

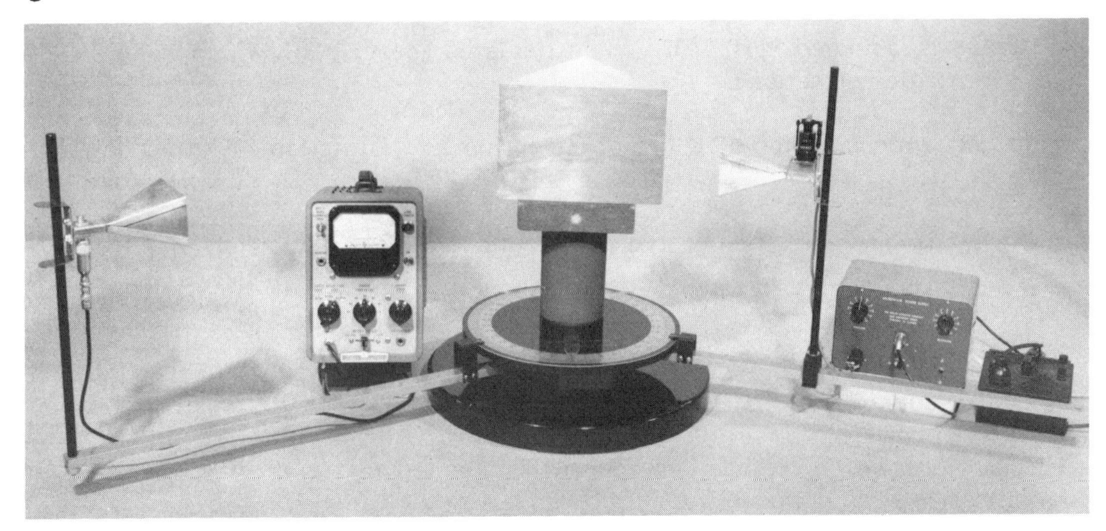

Fig. 13-8. Microwave Prism Spectrometer Including (from left to right): Receiving Horn, Detector, Plastic Prism on Spectrometer Table, Transmitting Horn and Klystron, Power Supply, Modulation Supply.*

*This apparatus was developed by Prof. Harry F. Meiners. It can be obtained from the Sargent Welch Scientific Corp., Chicago, Ill. For construction details of prism, see p. 1022, Harry F. Meiners, Ed., Physics Demonstration Experiments, Vol. II (Robert E. Krieger Publishing Co., Malabar, Fl. 32950, 1985); or same Vol. II (The Ronald Press Co., N.Y., N.Y., 1970 - out of press).

If a laser is available, measure the angle of minimum deviation of a prism with a laser beam. A spectrometer is unnecessary, since the beam is well collimated. Compute the index of refraction. Investigate different types of prisms; you may also want to observe internal reflections if a rectangular prism is used. DO NOT LOOK INTO THE LASER BEAM!

QUESTIONS

1. Determine the probably limit of error for the index of refraction of the prism found by use of yellow light. Use the least count of the spectrometer as the instrumental limit of error. Represent the error numerically.

2. Derive the equation for the index of refraction as a function of the angle of minimum deviation and the prism angle.

3. Show that the prism angle is one-half the angle through which the telescope was moved in the first part of the experiment.

4. How does the index of refraction vary with the dielectric constant of a material? Does this suggest another way to determine the index of refraction? Discuss.

5. Compute the dielectric constant of glass in the visible region.

6. Define dispersion. What is meant by anomalous dispersion?

7. Which color is deviated most? Which is deviated least? For which color has glass its largest index of refraction?

8. How do you expect a plot of angle of deviation as a function of angle of incidence to look?

9. Define what is meant by dispersive power and angular dispersion of a material.

10. Calculate both the dispersive power and the angular dispersion for the prism in your experiment.

11. What is the purpose of the collimator on the spectrometer?

12. How does the index of refraction vary with the density of the liquids used in the hollow prism?

13. Would you expect the index of refraction to be the same at microwave frequencies as in the visible region of the spectrum?

Experiment 13-4 INTERFERENCE AND DIFFRACTION

INTRODUCTION

If two coherent waves of the same frequency travel in approximately the same direction, they will interfere so that their energy is not distributed uniformly in space, but is a maximum at certain points and a minimum at others. The resultant interference phenomena can be easily observed for different types of waves. The demonstration of such interference effects for light first established the wave theory of light (Thomas Young in 1801).

Diffraction is the bending of light around an obstacle, such as the edge of a slit. For the detailed analysis of interference and diffraction the student is referred to physics textbooks. In the following experiments we investigate interference patterns both, qualitatively and quantitatively. From the data obtained, the wavelength of the interfering waves can be computed.

The interference phenomena are general and apply to all types of waves. Five different setups are suggested, all illustrating the same type of interference pattern. Students are not expected to perform all five parts of this experiment, but should be aware of the various ways in which interference and diffraction can be demonstrated.

PART A - Water Waves

A ripple tank is set up with two point-source ripplers. The resultant interference pattern is observed when the separation between sources is varied and when the frequency is changed. In some equipment the phase difference between the two sources may also be altered.

A plane wave incident on two adjacent slits will produce a similar interference pattern which should also be investigated in detail.

Repeat the previous part for a single slit, varying the slit width as well as the frequency.

PART B - Sound Waves

The equipment used in Experiment 12-4 may also be used to study interference of sound waves. A loudspeaker is mounted at one end of a 6-foot long aluminum tube lined with sound absorbing material. The speaker is driven by an audio oscillator through an amplifier. A microphone on a long movable boom is used as the detector; its output is put onto an oscilloscope.

A double slit is mounted at the open end of the tube. The sound intensity at the microphone can be measured and plotted as a function of the angle θ through which the boom is turned, from its initial zero position (parallel to the boom). The dimensions of the slits should be measured.

Plot the sound intensity as a function of sin θ and compute the wavelength from this data. Compare the result with the wavelength found by knowing the frequency of the oscillator.

The experiment should be carried out for different oscillator frequencies and for different sets of slits, varying the distance between slits. The effect of the width of the slits should also be investigated.

Repeat the experiment for single slits of different width.

PART C - Microwaves

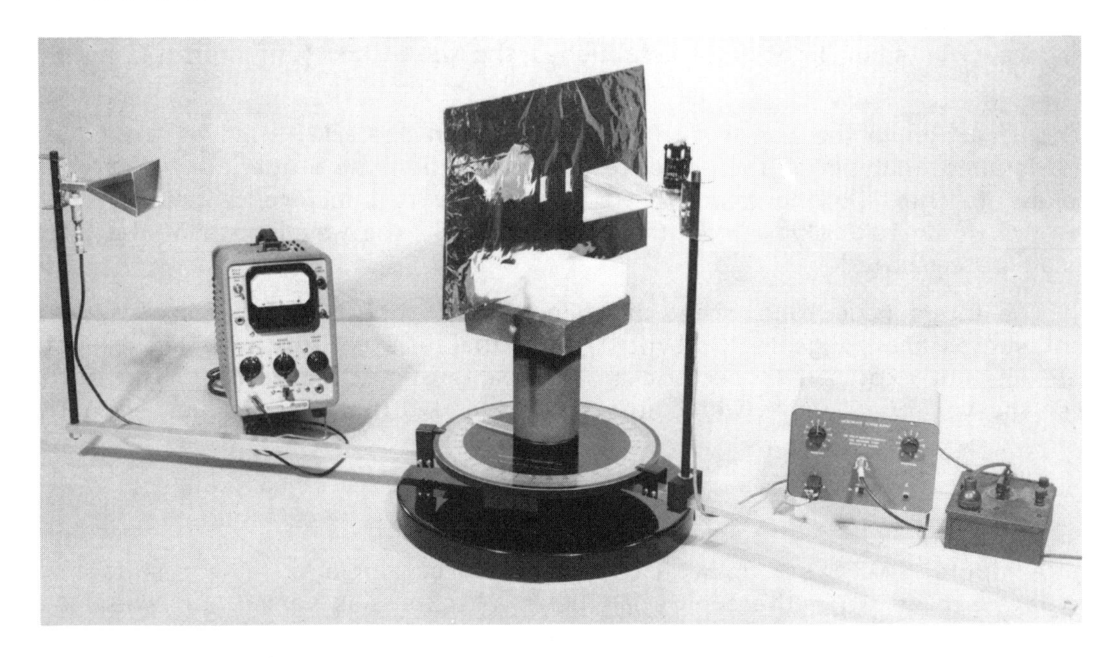

Fig. 13-9. Young's Experiment with Microwaves Showing (from left to right) the Receiving Horn, Detector, Spectrometer Table with Double Slit, Transmitter and Power Supplies.*

Microwaves may be used to study the two-slit interference pattern with a setup of the type shown in Fig. 13-9. Once again quantitative data should be taken for different sets of slits, varying the distance between slits. The effect of the slit width can again be investigated. In each case, plot the microwave intensity as a function of sin θ. The distance between slits (center to center) and the slit width must be measured.

If a recorder or an x-y plotter is available, it will speed up taking data.

Compute the wavelength of the microwaves and compare the result with the wavelength obtained by other methods (standing waves, or interferometer, for instance).

Repeat the experiment for single slits of different width.

---

*For construction details of slits, see Harry F. Meiners, Ed., Physics Demonstration Experiments, pp. 1020-1022.

PART D - Visible Light

Mount the double slit (rulings on a glass slide) on a spectrometer table and properly adjust the spectrometer. For a light source a straight filament light bulb is recommended; to get monochromatic light a filter should be used.

Sketch the pattern observed and measure the angular displacement between maxima. This should be done for different sets of double slits.

The distance between slits (center to center) and the width of each slit can be measured by means of a microscope.

From the data compute the wavelength of light used. Check the wavelength with the color on a spectral chart.

Repeat the experiment for single slits.

PART E - Laser Light and Microcomputer*

EQUIPMENT

The apparatus for this experiment consists of 4 major parts.

1. The laser, an intense coherent source of light.
2. The slits, creating the interference and diffraction patterns.
3. The detector, a movable photo cell to measure the light intensity at a given point of the interference or diffraction pattern.
4. The microcomputer which positions the detector, and the records and displays the intensity of the light pattern.

DO NOT LOOK INTO THE LASER BEAM!

---

*These directions are written for use with the microprocessor and its interface designed under the supervision of Dr. Thomas Shannon under a grant from the RCA Corporation. A detailed description may be obtained from Professor Walter Eppenstein, Physics Department, Rensselaer Polytechnic Institute, Troy, N.Y. 12180-3590.

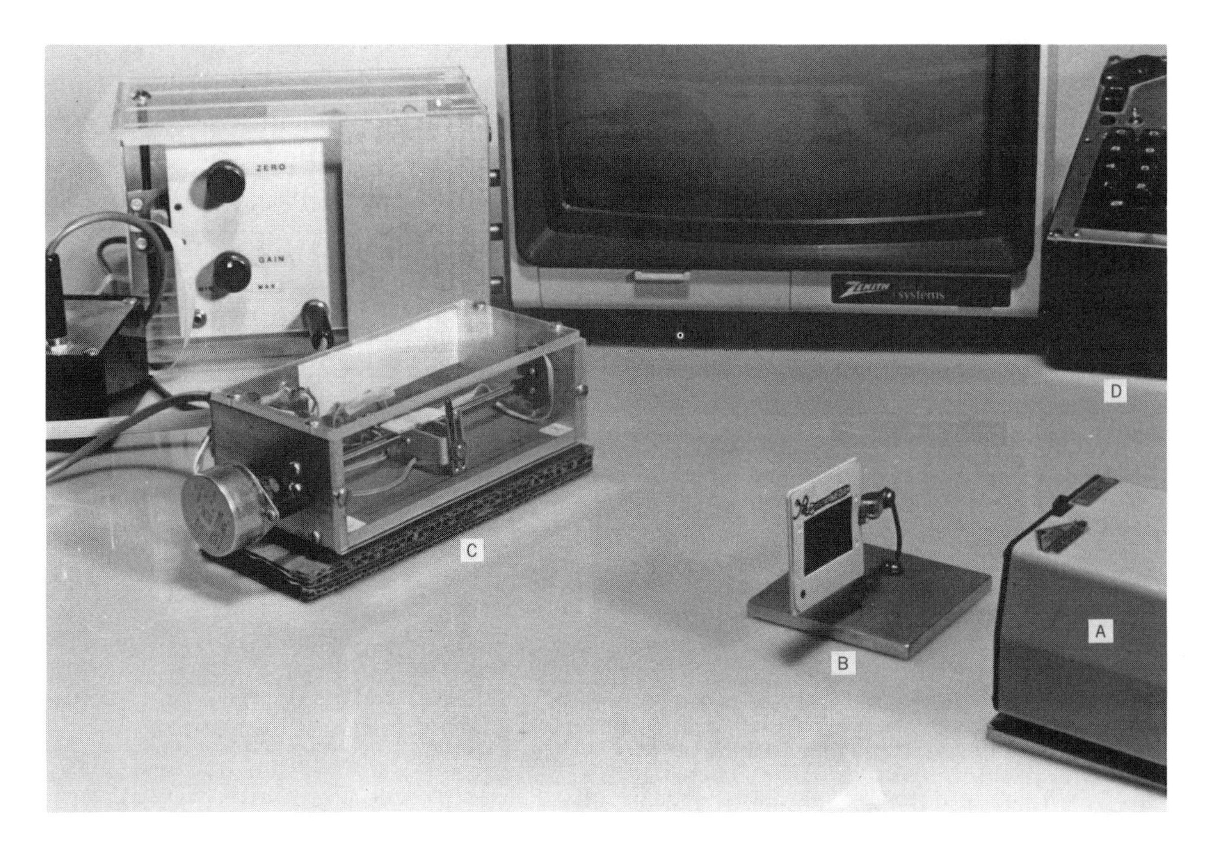

Fig. 13-10. The Laser A, Slits B, Detector C and Microcomputer D Used to Study Interference and Diffraction Patterns.

Some of the pertinent facts regarding this equipment are:

1. The wavelength λ of the laser light is 632.8 nm.
2. Full range of the detector is 256 steps.
3. There are 28 threads/inch on the drive screw of the detector box, thus each revolution of the stepper motor moves the photo cell 1/28".
4. It requires 48 pulses to the stepper motor to cause it to turn 1 revolution.
5. There are gain and zero offset controls on the micro interface card. It is suggested that the zero be set with the laser beam blocked or off and the gain set to give an intensity value of 200 on the central max. However, you might want a higher gain which would clip the high peak but make the smaller ones easier to observe. The controls may interact somewhat so that you will have to recheck one after changing the other one.

PROCEDURE

Line up the laser, slits and the detector box so that the photo cell will intercept the light pattern. It is very important that the pattern be exactly horizontal. Run the micro program several times to become familiar with the equipment. The slide contains 5 configurations of slits: a single slit, a double slit and 5 slits (narrow spacing) and a double and 5 slits (wide spacing). Observe the patterns produced by each of the 5 configurations. Then use the single slit to find the width of the slit; use the narrow double slit and find their separation. Use the same keys on the micro as were used to move the detector to move the cursor in the display mode.

QUESTIONS

1. From the pattern observed for the double-slit, determine the ratio of the distance between the slits to the width of each slit. Do this for each of the five parts for which data was taken.

2. Sketch the intensity curve expected from theory for the single slit and the double slit. Compare with the results of the experiment and try to explain the difference.

3. If the ratio of distance between slits to slit width is twice the value obtained in one of the experiments, sketch an intensity curve for the double slit.

4. What is the effect on the diffraction pattern from a double slit if (a) the distance between slits is kept unchanged and the slit width is varied, (b) the slit width is kept fixed and the distance between slits is varied?

5. What is meant by diffraction?

6. What is the difference between Fresnel and Fraunhofer diffraction?

7. Name all conditions required for obtaining an interference pattern.

8. Define what is meant by a missing order in Fraunhofer diffraction.

9. What is the relation between the slit width and slit separation so that missing orders can occur?

10. What is meant by coherent light? What is meant by monochromatic light? Give examples of each.

11. If a separate light source is used for each slit, would you observe an interference pattern? A diffraction pattern? Explain.

12. What important role did Young's experiment (double slit interference) play in the development of our present physical theories?

Experiment 13-5 DIFFRACTION GRATINGS

INTRODUCTION

A diffraction grating consists of a very large number of lines ruled on glass. Analysis of the interference pattern produced by a grating enables one to determine the spectrum of the radiation emitted by a light source. The theory of the diffraction grating is a logical extension of the theory of a two-slit interference pattern to the case of many slits (of the order of 15,000 per inch). The condition for the occurrence of a principle maximum is given by

$$d \sin \theta = m\lambda \qquad m = 0, 1, 2, \ldots \qquad 13\text{-}2$$

where d is the spacing between the centers of adjacent slits, θ is the angle between the normal to the grating and the direction of observation (see Fig. 13-11), m is called the order number and λ is the wavelength. One sees that this equation is identical with the one which locates the intensity maxima for the double slit.

Fig. 13-11 shows a schematic drawing of the experimental setup used. S is a light source which is assumed to emit a number of discrete wavelengths. C is a collimator which produces parallel light. g is the grating and t is the telescope used to observe the interference pattern.

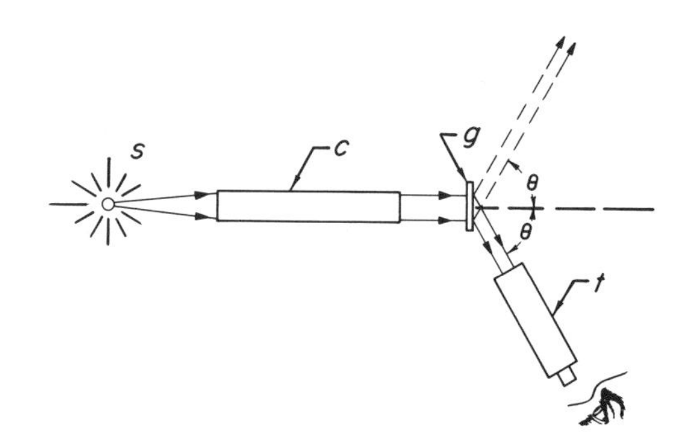

Fig. 13-11. A Simple Type of Grating Spectrometer.

PROCEDURE AND ANALYSIS

PART A - Grating Spectrometer

Read the instructions on the use of the spectrometer in Chapter 4. After the spectrometer has been properly adjusted, place the diffraction grating on the spectrometer table. Adjust the orientation of the grating until it appears to be normal to the path of the incident light. Then observe with the telescope the image of the slit which is formed by the undiffracted light. Record the angular position of the telescope. Move the telescope to the left and focus it on a convenient spectrum line. Again record the position of the telescope. Now move the telescope to the right until it is focused on the same spectrum line. Record the position of the telescope. Calculate the angular position of the spectrum line on both sides of the forward maximum. If they are not the same, the grating is not oriented perpendicular to the incident light. Repeat the above procedure until the grating is properly oriented.

Now observe and record the positions of all clearly visible spectrum lines of helium. Record the positions on both sides of the forward maximum. Also sketch the appearance of the spectrum lines.

Be sure to record the number of lines per inch for your grating. This is needed to calculate d, the slit spacing. Compute the wavelengths and draw a labelled sketch of the spectrum, indicating the color and wavelength of each visible line. Compare your results with published data.

Investigate the effect of changing the number of lines per inch by using different gratings.

PART B - Laser (DO NOT LOOK INTO THE LASER BEAM!)

If a laser is available, project patterns due to various gratings (do not use the spectrometer). Ronchi rulings (a course grating of opaque and clear spaces produced photographically) may be used; they are usually available in a wide variety of different spacings between lines. Investigate the effect of changing the number of lines per inch. Quantitative data can be taken by measuring the distances between lines in the projected image and also measuring the distance from the grating to the screen.

Investigate the pattern produced by two identical gratings with their lines at right angles to each other. Discuss the results.

Now use three identical gratings to produce a symmetric pattern. Add more gratings at various angles. Relate the resulting patterns produced to the type of pattern obtained in x-ray diffraction or electron diffraction (see Experiment 14-1).

PART C - Microwaves

If microwave equipment is available, investigate the effects of a "diffraction grating" for microwaves. What should be the distance between "lines" in such a grating?

QUESTIONS

1. Is the spectrum produced by a diffraction grating rational or irrational? Compare with the spectrum produced by a prism.

2. How many orders can be observed with the grating used? What determines the number of orders of spectra that can be used?

3. Explain why no spectrum is observed if the angle θ is zero - i.e., if the telescope is in line with the grating and slit. What is observed in this position?

4. What would be the effect on the results of the experiment if a grating with fewer lines had been used?

5. Distinguish between transmission and reflection gratings.

6. Derive the grating equation.

7. If the light incident on a diffraction grating makes an angle ψ with respect to the normal to the grating, show that the grating equation must be written as

$$d\ (\sin\psi + \sin\theta) = m\lambda\ , \qquad m = 0,\ 1,\ 2,\ \ldots$$

8. Using the Rayleigh criterion show that the resolving power of a grating is given by

$$R = Nm$$

where N is the total number of rulings and m is the order. Calculate the resolving power for your grating.

9. Define the dispersion D of a grating.

10. Show that the dispersion of a grating can be written as $D = \tan\theta/\lambda$.

11. Calculate the dispersion of your grating for one of the lines.

12. Sketch the light intensity as a function of $\sin\theta$ for an interference pattern due to (a) two slits (see Experiment 13-4), (b) three slits, (c) four slits, (d) 100 slits and (e) 15,000 slits. Use the same coordinates and compare the plots.

13. Discuss two-dimensional diffraction gratings and their interference patterns.

Experiment 13-6 THE MICHELSON INTERFEROMETER

INTRODUCTION

In the Michelson interferometer light from an extended source is divided into two parts by a beam splitter (a half-silvered mirror). The two beams of light are then reflected by two plane mirrors and brought together again to form interference fringes as shown in Fig. 13-12.

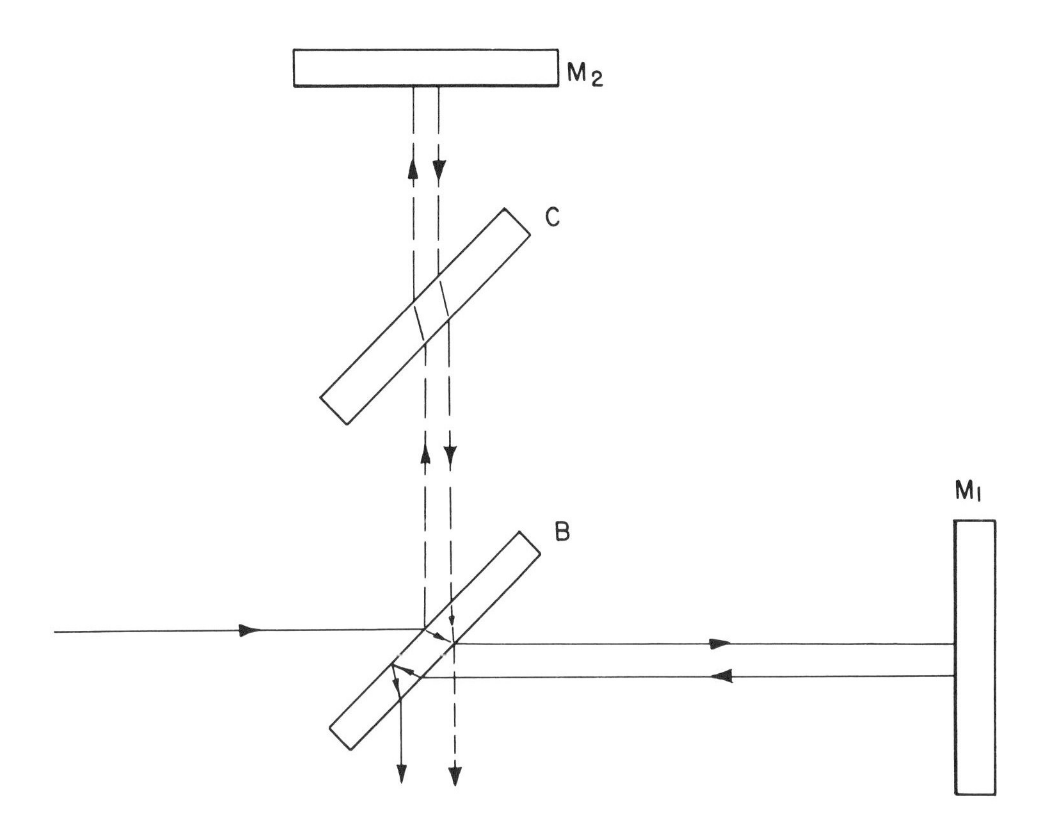

Fig. 13-12. Schematic Diagram of the Michelson Interferometer.

The main optical parts consist of two highly polished plane mirrors (M_1 and M_2) and two plane parallel plates (beam splitter B and compensator C). One side of the beam splitter B is lightly silvered so that the light coming from the source is divided into a reflected beam and a transmitted beam of equal intensities. The light reflected normally from mirror M_1 passes through the plate B for the second and third times and reaches the eye. The light reflected from mirror M_2 passes back through the plates C and B into the eye of the observer.

One of the mirrors is mounted on a carriage and can be moved through small distances by means of a micrometer screw.

ADJUSTMENTS (Consult manufacturer's directions if necessary.)

Measure the distance from the beam splitter B to mirror M_1, the fixed mirror. Adjust the carriage so that mirror M_2 is the same distance away from the beam splitter. All measurements should be made from the top center of the beam splitter frame to the nearest top edge of each mirror. These distances should be equal to within 1 millimeter for easiest adjustment.

The next step is to bring the mirrors into exact perpendicularity. This is another way of saying that one mirror should be parallel to the image of the other in the beam splitter. The adjustment can be accomplished as follows:

Turn on the light source and observe the focusing pin which is situated between the light source and the beam splitter. Two images of the pin will be seen; one coming from the reflection at the front surface of the beam splitter, the other from the reflection at its back surface.

Line up the vertical positions of the focusing pin. This can be done by means of the adjusting screw of mirror M_1. The adjusting screw of mirror M_2 will line up the focusing pin horizontally. When only one image of the pin is obtained, fringes should appear. To best observe these fringes, look straight into the back mirror from the front of the interferometer.

It takes a little practice to obtain the fringes. As stated previously, before touching the adjusting screws, make certain that the two mirrors are equally distant from the beam splitter. When the fringes first appear, they can be sharpened by very careful and minute adjustment of the screws. If the adjustment is accomplished in a room with a great deal of vibration or on an unsteady table, the fringes will soon disappear.

PROCEDURE AND ANALYSIS

PART A - Wavelength Measurement of Filtered Mercury Source

Use a mercury source with a filter for one of the mercury spectral lines to measure the wavelength of that line. After the interferometer has been properly adjusted and a good pattern of fringes has been obtained, take a reading of the micrometer head. By turning the head, the carriage can be moved slowly in either direction. Count the fringes as they pass by the focusing pin or as they appear or disappear in the bull's eye. For satisfactory precision count at least fifty or one hundred fringes. After counting them off, take a new reading of the micrometer head. From the number of fringes Δn passed over and the distance (d_1-d_2) traversed by the mirror, we may determine the wavelength of the monochromatic light by means of the equation

$$2(d_1 - d_2) = \Delta n\lambda \ . \qquad 13\text{-}3$$

The distance (d_1-d_2) traveled by the mirror in centimeters is given by

$$(d_1 - d_2) = K(D_1 - D_2),$$

where (D_1-D_2) is the change of the micrometer reading and K is the ratio of carriage movement to micrometer screw reading, a constant for any particular interferometer.*

If in doubt ask your instructor for the constant. From the data taken determine the wavelength of light emitted from the source used.

PART B - Wavelength Measurement of Laser Light

When using a laser, the pattern is not observed directly, but projected on a screen or the wall. DO NOT LOOK DIRECTLY INTO THE LASER BEAM! Because of the coherence of laser light, the adjustments of the interferometer are easier to make; the light path for the two beams does not have to be the same.

In order to produce a good fringe pattern, it is necessary to enlarge the laser beam so that the major portion of the interferometer beam splitter is filled. For this purpose a lens or a lens system is used.

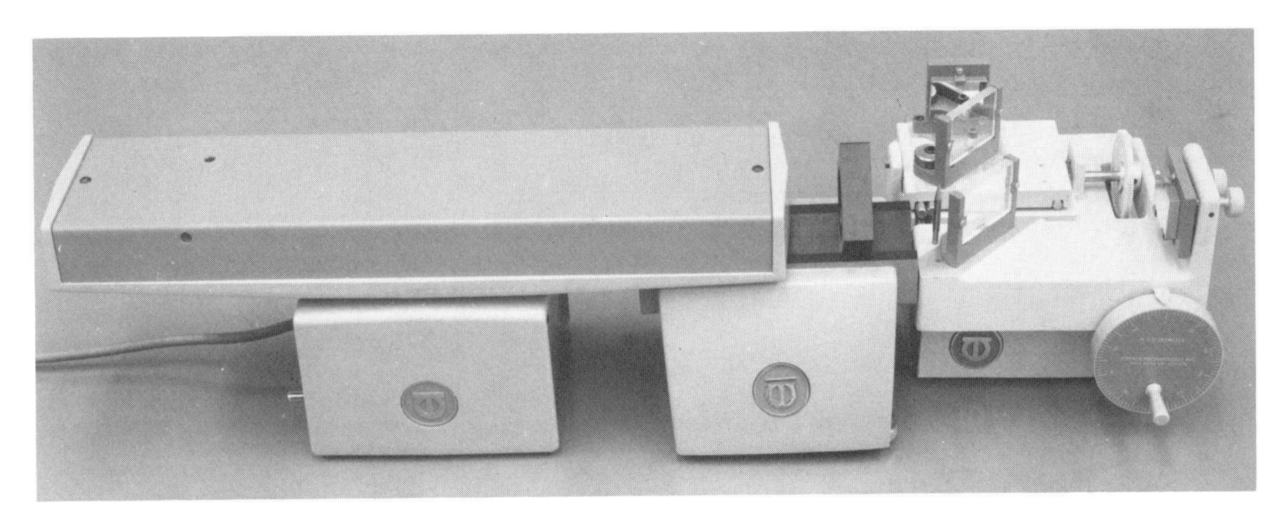

Fig. 13-13. Helium-Neon Gas Laser, Lens and Michelson Interferometer.

PART C - Index of Refraction of Glass Plate

A thin glass plate should be mounted in a holder on the interferometer between the beam splitter and the fixed arrow. The holder must be capable of a slow rotation through a measured angle.

The instrument is aligned to produce circular fringes of a monochromatic light. The glass plate is then rotated through an angle sufficient to produce a shift of about one hundred fringes. This shift in fringes is caused by the increase in optical path due to the index of refraction of the glass. The fringe shift (number of fringes) multiplied by the wavelength is equal to twice the increase in optical path length, $n\Delta x$. From the measurement of the angle through which the glass plate was rotated, the fringe shift and the thickness of the glass plate, its index of refraction can be calculated.

*For the Atomic Laboratories (CENCO) Interferometer the constant K = 0.050 for the M-3 model and 0.002 for the M-4 model. In the Optics Technology Model 176, the distance the mirror travels is read directly on the scale of the drum so that K = 1.

PART D - Index of Refraction of Air

Fig. 13-14 shows an interferometer equipped with a vacuum cell containing precision windows at both sides. Air is pumped out of the cell with a hand pump or a water aspirator. As air is slowly let back into the vacuum cell, there will be a shift of fringes. By counting the number of fringes that shift and measuring the length of the cell, the index of refraction of air can be computed.

The vacuum cell may also be used to measure the index of refraction of a gas.

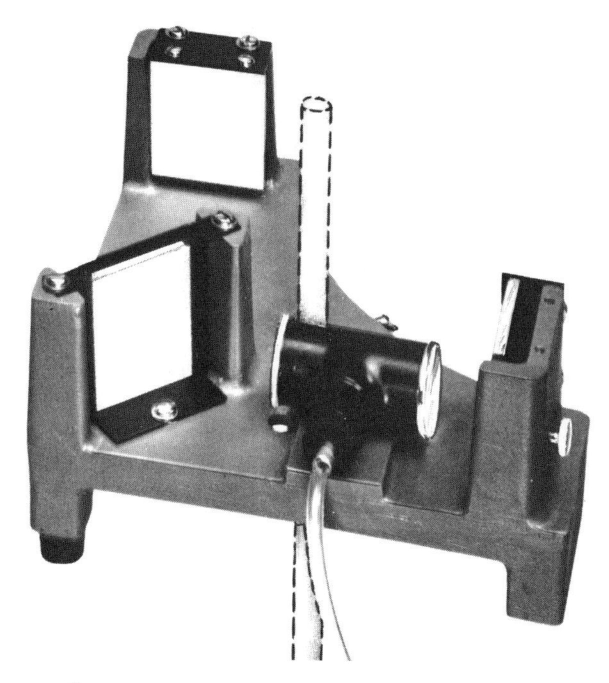

Fig. 13-14. Michelson Interferometer with a Vacuum Cell.

PART E - Measurement of Wavelength of Microwaves

A microwave interferometer can be set up as shown in Fig. 13-15. Determine the wavelength of the microwaves.

PART F - Other Measurements

There are a number of additional experiments possible; the student may want to perform some of these depending on his interest and the equipment available. Details can usually be found in experiment manuals supplied by the manufacturers of interferometers.

Among possible experiments are the following: Measurement of Sodium Doublet; Observation of White Light Fringes; Determination of Wavelength Differences for the Balmer Lines of Hydrogen and Deuterium.

QUESTIONS

1. Justify Equation 13-3.
2. Discuss possible sources of errors in this experiment.
3. Discuss the use of a Michelson interferometer for the calibration of a meter bar in terms of the wavelength of Krypton 86 used as a length standard.
4. How could the index of refraction of a transparent solid be determined by means of the Michelson interferometer?
5. How could the index of refraction of a gas be determined by means of the Michelson interferometer?

Fig. 13-15. Microwave Interferometer. The beam splitter is an aluminum grid.

6. Express the wavelength of light determined in nm, mμ, Å, cm, m, and inches.

7. Discuss the Michelson Morley experiment and the role it played in the developments preceding the special theory of relativity.

8. What other types of interferometers are used to measure the wavelength of light?

9. What other types of experiments, besides interferometry, can be used to measure the wavelength of light?

10. What is the purpose of the compensator between the half-silvered mirror (beam splitter) and one of the mirrors? Why is no compensator used when measuring the wavelength of microwaves?

11. Suppose instead of a monochromatic source you use a source consisting of two closely spaced wavelengths such as the sodium doublet. What would you observe as you move the carriage? To what does this correspond in the case of sound waves?

12. Why is the microwave interferometer easier to use than the optical interferometer?

13. Why is it easier to obtain fringes with a laser than with another light source?

14. If you have taken data for Part C, derive all equations used to calculate the index of refraction of a glass plate.

15. If you have used a vacuum cell (Part D), derive the equation used to calculate the index of refraction of air.

Experiment 13-7 POLARIZATION OF LIGHT

INTRODUCTION

For Parts A through F an ordinary light bulb (unpolarized light) may be used as a source. In Part E a filter of known wavelength must be put in front of the source if quantitative results are to be obtained.

If a laser is used it must be remembered that the beam may be plane polarized by virtue of the "Brewster windows," which are optical flats used at the polarizing angles at each end of the plasma tube. If no Brewster window is used in the construction of the laser, the beam is not polarized. This may be checked easily by using a piece of polaroid.

PART A - Polarization by "Polaroids"

Light is linearly polarized by means of a "polarizer." A second polarizer is used to look at the polarized light and is called an "analyzer." Both polarizer and analyzer are mounted on the optical bench together with a light source and a photronic photocell. The cell is connected to a galvanometer and is used to measure the light intensity transmitted through the analyzer as a function of the angle through which the analyzer is rotated. From this data a plot should be made which can be used to verify Malus' Law. It is left up to the student to choose the coordinates for this plot.

Insert and rotate an additional sheet of polaroid between two crossed polaroids and discuss the results.

PART B - Polarization by Reflection

For this part of the experiment, a round force table may be used. Let polarized light fall on the glass plates mounted on the center of the table and examine the reflected beam first with the naked eye and then with the telescope. The glass plates now act as "analyzer," and if the angle of incidence is made equal to the polarizing angle (Brewster's angle), the image can be extinguished by rotating the polarizer. From the polarizing angle determine the index of refraction of the glass plates.

Data may also be taken by using photocells to measure the light intensities of the reflected and the transmitted beam for different angles of incidence. What conclusions can you reach from this data?

Fig. 13-16. Apparatus for Polarization. Light source L, polarizer P, analyzer A and photocell C are mounted on optical bench. The output from the photocell is connected to a shunt R and a galvanometer G.

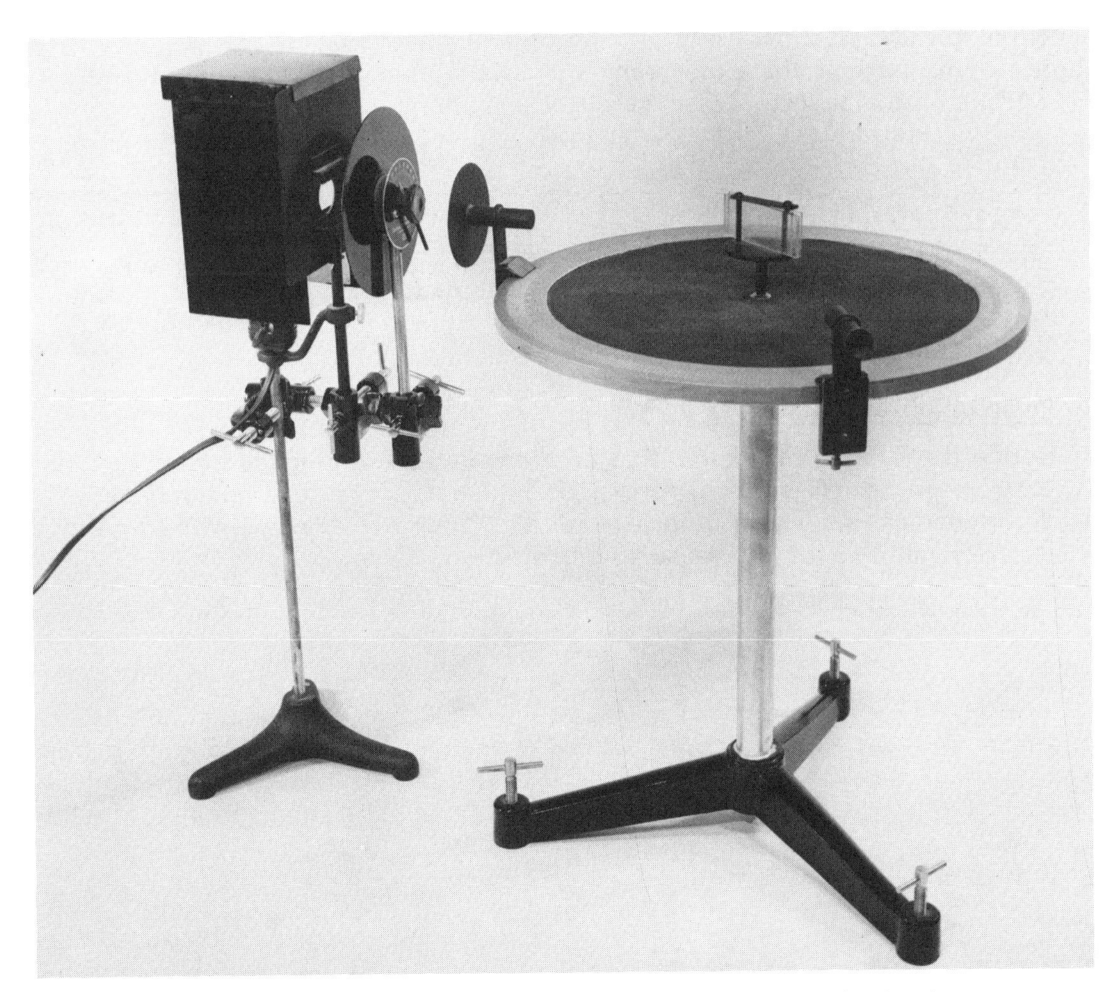

Fig. 13-17. Apparatus for Determining Brewster's Angle.

PART C - Polarization by Refraction

With the same apparatus used in Part II examine the refracted beam when the angle of incidence is equal to the polarizing angle. Is the refracted light polarized? If so, totally or partially?

PART D - Polarization by Scattering

Send polarized light into a tube containing small (invisible) particles of matter held in suspension in water. Light is scattered sideways by these particles. Look at the tube from the front and the top, each time rotating the polarizer. Discuss all observations.

PART E - Rotary Polarization

Certain substances rotate the direction of vibration of polarized light. To observe this phenomenon, set up the polarizer and the analyzer on the optical bench and rotate the polarizer until minimum light intensity is obtained through the analyzer. Now a tube containing a sugar solution is mounted between the polarizer and the analyzer. By rotating the analyzer to its new minimum position, the angle through which the plane of vibration has been rotated by the sugar solution is determined. This angle is proportional to the concentration of the sugar solution. It is also a function of the wavelength of the incident light. This part of the experiment should be tried with white light and with a red filter in front of the source. Discuss all observations.

PART F - Stress Analysis

Check various pieces of glassware, plastic rulers, etc., for stresses by viewing them between crossed polarizers. Discuss all observations.

PART G - Polarization of Microwaves

Investigate the state of polarization of microwaves (see Fig. 4-40 on pg. 107). Insert a grid between "polarizer" and "analyzer." Are the results what you expected from a simple mechanical slit model of polarization? Find the index of refraction of a piece of plastic for microwaves by measuring Brewster's angle.

QUESTIONS

1. Can all waves be polarized? Explain.

2. Can polarization of light be recognized by the unaided eye?

3. In order to produce polarization by reflection, a stack of glass plates are used. Could the same effect be obtained by means of a silvered mirror? Explain.

4. Unpolarized light is incident on a glass slab at the Brewster angle. Is the refracted beam completely polarized? Explain.

5. Unpolarized light is incident on two polarizing sheets oriented so that no light is transmitted. If a third polarizing sheet is placed between them, can light be transmitted? Explain.

6. A plane polarized beam from one light source is allowed to combine with a beam from another source polarized in a plane perpendicular to the first. Discuss the resultant combination.

7. Discuss Polaroid sunglasses and how they prevent glare. What is the direction of the polarization axis in these glasses?

8. Discuss the application of polarization to anti-glare headlights.

9. Unpolarized light passes through a stack of three polaroids: the two outer ones have parallel polarization axes; the center polaroid makes an angle of 45° with the others. What fraction of the incident light beam is transmitted through all three polaroids?

10. Repeat Problem 9 for the case where the two outer polaroids are crossed and the center one is still at an angle of 45°.

11. In this experiment you have observed polarization by scattering; do you think polarization can take place by absorption? Explain.

12. Discuss a method for using polarization observations of a liquid to investigate properties such as concentration of impurities.

13. In the microwave experiment (Part G) if the electric field vector is parallel to the grid wires, will the radiation be transmitted? Explain.

14. If the microwave transmitter is oriented at a 90° angle to the receiver, how will the response vary with the angle of rotation of a grid between the two?

Chapter 14

MODERN PHYSICS

Experiment 14-1 MILLIKAN'S OIL DROP EXPERIMENT

INTRODUCTION

One of the most important constants in physics is the value of the electrical charge carried by an electron. Around 1909 R. A. Millikan used the oil-drop experiment to make the first accurate measurement of the charge of the electron and thereby demonstrated the discreteness of electric charge.

In the Millikan oil-drop experiment fine droplets of oil are sprayed on top of two parallel plates. A few of the drops fall through a small hole in the upper plate of the parallel plate capacitor. A beam of light is directed between the plates, and a telescope with a horizontal scale permits the observation of a single drop and the measurement of of the vertical velocity of the drop. In the experiment, the oil drops are subject to three different forces: viscous, gravitational and electrical. By analyzing these various forces, an expression can be derived which will enable us to measure the charge carried by each oil drop and determine the charge on an electron.

First consider the case when there is no electric field present. The drop under observation will fall slowly, subject to the downward pull of gravity, and the upward force due to the viscous resistance of the air to the motion of the oil drop. The upward buoyant force given by Archimedes' Principle will be neglected because the density of air is very much smaller than the density of oil. The resistance offered by a viscous fluid to the steady motion of a sphere was investigated by Stokes. The result he obtained for a sphere of radius a moving with a constant velocity v through a fluid with a coefficient of viscosity η gives the retarding force acting on the sphere.

$$F_r = 6\pi a \eta v. \qquad 14\text{-}1$$

For an oil drop which has reached its terminal velocity v_o, the upward retarding force is balanced by the downward gravitational force, or

$$F_r = mg. \qquad 14\text{-}2$$

The mass of the oil drop is the density σ multiplied by the volume, or

$$F_r = mg = \frac{4}{3}\pi a^3 \sigma g = 6\pi a \eta v_o. \qquad 14\text{-}3$$

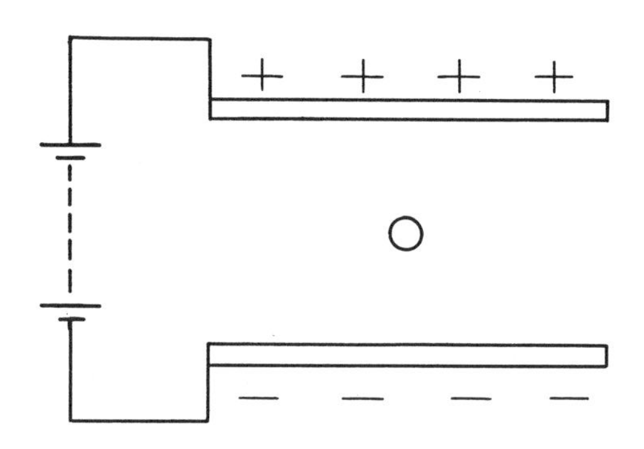

Fig. 14-1. The Plates in the Oil Drop Experiment.

Now let an electric field E be applied between the plates in such a direction as to make the drop move upward with a terminal velocity v_1. The viscous force again opposes the motion and thus acts downward in this case. If the oil drop has an electrical charge q when it reaches its terminal velocity, the forces acting on the drop are again in equilibrium resulting in the relation,

$$Eq - mg = 6\pi a \eta v_1. \qquad 14\text{-}4$$

Solving this last equation for mg and equating this to Eq. 14-3, we get as a result,

$$Eq = 6\pi a \eta (v_o + v_1). \qquad 14\text{-}5$$

Substituting the expression for the radius a of the drop found from Eq. 14-3 in the last expression, the charge of the oil drop is found to be:

$$q = \frac{6\pi}{E}\sqrt{\frac{9}{2}\frac{v_o}{\sigma g}\eta^3}\,(v_o + v_1). \qquad 14\text{-}6$$

The electric field E is obtained by applying a voltage V across the parallel plates separated by a distance d. The field therefore is:

$$E = V/d. \qquad 14\text{-}7$$

The charge is given by,

$$q = \frac{6\pi d}{V}\sqrt{\frac{9}{2}\frac{\eta^3 v_o}{\sigma g}}\,(v_o + v_1). \qquad 14\text{-}8$$

Millikan in his experiments found that the electron charge resulting from the measurements seemed to depend somewhat on the size of the particular oil drop used and on the air pressure. He suspected that the difficulty was inherent in Stokes' law which he found not to hold for very small drops. It is necessary to make a correction in Eq. 14-3 by dividing the velocity v_o by the factor $(1 + \frac{b}{pa})$, where p is the barometric pressure in mm Hg, a the radius of the drop in meters*, and b a constant with the numerical value of 6.71×10^{-6}.

The corrected charge on the oil drop is, therefore, given by

$$q = \frac{6\pi d}{V}\sqrt{\frac{9\eta^3 v_o}{2\sigma g}}\left(1 + \frac{b}{pa}\right)^{-3/2}(v_o + v_1). \qquad 14\text{-}9$$

In the experiment a given drop is timed for several upward and downward flights. Sudden changes sometimes occur because the drop may pick up or lose one or more electrons. Calculations for the charge picked up or lost may be made. The charge on the drop may be changed by bringing a radioactive material near the holes in the upper plate.

THE EXPERIMENT

In this experiment, it is important to follow the directions very closely. For this reason the apparatus available from two different manufacturers is described separately in some detail. For different equipment, the literature supplied by the manufacturers should be consulted.

PART A - Sargent-Welch Equipment

APPARATUS AND ADJUSTMENTS

The most important part of the oil-drop apparatus is the capacitor C. This will be found mounted rigidly on top of the central supporting rod S. It consists of two carefully faced metal plates separated by bakelite strips of definite and uniform thickness. There are several small holes in the top plate, under the cover, through which drops of oil sprayed from an atomizer A may fall into the region between the parallel plates. Two glass windows are provided to eliminate air currents between the capacitor plates and to allow observation of the drops of oil. A beam of light from bulb L is sent straight into the capacitor through the smaller window. It is then scattered by an oil drop and observed from the opposite side by a long focus microscope M. The plates are charged, one positive, the other negative, by power supply P connected through the reversing switch R mounted on the supporting base.

Connect the power supply (+ and - terminals) to the two bottom binding posts of the reversing switch. Connect the D.C. voltmeter V, reading 250 volts maximum, across these same binding posts or across the supply.

A wire from the binding post marked "Cond" should go to the binding post on top of the parallel plates. Be careful that it does not make electrical contact with the support. A terminal inside the switch and the lower plate are both grounded, thus making direct connection for the lower plate unnecessary. Be sure the set screw which holds the capacitor to the support is firmly tightened. Moving the switch handle to one side or the other charges the upper plate positive or negative, the lower plate taking the opposite polarity. When the switch handle is vertical, the parallel plate is not only disconnected from the

*This correction is small and the rough value of the radius a computed from Eq. 14-3 may be used to calculate it. Otherwise a rather complicated quadratic equation would arise.

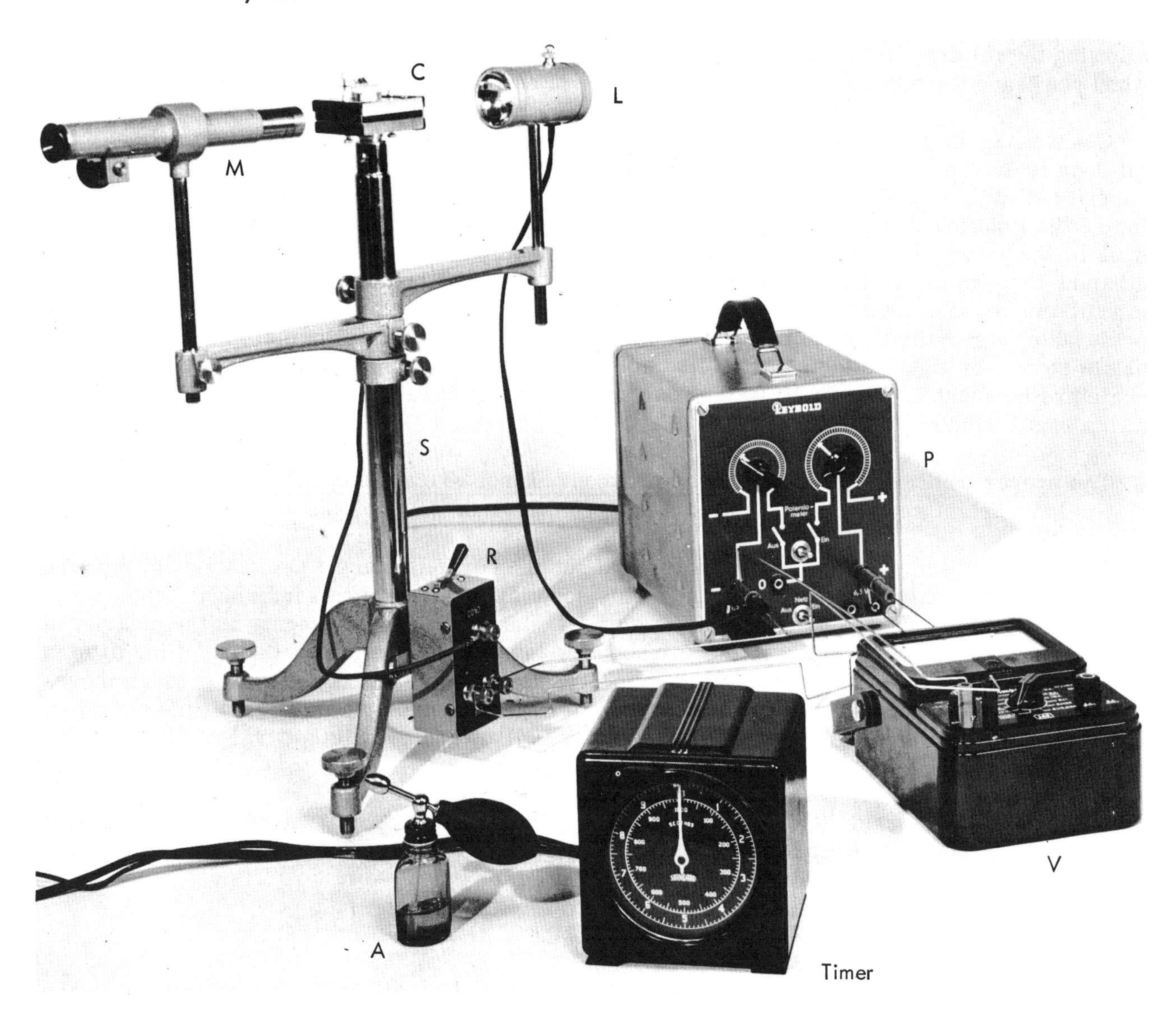

Fig. 14-2. The Millikan Oil Drop Apparatus.

battery but also short circuited, allowing the oil drop to fall freely under the attraction of gravity. Disconnect the power supply when you leave the apparatus in order to stop the current through the voltmeter.

Connect the light source to the 6.3 tap of the power supply. The proper illumination of the oil drop is extremely important. Turn the light source to one side and toward a sheet of white paper held at approximately the same distance as the center of the plates. Loosen the nut on top of the light and by sliding the lamp carrier forward or back, adjust the focus for a bright spot in the center of a faintly illuminated area. Rotate the lamp so that the beam passes straight through the small window and illuminates the upper and lower plates equally. Short-circuit the parallel plates, pass a small wire (to avoid a possible short, a thread of glass is preferable) vertically through the central hole in the top plate and focus the microscope on the bright streak of light reflected from its side. Rotate the microscope so that the lines of its scale are horizontal. See that the microscope is pointing toward the dark corner on one side of the small window.

The proper position of the lamp and microscope cannot be over-emphasized as it is imperative to have bright spots of light against a dark background. Remove the small wire.

Gently spray some oil toward the holes and put the cover in place. On looking through the microscope, a diffused light will be seen. This soon thins out and small individual bright spots (oil drops) may be seen. Select one of the slowest, and follow its movements as the polarity of the plates is changed. Note that the microscope inverts the motion so that a falling drop is seen to rise in the field of view and vice versa. It is advisable to choose a drop which moves slowly when the electrical field is applied as its charge is then only a small multiple of that on an electron. If the drop is too small, it will not travel up and down in a straight line but will waver back and forth with a Brownian movement.

If the drop tends to drift out of focus, move the microscope being careful not to alter the relative position of the eyepiece and objective. If this drifting is excessive, level the entire apparatus by means of the screws in the legs of the base.

PROCEDURE

1. Read the barometric pressure (p in mm of mercury) at the beginning and end of the experiment. Use the average value.

2. Follow a single drop as long as possible while recording the time for the drop to move up a given number of divisions. Occasionally record the voltage (V in volts) and the time for the drop to move down (up in the microscope) the same number of divisions. Use the averages of the latter in your calculations.

   Sudden changes sometimes occur in the time for the drop to move up under the electrical field. This is due to its picking up or losing one or more electrons. Calculations for the charge picked up or lost may be made by using differences between these average times.

3. Either before or after the experiment, remove the front glass plate and measure the distance d between the inner faces of the parallel plates. Similarly the microscope should be calibrated at a convenient time. To do this, raise it so that it is brought to bear on the 0.1-mm scale placed on top of the plates. Be careful not to change its focus. Move the 0.1-mm scale back and forth until it is clearly in focus and determine the distance corresponding to the number of divisions in the microscope eyepiece which you have used in observing the oil drop.

ANALYSIS

1. Divide the equivalent distance (in meters) of the microscope divisions by the averaged values of the times of rise or fall (in seconds). Obtain the velocities of rise v_1 and of free fall v_0.

2. Compute the approximate radius a of the oil drop from Eq. 14-3. The viscosity of the air η may be taken as 0.000182 poise or 0.0000182 N-s/m^2. (One poise is one dyne-sec/cm^2.) Use 9.8 m/sec^2 for the acceleration of gravity g and

0.92 grams/cm^3, or 920 kg/m^3, for the density of Nye's watch oil.*

3. Compute the electrical charge on the oil drop from Eq. 14-9. Note that most quantities in the equation remain constant during the time of the experiment and need to be computed only once. Calculate either the charges on the oil drop or the charges gained or lost. Using the MKS system throughout, the charge will come out in coulombs and will always be a multiple of the correct value of the charge of an electron.

4. Determine the number of electrons on the drop or the number gained or lost by a rough inspection of the data. This will be easy if the drop carried relatively few electrons. Hence, it is advisable to select a drop which moves upward in the electrical field at a slow rate.

5. Divide the charge on the drop by the number of electrons which it carried to obtain a value for the charge of an electron. These values may fluctuate significantly but their average should be within 2 per cent of the correct value, if the experiment has been correctly performed.

PART B - Pasco Scientific Equipment

APPARATUS AND ADJUSTMENTS

Fig. 14-3. Pasco Scientific Millikan Oil Drop Unit.

*Nye's watch oil, William F. Nye, Inc., New Bedford, Mass.

The condenser assembly consists of the following parts (Fig. 14-4):

A, Housing Cover
B, Housing
C, Droplet Hole Cover
D, Top Plate
E, Plastic Spacer.

The unit is leveled by placing a small level on the bottom brass plate and adjusting the three leveling screws on the bottom of the main chassis. NOTE: Reasonable care must be exercised when handling the brass plates to insure that no scratches or pits are formed on the plastic spacer. Any irregularity on the plane surfaces of these parts will make measurement of the plate separation extremely difficult.

The high voltage supply which is to be used for the experiment (400 - 600 volts) is connected to the binding posts of the main chassis in accordance with the POSitive and NEGative markings. (The PLATES switch, on the control box, should be in the OFF position.) The D.C. voltmeter for measuring plate potential should be connected across the binding posts, since a 130-megohm resistor is placed in series with the POSitive binding post and the positive plate to prevent accidental shock to the operator.

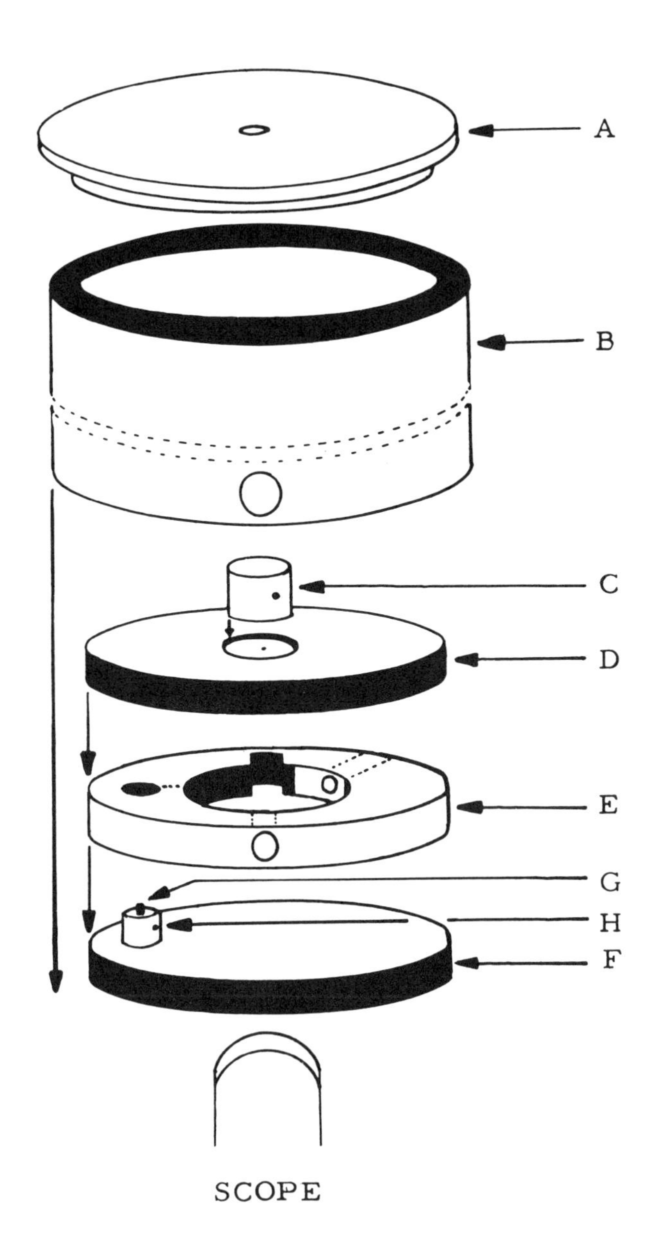

Fig. 14-4. Condenser Assembly.

The distance between the condenser plates is determined by measuring the thickness of the plastic spacer used to separate the plates. Two possible methods are suggested for making this measurement:

a. The thickness of the plastic spacer (Part E) is measured with a micrometer around the outer rim and the average value used.

b. A more accurate method is to place the plastic spacer between two small (2.5" x 2/5") pieces of plate glass and measure the combined thickness of the three. The point of measurement should be as close to the center of the plastic spacer as possible. The spacer is then removed and the thickness of the two glass plates measured. Subtracting the latter quantity from the former quantity gives the separation of the plates. This method has the advantage of taking into account any surface imperfections which may have been found.

NOTE: All surfaces involved in the measurement should be clean to prevent inaccurate readings.

The distance between the top of the upper reticule line may be given or it may be measured in two ways:

a. A microscope reticule, ruled in 0.01 mm, is placed on the bottom brass plate and the scope focused on the reticule.

b. The scope is removed and focused on the anvil and spindle of a micrometer.

This distance should be measured to the limit of the experimenter's ability since a measurement of 0.01 mm is still 1 part in 50 when the reticule line separation is about 0.5 mm.

The condenser housing, plates, spacer, and droplet hole cover should be cleaned, with particular attention to the droplet hole, the glass observation port covers on the condenser housing, and the droplet hole cover. If a solvent, such as carbon tetrachloride, is used, then all the solvent must evaporate from the parts before the condenser assembly is reassembled. The condenser assembly is reassembled except for the housing cover (Part A) and the droplet hole cover (Part C). The two glass ports on the housing (Part B) should line up with the ports of the plastic spacer. The spring contact (Part G) should make good electrical contact with the top plate (Part D). All parts are oriented as shown in Fig. 14-4.

To adjust the optical system a short piece of wire, or a pin (less than 0.010" dia.) is inserted through the hole in the center of the top plate and made to extend enough so as to touch the bottom plate when the condenser is assembled. The LIGHT SOURCE is turned ON. The two Focusing Screws on the observation scope mounting (Fig. 14-5) are loosened and the scope moved either forward or backward until the wire is in sharp focus. If the wire cannot be seen at all, check the orientation of the plastic spacer and the condenser housing. If the wire is still not visible, adjust the light source, in the manner described in the next step, until the wire is visible and focus the scope. Tightening the Focusing Screws locks the scope in position.

While viewing the wire, adjust the eyepiece, by turning it, until the reticule lines and wire are both in sharp focus. If the reticule lines are not horizontal, loosen the Reticule Adjustment Screws (Fig. 14-5), rotate the scope barrel until the lines are horizontal, and tighten the screws.

To adjust the light source, remove the light source housing and loosen the light adjustment screw. Move the light source laterally until the area of the wire observed in the scope, between the reticule lines, is at maximum illumination. Tighten the adjustment screw at this point and replace the housing.

Remove the focusing wire or pin from the top plate and place the condenser housing cover and droplet hole cover in position, as shown in Fig. 14-4. The apparatus is now ready to make measurements.

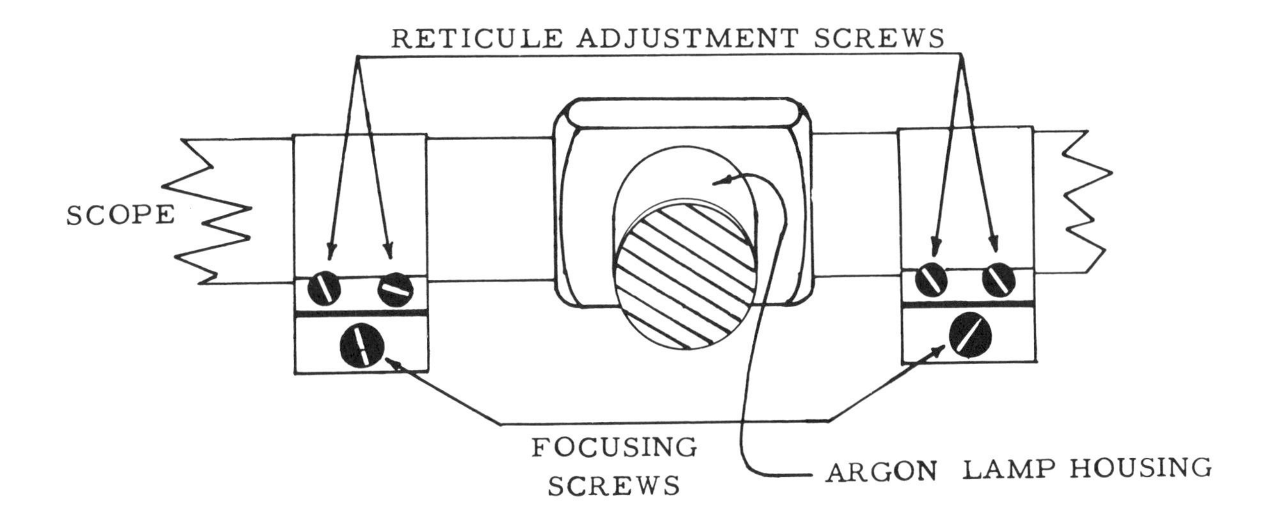

Fig. 14-5. Observation Scope.

PROCEDURE

The room should be made as dark as possible. The LIGHT SOURCE and RETICULE ILLUMINATION switches are turned ON, the PLATES switch is turned OFF, the POLARITY switch is on NORmal, and the RADIATION SOURCE lever is at the IN position. The RETICULE ILLUMination ADJustment should be set so that the reticule lines are just bright enough to be easily visible. Excessive illumination of these lines may make it difficult to observe very small droplets. Non-volatile oil, with a known density, is placed in the atomizer.

The nozzle of the atomizer is placed into the hole of the condenser housing cover. A few quick "squirts" of oil will fill the upper chamber of the condenser with drops and begin to force some drops into the viewing area. If no drops are seen, squeeze the atomizer bulb gently until drops appear in the viewing area. If repeated "squirts" of the atomizer fail to produce any drops in the viewing area, but rather a cloudy brightening of the field, the hole in the top plate is probably clogged, and should be cleaned.

The exact technique of introducing drops will have to be developed by the experimenter. The object is to get a small number of drops, not a large, bright cloud, from which a single drop can be chosen. It is important to remember that the drops are being forced into the viewing area by the pressure of the atomizer. Therefore, excessive use of the atomizer can cause too many drops to be forced into the viewing area and, more important, into the area between the condenser wall and the focal point of the scope. Drops in this area prevent observation of drops at the focal point of the scope.

NOTE: If the entire viewing area becomes filled with drops, so that no one drop can be selected, either wait three or four minutes until the drops settle out of view, or disassemble the condenser, thus removing the drops. When the amount of oil on the condenser parts becomes excessive, clean the assembly as explained previously. The less oil that is sprayed into the chamber, the fewer times the chamber must be cleaned.

From the drops in view the experimenter should select a slow falling, and when the plates are charged, a slow rising drop. Immediately after the drop is selected move the RADIATION SOURCE lever to the OUT position.

About 20 measurements of the rise and fall velocities of the drop should be made and its charge calculated. If the result of this first determination for the charge on a drop is greater than 5 excess electrons, then the student should use slower moving drops in subsequent determinations.

To change the charge of the drops proceed as follows: Drops are introduced into the viewing area and a new drop is selected. After about 20 measurements on this drop have been made, the drop is brought to the top of the field of view and allowed to fall with the RADIATION SOURCE lever at the IN position. A few seconds later the plates should be charged, and, if the rising velocity has changed, then the RADIATION SOURCE is moved to the OUT position and a new series of measurements taken. If, however, the charge has not changed, then turn the PLATES switch to OFF and allow the drop to continue falling. After a few seconds, again check for a change in the rising velocity. Continue this procedure until the drop has captured an ion.

If the drop captures an ion such that the drop moves rapidly downward, then reverse the polarity of the plates so that the drop can be made to rise.

Make about 20 measurements of the rising and falling velocity of the drop, and, if possible, change the charge again and repeat the measurements procedure.

It is desirable to observe as many changes of charge on a single drop as possible.

The plate potential is recorded for each determination; the density of the oil determined; the viscosity of air, at room temperature, found from a suitable handbook; and the barometric pressure recorded.

ANALYSIS

From the data taken compute the charge of an electron.

PART C - Plastic Spheres

The experiment can be simplified considerably by using microscopic latex spheres instead of oil drops as used by Millikan. The spheres have a diameter of approximately 1 micron and are suspended in water. One such Millikan apparatus using latex spheres is shown in Fig. 14-6; others are available.

QUESTIONS

1. Use Stokes' law and derive Eq. 14-9, showing all steps.

2. Discuss possible reasons for the failure of Stokes' law for very small drops.

3. How much difference does the correction for the velocity V_0 in Stokes' law make in the final result? Is it worth taking into account?

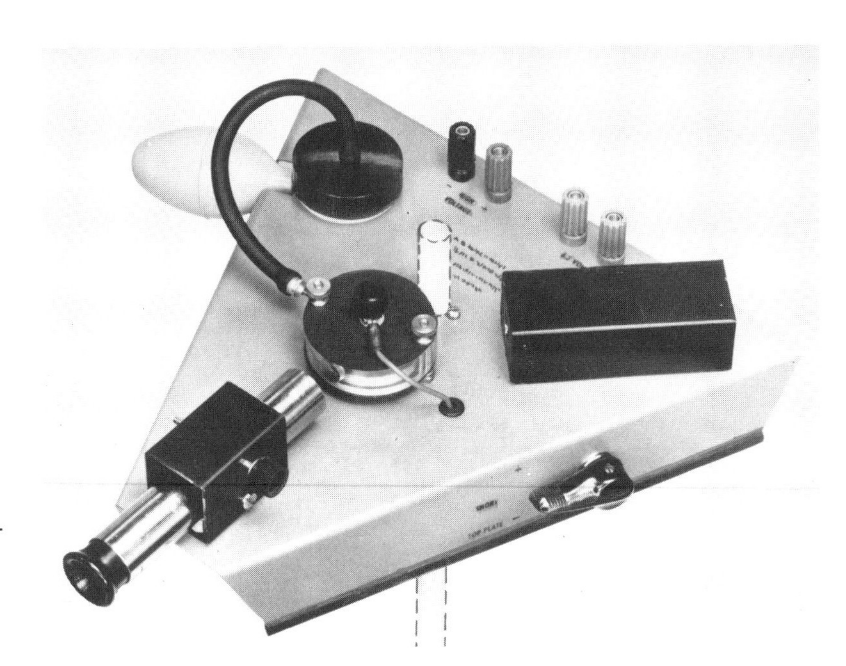

Fig. 14-6.
Millikan Apparatus Using Plastic Spheres.
(Courtesy of The Sargent-Welch Scientific Company)

4. Show that the velocity of the drop as a function of time is given by:

$$v = \frac{mg}{b}\left(1 - e^{-\frac{b}{m}t}\right), \text{ where}$$

$b = 6\pi a\eta$,
m = mass of the drop, and
g = acceleration due to gravity

Plot v as a function of t.

5. Using the equation derived in Question 4, estimate how long it takes the drop to achieve terminal velocity.

6. Does the result of Question 5 justify assuming the drop attains terminal velocity almost instantaneously?

7. How would you measure the viscosity of air? Of a liquid?

8. How do the drops acquire charge?

9. Compute the buoyant force acting on the oil drop and compare its magnitude with the gravitational, viscous and electrical forces.

10. How does the charge of an alpha particle compare with the electronic charge? A proton? A deuteron?

Experiment 14-2 THE PHOTOELECTRIC EFFECT

INTRODUCTION

One of the experiments that the classical picture of electromagnetic waves is unable to explain is the photoelectric effect. When light falls onto the surface of a metal it can, under certain circumstances, liberate electrons from the metal, forming a photoelectric current. When the intensity of the light falling on the cathode is increased, while keeping the wavelength constant, one would expect the greater intensity and corresponding greater electric and magnetic field strengths to give some electrons more energy than they had before. However, it is found that the kinetic energy of the emitted electrons remains the same as long as the wavelength of the incident light is not changed. Increasing the intensity of the light serves only to increase the photoelectric current.

The maximum kinetic energy of the photoelectrons emitted can be measured by plotting the photoelectric current as a function of the potential difference between the collector C and photosensitive surface S of a photoelectric cell (see Fig. 14-7).

The voltage at which the current vanishes is called the stopping potential or the cut-off voltage and is related to the maximum kinetic energy of the photoelectrons by the energy relation,

$$\frac{1}{2} mv_m^2 = eV_o \tag{14-10}$$

where v_m is the maximum velocity which any electron has when it leaves the photosensitive surface, V_o is the voltage at which the current is reduced to zero (stopping potential), and e and m are the charge and mass of the electron respectively.

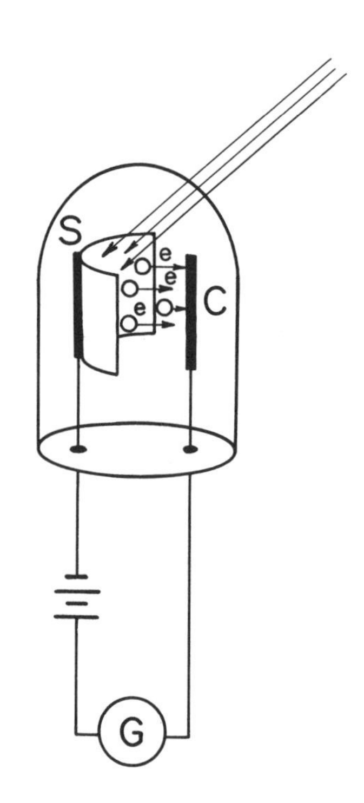

Fig. 14-7.
The Photoelectric Cell.

In this experiment students are asked to determine experimentally the relationship between photo current and voltage applied for various photocells and light sources, as well as the relationship between stopping potential and the frequency of the incident light. This last relationship was interpreted by Einstein in terms of the quantum or photon nature of light. Each photon has an energy $h\nu$, where h is Planck's constant and ν is the frequency of the incident light. In the photoelectric effect all or none of the energy $h\nu$ is transferred to an electron in the metal. Some of the energy the electron may acquire is used in escaping from the metal. This energy is called the work function and represented by ϕ. The rest of the energy supplied by the photon goes into kinetic energy of the electron. From these considerations it is seen that we can write the following relation:

$$h\nu = \frac{1}{2} mv_m^2 + \phi = eV_o + \phi. \tag{14-11}$$

Thus, a measurement of the stopping potential as a function of the frequency of the incident light allows us to determine Planck's constant.

PROCEDURE

PART I - Investigation of Characteristics of Photocells

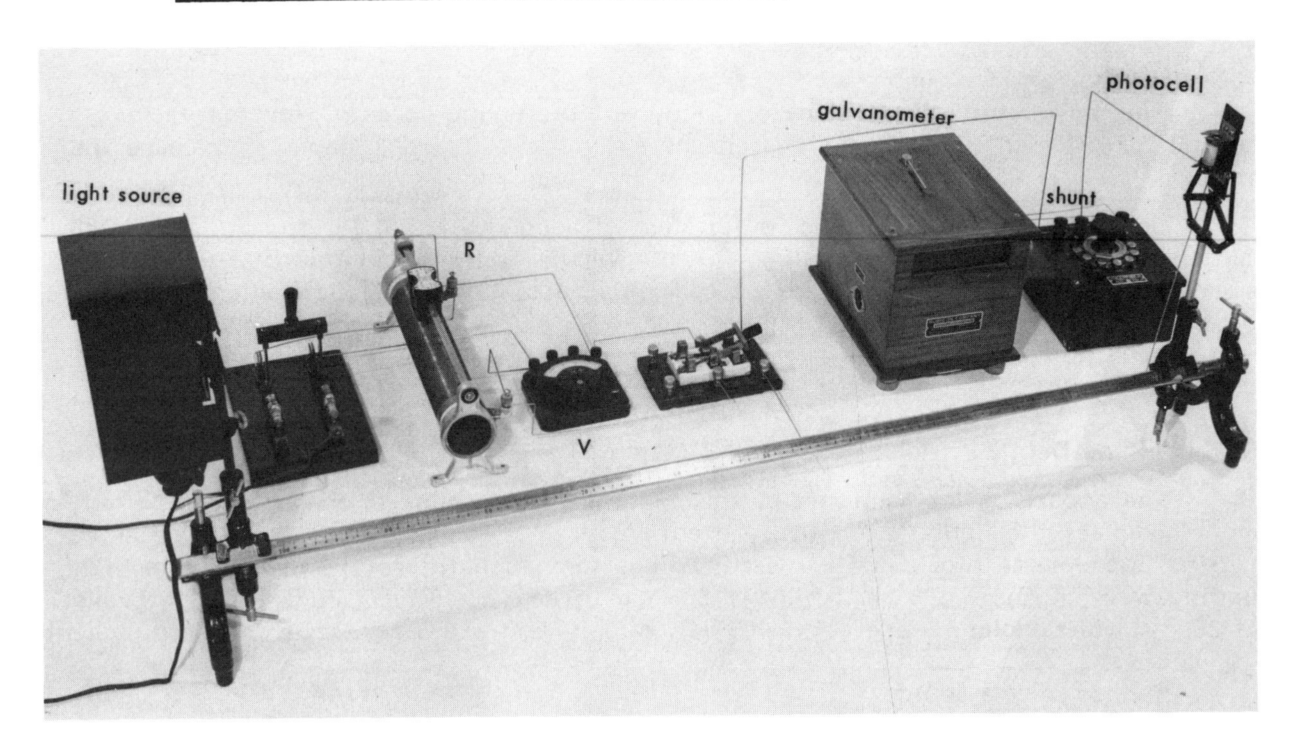

Fig. 14-8. Apparatus for Obtaining the Characteristics of Photocells.

Connect the apparatus according to the wiring diagram shown in Fig. 14-9.

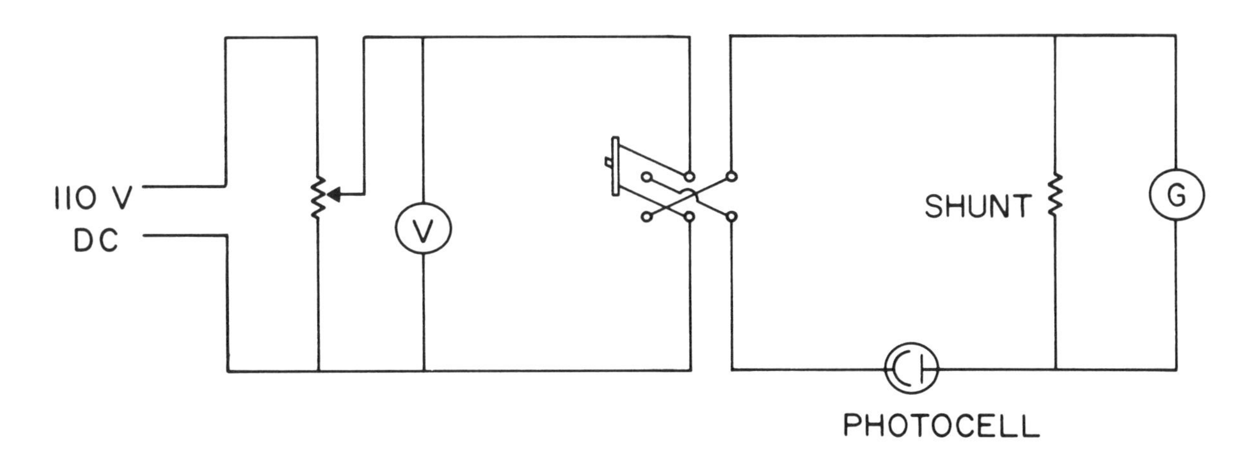

Fig. 14-9. Circuit for Obtaining the Characteristics of Photocells.

Use a vacuum cell (number 922, for instance) and measure the photocell current as a function of anode potential from +80 to -80 volts. This should be done for different values of light intensity. Experimental points should be spaced more closely when the current is changing rapidly.

Next, use a gas-filled cell (number 921) and measure the photocell current as a function of anode potential from +80 to -80 volts.

ANALYSIS

Plot the photocell current as a function of the anode potential for the vacuum photocell for different light intensities on the same sheet of graph paper. Compare the curves.

Plot the photocell current as a function of anode potential for the gas-filled cell. Compare this curve with those obtained for the vacuum cell. From this plot also determine the ionization potential of the gas in the cell. Use a table of ionization potentials to determine the gas used in the photocell.

PART II - The Determination of Planck's Constant

The apparatus for this part of the experiment is similar to the equipment used in Part I. Replace the light bulb with a mercury source. Monochromatic light is obtained by using appropriate filters. The mercury lines for which filters are provided have the following wavelengths: 578 nm (yellow), 546 nm (green), 436 nm (blue), 405 nm (violet), and 366 nm (ultraviolet). Use a 6-volt D.C. supply and a 935 photocell or its equivalent.

Instead of the equipment that has been described, the Planck's Constant apparatus shown in Fig. 14-10 may be used.* It consists of a phototube; two rheostats for varying the voltage across the tube; and connecting terminals all mounted on a housing. Three filters for a mercury source are on a card that fits into a slot in front of the phototube.

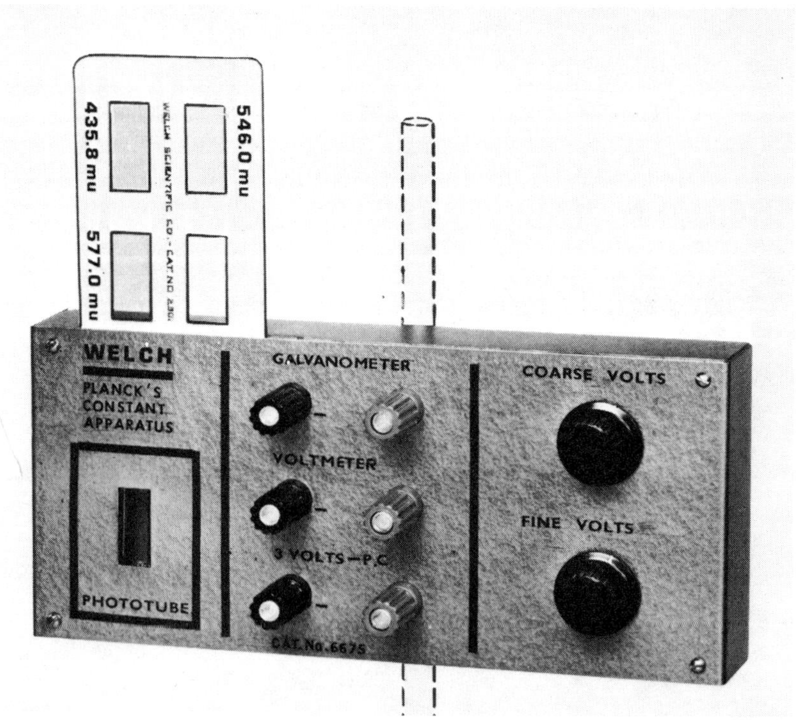

*Available from the Sargent-Welch Scientific Company, No. 2120.

Fig. 14-10. Planck's Constant Apparatus. (Courtesy of the Sargent-Welch Scientific Company)

Take all data necessary to plot the photocell current as a function of anode potential for each wavelength used. Determine the stopping potentials from the current-voltage graphs and plot stopping potential as a function of frequency. Discuss the resulting graph and determine Planck's constant.

The errors to be expected in this experiment, using the equipment provided, are high. Try to discuss some of the sources for these rather large errors.

Optional Part: Since the errors to be expected in determining Planck's constant are very large, it is desirable to find the equation of the straight line which best fits the observed data. By using the method of "least squares" find the straight line which best approximates your data. Determine Planck's constant and compare with your previous result.

QUESTIONS

1. What is meant by "ionization potential"?
2. Explain the difference in shape between the characteristic curves of gas-filled photocells and vacuum photocells.
3. Discuss some practical applications of photocells.
4. What is the difference between photocell and photronic cell?
5. Does the photo current vary with the intensity of the light source? Does this agree with the wave theory of light? Explain.
6. How can the "stopping potential" be determined from the characteristic plot of a photocell?
7. Does the stopping potential vary with the intensity of the light source? Does this agree with the wave theory of light? Explain.
8. Given a plot of photo current vs. retarding potential of a photo tube, what is the physical significance of the slope of this curve?
9. From the plot of stopping potential vs. frequency of incident light in Part II, could you determine the work function of the metal?
10. What is the threshold frequency of a metal? How is the threshold frequency related to the work function?
11. Look up the characteristics of the photocell used in Part II; determine the work function. Express the work function in electron volts and in joules.
12. Knowing the work function for the photocell in Part II, compute the threshold frequency and wavelength.
13. Compute the maximum velocity with which the electrons escape from the metal for each wavelength of light used in Part II. Should relativistic equations be used for calculating these velocities? Explain.
14. Why do many electrons arrive outside the metal with a velocity less than their maximum value? Can you use this to explain why the current decreases smoothly to zero when the anode voltage becomes negative (retarding) in the photocell characteristic plot? Explain.
15. Do you think electrons could be emitted from a metal if, instead of light, high energy particles bombarded the metal? Explain.

16. Discuss the significance of the photoelectric effect in establishing the quantum ideas and the dual nature of light.
17. Name several experiments performed in this laboratory proving the wave nature of light.
18. Give some suggestions for reducing the experimental error in the determination of Planck's constant.

Experiment 14-3 ANALYSIS OF SPECTRA

INTRODUCTION

When a beam of white light is sent through the prism or the grating of a spectrometer, it is dispersed into a spectrum. The prism or grating thus breaks up the light into its component colors. The character of the spectrum observed depends upon the source. An incandescent light bulb, for instance, gives off a continuous spectrum. A gas, on the other hand, will cause a line spectrum which is characteristic of the gas being used. The study of these spectra has played an extremely important role in the development of atomic physics. The emission of the line spectrum was one of the experimental facts which could not be explained by classical physics. Only the atomic structure proposed by Bohr, using the quantum theory, is able adequately to explain line spectra.

PROCEDURE

Thoroughly understand the spectrometer before using it. The instrument section should be consulted for details. The adjustments include the focusing of the crosshairs and the telescope, the elimination of parallax, the proper width of the slit, and the proper alignment of all parts of the instrument.

Mount the prism or grating and observe the continuous spectrum of a light bulb. Make a rough sketch, indicating the colors observed. What is the difference between the spectrum produced by a prism and that due to a grating?

Use the helium discharge tube as a light source. If a prism is used, it should be set for minimum deviation of the yellow line of helium. This is done by rotating the prism table until the image of the slit comes to a standstill, and then reverses its motion as rotation of the prism is continued. Clamp the prism table in this position and make sure the prism is not moved during the rest of the experiment. A grating must be perpendicular to the incident light.

As the telescope is moved, other slit images will appear, corresponding to different colors (or wavelengths). These images form a line spectrum; the wavelengths of the different lines in the visible region of helium are shown in Fig. 14-11. Dotted lines in the figure indicate lines that are faint. Identify in the spectrum an image corresponding to each solid line in the diagram. Now set the crosshairs on each image in turn, read and record the corresponding scale position.

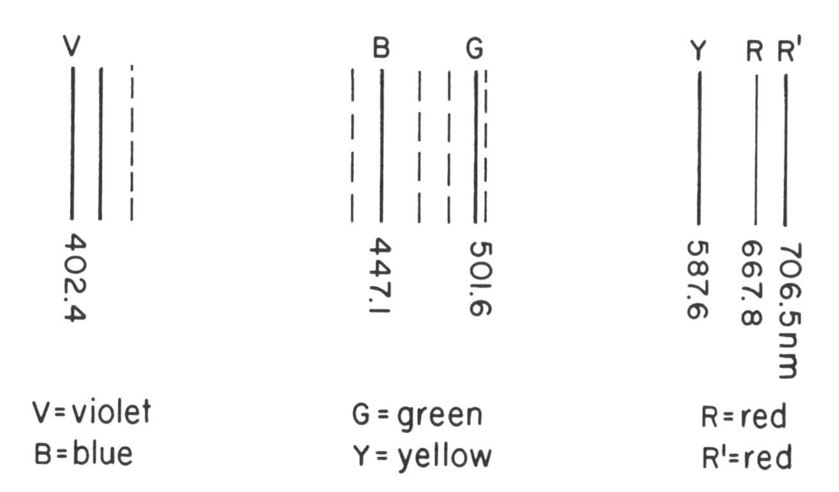

Fig. 14-11. The Line Spectrum of Helium. The wavelengths are in nm.

If a prism is used, plot the scale reading as a function of wavelength. This is the calibration curve for the spectrometer for one particular prism setting. By reading the scale position for any other line, the wavelength can now be found from this calibration curve.

If a grating is used, the wavelength can be computed directly knowing the number of lines on the grating.

Remove the helium discharge tube and put a hydrogen discharge tube in its place. Obtain scale readings for the strong red line and the intense line in the blue. Find the wavelength of these two lines. They are the first two members of the Balmer series of hydrogen. Substitute these wavelengths into the Balmer series together with the Rydberg constant, and compute the values of "n" for the red and blue lines.

$$\frac{1}{\lambda} = R\left(\frac{1}{2^2} - \frac{1}{n^2}\right) \qquad n = 3, 4, 5, \ldots \qquad 14\text{-}12$$

$$R(\text{Rydberg constant}) = 1.097 \times 10^7 \text{ m}^{-1}$$

Discuss the significance of the two values of n that were obtained.

If a fluorescent light is available, investigate its spectrum. From a measurement of the wavelength of the lines determine the gas inside the tube.

If the overhead lights are fluorescent, use a 90° prism to bring the light into your spectrometer. Also investigate the spectrum of the sun.

Insert other discharge tubes or use spectral lamps to investigate the spectrum emitted by various gases. Sodium and mercury should be looked at because their characteristic wavelengths are used in other experiments. The spectrum of neon might be looked at because of the use of helium and neon in the continuous gas laser.

Various special tubes are available for investigations of spectral lines. Students may, for instance, want to look at the spectrum due to the discharge in air at low pressures (Experiment 10-6).

Special tubes containing mercury and neon are available to illustrate the Franck-Hertz experiment. The pressure of the gas mixture is pre-adjusted so that the maxima and minima due to inelastic collisions of electrons with gas atoms may be studied visually by means of characteristic color changes, spectroscopically by observing the bright lines at specific voltages, as well as electrically by recording the voltages at which the current dips and peaks occur. Both, mercury and neon lines are visible at certain voltages. For the details the student is referred to the instructions supplied by the manufacturer.

QUESTIONS

1. Define what is meant by emission spectra and absorption spectra. Give examples of each.

2. Define what is meant by line, continuous, and band spectra. Give examples of each.

3. Is the spectrum observed through the prism a rational or an irrational spectrum? What about the spectrum observed through a grating?

4. What causes the spectra of different elements to differ?

5. Can this spectrometer be used to observe lines of series other than the Balmer series? (Lyman, Paschen, Bracket). Explain.

6. What determines the "breadth" of spectral lines?

7. Use Bohr's postulates to derive the expression for the Balmer series of hydrogen. (Eq. 14-12)

8. What is the relationship between the wavelength of a particular transition in singly ionized helium and the wavelength for the same transition in hydrogen? What is the relationship between doubly ionized lithium and hydrogen?

9. Explain the principle of a discharge tube.

Experiment 14-4 BRAGG DIFFRACTION WITH MICROWAVES

INTRODUCTION

In 1912 Bragg* derived a relation which explained x-ray diffraction effects in terms of reflections from a family of atomic planes within a crystal. Several conditions are necessary for Bragg diffraction.

1. No matter what the value of the incident angle, each individual plane consisting of an ordered array of diffracting centers in the family acts as a plane mirror. The reflected waves reinforce each other to produce maximum intensity when the angle of incidence equals the angle of reflection. In the case of reflections from atomic planes, the angle θ is the angle between the incident or reflected beam and the plane rather than the angle between the beam and the normal to the plane, as is customary in optics.
2. When a beam of radiation strikes a family of parallel planes, each plane will reflect part of the energy. If the reflected waves from O and Q, as indicated in Fig. 14-12 are to be in phase (interfere constructively), the path difference

$$PQ + QR = 2d \sin \theta$$

must equal an integral number of wavelengths or

$$2d \sin \theta = n\lambda \qquad n = 1, 2, 3, 4 \ldots \tag{14-13}$$

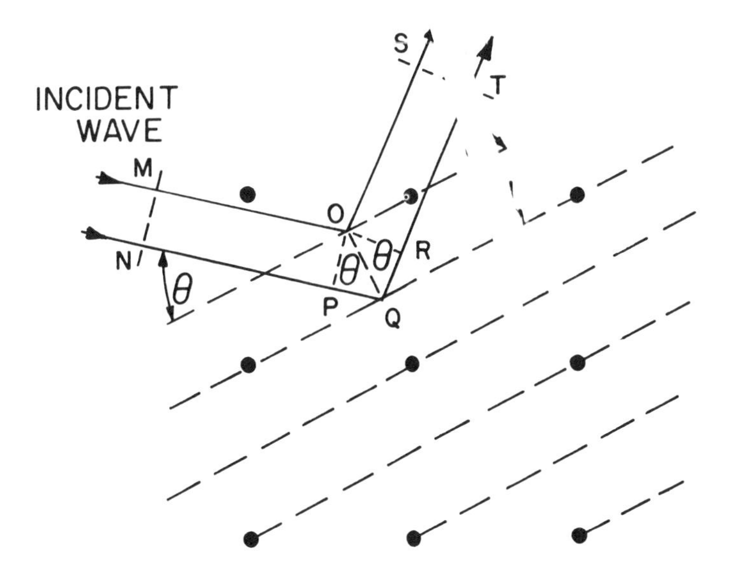

Fig. 14-12.
Incident waves falling on a family of planes are reflected and reinforce each other (interfere constructively) to produce strong reflections (maximum intensity) only when Bragg's law is satisfied. Dotted lines correspond to the (210) family of planes in a simple cubic crystal.

*W. L. Bragg, Proc. Cambridge Phil Soc. 17, 43 (1912).

The path length NQT is an integral number of wavelengths longer than the length MOS.

Equation 14-13 is Bragg's law and governs the diffraction of waves from parallel planes in our lattice. As contrasted to the mirror behavior of a single plane which gives reflections for any angle θ, only particular values of θ will satisfy Bragg's law and allow constructive interference. For other angles there is no reflected beam because of destructive interference.

The perpendicular distance between adjacent parallel planes is d. As shown in Fig. 14-13, other families of planes with different d values are present in the same crystal. There is a decrease in the number of atoms per plane associated with a decrease in d for a simple cubic structure which has the same type of atom at each lattice site. Reflections, therefore, become weaker as d decreases. This is not, in general, true for more complex structures.

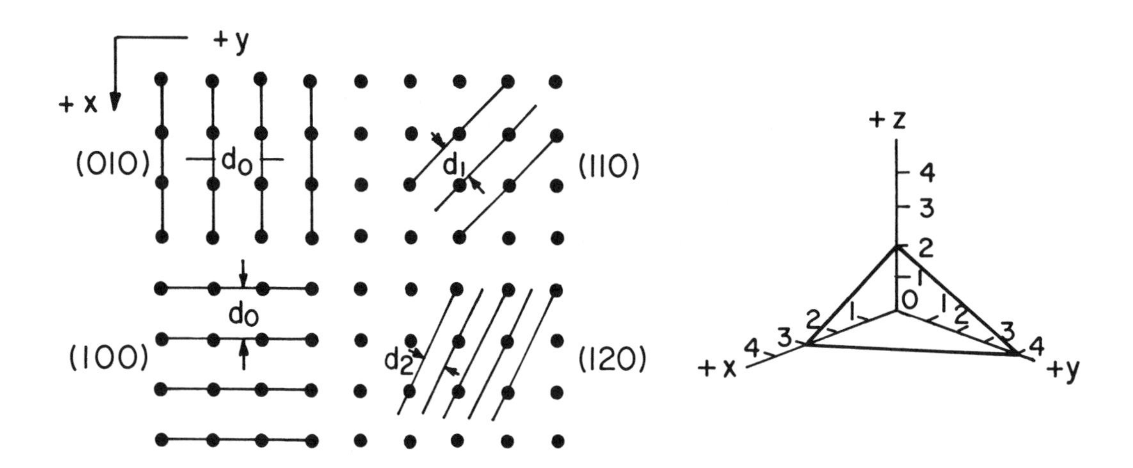

Fig. 14-13. Several families of planes in a simple cubic crystal. As the interplanar spacing d decreases, the intensity of the diffracted waves is reduced since the number of diffracting centers per unit area in a plane decreases. The reference line connecting lattice points occupied by atoms is of the same unit length for each family of planes.

When we analyze a crystal at a specific orientation with monochromatic radiation, a spectrum of reflections is obtained as a function of the angle θ. If the value of θ corresponding to the strongest reflection peak is used in Bragg's relationship, the interplanar spacing d can be calculated for the family of planes contributing to the maximum peak(s). Reflections are produced by more than one family of planes providing Bragg's law is satisfied. These weaker reflections are observed as background effects in a plot of relative reflection intensity versus θ (R.E. vs θ). Depending on which family of planes is investigated with monochromatic radiation, various orders of diffraction are obtained.

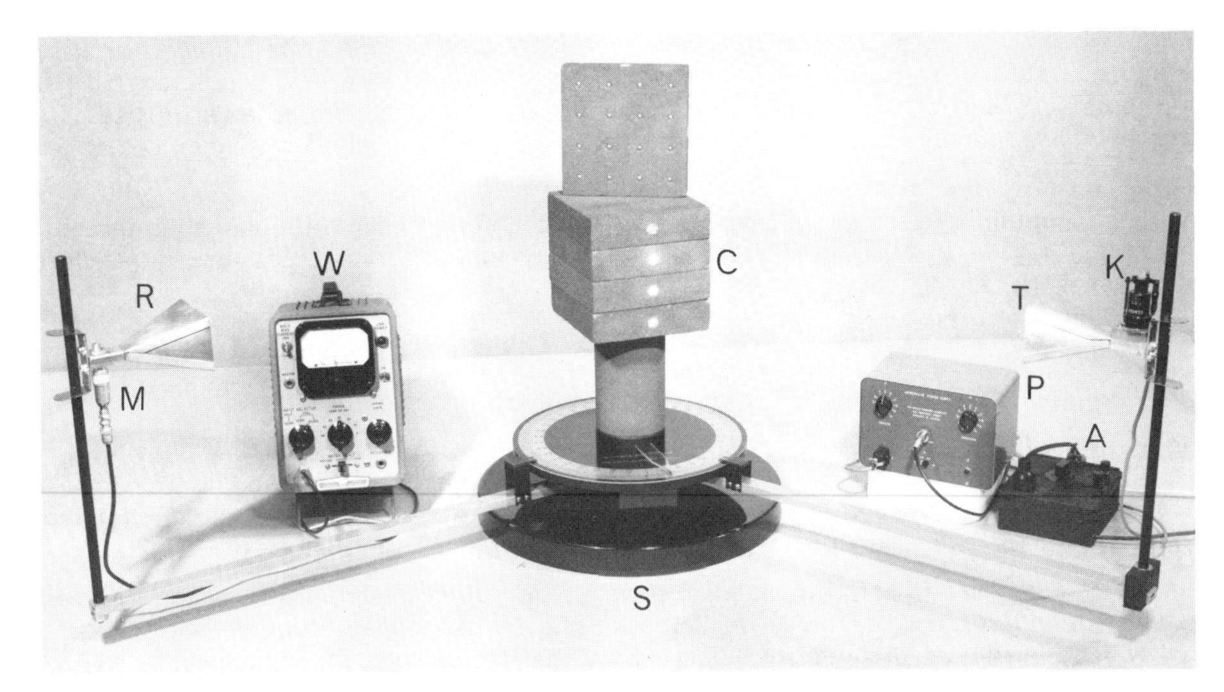

Fig. 14-14. Bragg Diffraction Apparatus (PATENTED*). Three-cm microwaves generated by a reflex klystron K supported on a transmitter horn T are used to investigate the crystal C which contains a lattice of 10-mm aluminum spheres in an 18 x 18 x 18-cm foam plastic matrix. The crystal rests on a column of foam plastic at the center of the spectrometer S. The spectrometer is provided with a boom system which permits independent adjustment and measurement of the relative angular positions of the transmitter horn, lattice, and receiver horn R to 1°. M is a crystal mount for a 1N23B silicon diode. A standing wave indicator W is attached containing the diode. The regulated D.C. power supply P has an output jack O which connects to the transmitter horn. Input jack I is used for external modulation by autotransformer A. Crystal C and its support column are transparent to microwave radiation.

Instead of x-rays used in diffraction studies of crystalline solids, the Bragg diffraction apparatus uses 3-cm microwaves. They are produced by a 2K25 or 723 A/B reflex klystron which is a resonant cavity capable of sustaining oscillations at frequencies between 8×10^9 and 10×10^9 cycles per second. Our investigations in this experiment are concerned with using microwaves to determine the separation d between several families of planes in a simple cubic lattice. For example if $\lambda = 3.00$ cm and $d = 4.00$ cm, then diffracted waves are possible at $\theta \cong 22°$ ($n = 1$) and $\theta \cong 48°35'$ ($n = 2$). Higher order reflections ($n = 3, 4, 5, \ldots$) can not be obtained because the sin θ can not be greater than unity.

* PATENT assigned to Sargent-Welch Scientific Co. by Professor Harry F. Meiners.

In order to identify the various crystal faces or planes, the Miller[1,2] indices method of notation is used. The indices for a particular plane are calculated from the intercepts on a set of three axes. Customarily, a right-hand set of axes is used. The intercepts are measured in terms of multiples or fractions of the unit distances on each axes.

To obtain the Miller indices for the plane shown in Fig. 14-13 with intercepts $x = 3$, $y = 4$, $z = 2$, take the reciprocal of each value, clear of fractions and enclose in parentheses $(hk\ell)$ as follows:

$$\frac{1}{3}\ \frac{1}{4}\ \frac{1}{2} = \frac{4}{12}\ \frac{3}{12}\ \frac{6}{12} = (436)$$

The Miller indices of the plane are (436). The plane with intercepts $x = 1$, $y = 1$, $z = \infty$ has Miller indices (110). The indices of the other planes can be found in the same way. Because the booms supporting the transmitter horn and receiver horn can be rotated only about the z axis, we limit our investigations to the planes parallel to the z axis.

The wavelength of the microwaves must be determined before we can proceed to analyze our macroscopic crystal. A method different from that often used to measure x-rays will be employed because we assume that the lattice constant d_0 is not known. In Project I of this experiment the wavelength of the microwaves is measured by the method of standing waves already discussed in Experiment 12-1, Part C; the setup of the apparatus is shown in Figures 12-4 and 14-15.

In summary we are concerned in this experiment only with the simpler analysis of diffraction by a space lattice. As shown in Fig. 14-12, we considered Bragg reflections from parallel planes passing through atoms. Incident waves are reflected just as if the planes were a stack of mirrors. For a depth treatment of the phenomenon of x-ray diffraction by crystals, other references should be consulted. Actually x-rays are scattered by the electrons of the atoms. Analysis of the resulting diffraction patterns provides information about the unit cell of the crystal. Measurements of the intensities of the diffraction patterns permits location of the atoms within a crystal.

PROJECT I - Determination of the Wavelength of Microwaves

PROCEDURE

1. Arrange the apparatus as shown in Fig. 14-15. Set the reflector control at position 4 or 5 and the resonator control at 10. The reflector control regulates the negative reflector voltage which affects slightly the density of the electron beam in the reflex klystron. Variation of the reflector voltage changes the power output of the klystron. The lower (less negative) the reflector voltage, the less power produced for a particular mode of operation. The resonator control is used to set the electron beam voltage. It regulates the positive voltage applied to the accelerator grid within the klystron.

---

[1] C. S. Barrett, Structure of Metals, McGraw-Hill Book Company, Inc., New York (1943) Chapter I and IV.

[2] H. P. Klug and L. E. Alexander, X-Ray Diffraction Procedures, John Wiley and Sons, Inc., New York (1954) Chapter I.

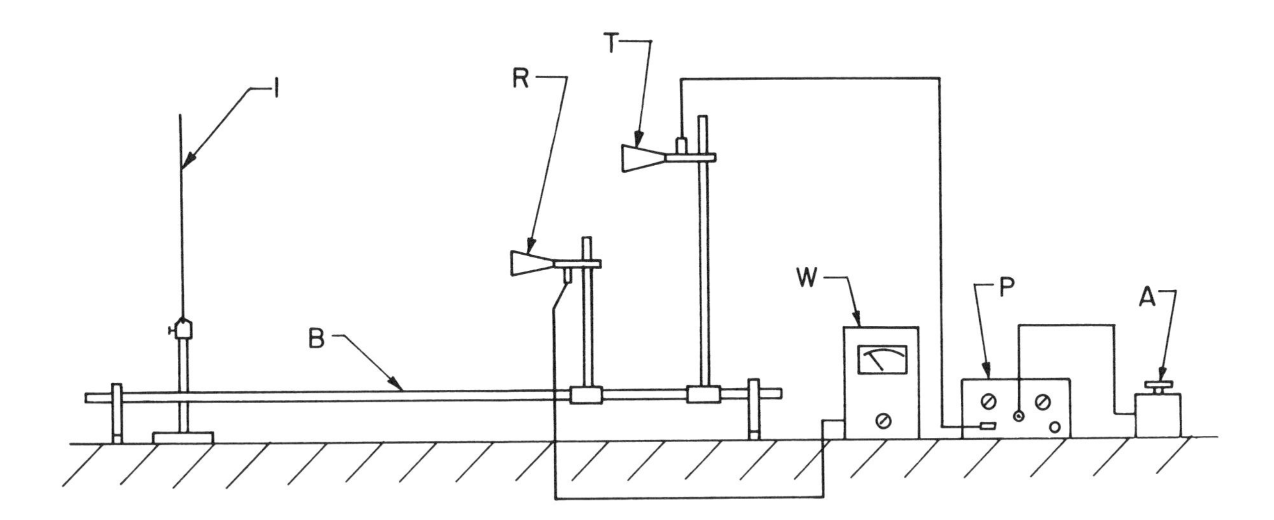

Fig. 14-15. Apparatus Used to Measure Wavelength of Microwaves. To study standing waves set up between a metal 30 x 30-cm mirror I and the transmitter horn T, receiver horn R is moved slowly on the optical bench B toward the mirror. Positions of nodes and antinodes are observed on the indicating meter W. (Photograph of equipment is shown in Fig. 12-4 on Page 543).

2. Before attempting to take data, experiment with the apparatus. Make certain that you know how to read the indicating meter and to distinguish between nodes and antinodes. Would you expect to find a node or antinode at the mirror; at the transmitter horn? Note the intensity registered on the meter and try moving the receiver horn toward the mirror from several different positions so as to obtain the best readings. Should you also try several different spacing arrangements between the mirror and the transmitter horn?

ANALYSIS

Since nodes can be observed more easily than antinodes, record as accurately as possible the position on the optical bench of one node N_i. Move the receiver slowly toward the mirror until 20 antinodes are passed. Record carefully the position of the last node N_f and calculate the wavelength λ. Repeat the measurements four times. Record all the measurements and their mean in your data table.

QUESTIONS

1. Determine the scale accuracy (P.I. - limit of error) of the optical bench and calculate the maximum numerical error and the relative (per cent) error for the wavelength λ. What is the probable and relative (per cent) probable error?

2. Discuss other more accurate methods which could be used to find the wavelength of the microwaves.

3. What type of radiation is produced by a reflex klystron? Compare the optical properties of electromagnetic radiation with those of light and discuss the differences.

4. Show in cross-sectional views of width, length and depth of a waveguide the wave patterns of the electric and magnetic fields.

5. When standing waves are set up between the transmitter horn and mirror, is there a node or antinode of the electric field at the mirror? At the horn? Is there a node or antinode of the magnetic field at the mirror? At the horn? Where would the nodes or antinodes for the electric and magnetic fields be located with respect to the transmitter horn; in front of the horn or at the end of the short section containing the exciting antenna?

6. Do waves of the electric field undergo a phase change at the mirror? Do waves of the magnetic field experience a similar phase change? Explain.

7. If the free space wavelength of the microwave is 3.2 cm and the width of a waveguide is 2.0 cm, calculate the guide wavelength. Why is there a difference in length between the free space and the guide wavelengths?

8. Using 3.2 cm for the free space wavelength and 2.0 cm for the width of the waveguide, find the phase speed v_{ph}, and the group speed v_{gr} of the electromagnetic radiation in the guide. What is the speed v of the electromagnetic waves in free space? Explain the differences between the various wave speeds.

9. For a free space wavelength of 3.2 cm and a waveguide width of 2.0 cm, determine the cutoff frequency.

PROJECT II - Quantitative Investigation of a Macroscopic Crystal with Microwaves

ADJUSTMENTS AND PROCEDURE

1. The power supply P and the indicating meter W near the Bragg spectrometer in Fig. 14-14 must be placed at a level below the apparatus to reduce the effects of background reflections. For the same reason students when taking data should stand behind the transmitter and receiver horns.

2. The plastic cover, which is transparent to the 3-cm electromagnetic radiation, will be removed from the crystal by the instructor after your analysis is completed.

3. During the investigation, the crystal is to be considered an unknown structure. The reference pointer at the base of the plastic column supporting the crystal has been set at an angle which permits a particular family of planes within the crystal to be analyzed. Do not change the position of the pointer.

4. The transmitter and receiver horns should be directed toward the central layer of the crystal.

5. Set the reflector control at 4 or 5 and the resonator control at 10. To become familiar with the apparatus, move the transmitter and receiver booms slowly toward each other at the same rate. Observe the relative intensity readings on the detector W. If necessary, change the sensitivity of the detector. Do not change the settings on the power supply. What is the significance of the meter variations?

6. Take data every 1°. Record all measurements. After data is recorded for a particular lattice orientation, ask the instructor if another family of planes is to be investigated. If so, the reference pointer must be set at a new angle. Important! Plastic cover is not to be removed until the analysis of the results of all the curves is completed.

ANALYSIS

1. Plot a graph of relative intensity versus θ for each set of data.

2. Using values from your graphs, calculate the interplanar spacing d for each family of planes which was investigated. Do you know the wavelength of the microwaves?

3. Open the lattice and measure carefully the spacings between the balls. Take a group of measurements of the distance between different sets of balls so that a reasonable mean value can be obtained. What is the lattice constant d_o?

4. From a knowledge of the geometry of the lattice, calculate the interplanar spacings between the (110) and (210) families of planes.

5. Compare the values for the interplanar spacings obtained from your graphs (paragraph 2) with the actual spacings computed from the lattice geometry (paragraph 4). Discuss.

6. How could you determine the wavelength of the microwaves from an analysis of a plot of relative intensity versus θ? Use data obtained from your graph(s) and calculate λ. Compare your value with the known wavelength obtained in Project I. Discuss.

QUESTIONS

1. Derive the Miller indices for the (100) and (120) families of planes in Figure 14-13. What is the relationship between the (010) and (100) planes; the (120) and (210) planes?

2. Could x-rays be used to investigate the crystal instead of microwaves? Could radiowaves be used? Explain fully.

3. What would be the shape of the relative intensity versus θ plot for a particular family of planes if the population of diffracting centers (balls) per plane is increased?

4. Since Miller indices containing a common factor are used to designate a higher order reflection from a particular plane, how would the second-order reflections from the (100) planes be represented? If a different wavelength of microwaves was used so that higher order reflections were possible from the (110) and (210) planes, how could the higher orders be specified in terms of Miller indices?

5. If a plane were to cut the y-axis on the negative side of the origin, how could this be shown using the Miller indices $(hk\ell)$?

6. List the different crystal systems and their characteristics. Make up a table which compares the properties of each system.

7. What is a unit cell? What are the dimensions of a unit cell for the following lattices?
   (a) simple cubic
   (b) face-centered cubic
   (c) body-centered cubic
   (d) hexagonal

8. Calculate the number(s) of atoms per unit cell for the following lattices:
   (a) simple cubic (b) face-centered cubic (c) body-centered cubic.

9. The general rule for the spacings in a simple cubic lattice is

$$d_{(hk\ell)} = \frac{a_o}{\sqrt{h^2 + k^2 + \ell^2}} .$$

   In this formula d is the interplanar spacing; a_o is the edge of the unit cell in our macroscopic lattice which is also equal to the interplanar spacing d_o between the (100) planes or 4.35 cm; $(hk\ell)$ are Miller indices. Use the Miller indices (110) and (210) and calculate the interplanar spacings d_1 and d_2. Compare the calculated values with those obtained from your plots of relative intensity versus θ. For extra credit derive the general rule for the spacings in a simple cubic lattice.

10. The wavelength of a reflex klystron is measured by precision methods and found to be $\lambda = 3.25 \pm 0.02$ cm. When the microwaves are used to analyze a crystal and the results plotted on a graph, second-order reflections are observed at an angle $\theta = 50 \pm 1°$. Find the maximum numerical error and the relative (per cent) error for the interplanar spacing d_o.

11. Calculate the lattice constant d_o for face-centered cubic aluminum.

12. Review your plots of relative intensity vs. θ for the (100), (110) and (210) families of planes. Compare the relative intensity of the reflected waves from each family of planes and discuss the reasons for any unexpected variations.

Experiment 14-5 ELECTRON DIFFRACTION

INTRODUCTION

In 1924 de Broglie[1] predicted that the wavelength of matter waves could be found by using the same relationship that held for light, namely $\lambda = h/p$, where λ is the wavelength of a light wave, h is Planck's constant and p is the momentum of the photons. De Broglie proposed that matter, as well as light, has a dual character behaving in some circumstances like particles and in others like waves. He suggested that a particle of matter having a momentum p would have an associated wavelength λ.

The first experimental evidence of the existence of matter waves was obtained by Davisson and Germer[2] in 1927. They "reflected" slow electrons from a single crystal of nickel and applied de Broglie's relationship (See Eq. 14-16). The wavelength of the electrons was determined and compared with that calculated from Bragg's[3] expression (See Eq. 14-17). Excellent agreement was obtained.

The wavelength of a beam of electrons, as indicated above, is given by:

$$\lambda = \frac{h}{p} = \frac{h}{mv}. \qquad 14\text{-}14$$

where mv is the momentum of a moving electron. The wavelength is therefore inversely proportional to the velocity v of the electrons. This velocity can be obtained directly from the accelerating potential V in the vacuum tube. Since the non-relativistic kinetic energy of the electrons is given by:

$$\frac{1}{2} mv^2 = eV \qquad 14\text{-}15$$

the wavelength

$$\lambda = \frac{h}{\sqrt{2meV}}$$

where m is the mass and e is the charge of the electron. When the values of h, m, and e are substituted, this becomes

$$\lambda = \sqrt{\frac{150}{V}} \qquad 14\text{-}16$$

if λ is expressed in angstroms (1 Å = 0.1 nm = 10^{-10} m) and V in volts. Since our apparatus is operated in general below 10 KV, electron energies are non-relativistic and no relativistic correction factor is needed.

[1] L. de Broglie, Phil. Mag. 47, 446 (1924).
[2] C. J. Davisson and L. H. Germer, Phys. Rev. 30, 705 (1927).
[3] W. L. Bragg, Proc. Camb. Phil. Soc. 17, 43 (1913).

The results of Thomson's[4] experiments with fast electrons supplied additional evidence about the behavior of electrons. Thomson analyzed photographically diffraction patterns produced by electron beams passing through thin films of gold, aluminum and other materials. From measurements of the size of the electron diffraction rings on a fluorescent screen, the wavelength of the electrons was found and again agreed with that predicted by de Broglie's equation. Other experimental work has shown that all particles have dual properties.

This experiment uses Thomson's method of transmitting electrons through a thin film of randomly oriented crystals to investigate a ring diffraction pattern with the aid of Bragg's law,

$$2d \sin \theta = n\lambda^*. \qquad 14\text{-}17$$

In this equation d is the separation between lattice planes, λ^* is the wavelength of the electrons, θ is the ordinary angle between the incident beam and the reflecting plane, which is the same as the angle between the reflected beam and the reflecting plane, and n is the order of the reflection. The angle between the direct and diffracted beams is 2θ, as shown in Fig. 14-16.

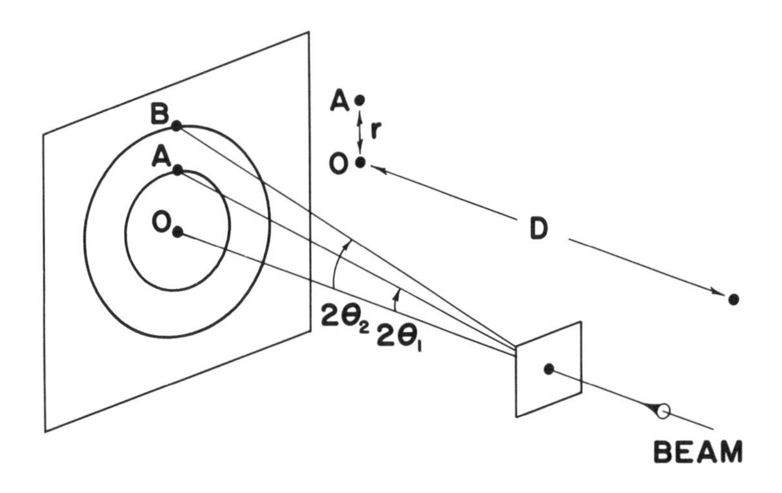

Fig. 14-16.
Ring Diffraction Pattern produced by the constructive interference of the electron waves diffracted from the various families of planes within the randomly oriented crystals in the thin film aluminum target.

1. $$2\theta = \frac{r}{D}$$

2. $$2a^* \sin\theta = (h^2 + k^2 + l^2)^{\frac{1}{2}} \lambda^* = 2a^*\theta = \frac{a^* r}{D}$$

3. $$\lambda^* = \frac{a^* r}{D} \frac{1}{(h^2 + k^2 + l^2)^{\frac{1}{2}}}$$

[4] G. P. Thomson, Proc. Roy. Soc. 117, 600(1928); 119, 652(1928).

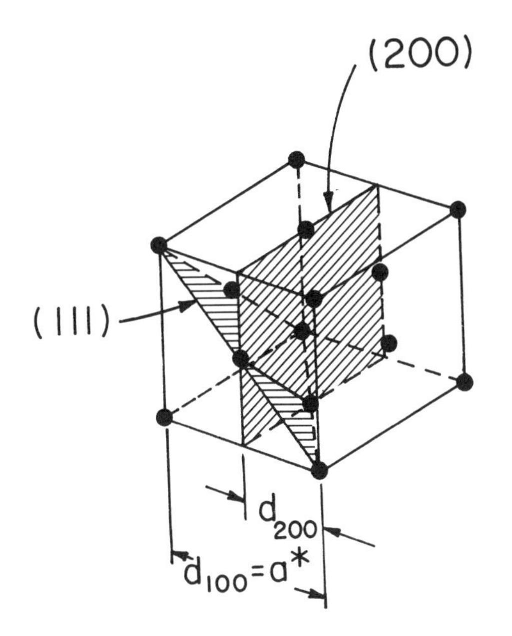

Fig. 14-17.
The Face-Centered Cubic Lattice of Aluminum. Two possible reflecting planes are shown. The separation between the (100) planes is called the lattice constant ($d_{100} = a^*$).

The separation d between reflection planes[5] for a face-centered cubic structure (FCC), such as aluminum, in terms of the Miller indices (HKL) is

$$d = \frac{a^*}{\sqrt{H^2 + K^2 + L^2}} , \tag{14-18}$$

where a* is the length of the edge of the unit cell (Fig. 14-17). When the Miller indices are multiplied by the order of the reflection n, any order n of Bragg reflection from planes (HKL) is considered to be first order Bragg reflection from planes [$hk\ell$]. Therefore,

$$d = \frac{na^*}{\sqrt{h^2 + k^2 + \ell^2}} , \tag{14-19}$$

where $nH = h$, $nK = k$ and $nL = \ell$. Substituting into Eq. 14-17, we obtain

$$\lambda^* = \frac{2a^* \sin\theta}{\sqrt{h^2 + k^2 + \ell^2}} . \tag{14-20}$$

When the angle θ in Eq. 14-20 is small, the $\sin\theta$ can be replaced with

$$\theta = \frac{r}{2D} ,$$

[5] G. P. Harnwell and J. J. Livingood, Experimental Atomic Physics, McGraw-Hill Book Co. (1961).

and Eq. 14-20 becomes

$$\lambda^* = \frac{a^* r}{D} \cdot \frac{1}{\sqrt{h^2 + k^2 + \ell^2}}, \qquad 14\text{-}21$$

where D is the distance from target to screen and r is the radii of rings. a* is known from x-ray measurements. D and r are obtained by direct measurements.

When comparing the wavelength calculated from de Broglie's relationship with that calculated from Bragg's expression, it is helpful to designate the two values as λ and λ^*, respectively, so that they can be tabulated without confusion. λ and λ^* represent the wavelength of the same electrons.

The observed diffraction pattern consisting of rings of various radii is produced by the constructive interference of the electron waves diffracted from the various families of planes within the randomly oriented crystals in the thin film target. The intensity of a reflection in the diffraction pattern is proportional to the square of the corresponding structure factor, i.e.,

$$I_{(hk\ell)} \propto [F_{(hk\ell)}]^2 .$$

For a face-centered cubic structure,[6]

$$F_{(hk\ell)} \backsim 1 + e^{i\pi(h+k)} + e^{i\pi(k+\ell)} + e^{i\pi(\ell+h)}. \qquad 14\text{-}22$$

The structure factor actually takes into consideration the coordinates and differences in scattering power of the individual atoms, the Miller indices $hk\ell$, and the addition of sine waves of different amplitude and phase but of the same wavelength. When squared, the structure factor vanishes unless h, k, ℓ are all odd or even, in which case $F \backsim 4$. (See Table I for allowed FCC reflections for aluminum.) Therefore, rings will occur for which h, k, ℓ are all odd or even.

TABLE I

Allowed FCC Reflections for Aluminum

$hk\ell$	$(h^2 + k^2 + \ell^2)$	$(h^2 + k^2 + \ell^2)^{\frac{1}{2}}$
111	3	1.732
200	4	2.000
220	8	2.828
311	11	3.316
222	12	3.464
400	16	4.000
331	19	4.358
420	20	4.472
422	24	4.898
511,333	27	5.196
440	32	5.656

[6] C. Kittel, Introduction to Solid State Physics, 6th, John Wiley and Sons, Inc. (1986).

Using methods first applied by Ewald[7] and von Laue[8], spot patterns from hexagonal pyrolytic graphite will be investigated. Measurements are made directly from the tube screen (see Fig. 14-18).

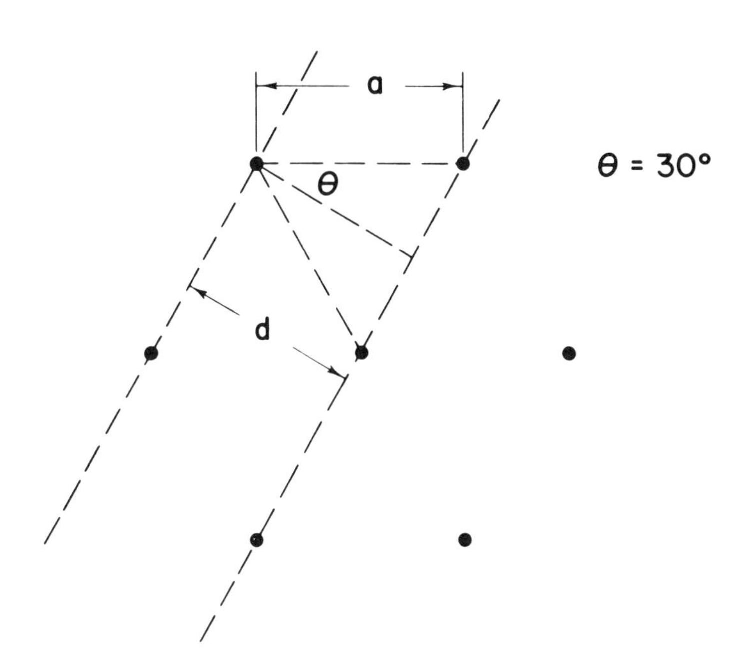

Fig. 14-18.
Analysis of Hexagonal Spot Diffraction Pattern. d is the separation between planes. The distance r from the central spot to each of the spots on the hexagon is measured. a is the calculated lattice constant which is compared to the known a* found by x-ray measurements. From the geometry it can be seen that $a = \lambda D/r \cos\theta$.

To obtain the separation d between planes[9], first calculate the wavelength λ of the electrons from de Broglie's relationship (Eq. 14-16) and substitute in Bragg's relationship, λ for λ*. For small angles the latter becomes

$$d = \frac{\lambda D}{r}, \qquad 14\text{-}22$$

where D is the distance from target to screen.

Once d is determined, the lattice constant a for hexagonal pyrolytic graphite can be calculated and compared with the known constant a* obtained from x-ray measurements. Conversely, given a*, the wavelength λ* can be calculated and compared with λ computed from the voltage.

[7] P. P. Ewald, Z. Krist. 56, 129(1921).

[8] M. von Laue, Ann. d. Phys. 29, 211, (1937).

[9] See G. Thomas, Transmission Electron Microscopy of Metals, John Wiley and Sons, New York, p.30 or Harnwell and Livingood, p.167.

OPERATING PROCEDURE

WARNING: Do not touch connecting wires at the rear of the power supply or tube. High voltage is dangerous. If for any reason the tube must be moved or wires disconnected, call your instructor. The tube is evacuated to a pressure of 10^{-8} mm Hg and should be handled with caution to avoid implosion.

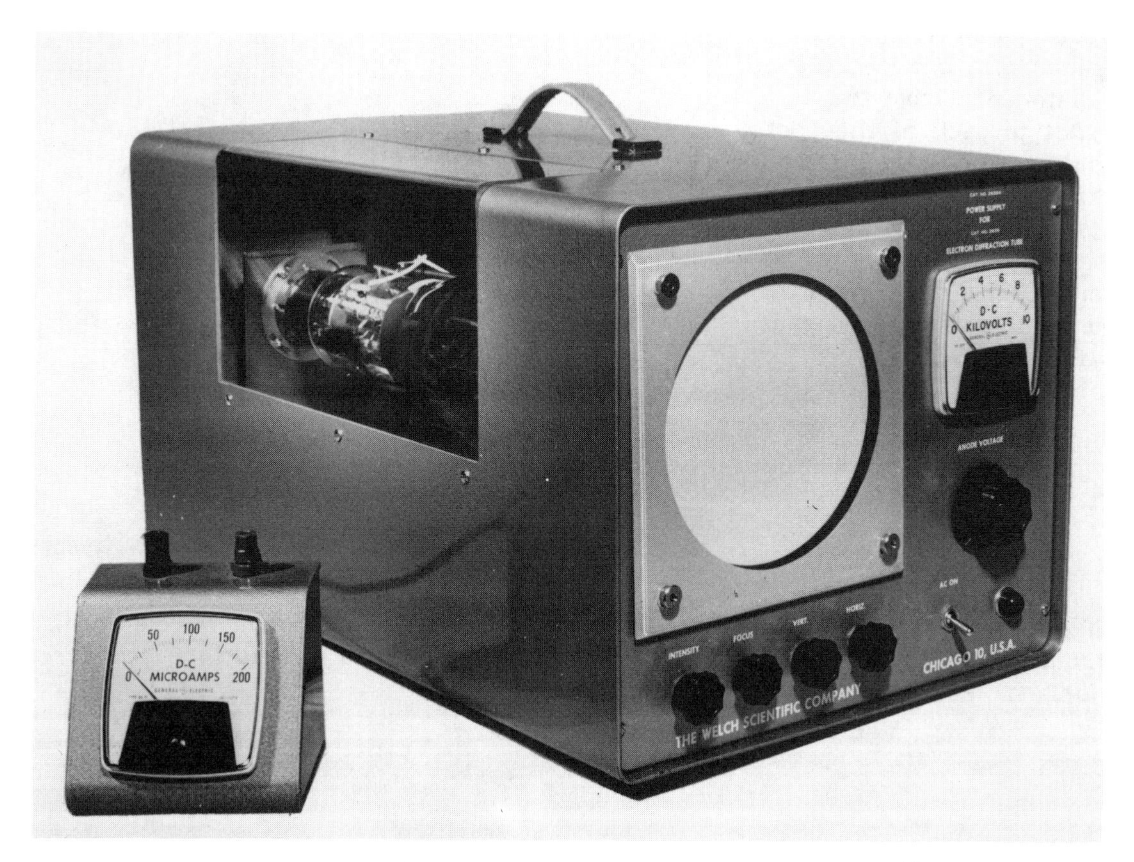

Fig. 14-19. Electron Diffraction Tube (PATENTED[10]).

1. A microammeter should be connected to the external meter jack on the power supply and used at all times to measure target currents. Target current is usually 5 to 10 microamperes and should not exceed 10 microamperes. Gun filament and target are designed to have a long life when operated under normal laboratory conditions with subdued illumination.

2. Before turning on the power supply, turn the intensity (bias) and voltage controls to "off". Turn the A.C. line switch on. Allow a few minutes for the power supply and the tube filament to warm up. Then turn the high voltage control to the desired value.

[10] PATENT assigned to the Sargent-Welch Scientific Co. by Professor Harry F. Meiners of Rensselaer Polytechnic Institute and Dr. Stanley A. Williams of Iowa State University, Ames, Iowa.

3. To avoid burning the phosphor screen, turn the intensity (or bias) control on slowly. Do not exceed current ratings. Do not at any time increase the intensity beyond the point where the power supply overloads and the high voltage meter dips. Watch the central beam spot for possible burning of the screen due to prolonged high intensity at one position. This will not occur if the centering controls are used to move the beam slowly while searching for diffraction patterns. If the suggested target current ratings are followed, the screen will not be burned when the beam is focused on a particular target.

4. Always adjust the focus control until the spot size on the screen is as small as possible for each selected high voltage reading.

5. Turn the intensity and voltage controls to the "off" position before switching off A.C. line voltage.

NOTE: Always work with the minimum current needed to give good patterns for the purpose desired. Keep the spot moving slowly while searching for a good target. When the beam is focused on a target, the intensity can be increased without damage to the screen. Correct the voltage after using the centering and focusing controls since there is some interaction between controls.

PART I - Analysis of the Electron Diffraction Pattern of Polycrystalline Aluminum

1. Look up the value of the lattice constant a^* which is known from x-ray measurements.

2. Using a series of voltages, measure the radii of all observed circular rings at each voltage. So that the effects of ring distortion may be minimized, obtain an average of four measurements of the radius of each ring by measuring across four different diameters. Check the voltage continuously while making measurements or correcting focus. A.C. line voltage may vary and affect meter readings. Record all your measurements, including the voltages.

3. Bring a small magnet near the face of the tube. The diffraction pattern is deflected with little distortion demonstrating that the pattern is formed by electrons and not electromagnetic radiation. Use the magnet carefully. If the electron beam is deflected greatly, internal arcing might result.

4. Apply Bragg's relationship and calculate the wavelength λ^* of the electrons for all permitted reflections (see Table I) at each voltage selection in Paragraph 2. For the same voltages compute from de Broglie's equation the wavelength λ. Record all reflections and the corresponding values of λ^* and λ.

   Since some allowed reflections cannot be observed because of their low intensity, check carefully for an obvious lack of agreement between λ^* and λ. For example, if the choice of the Miller indices for a particular measurement of the radius r of a ring pattern results in an unusual spread between λ^* and λ, then try other combinations of the indices until good agreement between λ^* and λ is obtained.

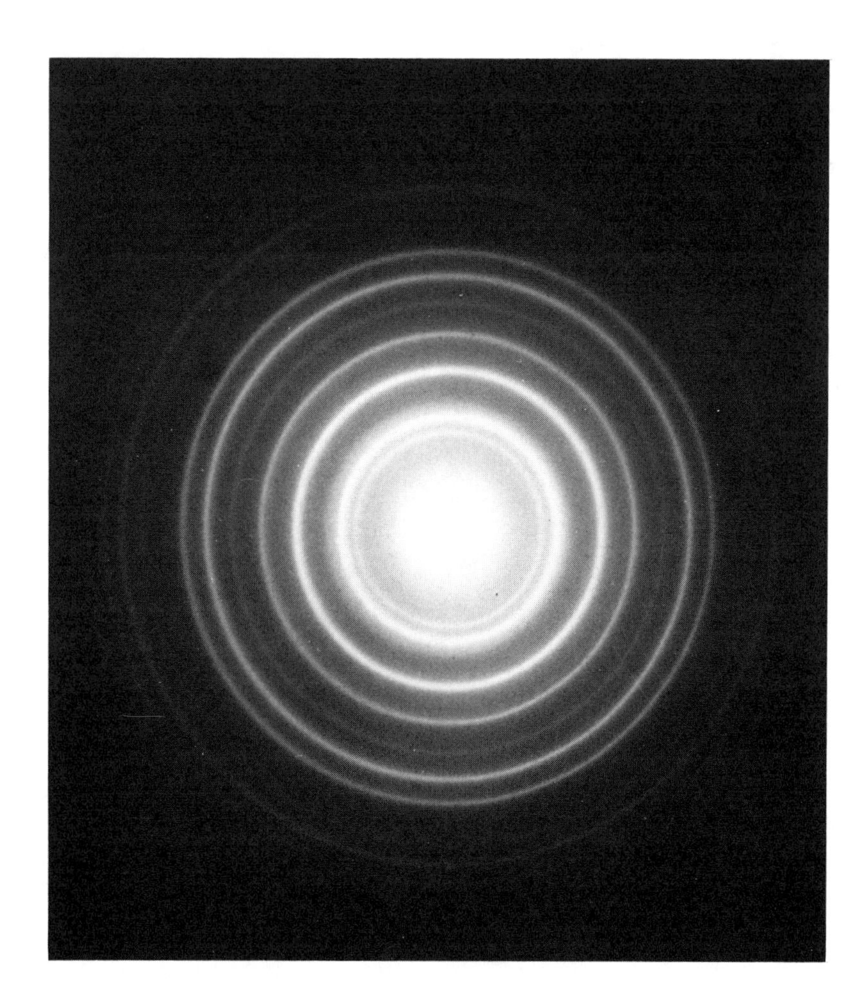

Fig. 14-20.
Diffraction Pattern due to Polycrystalline Aluminum.

5. Discuss your results. Explain why the ring diameters increase with decrease in voltage.

Suggested Data Table for Part I

Lattice constant a*, _________ from x-ray data.
Target to screen distance D = _________ marked on the tube.

Accelerating Potential (Volts)	Reflection Plane	$(h^2 + k^2 + \ell^2)^{\frac{1}{2}}$	$\bar{r}$ (cm)	λ (A)	λ*(A)	%
	111	1.732				
	200	2.000				
	220	2.828				
	311	3.316				

Note. Repeat the table for each voltage selected. Include under $\bar{r}$ the four measurements of the radius of each ring and the mean $\bar{r}$.

PART II - Calculation of the Lattice Constant of Aluminum

1. Determine the number of atoms associated with a unit cell for a face-centered cubic lattice. Use the molecular weight of aluminum M, its density ρ, Avogadro's number N_0, the number of atoms in the unit cell, and calculate the lattice constant a^+.

2. Substitute for λ* in Bragg's relationship, each λ obtained from de Broglie's relation for the various voltages used in Part I. Calculate the lattice constant a for the allowed combinations of reflections. Record all values of a and the mean $\bar{a}$. Also include the corresponding voltages.

3. Compare the known value of the lattice constant a*, obtained from x-ray measurements, with the mean $\bar{a}$, found from Bragg's relationship, and a^+, calculated above in Paragraph 1. Discuss.

Suggested Data Table for Part II

Lattice constant a^+ ________ directly from unit cell.
Lattice constant a* ________ from x-ray data.
Target to screen distance D ________.

Accelerating Potential (Volts)	Reflection Plane	$(h^2 + k^2 + \ell^2)^{\frac{1}{2}}$	r(cm)	λ(A)	a(A)
	111	1.732	1____ 3____ 2____ 4____ $\bar{r}$ = ________		
	200	2.000	1____ 3____ 2____ 4____ $\bar{r}$ = ________		
	220	2.828	1____ 3____ 2____ 4____ $\bar{r}$ = ________		
	311	3.316	1____ 3____ 2____ 4____ $\bar{r}$ = ________		

Note. Repeat table for each voltage selected.

PART III - Analysis of the Spot Electron Diffraction Patterns of Pyrolytic Graphite

1. Derive from Bragg's relationship an expression which can be used to calculate the wavelength λ^* of the electrons. The equation should include the known lattice constant a^* of hexagonal pyrolytic graphite which is 2.456 A obtained from x-ray measurements.

Fig. 14-21. Diffraction Pattern Due to Pyrolytic Graphite.

2. For a series of voltages measure the distance r from the central beam spot to each of the spots in the hexagonal diffraction pattern. Check the voltage continuously while making measurements or correcting focus. Record each measurement r and the mean $\bar{r}$ for the corresponding voltages.

3. Use the mean $\bar{r}$ of the spot measurements to calculate the wavelength λ^* of the electrons. Also compute the wavelength λ by applying de Broglie's relationship for each voltage. Record λ^* and λ.

4. Does the agreement between λ^* and λ fall within the limits of accuracy of the apparatus? Discuss.

Suggested Data Table for Part III

Lattice constant a*_________ from x-ray data.
Target to screen distance D _________.

Accelerating Potential (Volts)	r(cm)	d(a)	λ^*(A)	λ(A)	%
	1____ 4____ 2____ 5____ 3____ 6____ $\bar{r}$ = ______				

Note. Repeat table for each voltage selected.

PART IV - Calculation of the Lattice Constant of Pyrolytic Graphite

1. Substitute for the wavelength λ^* in Bragg's relationship the value of λ that was obtained from de Broglie's equation. Compute the lattice constant a. Do this for each voltage selected in Part III or for a series of voltages recommended by your instructor. If voltages other than those used in Part III are selected, measure the distance r from the central beam spot to each of the spots in the hexagonal diffraction pattern. Be sure to check the voltage continuously while making measurements or correcting the focus. Record each measurement r and the mean $\bar{r}$ for the corresponding voltages. Also record the calculated lattice constant a for each voltage.

2. Compare the value of the lattice constant a* known from x-ray measurements with the mean $\bar{a}$ of your calculated a values. Discuss your results.

Suggested Data Table for Part IV

Lattice constant a* _________ from x-ray data.
Target to screen distance D _____________.

Accelerating Potential (Volts)	r(cm)	d(A)	λ(A)	a(A)	%
	1____ 4____ 2____ 5____ 3____ 6____ $\bar{r}$ = ______				

Note. Repeat table for each voltage selected.

QUESTIONS

1. Prove that $\lambda(A) = \sqrt{\frac{150}{V}}$.

2. Would the above equation be valid for a 10-Mev electron beam? If not, why not? Derive an equation which would apply.

3. How is the wavelength related to the electron's kinetic energy?

4. Assume that the electron beam is replaced with a beam of positively charged particles. Could you carry out the same analysis if the particles were positrons? What would be the result if the particles were protons?

5. Plot a graph of λ^2 vs 1/V, where λ is in angstroms and V is in volts. Use the data obtained from de Broglie's relationship. What is the significance of the slope?

6. Derive an expression for the separation d between planes for a simple cubic lattice.

7. Why is it necessary for the inner wall of the tube to be covered almost completely with a conducting coating, i.e., graphite?

8. Why does the diffraction pattern consist of concentric rings? Would a single crystal produce the same pattern? Explain.

9. Why are the target, screen, and conducting coating on the inner wall of the tube connected to ground?

10. Assuming that Helmholtz coils are placed on each side of the tube, derive an expression based on pattern deflection which could be used to calculate the speed v of the electrons. Also derive in terms of the accelerating voltage an alternate equation for the speed v of the electrons. Could the ratio of e/m for the electrons be found in this manner? Explain.

11. Explain why an understanding of matter waves is important in the study of wave mechanics.

Experiment 14-6 ABSORPTION OF GAMMA AND BETA RAYS

INTRODUCTION

If we pass gamma-rays through matter, a certain number of the incoming photons will be absorbed and only a fraction of the initial radiation will pass through the absorber. There are three fundamental processes by which photons are absorbed: the photoelectric effect, the Compton effect, and pair production. (Strictly speaking the Compton effect is a scattering, not an absorption process.) All of the above processes are "one-shot" processes in the sense that if they occur a photon is removed from the incident direction. It is this property of these processes which accounts for the fact that photons do not have a well-defined "range" in matter. Photons obey an "Exponential Law of Absorption," which we now derive.

Let N be the number of photons that pass through a thickness X of absorber, and let N_0 be the initial number of photons hitting the absorber. The number of photons that are absorbed in an additional absorber thickness dx is -dN. We can write

$$dN = -\mu N\,dx \qquad 14\text{-}23$$

where μ is a constant depending on the material of the absorber and is called the "absorption coefficient." Re-arranging Eq. 14-23 and integrating, we have

$$\int_{N_0}^{N} \frac{dN}{N} = -\mu \int_{0}^{x} dx$$

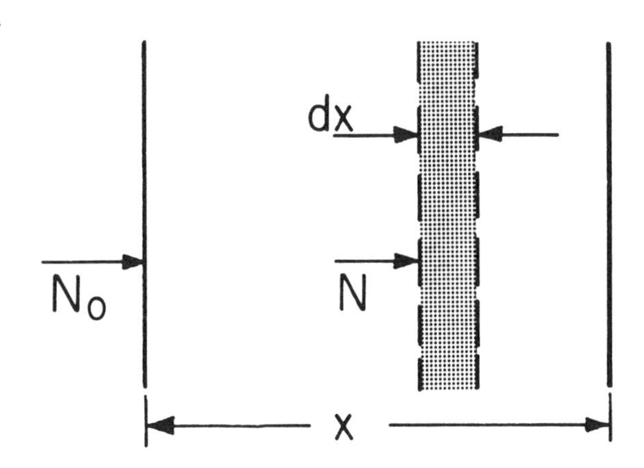

Fig. 14-22. Absorption of Gamma Rays.

or

$$\ln\left(\frac{N}{N_0}\right) = -\mu x$$

giving us

$$N = N_0\, e^{-\mu x}. \qquad 14\text{-}24$$

Eq. 14-24 applies not only to gamma rays, but to other kinds of radiation in certain situations.

Beta-rays are also absorbed when they are passed through matter, but the nature of the process is quite different from that of gamma-rays. Beta-rays do not obey an "Exponential Law of Absorption." This is because the processes involved in beta-ray absorption do not have the "one-shot" nature of the processes involved in gamma ray absorption. The processes involved in beta-ray absorption cause the electrons (or positrons) to lose energy in small amounts. Low energy beta-rays lose energy primarily by exciting or ionizing atoms; that is, by collision processes. High energy electrons can also lose energy through the mechanism of "Bremsstrahlung." The nature of these processes leads to beta-rays having a definite "range" in a particular material, which is proportional to the maximum energy of the beta-rays. This range can be measured by determining how much material (in our case aluminum) is necessary to stop all electrons. It must be remembered that even if all the beta-rays are stopped, gamma rays, which may be given off by the source, may still pass through the absorber.

EXPERIMENTAL ERRORS

The problem of experimental errors is somewhat different from the situations met in previous experiments. Radioactive decay is a pure chance phenomenon; the laws of probabilities therefore apply. If we put a radioactive source near a Geiger counter, a definite number of counts n will be recorded in a time t. If we repeat the experiment, we do not, in general find the same number n for the same time t. This indefiniteness in n comes from the random nature of the decay process. Obviously it is impossible to define a maximum error or error limit. All one can state is that a certain "safe error" is obtained only with a certain probability. There are different such "safe errors" used for different purposes. In the counting of nuclear events the "standard deviation" is commonly used. The probability is about 68% that the actual error does not exceed the standard deviation. The theory of probability shows that the standard deviation σ in our case is given by

$$\sigma = \sqrt{n} \,,$$

where n is the total number of counts.

PROCEDURE AND ANALYSIS

PART I - <u>Absorption of Gamma-Rays</u>

Before starting the experiment, read Section 4-11 on the operation of the nuclear detector. Take a five-minute background reading.

Mount the gamma-ray source in the source holder and take a one-minute reading with no absorber between the source and the counter. Take eight to ten readings for different thicknesses of aluminum between source and counter and repeat the same procedure with a different absorber, such as lead. Use a micrometer to measure the thickness of the aluminum and lead plates.

Plot the data on semi-logarithmic graph paper. Indicate the standard deviations on the plots. From the curves determine the absorption coefficients of aluminum and lead and compare the two.

PART II - <u>Absorption of Beta-Rays</u>

Before starting the experiment, read Section 4-11 on the operation of the nuclear detector. Take a five-minute background reading. Mount the beta-ray source in the source holder and take a one-minute reading with no absorber between the source and the counter. Put five sheets of aluminum foil on the tray in the source holder, and take another one-minute count. Keep on adding five sheets at a time, each time taking a one-minute count, until the readings do not change appreciably. Take about five counts after the readings have leveled off.

Since the aluminum foil is too thin to measure its thickness accurately with a micrometer, weigh a piece on the analytical balance, and measure its area. The mass per unit area (mg/cm^2) gives the areal density, which is very often used instead of the thickness.

Plot the log of the counting rate vs. areal thickness of aluminum in mg/cm^2; (Use semi-logarithmic graph paper.) From this plot determine the range of the beta-rays, or the thickness of aluminum (in mg/cm^2) that will absorb all electrons. The counter wall itself absorbs electrons; therefore its thickness[1] should be added to the absorber thickness determined from the graph. From the range of the electrons in aluminum the maximum energy of the beta-rays can be determined by using the empirical relation

$$R = 0.542\, E_m - 0.133 ,$$

where R is the total range in g/cm^2, and E_m is the maximum energy of the electrons in Mev (million electron volts). Discuss the shape of the curve as well as the result for the maximum range of beta-rays. This value may be compared with those given on charts or in tables for the isotope used.

QUESTIONS

1. Distinguish between alpha, beta and gamma rays.

2. Define the term cross section. How is the total cross section for photoprocesses related to the absorption coefficient for gamma rays?

3. How is the mean free path for a photon related to the absorption coefficient for gamma rays? (Hint: Calculate the average distance a photon will travel in the material.)

4. Discuss the physical significance of the absorption coefficient for gamma rays.

5. Compare the absorption coefficients for gamma rays found in the experiment with those given in the tables and discuss some of the reasons for any discrepancies.

6. For the given source of gamma rays, determine the total cross section per atom for lead and aluminum. Do you expect these cross sections to be energy dependent? Explain.

7. Determine the mean free paths for the given gamma rays in lead and aluminum. Do you expect these quantities to be energy dependent? Explain.

8. Describe the photo-processes which make up the total absorption coefficient for gamma rays. Sketch the absorption coefficient as a function of photon energy. Indicate the energy range over which each process is predominant.

9. Discuss why gamma rays are monenergetic and beta rays in general have a continuous energy distribution.

[1]This value will vary from a few mg/cm^2 for end-window counters to 30 mg/cm^2 for side-wall G-M tubes. The specifications of the manufacturer should be consulted.

10. Use an isotope chart or tables to list all properties of the radioactive isotope used in Part I in the experiment, such as the half-life, the particles given off, their energies, etc.

11. Why is semi-logarithmic graph paper used to plot the results of this experiment? Show how the curves in Part I would look on ordinary graph paper. Do the same for the curves in Part II.

12. Discuss at least one industrial application of the absorption of gamma rays.

13. How many counts have to be taken to keep the standard deviation in Part I below 10%? Below 1%? Do the same for Part II.

14. Explain why elements with high Z (atomic numbers) make the best γ-ray shields.

15. In Part II the thickness of the aluminum foil was found in grams per square centimeter. How much would this be if expressed in millimeters? Also express the thickness of aluminum foil in atoms/cm^2.

16. What radioactive isotope was used in Part II of this experiment? Use the isotope chart to look up the energy of all beta and gamma rays emitted by this isotope. What other information can be obtained from the chart regarding the isotope used?

17. Discuss why the range for electrons is less precisely defined than for heavier charged particles.

18. (a) Explain why the range is nearly independent of the nature of the absorber if the amount of absorber is expressed as the weight per unit area.
    (b) Explain why the ability of an element to stop β particles depends predominately on the Z/A ratio (atomic number/mass number).

19. When plotted on semi-logarithmic graph paper, the absorption curve for gamma-rays should be a straight line, while that for beta-rays does not come out to be a straight line. Discuss the reason for this difference between beta- and gamma-rays.

20. Discuss a few possible applications making use of the absorption characteristics of beta-rays.

21. Discuss the role played by the antineutrino in beta-decay.

22. Write the equations for the nuclear reactions for the radioactive sources used in the experiment.

Experiment 14-7 HALF-LIFE OF RADIOACTIVE SOURCES

INTRODUCTION

The "activity" of any radioactive source decreases with time. In some isotopes the activity falls off very rapidly within a few seconds, while in others it decreases very slowly over a period of years. From experimental evidence we know that for a given radioactive material the number of disintegrations per unit time, DN/dt, depends only on the total number N of radioactive atoms present in the sample, and is proportional to this number. We can therefore write

$$\frac{dN}{dt} = -\lambda N \qquad 14\text{-}25$$

where λ is called the decay constant and depends on the material. The negative sign is introduced because dN/dt is the rate at which N is decreasing. If we let N be the number of atoms initially present when t = 0, N the number remaining at time t, we have

$$\int_{N_o}^{N} \frac{dN}{N} = -\int_0^t \lambda\, dt,$$

$$\ln \frac{N}{N_o} = -\lambda t,$$

and

$$N = N_o e^{-\lambda t}. \qquad 14\text{-}26$$

The half-life T is defined as the time required for the activity of a source to decrease to one-half of its initial value. Substituting this half-life T in Equation 14-26 we find that

$$\frac{1}{2} N_o = N_o e^{-\lambda T},$$

$$2 = e^{\lambda T},$$

$$\ln 2 = \lambda T,$$

$$\lambda T = 0.693,$$

and

$$T = \frac{0.693}{\lambda}. \qquad 14\text{-}27$$

Equation 14-27 gives a relation between the half-life and the decay constant. Experimentally the half-life time can be determined very easily by determining how long it takes the counting rate to drop to one-half its value. The decay constant can then be determined from Equation 14-27. It can also be found from a plot of ln N vs. t.

The actual number of surviving atoms N cannot be determined by direct observation. What one actually observes is the activity of the sample. The activity is the number of decays per unit time, dN/dt. As indicated in Equation 14-25, the activity is directly proportional to the number of surviving atoms, and can therefore be used as a measure of N.

PART A - Half-life of an Artificial Radiation Source Using a G-M Counter and a Scalar[1]

INTRODUCTION

In this part of the experiment we determine the half-life of a radioactive source by determining how long it takes the counting rate to drop to one-half its initial value. The decay constant can then be found from Equation 14-27 or from a plot of ln N vs. t.

As a source we use an artificial radioactive substance with a short half-life. There are many different possibilities. A sheet of indium or a piece of silver, for instance, may be placed next to a neutron source to induce radioactivity by neutron bombardment. Different neutron sources are available for this purpose, such as a polonium-beryllium source, in which alpha particles given off by polonium cause a nuclear reaction in the beryllium resulting in neutrons being emitted.

PROCEDURE AND ANALYSIS

Before starting the experiment, read Section 4-11 on the operation of the counter. Take a five-minute background reading. Mount the radioactive source in the source holder and measure counting rates at regular intervals. The time intervals to be used depend on the half-life of the source. For indium, for instance, 1-minute counts every five minutes are suggested.

Make certain to subtract the background reading before plotting the log of the counting rate as a function of time. Use semi-logarithmic graph paper. From this plot determine the half-life time and the decay constant of the radioactive substance.

PART B - Half-life of a Radioactive Source Using a G-M Tube and a Microcomputer[2]

INTRODUCTION

There are two major points we would like to make in this experiment: (1) is the fact that collecting more information will give better precision and (2) is the experimental result that radioactive decay follows an exponential function (see Fig. 14-23).

---

[1] See Section 4-11 for a description of a G-M tube and scalar.

[2] These directions are written for use with the microprocessor and its interface designed under the supervision of Dr. Thomas Shannon under a grant from the RCA Corporation. A detailed description may be obtained from Professor Walter Eppenstein, Physics Department, Rensselaer Polytechnic Institute, Troy, N.Y. 12180-3590.

Fig. 14-23. Using a Geiger-Mueller tube (lower right) with a Microcomputer Taking the Place of a Scalar.

Radioactive decay is a random process which means that we cannot predict when a particular atom will decay or even when we will get a decay from a large number of atoms. But we can measure the time at which the decay rate (number of counts/time) has decreased to one-half its original value. If this time (half-life) is very long compared to the time that you are making measurements, then you can assume that the decay rate is essentially constant. In the first part of this experiment you will measure such a "constant" decay rate. The radioactive material that you are given is Cesium 137. It decays to Barium 137 with a half-life of about 30 years (very long compared to a laboratory period).

The Barium 137 decays with a half-life of a couple of minutes (short compared to a laboratory period). In the second part of this experiment you will observe the decay rate of Barium 137 as a function of time and measure its half-time. With any exponential function there is always a time constant τ associated, the time to go to 1/e of its initial (or final) value. However, historically the half-life $T_{\frac{1}{2}}$ (time to go to $\frac{1}{2}$ of the initial value) has been used for radioactive decay. These two times are related by $T_{\frac{1}{2}} = .693\tau$.

PROCEDURE

1. Run the program with no sample using 1 sec counting intervals. Note that there is a "back-ground count." We will not use this in our measurements; however, remember that it is always there. Put the plastic tray in the third pair of slots from the bottom. Place the blue plastic cylinder containing the Cesium 137 onto

the tray. Run using 1 sec counting intervals. It can be shown that the uncertainty for such a measurement is $\sqrt{N}$, where N is the number of counts. Thus if you got 36 counts, the rate would be reported as 36 ± 6 counts/sec. Plot in your lab book the first 10 readings. Calculate the uncertainty for each and put error bars on each point to show this uncertainty. Also calculate the average of the ten values and draw a horizontal line on your plot representing this average. Note that approximately 7 of the 10 values should include the average value within their error bars. Calculate a rough percent error for these measurements (e.g., 36 ± 6 has a percent error of 6/36 × 100 = 17%). Now repeat the same procedure except this time using 100 sec counting interval. Since the rate is the same, the count should be approximately 100 times the previous value and since we have collected more information (at the expense of time) the percent error should be smaller (about one tenth).

2. In this part we will separate the Barium 137 from the Cesium 137 by means of very dilute HCl so that we may observe the decay of Barium 137. Remove the caps from the blue plastic cylinder and from the spout of the plastic bottle. Place the tip of the bottle firmly into the larger opening of the cylinder. The tip will "snap" into place. Holding the smaller opening over the metal plachet gently squeeze the plastic bottle to slowly collect about 6 drops of liquid in the plachet. Place the plachet on the plastic tray which should be in the topmost position on the counter. Run the program using an appropriate counting time interval. (Remember the half-life of Barium 137 is a couple of minutes and you can collect 32 data points.) At the conclusion of the run plot 10 of the data points on semi-log paper and find the slope (τ). <u>Caution</u>: You want $[\ln \text{Data}_1 - \ln \text{Data}_2]/T_1 - T_2$, not $[\text{Data}_1 - \text{Data}_2]/T_1 - T_2$. Multiply by .693 to get half-life $T_{\frac{1}{2}}$. Finally put error bars on the first and last data points that you plotted.

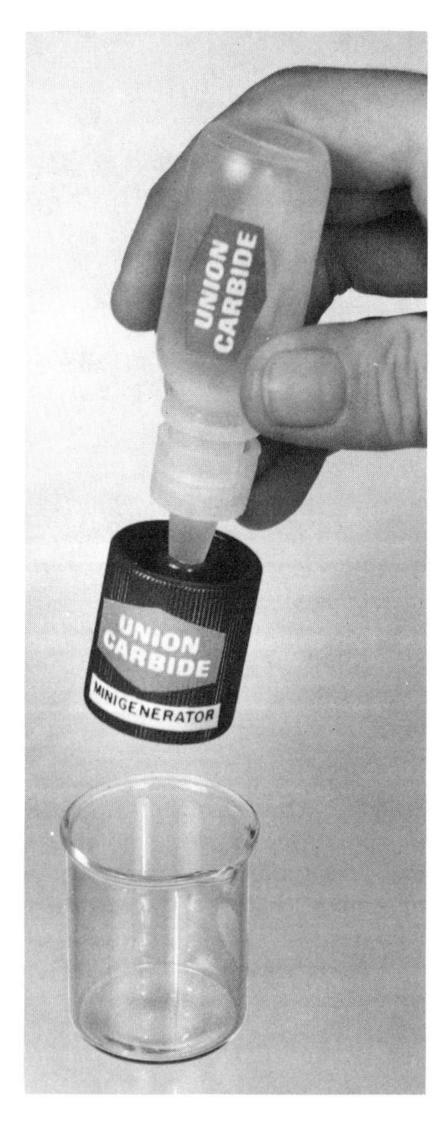

Fig. 14-24.
Eluting of the Minigenerator. (Courtesy of the Sargent-Welch Scientific Company)

PART C - Using the Emanation Electroscope

INTRODUCTION

In this part of the experiment we measure the activity in terms of the rate of discharge of an electroscope.

The emanation electroscope consists of three aluminum chambers mounted on a wooden base as shown in Fig. 14-25. In the bottom chamber are the storage bottles B containing the radioactive material (in this case thorium oxide) together with the necessary filtering and drying materials. The top chamber is the electroscope. The aluminum leaf L of the electroscope is mounted on an aluminum stem which extends downward into the middle chamber and is supported by a large polystyrene disk D. A contact button C and binding post on top of the instrument connect the leaf to the charging potential. The middle chamber is the ionization chamber into which the radioactive gas is pumped.

A small measuring microscope is attached to the top to measure the rate of discharge of the leaf. An electrostatic charger E is attached to the base. This may be used to charge the leaf to the desired potential.

The radioactive gas (an isotope of radon ^{220}Rn) is pumped into the ionization chamber from the storage bottles by means of a double-valved rubber bulb R. The radon 220 is obtained after a series of radioactive decays from the thorium ^{232}Th in thorium oxide.

ADJUSTMENTS

Charge the leaf with the charger as follows: Place the two-piece charging rod in the metal sleeve with the metal end down. Move the rod up and down several times. With the rod down depress the contact button on the top chamber and raise the charging rod slowly. The leaf should deflect. When the leaf is deflected, release the button and lower the rod. The leaf should remain deflected. If it does not, there is electrical leakage which must be eliminated. To do this, it is usually sufficient to clean with alcohol the polystyrene disk which supports the electroscope. The chamber may also have been contaminated with radioactive material. Consult your instructor.

Move the eyepiece of the microscope in or out until the graduated reticle is in sharp focus. Move the microscope in or out until the nearer edge of the leaf is in focus. The microscope mount may be rotated about a vertical axis to position the image of the leaf properly with respect to the graduations and the microscope may be rotated about its own axis to make the graduations parallel to the edge of the leaf.

Charge and discharge the leaf a number of times while viewing it through the microscope. Its movement should be smooth and without jerks, the latter being due to kinks or bends in the leaf. The edge of the leaf should be sharp and distinct so that its position is easily determined on the graduated reticle.

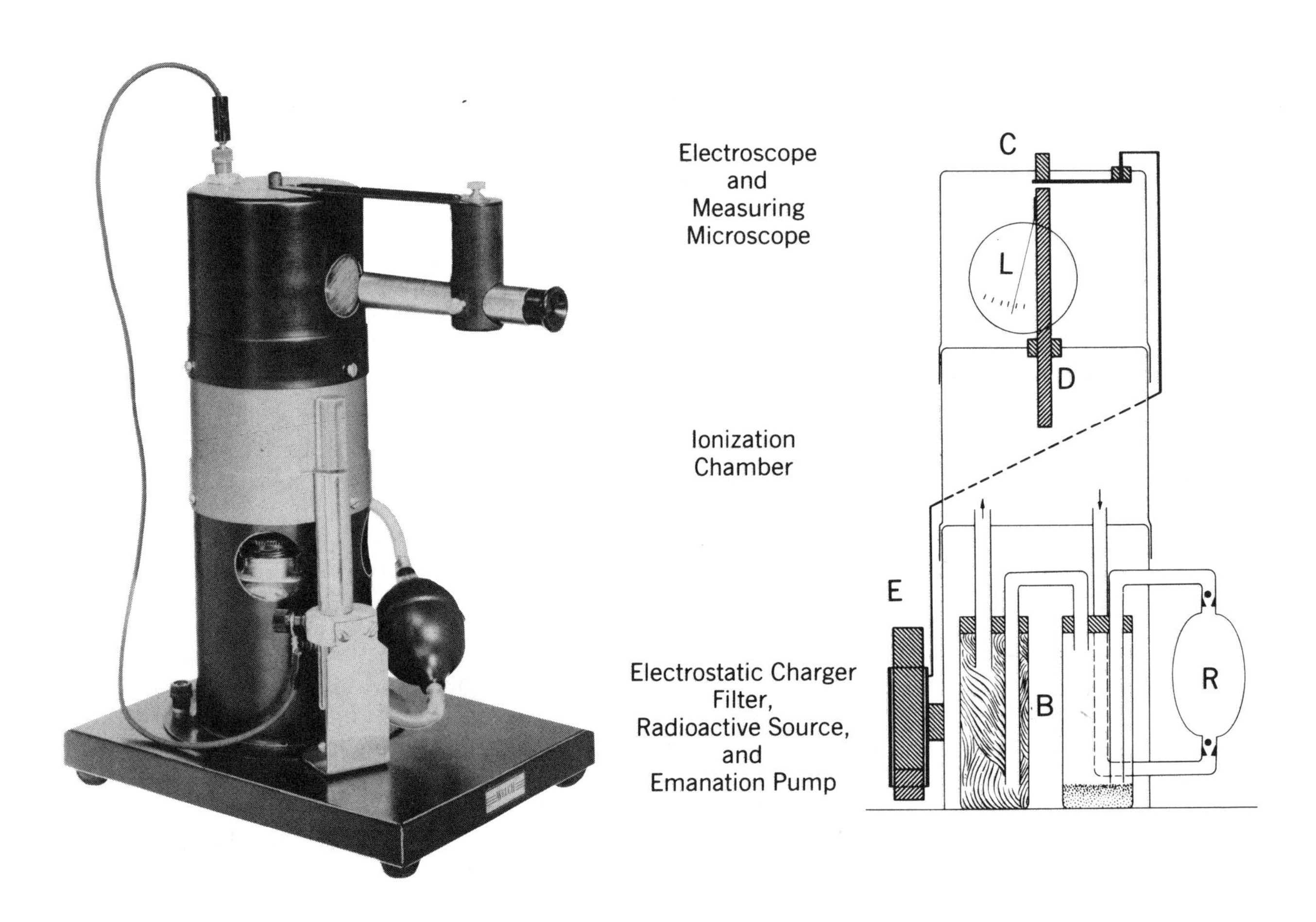

Fig. 14-25. The Emanation Electroscope.
(Courtesy of The Sargent-Welch Scientific Co.)

PROCEDURE AND ANALYSIS

In this experiment we measure the interval of time t for the leaf to fall the distance between any two convenient lines on the reticle. The rate of discharge 1/t is a measure of the activity. The radiation from the decay products of radon 220 may be neglected in this experiment.

Since there is always some residual activity in the ionizing chamber, this activity should be measured. First, charge the leaf without introducing any gas and measure the time required for it to fall the given distance. Call this time t_b. Then $1/t_b$ is a measure of background activity, and also any electrical leakage due to imperfect insulation. If t_b is less than 250 seconds, subtract $1/t_b$ from each subsequently observed value of activity, 1/t.

Data is most easily obtained by two students working together. One student should observe the leaf and measure the time of discharge for a predetermined distance. The partner should use another timer (a wall clock will do) to measure the total running and should record all the data.

First make a trial run to develop technique. Pump a charge of gas into the chamber, charge the leaf, and measure the time of discharge t_1. Without pausing or pressing the bulb again, measure t_2, t_3, etc., until t equals about 100 sec. (Charge the leaf between each measurement.) The partner should record these values of t, and at the same time observe and record the running times s_1, s_2, s_3, etc. After developing the technique, make a serious run and record the data. Data and results should be recorded in tabular form.

Plot a graph of 1/t against (s + t/2) on both rectangular and semi-log graph paper. Calculate the half-life T. Compute λ. Compare T and λ with published values and discuss possible sources of errors.

QUESTIONS

1. Are alpha, beta and gamma rays equally effective in ionizing a gas? Discuss. Can neutrons ionize a gas? Explain.

2. How can you determine the type of particle given off in a nuclear decay? As specific examples consider alpha-particles and neutrons.

3. How are the mean life, half-life, and disintegration constant related? Discuss the physical significance of each.

4. Why is the activity defined as the negative time derivative of the number of atoms surviving?

5. Define the term "cross section" as used in connection with nuclear reactions.

6. Write the equation for the nuclear reaction which produces the neutrons in the neutron source, such as a polonium-beryllium source.

7. What determines the half-life of the neutron source? Use an isotope chart or tables to determine the half-life of the polonium-beryllium source.

8. Compare the half-life found for indium or silver with the corresponding value given in charts or tables. Discuss any differences.

9. Do indium or silver have any half-lives? If so, why were they not detected?

10. Write a nuclear disintegration equation for the indium or silver as it decays.

11. What is a curie? Express the activities measured in Part I of the experiment in curies.

12. If a radionuclide "cow" was used, write all nuclear reactions associated with the parent and daughter substances.

13. How does the electroscope in Part II work? Must it be modified to determine the rate of beta decay? Gamma decay? How?

14. Using the chart of the nuclides, determine the decay scheme from ^{232}Th to ^{220}Rn. Indicate the half-life for each decay.

15. To what natural radioactive series does the decay scheme in 14 belong?

Experiment 14-8 NUCLEAR AND HIGH ENERGY PARTICLES

PART A - The Diffusion Cloud Chamber

INTRODUCTION

The cloud chamber is one of the earliest, simplest and most useful detecting instruments in nuclear physics. It is based on the behavior of supersaturated vapor. An ionizing particle passing through such a supersaturated vapor leaves a trail of ions along its path about which condensation takes place. The tracks of the liquid droplets may easily be seen and photographed.

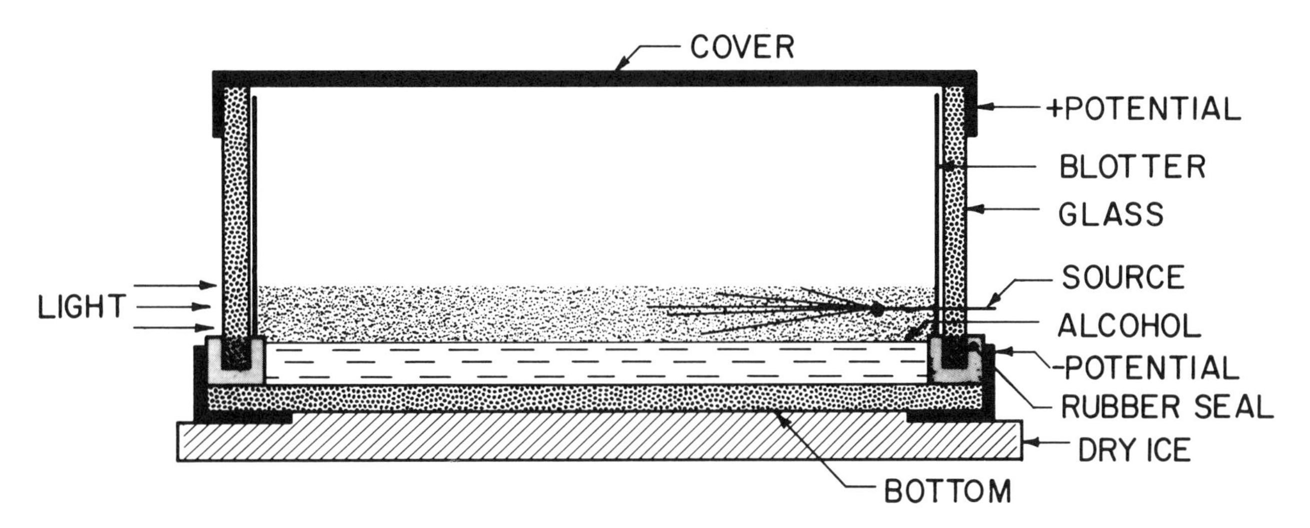

Fig. 14-26. Cross-section of a Diffusion Cloud Chamber.

In the diffusion cloud chamber the bottom of the container (a glass or plastic cylinder on a shallow metal pan) is maintained at a considerable lower temperature than the top. This is usually accomplished by placing the cloud chamber on a piece of dry ice (solid CO_2). The bottom pan contains some alcohol and a blotter around the inside wall of the cylinder rests with its lower edge in the alcohol. Evaporated alcohol around the warm upper edge of the blotter mixes with the air. As the vapor diffuses downward, it is cooled and becomes supersaturated. The chamber is continuously sensitive in the region of the supersaturated vapor and will show tracks as small droplets form on the ions present.

A potential difference between the bottom and the top of the chamber will clear the field of ions so that the newly formed tracks are easily visible. A strong light source is used through the side of the cylinder; an opening for the light has to be left in the blotter.

OPERATION OF THE CLOUD CHAMBER

Fill the bottom of the chamber with methyl alcohol containing a black dye to a depth of about 1/8 of an inch. For proper viewing, alcohol must be jet black. It is only against a perfectly black background that the white tracks really stand out.

Make certain that the level of the alcohol in the chamber is high enough to touch the bottom edge of the black blotter which lines the inside of the cloud chamber cylinder. The alcohol will then rise by capillary action until the entire blotter is saturated. At this time it may be necessary to add a little more alcohol to maintain the proper level of alcohol in the bottom tray. The optimum depth is 1/8 of an inch. Too much alcohol acts as a heat insulator and tends to impair cloud chamber operation. Too little alcohol will weaken the black background.

Once the blackened alcohol is at the proper level in the chamber, the next step is to replace the cover glass and fit the aluminum top ring upon it. Be sure it fits snugly because the colloidal silver electrode must touch the aluminum top ring for proper operation of the clearing field.

The chamber is now ready to be placed on dry ice. A slab about 1 inch thick and 7 inches square is about the right size. In any case, the slab should be large enough for the entire bottom area of the cloud chamber to rest upon it evenly.

The radioactive alpha and beta sources are identified by name and mounted on the heads of needles which are embedded in rubber stoppers. Insert the sources in the hole in the side of the cloud chamber, pushing the rubber stopper tightly into the hole to prevent air leaks. (Air leaks will set up convection currents that are extremely detrimental to cloud chamber operation.) Sources may be interchanged at any time. A plain rubber stopper should be used for observation of cosmic ray tracks (CAUTION: alpha source contains radium; beta source contains strontium 90. Both sources contain such small amounts of radioactive material that they are absolutely safe. After direct contact, however, wash your hands.)

A 45-volt battery is provided for use as a clearing field power supply. The purpose of the clearing field is to remove as many stray ions as possible from the sensitive volume in which the tracks appear. Without a clearing field these ions will drift to the bottom, forming droplets. Too many of these droplets in the cloud chamber result in a fogginess which not only tends to obscure tracks, but diminishes the amount of vapor available for track formation.

The positive terminal of the battery leads should be attached to the bottom tray (which serves as the bottom electrode). The negative terminal should be attached to the top ring which makes contact with the electrode fastened to the underside of the cover glass. Thus, positive ions are attracted to the top and negative ions and electrons are attracted to the bottom. A switch is provided to turn the clearning field off and on. It is not necessary to keep the clearing field on all the time. In fact, some of the best viewing comes during the few seconds after turning it off.

The light source is placed about 6" from the open space in the cloud chamber side wall and directed so that it skims the top of the alcohol surface in the cloud chamber. With a very strong light source the tracks may be viewed directly from above at an angle of 90° to the direction of the light. This is highly desirable for good, convenient viewing. With a weaker light source it may be necessary to direct your gaze toward the light at an angle of about 45° to its direction.

It takes about 5 minutes for the cloud chamber to cool sufficiently for proper operation after it has been placed upon the dry ice. However, almost from the start a light precipitation of droplets is to be observed. Gradually this precipitation begins to subside and, if you observe the bottom area very carefully, cosmic ray tracks will begin to appear. If the radioactive alpha and beta sources are now inserted, alpha and beta tracks will appear.

After the chamber has been in operation for about half an hour, the cover glass will become so cold that water from the air will condense on it. This may be removed by wiping with a clean handkerchief. An alternate method is to remove the cover glass and rinse in warm tap water. Then dry carefully, replace and use as before. Whenever the cover glass is removed, however, the operation of the chamber will be interrupted for about 10 minutes. The impure air of the room will enter the chamber and it will require a little time for the condensation nuclei (dust particles, etc.) to be precipitated out.

When you are through with the cloud chamber for the day, disassemble it and rinse it with tap water. It is best to disassemble it over a sink, since a little of the blackened alcohol is usually spilled. The black alcohol should be saved. Never leave alcohol in the bottom tray as it deteriorates the metal.

Any foreign matter - grease, dirt, odorous substances - may prevent the cloud chamber from operating satisfactorily.

PROCEDURE AND ANALYSIS

Put the cloud chamber into operation as explained above. Become familiar with the operation of the chamber. Observe the tracks with respect to (a) cosmic rays, (b) beta particles, and (c) alpha particles. For each case make a sketch of the appearance of the tracks. Compare the tracks. You should use a celluloid ruler and measure the approximate range of the alpha particles. Ranges will seem to vary; actually all the alphas have the same energy, but some are emitted from underneath the surface of the source and must expend some of their energy in order to reach the surface.

Knowing the range of alpha particles in air, we can compute the approximate half-life from the Geiger-Nuttall Law:

$$\log_{10} R = 0.707 + 0.0174 \log_{10} \lambda$$

where R = range, in air, in cm

λ = decay constant in $\sec^{-1} = \dfrac{.693}{T_{\frac{1}{2}}}$

$T_{\frac{1}{2}}$ = half-life

The alpha source is a ^{226}Ra compound. Compare your computed half-life with its true value.

If a magnet is available, place the cloud chamber into a magnetic field. Compare the curvature of alpha and beta particles. Try to estimate the radii of curvature. What information can you obtain if the magnetic field strength of the magnet is known? Check on the direction of curvature to find out if the particles are positively or negatively charged.

PART B - Elementary Particles

INTRODUCTION

It is not possible to go into the various aspects of high energy physics and the analysis of nuclear tracks in this experiment. A brief summary of some of the ideas is given below. It is assumed that students performing this experiment have consulted some of the available references and have a copy of the directions supplied with the bubble chamber pictures used in this experiment.

A number of different sets of bubble chamber photos are available commercially[1] and some have been published in journals[2]; each contains detailed information.

ELEMENTARY PARTICLES

The term "elementary particles" refers to those particles which cannot at this time be described as composites of other particles. A large number are known today, and most of them are unstable; that is, they spontaneously decay into two or more other particles. The more interesting particles are relatively stable, with an average lifetime of 10^{-16} sec or more. There are about thirty of these, and a list of their properties can be found in many books. Other particles decay in about 10^{-23} sec and are often called resonances to distinguish them from their more stable companions.

Some of the elementary particles are well known, such as electrons, protons, and neutrons. Others, positrons and neutrinos, for instance, are emitted by nuclei in radioactive decay processes. Pions and other mesons seem to be the carriers of the strong force that holds protons and neutrons together in atomic nuclei. Muons and a few other particles were first observed in cosmic rays, but the remaining particles on the list were not observed until the advent of high energy accelerators in the 1950's. It is now customary to group the elementary particles into four categories, in order of increasing mass. These are: photons, leptons, mesons, and baryons.

[1] "Analysis of Bubble Chamber Photographs," 38 different slides selected from exposures at Brookhaven National Laboratory, distributed together with projection apparatus and instructions by The Sargent-Welch Scientific Company.

[2] H. Whiteside, J. N. Palmier, and R. A. Burnstein, "An Experiment in Elementary Particle Physics for the Elementary Laboratory," includes two large and four small bubble chamber pictures and a curvature template picture, Am. J. Phys. 34, 1005 (1966).

THE BUBBLE CHAMBER

The bubble chamber, invented in 1952 by D. H. Glaser, has become one of the most valuable instruments for studying elementary particles and high energy events. In the cloud chamber used in Part A, fogdrops form on the ions produced by charged particles traveling through the gas-filled chamber. In the bubble chamber, the ions formed by charged particles travel through a liquid and form local heat centers in which tiny gas bubbles develop and grow. A hydrogen bubble chamber, is a tank filled with liquid hydrogen under a pressure of several atmospheres. If the pressure is suddenly released, the hydrogen will begin to boil, with the first bubbles forming around any ions in the liquid. Since a charged particle passing through any material ionizes some of the molecules along its path, a reduction of the chamber pressure immediately after the passage of a charged particle will produce a trail of bubbles along its trajectory. If the interior of the bubble chamber is photographed immediately from two different angles, the trajectory of the particle in three dimensions can be determined. Uncharged particles produce no ions and cannot be observed in this way.

A bubble chamber is normally operated in a uniform magnetic field so that the momenta of charged particles can be determined from the curvature of their tracks. Knowledge of particle speed can be obtained from the density of bubbles along the track, since there is a monotonic relation between these quantities for hydrogen in the energy region where a bubble density can be measured.

RELATIVISTIC MECHANICS

Because of the high energy and velocity of the particles involved, relativistic equations have to be used in all calculations. Derivation of the relevant equations can be found in most texts on relativity. A particle of rest mass m_0 and velocity v has a momentum of

$$p = \frac{m_0 v}{\sqrt{1 - v^2/c^2}} , \qquad 14\text{-}28$$

where c is the speed of light in vacuum. The total energy E of the particle (rest energy plus kinetic energy) is given by the expression

$$E^2 = (pc)^2 + (m_0 c^2)^2 . \qquad 14\text{-}29$$

Note that E, pc and m_0c^2 have the dimensions of energy; it is customary in particle physics to express each of them in units of Mev (one million electron volts). The momentum p is therefore expressed in Mev/c.

A particle with charge q moving at constant speed in a direction perpendicular to a magnetic field B follows a circular path, with the radius of curvature R related to the momentum of the particle by the equation

$$p = qBR \qquad 14\text{-}30$$

in the MKS system of units. Usually B is expressed in gauss (10^4 gauss = 1 tesla) resulting in the relation

$$p = 3 \cdot 10^{-2}\ BR\ \text{Mev/c}, \qquad 14\text{-}31$$

where R is in meters, p in Mev/c, and the particle has a charge e (electronic charge).

CONSERVATION LAWS

In all events observed in the bubble chamber certain conservation laws are valid and can be used to identify unknown particles. Both the relativistic energy (Equation 14-27) and the momentum of a system of particles must be conserved. The total charge of a system of particles must also be conserved in all interactions. It should be remembered, however, that uncharged particles do not leave tracks in the bubble chamber.

The total angular momentum, including the spin contribution, of a system of particles is always conserved in interactions.

In any interaction the total lepton number and the total baryon number must be conserved. The strangeness of a system is conserved as long as only strong nuclear or electromagnetic forces are involved in the interaction.

MEASUREMENTS AND ANALYSIS

Students are presented with photographs or slides taken in bubble chambers near accelerators. The detailed analysis of the events depends on the photographs used. It must be remembered that the tracks in the bubble chamber are three-dimensional, while our photographs show only two dimensions. Very rarely do we have interactions where all particles involved move in two dimensions only; there are, however, a few pictures available and these lend themselves very easily for an analysis.

With the pictures to be analyzed students must receive some information about the tracks. First of all the type of incoming particle is usually known. The strength of the magnetic field across the bubble chamber should also be given. By measuring the radii of curvature, the momenta of the particles can be computed from either Equation 14-30 or 14-31, if the charge is known. Radii of curvature can best be measured by the use of a transparent template specially designed for this purpose. The scale factor of the pictures has to be considered if an actual value of momentum is to be obtained. Knowing the momentum, the mass of the particle can be determined.

An estimate of the time of flight of a particle can be made by assuming that it travels at constant speed from one event to another. The mean life-time of the particle can therefore be computed.

By applying the conservation laws to the event, a considerable amount of information can be obtained. For further details the student is referred to the instruction manual or general information sheet supplied with the bubble chamber pictures.

QUESTIONS

1. Name as many different methods of detecting nuclear particles as you can.

2. How does the diffusion cloud chamber differ from the Wilson cloud chamber? What are its advantages?

3. Describe in detail how tracks of charged particles are made visible in the cloud chamber used.

4. How do alpha and beta particles lose their energy in the cloud chamber?

5. How does a bubble chamber differ from a cloud chamber? What are its advantages?

6. What is a spark chamber? What are its advantages?

7. If an alpha and a beta particle are subjected to the same magnetic field in a cloud chamber, compare their radii of curvature. Which one is larger and by how much?

8. Can you ever see γ-rays, neutrons or neutrinos in a cloud or bubble chamber? Explain the reason for your answer.

9. Are alpha-particles emitted by a natural radioactive source monoenergetic? Do they all have the same range?

10. Are beta-rays emitted by a radioactive source monoenergetic? Explain.

11. Write the nuclear reactions that occur in the alpha and beta sources used in the cloud chamber. What are their half-lives?

12. Write the reactions for all bubble chamber events studied in Part B of the experiment.

13. How could you take into account the fact that the actual bubble chamber tracks are three-dimensional, but our photographs are two-dimensional?

14. Derive Equations 14-28, 14-29 and 14-30.

Appendix 1 SYMBOLS

Symbol	Meaning
A	area
A	amplitude
A	Angstrom (10^{-10} m)
a	acceleration
a_R	radial acceleration
a.c.	alternating current
A.D.	a.d./$\sqrt{n}$
a.d.	average deviation
B	magnetic field
C	capacitance
c	specific heat capacity
cm	centimeter
D	angle of minimum deviation
D	diameter
d	deviation
d	distance
d.c.	direct current
E	electric field
E_m	maximum energy of beta-rays
F	force
F	farad
f	focal length
f	frequency
G	gravitational constant
g	acceleration of gravity
H	heat flow
h,H	height
I,i	current
I	rotational inertia
I	image size
ILE	instrumental limit of error
J	Joule's equivalent of heat
K,k	constants
k	thermal conductivity
KE	kinetic energy
L	latent heat of fusion
L	length
L	inductance
LE	limit of error
ln	logarithm - base e
log	logarithm - base 10
M	mean value
m	magnification
m	mass
m	meter
mg	weight
mm	millimeter
n	index of refraction
n	number of readings
n	principle quantum number
nm	nanometer
NA	numerical aperture
O	object size
P	power
P	pressure
PE	potential energy
PE	probable error
pF	picofarad
PI	precision index
Q,q	charge
Q	quantity of heat
R,r	radius
R,r	resistance
R	range of beta-rays
R	Rydberg constant
RE	random error
RMS	root mean square
S	object distance
S'	image distance
S.D.	standard deviation
T	half-life time
T	period
T	tension
T	temperature
T	tesla
t	time
V	potential drop
V	volume
v	velocity
v_T	tangential velocity
w	weber
X	reactance
X_C	capacitive reactance

X_L	inductive reactance
x	displacement
Y	Young's Modulus
Z	impedance
α	coefficient of linear expansion
γ	ratio of specific heats of a gas
Δ	change in
ε	electromotive force (emf)
λ	radioactive decay constant
λ	wavelength
μ	absorption coefficent
μ	coefficient of friction
μ	linear density
μF	microfarad
μ_o	permeability of free space
μ	mean value of a large set of measurements
ρ	density
Σ	summation
σ	standard deviation of a large group of measurements from their mean
τ	torque
ϕ	flux
ϕ	phase angle
Ω, ω	Ohm

Appendix 2 DENSITY OF AIR in gm/cm³

Temp.	Atmospheric Pressure (P_o) in cm of Mercury				
°C	73.0	74.0	75.0	76.0	77.0
16	.001173	.001189	.001205	.001221	.001238
18	.001165	.001181	.001197	.001213	.001229
20	.001157	.001173	.001189	.001205	.001221
22	.001149	.001165	.001181	.001197	.001212
24	.001142	.001157	.001173	.001189	.001204
26	.001134	.001149	.001165	.001181	.001196
28	.001126	.001142	.001157	.001173	.001188
30	.001119	.001134	.001150	.001165	.001180

Appendix 3 SOME SI PREFIXES COMMONLY USED IN THE LABORATORY

Factor	Prefix	Symbol	Examples
10^{-12}	pico	p	pF (capacitance)
10^{-9}	nano	n	nm (wavelength)
10^{-6}	micro	μ	μF (capacitance)
10^{-3}	milli	m	mm (distance)
10^{-2}	centi	c	cm (distance)
10^{3}	kilo	k	km (distance), kg (mass)
10^{6}	mega	M	MΩ (resistance), MHZ (frequency)
10^{9}	giga	G	GV (voltage)

Appendix 4 SOME CONVERSION FACTORS COMMONLY USED IN THE LABORATORY

1 radian	= 180/π or 57.3 degrees	1 lb/in^2	= $6.895 \cdot 10^3$ Pa
1 cm	= 10^{-2} m (meter)	1 BTU	= 1055 J (Joule)
1 mm	= 10^{-3} m	1 erg	= 10^{-7} J
1 km	= 10^3	1 ft-lb	= 1.356 J
1 inch	= .0254 m	1 HP-hr	= $2.685 \cdot 10^6$ J
1 nm	= 10^{-9} m	1 cal	= 4.187 J
1 Å	= 10^{-10} m	1 KW-hr	= $3.6 \cdot 10^6$ J
1 gram	= 10^{-3} kg (kilogram)	1 eV	= $1.6 \cdot 10^{-19}$ J
1 ft/s	= 0.305 m/s (meter per second)	1 HP	= 745.7 W (Watt)
1 mi/hr	= 0.447 m/s	1 cal/s	= 4.187 W
1 dyne	= 10^{-5} N (Newton)	1 μF	= 10^{-6} F (Farad)
1 pound	= 4.45 N	1 pF	= 10^{-12} F
1 atmosphere	= $1.013 \cdot 10^5$ Pa (Pascal)	1 mH	= 10^{-3} H (Henry)
1 cm of Hg (at 0°C)	= 1333 Pa	1 gauss	= 10^{-4} T (Tesla)

INDEX